Graphene and 2D Materials in Heat Transfer

Micro and Nano Technologies

Graphene and 2D Materials in Heat Transfer

Fundamentals and Applications

Edited by

Mohammad Khalid
Professor, Materials for Green Energy Technologies,
James Watt School of Engineering, University of Glasgow,
Glasgow, United Kingdom

Kazi Salim Newaz
Associate Professor, Department of Mechanical Engineering,
Faculty of Engineering, Universiti Malaya, Kuala Lumpur,
Malaysia

Abdul Khaliq Rasheed
Assistant Professor, Department of New Energy Science and Engineering,
School of Energy and Chemical Engineering,
Xiamen University Malaysia,
Sepang, Selangor, Malaysia

Rashmi Walvekar
Lecturer, Department of Chemical and Process Engineering,
University of Strathclyde,
Glasgow, United Kingdom

ELSEVIER

Elsevier
Radarweg 29, PO Box 211, 1000 AE Amsterdam, Netherlands
50 Hampshire Street, 5th Floor, Cambridge, MA 02139, United States

ISBN: 978-0-12-821964-5

For Information on all Elsevier publications visit our website at https://www.elsevier.com/books-and-journals

Publisher: Matthew Deans
Acquisitions Editor: Ana Claudia Garcia
Editorial Project Manager: Manisha Rana
Production Project Manager: Maria Bernard
Cover Designer: Raman Kumar

Typeset by Aptara, New Delhi, India

Contents

Contributors

Khouloud Abid
Georesources Materials Environment and Global Changes Laboratory (GEOGLOB), Faculty of Sciences of Sfax, University of Sfax, Sfax, Tunisia; Institute of Advanced Technologies for Energy "Nicola Giordano", Messina, Italy

Elham Abohamzeh
Department of Energy, Materials and Energy Research Center (MERC), Karaj, Iran

Bilal Akbar
Department of Mechanical Engineering, Brunel University of London, Uxbridge, London, United Kingdom

Naveed Akram
Department of Mechanical Engineering, Mirpur University of Science and Technology (MUST), Mirpur, Azad Jammu and Kashmir, Pakistan; Department of Mechanical Engineering, Faculty of Engineering, Universiti Malaya, Kuala Lumpur, Malaysia

O.A. Alawi
Department of Thermofluids, School of Mechanical Engineering, Universiti Teknologi Malaysia, Johor Bahru, Malaysia

Jorge Luis García Alcaraz
Department of Industrial Engineering and Manufacturing, Autonomous University of Ciudad Juárez, Juárez, Chihuahua, México

Alok Kumar Ansu
Integrated Centre for Energy and Sustainable Technologies, Manipal University Jaipur, Jaipur, Rajasthan, India

Yashar Aryanfar
Department of Electric Engineering and Computation, Autonomous University of Ciudad Juárez, Juárez, Chihuahua, México

Muhammad Asim
School of Professional Education and Executive Development, The Hong Kong Polytechnic University, Hong Kong, China

Mamdouh El Haj Assad
Department of Sustainable and Renewable Energy Engineering, University of Sharjah, Sharjah, United Arab Emirates

Hafiz Taimoor Ahmed Awan
James Watt School of Engineering, University of Glasgow, Glasgow, United Kingdom

Seitkhan Azat
Laboratory of Engineering Profile, Satbayev University, Almaty, Kazakhstan

Nour Hamza Belkhir
Georesources Materials Environment and Global Changes Laboratory (GEOGLOB), Faculty of Sciences of Sfax, University of Sfax, Sfax, Tunisia; Physics Department, Faculty of Sciences, University of Hafr Al Batin, Hafar Al Batin, Kingdom of Saudi Arabia

Ram Prakash Bharti
Department of Chemical Engineering, Indian Institute of Technology Roorkee, Roorkee, Uttarakhand, India

C. Bhinitha
Department of Mechanical Engineering, Faculty of Engineering, Universiti Malaya, Kuala Lumpur, Malaysia

Chitra Boobalan
Department of Chemical Engineering, Sri Sivasubramaniya Nadar College of Engineering, Kalavakkam, Chennai, Tamil Nadu, India

Swapnil Dharaskar
Department of Chemical Engineering, School of Energy Technology, Pandit Deendayal Energy University, Gandhinagar, Gujarat, India

Shobana Gabriel
Department of Biomedical Sciences, Faculty of Medical and Life Sciences, Sir Jeffrey Cheah Sunway Medical School, Sunway University, Bandar Sunway, Selangor, Malaysia

Krunal M. Gangawane
Department of Chemical Engineering, National Institute of Technology Rourkela, Rourkela, Odisha, India; Department of Chemical Engineering, Indian Institute of Technology Jodhpur, Karwar, Jodhpur, Rajasthan, India

Syed Asad Hussain
Life Cycle Management Laboratory (LCML), Faculty of Applied Science, School of Engineering, The University of British Columbia, Vancouver, BC, Canada

Kashif Ishfaq
Department of Industrial and Manufacturing Engineering, University of Engineering and Technology, Lahore, Pakistan

Afrin Jahan
Department of Mechanical Engineering, Faculty of Engineering, Nanotechnology & Catalysis Research Centre (NANOCAT), Institute for Advanced Studies, Universiti Malaya, Kuala Lumpur, Malaysia

Imdad Hussain Kalhoro
Department of Electrical Engineering Quaid-e-Awam University of Engineering, Science & Technology (QUEST), Nawabshah, Sindh, Pakistan

H.M. Kamar
Department of Thermofluids, School of Mechanical Engineering, Universiti Teknologi Malaysia, Johor Bahru, Malaysia

Sathishkumar Kannaiyan
Department of Chemical Engineering, Sri Sivasubramaniya Nadar College of Engineering, Kalavakkam, Chennai, Tamil Nadu, India

Mohammad Khalid
James Watt School of Engineering, University of Glasgow, Glasgow, United Kingdom

Hasan Koten
Faculty of Engineering and Pure Sciences, Istanbul Medeniyet University, Istanbul, Turkey

Laveet Kumar
Department of Mechanical Engineering, Mehran University of Engineering and Technology, Jamshoro, Pakistan

L. Harish Kumar
Department of Mechanical Engineering, NITTE Meenakshi Institute of Technology, NITTE (Deemed to be University), Bengaluru, Karnataka, India

Priyanka Mahajan
University Centre for Research & Development, Chandigarh University, Mohali, Punjab, India

A.R. Mallah
Department of Mechanical Engineering, Faculty of Engineering, Universiti Malaya, Kuala Lumpur, Malaysia

H.H. Masjuki
Department of Mechanical Engineering, Faculty of Engineering, International Islamic University Malaysia, Kuala Lumpur, Malaysia

Ramzi Maâlej
Georesources Materials Environment and Global Changes Laboratory (GEOGLOB), Faculty of Sciences of Sfax, University of Sfax, Sfax, Tunisia; Fluoink Nanotechnologies, Pouderiere Industrial Zone, Sfax, Tunisia

Faaiza Naaz
Department of Biochemistry, Auxilium College, Vellore, Tamil Nadu, India

Fedal Castro Nagarajan
Department of Mechanical Engineering, Aarupadai Veedu Institute of Technology, Paiyanoor, Chennai, Tamil Nadu, India

Thachnatharen Nagarajan
Department of Chemical & Petroleum Engineering, Faculty of Engineering, Technology & Built Environment, UCSI University, Kuala Lumpur, Malaysia

Swapnil Narkhede
Department of Mechanical Engineering, Dr. D. Y. Patil Institute of Technology, Pune, Maharashtra, India

Kazi Salim Newaz
Department of Mechanical Engineering, Faculty of Engineering, Universiti Malaya, Kuala Lumpur, Malaysia

Debabrata Panda
Department of Chemical Engineering, National Institute of Technology Rourkela, Rourkela, Odisha, India

Surya Pattnaik
Department of Chemical Engineering, National Institute of Technology Rourkela, Rourkela, Odisha, India

Abdul Khaliq Rasheed
Department of New Energy Science and Engineering, School of Energy and Chemical Engineering, Xiamen University Malaysia, Sepang, Selangor, Malaysia

Shahid Rasul
Mechanical and Construction Engineering, Northumbria University, Newcastle upon Tyne, United Kingdom

Adiba Sharif
Department of Mechanical Engineering, Mirpur University of Science and Technology (MUST), Mirpur, Azad Jammu and Kashmir, Pakistan

Ravi Kumar Sharma
Department of Mechanical Engineering, Manipal University Jaipur, Jaipur, Rajasthan, India; Integrated Centre for Energy and Sustainable Technologies, Manipal University Jaipur, Jaipur, Rajasthan, India

Yashvi Sheth
Centre of Nanotechnology, Indian Institute of Technology Bombay, Mumbai, Maharashtra, India

Samson Eugin Simon
Department of Hematology and Oncology, College of Medicine, Cancer Research Center, University of Tennessee Health Science Center, Memphis, TN, United States

Kamaljit Singh
Department of Mechanical Engineering, School of Engineering and Technology, CGC University, Mohali, Punjab, India

Pooja Singh
Department of Mechanical Engineering, Manipal University Jaipur, Jaipur, Rajasthan, India

Anirban Sur
Symbiosis Institute of Technology (SIT), Symbiosis International (Deemed University), Pune, Maharashtra, India

Ratiba Wali
Georesources Materials Environment and Global Changes Laboratory (GEOGLOB), Faculty of Sciences of Sfax, University of Sfax, Sfax, Tunisia

Rashmi Walvekar
Department of Chemical and Process Engineering, University of Strathclyde, Glasgow, United Kingdom

Weng Pin Wong
Sunway Centre for Electrochemical Energy and Sustainable Technology (SCEEST), Faculty of Engineering and Technology, Sunway University, Bandar Sunway, Selangor Darul Ehsan, Malaysia

Abdul Wasy Zia
Institute of Mechanical, Process and Energy Engineering (IMPEE), School of Engineering and Physical Sciences, Heriot-Watt University, Edinburgh, United Kingdom

M.N.M. Zubir
Department of Mechanical Engineering, Faculty of Engineering, Universiti Malaya, Kuala Lumpur, Malaysia

CHAPTER

2D materials and heat transfer

1

Elham Abohamzeh

Department of Energy, Materials and Energy Research Center (MERC), Karaj, Iran

1.1 Evolution of graphene and two-dimensional materials

Understanding the properties of materials and the factors influencing these properties is crucial for technological advancement. Different materials possess specific properties that make them suitable for various applications. For instance, copper's high electrical conductivity makes it ideal for use in electrical circuits, whereas concrete's compressive strength is valuable for constructing skyscrapers. However, material properties are not solely determined by their composition but also by their internal structures. Metallic bonds in metals, for example, enable them to conduct electricity by allowing electrons to move freely within the material when an electric field is applied. Another influential factor is the dimensionality of materials, particularly for nanoscale materials (sized a few hundred nanometers or smaller), which can exhibit unique mechanical, chemical, electrical, magnetic, and optical characteristics. The advancements in nanomaterials science have revealed unexpected features, paving the way for future technologies based on material size and bulk properties. Nanomaterials are categorized based on their number of nanoscale dimensions, including one-dimensional (1D), two-dimensional (2D), and zero-dimensional (0D) materials. Materials with three nanosized dimensions are 0D, including quantum dots and fullerenes as the first examples of this category. Fullerenes, discovered in 1985, are carbon allotropes with molecules consisting of carbon atoms that are held together by double and single bonds, forming a closed mesh. 1D materials, including nanowires, nanorods, and nanotubes, have two nanosized dimensions, and 1D is outside the nanoscale. Carbon nanotubes (CNTs), discovered in 1991, are molecular size structures composed of carbon atoms placed in one or two rolled-up layers, connected by covalent bonds in a hexagonal tiled pattern in each layer, forming a hollow tube with a diameter of about a few hundred nanometers. 2D materials with plate-like shapes have only one nanosized dimension. Nanocoatings, nanolayers, nanofilms, and graphene are examples of 2D materials. The carbon-based nanomaterials are depicted in Fig. 1.1.

2D materials have gained much attention lately because of their exceptional properties, surpassing those of bulk materials. The discovery of graphene [1] led scientists to explore the possibility of producing new 2D materials. Over the years, many 2D materials comprising carbon and other elements have been developed. These materials possess excellent mechanical flexibility, high carrier motilities, very low weight, significant ultraviolet and optical absorption, and also high thermal conductivity [2]. The superior thermal properties of 2D materials have provided an excellent opportunity for their possible applications in energy-related devices. This can be extremely helpful considering the substantial effect

Graphene and 2D Materials in Heat Transfer: Fundamentals and Applications. DOI: https://doi.org/10.1016/B978-0-12-821964-5.00017-5

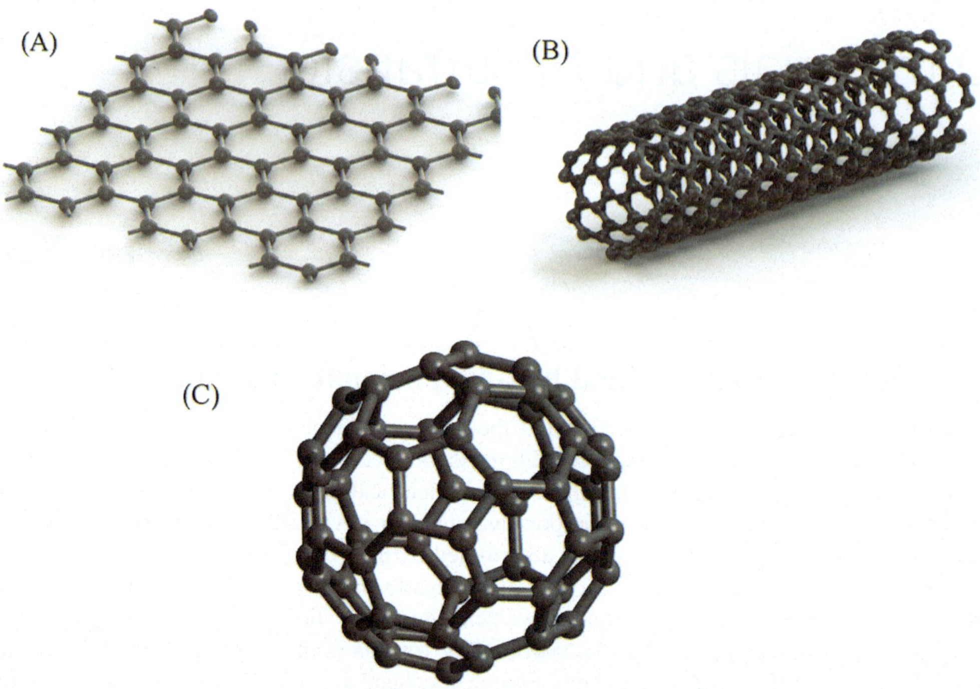

FIGURE 1.1

(A) 2D graphene sheet, (B) 1D carbon nanotube, and (C) 0D fullerene. *0D*, Zero-dimensional; *1D*, one-dimensional; *2D*, two-dimensional.

of intensifying heat transfer on improving the efficiency of engineering devices, reducing their size, and decreasing energy consumption in industry sectors.

The discovery of graphene ignited a revolution in 2D materials research [3,4]. The term "graphene" was coined in 1986 by Boehm et al. [5], derived from a combination of a suffix associated with polycyclic aromatic hydrocarbons and "graphite." This term initially described a single layer of graphite within larger configurations, such as graphite interaction compounds. However, the concept was first proposed theoretically by Phil Wallace in 1947 as an initial step toward understanding the electronic properties of 3D graphite [6]. Despite early theoretical groundwork the existence of 2D materials was deemed impossible because of stability issues, according to the Mermin–Wagner theorem in statistical mechanics and quantum field theory. This theorem hindered widespread recognition of graphene until its isolation in 2004, despite earlier production, such as the synthesis of graphene oxide (GO) by Brodie [7]. Brodie's treatment of graphite with acids led to the discovery of what he termed "carbonic acid," believing it to be a novel form of carbon known as "graphon." Subsequent research revealed that this substance was actually a suspension of tiny GO crystals. Further investigations into graphite oxide led to the understanding of its layered structure. A significant milestone occurred in 1984 when Ruess and Vogt observed creased GO flakes less than a few nanometers thick using transmission electron microscopy (TEM) [8]. In 1962 Boehm attempted to identify the thinnest pieces of reduced graphite oxide and noted some monolayers [9]. However, little attention was paid to this observation until 2009–10. It is worth noting that their 1962 study referenced TEM contrast, a method that is not foundational in current research because of its dependency on focusing conditions. Nevertheless, their work is now recognized as the initial observation of graphene sheets using TEM [10]. By the 1970s single layers of graphene were grown over other materials through epitaxial growth, forming a hexagonal lattice of carbon atoms bonded in sp^2 configuration with atomic-scale thinness. Notably, some instances witnessed significant charge transfer from the substrate to graphene, leading to a hybridization of graphene's π orbitals and the d-orbitals of substrate atoms, thereby altering the electronic structure significantly compared to freestanding graphene. From the early 1990s numerous attempts were made to produce extremely thin graphite films through methods like micromechanical lamination, yet until 2004, no carbon material thinner than 50–100 layers had been achieved.

Prior to 2004 the prevailing assumption was that isolating single-layer graphene was impossible because of its perceived inherent instability. Attempts at micromechanical exfoliation, like Kim's "nanopencil" method in 2000, yielded thick materials of tens of atomic layers, confirming this belief [11]. However, in a groundbreaking discovery in 2004, Andre Geim and Konstantin Novoselov at the University of Manchester demonstrated a novel method for isolating graphene. Their technique involved micromechanical exfoliation using adhesive tape to peel off single atomic layers from graphite and transfer them onto a silicon dioxide substrate. This achievement marked the birth of graphene, the first recognized 2D material [1,12].

In graphene, carbon atoms are interconnected by strong sigma bonds within a single-atom-thick layer (approximately 0.14 nm). Each graphene layer consists of hexagonal carbon rings arranged in a honeycomb lattice structure, presenting a simple yet exceptionally robust carbon configuration. The unique properties of graphene are highlighted as follows:

1. **Strength and flexibility:** Graphene exhibits remarkable strength and flexibility. It can withstand significant deformation (up to 20%–25% of its size) without breaking its atomic bonds, making it the material with the highest known tensile strength.
2. **Thin and lightweight:** With a thickness of only one atom, graphene is incredibly light and thin, contributing to its versatility in various applications.
3. **Outstanding thermal conductivity:** Graphene possesses an unparalleled thermal conductivity, surpassing that of silver, copper, graphite, and even diamond. Integrating graphene into different materials enhances their thermal properties, offering opportunities for diverse applications.
4. **Excellent electrical conductivity:** The flat hexagonal structure of graphene allows electrons to move at extremely high speeds (approximately 1/300 of the speed of light), resulting in outstanding electrical conductivity. Graphene maintains excellent conductivity even at room temperature, unlike superconductors that require very low temperatures.
5. **Optical transparency:** Because of its minimal thickness, graphene transmits a significant amount of light. With fewer atoms obstructing photon transmission, graphene transmits approximately 97.98% of the incident light, surpassing the transparency of standard glass (80%–90%).

Given the unique properties of graphene and its potential applications in photonics, electronics, and various other fields, it has garnered significant attention [13,14]. The discovery of graphene has spurred international research efforts into materials and methods for producing them with high quality [15,16]. Notably the library of 2D materials continues to expand and now includes over 150 families. Research interest in these materials is growing annually, with several new materials being added to this list each year [17]. The classification of 2D materials is generally based on their structural characteristics.

Examples of 2D materials include graphene, transition metal dichalcogenides (TMDCs) [18], layered double hydroxides, elemental monocompounds (Xenes), hexagonal boron nitride (h-BN) [19,20], five-layered V–VI trichalcogenides, black phosphorus, transition metal halides, Laponite clay, metal oxides, 2D polymers [21], perovskite-type oxides [22], transition metal oxides [23], metal nitrides/carbides (MXenes), and graphitic carbon nitride (g-C_3N_4) [24–27]. Many of these materials exhibit properties comparable to or surpassing those of graphene. For example, although pristine graphene lacks a bandgap, the direct bandgap present in certain monolayer TMDCs makes them particularly appealing for optoelectronic and electronic applications [28,29]. The development of new 2D materials has shown their remarkable applications in different fields such as LEDs, sensors, biomedicine, catalysis, and physical sciences [30]. The drive for higher efficiency electric devices with reduced energy consumption has spurred research into 2D material-enabled electronic devices. The significance of 2D materials lies in their potential applications in atomic- and nanoscale devices. Electronic devices based on 2D materials offer enhanced effectiveness because of their smaller size compared to traditional silicon-based counterparts. The growth of thin sheets of 2D materials over various substrates enables the synthesis of free-standing sheets and facilitates device applications. Following the discovery of graphene, numerous other 2D layered crystals have been identified, such as metal halides, h-BN, and MXenes. These materials, when exfoliated, expose nearly all their atoms on the surface, significantly increasing the surface area of 2D materials. This enhanced surface area substantially enhances their physical and chemical reactivities and impacts the 2D wave function because of quantum confinement effects. Consequently, 2D materials exhibit exceptional magnetic, photonic, electronic, and catalytic characteristics distinct from their bulk forms, rendering them suitable for a wide range of applications [31–33].

1.2 Introduction to macro-, micro-, and nanoscale heat transfer

Heat transfer is the heat energy exchange between two or more physical objects or systems and occurs in three ways: conduction, convection, and radiation. This section will first review the main concepts behind the conventional macroscopic definition of these mechanisms. Then, length and time scales will be discussed as parameters that characterize the mentioned heat transfer models. Subsequently the validity field of these mechanisms, considering the fundamental assumptions in these models, will be studied, and the discussion on these phenomena beyond the validity of the macroscopic principles will be presented.

1.2.1 Conduction heat transfer

The Fourier's law relates the flux to the temperature gradient and is applied to heat conduction in media [34,35]:

$$\phi = -k\nabla T \tag{1.1}$$

The following formula presents energy conservation [34,35]:

$$\nabla.\phi = -\frac{\partial T}{\partial t}c_p\rho \tag{1.2}$$

In the above equations, k is the thermal conductivity, and c_p is the specific heat capacity. A diffusion equation can be derived for the thermal field by inserting Fourier's law into Eq. (1.2) as well as with the assumption of homogeneous thermal conductivity [34,35]:

$$\rho c_p \frac{\partial T}{\partial t} = k\nabla^2 T \tag{1.3}$$

Considering thermal diffusivity, which is the rate of heat transfer through the material, is defined as $a = \frac{k}{\rho c_p}$.

Therefore Eq. (1.3) can be rearranged as

$$\nabla^2 T = \frac{1}{a}\frac{\partial T}{\partial t} \tag{1.4}$$

1.2.1.1 Characteristic time and length scales in conduction heat transfer

By the consideration of $x = x^+ L$ and $t = t^+ t_0$ the dimensionless form of the heat equation is achieved in which t_0 is characteristic time [34,35]:

$$\nabla_+^2 T = \frac{L^2}{a t_0}\frac{\partial T}{\partial t^+} \tag{1.5}$$

where $\frac{a t_0}{L^2}$ is a dimensionless quantity, recognized as the Fourier number. Defining the solution to the problem outlined by Eq. (1.6) that is associated with the field of temperature caused by a heat pulse at $r = 0$ as well as $t = 0$ helps to visualize the diffusive mechanism behavior [34,35]:

$$\nabla^2 T - \frac{1}{a}\frac{\partial T}{\partial t} = S\delta(t)\delta(r) \tag{1.6}$$

The solution can be specified with the definition of boundary conditions for an infinite environment where the temperature at infinity reduces to zero [34,35]:

$$T(r, t) = \frac{S}{(4\pi a t)^{3/2}} \exp\left(-\frac{r^2}{4at}\right) \tag{1.7}$$

Which represents a Gaussian energy distribution in time and space, having a characteristic width of $2\sqrt{at}$. It can be observed that heat diffuses through a sphere having the radius of $2\sqrt{at}$ following a time, t, has passed. If we consider an object with a radius of R, which is heated at the surface, the time needed for the object to thermalize is the required time for diffusing the heat throughout the object volume, for instance $t_{cd} = R^2/a$.

When heating is so that it changes in time in a sinusoidal manner $Q\left[1 + \sin \frac{2\pi t}{T}\right]$, the behavior would be dependent on the value of T. When the value of T is larger than conduction time, the object turns out to be isothermal, and the system thermalizes progressively and follows imposed variations, whereas for T values that are smaller than the conduction time, it would not be possible for the system to follow the variation imposed upon it.

When the radius of the object reduces to 1 nm, the diffusion time would be calculated as equal to 1 picosecond, and the mentioned arguments cannot be applied anymore. Here are the limitations of the macroscopic heat transfer model's validity, and a different physical problem's characteristic time should be encountered.

1.2.1.1.1 Short-time limit of the diffusion equation

The challenge here arises from the typical picosecond-order relaxation time of phonons, which presents a significant difficulty in the context of system relaxation within the diffusion equation at small time scales. Considering the solution in response to a heat pulse, it becomes evident that the temperature field at each spatial point is instantaneously modified upon the pulse's activation. However, at any given time t, it is physically impossible for this temperature modification to propagate over a distance greater than ct; otherwise the pulse effect would exceed the speed of light. Therefore a modification of the diffusion equation becomes necessary to address this issue. As Fourier's law describes, the fundamental problem originates from the immediate response to excitation, which neglects the system's response time. Instantaneous response to an applied temperature profile is unrealistic, particularly in linear systems. An appropriate time scale for linear systems can be defined as the response time to excitation, characterized by the time required for equilibrium to be reestablished, such as the relaxation time governed by microscopic collision processes. The relationship between responses to a Dirac delta function and a step function is revealed through differentiation for each linear system. Hence the relaxation time serves as the natural time scale of the physical problem, specifying the domain in which Fourier's law remains valid. Modifications to Fourier's law can therefore be derived from considerations within the Boltzmann equation [36]:

$$\tau \frac{\partial \phi}{\partial t} + \phi = -k\nabla T \tag{1.8}$$

where τ is the average relaxation time. The obtained formula includes an additional term in proportion to velocity propagation at the velocity of $v = \sqrt{\frac{a}{\tau}}$. The validity of Fourier's law depends on the relationship between the relaxation time and the characteristic time scale. Fourier's law holds that when the relaxation time is much shorter than the characteristic time scale, it becomes invalid when the characteristic time scale is smaller than the relaxation time. Corrections involving the dispersivity of Fourier's law should be considered under these conditions. Similarly, comparing a characteristic length scale to the average free path determines the applicability of local Fourier's law. If the characteristic length is significantly larger than the average free path, local Fourier's law is valid. However, nonlocal effects must be considered if the characteristic length is shorter than the average free path. This implies that the energy flux at a position depends on the temperature profile in the vicinity of that position. The limitations of the macroscopic approach to heat conduction relate to the length scale compared to the phonon mean free path (MFP) and the time scale compared to the phonon relaxation time. In addition, considering geometrical dimensions in comparison with the phonon wavelength presents another constraint.

1.2.1.2 Harmonic conduction regime

Think about the temperature field distribution as a result of applying a modulated heat source with the angular frequency of ω. To find a solution having the form of [34–36]

$$T(r,t) = e^{(-i\omega t)}\theta(r) + T_0(r) \tag{1.9}$$

The time-dependent part can be achieved as [34–36]

$$\nabla^2\theta + \frac{i\omega}{a}\theta = 0 \tag{1.10}$$

The above formula has a structure similar to the Helmholtz equation. A complex wave vector presented in Eq. (1.11) can be defined if we consider a solution in the form of exp (iKx) [34–36].

$$K = \frac{1+i}{\sqrt{2}}\sqrt{\frac{\omega}{a}} \tag{1.11}$$

During the wave propagation, its attenuation is described by the imaginary part of K. As mentioned earlier, responding to quick changes is not possible for the system. This can also be noticed in damping waves with high frequency in extremely small spaces. The magnitude order of the attenuation distance or the wavelength can be established. With the consideration of the Fourier number and the characteristics time, $l = \sqrt{aT}$ is obtained. The amplitude is commonly presented as $\exp \frac{i2\pi x}{\lambda} \exp\left(-\frac{x}{\delta}\right)$.

If we consider the modified form of the Fourier law, which is expressed by Eq. (1.8), it can be noticed that conductivity is dependent upon ω. In the regime of monochromatic [34–36]:

$$\phi = -\frac{k}{1 - i\omega\tau}\nabla\theta \tag{1.12}$$

where τ, as the time constant, is associated with the phonon relaxation time with a range of picoseconds; for frequencies lower than $1/\tau$ there is no need to consider this impact. So, it can be found that at low frequencies, k is nondispersive. It is the same as electrical conductivity, where it is possible to neglect dispersion impacts up to a few GHz.

1.2.1.3 *Short-scale transfer*

In summation, time and length scales comparable to the relaxation time of phonons and the MFP of phonons are primary limitations of the macroscopic conduction approach. Another limitation is the phonon wavelength, which is comparable with geometrical dimensions. In such a case the impact of mode quantization becomes relevant. Think about a wire that has a diameter of a few nanometers. So, there would be discrete numbers of modes that could propagate, leading to conductance quantization, a phenomenon renowned in electrical conduction [37].

1.2.2 Convection

In a stationary fluid, heat transfer occurs primarily through conduction, according to Fourier's law. However, when gravity is present, density changes due to temperature variations create forces that induce fluid motion. This phenomenon is referred to as natural convection. The heat flux per unit area between a fluid and a surface can be presented as follows [34–36]:

$$\phi = h(T_s - T_f) \tag{1.13}$$

In the above equation, T_f as well as T_s present fluid and solid temperatures, respectively. Because the speed at the surface is equal to zero, throughout the surface, there is no fluid movement, and no convection exists there; in other words, there is no energy transfer related to the transfer of mass. Conduction in the boundary layer is the only energy transfer mechanism. So the heat flux at the surface can be expressed as [34–36]

$$\phi = -k_f\frac{dT}{dx} \tag{1.14}$$

In which k_f is fluid thermal conductivity. By the consideration of a nondimensional parameter $x = x^+ L$ and the nondimensional temperature, [34–36]

$$[T_s - T(x)] = T^+(x)[T_s - T_f] \tag{1.15}$$

The above equation results in:

$$\phi = -\frac{(T_f - T_s)}{L} k_f \left(-\frac{dT^+}{dx^+}\right) = h(T_s - T_f) \tag{1.16}$$

In the above equation $h \frac{k_f Nu}{L}$ represents the exchange equation, in which nondimensional Nu expresses the Nusselt number (the nondimensional temperature profile in the boundary layer $Nu = dT^+/dx^+$).

1.2.2.1 Characteristic convection time

A characteristic time scale can also be defined for convective heat transfer. The energy balance formula for a spherical object with an area of S and a volume of V is defined as [34–36]

$$\rho c_P V \frac{\partial T}{\partial t} = hS(T - T_f) \tag{1.17}$$

If we assume a larger time scale compared to the conduction time, the sphere object can be considered isothermal. With the above solving equation the convection time τ_{cv} can be achieved as follows:

$$\tau_{cv} = \frac{\rho c_P V}{hS} \tag{1.18}$$

1.2.2.2 Ballistic transport (short-scale transfer)

When considering heat transfer between two parallel surfaces with different temperatures and a separation distance of d the average free path of the molecules becomes a crucial parameter. This average free path represents the mean distance traveled between two consecutive collisions. If the distance between the surfaces is smaller than the average free path, a phenomenon known as ballistic transport dominates. In this regime, molecules leaving one surface without colliding directly reach the opposite surface, leading to a much higher transfer efficiency compared to diffusive transport. This is because each molecule directly participates in energy transfer without undergoing collisions or backscattering.

In contrast, diffusive transport involves multiple collisions, causing molecules to spread forward and backwards, reducing the overall energy transfer efficiency. This regime is characterized by two timescales: the time-of-flight between the plates and the collision time. When the collision time is longer than the time-of-flight, the transport is considered ballistic. Therefore the traditional method for conduction, based on Fourier's law, is valid only when the timescales are much larger than the collision time and the distances are much greater than the average free path. In situations where these conditions are not met, ballistic transport becomes the dominant mechanism, requiring a different approach to model the heat transfer process accurately.

1.2.3 Radiation

Considering a surface with an area of dS a radiative flux passing the surface in the u direction with a frequency range of $[\nu, \nu + d\nu]$ and a solid angle of $d\Omega$ and can be defined as [34–36]

$$d\phi_\nu = L_\nu(u, r)dS.\cos\theta.d\Omega.d\nu \tag{1.19}$$

The specific intensity, L, is related to the considered point, the orientation, and frequency. It could be viewed as the product of photons' number $n_\nu(u, r)$ in volume unit having the energy $h\nu$, also the propagation velocity of c, and spreading in u-direction.

1.2.3.1 Equilibrium-specific intensity

In thermal equilibration, any given mode is filled by photon quantities specified by the distribution of the Bose–Einstein. $n_{\nu,}$ which was mentioned before, can be defined as [34–36]

$$n_\nu(u, r) = \frac{1}{\exp\left(h\nu/kT\right) - 1} \tag{1.20}$$

The number of modes in a solid angle of $d\Omega$ per unit volume can be specified. Considering a finite-sized cavity having the side of L and leaving L to go to infinity the quantities of the electromagnetic modes can be counted. The expression for black body-specific intensity is as follows [34–36]:

$$L_\nu^0(T) = \frac{2h\nu^3}{c^2}\frac{1}{\exp\left(h\nu/kT\right) - 1} \tag{1.21}$$

It is also noteworthy to emphasize that plane waves spreading in a vacuum are the only solutions to Maxwell's equations, which are considered in the above explanation. Close to an interface within a cavity with a dimension that is assimilated with the wavelength, Maxwell's equations solutions are dissimilar. When the physical system is different, the modes also change. Bose–Einstein statistics give the average occupation number for every mode under equilibrium. On the other hand the number of modes per unit volume cannot be more than $2\nu^3/c^3$ that and changes considerably with the change in the system.

1.2.3.2 Absorbed and emitted flux

L_ν^0 represents the specific intensity under equilibration. When specific intensities reflected and emitted from a plate add together, the specific intensity that leaves a plate can be calculated. The emitted specific intensity can be specified based on the following formula [37]:

$$L_\nu(T) = \epsilon_\nu L_\nu^0(T) \tag{1.22}$$

where ϵ_ν is emissivity. The proportion of absorbed and incident power (both per surface unit) defines absorptivity α_ν. The polarized directive monochromatic emissivity is equal to the polarized directive monochromatic absorptance, based on the reciprocity theorem [38]:

$$\alpha_\nu(u) = \epsilon_\nu(u) \tag{1.23}$$

1.2.3.3 Characteristic length and time scales

In radiative problems, different length scales, such as the coherence length, the skin depth, and the wavelength, can appear.

1.2.3.3.1 Wavelength

The wavelength is a fundamental parameter that governs various behaviors in electromagnetic systems. When distances involved are much smaller than the wavelength, retardation effects become negligible, allowing the problem to be treated as electrostatics, particularly in terms of spatial field variation. It is crucial to note that certain medium properties, such as permittivity and permeability, vary with frequency. The wavelength also sets the scale for the validity of geometric optics; when structures approach or exceed the wavelength size, the assumption of light rays propagating in straight lines becomes unreliable.

The wavelength plays a critical role in transmission modes between objects with dimensions smaller than itself. Structures shorter than the wavelength can have a behavior that resembles that of a radiative dipole. Concepts such as absorptance and emissivity may not apply appropriately in such scenarios. Instead, energy transport is influenced by van der Waals forces, involving both momentum and energy transfer. This type of energy exchange occurs in the near-field of electromagnetic interactions and is called near-field radiative transfer. In molecular systems, this transfer mode is often referred to as nonradiative transfer, as it does not strictly involve photon absorption and emission according to quantum mechanics. Nonetheless, it represents a significant energy transfer mechanism mediated by electromagnetic interactions.

1.2.3.3.2 Skin depth

The skin depth in an opaque medium represents the depth at which an electromagnetic field can penetrate. Power absorption occurs predominantly within this depth from the surface layer, where field emission also originates. The concept of emissivity applies effectively under conditions of uniform temperature throughout the skin depth. However, additional considerations are necessary in cases of nonuniform temperature distribution, particularly in semitransparent media.

1.2.3.3.3 Coherence length

In an emission phenomenon, there is another significant factor, which is spatial coherence length. Contributions from various spots on the plane are added together if the field is not coherent. When the field is coherent, the interference of various plane points is possible, resulting in totally different outputs. It is possible to demonstrate that the wavelength is much longer than the coherence length in case surface waves are not carried by the medium [5]. If it is not, an extremely coherent field would be created at the interface as a result of waves propagating along the surface.

This results in massive anomalies in the absorptance as well as emissivity that could be used to create an extreme correction to the surface's radiating characteristics. Based on the explanations above, nano- or microheat transfer points out phenomena that exceed conventional macroscopic laws. The limit here can be defined by two parameters: the time scale (relaxation time) and length scale (average free path) of the considered system. Effective heat transfer relies on collisions between energy carriers. The average distance an energy carrier travels per collision across numerous collisions is termed the average free path, whereas the average time taken by energy carriers between collisions is referred to as the mean free time. In lattice structures the order of the MFP typically ranges from 10 to 100 nm. For phonon–phonon collisions in insulators, dielectric crystals, and semiconductors, the mean free time is on the order of 10 picoseconds (ps), whereas for electron–phonon interactions in metals, it is around 100 ps, and for electron–electron collisions, it is about 100 femtoseconds (fs).

When the system size decreases to dimensions comparable to or smaller than the average free path of energy carriers, or when the time scale of processes approaches the range of the mean free time, distinct and statistically meaningful behaviors of energy carriers become evident. Microscale heat transfer phenomena differ significantly from macroscale heat transfer, which relies on averaging over hundreds of thousands of collisions in time and particles in space.

Microscale heat transfer involves various physical mechanisms depending on the type of energy carriers present. In semiconductors, insulators, and dielectric films, phonon scattering governs heat transfer, whereas phonon–electron interactions play a crucial role in metallic films. Conventional macroscopic approaches are inadequate for addressing microscale heat transfer problems, as they are designed to describe averaged macroscopic behavior across large numbers of particles [37,39,40].

1.3 Overview of graphene in heat transfer applications

Recently, researchers have revealed the extraordinary properties of graphene and announced different applications of graphene in proton-conducting [41], weed barriers for molecules [42], solid-phase microextraction [43], electrochemical catalysts [44], chemical synthesis [45], and electric energy conversion industry [46]. Moreover, graphene has been reported to have wide applications in the energy industry and energy-related aspects [47–51]. Identifying the possible applications of graphene for energy-related problems has led scientists to discover more about utilizing those excellent properties for heat transfer as well as thermal conductivity and storage, and much attention has been paid to finding novel applications of graphene in heat and energy transfer [52].

1.3.1 Nanographene

Nanofluids consisting of a base fluid and nanoparticles with excellent thermal conductivities are widely applied in different industries aiming to improve convective heat transfer [53,54]. The most significant variables that affect the heat transfer coefficient of nanomaterials are temperature, shape, and type of particles. The other effective factors include the type of base fluid, volume fraction of nanoparticles, morphology, additives, clustering, and acidity [55]. Graphene can be easily prepared, and it has numerous benefits, namely, high dispersion stability, greater thermal conductivity, less corrosion and erosion, and a large ratio of surface area to volume [56,57]. Thus graphene can be considered a great additive in preparing nanofluids. The novel heat transfer nanofluids, which are based on nanographene, are heat transfer systems in which the effectiveness of latent heat generated by the phase change materials' (PCMs) thermal conductivity is mainly specified by nanographene. Some examples of utilization of nanographene for enhancing heat transfers are graphene nanoribbons under great heat flux [58], paraffin/expanded perlite phase change composites with graphene nanoplatelets (GNPs) [59], and GNP–silver hybrid nanofluids [55]. Outputs of previous research studies revealed that thermal transport characteristics of graphene-based nanofluids could be influenced by filling fraction, size, temperature, etc. Esfahani et al. [60] prepared GO nanofluids at various mass fractions

(0.01%, 0.05%, 0.1%, and 0.5%) in water. According to the outputs, two parameters that govern the thermal conductivity of this GO nanofluid are viscosity and particle size distribution. Moreover, they observed that nanofluids filled with GO have higher thermal conductivity rather than water. With increasing the mass fraction of GO, thermal conductivity enhances initially, but further increase in the mass fraction speeds up agglomeration and leads to decreasing the thermal conductivity. Therefore there is an optimum value for mass fraction. The rising thermal conductivity of nanofluids could be discussed considering nanoparticles' stochastic motions. At high temperatures, violent Brownian-like motions enhance thermal conductivity. Applying surface functionalization might improve the transport properties [61,62]. Naddaf et al. [63] dispersed multiwalled CNT (MWCNT) and GNPs functionalized with hexylamine in diesel oil. Their experimental studies revealed that thermal conductivity was enhanced under constant temperature and at all considered mass fractions compared with pure diesel oil. To investigate heat transfer in armchair graphene nanoribbons under large heat flux, Zheng et al. [58] conducted nonequilibrium molecular dynamics (MD) simulations. They observed two energy transfer paths: anomalous wave-like kinetic transport of energy, constituting a great value of total energy, and energy transfer through heat conduction based on Fourier's law. This indicates that by wave kinetics, the efficiency of heat transfer in graphene is enhanced significantly. Based on the outputs of the investigation conducted by Ramakrishnan et al. [59], with the addition of 1% nanographene to the paraffin/expanded, the thermal conductivity increases by 49%, whereas the duration of the heat storage/release decreases to 33%. In the experiment of Yarmand et al. [55] a synthesis approach was presented for decorating Ag on the GNPs and nanofluid preparation. The outputs of experimental tests on GNP–Ag nanofluids revealed that the thermophysical characteristics of hybrid nanofluids are significantly enhanced, and a considerable improvement in heat transfer was noticed. Sadeghinezhad et al. [64] scrutinized the turbulent heat transfer properties and pressure drop of GNP nanofluids numerically and experimentally. The outer surface of the tube containing GNP nanofluids is exposed to a uniform heat flux. The influences of nanoparticle concentrations and heat flux were investigated. The synergistic impact of graphene–MWCNT composite on nanofluid thermal conductivity was studied by Aravind et al. [65]. The increase of 9.2% and 10.5% in the thermal conductivity of deionized H_2O with graphene–MWCNT and graphene was observed. The extraordinary thermophysical characteristics of graphene–MWCNT composite nanofluids can be explained by special complementary structure between MWCNT and graphene, excellent thermal transport of MWCNT and graphene, and also great aspect ratio. The thermal conductivity of nanofluids can enhance considerably with graphene. It is efficient to utilize nanographene, which has great heat transfer properties in oscillating heat pipes, heat pipe heat exchangers [66,67], and other heat transfer devices [68].

1.3.2 Graphene coating

Applying graphene monolayer or multilayer coating has provided effective heat transfer. Some applications reported in the literature are coating on composite anti-/deicing components [69], volatile organic compounds [70], and graphene/h-BN multilayers [71].

1.3.2.1 Convective boiling

In the field of heat transfer, boiling is regarded as one of the important subjects. This phenomenon is a convection process from a surface into a fluid, resulting in the boiling and vaporizing of the liquid and dissipating a huge amount of heat through small areas via transpiration and evaporation. Jaikumar et al. [72] conducted tests to study the influence of graphene coating on the pool boiling efficiency of copper chips. They observed that critical heat flux (CHF) and the heat transfer coefficient were significantly improved compared with the uncoated case. The increase in heat transfer coefficient was 47%, and also 42% augmentation in the CHF was perceived. To understand the mechanism of heat transfer improvement, they considered several parameters such as contact angle hysteresis, graphene film thermal conductivity, the degree of wetting, and influences of morphological scale. It was concluded that wettability and thermal conductivity were not contributive parameters. The key factors that affected heat transfer were roughness and contact angle hysteresis [73]. Kim et al. [74] examined the improvement of nucleate boiling heat transfer resulting from modifying surfaces with graphene coating. They observed that coating with 2D or 3D graphene and changing the surface orientation angle from 0° to 170° improved the boiling heat transfer coefficient and CHF compared with the bare surface. The study reported a 40% enhancement in boiling heat transfer coefficient (BHTC) for a 2D-graphene-coated surface in the downward-facing condition. The CHF enhancement ratio was 40% for the 2D-graphene-coated surface and 20–25% for the 3D-graphene-coated surface. The boiling structures are illustrated in Fig. 1.2 [74]. Under downward orientation the 2D graphene surface exhibited improved nucleate site density, leading to a 40% enhancement in BHTC compared to the bare surface, whereas such enhancement was not observed for the 3D surface. Boiling experiments were conducted to investigate heat transfer properties and establish essential contact between thermal reduction and CHF improvement [75], aiding in thermal conductivity enhancement. The study noted that increasing the thickness of the GO film increased CHF. To improve the performance of nucleate boiling, Kim et al. [76] created composites featuring porous graphene networks and micropillar arrays. They concluded that only micropillars' heads should be coated with porous graphene to optimize the composite boiling characteristics.

1.3.2.2 Near-field radiative thermal transport

The recent investigation of near-field radiation utilizing graphene-covered coating has gained much attention based on the exact scattering theory [77]. Radiative heat transfer, classified among basic heat transfer mechanisms, has wide applications in different areas, such as thermal management and energy harvesting. The results of the study by Zhao et al. [71] revealed that layer numbers in the stack do not affect the near-field heat transfer within multilayer structures, which makes it possible to control the rate of heat transfer by graphene layer numbers in the multilayer structure. Graphene coating can be applied to enhance radiative heat flux among corrugated polar materials and mitigate performance sensitivity [78]. Conducting different research studies on graphene and its effects on heat transfer revealed the practical benefits [79] of graphene employment for enhancing thermal conductivity. As measurement techniques for determining graphene conductivity have been developed [80–82], different research studies have approved that the graphene layer can improve heat transfer rate [68,83–87].

1.3.3 Graphene thermal interface materials

Thermal interface materials (TIMs) are utilized between two solid contact surfaces to enhance thermal transport efficiency. The escalating power density in electronic and electrical devices has spurred the urgent need for TIMs with high through-plane thermal conductivity to address thermal management

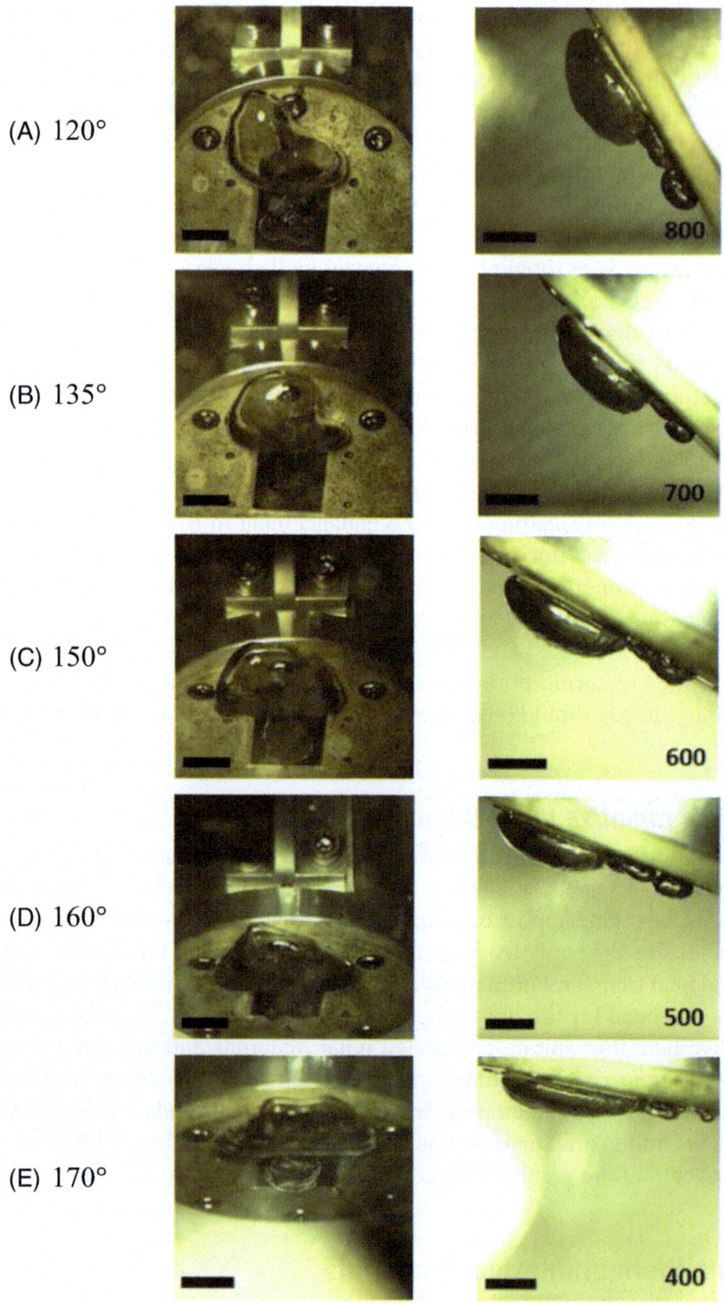

FIGURE 1.2

Front and side-view images of slug behavior in the fully developed nucleate boiling regime at different surface orientation angles: (A) 120°, (B) 135°, (C) 150°, (D) 160°, and (E) 170° [74].

challenges. Various types of TIMs have emerged, including carbon-based materials, liquid metals, and particle-filled polymers. Graphene-based TIMs have garnered significant attention because of graphene's large specific surface area and excellent thermal conductivity [88]. The interface thermal resistance is an important factor influencing the thermophysical properties of graphene-based TIMs [3,89,90]. Establishing a network structure with high heat conductivity is a fundamental concept [91]. Achieving "end-to-end" linking of graphene through directional alignment technology can create a thermal conducting network within the matrix. Wang et al. [92] investigated the change in paraffin-graphene interfacial thermal resistance through covalent functionalization of graphene using MD simulations. Their findings suggest that an effective strategy to reduce interface thermal resistance is the modification of graphene with functional groups, depending on the coverage and type of functional groups. However, Shen et al. [26] demonstrated that the effectiveness of modifying graphene with functional groups is limited by graphene size. Although efficient for small-sized graphene, this approach is less effective for larger graphene sheets.

1.3.4 Graphene-polymer composites

Thermal conductive polymer composites exhibit excellent heat transfer properties and corrosion resistance while maintaining ease of processing, making them highly versatile across aerospace, microelectronics, and electrical industries. The enhanced thermal conductivity of these composites largely depends on the function of the fillers used. Graphene, owing to its exceptional strength, large surface area, high thermal conductivity, and rapid carrier mobility, proves to be an effective enhancer of polymer thermal properties. Yu et al. [93] demonstrated significant improvements in the thermal conductivity of nylon 6 composites by incorporating graphene nanosheets as fillers via mechanical blending. Results showed a remarkable increase in thermal conductivity, with a 42.6% improvement at a 1.0% volume fraction of graphene and over 15 times enhancement at 20% volume fraction compared to pure nylon 6, highlighting its potential for effective heat management applications. The properties of the filler, including thermal conductivity, surface characteristics, structure, and particle size, critically influence the heat transfer characteristics of particle-loaded composite materials. Graphene-based thermal conductive polymers offer efficient thermal transport, characterized by low phonon scattering, reduced transmission resistance, and exceptional thermal conductivity. Moreover, the planar 2D structure of graphene facilitates the formation of a thermal conduction network because of its large surface area. Fig. 1.3 illustrates the modification of GO sheets via polyvinylpyrrolidone noncovalent surface treatment, leading to homogeneous dispersion within the styrene–butadiene rubber matrix. Experimental studies revealed a 30% increase in thermal conductivity with just five parts per hundred rubber of modified GO in the nanocomposite. This research underscores the promising role of graphene-based fillers in advancing the thermal properties of polymer composites for practical heat management solutions [94].

1.3.5 Graphene-enhanced hybrid phase change materials

PCMs are materials that are able to absorb and release heat through phase transition for the purpose of energy storage. They have wide applications in a variety of fields. For practical applications, PCMs are desired to have some properties such as high latent heat capacity, great thermal conductivity and stable

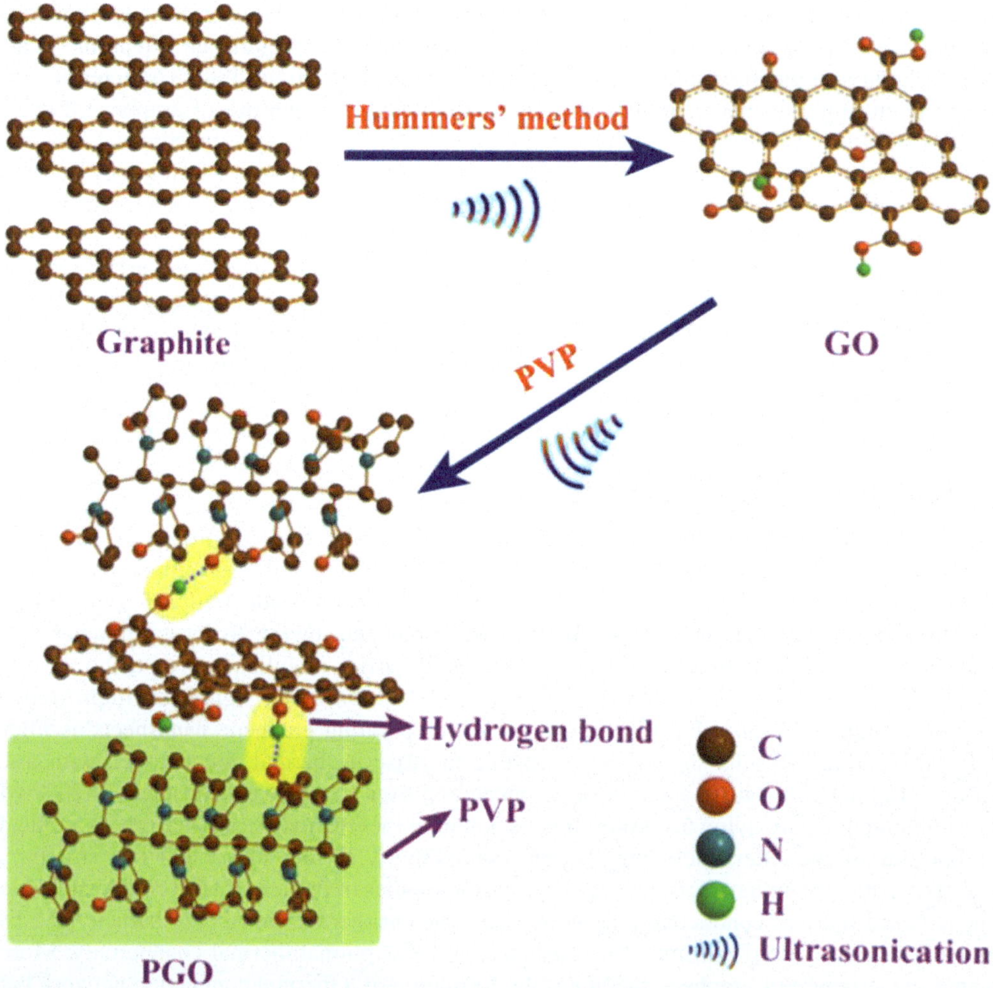

FIGURE 1.3

Hydrogen bonds between PVP and GO [94]. *GO*, Graphene oxide; *PVP*, polyvinylpyrrolidone.

thermal characteristics, low-cost and easy access, low volume expansion coefficient, environmentally friendly, noncorrosiveness, and nontoxicity. Although organic PCMs, such as polybasic alcohol, fatty acid, and paraffin, have most of the mentioned properties, low thermal performance is regarded as the main drawback of these materials. Finding solutions for improving thermal conductivity leads to intensifying the releasing and absorbing heat processes and improving heat storage efficiency. The thermal properties of organic PCMs can be enhanced by applying methods such as filling additives with great thermal conductivity. The unique properties of graphene, such as low density, excellent thermal conductivity, and large specific surface area, make it a great additive for organic PCMs. Achieving organic PCMs with high heat transfer properties has remained a considerable challenge, so quite a bit of research has been conducted recently on the addition of high thermal conductivity graphene for enhancing the thermal characteristics of organic PCMs. By high-temperature annealing, graphene can considerably decrease defects. Furthermore, it has been observed that incorporating annealed graphene (10 wt%) into PCMs leads to a significant enhancement in thermal conductivity. Specifically, PCMs filled with annealed graphene exhibit a sixfold improvement in thermal conductivity compared to graphene-based PCMs without annealing and a remarkable 16-fold improvement compared to pristine PCMs. Encapsulated PCMs typically consist of a core–shell structure achieved through methods such as interfacial polymerization, miniemulsion polymerization, or emulsion polymerization. The PCM core within these capsules functions to absorb, store, and release latent thermal energy during phase change processes. Encapsulation in spherical capsules isolates PCMs from the external environment, addressing challenges such as phase separation, stability, and corrosion. This core–shell configuration extends the lifetime of PCMs and enhances thermal transport performance by increasing the contact area of the PCM. For example, stearic acid/graphene composite microcapsules were produced using latex technology for thermal energy applications [95]. The stearic acid core demonstrated a stable structure and distinct phase transition behavior. Results indicated the composite formulation enhanced both thermal conductivity and latent heat capacity relative to pristine PCMs [96].

1.4 Phonon transport and heat conductivity

Measuring the lattice thermal conductivity of a specific material is the primary experimental method for investigating the transport of acoustic phonons in these materials. The most important properties of heat conduction are discussed as follows. Based on Fourier's law for conduction, heat transfer, heat flux ϕ, and temperature gradients ∇T are related together, as was shown in Eq. (1.1). In this formula $K = (K_{ij})$ is the thermal conductivity tensor. K is considered a constant, and thermal conductivity is not dependent on heat flow direction for the isotropic media. Thermal conductivity can be viewed as a constant in cases with small temperature changes. Otherwise, K is regarded as a temperature function $K = K(T)$. Generally, electrons and phonons are heat carriers in solid materials. Therefore thermal conductivity is defined as $K = K_p + K_e$.where K_p and K_e are contributions of phonons as well as electrons, because of the great magnitude of free carrier density, K_e is dominant in metals and degenerate semiconductors. Applying the law of Wiedemann–Franz the magnitude of K_e can be specified by measuring the electrical conductivity, σ [97]:

$$\frac{K_e}{\sigma T} = \frac{\pi^2 k_B^2}{3e^2} \tag{1.24}$$

In the above equation, k_B is Boltzmann's constant, and e is the electron charge. Within carbon-based materials, phonons play a primary role in conductive thermal transport mechanisms. For graphite and other layered carbons exhibiting metal-like properties, acoustic phonons are the dominant heat carriers [98]. This is because of the powerful covalent sp^2 bonding, resulting in great phonon group speeds as well as small crystal lattice anharmonicity for vibrations. The thermal conductivity of a phonon could be defined by applying the following formula [96]:

$$K_p = \sum_j \int C_j(\omega) v_{x,j}(\omega) \tau_j(\omega) v_{x,j}(\omega) d\omega \tag{1.25}$$

where τ_j is the relaxation time of the phonon. There are two transverse and one longitudinal acoustic (LA) branch in phonon polarization branches, j, over which the summation is done.

The phonon group velocity of the jth can be specified by the following equation [96]:

$$\vec{v}_j = \frac{d\omega_j}{d\vec{q}} \tag{1.26}$$

It can be estimated with the sound speed in lots of solids. $v_{x,j}$ is the projection of v_j onto the x-axis. C_j in Eq. (1.25) expresses the contribution of the jth branch in heat capacity and can be defined as [96]

$$C_j = \hbar\omega_j \partial N_0\left(\hbar\omega_J/k_B T\right)/\partial T \tag{1.27}$$

where the Bose–Einstein distributions of phonons function is $N_0\left(\frac{\hbar\omega_j}{K_B T}\right) = \left[\exp\left(\frac{\hbar\omega_j}{K_B T}\right) - 1\right]^{-1}$. The relation between the relaxation time and the MFP of phonon,Λ, is defined by applying the following expression [96]:

$$\Lambda = \tau v \tag{1.28}$$

Different scattering mechanisms are often taken into consideration for the relaxation time approximation. Acoustic phonons as heat carriers in common solids are scattered by interfaces, conduction electrons, impurities, lattice defects, and other phonons [99–102]. The phonon–phonon scattering restricts phonon MFP in ideal crystals that do not contain lattice defects or rough boundaries. Thermal conductivity, in this case, is said to be intrinsic. On the other hand, when extrinsic impacts like phonon–defect scattering or phonon–rough boundary limit thermal conductivity, they would be referred to as extrinsic. Umklapp processes describe the anharmonic interaction of phonons, which cause finite 3D thermal conductivity [99]. The lattice anharmonicity degree is specified by the Gruneisen parameter γ, which is a parameter upon which the Umklapp scattering rates are dependent [100]. The phonon energy spectra in nanostructures are quantized because of the acoustic phonons' spatial confinement, which leads to a decrease in phonon group velocity. Nanostructures' thermal conductivity is affected by modifying phonon energies, phonon scattering from boundaries, and group velocities and density of states. In many cases the thermal conductivity of phonons decreases because of the acoustic phonons' spatial confinement [103,104]. Nevertheless, for nanostructures embedded into the acoustically hard barrier layers, the acoustic phonons' spatial confinement is anticipated to enhance the thermal conductivity [105–108]. The scattering of the phonon–boundary can be estimated as [102]

$$\frac{1}{\tau_{B,j}} = \frac{1-p}{1+p}\frac{v_{x,j}}{D} \tag{1.29}$$

The specularity parameter, p, is determined as the specular scattering probability across boundaries, and D is the size of the nanostructure. For a smooth interface ($p = 1$), the specular scattering conserves momentum and is not added to thermal resistance. So, only diffuse phonon scattering at rough boundaries ($p \to 0$), which differs from the phonon momentum, limits the MFP of the phonon. The phonon specularity could be expressed as [102,109,110]

$$p(\lambda) = \exp\left(-\frac{16\pi^2\eta^2}{\lambda^2}\right) \tag{1.30}$$

In the above equation, λ is the incident phonon wavelength, and η is the root mean square divergence of the surface height from the base plane. With the assumption of the dominance of the phonon–boundary scattering, thermal conductivity scales with the size of the nanostructure, D as $K_p \sim C_p v \Lambda \sim C_p v^2 \tau_B \sim C_p v D$. When D is too small, the dependency of the thermal conductivity on the structure dimension would be more complex because of the strong quantization of the phonon energy spectra [103,105,108]. C_p is dependent on the state's phonon density, resulting in different $C_p(T)$ dependencies for 1D, 2D, and 3D systems [99,102]. Under low temperatures for bulk systems $K(T) \sim T^3$, but for 2D systems $K(T) \sim T^2$.

The capability of a specific material to conduct heat is defined by thermal conductivity. The rate of conducting heat in a material is measured by another property, thermal diffusivity, which can be specified as follows [7]:

$$\alpha = \frac{k}{C_p \rho_m} \tag{1.31}$$

where ρ_m is the density. It must be mentioned that thermal diffusivity is the parameter mostly measured by experimental techniques [111].

Over the recent years, several theoretical approaches have been developed for investigating the heat transport mechanism in 2D materials, including the Boltzmann transport equation (BTE) [112, 113], nonequilibrium Green's function method [114, 115], and MD simulations [116, 117]. On the other hand, there have been limited experimental studies on these materials because of difficulties in measuring the temperature distribution on the nano/microscale and the suspension of nanomaterials appropriate for thermal measurements. Despite these challenges, measuring temperature gradients in nanoscale materials is possible by applying modern nanofabrication technologies. Different techniques are available for measuring intrinsic thermal conductance of low-dimensional materials, such as the thermal flash method [118], scanning thermal microscope [119,120], time-domain thermoreflectance method [121], 3ω method [122], thermal bridge method [123,124], and confocal micro-Raman method [125]. The latter two are two popular methods [126].

1.4.1 The Boltzmann equation solution for phonon transport

When the dimension of the object is in the scale of the submicron (10–100 nm), the Fourier law cannot correctly describe the heat conduction. Here the fundamental physics of phenomena should be considered, which is the physics of vibration in crystal lattices in the case of nonmetallic materials. In this context, phonons are the main energy carriers. They are acceptably similar to photons for justifying the definition of a phonon intensity and a transfer equation that is similar, considering particular assumptions, to the one utilized for thermal radiation in emitting–absorbing medium. Consequently, different computational approaches acquired from radiative transfer are applied to make nanoscale conduction predictions. Two of these very popular methods are P1 and discrete ordinate methods. The P1 method is more concise. In this method the intensity is not used directly as the variable. The intensity is a quantity that depends on the propagation direction. Alternatively an integrated quantity equivalent to the internal energy is used. With the application of this method an uncomplicated second-order partial differential equation is obtained, which is easily solved with standard numerical approaches. Nevertheless, this conciseness is achieved at the expense of the ballistic component of transfer (as a result of the bounding surfaces). Therefore the P1 method can be applied only when the media sizes are considerably larger than the phonon's MFP. On the contrary the discrete ordinate method completely considers the intensity-directional aspect. The transfer equation is solved in a limited number of directions, and the integrated quantities are calculated, including the flux or the internal energy, with quadrature formulas. The discrete ordinate method is known to be accurate and economical regarding computational time. However, a reassessment of general considerations is needed in each case. This method has gained a reputation as the most popular method within the framework of thermal radiation. These two methods are explained briefly in this section.

1.4.2 P1 method

Modeling phonon radiative transfer poses a significant challenge because of the directional nature of energy propagation. To address this, researchers have explored approximate methods aimed at efficiently handling directional aspects to reduce computational time. One example is the Rosseland model, which operates within a diffusive regime. Another approach involves using the P1 method, which allows for circumventing directional calculations while yielding relatively accurate results. Originally developed for stellar radiation modeling in astrophysics [127], the P1 method later found applications in neutronics, alongside Monte Carlo and discrete ordinate approaches. The method falls under the category of PN methods, named for the intensity expansion in terms of orthogonal functions truncated at order N. Through empirical studies, it has been found that satisfactory results can often be achieved using first-order expansion, as higher order calculations tend to substantially increase complexity without offering significant accuracy improvements. Modest [128] provides a comprehensive explanation of the P1 method in the context of thermal radiation. The P1 method provides several advantages, including its applicability to linear problems, reduced computational time, straightforward formulation, and the solution of a single equation. However, limitations such as the approximate handling of intensity and fluxes at bounding walls represent notable drawbacks associated with this method [129].

1.4.3 Discrete ordinate method

Numerical quadrature is used in the discrete ordinate method for calculating integrals along solid angles arising in calculating radiative fluxes and internal energy. The first step is discretizing the angle space in a limited number of directions and solving the radiative transfer equation to produce different intensity values at each point. The values of q''_ω and e_ω are calculated locally with the weighted summation of these intensities. This method was developed by Chandrasekhar [130]. However, Carlson and Lathrop [131] then largely developed this method for application to neutronics. It has been mainly developed by Fiveland [132] to apply discrete ordinates to thermal radiation. Moreover, Modest has described this method very well in his book [128]. This method has been used for the phonon transfer afterward [133–135].

Because of its efficiency the discrete ordinate method is primarily utilized for thermal radiation calculations. Unlike the Monte Carlo method, which offers high accuracy but requires significantly longer computational time, the discrete ordinate method provides a computationally fast and explicit solution. This method involves solving the transfer equation for radiation in discrete directions of the quadrature without inverting matrices. It efficiently accounts for directional aspects of radiation transfer and handles nonisotropic boundary conditions effectively. The precision achieved relative to computational time is generally excellent. However, there are limitations, such as challenges with nonrectangular meshes, arbitrary geometries, and essentially diffusive media. In addition, two specific numerical issues may impact the method's accuracy [129].

1.5 Phonon modes and spectra in two-dimensional materials

All physical phenomena in solids are affected by phonons [136]. Phonons limit the movements of electrons under ambient temperature. They also influence the optical characteristics of crystals. The primary heat carriers in semiconductors and insulators are acoustic phonons. The characterization of phonons is through their dispersion ω (q).

Here, q is the phonon wave vector, and ω is the angular frequency. For a crystal with N atoms in the primitive unit cell, 3N phonon dispersion branches exist for each phonon wave vector q. These include three acoustic branches, which describe collective vibrations of the unit cell, and 3N − 3 optical branches, which describe relative vibrations of atoms within the unit cell. In graphene, there are two atoms in the primitive unit cell; therefore, six phonon branches exist: longitudinal acoustic (LA), transverse acoustic (TA), flexural acoustic or out-of-plane acoustic (ZA), longitudinal optical (LO), transverse optical (TO), and out-of-plane optical (ZO). The branches of acoustic polarization are usually mentioned as LA and transverse acoustic (TA). z-Axis acoustic (ZA) phonons in graphene denote the out-of-plane transverse vibrations. An almost linear dispersion can be observed in the bulk crystal from acoustic phonons in the long-wavelength limit. This could be expressed as$\omega = V_s q$. Here,$V_s$is the speed of the sound. On the other hand the dispersion of optical phonons is less, and their group velocity $V_G = d\omega/dq$ is small. In graphene the primary heat carriers are acoustic phonons, whereas optical phonons can be utilized to determine the atomic planes via Raman spectroscopy in few-layer graphene. The extraordinary thermal conductivity of 2D materials can be interpreted by the unique phonon transport in these materials [137].

For the 2D materials with planar structures such as graphene and boron nitride, there is no coupling between the in-plane and out-of-plane displacements of atoms. Therefore the phonon modes can be classified into the in-plane modes, which consist of longitudinal optical (LO), LA, transverse optical (TO), TA, and also flexural out-of-plane modes comprising flexural optical (ZO) and flexural acoustic (ZA) modes. However, the structure of the plane makes the constants of third-order anharmonic force zero. The third-order derivative of the crystal energy, E, in the matter of atom motions involving the movement of i, j, and k atoms through α, β, and γ directions. It implies that, at the same time, there is no (anharmonic) phonon–phonon scattering resulting from the displacement of the three components in accordance with these zero constants. If an odd number of flexural modes consisting of optical (ZO) and acoustic (ZA) modes are included in the three-phonon scatterings, all the third force constants are zero.

Thus the three-phonon scattering processes are impossible (i.e., the scattering rate is zero). As a result the flexural acoustic phonon modes have greater relaxation times [112]. One significant cause of the great thermal conductivity observed in both boron nitride and single-layer graphene is this special choice regulation for planar symmetry structures [112, 138]. In 2D materials lacking planar crystals, such as 2H-phase transition metal dichalcogenides (TMDs), the absence of pure out-of-plane vibrations from flexural phonon modes disrupts the conventional symmetry selection rules. This disruption enables processes like three-photon scattering involving odd numbers of flexural phonons. Specifically, for triplane symmetric structures where metallic atoms in the middle layer are represented by indices i, j, and k, the quantity, $\psi_{i,j,k}^{\alpha\beta\gamma}$ is confirmed to be zero. This indicates that the relative movement of these metal atoms along the z-direction does not induce anharmonicity in such systems. Therefore the scattering ratios of these ZA phonon modes are extremely low because of the dominance of ZA phonons' atomic motion, which is through the z-direction [139]. Let's consider other 2D crystals, like 1T TMDs as well as buckled honeycomb crystals, in their crystals. No symmetry is observed, leading to greater phonon scattering in comparison with 2H TMDs and planar crystals [140, 141]. In addition to the crystal structure the atomic masses are another significant parameter that explains the thermal conductivity difference in the materials having an equivalent crystal structure. Metal atoms' mass for a small number of TMDs is far greater than the chalcogen atoms. The great frequency gap among the acoustic and optical phonon branches results from the great variation of the basis atoms' mass [140]. For instance the frequency gap of the acoustic-optical of WS_2 is as high as 110 cm^{-1}, which is near the acoustic phonon range for WS_2 (0–178 cm^{-1}), whereas 45 cm^{-1} is the gap for MoS_2, which is too lower than the range of acoustic phonon frequency (0–230 cm^{-1}). Because of the three-photon scattering energy conservation demand, as well as the great phonon frequency gap of WS_2, a significant phonon scattering channel (the process of annihilation of two acoustic phonon modes to one optical mode) is powerfully suppressed. Consequently the ratio of phonon–phonon scattering in WS_2 is weaker, leading to an extremely greater thermal conductivity in WS_2 compared to MoS_2 [142].

1.6 Thermal conductivity because of phonons in graphene

The exceptional nature of 2D phonon transport in graphene and relevant materials leads to unusual thermal transport. Investigating the phonons (lattice vibrational modes) of graphene is the first step for understanding its thermal properties, considering that phonon properties significantly influence thermal characteristics of graphene (thermal conductivity coefficient, thermal expansion coefficient, and specific heat). Moreover, the 2D structure of graphene determines phonon spectrum properties. Six phonon branches appear in graphene because of its hexagonal structure having two carbon atoms in every unit cell. The six phonon branches include three acoustics branches, which are LA, TA, and out-of-plane acoustic (ZA), and also three optical branches, namely, LO, TO, and out-of-plane optical (ZO). The displacement vector of ZA and ZO is along the z-axis. with the corrected version: "LO and LA are longitudinal in-plane phonon modes, in which atomic displacements are parallel to the wave propagation direction, whereas TO and TA are transverse in-plane phonon modes, in which atomic displacements are perpendicular to the wave propagation direction [143,144]. Transverse modes in common 3D solids possess two corresponding polarizations, whereas the extraordinary 2D structure of graphene enables out-of-plane displacements, which are recognized as flexural (Z) phonons. In graphene the main heat carriers at room temperature are acoustic, whereas optical phonons are necessary for understanding Raman spectra. Based on theoretical and experimental studies, it has been revealed that phonons' transport properties, such as scattering rates and energy dispersion in quasi-2D systems like graphene, have significant differences with those of basal planes in 3D bulk crystals or graphite.

The relation between frequency (ω) and phonon wave vector (q) is known as dispersion. At low q the frequencies of LA and TA have linear dispersions close to the Brillouin zone center [143,145]. At, $\omega_{LA} \approx v_{LA}q$ and $\omega_{TA} \approx v_{TA}q$, $v_{LA} \approx 21.3 km/s$ and $v_{TA} \approx 13.6 km/s$. On the other hand, there is an approximately quadratic dispersion for flexural ZA modes [8,9], $\omega_{ZA} = \alpha q^2$ where $\alpha \approx 6.2\times 10^{-7} m^2/s$.

It must be mentioned that the reason for many unique graphene thermal properties is the presence and modifications of these ZA modes.

BTE determines the thermal conductivity of phonon as [146,147]

$$\frac{\partial n_s(\vec{q})}{\partial t} + v.\frac{\partial n_s(\vec{q})}{\partial x} = \frac{\partial n_s(\vec{q})}{\partial t}|_{\text{coll}} \tag{1.32}$$

$n_s(\vec{q})$ is the phonon distribution function of polarization, s, and $\frac{\partial n_s(\vec{q})}{\partial t}|_{\text{coll}}$ is the collision integral.

In the phonon gas, quasitotal phonon momentum and the total number of phonons are not conserved. The conservation of the whole quasimomentum does not occur with the consideration of Umklapp processes, where for the reciprocal lattice vector's value, the general quasimomentum can differ [122,123]:

$$\sum_i \vec{q}_i = \sum_i \vec{q}_i' + \vec{b} \tag{1.33}$$

The total quasimomentum is conserved in normal N-processes [146,147]. To solve the Boltzmann equation the common method includes linearizing the collision integral around the phonon distribution function equilibrium. The assumptions for applying this approach are that the phonon distribution function is near equilibrium, small temperature gradients, and slow temperature changes determine the distribution function throughout the graphene sheet.

For the definition of heat flux the following expression is applied [122,123]:

$$\vec{W} = -k\nabla TV;\ W_\alpha = -\sum_\beta k_{\alpha\beta}\frac{\partial T}{\partial x_\beta}V \tag{1.34}$$

In the above equation, V is the graphene sheet volume.

The left-hand side of the Boltzmann equation for a stationary situation, $\frac{\partial n_s^0(\vec{q})}{\partial t} = 0$, turns into [122,123]:

$$\frac{\partial n_s(\vec{q})}{\partial x}.v_s(\vec{q}) = \frac{\partial n_s(\vec{q})}{\partial T}\frac{\partial T}{\partial x}.v_s(\vec{q}) = \frac{\partial n_s(\vec{q})}{\partial T}\nabla_X T.v_S(\vec{q}) \tag{1.35}$$

Linearizing the collision integral close to the equilibrium distribution [122,123]:

$$n_s(\vec{q}) = n_s^0(\vec{q}) + n_s^1(\vec{q}) \tag{1.36}$$

$n_s^0(\vec{q})$ is phonon equilibrium distribution function of polarization and $n_s^1(\vec{q})$ is the equivalent deviation from near-equilibrium, and also linearizing is applied in the relaxation times approximation [122,123]:

$$\frac{\partial n_s}{\partial t}|_{\text{coll}} = -\frac{\left[n_s(\vec{q}) - n_s^0(\vec{q})\right]}{\tau_s(\vec{q})} = -\frac{n_s^1(\vec{q})}{\tau_s(\vec{q})} \tag{1.37}$$

where $\tau_s(\vec{q})$ is the phonons of polarization relaxation times.

Combining the stationary form of Eq. (1.32) with Eqs. (1.35)–(1.37), the Boltzmann equation becomes [122,123]

$$-\frac{n_s^1(\vec{q})}{\tau_s(\vec{q})} = \frac{\partial n_s(\vec{q})}{\partial T}\nabla xT.v_s(\vec{q}) \tag{1.38}$$

The expression of heat flux is as follows [122,123]:

$$W_\alpha = \sum_\alpha \sum_\beta n_s^1(\vec{q})\hbar\omega_{s\beta}(\vec{q})v_{s\beta}(\vec{q})V \tag{1.39}$$

The distribution function of phonons is a near-equilibrium distribution function. As a result, $\frac{\partial n_s(\vec{q})}{\partial T} \approx \frac{\partial n_s^0(\vec{q})}{\partial T}$ can be applied, and Eq. (1.38) can be rewritten as [122,123]

$$n_s^1(\vec{q}) = -\tau_s(\vec{q})\frac{\partial n_s^0(\vec{q})}{\partial T}\nabla_x T.v_s(\vec{q}) \tag{1.40}$$

Based on Eqs. (1.39) and (1.40), heat flux could be defined as [122,123]

$$W_\alpha = -\sum_s \sum_\beta \tau_s(\vec{q})\frac{\partial n_s^0(\vec{q})}{\partial T}\nabla xT.v_{s\alpha}(\vec{q})\hbar\omega_{s\beta}(\vec{q})v_{s\beta}(\vec{q})V \tag{1.41}$$

$$W_\alpha = -\sum_\beta (\nabla xT)_\beta \sum_s \tau_s(\vec{q})\frac{\partial n_s^0(\vec{q})}{\partial T}.v_{s\alpha}(\vec{q})\hbar\omega_{s\beta}(\vec{q})v_{s\beta}(\vec{q})V \tag{1.42}$$

$$W_\alpha = -\kappa_{\alpha\beta}(\nabla T)_\beta L_x L_y L_z;\ \kappa_{\alpha\beta} = \kappa\delta_{\alpha\beta}. \tag{1.43}$$

The diagonal component of thermal conductivity tensor is defined as [122, 123]

$$\begin{aligned}\kappa_{\alpha\alpha} \equiv \kappa &= \frac{1}{L_x L_y L_z}\sum_s \sum_{\vec{q}} \tau_s(\vec{q})\frac{\partial n_s^0(\vec{q})}{\partial T}\hbar\omega_{s\alpha}(\vec{q})v_{s\alpha}^2(\vec{q})\cos^2\varphi \\ &= \frac{1}{L_x L_y L_z}\sum_s \sum_{\vec{q}} \tau_s(\vec{q})\frac{\partial n_s^0(\vec{q})}{\partial T}\hbar\omega_s(\vec{q})v_s^2(\vec{q})\cos^2\varphi\end{aligned} \tag{1.44}$$

where φ is the angle between the heat propagation direction and graphene direction (x-axis). Converting summation into an integration: $\sum_{\vec{q}} \rightarrow \int d^2\vec{q} \rightarrow \int g_{2D}(\omega)$[146]. The phonon density of state for the 2D case is determined as [122,123]

$$g_{2D} = \frac{L_x L_y}{2\pi}\frac{\omega_s(\vec{q})}{v_s(\vec{q})u_s(\vec{q})} \tag{1.45}$$

where $\vec{v}$ is group velocity, and $\vec{u}$ is phase velocity.

Finally, the thermal conductivity of graphene can be obtained by converting the summation in Eq. (1.44) into an integral using the two-dimensional phonon density of states in Eq. (1.45) and by averaging over the angle φ [122,123]:

$$\kappa = \frac{1}{4\pi L_z k_B T^2} \sum_S \int_{\omega_{\min}}^{\omega_{\max}} (\hbar\omega_s)^2 \frac{e^{\frac{\hbar\omega_S}{k_B T}}}{\left(e^{\frac{\hbar\omega_S}{k_B T}} - 1\right)^2} \tau(\omega_S) \frac{v}{u} \omega_S d\omega_S \tag{1.46}$$

The initial measurements of graphene's thermal conductivity were conducted in 2007 at UC Riverside using optothermal Raman techniques to study phonon transport [148–151]. These experiments utilized layers of suspended graphene. The results yielded a wide range of thermal conductivity values, with some measurements exceeding 2000 W/mK, comparable to bulk graphite's thermal conductivity at room temperature. In addition, it was observed that phonons significantly contribute to thermal conduction in ungated graphene compared to electrons near room temperature. The phonon MFP was estimated to be around 800 nm in graphene at room temperature [125].

The Raman optothermal technique has been employed in subsequent studies with various modifications, including incorporating a power meter beneath the graphene-suspended region. Notably, graphene suspended via chemical vapor deposition has achieved thermal conductivity values exceeding ~2500 W/mK at 350 K and approximately 1400 W/mK at 500 K [151]. The measured thermal conductivity of suspended graphene closely approaches its intrinsic thermal conductivity because of reduced scattering from substrate impurities and minimized thermal coupling with substrates. In contrast, thermal conductivity values for fully supported graphene have been reported to be lower. Based on the experimental study conducted by Seol et al. [152] the thermal conductivity of 600 W/mK has been measured for exfoliated graphene on SiO_2/Si support. They solved the BTE, compared the results with the experimental outputs, and concluded that free graphene should have a thermal conductivity of about 3000 W/mK near room temperature.

Although different amounts have been reported for the graphene thermal conductivity, it can be concluded that the amount is very high compared to important materials for electronic applications such as bulk copper ($K = 400$ W/mK at RT) or bulk silicon ($K = 145$ W/mK at RT). Some factors that can be considered the cause of the differences in graphene thermal conductivity are the thickness nonuniformity, changes in lateral sizes of the graphene sample, the quality of material, strain distributions, and also the orientation and size of the grain. In previous research studies, the effects of different parameters on graphene thermal conductivity have been investigated, such as edge configurations [115,153], vacancy [154,155], strain [156,157], surface functionalization [92,158], isotopic doping [159], substrate effect [117,160], and size effect [111,161,162].

1.7 Challenges and future prospects

Considering their extraordinary properties, 2D materials have exhibited great potential for heat transfer-related applications. Right now, there is a deeper understanding of the underlying physical mechanism of thermal transport in 2D materials compared to when graphene, the first 2D material, was discovered. However, several challenges and questions should be addressed for further development of these materials and the expansion of their applications. For instance, there are difficulties concerning the experimental measurement of the thermal contact resistance and detecting the temperature distributions at the nano- and microscales with high sensitivity. Some problems have been solved by applying novel experimental techniques [163,164]. Another challenge is the physical fundamentals of novel thermal conduction, such as phonon–phonon scattering in 2D sheets with micro-/nanoscale electron–phonon coupling on the contact and mode contributions [126]. The suspension of clean 2D materials can be another challenge because the intrinsic thermal conduction of phonons is obscured in the presence of residues on the surfaces. Several studies have been performed to decrease the contamination of graphene surfaces, including applying surface treatments with O_2 or ozone plasma and elevated-temperature vacuum annealing. Moreover, developing cost-efficient approaches for providing 2D material-based products is essential. To achieve a widespread market for 2D materials the most critical measures that should be taken are providing reliable and repeatable fabrication approaches; accurate characterization methods based on 2D materials; large production; and achieving low contact resistance [17].

1.8 Conclusion

Addressing the international energy crisis through more effective heat transfer solutions is crucial as global energy demand increases along with advancing technology. Materials with enhanced thermal properties offer promising direct and indirect energy savings opportunities. 2D materials, in particular, have opened new avenues for research because of their unique properties. Their thermal characteristics make 2D materials, notably graphene, well-suited for applications in bioenergy, solar power, electronics like sensors, and portable devices like batteries. This chapter provides a comprehensive discussion of 2D material thermal transport fundamentals, from the underlying phonon dynamics to various experimental and theoretical techniques used to measure thermal conductivity. A review of graphene applications in heat transfer contexts is also presented. Graphene and other 2D nanomaterials are poised to play an important future role as flexible, portable thermal engineering solutions. Their potential extends across sectors, including energy production, conversion, and storage. Understanding 2D material thermal transport mechanisms at microscopic and macroscopic scales remains important as these materials are developed for practical heat transfer applications. The continuing exploration of 2D material thermal properties aims to advance progress on effective energy solutions, from molecular phonon interactions to full system-level heat transfer behaviors and energy technologies.

References

[1] K.S. Novoselov, A.K. Geim, S.V. Morozov, D. Jiang, Y. Zhang, S.V. Dubonos, I.V. Grigorieva, A.A. Firsov, Electric field effect in atomically thin carbon films, Science 306 (5696) (2004) 666–669.

[2] N. Mounet, M. Gibertini, P. Schwaller, D. Campi, A. Merkys, A. Marrazzo, T. Sohier, I.E. Castelli, A. Cepellotti, G. Pizzi, Two-dimensional materials from high-throughput computational exfoliation of experimentally known compounds, Nat. Nanotechnol. 13 (3) (2018) 246–252.

[3] M. Qiu, W.X. Ren, T. Jeong, M. Won, G.Y. Park, D.K. Sang, L.-P. Liu, H. Zhang, J.S. Kim, Omnipotent phosphorene: a next-generation, two-dimensional nanoplatform for multidisciplinary biomedical applications, Chem. Soc. Rev. 47 (15) (2018) 5588–5601.

[4] W. Tao, N. Kong, X. Ji, Y. Zhang, A. Sharma, J. Ouyang, B. Qi, J. Wang, N. Xie, C. Kang, Emerging two-dimensional monoelemental materials (Xenes) for biomedical applications, Chem. Soc. Rev. 48 (11) (2019) 2891–2912.

[5] H.P. Boehm, R. Setton, E. Stumpp, Nomenclature and terminology of graphite intercalation compounds, in: Carbon, 24, 2nd, Pergamon, 1986, pp. 241–245. https://doi.org/10.1016/0008-6223(86)90126-0.
[6] P.R. Wallace, The band theory of graphite, Phys. Rev. 71 (9) (1947) 622.
[7] B.C. Brodie, XIII. On the atomic weight of graphite, Phil. Trans. R. Soc. Lond. 149 (1859) 249–259.
[8] G. Ruess, F. Vogt, Höchstlamellarer Kohlenstoff aus Graphitoxyhydroxyd, Monatshefte für Chemie und verwandte Teile anderer Wissenschaften 78 (3–4) (1948) 222–242.
[9] H.-P. Boehm, A. Clauss, G.O. Fischer, U. Hofmann, Das adsorptionsverhalten sehr dünner kohlenstoff-folien, Zeitschrift für anorganische und allgemeine Chemie 316 (3–4) (1962) 119–127.
[10] J.C. Meyer, A.K. Geim, M.I. Katsnelson, K.S. Novoselov, D. Obergfell, S. Roth, C. Girit, A. Zettl, On the roughness of single-and bi-layer graphene membranes, Solid State Commun. 143 (1–2) (2007) 101–109.
[11] Y. Zhang, Y.-W. Tan, H.L. Stormer, P. Kim, Experimental observation of the quantum Hall effect and Berry's phase in graphene, Nature 438 (7065) (2005) 201–204.
[12] K.S. Novoselov, A.K. Geim, S.V. Morozov, D. Jiang, M.I. Katsnelson, I. Grigorieva, S. Dubonos, A.A. Firsov, Two-dimensional gas of massless Dirac fermions in graphene, Nature 438 (7065) (2005) 197–200.
[13] A.K. Geim, K.S. Novoselov, The rise of graphene, Nature Materials 6 (3) (2010) 183–189, doi:10.1038/nmat1849.
[14] A.H.C. Neto, F. Guinea, N.M.R. Peres, K.S. Novoselov, A.K. Geim, The electronic properties of graphene, Rev. Mod. Phys. 81 (1) (2009) 109.
[15] M. Xu, T. Liang, M. Shi, H. Chen, Graphene-like two-dimensional materials, Chem. Rev. 113 (5) (2013) 3766–3798.
[16] S.Z. Butler, S.M. Hollen, L. Cao, Y. Cui, J.A. Gupta, H.R. Gutiérrez, T.F. Heinz, S.S. Hong, J. Huang, A.F. Ismach, Progress, challenges, and opportunities in two-dimensional materials beyond graphene, ACS Nano 7 (4) (2013) 2898–2926.
[17] K. Khan, A.K. Tareen, M. Aslam, R. Wang, Y. Zhang, A. Mahmood, Z. Ouyang, H. Zhang, Z. Guo, Recent developments in emerging two-dimensional materials and their applications, J. Mater. Chem. C 8 (2) (2020) 387–440.
[18] F. Torrisi, J.N. Coleman, Electrifying inks with 2D materials, Nat. Nanotechnol. 9 (10) (2014) 738–739.
[19] L. Liu, J. Park, D.A. Siegel, K.F. McCarty, K.W. Clark, W. Deng, L. Basile, J.C. Idrobo, A.-P. Li, G. Gu, Heteroepitaxial growth of two-dimensional hexagonal boron nitride templated by graphene edges, Science 343 (6167) (2014) 163–167.
[20] C.C. Coleman, H. Goldwhite, W. Tikkanen, A review of intercalation in heavy metal iodides, Chem. Mater. 10 (10) (1998) 2794–2800.
[21] C.E. Boott, A. Nazemi, I. Manners, Synthetic covalent and non-covalent 2D materials, Angew. Chem. Int. Ed. 54 (47) (2015) 13876–13894.
[22] S. Ida, C. Ogata, M. Eguchi, W.J. Youngblood, T.E. Mallouk, Y. Matsumoto, Photoluminescence of perovskite nanosheets prepared by exfoliation of layered oxides, K2Ln2Ti3O10, KLnNb2O7, and RbLnTa2O7 (Ln: lanthanide ion), J. Am. Chem. Soc. 130 (22) (2008) 7052–7059.
[23] Z. Sun, T. Liao, Y. Dou, S.M. Hwang, M.-S. Park, L. Jiang, J.H. Kim, S.X. Dou, Generalized self-assembly of scalable two-dimensional transition metal oxide nanosheets, Nat. Commun. 5 (1) (2014) 1–9.
[24] X. Bao, Q. Ou, Z.Q. Xu, Y. Zhang, Q. Bao, H. Zhang, Band structure engineering in 2D materials for optoelectronic applications, Adv. Mater. Technol. 3 (11) (2018) 1800072.
[25] K. Khan, A.K. Tareen, M. Aslam, K.H. Thebo, U. Khan, R. Wang, S.S. Shams, Z. Han, Z. Ouyang, A comprehensive review on synthesis of pristine and doped inorganic room temperature stable mayenite electride,[Ca24Al28O64] 4+(e−) 4 and its applications as a catalyst, Prog. Solid State Chem. 54 (2019) 1–19.

[26] X. Shen, Z. Wang, Y. Wu, X. Liu, J.-K. Kim, Effect of functionalization on thermal conductivities of graphene/epoxy composites, Carbon 108 (2016) 412–422.

[27] L. Li, Y. Yu, G.J. Ye, Q. Ge, X. Ou, H. Wu, D. Feng, X.H. Chen, Y. Zhang, Black phosphorus field-effect transistors, Nat. Nanotechnol. 9 (5) (2014) 372.

[28] Q.H. Wang, K. Kalantar-Zadeh, A. Kis, J.N. Coleman, M.S. Strano, Electronics and optoelectronics of two-dimensional transition metal dichalcogenides, Nat. Nanotechnol. 7 (11) (2012) 699–712.

[29] M. Chhowalla, H.S. Shin, G. Eda, L.-J. Li, K.P. Loh, H. Zhang, The chemistry of two-dimensional layered transition metal dichalcogenide nanosheets, Nat. Chem. 5 (4) (2013) 263–275.

[30] K. Khan, A.K. Tareen, M. Aslam, Q. Khan, R. Wang, S.A. Khan, A.S. Saleemi, Q.U. Khan, Y. Zhang, Z. Ouyang, Novel two dimensional carbon-chromium nitride based composite as an electrocatalyst for oxygen reduction reaction, Front. Chem. 7 (2019) 738.

[31] Q. Wu, Y.J. Song, The environmental stability of large-size and single-crystalline antimony flakes grown by chemical vapor deposition on SiO_2 substrates, Chem. Commun. 54 (69) (2018) 9671–9674.

[32] G. Wang, R. Pandey, S.P. Karna, Atomically thin group V elemental films: theoretical investigations of antimonene allotropes, ACS Appl. Mater. Interfaces 7 (21) (2015) 11490–11496.

[33] A. Nourbakhsh, L. Yu, Y. Lin, M. Hempel, R.-J. Shiue, D. Englund, T. Palacios, Heterogeneous integration of 2D materials and devices on a Si platform, in: R.O. Topaloglu, H.-S.P. Wong (Eds.), Beyond-CMOS Technologies for Next Generation Computer Design, Springer, Cham, Switzerland, 2019, pp. 43–84.

[34] G.P. Srivastava, The Physics of Phonons, Adam Hilger, Bristol, UK, 1990.

[35] A.F. Mills, Heat and Mass Transfer, Richard D Irwin, Boston, 1995.

[36] S.G. Volz, Thermal insulating behavior in crystals at high frequencies, Phys. Rev. Lett. 87 (7) (2001) 074301.

[37] S. Volz, R. Carminati, P. Chantrenne, S. Dilhaire, S. Gomez, N. Trannoy, G. Tessier, Microscale and Nanoscale Heat Transfer, Topics in Applied Physics, 107, Springer, Berlin/Heidelberg, Germany, 2007.

[38] J.-J. Greffet, M. Nieto-Vesperinas, Field theory for generalized bidirectional reflectivity: derivation of Helmholtz's reciprocity principle and Kirchhoff's law, JOSA A 15 (10) (1998) 2735–2744.

[39] J. Darabi, Micro-and nanoscale heat transfer: challenges and opportunities, Heat Transfer Eng. 23 (2) (2002) 1–2.

[40] D.Y. Tzou, Macro-to Microscale Heat Transfer: The Lagging Behavior, 2nd, John Wiley & Sons Ltd., Chichester, UK, 2014.

[41] Z. Jiang, Y. Shi, Z.-J. Jiang, X. Tian, L. Luo, W. Chen, High performance of a free-standing sulfonic acid functionalized holey graphene oxide paper as a proton conducting polymer electrolyte for air-breathing direct methanol fuel cells, J. Mater. Chem. A 2 (18) (2014) 6494–6503.

[42] M.L. Gethers, J.C. Thomas, S. Jiang, N.O. Weiss, X. Duan, W.A. Goddard Iii, P.S. Weiss, Holey graphene as a weed barrier for molecules, ACS Nano 9 (11) (2015) 10909–10915.

[43] Y.-B. Luo, B.-F. Yuan, Q.-W. Yu, Y.-Q. Feng, Substrateless graphene fiber: a sorbent for solid-phase microextraction, J. Chromatogr. A 1268 (2012) 9–15.

[44] D. Yu, L. Wei, W. Jiang, H. Wang, B. Sun, Q. Zhang, K. Goh, R. Si, Y. Chen, Nitrogen doped holey graphene as an efficient metal-free multifunctional electrochemical catalyst for hydrazine oxidation and oxygen reduction, Nanoscale 5 (8) (2013) 3457–3464.

[45] F. Meng, R. Li, Q. Li, W. Lu, T.-W. Chou, Synthesis and failure behavior of super-aligned carbon nanotube film wrapped graphene fibers, Carbon 72 (2014) 250–256.

[46] S.-c. Yan, B.-c. Zheng, J.-h. Chen, F. Xu, Y.-q. Lu, Optical electrical current sensor utilizing a graphene-microfiber-integrated coil resonator, Appl. Phys. Lett. 107 (5) (2015) 053502.

[47] X.H. Li, M. Antonietti, Polycondensation of boron-and nitrogen-codoped holey graphene monoliths from molecules: carbocatalysts for selective oxidation, Angew. Chem. 125 (17) (2013) 4670–4674.

[48] A. Kumar, A.L.M. Reddy, A. Mukherjee, M. Dubey, X. Zhan, N. Singh, L. Ci, W.E. Billups, J. Nagurny, G. Mital, Direct synthesis of lithium-intercalated graphene for electrochemical energy storage application, ACS nano 5 (6) (2011) 4345–4349.

[49] X. Huang, B. Sun, D. Su, D. Zhao, G. Wang, Soft-template synthesis of 3D porous graphene foams with tunable architectures for lithium–O_2 batteries and oil adsorption applications, J. Mater. Chem. A 2 (21) (2014) 7973–7979.

[50] H. Chen, M.B. Müller, K.J. Gilmore, G.G. Wallace, D. Li, Mechanically strong, electrically conductive, and biocompatible graphene paper, Adv. Mater. 20 (18) (2008) 3557–3561.

[51] B. Bera, D. Mandal, M.D. Sarkar, Sensor made of PVDF/graphene electrospinning fiber and comparison between electrospinning PVDF fiber and PVDF/graphene fiber, Imperial J. Interdiscip. Res. 2 (5) (2016) 1411–1413p.

[52] D.A.C. Brownson, D.K. Kampouris, C.E. Banks, An overview of graphene in energy production and storage applications, J. Power Sources 196 (11) (2011) 4873–4885.

[53] E. Sadeghinezhad, M. Mehrali, R. Saidur, M. Mehrali, S.T. Latibari, A.R. Akhiani, H.S.C. Metselaar, A comprehensive review on graphene nanofluids: recent research, development and applications, Energy Convers. Manage. 111 (2016) 466–487.

[54] W. Yu, H. Xie, W. Chen, Experimental investigation on thermal conductivity of nanofluids containing graphene oxide nanosheets, J. Appl. Phys. 107 (9) (2010) 094317.

[55] H. Yarmand, S. Gharehkhani, G. Ahmadi, S.F.S. Shirazi, S. Baradaran, E. Montazer, M.N.M. Zubir, M.S. Alehashem, S.N. Kazi, M. Dahari, Graphene nanoplatelets–silver hybrid nanofluids for enhanced heat transfer, Energy Convers. Manage. 100 (2015) 419–428.

[56] M. Kole, T.K. Dey, Investigation of thermal conductivity, viscosity, and electrical conductivity of graphene based nanofluids, J. Appl. Phys. 113 (8) (2013) 084307.

[57] A. Ghozatloo, M. Shariaty-Niasar, A.M. Rashidi, Preparation of nanofluids from functionalized graphene by new alkaline method and study on the thermal conductivity and stability, Int. Commun. Heat Mass Transfer 42 (2013) 89–94.

[58] K. Zheng, L. Wang, S. Bai, J. Yu, Z. Tang, Z. Huang, An anomalous wave-like kinetic energy transport in graphene nanoribbons at high heat flux, Physica B 434 (2014) 64–68.

[59] S. Ramakrishnan, X. Wang, J. Sanjayan, J. Wilson, Heat transfer performance enhancement of paraffin/expanded perlite phase change composites with graphene nano-platelets, Energy Procedia 105 (2017) 4866–4871.

[60] M.R. Esfahani, E.M. Languri, M.R. Nunna, Effect of particle size and viscosity on thermal conductivity enhancement of graphene oxide nanofluid, Int. Commun. Heat Mass Transfer 76 (2016) 308–315.

[61] T.T. Baby, R. Sundara, Synthesis and transport properties of metal oxide decorated graphene dispersed nanofluids, J. Phys. Chem. C 115 (17) (2011) 8527–8533.

[62] C. Hermida-Merino, M. Perez-Rodriguez, A.B. Pereiro, M.M. Piñeiro, M.J. Pastoriza-Gallego, Tailoring nanofluid thermophysical profile through graphene nanoplatelets surface functionalization, ACS Omega 3 (1) (2018) 744–752.

[63] A. Naddaf, S.Z. Heris, Experimental study on thermal conductivity and electrical conductivity of diesel oil-based nanofluids of graphene nanoplatelets and carbon nanotubes, Int. Commun. Heat Mass Transfer 95 (2018) 116–122.

[64] E. Sadeghinezhad, H. Togun, M. Mehrali, P.S. Nejad, S.T. Latibari, T. Abdulrazzaq, S.N. Kazi, H.S.C. Metselaar, An experimental and numerical investigation of heat transfer enhancement for graphene nanoplatelets nanofluids in turbulent flow conditions, Int. J. Heat Mass Transfer 81 (2015) 41–51.

[65] S.S.J. Aravind, S. Ramaprabhu, Graphene–multiwalled carbon nanotube-based nanofluids for improved heat dissipation, RSC Adv. 3 (13) (2013) 4199–4206.

[66] E. Jiaqiang, X. Zhao, H. Liu, J. Chen, W. Zuo, Q. Peng, Field synergy analysis for enhancing heat transfer capability of a novel narrow-tube closed oscillating heat pipe, Appl. Energy 175 (2016) 218–228.

[67] E. Jiaqiang, X. Zhao, Y. Deng, H. Zhu, Pressure distribution and flow characteristics of closed oscillating heat pipe during the starting process at different vacuum degrees, Appl. Therm. Eng. 93 (2016) 166–173.

[68] X. Zhao, E. Jiaqiang, G. Wu, Y. Deng, D. Han, B. Zhang, Z. Zhang, A review of studies using graphenes in energy conversion, energy storage and heat transfer development, Energy Convers. Manage. 184 (2019) 581–599.

[69] L. Chen, Y. Zhang, Q. Wu, Effect of graphene coating on the heat transfer performance of a composite anti-/deicing component, Coatings 7 (10) (2017) 158.

[70] S.K. Georgantzinos, G.I. Giannopoulos, A. Fatsis, N.V. Vlachakis, Analytical expressions for electrostatics of graphene structures, Physica E 84 (2016) 27–36.

[71] B. Zhao, B. Guizal, Z.M. Zhang, S. Fan, M. Antezza, Near-field heat transfer between graphene/hBN multilayers, Phys. Rev. B 95 (24) (2017) 245437.

[72] A. Jaikumar, S.G. Kandlikar, A. Gupta, Pool boiling enhancement through graphene and graphene oxide coatings, Heat Transfer Eng. 38 (14–15) (2017) 1274–1284.

[73] A. Jaikumar, A. Gupta, S.G. Kandlikar, C.-Y. Yang, C.-Y. Su, Scale effects of graphene and graphene oxide coatings on pool boiling enhancement mechanisms, Int. J. Heat Mass Transfer 109 (2017) 357–366.

[74] T. Kim, J.M. Kim, J.H. Kim, S.C. Park, H.S. Ahn, Orientation effects on bubble dynamics and nucleate pool boiling heat transfer of graphene-modified surface, Int. J. Heat Mass Transfer 108 (2017) 1393–1405.

[75] J.M. Kim, T. Kim, J. Kim, M.H. Kim, H.S. Ahn, Effect of a graphene oxide coating layer on critical heat flux enhancement under pool boiling, Int. J. Heat Mass Transfer 77 (2014) 919–927.

[76] J.M. Kim, S.C. Park, B. Kong, H.S. Ahn, Effect of porous graphene networks and micropillar arrays on boiling heat transfer performance, Exp. Therm. Fluid Sci. 93 (2018) 153–164.

[77] Q. Zhao, T. Zhou, T. Wang, W. Liu, J. Liu, T. Yu, Q. Liao, N. Liu, Active control of near-field radiative heat transfer between graphene-covered metamaterials, J. Phys. D Appl. Phys. 50 (14) (2017) 145101.

[78] X.L. Liu, Z.M. Zhang, Graphene-assisted near-field radiative heat transfer between corrugated polar materials, Appl. Phys. Lett. 104 (25) (2014) 251911.

[79] D. Drosdoff, A.D. Phan, L.M. Woods, Transverse electric mode for near-field radiative heat transfer in graphene–metamaterial systems, Adv. Opt. Mater. 2 (11) (2014) 1038–1042.

[80] K.F. Mak, M.Y. Sfeir, Y. Wu, C.H. Lui, J.A. Misewich, T.F. Heinz, Measurement of the optical conductivity of graphene, Phys. Rev. Lett. 101 (19) (2008) 196405.

[81] J. Yang, X. Li, S. Han, Y. Zhang, P. Min, N. Koratkar, Z.-Z. Yu, Air-dried, high-density graphene hybrid aerogels for phase change composites with exceptional thermal conductivity and shape stability, J. Mater. Chem. A 4 (46) (2016) 18067–18074.

[82] B. Mortazavi, O. Rahaman, T. Rabczuk, L.F.C. Pereira, Thermal conductivity and mechanical properties of nitrogenated holey graphene, Carbon 106 (2016) 1–8.

[83] Z.-S. Wu, G. Zhou, L.-C. Yin, W. Ren, F. Li, H.-M. Cheng, Graphene/metal oxide composite electrode materials for energy storage, Nano Energy 1 (1) (2012) 107–131.

[84] M. Zhou, T. Lin, F. Huang, Y. Zhong, Z. Wang, Y. Tang, H. Bi, D. Wan, J. Lin, Highly conductive porous graphene/ceramic composites for heat transfer and thermal energy storage, Adv. Funct. Mater. 23 (18) (2013) 2263–2269.

[85] G. Zhang, X. Guo, S. Wang, X. Wang, Y. Zhou, H. Xu, New graphene fiber coating for volatile organic compounds analysis, J. Chromatogr. B 969 (2014) 128–131.

[86] P. Yu, S. Yu, W. Zhou, Evaluation of thermal performance of graphene overcoat on multi-layered structure subject to laser heating, Int. Commun. Heat Mass Transfer 68 (2015) 27–31.

[87] A. Tewari, S. Böhm, S. Gandla, C.R. McNeill, D. Gupta, Graphene-MWNTs composite coatings with enhanced electrical conductivity, FlatChem 4 (2017) 33–41.

[88] T. Cui, Q. Li, Y. Xuan, P. Zhang, Preparation and thermal properties of the graphene–polyolefin adhesive composites: application in thermal interface materials, Microelectron. Reliab. 55 (12) (2015) 2569–2574.

[89] L. Zhao, H. Liu, X. Chen, S. Chu, H. Liu, Z. Lin, Q. Li, G. Chu, H. Zhang, Liquid metal nano/micro-channels as thermal interface materials for efficient energy saving, J. Mater. Chem. C 6 (39) (2018) 10611–10617.

[90] K.M. Razeeb, E. Dalton, G.L.W. Cross, A.J. Robinson, Present and future thermal interface materials for electronic devices, Int. Mater. Rev. 63 (1) (2018) 1–21.
[91] J. Gu, C. Liang, X. Zhao, B. Gan, H. Qiu, Y. Guo, X. Yang, Q. Zhang, D.-Y. Wang, Highly thermally conductive flame-retardant epoxy nanocomposites with reduced ignitability and excellent electrical conductivities, Compos. Sci. Technol. 139 (2017) 83–89.
[92] Y. Wang, H.F. Zhan, Y. Xiang, C. Yang, C.M. Wang, Y.Y. Zhang, Effect of covalent functionalization on thermal transport across graphene–polymer interfaces, J. Phys. Chem. C 119 (22) (2015) 12731–12738.
[93] W. Yu, H. Xie, L. Chen, Graphene nanoplatelets/nylon 6 composites with high thermal conductivity, J. Eng. Thermophys. 34 (9) (2013) 1749–1751.
[94] B. Yin, J. Wang, H. Jia, J. He, X. Zhang, Z. Xu, Enhanced mechanical properties and thermal conductivity of styrene–butadiene rubber reinforced with polyvinylpyrrolidone-modified graphene oxide, J. Mater. Sci. 51 (12) (2016) 5724–5737.
[95] T.D. Dao, H.M. Jeong, A Pickering emulsion route to a stearic acid/graphene core–shell composite phase change material, Carbon 99 (2016) 49–57.
[96] C. Liu, M. Chen, W. Yu, Y. He, Recent advance on graphene in heat transfer enhancement of composites, ES Energy Environ. 2 (2018) 31–42.
[97] J.M. Ziman, Principles of the Theory of Solids, 2nd, Cambridge University Press, Cambridge, UK, 1972.
[98] P.G. Klemens, Theory of the a-plane thermal conductivity of graphite, J. Wide Bandgap Mater. 7 (4) (2000) 332–339.
[99] P.G. Klemens, Thermal conductivity and lattice vibrational modes, Solid state phys. 7 (1958) 1–98.
[100] J. Callaway, Model for lattice thermal conductivity at low temperatures, Phys. Rev. 113 (4) (1959) 1046.
[101] J.E. Parrott, A.D. Stuckes, Thermal Conductivity of Solids, Methuen, New York, 1975.
[102] J.M. Ziman, Electrons and Phonons: The Theory of Transport Phenomena in Solids, Oxford University Press, Oxford, UK, 2001.
[103] A. Balandin, K.L. Wang, Significant decrease of the lattice thermal conductivity due to phonon confinement in a free-standing semiconductor quantum well, Phys. Rev. B 58 (3) (1998) 1544.
[104] J. Zou, A. Balandin, Phonon heat conduction in a semiconductor nanowire, J. Appl. Phys. 89 (5) (2001) 2932–2938.
[105] A.A. Balandin, E.P. Pokatilov, D.L. Nika, Phonon engineering in hetero-and nanostructures, J. Nanoelectron. Optoelectron. 2 (2) (2007) 140–170.
[106] E.P. Pokatilov, D.L. Nika, A.A. Balandin, Acoustic-phonon propagation in rectangular semiconductor nanowires with elastically dissimilar barriers, Phys. Rev. B 72 (11) (2005) 113311.
[107] D.L. Nika, N.D. Zincenco, E.P. Pokatilov, Engineering of thermal fluxes in phonon mismatched heterostructures, J. Nanoelectron. Optoelectron. 4 (1) (2009) 180–185.
[108] A.A. Balandin, Nanophononics: phonon engineering in nanostructures and nanodevices, J. Nanosci. Nanotechnol. 5 (7) (2005) 1015–1022.
[109] Z. Aksamija, I. Knezevic, Anisotropy and boundary scattering in the lattice thermal conductivity of silicon nanomembranes, Phys. Rev. B 82 (4) (2010) 045319.
[110] S.B. Soffer, Statistical model for the size effect in electrical conduction, J. Appl. Phys. 38 (4) (1967) 1710–1715.
[111] D.L. Nika, A.A. Balandin, Two-dimensional phonon transport in graphene, J. Phys. Condens. Matter 24 (23) (2012) 233203.
[112] L. Lindsay, D.A. Broido, N. Mingo, Flexural phonons and thermal transport in graphene, Phys. Rev. B 82 (11) (2010) 115427.
[113] D. Singh, J.Y. Murthy, T.S. Fisher, Mechanism of thermal conductivity reduction in few-layer graphene, J. Appl. Phys. 110 (4) (2011) 044317.
[114] Y. Cai, J. Lan, G. Zhang, Y.-W. Zhang, Lattice vibrational modes and phonon thermal conductivity of monolayer MoS_2, Phys. Rev. B 89 (3) (2014) 035438.

[115] Z.-X. Xie, K.-Q. Chen, W. Duan, Thermal transport by phonons in zigzag graphene nanoribbons with structural defects, J. Phys. Condens. Matter 23 (31) (2011) 315302.
[116] J. Chen, J.H. Walther, P. Koumoutsakos, Strain engineering of Kapitza resistance in few-layer graphene, Nano Lett. 14 (2) (2014) 819–825.
[117] Z.-Y. Ong, E. Pop, Effect of substrate modes on thermal transport in supported graphene, Phys. Rev. B 84 (7) (2011) 075471.
[118] M. Bozlar, D. He, J. Bai, Y. Chalopin, N. Mingo, S. Volz, Carbon nanotube microarchitectures for enhanced thermal conduction at ultralow mass fraction in polymer composites, Adv. Mater. 22 (14) (2010) 1654–1658.
[119] A. Majumdar, Scanning thermal microscopy, Annu. Rev. Mater. Sci. 29 (1) (1999) 505–585.
[120] C.C. Williams, H.K. Wickramasinghe, Scanning thermal profiler, Microelectron. Eng. 5 (1–4) (1986) 509–513.
[121] D.G. Cahill, Analysis of heat flow in layered structures for time-domain thermoreflectance, Rev. Sci. Instrum. 75 (12) (2004) 5119–5122.
[122] D.G. Cahill, Thermal conductivity measurement from 30 to 750 K: the 3ω method, Rev. Sci. Instrum. 61 (2) (1990) 802–808.
[123] M.C. Wingert, Z.C.Y. Chen, S. Kwon, J. Xiang, R. Chen, Ultra-sensitive thermal conductance measurement of one-dimensional nanostructures enhanced by differential bridge, Rev. Sci. Instrum. 83 (2) (2012) 024901.
[124] J. Zheng, M.C. Wingert, E. Dechaumphai, R. Chen, Sub-picowatt/kelvin resistive thermometry for probing nanoscale thermal transport, Rev. Sci. Instrum. 84 (11) (2013) 114901.
[125] A.A. Balandin, S. Ghosh, D. Teweldebrhan, I. Calizo, W. Bao, F. Miao, C.N. Lau, Extremely high thermal conductivity of graphene: Prospects for thermal management applications in silicon nanoelectronics. https://doi.org/10.1109/SNW.2008.5418404.
[126] X. Xu, J. Chen, B. Li, Phonon thermal conduction in novel 2D materials, J. Phys. Condens. Matter 28 (48) (2016) 483001.
[127] J.H. Jeans, Stars, gaseous, radiative transfer of energy, Mon. Not. R. Astron. Soc. 78 (1917) 28–36.
[128] M.F. Modest, Radiative Heat Transfer, 3rd, Academic Press, New York, NY, USA, 2013.
[129] D. Lemonnier, Solution of the Boltzmann equation for phonon transport, in: S. Volz (Ed.), Microscale and Nanoscale Heat Transfer, 107, Springer Nature, Berlin/Heidelberg, Germany, 2007, pp. 77–106. https://doi.org/10.1007/11767862_5.
[130] S. Chandrasekar, Radiative Transfer, Dover Publications, New York, 1960.
[131] B.G. Carlson, K.D. Lathrop, Transport theory—the method of discrete ordinates, in: H. Greenspan, C.N. Kelber, D. Okrent (Eds.), Computing Methods in Reactor Physics, Gordon and Breach, New York, 1968.
[132] W.A. Fiveland, Three-dimensional radiative heat-transfer solutions by the discrete-ordinates method, J. Thermophys. Heat Transfer 2 (4) (1988) 309–316.
[133] J.D. Chung, M. Kaviany, Effects of phonon pore scattering and pore randomness on effective conductivity of porous silicon, Int. J. Heat Mass Transfer 43 (4) (2000) 521–538.
[134] S. Volz, D. Lemonnier, J.-B. Saulnier, Clamped nanowire thermal conductivity based on phonon transport equation, Microscale Thermophys. Eng. 5 (3) (2001) 191–207.
[135] R. Koch, R. Becker, Evaluation of quadrature schemes for the discrete ordinates method, J. Quant. Spectrosc. Radiat. Transfer 84 (4) (2004) 423–435.
[136] M.A. Stroscio, M. Dutta, Phonons in Nanostructures, Cambridge University Press, Cambridge, UK, 2001.
[137] A.A. Balandin, D.L. Nika, Phononics in low-dimensional materials, Mater. Today 15 (6) (2012) 266–275.
[138] L. Lindsay, D.A. Broido, Theory of thermal transport in multilayer hexagonal boron nitride and nanotubes, Phys. Rev. B 85 (3) (2012) 035436.
[139] X. Gu, R. Yang, Phonon transport in single-layer $Mo_{1-x}W_xS_2$ alloy embedded with WS_2 nanodomains, Phys. Rev. B 94 (7) (2016) 075308.

[140] X. Gu, R. Yang, Phonon transport in single-layer transition metal dichalcogenides: a first-principles study, Appl. Phys. Lett. 105 (13) (2014) 131903.

[141] X. Gu, R. Yang, First-principles prediction of phononic thermal conductivity of silicene: a comparison with graphene, J. Appl. Phys. 117 (2) (2015) 025102.

[142] X. Gu, Y. Wei, X. Yin, B. Li, R. Yang, Colloquium: phononic thermal properties of two-dimensional materials, Rev. Mod. Phys. 90 (2018) 041002.

[143] D.L. Nika, E.P. Pokatilov, A.S. Askerov, A.A. Balandin, Phonon thermal conduction in graphene: role of Umklapp and edge roughness scattering, Phys. Rev. B 79 (15) (2009) 155413.

[144] A. Alofi, G.P. Srivastava, Thermal conductivity of graphene and graphite, Phys. Rev. B 87 (11) (2013) 115421.

[145] N. Mingo, D.A. Broido, Carbon nanotube ballistic thermal conductance and its limits, Phys. Rev. Lett. 95 (9) (2005) 096105.

[146] V. Adamyan, V. Zavalniuk, Lattice thermal conductivity of graphene with conventionally isotopic defects, J. Phys. Condens. Matter 24 (41) (2012) 415401.

[147] V. Adamyan, V. Zavalniuk, Phonons in graphene with point defects, J. Phys. Condens. Matter 23 (1) (2010) 015402.

[148] A.A. Balandin, S. Ghosh, W. Bao, I. Calizo, D. Teweldebrhan, F. Miao, C.N. Lau, Superior thermal conductivity of single-layer graphene, Nano Lett. 8 (3) (2008) 902–907.

[149] A.A. Balandin, S. Ghosh, D.L. Nika, E.P. Pokatilov, Extraordinary thermal conductivity of graphene: possible applications in thermal management, ECS Trans. 28 (5) (2010) 63.

[150] A.A. Balandin, S. Ghosh, D.L. Nika, E.P. Pokatilov, Thermal conduction in suspended graphene layers, Fullerenes Nanotub. Carbon Nanostruct. 18 (4–6) (2010) 474–486.

[151] W. Cai, A.L. Moore, Y. Zhu, X. Li, S. Chen, L. Shi, R.S. Ruoff, Thermal transport in suspended and supported monolayer graphene grown by chemical vapor deposition, Nano Lett. 10 (5) (2010) 1645–1651.

[152] J.H. Seol, I. Jo, A.L. Moore, L. Lindsay, Z.H. Aitken, M.T. Pettes, X. Li, Z. Yao, R. Huang, D. Broido, Two-dimensional phonon transport in supported graphene, Science 328 (5975) (2010) 213–216.

[153] J. Lan, Y. Cai, G. Zhang, J.-S. Wang, Y.-W. Zhang, Topological symmetry-induced width dependence of thermal conductance of edge-reconstructed graphene nanoribbons, J. Phys. D Appl. Phys. 47 (26) (2014) 265303.

[154] G. Xie, Y. Shen, X. Wei, L. Yang, H. Xiao, J. Zhong, G. Zhang, A bond-order theory on the phonon scattering by vacancies in two-dimensional materials, Sci. Rep. 4 (1) (2014) 1–6.

[155] H. Zhan, Y. Zhang, J.M. Bell, Y. Gu, Suppressed thermal conductivity of bilayer graphene with vacancy-initiated linkages, J. Phys. Chem. C 119 (4) (2015) 1748–1752.

[156] N. Wei, L. Xu, H.-Q. Wang, J.-C. Zheng, Strain engineering of thermal conductivity in graphene sheets and nanoribbons: a demonstration of magic flexibility, Nanotechnology 22 (10) (2011) 105705.

[157] H.F. Zhan, G. Zhang, J.M. Bell, Y. Gu, Thermal conductivity of configurable two-dimensional carbon nanotube architecture and strain modulation, Appl. Phys. Lett. 105 (15) (2014) 153105.

[158] X. Liu, G. Zhang, Y.-W. Zhang, Thermal conduction across graphene cross-linkers, J. Phys. Chem. C 118 (23) (2014) 12541–12547.

[159] X. Li, J. Chen, C. Yu, G. Zhang, Comparison of isotope effects on thermal conductivity of graphene nanoribbons and carbon nanotubes, Appl. Phys. Lett. 103 (1) (2013) 013111.

[160] J. Chen, G. Zhang, B. Li, Substrate coupling suppresses size dependence of thermal conductivity in supported graphene, Nanoscale 5 (2) (2013) 532–536.

[161] D.L. Nika, A.S. Askerov, A.A. Balandin, Anomalous size dependence of the thermal conductivity of graphene ribbons, Nano Lett. 12 (6) (2012) 3238–3244.

[162] X. Xu, L.F.C. Pereira, Y. Wang, J. Wu, K. Zhang, X. Zhao, S. Bae, C.T. Bui, R. Xie, J.T.L. Thong, Length-dependent thermal conductivity in suspended single-layer graphene, Nat. Commun. 5 (1) (2014) 1–6.

[163] J.S. Reparaz, E. Chavez-Angel, M.R. Wagner, B. Graczykowski, J. Gomis-Bresco, F. Alzina, C.M. Sotomayor Torres, A novel contactless technique for thermal field mapping and thermal conductivity determination: two-laser Raman thermometry, Rev. Sci. Instrum. 85 (3) (2014) 034901.

[164] D. Liu, R. Xie, N. Yang, B. Li, J.T.L. Thong, Profiling nanowire thermal resistance with a spatial resolution of nanometers, Nano Lett. 14 (2) (2014) 806–812.

CHAPTER

Current trends in graphene and two-dimensional material synthesis and characterization

2

Swapnil Dharaskar[a], Yashvi Sheth[b] and Surya Pattnaik[c]

[a]*Department of Chemical Engineering, School of Energy Technology, Pandit Deendayal Energy University, Gandhinagar, Gujarat, India,* [b]*Centre of Nanotechnology, Indian Institute of Technology, Bombay, Mumbai, Maharashtra, India,* [c]*Department of Chemical Engineering, National Institute of Technology Rourkela, Rourkela, Odisha, India*

2.1 Introduction

Nanotechnology and nanomaterials have made revolutionary changes because of their applications in almost every field. The increasing importance increases because of their unique properties at the nanosizes, such as improved electrical conductance, enlarged surface areas, high surface-to-volume ratio, and increased reactivity, because the size of the particles has a significant effect on these properties. Nanomaterials can be classified as two-dimensional (2D) nanoparticles or layered nanoparticles whose one of the three dimensions is in nanoscale, one-dimensional particles whose two dimensions are in nanoscale, and zero-dimensional whose all the dimensions are in nanoscale [1].

As the discovery of the first 2D graphene nanosheet structures in 2004, significant research has been directed toward 2D nanomaterials. Numerous 2D nanoparticles, commonly referred to as graphene analogs (Fig. 2.1), have been explored since 2004, including molybdenum disulfide, boron nitride, MXene, carbon nitrides, and metal halides, among others, owing to their attractive and unique properties. These 2D materials are layered, appearing as sheets with interlayer spacings ranging from 0 to 100 nm. The pioneering research on graphene described it as "a rising star on the horizon of condensed matter physics and materials science." Over the past decade, more than a dozen different 2D nanomaterials have been successfully isolated [2]. Because of their unique properties, they have found promising applications in a wide range of applications.

This chapter presents a detailed exploration of the synthesis and characterization of graphene and its analogs, including MXenes, transition metal dichalcogenides (TMDs), metal halides, and boron-based materials. The synthesis methods specific to each material family are emphasized, providing insights into the diverse approaches to producing these nanomaterials. In addition, various characterization techniques, such as scanning electron microscopy (SEM), transmission electron microscopy (TEM), Raman spectroscopy, and X-ray diffraction (XRD), are discussed in relation to graphene and its analogs. This comprehensive overview aims to elucidate the synthesis routes and analytical tools essential for advancing the understanding and application of these intriguing nanomaterials.

Graphene and 2D Materials in Heat Transfer: Fundamentals and Applications. DOI: https://doi.org/10.1016/B978-0-12-821964-5.00003-5

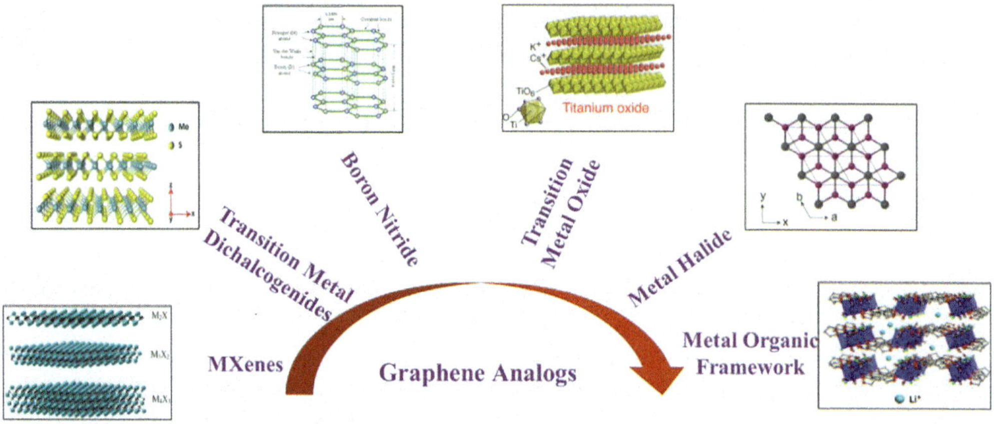

FIGURE 2.1

Analogs of graphene.

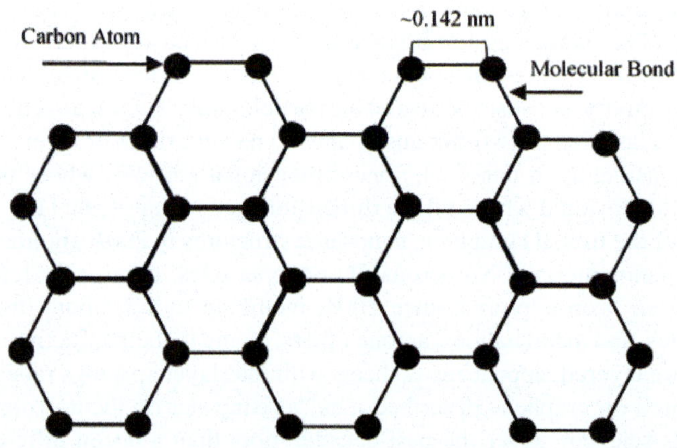

FIGURE 2.2

Structure of graphene [4].

2.2 Graphene

The word "graphene" is composed of "graph," which is derived from graphite, and the suffix "ene," which represents the carbon/carbon double bonds having bond lengths equal to 0.142 nm [3]. Graphene is formed by a special arrangement of carbon atoms through sp^2 hybridization, leading to the formation of a 2D honeycomb-like lattice structure. The atomic structure of graphene is shown in Fig. 2.2. It has received tremendous attention in research and practical applications because of its attractive, unique, and versatile properties. Because of this structure, graphene has several unique structural, optical, and

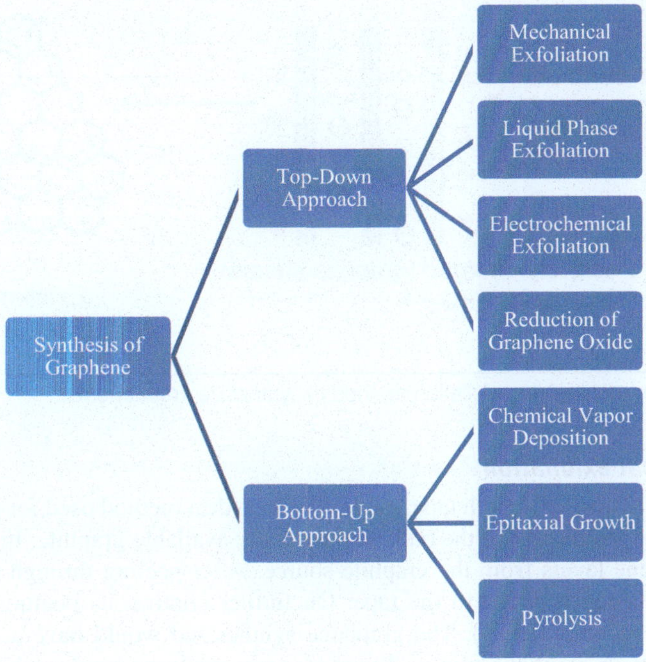

FIGURE 2.3

Flowchart for graphene synthesis through various methods.

electronic properties. It is an allotrope of carbon. The graphene monolayer is a flat sheet of atoms bonded through the van der Waals thickness of ~0.34 nm. In physical structure, graphene is the basic structure of various carbon nanomaterials, such as one-dimensional nanomaterials like carbon nanotubes and zero-dimensional nanomaterials like fullerenes. This monolayer represents a building block for all other graphitic materials. Graphene can exist as a single sheet or even stacked as multiple layers of graphene sheets. Because of the sp^2 hybridization or π bonds between the carbon atoms the delocalization of electrons is possible, making graphene an excellent electrical conductor.

2.2.1 Synthesis of two-dimensional graphene

Several methods have been exploited for the synthesis of 2D graphene. These methods can be classified into bottom-up methods and top-down methods. Fig. 2.3 shows the classification of various top-down and bottom-up techniques used for graphene synthesis. Mechanical exfoliation is the most common top-down method, which has been exploited. The synthesis of the 2D material graphene was done through mechanical exfoliation when it was discovered for the first time in 2004. As then the application of graphene 2D nanoparticles has grown exponentially in diverse fields because of their unique properties.

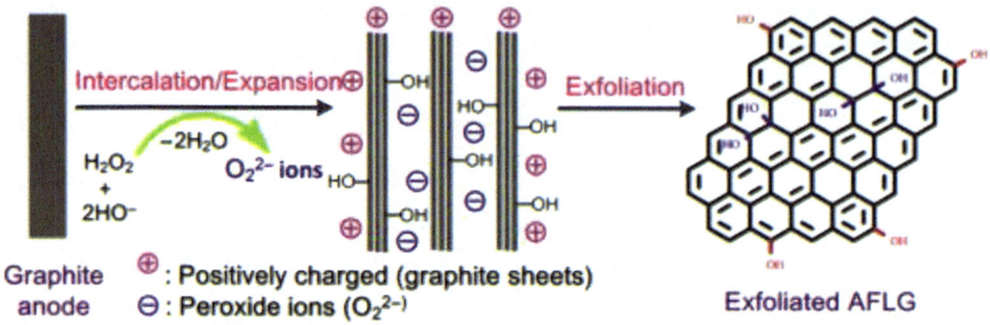

FIGURE 2.4

Schematics for the electrochemical exfoliation method to synthesize graphene [8].

2.2.1.1 Mechanical exfoliation

The mechanical exfoliation, or Scotch tape, method is the oldest method used for graphene synthesis. It is a simple method of peeling out of the bulk commercially available graphite. In this method a Scotch tape peels off graphene layers from the graphite source. After peeling through the tape multilayered graphene is obtained, which sticks to the tape. On further continuous peeling, few layers of thick graphene flakes can be obtained [5]. The graphene synthesized would have a single layer or a few layers' thick structure. A graphene film of 20 μm size can easily be synthesized through this method [6]. However, it takes a longer time for synthesis, and the uniformity is not maintained in the thickness of the graphene nanosheets. The solvents used for the synthesis usually have high boiling temperatures, which makes the process costly and difficult to recover. Hence applying this synthesis technique for large-scale production is impossible [7].

2.2.1.2 Electrochemical exfoliation

Electrochemical exfoliation, as the name suggests, utilizes an electric current in the presence of electrolytes within an electrolysis cell. As illustrated in Fig. 2.4, this method involves applying voltage to bulk graphite immersed in a salt electrolyte solution. The applied voltage generates ionic species from the electrolyte, which intercalate into the graphite structure, thereby increasing the interlayer spacing. The electrolytes typically used in this process include mixtures of deionized water with sodium chloride, sodium hydroxide, and dimethyl sulfoxide [8]. The generation of gaseous molecules during this stage leads to their expansion and subsequent bonding, forming graphene nanosheets [9]. This technique is particularly promising because of its use of nonhazardous materials and its ability to produce high-quality graphene nanosheets. Consequently, electrochemical exfoliation is well-suited for large-scale synthesis, offering advantages over methods like mechanical exfoliation [10].

2.2.1.3 Liquid-phase exfoliation

Liquid exfoliation is a prominent top-down method for synthesizing graphene nanosheets from bulk graphite. This technique involves preparing a solution with a solvent, which is subsequently subjected to ultrasonication, resulting in the formation of graphene nanosheets. Commonly, organic solvents such

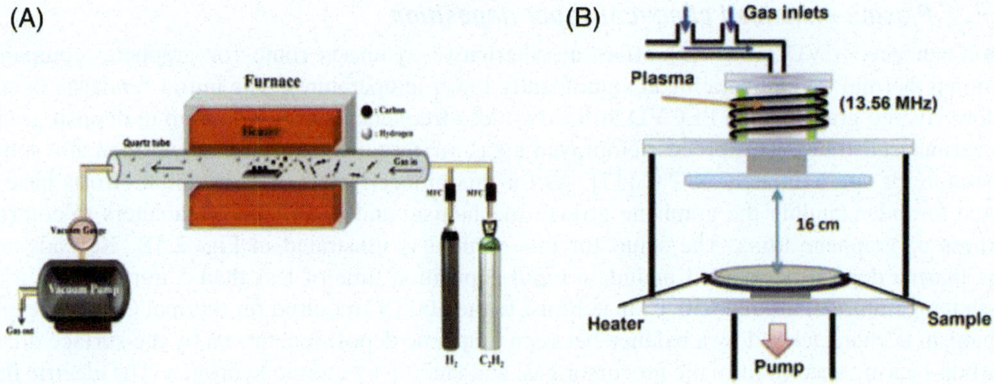

FIGURE 2.5

Setup for (A) thermal chemical vapor deposition [8] and (B) plasma-enhanced chemical vapor deposition [16].

as *N*-methyl-2-pyrrolidone are employed in this process. Liquid exfoliation is capable of producing a mixture of single- and multilayer graphene nanosheets [11].

2.2.1.4 Chemical vapor deposition

The chemical vapor deposition (CVD) process is a widely recognized technique for synthesizing 2D nanoparticles, particularly graphene nanosheets. In the CVD process, 2D materials grow on a substrate, typically a metal surface, through activated chemical reactions of precursors in a controlled environment [12]. Fig. 2.5A illustrates the CVD setup for graphene synthesis. This method has significantly advanced graphene production, allowing for efficient synthesis through a bottom-up approach. Subsequently, various combinations of metal substrates have been explored for graphene synthesis [13]. Upon completing the chemical etching of the metal substrate the graphene layers separate from the substrate and can be transferred to another substrate, yielding high-quality graphene without requiring mechanical or chemical treatments. The growth mechanism is straightforward when carbon atoms have good solubility in the metal substrate. In such cases, carbon atoms dissolve into the metal substrate at high temperatures and precipitate upon cooling to form graphene. The cooling rate can control the number of graphene layers, with slower cooling resulting in additional layers. For substrates with low carbon solubility, reactant molecules are typically transformed into a gaseous state by heating. These molecules then react or decompose on the substrate surface to form the desired film [5]. Several factors influence the final product, including the choice of precursor, reaction conditions, atmosphere, and substrate surface. For graphene synthesis, hydrocarbon precursors are generally used with metal substrates such as nickel (Ni), palladium (Pd), ruthenium (Ru), iridium (Ir), and copper (Cu). Most studies focus on Ni and Cu, with Ni offering good carbon solubility and Cu offering very low carbon solubility [14]. Because of its low carbon solubility, Cu is frequently used, resulting in a higher yield and purity of graphene nanosheets [15]. The CVD process is currently being scaled up for large-scale industrial production of high-quality graphene nanosheets.

2.2.1.5 Plasma-enhanced chemical vapor deposition

Plasma-enhanced CVD (PECVD) offers an alternative synthesis route for graphene compared to traditional thermal CVD, operating at significantly lower temperatures. The initial synthesis of single- and few-layered graphene via PECVD utilized a radio frequency PECVD system to deposit graphene onto various substrates. This process employed a gas mixture of methane and hydrogen at a substrate temperature of approximately 680°C [17]. As for this pioneering work, substantial efforts have been devoted to understanding the graphene growth mechanism and optimizing parameters to control the thickness of graphene films. The setup for this method is illustrated in Fig. 2.5B. Key advantages of the plasma deposition method include a rapid deposition time of less than 5 minutes and a lower growth temperature of around 650°C, in contrast to the 1000°C required for thermal CVD. The growth mechanism is characterized by a balance between graphene deposition, driven by the surface diffusion of carbon-bearing species from the precursor gas, and etching by atomic hydrogen. The electric field in the plasma directs the synthesis, promoting the vertical alignment of the graphene sheets [18].

2.2.1.6 Reduction of graphene oxide

The synthesis of graphene nanosheets through the reduction of graphene oxide is a prominent top-down method suitable for large-scale production. This technique, first utilized in 1962 to synthesize graphene nanoplatelets, involves using various reducing agents to remove oxygen from graphene oxide particles [5]. Typical reducing agents include hydrazine and sodium borohydride. Initially, graphene oxide is dispersed in water, creating a brown-colored solution. After introducing a reducing agent, oxygen atoms are removed, converting the color of the solution from brown to black, which signifies the formation of graphene. Because of the insolubility of graphene in water, it precipitates as black particles that can be subsequently recovered.

2.2.1.7 Comparison of different synthesis methods

The characteristics and properties of graphene synthesized through top-down and bottom-up approaches differ because of the distinct mechanisms involved in each method. This section compares mechanical exfoliation and CVD, as they are the most widely studied and utilized methods for industrial-scale production. Research indicates that the CVD method yields higher quality graphene with larger surface areas and fewer defects. In addition, the overall cost of synthesis is lower for CVD, making it a more advantageous option [2]. In contrast, mechanical exfoliation from bulk materials produces nanoparticles with smaller sheet sizes. However, the quality of the graphene obtained via mechanical exfoliation is generally low, and controlling the particle size is challenging [12].

2.2.2 Characterization

Graphene is characterized by various advanced techniques crucial for advancing graphene research. These characterization methods enable the precise determination of synthesis quality, identification of structural defects, assessment of layer structure and number of layers, and evaluation of material properties. Key techniques include TEM, SEM, atomic force microscopy (AFM), Raman spectroscopy, and XRD. Each technique provides valuable insights into graphene's characteristics, contributing to a comprehensive understanding of its properties and potential applications.

2.2.2.1 Optical imaging

A variety of imaging techniques can be employed to visualize single-, bi-, and few-layer graphene at the nanoscale. Optical microscopy, AFM, high-resolution TEM (HRTEM), and SEM all enable the characterization of graphene's morphology and structure. Historically, optical microscopy was commonly utilized, given its low cost and accessibility. However, SEM, TEM, and AFM have now become the preferred methods, as they offer higher resolution images and a clearer view of graphene's true structure at the nanoscale. A multitechnique analytical approach often provides the most robust structural examination, leveraging each technology's strengths. For example, optical microscopy serves as an initial screening tool, whereas AFM can confirm the number of layers. SEM reveals surface topography and defects. HRTEM is particularly powerful because of its ability to resolve individual carbon atoms and identify defects. The combination of these advanced microscopic techniques, such as correlative AFM–TEM, has enabled breakthrough discoveries in graphene research by revealing insights not feasible using any single method alone. Their application remains crucial for ongoing progress in synthesizing, manipulating, and practically applying this novel 2D material.

2.2.2.2 Atomic force microscopy

AFM has emerged as a powerful and widely applied characterization technique in graphene research. AFM uses a cantilever probe to track surface topography at the nanoscale, providing insights beyond what is visible using traditional optical or electron microscopy alone. Through measurements of cantilever deflection across graphene nanosheets and substrate surfaces, AFM enables high-resolution analysis of key properties such as topological features, sheet dimensions, and layer thickness. Modalities such as conductive and piezoresponse force microscopy further allow mechanical and electronic properties to be probed with nanoscale precision. Unlike electron microscopy methods, AFM can also generate images of graphene and composites immersed in liquid environments. Together, these capabilities have made AFM a mainstay for quantitatively measuring graphene surface structure in three dimensions directly from layer counting and thickness measurements. Moreover, AFM returns morphological and property data simultaneously, advancing understanding of synthesis–structure–property relationships. The technique continues to advance graphene characterization abilities with emerging new modes and enables insights that further the fields of graphene synthesis optimization and application development. Graphene composites with biomolecules, where protein or DNA strongly adsorbs to the graphene layers, are usually visualized by AFM to get high-quality images and perform height analysis.

2.2.2.3 Scanning electron microscopy

SEM is instrumental in identifying and interpreting the morphology and layered structure of graphene. It offers advantages in detecting impurities, folds, and discontinuities in synthesized graphene across various substrates. However, SEM's lower resolution limits its effectiveness for imaging ultrathin graphene [19]. SEM analysis can identify the layered structure and examine the number of graphene layers. Fig. 2.6 presents SEM images of layered graphene on an alumina substrate, showcasing different layer configurations. The underlying alumina structure is visible in SEM images for single- or bilayer graphene, as illustrated in Fig. 2.5. Conversely, this alumina structure is obscured in SEM images of graphene with more layers.

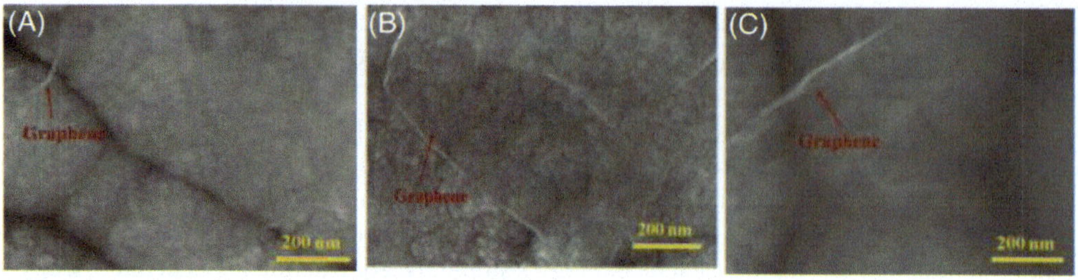

FIGURE 2.6

Scanning electron microscopy (SEM) images of graphene (A) single-layered, (B) bi-layered, and (C) tri-layered [19].

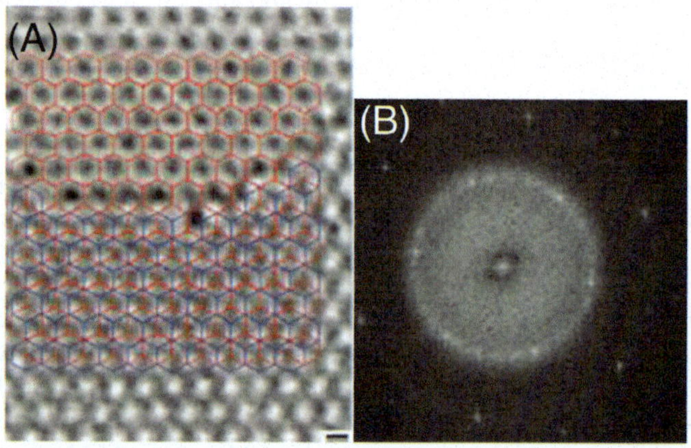

FIGURE 2.7

Transmission electron microscopy images of grapheme: (A) view of layers and (B) atomic structure.

2.2.2.4 Transmission electron microscopy

In general, TEM is widely used to obtain high-resolution images of nanoscale materials. In TEM an electron beam is transmitted through a thin sample, and the resulting interactions are detected to form an image. This technique is particularly effective for imaging graphene and reduced graphene oxide because of their atomic thickness, allowing for images with exceptional resolution. However, conventional TEM systems are not ideal for this purpose. They often produce poor-quality images at low operating voltages, whereas higher voltages can damage the monolayer structure. Recently, researchers have adopted advanced TEM systems featuring aberration correction and monochromators, which achieve a resolution of 1 Å at an acceleration voltage of only 80 kV. Fig. 2.7A presents a TEM image of graphene, where the distinct layers are marked with red connections for the first layer and blue for the second layer. Fig. 2.7B illustrates the hexagonal arrangement of carbon atoms within the graphene sheet, highlighting its atomic structure.

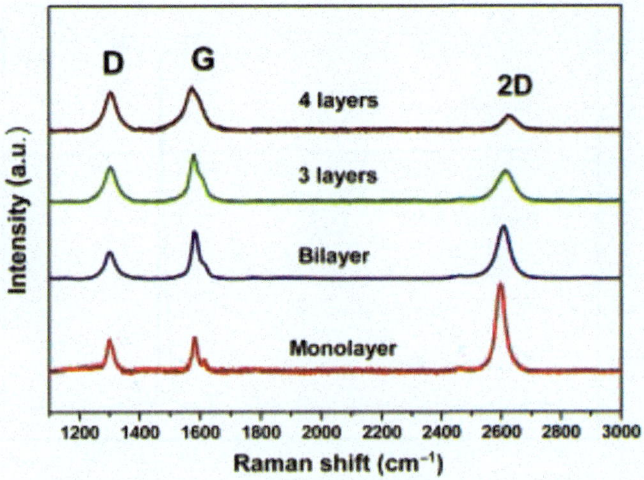

FIGURE 2.8

Raman analysis spectrum for single- and multilayers of graphene [25].

2.2.2.5 ***Raman spectroscopy***

Raman spectroscopy is an effective technique for imaging graphene-based nanomaterials, leveraging their intrinsic structures and properties. Widely adopted to analyze the electronic properties and microstructure of graphite materials, Raman spectroscopy is particularly useful for assessing the quality and thickness of graphene nanosheets.

This technique involves directing a monochromatic laser light onto the material, where the interaction with electrons causes a shift in laser energies because of light scattering. Three primary peaks in the Raman spectrum correspond to graphene's structure: the D peak at 1350 cm^{-1}, the G peak at 1580 cm^{-1}, and the 2D peak at 2700 cm^{-1}. The D peak indicates disorder in sp^2 hybridized carbons from photon backscattering, whereas the G peak represents lattice vibrations [20,21]. The 2D peak, at a Raman shift of 2700 cm^{-1}, results from the second-order Raman scattering effect [22]. The ratio of the peak intensities (I_D/I_G and I_{2D}/I_G) helps in identifying defects and the number of graphene layers [22]. A higher I_D/I_G ratio indicates a greater degree of disorder and an amorphous graphene structure, as more defects lead to increased elastic light scattering. The I_{2D}/I_G ratio, along with the shape and location of the 2D peak, is useful for determining the number of graphene layers. The intensity of the G peak increases with the number of layers, whereas the 2D peak becomes wider and smaller [23,24]. Fig. 2.8 illustrates the Raman spectrum of single-, bi-, tri-, and four-layered graphene on Ni interdigitated electrodes deposited on alumina substrates [19]. The 2D peak of monolayer graphene is notably sharper and more intense compared to the 2D peaks of multilayered graphene. In addition, a slight frequency shift and an increase in the G peak intensity are observed.

2.2.2.6 ***X-ray diffraction***

XRD is one of the essential analytical methods for the identification of phases of materials based on the cell dimensions. Fig. 2.9 shows the XRD image of graphite, grapheme, and graphene oxide [25].

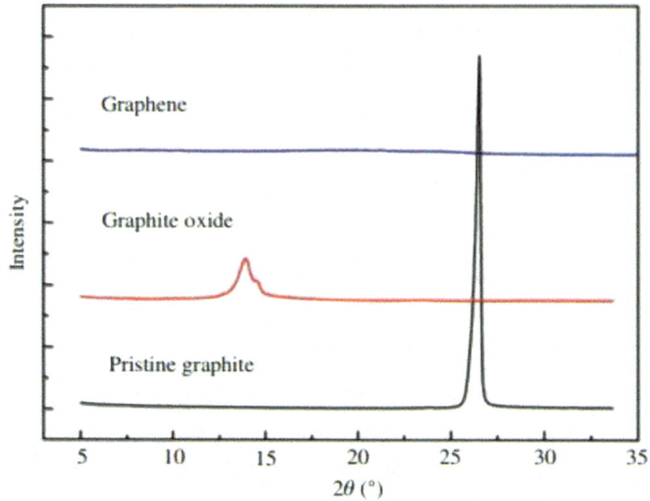

FIGURE 2.9

X-ray diffraction image of graphene [25].

Pristine graphite displays a sharp, narrow, high-intensity peak at a 2θ angle of 26.6 degrees, indicative of its crystalline structure. In contrast, graphene oxide exhibits a peak at 13.9 degrees, attributed to the presence of oxygen-containing groups. Graphene, ideally, shows no significant peak, indicating the absence of elements other than carbon. This confirms the successful synthesis of pure graphene.

2.3 MXenes

MXenes belong to a growing family of 2D transition metal carbides and nitrides. These materials are synthesized using a top-down approach by selectively etching a specific element (A) from layered MAX phases. MAX phases possess a unique structure with strong bonds between metal (M) and element X (carbon and/or nitrogen), whereas the M–A bonds are weaker. This allows for the selective removal of A atoms with appropriate etching techniques, resulting in the formation of MXene sheets with a characteristic layered structure [26]. MXenes have a general formula of $M_{n+1}AX_n$, where M is an early transition metal (e.g., Sc, Ti, V, Cr, and Mo), A is primarily a Group IIIA or IVA element (e.g., Al, Si, Sn, and In), X is carbon and/or nitrogen, and n can be 1, 2, or 3. Common MXenes include M_2X, M_3X_2, and M_4X_3. Their properties can be tailored by modifying their structure, composition, surface terminal groups, and interlayer spacing. MXenes exhibit a range of unique properties, such as good metallic conductivity and hydrophilicity, which are uncommon in other 2D materials [27]. Titanium carbide (Ti_3C_2) was the first MXene reported in the literature in 2011. It was synthesized by selectively etching aluminum from the titanium aluminum carbide (Ti_3AlC_2) MAX phase using an Hydrofluoric Acid (HF) solution under ambient conditions. Since then, numerous other MXenes have been developed, including Ti_2C, $(Ti_{0.5}, Nb_{0.5})_2C$, Nb_2C, Mo_2C, V_2C, $(V_{0.5}, Cr_{0.5})_3C_2$, Ti_3CN, and Nb_4C_3. Notably, Mo_2CT_x was the first MXene synthesized from a non-MAX phase, specifically from Mo_2Ga_2C [28].

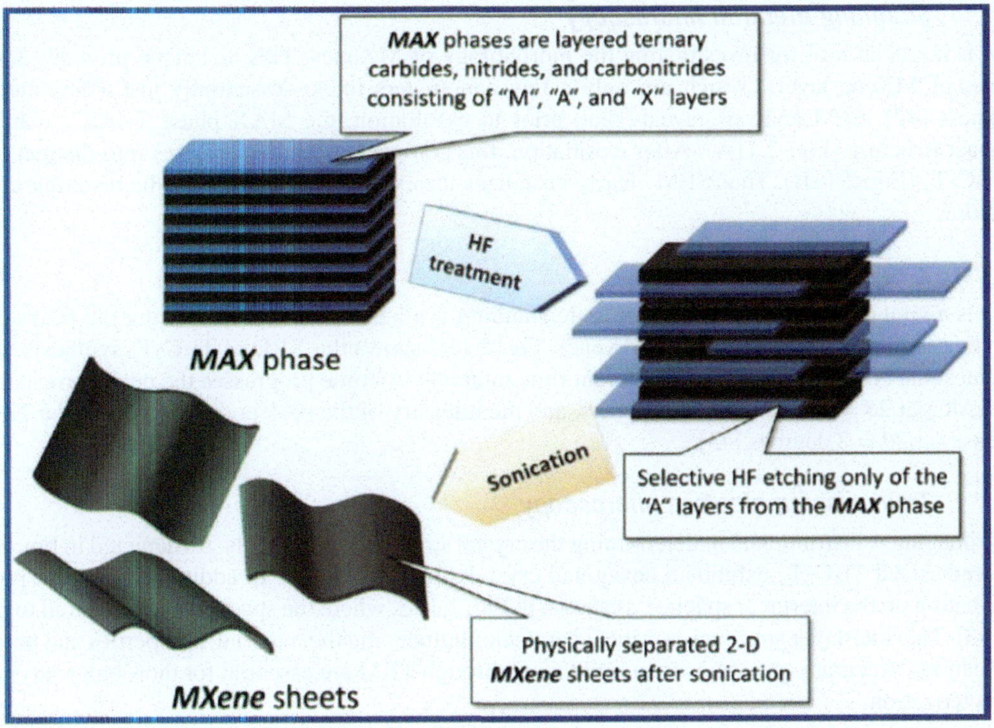

FIGURE 2.10

Synthesis of MXene from MAX phase using HF solution by selective etching method [41].

The most common method for MXene synthesis remains selective etching with HF solution, as depicted in Fig. 2.10. Alternative etching methods using LiF + HCl [29] and NH_4HF_2[30–32] have been developed to mitigate the toxicity of HF. In addition, hydrothermal synthesis at high temperatures (approximately 180°C) using nontoxic NaOH solutions has been successfully reported [33]. The tunable physicochemical properties of MXenes make them suitable for a wide range of applications, including supercapacitors, electrode materials, hybrid devices, energy storage, water purification (adsorption, photodegradation, membrane separations), gas sensors, optoelectronics, thermoelectric materials, and insulators [34–40].

2.3.1 Characterization of MXenes

For a comprehensive understanding of the successful synthesis of MXenes, SEM, TEM, and XRD analyses are commonly employed. These techniques provide crucial insights into morphology and reaction completeness. Given its widespread utilization, this section focuses on SEM, TEM, and XRD analyses of $Ti_3C_2T_x$ as a representative MXene.

2.3.1.1 Scanning electron microscopy

SEM is a crucial tool for investigating the morphology of MXenes. This technique provides detailed images of MXene layers, which are only a few nanometers thick—essentially just a few atoms in thickness [42]. SEM analysis reveals that, prior to exfoliation, the MAX phase Ti_3AlC_2 exhibits a compact structure (Fig. 2.11A). After exfoliation, this compact structure separates into distinct layers of $Ti_3C_2T_x$ (Fig. 2.11B). Thus SEM clearly visualizes the exfoliation process and the resulting layered structure.

2.3.1.2 X-ray diffraction

XRD is a vital characterization tool for understanding synthesis progress by viewing the peaks corresponding to the MAX phase and the MXenes. Fig. 2.11C shows the XRD of $Ti_3C_2T_x$ synthesis, which indicates that XRD is compared at different time intervals. As time progresses the peak corresponding to Ti_3AlC_2 at $2\theta = 35.9$ degrees disappears, and the intensity of the peak corresponding to the $Ti_3C_2T_x$ increases at $2\theta = 8$ degrees [43].

2.3.1.3 Transmission electron microscopy

TEM imaging is instrumental in determining the crystal structure of materials. As depicted in Fig. 2.11D the synthesized $Ti_3C_2T_x$ exhibits a hexagonal crystal structure pattern. In addition, TEM can provide information on the interlayer spacing, as shown in Fig. 2.11E, where the spacing was measured to be 1.2 nm [43]. This interlayer spacing is a critical parameter influencing the material's properties and potential applications. Accurate spacing estimates obtained through TEM are essential for thorough analysis and characterization.

2.4 Transition metal dichalcogenides

TMDs represent another class of graphene analogs characterized by a layered structure similar to graphene. These compounds have a general chemical formula of MX_2, where M denotes a central transition metal such as Mo, W, V, or Re, which forms a single metal layer. This metal layer is flanked by two layers of chalcogen elements (X), such as Se, O, or S [44]. TMD layers are bonded covalently between the metal and chalcogen elements, forming a hexagonal crystal lattice, whereas weak van der Waals forces hold adjacent layers together. Among the various TMDs studied, compounds, such as $MoSe_2$, MoS_2, WSe_2, and WS_2, are noteworthy because of their broad applicability [45]. Extensive research focuses on TMDs because of their unique synthesis methods and significant potential in diverse applications, including batteries, supercapacitors, material sciences, and various energy storage devices [46,47].

2.4.1 Synthesis

2.4.1.1 Hydrothermal/Solvothermal

Hydrothermal and solvothermal methods have garnered significant attention for synthesizing TMDs because of their scalability, rapid synthesis, and cost-effectiveness. In these methods, metal salts, such as molybdenum (Mo), tungsten (W), and vanadium (V), serve as precursors for the transition metal

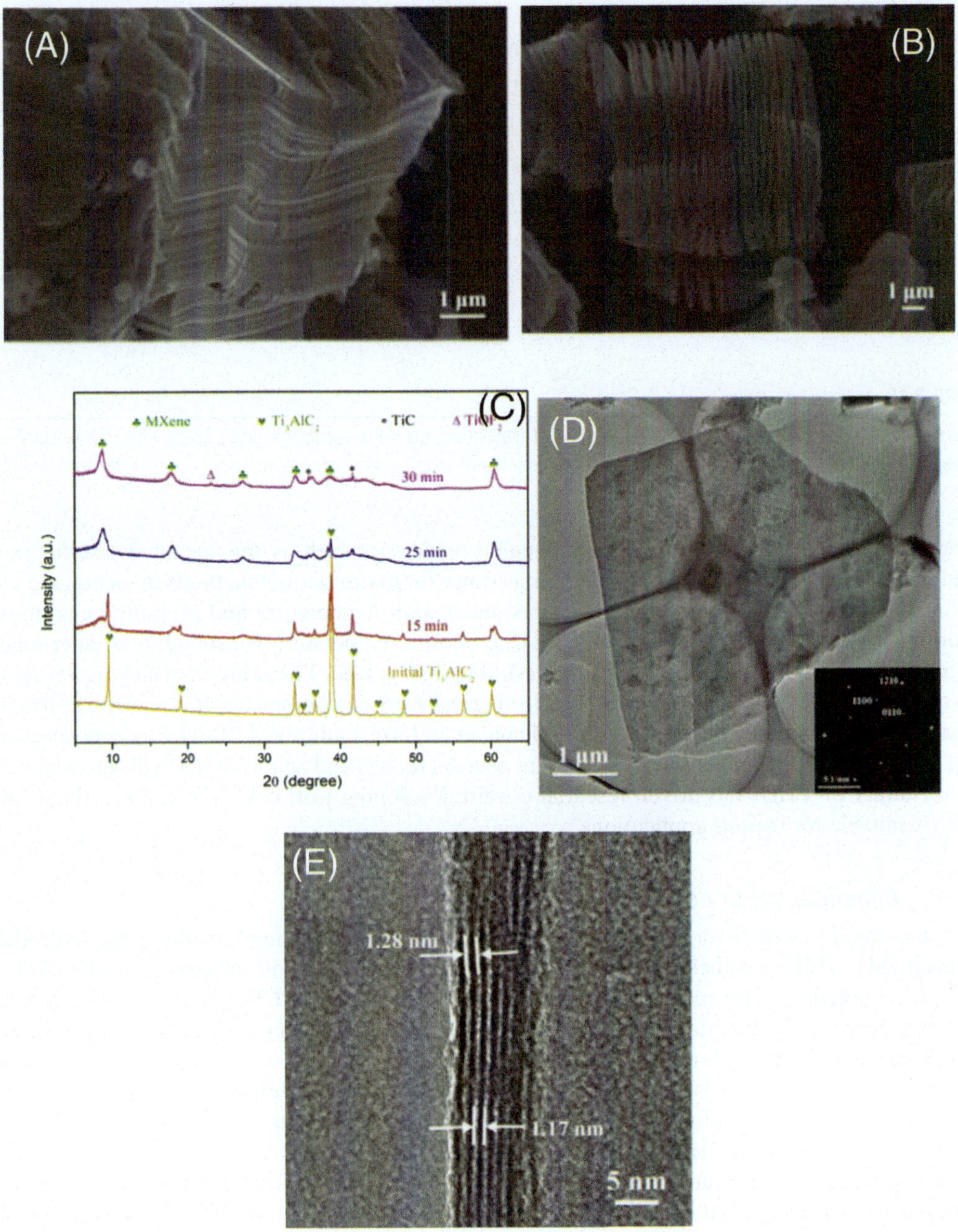

FIGURE 2.11

(A) SEM image of Ti_3AlC_2 and (B) SEM image of exfoliated $Ti_3C_2T_x$, (C) XRD image of Ti_3AlC_2 and $Ti_3C_2T_x$, and (D and E) TEM image of $Ti_3C_2T_x$ [43]. *SEM*, Scanning electron microscopy; *TEM*, transmission electron microscopy; *XRD*, X-ray diffraction.

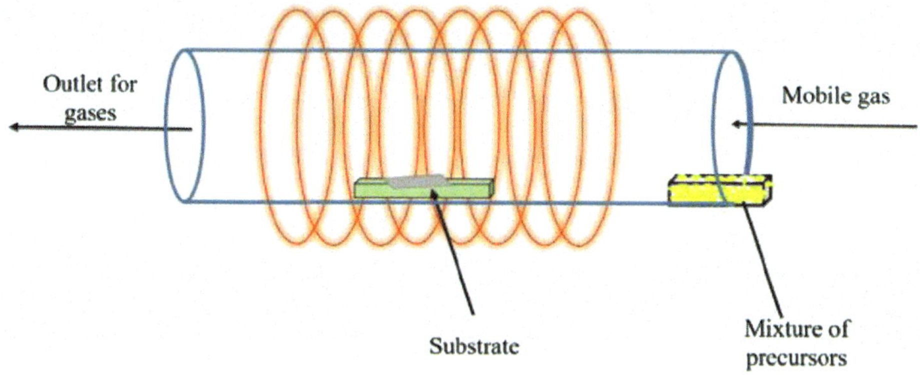

FIGURE 2.12

Schematics for the CVD setup for the synthesis of transition metal dichalcogenides [52]. *CVD*, Chemical vapor deposition.

atoms. These metal salts react with chalcogenide precursors such as thioacetamide, sulfur powder, thiourea, and L-cysteine. The reaction medium is water for hydrothermal methods or an organic solvent for solvothermal methods. This mixture reacts under high temperature and pressure conditions in a Teflon-lined hydrothermal autoclave reactor. The chosen temperature is crucial as it determines the operating pressure and the properties of the resulting TMDs [48]. For example, Chakravarty and Late [49] synthesized WSe_2 nanorods using a hydrothermal method combined with microwave irradiation for capacitor applications. In another study, Huang et al. [50] synthesized 2D $MoSe_2$ nanosheets on Ni foam using sodium molybdate and Se powder as sources for Mo and Se through a hydrothermal method. The versatility of TMDs has driven research toward developing fast, low-cost, and scalable synthesis methods suitable for various applications.

2.4.1.2 Chemical vapor deposition

CVD is a crucial method for the large-scale synthesis of TMDs, complementing the hydrothermal approach [51]. TMDs synthesized via CVD exhibit superior electronic properties, minimal defects, high-quality outcomes, and customizable thickness. The conventional CVD process has been modified to achieve the desired properties and scalability. Key steps in the CVD process include vapor generation and subsequent deposition on a substrate. For TMD synthesis, vapor generation typically occurs at high temperatures (500°C–900°C) and lower pressures (~100 Torr), which helps reduce the boiling point and produce sufficient vapors. This vapor generation relies on the volatilization and thermolysis of precursor materials, where transition metal salts react with a chalcogen source in the gas phase to form TMDs [52]. The generalized schematic for the CVD process of TMDs is illustrated in Fig. 2.12. For example, Zhao et al. [53] developed a novel CVD method for synthesizing $MoSe_2$ on large-scale silica substrates (SiO_2/Si), achieving uniform and highly crystalline products. Similarly, Hyun et al. [54] synthesized 2D MoS_2-layered crystals using a CVD method that employed MoO_3 and sulfur powder as precursors, detailing the synthesis mechanism. Mahyavanshi et al. [55] also utilized CVD to synthesize MoS_2 with a nanoribbon morphology.

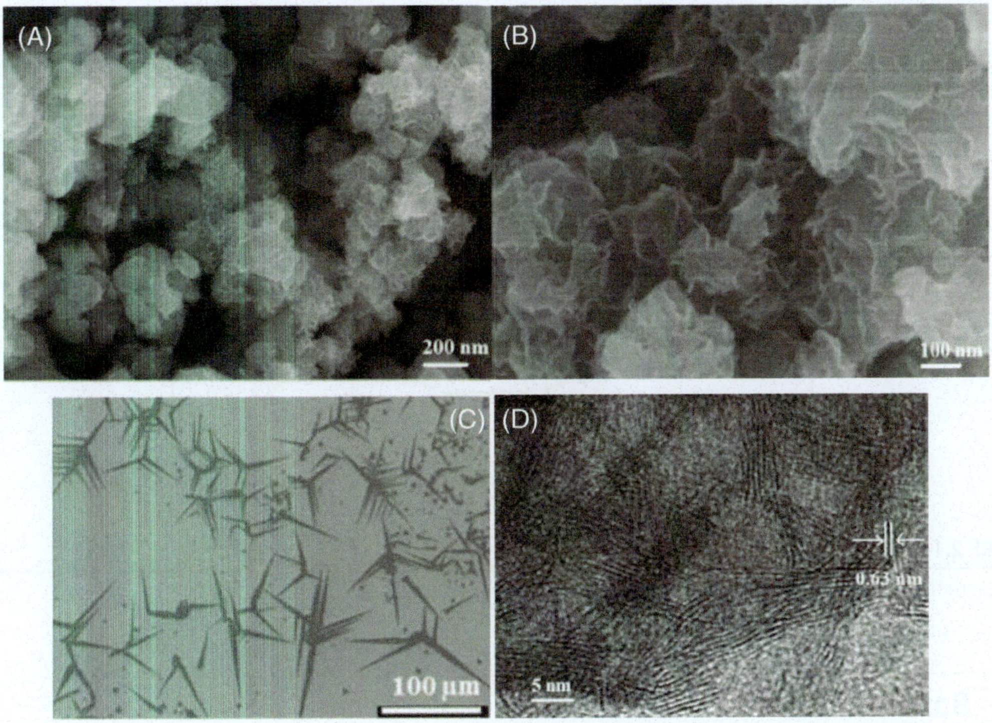

FIGURE 2.13

(A and B) SEM images of MoS_2 having nanoflower morphology [56]; (C) SEM image of MoS_2 having nanoribbon morphology [55]; and (D) TEM image of MoS_2 which has nanoflower morphology [56]. *SEM*, Scanning electron microscopy; *TEM*, transmission electron microscopy.

2.4.2 Characterization

2.4.2.1 *Scanning electron microscopy*

Fig. 2.13A and B display SEM images of MoS_2 nanosheets synthesized via the hydrothermal method, revealing that the nanosheets aggregate to form nanoflower structures [56]. In contrast, Fig. 2.13C shows the nanoribbon morphology of MoS_2 when synthesized using the CVD technique [55]. These observations demonstrate that the synthesis method and process parameters significantly influence the resulting structure, which may be critical depending on the specific application.

2.4.2.2 *Transmission electron microscopy*

TEM enables visualization of the actual layered structure and interlayer spacing of nanoparticles, even in potentially agglomerated structures. For example, SEM analysis revealed a flower-like morphology of MoS_2. TEM analysis of this sample confirmed the layered structure, with an interlayer spacing of 0.63 nm (Fig. 2.13D and E) [56].

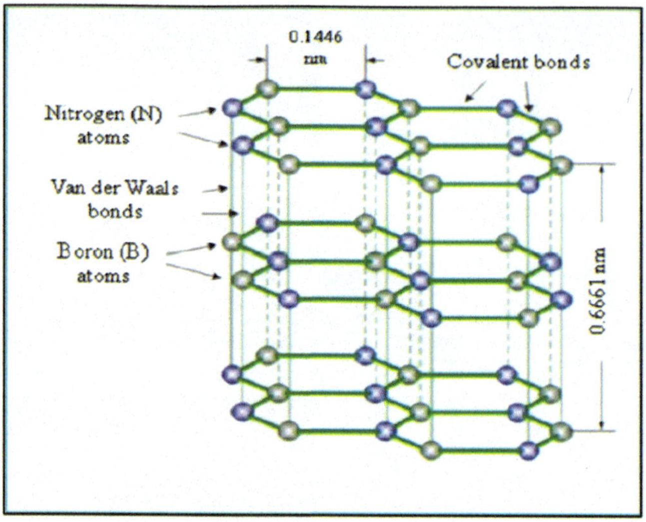

FIGURE 2.14

Structure of hexagonal boron nitride [57].

2.5 Boron nitride

Hexagonal boron nitride (h-BN) consists of boron and nitrogen atoms in a 1:1 ratio, forming a hexagonal Bernal structure. In this arrangement, boron and nitrogen atoms stack in alternating layers, as depicted in Fig. 2.14. Because of the differing electronegativities of boron and nitrogen, h-BN exhibits a net ionic charge. Structurally analogous to graphene, single-layer boron nitride is often called "white graphene." h-BN can exhibit either zigzag or armchair configurations, with weak van der Waals forces acting between the layers. Boron nitrides have significant potential applications across various fields, including electronic devices, energy sectors, and water treatment as photocatalysts. Despite their versatility, challenges remain in achieving high-quality, large-scale synthesis, highlighting an area for further research.

2.5.1 Synthesis

2.5.1.1 Synthesis using the top-down method

The top-down method for hexagonal boron nitride is mechanically breaking the van der Waal's force of attraction between boron nitride crystals. However, this method is now obsolete. The more preferred ones are CVD and hydrothermal.

2.5.1.2 Synthesis using chemical vapor deposition

CVD is a crucial technique for synthesizing high-quality boron nitride with desired properties. Precursors, such as borazine and hexachloroborazine ($B_3N_3Cl_6$), serve as sources of boron and nitrogen. The in situ method is advantageous for synthesizing metal oxide and boron nitride composites, using boron nitride as a template for metal oxide growth [58]. Wang et al. [59] developed a two-step synthesis process for boron nitride nanosheets. A citrate–nitrate mixture was combusted in the first step to produce boron-powder-coated cobalt oxide. In the second step, catalytic CVD of the boron precursor was performed. The resulting boron nitride nanosheets ranged in size from 20 to 80 nm.

2.5.1.3 Synthesis by hydrothermal method

Boron nitride is commonly synthesized using the hydrothermal process. For example, Hao et al. [60] successfully synthesized hexagonal boron nitride nanocrystals in an autoclave at 300°C for 24 hours, using boric acid as the boron precursor and trimethylamine as the nitrogen source. In addition, Chen et al. [61] employed hydrothermal technology to synthesize modified hexagonal boron nitride using carbon nitride, which was then applied in photocatalysis. They explored the impact of different solvents, such as ethanol, on the resulting structure. Although the hydrothermal method is predominantly used for synthesizing hexagonal boron nitride, other techniques like laser ablation are employed less frequently.

2.5.2 Characterization

2.5.2.1 Scanning electron microscopy analysis

The synthesized boron nitride nanoparticles exhibit various morphologies, which can be effectively identified using SEM analysis. For example, boron nitride synthesized via CVD at high temperatures (1400°C) under an argon atmosphere resulted in a nanotube structure, as illustrated in Fig. 2.15A and B. These nanotubes displayed a uniform diameter of approximately 80 nm and lengths extending to several micrometers [59].

2.5.2.2 Transmission electron microscopy analysis

To investigate the internal morphology of boron nitride synthesized via CVD at 1400°C under an argon atmosphere, TEM analysis was conducted. Fig. 2.15C shows that the parallel lattice fringes exhibit an interlayer spacing of 0.335 nm [59]. In addition, the hexagonal arrangement of boron nitride is evident from the TEM analysis. Fig. 2.16 illustrates the TEM analysis of boron nitride sheets produced through ultrasonication, confirming the hexagonal arrangement of boron and nitrogen atoms [58].

2.6 Metal halides

Metal halides are another class of 2D layered materials composed of metal ions (M) such as Mg, Ca, Ge, Cd, Pb, and Zn and halogen atoms (H) such as Cl, Br, or I with composition MH_2 (M:H = 1:2). These materials have semiconducting properties with the bandgap falling in wide range of visible to UV. In comparison with all the graphene analogs, research on metal hydrides is still in progress. Metal hydrides can be used in solar cells, catalytic fields, and photodetectors because of their semiconducting properties [62–64].

Among the metal halides, lead iodide is the most popular one. Lead iodide (PbI_2) is a high-anisotropic intrinsic semiconductor material with a direct bandgap of 2.28–2.5 eV [65]. The layers

FIGURE 2.15

FESEM images of boron nitride synthesized through CVD method: (A) low magnification, (B) high magnification, and (C) TEM images of boron nitride [59]. *CVD*, Chemical vapor deposition; *TEM*, transmission electron microscopy.

of PbI_2 constitute iodine and lead atoms arranged in a hexagonally close-packed manner loaded along the *c*-axis. Strong covalent bonding exists between atoms, whereas interlayers are bonded through a weak van der Waals force of attraction [66].

2.6.1 Synthesis

PbI_2 can usually be synthesized through mechanical exfoliation, solution method, and physical vapor deposition (PVD) [66–70]. However, the solution method is limited by low yield and poor product quality because of impurities degrading crystal quality. In contrast, PVD ensures a high yield, excellent crystallinity, and superior quality [70].

Using PVD under low-pressure conditions and temperature gradients, PbI_2 flakes can be synthesized efficiently. In this process the precursor material, typically in bulk form, is placed at the center of a furnace, whereas another substrate is positioned at a distance to facilitate the growth of 2D material layers. The furnace operates under low pressure, with the center maintained at a higher temperature to vaporize the precursor. This vapor then migrates under the temperature gradient and condenses on the

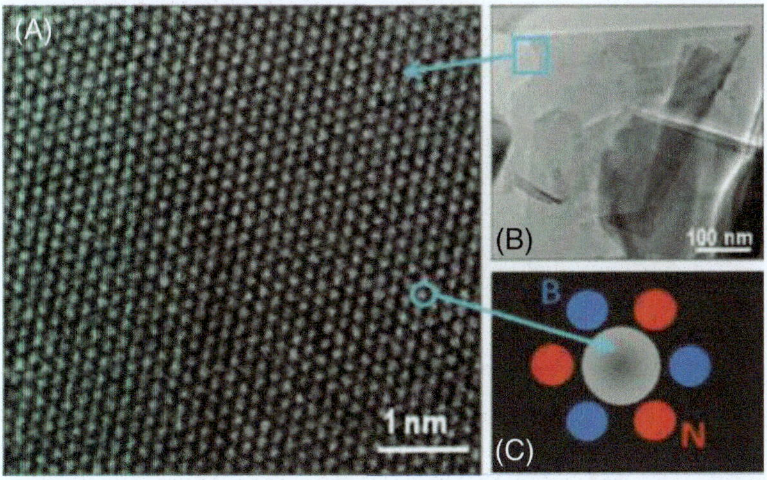

FIGURE 2.16

(A) HRTEM images of boron nitride nanosheets a thin few-layer-thick boron nitride sheet, and an image taken from the edge of the thinnest sheet; (B) HRTEM for bond distance; and (C) a bright spot of HRTEM shows the center for the boron and nitrogen atoms that are arranged in lattices of a hexagon [58]. *HRTEM*, High-resolution transmission electron microscopy.

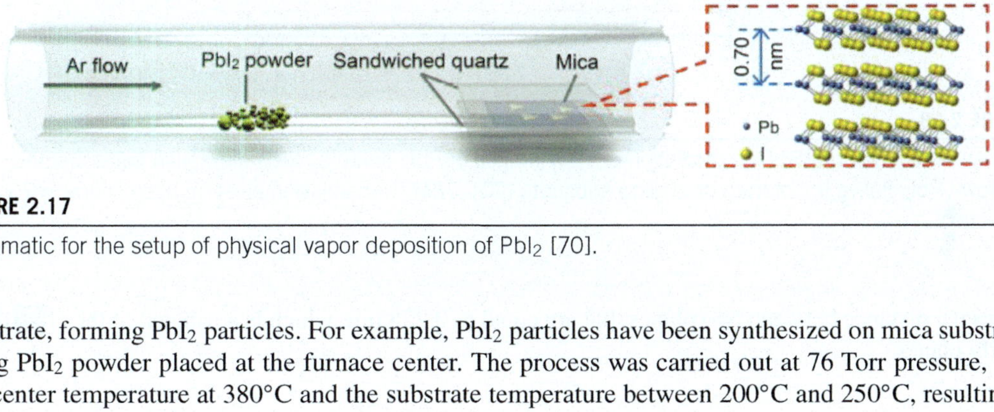

FIGURE 2.17

Schematic for the setup of physical vapor deposition of PbI_2 [70].

substrate, forming PbI_2 particles. For example, PbI_2 particles have been synthesized on mica substrates using PbI_2 powder placed at the furnace center. The process was carried out at 76 Torr pressure, with the center temperature at 380°C and the substrate temperature between 200°C and 250°C, resulting in the successful growth of PbI_2 particles on the mica substrate [70]. A schematic of this arrangement is shown in Fig. 2.17.

2.6.2 Characterization

2.6.2.1 Transmission electron microscopy analysis

TEM is adopted to confirm the crystalline structure and the arrangement of atoms of PbI_2. Fig. 2.18A shows the low-resolution TEM image of PbI_2 composite. An HRTEM image of a PbI_2 flake (Fig. 2.18B)

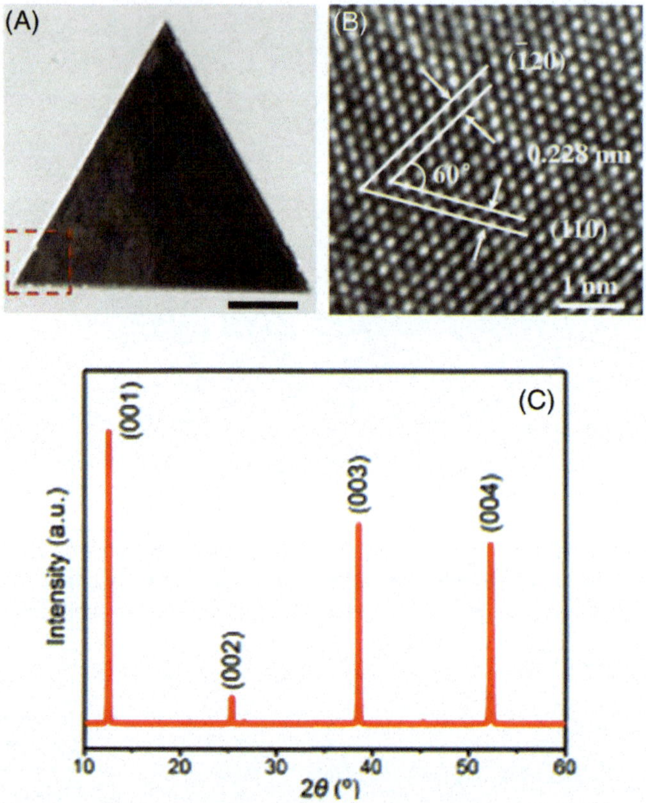

FIGURE 2.18

(A) Low resolution and (B) high resolution of TEM images of PbI_2 flake (C) XRD diffraction pattern of the as-grown PbI_2 flakes transferred onto glass substrate [70]. *TEM*, Transmission electron microscopy; *XRD*, X-ray diffraction.

shows the distance between two planes that are equal to 0.228 nm, which is attributed to the (1[redacted]1[redacted]0) family planes.

2.6.2.2 X-ray diffraction analysis

The crystallinity and phase structure of the synthesized PbI_2 flakes were evaluated using XRD analysis. As shown in Fig. 2.18C the XRD pattern closely matched that of standard hexagonal PbI_2, with only the characteristic (0[redacted]0[redacted]1) diffraction peaks observed. This confirmed the successful preparation of phase-pure PbI_2 with the expected crystal structure. Throughout this chapter, various critical methods for synthesizing and characterizing graphene and its analogs have been discussed. Table 2.1 summarizes these significant methods and key nanoparticles, providing a comprehensive overview of the techniques and materials explored.

Table 2.1 Summary of key synthesis method and important nanoparticles of various 2D nanoparticles.

2D nanoparticle	Important nanoparticles	Key synthesis method
Graphene	–	CVD, mechanical exfoliation
MXenes	$Ti_3C_2T_x$	Selective etching
Transition metal dichalcogenides	MoS_2, $MoSe_2$, WS_2	CVD, hydrothermal
Metal halide	PbI_2	PVD
Boron family	Hexagonal boron nitride	CVD, hydrothermal

2D, *two-dimensional;* CVD, *chemical vapor deposition;* PVD, *physical vapor deposition.*

2.7 Conclusion

This chapter provided an overview of prominent 2D layered nanomaterials, with a focus on graphene and analogs. Both top-down and bottom-up synthesis routes were examined, including mechanical exfoliation, electrochemical exfoliation, CVD, reduced graphene oxide, and hydrothermal and selective etching approaches. Key findings were that CVD, hydrothermal synthesis, and selective etching allow the production of high-quality graphene and related materials at scale. Various characterization techniques that are crucial for analyzing structure were also reviewed, such as optical imaging, XRD, SEN/TEM, and Raman spectroscopy. Although significant advancements have been made, widespread industrial application of 2D nanomaterials has yet to be fully realized. Further optimization of scalable synthesis methods may be needed to fully address challenges like control over number of layers and defect reduction. Continued exploration of novel 2D materials is equally important. With additional progress in materials engineering and understanding structure–property links, 2D nanomaterials show great promise to enable transformational technologies across multiple sectors through their unique nanoscale attributes. With continued progress, these unique 2D nanomaterials may eventually be incorporated into real-world technologies, enabling revolutionary performance improvements across many industries.

References

[1] R. Mas-Ballesté, C. Gómez-Navarro, J. Gómez-Herrero, F. Zamora, 2D materials: to graphene and beyond, Nanoscale 3 (2011) 20–30. https://doi.org/10.1039/c0nr00323a.

[2] A. Gupta, T. Sakthivel, S. Seal, Recent development in 2D materials beyond graphene, Prog. Mater. Sci. 73 (2015) 44–126. https://doi.org/10.1016/j.pmatsci.2015.02.002.

[3] O.V. Penkov, Introduction to graphene, Tribol. Graphene (2020) 1–10. https://doi.org/10.1016/b978-0-12-818641-1.00001-0.

[4] C.B. Clemons, M.W. Roberts, J.P. Wilber, G.W. Young, A. Buldum, D.D. Quinn, Continuum plate theory and atomistic modeling to find the flexural rigidity of a graphene sheet interacting with a substrate, J. Nanotechnol. (2010). https://doi.org/10.1155/2010/868492.

[5] M.S.A. Bhuyan, M.N. Uddin, M.M. Islam, F.A. Bipasha, S.S. Hossain, Synthesis of graphene, Int. Nano Lett. 6 (2016) 65–83. https://doi.org/10.1007/s40089-015-0176-1.

[6] L.Y. Lin, D.E. Kim, W.K. Kim, S.C. Jun, Friction and wear characteristics of multilayer graphene films investigated by atomic force microscopy, Surf. Coatings Technol. 205 (2011) 4864–4869. https://doi.org/10.1016/j.surfcoat.2011.04.092.

[7] R. Bhoria, Enhancing liquid phase exfoliation of graphene in organic solvents with additives, in: I. Ahmad, F.I. Ezema (Eds.), Graphene and Its Derivatives—Synthesis and Applications, 2019, pp. 1–15. [Working Title]. https://doi.org/10.5772/intechopen.81462 .

[8] Y. Seekaew, O. Arayawut, K. Timsorn, C. Wongchoosuk, Synthesis, Characterization, and Applications of Graphene and Derivatives, Elsevier Inc., 2018. https://doi.org/10.1016/B978-0-12-813248-7.00009-2.

[9] T.C. Achee, W. Sun, J.T. Hope, S.G. Quitzau, C.B. Sweeney, S.A. Shah, T. Habib, M.J. Green, High-yield scalable graphene nanosheet production from compressed graphite using electrochemical exfoliation, Sci. Rep. 8 (2018) 1–8. https://doi.org/10.1038/s41598-018-32741-3.

[10] P. Yu, S.E. Lowe, G.P. Simon, Y.L. Zhong, Electrochemical exfoliation of graphite and production of functional graphene, Curr. Opin. Colloid Interface Sci. 20 (2015) 329–338. https://doi.org/10.1016/j.cocis.2015.10.007.

[11] C. Huo, Z. Yan, X. Song, H. Zeng, 2D materials via liquid exfoliation: a review on fabrication and applications, Sci. Bull. 60 (2015) 1994–2008. https://doi.org/10.1007/s11434-015-0936-3.

[12] J. Yu, J. Li, W. Zhang, H. Chang, Synthesis of high quality two-dimensional materials via chemical vapor deposition, Chem. Sci. 6 (2015) 6705–6716. https://doi.org/10.1039/c5sc01941a.

[13] S. Stankovich, D.A. Dikin, R.D. Piner, K.A. Kohlhaas, A. Kleinhammes, Y. Jia, Y. Wu, S.B.T. Nguyen, R.S. Ruoff, Synthesis of graphene-based nanosheets via chemical reduction of exfoliated graphite oxide, Carbon N. Y. 45 (2007) 1558–1565. https://doi.org/10.1016/j.carbon.2007.02.034.

[14] V. Singh, D. Joung, L. Zhai, S. Das, S.I. Khondaker, S. Seal, Graphene based materials: past, present and future, Prog. Mater. Sci. 56 (2011) 1178–1271. https://doi.org/10.1016/j.pmatsci.2011.03.003.

[15] X. Li, L. Colombo, R.S. Ruoff, Synthesis of graphene films on copper foils by chemical vapor deposition, Adv. Mater. 28 (2016) 6247–6252. https://doi.org/10.1002/adma.201504760.

[16] L. Cheng, K. Yun, A. Lucero, J. Huang, X. Meng, G. Lian, H.S. Nam, R.M. Wallace, M. Kim, A. Venugopal, L. Colombo, J. Kim, Low temperature synthesis of graphite on Ni films using inductively coupled plasma enhanced CVD, J. Mater. Chem. C 3 (2015) 5192–5198. https://doi.org/10.1039/c5tc00635j.

[17] J.J. Wang, M.Y. Zhu, R.A. Outlaw, X. Zhao, D.M. Manos, B.C. Holloway, V.P. Mammana, Free-standing subnanometer qraphite sheets, Appl. Phys. Lett. 85 (2004) 1265–1267. https://doi.org/10.1063/1.1782253.

[18] M. Zhu, J. Wang, B.C. Holloway, R.A. Outlaw, X. Zhao, K. Hou, V. Shutthanandan, D.M. Manos, A mechanism for carbon nanosheet formation, Carbon N. Y. 45 (2007) 2229–2234. https://doi.org/10.1016/j.carbon.2007.06.017.

[19] Y. Seekaew, D. Phokharatkul, A. Wisitsoraat, C. Wongchoosuk, Highly sensitive and selective room-temperature NO_2 gas sensor based on bilayer transferred chemical vapor deposited graphene, Appl. Surf. Sci. 404 (2017) 357–363. https://doi.org/10.1016/j.apsusc.2017.01.286.

[20] J.M. Allen, T.C. Vincent, K.B. Richard, Honeycomb carbon: a review of graphene, Chem. Rev. 110 (2010) 132–145.

[21] P.C. Eklund, J.M. Holden, R.A. Jishi, Vibrational modes of carbon nanotubes; spectroscopy and theory, Carbon N. Y. 33 (1995) 959–972. https://doi.org/10.1016/0008-6223(95)00035-C.

[22] Y.Y. Wang, Z.H. Ni, T. Yu, Z.X. Shen, H.M. Wang, Y.H. Wu, W. Chen, A.T.S. Wee, Raman studies of monolayer graphene: the substrate effect, J. Phys. Chem. C 112 (2008) 10637–10640. https://doi.org/10.1021/jp8008404.

[23] A.C. Ferrari, J.C. Meyer, V. Scardaci, C. Casiraghi, M. Lazzeri, F. Mauri, S. Piscanec, D. Jiang, K.S. Novoselov, S. Roth, A.K. Geim, Raman spectrum of graphene and graphene layers, Phys. Rev. Lett. 97 (2006) 1–4. https://doi.org/10.1103/PhysRevLett.97.187401.

[24] Z. Ni, Y. Wang, T. Yu, Z. Shen, Raman spectroscopy and imaging of graphene, Nano Res. 1 (2008) 273–291. https://doi.org/10.1007/s12274-008-8036-1.

[25] H.Bin Zhang, W.G. Zheng, Q. Yan, Y. Yang, J.W. Wang, Z.H. Lu, G.Y. Ji, Z.Z. Yu, Electrically conductive polyethylene terephthalate/graphene nanocomposites prepared by melt compounding, Polymer (Guildf) 51 (2010) 1191–1196. https://doi.org/10.1016/j.polymer.2010.01.027.

[26] M. Naguib, V.N. Mochalin, M.W. Barsoum, Y. Gogotsi, 25th anniversary article: MXenes: a new family of two-dimensional materials, Adv. Mater. 26 (2014) 992–1005. https://doi.org/10.1002/adma.201304138.

[27] B. Anasori, Y. Xie, M. Beidaghi, J. Lu, B.C. Hosler, L. Hultman, P.R.C. Kent, Y. Gogotsi, M.W. Barsoum, Two-dimensional, ordered, double transition metals carbides (MXenes), ACS Nano 9 (2015) 9507–9516. https://doi.org/10.1021/acsnano.5b03591.

[28] R. Meshkian, L.-Å. Näslund, J. Halim, J. Lu, M.W. Barsoum, J. Rosen, Synthesis of two-dimensional molybdenum carbide, Mo_2C, from the gallium based atomic laminate Mo_2Ga_2C, Scr. Mater. 108 (2015) 147–150. https://doi.org/10.1016/j.scriptamat.2015.07.003.

[29] A. Lipatov, M. Alhabeb, M.R. Lukatskaya, A. Boson, Y. Gogotsi, A. Sinitskii, Effect of synthesis on quality, electronic properties and environmental stability of individual monolayer Ti_3C_2 MXene flakes, Adv. Electron. Mater. 2 (2016). https://doi.org/10.1002/aelm.201600255.

[30] A. Feng, Y. Yu, F. Jiang, Y. Wang, L. Mi, Y. Yu, L. Song, Fabrication and thermal stability of NH4HF2-etched Ti_3C_2 MXene, Ceram. Int. 43 (2017) 6322–6328. https://doi.org/10.1016/j.ceramint.2017.02.039.

[31] A. Feng, Y. Yu, Y. Wang, F. Jiang, Y. Yu, L. Mi, L. Song, Two-dimensional MXene Ti_3C_2 produced by exfoliation of Ti_3AlC_2, Mater. Des. 114 (2017) 161–166. https://doi.org/10.1016/j.matdes.2016.10.053.

[32] J. Halim, M.R. Lukatskaya, K.M. Cook, J. Lu, C.R. Smith, L.Å. Näslund, S.J. May, L. Hultman, Y. Gogotsi, P. Eklund, M.W. Barsoum, Transparent conductive two-dimensional titanium carbide epitaxial thin films, Chem. Mater. 26 (2014) 2374–2381. https://doi.org/10.1021/cm500641a.

[33] H. Wang, H. Cui, X. Song, R. Xu, N. Wei, J. Tian, H. Niu, Facile synthesis of heterojunction of MXenes/TiO_2 nanoparticles towards enhanced hexavalent chromium, J. Colloid Interface Sci. (2019). https://doi.org/10.1016/j.jcis.2019.11.120.

[34] Q. Tang, Z. Zhou, P. Shen, Are MXenes promising anode materials for li ion batteries? Computational studies on electronic properties and Li storage capability of Ti_3C_2 and $Ti_3C_2 \times 2$ (X = F, OH) Monolayer, J. Am. Chem. Soc. 134 (2012) 16909–16916. https://doi.org/10.1021/ja308463r.

[35] Y. Dall'Agnese, P.-L. Taberna, Y. Gogotsi, P. Simon, Two-dimensional vanadium carbide (MXene) as positive electrode for sodium-ion capacitors, J. Phys. Chem. Lett. 6 (2015) 2305–2309. https://doi.org/10.1021/acs.jpclett.5b00868.

[36] C. Eames, M.S. Islam, Ion intercalation into two-dimensional transition-metal carbides: global screening for new high-capacity battery materials, J. Am. Chem. Soc. 136 (2014) 16270–16276. https://doi.org/10.1021/ja508154e.

[37] Y. Xie, M. Naguib, V.N. Mochalin, M.W. Barsoum, Y. Gogotsi, X. Yu, K.-W. Nam, X.-Q. Yang, A.I. Kolesnikov, P.R.C. Kent, Role of surface structure on Li-ion energy storage capacity of two-dimensional transition-metal carbides, J. Am. Chem. Soc. 136 (2014) 6385–6394. https://doi.org/10.1021/ja501520b.

[38] Y. Xie, Y. Dall'Agnese, M. Naguib, Y. Gogotsi, M.W. Barsoum, H.L. Zhuang, P.R.C. Kent, Prediction and characterization of MXene nanosheet anodes for non-lithium-ion batteries, ACS Nano 8 (2014) 9606–9615. https://doi.org/10.1021/nn503921j.

[39] H. Weng, A. Ranjbar, Y. Liang, Z. Song, M. Khazaei, S. Yunoki, M. Arai, Y. Kawazoe, Z. Fang, X. Dai, Large-gap two-dimensional topological insulator in oxygen functionalized MXene, Phys. Rev. B. 92 (2015) 75436. https://doi.org/10.1103/PhysRevB.92.075436.

[40] H. Liu, C. Duan, C. Yang, W. Shen, F. Wang, Z. Zhu, A novel nitrite biosensor based on the direct electrochemistry of hemoglobin immobilized on MXene-Ti_3C_2, Sensors Actuators, B: Chem 218 (2015) 60–66. https://doi.org/10.1016/j.snb.2015.04.090.

[41] M. Naguib, O. Mashtalir, J. Carle, V. Presser, J. Lu, L. Hultman, Y. Gogotsi, M.W. Barsoum, Two-dimensional transition metal carbides, ACS Nano 6 (2012) 1322–1331. https://doi.org/10.1021/nn204153h.

[42] A. Srivastava, C.S.P. Tripathi, V.K. Singh, R.R. Srivastava, S.K. Pandey, S. Rai, R. Dutt, A.K. Patel, Characterizations of nanoscale two-dimensional materials and heterostructures, in: S. Jit, S. Das (Eds.), 2D Nanoscale Heterostructured Materials: Synthesis, Properties, and Applications, Elsevier, 2020. https://doi.org/10.1016/b978-0-12-817678-8.00003-8.

[43] S. Hu, S. Li, W. Xu, J. Zhang, Y. Zhou, Z. Cheng, Rapid preparation, thermal stability and electromagnetic interference shielding properties of two-dimensional Ti_3C_2 MXene, Ceram. Int. 45 (2019) 19902–19909. https://doi.org/10.1016/j.ceramint.2019.06.246.

[44] D. Jariwala, V.K. Sangwan, L.J. Lauhon, T.J. Marks, M.C. Hersam, Emerging device applications for semiconducting two-dimensional transition metal dichalcogenides, ACS Nano 8 (2014) 1102–1120. http://www.ncbi.nlm.nih.gov/pubmed/24476095.

[45] S. Manzeli, D. Ovchinnikov, D. Pasquier, O.V. Yazyev, A. Kis, 2D transition metal dichalcogenides, Nat. Rev. Mater. 2 (2017) 17033. https://doi.org/10.1038/natrevmats.2017.33.

[46] W. Choi, N. Choudhary, G.H. Han, J. Park, D. Akinwande, Y.H. Lee, Recent development of two-dimensional transition metal dichalcogenides and their applications, Mater. Today 20 (2017) 116–130. https://doi.org/10.1016/j.mattod.2016.10.002.

[47] R. Lv, J.A. Robinson, R.E. Schaak, D. Sun, Y. Sun, T.E. Mallouk, M. Terrones, Transition metal dichalcogenides and beyond: synthesis, properties, and applications of single- and few-layer nanosheets, Acc. Chem. Res. 48 (2015) 56–64. https://doi.org/10.1021/ar5002846.

[48] H. Song, A. Tang, G. Xu, L. Liu, M. Yin, Y. Pan, One-step convenient hydrothermal synthesis of MoS_2/RGO as a high-performance anode for sodium-ion batteries, Int. J. Electrochem. Sci. 13 (2018) 4720–4730. https://doi.org/10.20964/2018.05.29.

[49] D. Chakravarty, D.J. Late, Microwave and hydrothermal syntheses of WSe_2 micro/nanorods and their application in supercapacitors, RSC Adv. 5 (2015) 21700–21709. https://doi.org/10.1039/c4ra12599a.

[50] K.J. Huang, J.Z. Zhang, Y. Fan, Preparation of layered $MoSe_2$ nanosheets on Ni-foam substrate with enhanced supercapacitor performance, Mater. Lett. 152 (2015) 244–247. https://doi.org/10.1016/j.matlet.2015.03.130.

[51] Y.-H. Chang, W. Zhang, Y. Zhu, Y. Han, Monolayer $MoSe_2$ grown by chemical vapor deposition for fast photodetection, ACS Nano 8 (2019) 8582–8590.

[52] U. Gupta, D. Gautam, V. Jain, B. Mukherjee, Design and synthesis of two-dimensional materials and their heterostructures, in: S. Jit, S. Das (Eds.), 2D Nanoscale Heterostructured Materials: Synthesis, Properties, and Applications, Elsevier, 2020. https://doi.org/10.1016/b978-0-12-817678-8.00002-6.

[53] Y. Zhao, H. Lee, W. Choi, W. Fei, C.J. Lee, Large-area synthesis of monolayer $MoSe_2$ films on SiO_2/Si substrates by atmospheric pressure chemical vapor deposition, RSC Adv. 7 (2017) 27969–27973. https://doi.org/10.1039/c7ra03642f.

[54] C.M. Hyun, J.H. Choi, S.W. Lee, J.H. Park, K.T. Lee, J.H. Ahn, Synthesis mechanism of MoS_2 layered crystals by chemical vapor deposition using MoO_3 and sulfur powders, J. Alloys Compd. 765 (2018) 380–384. https://doi.org/10.1016/j.jallcom.2018.06.183.

[55] R.D. Mahyavanshi, G. Kalita, K.P. Sharma, M. Kondo, T. Dewa, T. Kawahara, M. Tanemura, Synthesis of MoS_2 ribbons and their branched structures by chemical vapor deposition in sulfur-enriched environment, Appl. Surf. Sci. 409 (2017) 396–402. https://doi.org/10.1016/j.apsusc.2017.03.074.

[56] K.J. Huang, J.Z. Zhang, G.W. Shi, Y.M. Liu, Hydrothermal synthesis of molybdenum disulfide nanosheets as supercapacitors electrode material, Electrochim. Acta 132 (2014) 397–403. https://doi.org/10.1016/j.electacta.2014.04.007.

[57] S. Majety, X.K. Cao, R. Dahal, B.N. Pantha, J. Li, J.Y. Lin, H.X. Jiang, Semiconducting hexagonal boron nitride for deep ultraviolet photonics, Quantum Sens. Nanophotonic Devices IX 8268 (2012) 82682R. https://doi.org/10.1117/12.914084.

[58] D. Golberg, Y. Bando, Y. Huang, T. Terao, M. Mitome, C. Tang, C. Zhi, Boron nitride nanotubes and nanosheets, ACS Nano 4 (2002) 2979–2993. http://www.nanoarchive.org/3950/.

[59] H. Wang, W. Wang, H. Wang, F. Zhang, Y. Li, Z. Fu, Synthesis of boron nitride nanotubes by combining citrate-nitrate combustion reaction and catalytic chemical vapor deposition, Ceram. Int. 44 (2018) 13959–13966. https://doi.org/10.1016/j.ceramint.2018.04.245.

[60] X. Hao, S. Dong, W. Fang, J. Zhan, L. Li, X. Xu, M. Jiang, A novel hydrothermal route to synthesize boron nitride nanocrystals, Inorg. Chem. Commun. 7 (2004) 592–594. https://doi.org/10.1016/j.inoche.2004.02.026.

[61] T. Chen, Q. Zhang, Z. Xie, C. Tan, P. Chen, Y. Zeng, F. Wang, H. Liu, Y. Liu, G. Liu, W. Lv, Carbon nitride modified hexagonal boron nitride interface as highly efficient blue LED light-driven photocatalyst, Appl. Catal., B: Environ. 238 (2018) 410–421. https://doi.org/10.1016/j.apcatb.2018.07.053.

[62] J. Xu, F. Wu, Q. Jiang, J.K. Shang, Y.X. Li, Metal halides supported on mesoporous carbon nitride as efficient heterogeneous catalysts for the cycloaddition of CO_2, J. Mol. Catal., A: Chem. 403 (2015) 77–83. https://doi.org/10.1016/j.molcata.2015.03.024.

[63] W.S. Yang, B.W. Park, E.H. Jung, N.J. Jeon, Y.C. Kim, D.U. Lee, S.S. Shin, J. Seo, E.K. Kim, J.H. Noh, S.Il Seok, Iodide management in formamidinium-lead-halide-based perovskite layers for efficient solar cells, Science 356 (2017) 1376–1379. https://doi.org/10.1126/science.aan2301.

[64] C. Bao, J. Yang, S. Bai, W. Xu, Z. Yan, Q. Xu, J. Liu, W. Zhang, F. Gao, High performance and stable all-inorganic metal halide perovskite-based photodetectors for optical communication applications, Adv. Mater. 30 (2018) 1–8. https://doi.org/10.1002/adma.201803422.

[65] M. Han, J. Sun, L. Bian, Z. Wang, L. Zhang, Y. Yin, Z. Gao, F. Li, Q. Xin, L. He, N. Han, A. Song, Z. Yang, Two-step vapor deposition of self-catalyzed large-size PbI_2 nanobelts for high-performance photodetectors, J. Mater. Chem. C 6 (2018) 5746–5753. https://doi.org/10.1039/C8TC01180J.

[66] G. Zhu, P. Liu, M. Hojamberdiev, J. Zhou, X. Huang, B. Feng, R. Yang, Controllable synthesis of PbI_2 nanocrystals via a surfactant-assisted hydrothermal route, Appl. Phys. A. 98 (2009) 299. https://doi.org/10.1007/s00339-009-5412-y.

[67] H. Sun, X. Zhu, D. Yang, J. Yang, X. Gao, X. Li, Morphological and structural evolution during thermally physical vapor phase growth of PbI_2 polycrystalline thin films, J. Cryst. Growth 405 (2014) 29–34. https://doi.org/10.1016/j.jcrysgro.2014.07.043.

[68] J. Liu, Y. Xue, Z. Wang, Z.-Q. Xu, C. Zheng, B. Weber, J. Song, Y. Wang, Y. Lu, Y. Zhang, Q. Bao, Two-dimensional CH3NH3PbI3 perovskite: Synthesis and optoelectronic application, ACS Nano 10 (2016) 3536–3542. https://doi.org/10.1021/acsnano.5b07791.

[69] Z. Cheng, M. Chen, L. Xie, L. Peng, M. Yang, M. Li, Bioaugmentation of a sequencing batch biofilm reactor with *Comamonas testosteroni* and *Bacillus cereus* and their impact on reactor bacterial communities, Biotechnol. Lett. 37 (2015) 367–373. https://doi.org/10.1007/s10529-014-1684-1.

[70] P. Wangyang, H. Sun, X. Zhu, D. Yang, X. Gao, Mechanical exfoliation and Raman spectra of ultrathin PbI_2 single crystal, Mater. Lett. 168 (2016) 68–71. https://doi.org/10.1016/j.matlet.2016.01.034.

CHAPTER

Structure and properties of 2D materials

3

Khouloud Abid[a,b], Nour Hamza Belkhir[a,c], Ratiba Wali[a] and Ramzi Maâlej[a,d]

[a]*Georesources Materials Environment and Global Changes Laboratory (GEOGLOB), Faculty of Sciences of Sfax, University of Sfax, Sfax, Tunisia,* [b]*Institute of Advanced Technologies for Energy "Nicola Giordano", Messina, Italy,* [c]*Physics Department, Faculty of Sciences, University of Hafr Al Batin, Hafar Al Batin, Kingdom of Saudi Arabia,* [d]*Fluoink Nanotechnologies, Pouderiere Industrial Zone, Sfax, Tunisia*

3.1 Introduction

3.1.1 Atomic structure

Two-dimensional (2D) materials have a unique structure compared with their bulk compounds. They are layered materials with weak van der Waals bonds between layers but strong covalent bonds within interlayers.

In this section, we investigate the atomic structure of the most studied 2D materials such as graphene, transition metal dichalcogenides (TMDCs), transition metal (TM) carbides, and/or carbonitrides (MXenes), hexagonal boron nitride (h-BN), and black phosphorus (BP).

3.1.1.1 Graphene

Graphene has a honeycomb grid structure organized into a planar hexagonal network of carbon atoms (Fig. 3.1A) with D_{6h} as a space group [1]. A strong covalent bond interconnects these atoms with hybridization sp^2. The C–C bond length is equal to 1.42 Å, and the lattice constant is $a = \sqrt{3}\, d \approx 2.46$ Å (Fig. 3.1B) [2]. Moreover, the thickness of 1L-graphene is 0.335 nm [3]. The unit cell of graphene contains two nonequivalent carbon atoms labeled A and B that are separated by an angle $\beta = 120$ degrees (Fig. 3.1B) [1]. In fact, this nonequivalence is because of the inverse symmetry that exists in D_{6h} point group.

In addition to graphene, two notable derivatives that have garnered considerable attention within the research community are graphene oxide (GO) and reduced GO (rGO). These derivatives, GO and rGO, have exhibited significant promise in various scientific investigations. The first one has a 2D arrangement similar to graphene where the carbon monolayer is functionalized covalently with oxygen-enclosing groups such as carbonyl groups (C=O), oxygen epoxides (bridging oxygen atoms), phenols (C_6H_6O), hydroxyl (–OH), and organo-sulfate groups [4]. It is a nonstoichiometric chemical compound consisting of a variable proportion of carbon, oxygen, and hydrogen [4].

Graphene and 2D Materials in Heat Transfer: Fundamentals and Applications. DOI: https://doi.org/10.1016/B978-0-12-821964-5.00002-3

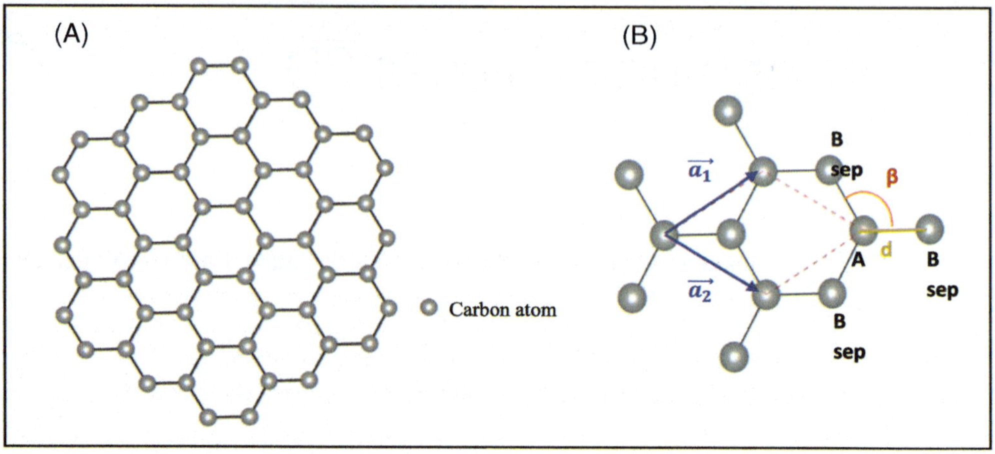

FIGURE 3.1

(A) Hexagonal honeycomb lattice of graphene and (B) unit cell of graphene where $\vec{a_1}$ and $\vec{a_2}$ are the primitive unit vectors.

The second derivative (rGO) has a heterogeneous structure composed of a graphene-like basal plane with structural defects and oxidized chemical groups. This material is produced by a reductive exfoliation treatment of GO [5].

3.1.1.2 Transition metal carbides and/or carbonitrides (MXenes)

The MXene family refers to 2D TM carbides, nitrides, and carbonitrides with a general formula of $M_{1.33}XT_z$ or $M_{n+1}X_nT_z$, where M represents a TM (Sc, V, Mo, Ti, Cr, etc.), X can be either a nitrogen atom or carbon atom, and T_z is a terminal group (–F, –OH, =O, or –Cl) (see Fig. 3.2). The index z denotes the surface functionalities number, whereas n is a stoichiometric variable, with a value varying from 1 to 3 [6,7]. According to Champagne et al., the MXene's single layer has a thickness of around 1 nm, which is sensitive to the nature of the terminal groups as well as the index value "n" [6].

This family is divided into four groups [8]. The first one is the monometal elements with the formula $M_{n+1}X_n$, where M-atoms have a compact hexagonal structure, and X-atoms are inserted into octahedral sites [6,7]. Among this group, there are Ti_2C, V_2C, and Nb_4C_3. The stoichiometric variable "n" can influence the configuration of atoms [7]. For $n = 1$ (M_2X), MXens materials have a hexagonal structure that becomes cubic centered when $n = 2$ or 3 with chemical formulas M_3X_2 and M_4X_3, respectively [7–9]. For instance the lattice parameter of Ti_2C is equal to 3.3338 Å [9].

The second is named solid solution double-metal with formula $(M^1, M^2)_3X_2$ such as $(Ti, V)_3C_2$ and $(Cr, V)_3C_2$[7–9]. The next group is ordered double-metal elements, divided into two classes [8]. The first one is the in-plane double-M (i-MAX phase), such as $(Mo_{2/3}Y_{1/3})_2CT_z$, whereas the second one is the out-of-plane double-M (o-MAX phase), such as $Mo_2TiC_2T_z$ [6,7]. The crystal structure of the in-plane double-M phase can be described by a monoclinic structure (C2/c) [10].

The last MXenes group is the vacancies group, classified into two types of vacancies. These vacancies can be either ordered distributed ($Mo_{1.33}CT_z$ and $W_{1.33}CT_z$) or randomly distributed as $Nb_{1.33}CT_z$ [6,7].

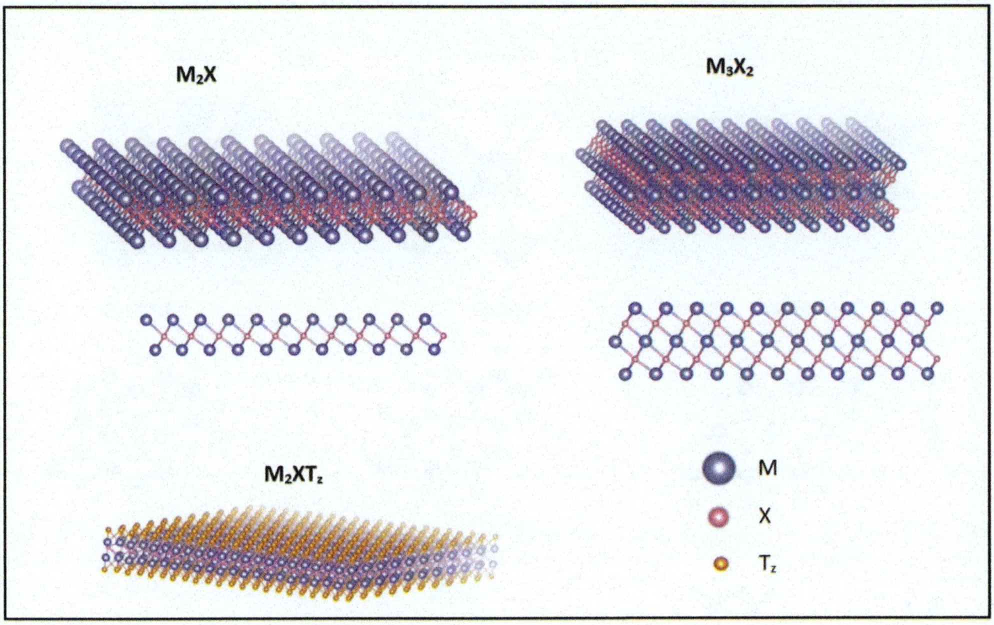

FIGURE 3.2

Structures of typical M_2X, M_3X_2, and M_2XT_z.

3.1.1.3 Transition metal dichalcogenides (TMDCs)

TM dichalcogenides (TMDCs) are an interesting class of layered 2D materials. Usually, TMDCs have a hexagonal structure with a general chemical formula MX_2, where M is a TM such as tungsten (W), molybdenum (Mo), or niobium (Nb), and X represents a chalcogen such as selenium (Se), sulfur (S), or tellurium (Te) [11].

For this family, each MX_2 sheet presents three planes of atoms in a pentagonal or a hexagonal lattice where the metal atoms plane is intercalated between the two chalcogen atoms planes (X–M–X) (see Fig. 3.3A) [11]. The metal and chalcogen layers are connected by a strong covalent bond, whereas the chalcogen layer links are of a van der Waals (VdW) type [12].

The TM dichalcogenides (TMDs) family encompasses a diverse set of 40 stable compounds formed through various combinations of constituent elements [11]. Within this family the bonding between the metal and chalcogen atoms is characterized by the presence of two tetrahedrons oriented in opposing directions [12,13]. Consequently the arrangement of these tetrahedra gives rise to two distinct phases denoted as T and H, which exhibit different stacking polytypes [12]. In the H-phase the two tetrahedra (upper and lower) are symmetrically arranged, resulting in a trigonal prismatic configuration (Fig. 3.3B). For the T-phase the tetrahedrons are rotated to each other by 180 degrees, creating a distorted octahedral structure (Fig. 3.3C) [12].

The most frequently studied polytypes are 1T, 1T′ (distorted 1-H), 1H, 2H, and 3R. The digit is assigned to the number of layers X–M–X inside a single unit cell, whereas T, H, and R illustrate tetragonal, hexagonal, and rhombohedral crystal symmetry, respectively [12,14].

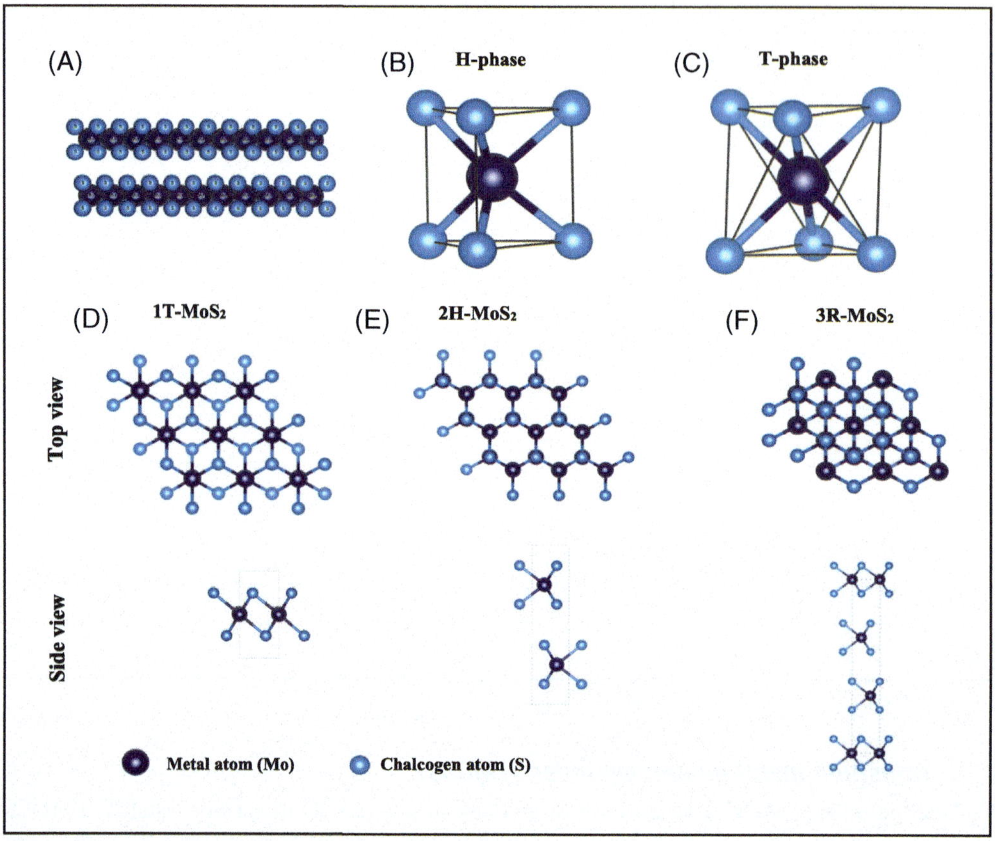

FIGURE 3.3

Atomic structure of TMDCs: (A) M-atom's layer sandwiched between two chalcogen atom's layers (B) T-phase (C) H-phase (D) structure of 2H-MoS_2 in top view and side view (E) structure of 3R-MoS_2 in top view and side view (F) structure of 1H-MoS_2 in top view and side view.

Table 3.1 summarizes the crystal system, point group, and symmetry group of different polytypes of TMDs. For instance, MoS_2 can exist under several polytypes, such as 1T, 2H, and 3R [15]. Moreover, molybdenum disulfide (MoS_2) possesses two stable polytypes. The first one is 2H-MoS_2, which is the most stable polytype with a lattice parameter $a = 3.15$ Å and $c = 12.30$ Å (Fig. 3.3D). The second stable polytype is 3R-MoS_2 with parameters lattice $a = 3.17$ Å and $c = 18.38$ Å (Fig. 3.3E) [14,15]. Finally, 1T-MoS_2 is a metastable phase of MoS_2 with a lattice parameter $a = 5.60$ Å and $c = 5.99$ Å (Fig. 3.3F) [14,15].

3.1.1.4 Hexagonal boron nitride (h-BN)

Hexagonal boron nitride (h-BN) is a layered structure. It is an inorganic material similar to graphene, also designated "white graphene" because of its white color [11]. The h-BN is a honeycomb hexagonal

Table 3.1 Different polytypes of TMDCs.

	T polytype		H polytype		
Polytypes	**1T′**	**1T**	**1H**	**2H**	**3R**
Cristal system	Monoclinic	Trigonal	Hexagonal		Rhombohedral
Point group	C2v	D6d	D6h		C3v
Symmetry's group	$P2_1$/m	$P\bar{3}$ m1	$P\bar{6}$ m2	$P6_3$/mmc	*R*3m
References	[12,16]	[12,14,16]	[12]	[12,14,16]	[12,14]

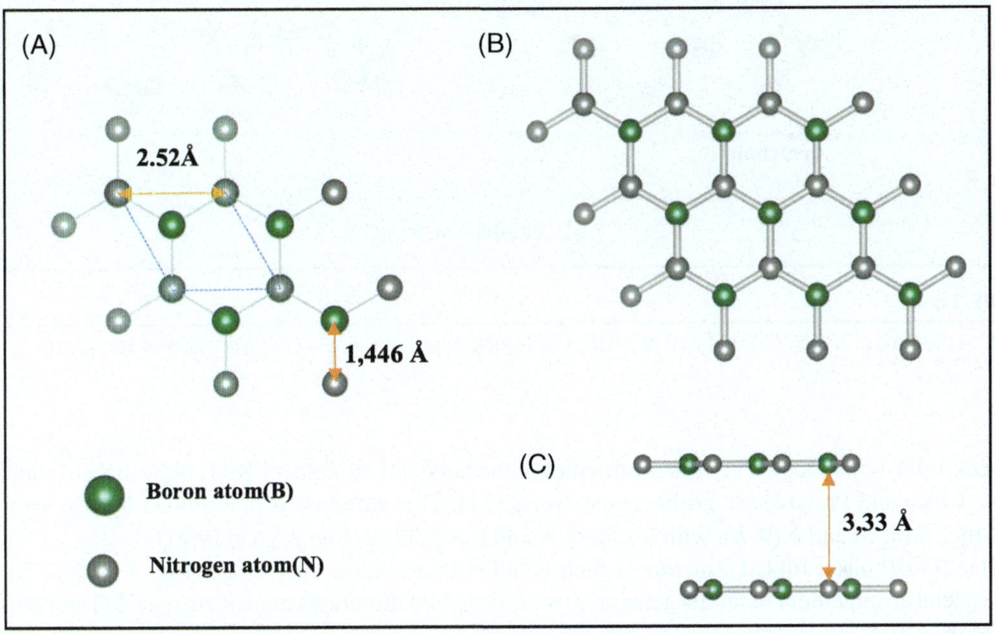

FIGURE 3.4

Atomic structure of h-BN: (A) unit cell of h-BN and (B) interlayer distance.

ring, where boron (B) and nitrogen (N) atoms are alternately arranged in a 2D thin sheet (Fig. 3.3B). The lattice constant of h-BN is equal to 2.52 Å (Fig. 3.4A) [17].

Nitrogen atoms are bonded with three neighboring boron atoms where the distance between an N atom and a neighboring B atom is 1446 Å (Fig. 3.4A) [17]. They are linked in the plane by a strong covalent force with sp^2 hybridization [17]. However, the bonding strengths of interlayers are VdW weak interactions [11]. The distance between the two plans is 333 Å (Fig. 3.4B) [18].

3.1.1.5 Black phosphorus

BP is a layered hexagonal compound with an sp^3 hybridization and a nonplanar structure (Fig. 3.5A) [19]. Inside these layers, P-atoms are linked by a strong covalent bond, whereas interlayers are connected

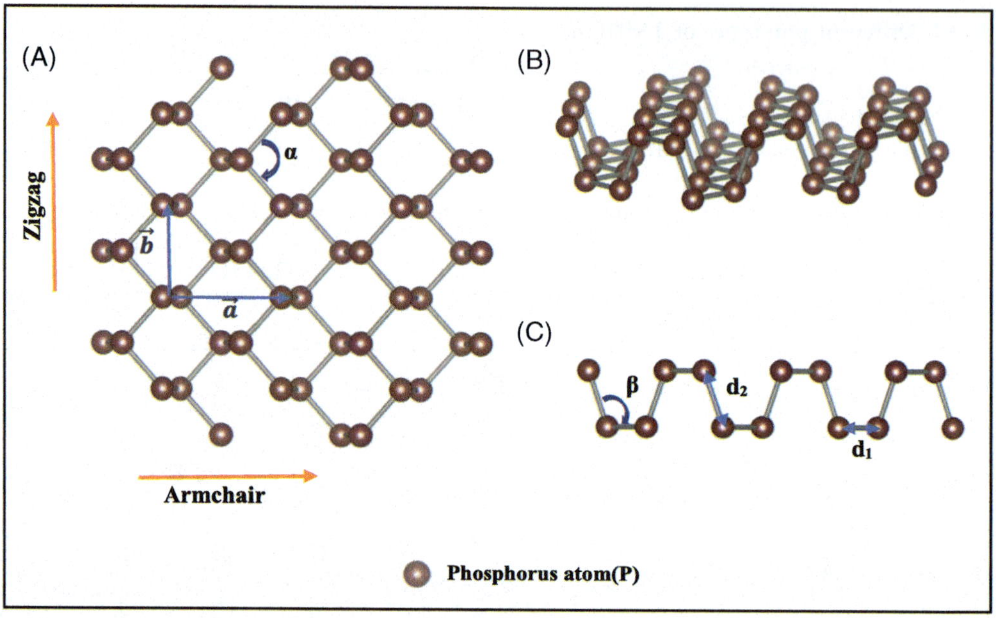

FIGURE 3.5

Atomic structure of h-BN: (A) unit cell of h-BN, (B) honeycomb lattice, and (C) side view of two layers.

by weak VdW bonds [20]. The crystal structure of the bulk and monolayer BP is orthorhombic, and they have a Cmca and Pmna space group, respectively [21]. This structure is determined by two vectors in the lattice $\vec{a}$ $(a, 0)$ and $\vec{b}$ $(0, b)$, with $a = 4.58$ Å and $b = 3.32$ Å (Fig. 3.5A) [19,21].

The 2D-BP has a folded structure, which is holed in a honeycomb lattice (Fig. 3.5A and B) [22]. This special arrangement of atoms generates two orthogonal directions named zigzag (ZZ) and armchair [19] (Fig. 3.5A).

The folded monolayer is composed of two atomic rows that are arranged in parallel. Within each atomic row, adjacent atoms are connected by bonds with a length of $d_1 = 2.24$ Å and an angle of $\alpha = 96$ degrees. Similarly the bonds between the two rows have a length of $d_2 = 2.28$ Å and an angle of $\beta = 103.51$ degrees (Fig. 3.5C) [19]. This anisotropic structure confers the BP's prominent features [19].

3.1.2 Features of two-dimensional materials

The 2D materials have gained prominent electrical, magnetic, and optical features, making them potential candidates for spintronic as well as optoelectronic applications owing to their exceptional atomic structure [23].

3.1.2.1 Mechanical properties

The unique structure of 2D materials generates remarkably distinct properties compared with their bulk counterparts. Among these properties, there are mechanical characteristics such as high Young's modulus (E), high breaking strength (σ_m), and toughness in the plan [24].

2D materials are the most suitable candidates for flexible, transparent electronic and composite applications [25]. As a result of the influence of their nanostructure on these properties, there are two deformation modes: stretching in the plane and bending out of the plane [26]. The 2D crystals exhibit exceptional characteristics such as low mass and exceptional carrier mobility, as well as high Young's modulus and notable mechanical property anisotropy both within the plane and between layers. These features make them highly promising for various scientific and technological applications. The lightweight nature of these crystals allows for efficient manipulation and integration into devices, whereas their exceptional carrier mobility facilitates rapid and efficient charge transport. In addition, their high Young's modulus indicates strong mechanical strength, enabling them to withstand substantial external forces. Furthermore the pronounced anisotropy of their mechanical properties offers distinct behavior depending on the direction of applied forces, providing opportunities for tailored material performance in specific applications. Overall the unique combination of these characteristics positions 2D crystals as compelling candidates for further exploration and utilization in diverse scientific and engineering fields [26].

3.1.2.1.1 Young's modulus

Young's modulus (E) is also called the modulus of elasticity [27]. It is the ratio of tensile stress (σ: force per unit area) to axial strain (ε: proportional deformation) in the material's linear elastic region; in fact, Young's modulus informs on the tensile stiffness of a solid substance, and it is determined using Eq. (3.1)[27]:

$$E = \frac{\sigma}{\varepsilon} \tag{3.1}$$

The 2D materials have a higher Young's modulus compared with their bulk counterparts [26]. This value and the important resistance of 2D materials result from covalent bonds between atoms. However, the easy slippage between them is ensured by the VdW force [3].

Graphene has the strongest mechanical properties because of the stronger in-plane covalent carbon atoms bonds [26]. Zhang et al. suggest that graphene's superhigh strength and stability make it an ideal candidate for various applications, such as flexible electronic sensors and composites [25,28]. This type of material is the most suitable candidate for flexible and transparent electronic and composite applications [25].

Table 3.2 summarizes the different mechanical properties of various 2D materials. In fact, graphene and h-BN have higher Young's modulus, about five and four times that of steel, respectively [11]. Moreover, graphene has 100 times higher breaking strength than steel and 70 times more than 1L-MoS_2 [11,29]. This feature makes them an ideal candidate for various applications, such as manufacturing reinforcing components or flexible electronic sensors and appliances [28].

Recent studies show WSe_2 has a lower modulus than MoS_2 and WS_2[25]. This finding is because of an enhancement in the network constant and a reduction in the load transfer, causing weakening of the bond between chalcogen and metal when switching from Mo to W and S to Se [25]. The various

Table 3.2 Mechanical properties of some 2D materials.

	Young's modulus (GPa)	Poisson's ratio	Breaking strength (GPa)
Steel	207 [30]	0.3 [30]	1.77 [31]
Graphene	1000 [6,27,32]	0.186 [27]	130 [25]
TMDCs	270 [6,32]		23 [25]
MoS_2	272 [27]	0.3 [24]	12.4 [29]
WS_2	170 [32]		5.6 [29]
WSe_2	110 [29]		
$MoTe_2$			
h-BN	870 [6]	0.22 [32]	70.5 [29]
BP	106.4 (ZZ) 41.3 (Ar) [25]	0.064 ~ 0.703 [24]	18 (ZZ) 8 (Ar) [25]
MXene	330 [6]	0.227 [34]	17.3 [25]
$Ti_3C_2T_x$	312 [33]		
Mo_2C			
Graphene oxide (GO)	207.6 [32]	0.165 [35]	12 [36]
Reduced GO	250 [32]	–	–

2D, *two-dimensional;* BP, *black phosphorus.*

crystalline orientations reveal high anisotropy in Young's modulus of BP. According to ZZ and armchair directions the elasticity moduli of the BP monolayer are 106.4 and 41.3 GPa, respectively.

Among the MXene family, Young's modulus of 1L-$Ti_3C_2T_z$ is three times that of graphene and two times that of h-BN. Moreover, it is almost similar to that of MoS_2 and GO. Note that nitride-based MXenes have higher values than carbide-based MXenes [6]. Although the MXenes have a lower Young's modulus than other 2D materials, their higher bending stiffness and other properties make them promising competitors for composites with polymers, carbon nanotubes, or oxides [6].

The mechanical properties of 2D materials are affected by the number of layers; as the number of layers increases, Young's modulus and fracture strength drop [25]. This variation is mainly induced by stacking errors between layers [25].

3.1.2.1.2 Breaking strength

Another important property of 2D materials is the breaking strength or maximum tensile strength (σ_m). It is defined as the maximum force that the material can withstand before breaking when it is stretched [27]. For instance, pristine graphene has a breaking strength of 130 GPa [27].

In general the breaking strength of 2D materials is considerably higher [28]. Notably, graphene demonstrates an exceptional breaking strength approximately 100 times higher than that of steel, with a maximum strength of approximately 100 GPa at a breaking strain close to 25%[26]. Furthermore, investigations conducted by Ng et al. reveal that the monolayer of MoS_2 exhibits a notable breaking strength of up to 22 GPa, surpassing steel by a factor of 30 [11]. These findings underscore the remarkable mechanical properties possessed by these 2D materials, suggesting their potential for various applications requiring high-strength materials.

For the TMDC family, the tensile strength of the WX_2 is higher than the MoX_2. Nonetheless, in the case of a similar TM, the most resistant are sulfides (MS_2), and the weakest are tellurides (MTe_2) [25].

For BP an anisotropic Young's modulus produces a breaking strength exhibiting anisotropy in the armchair and ZZ direction [26]. Most 2D materials can be used for flexible electronic devices, except 2D dielectrics such as mica and h-BN, at their breaking strain greater than 7% compared to that of current substrates, notably polydimethylsiloxane or polyimide [26]. In addition, BP's single layer is very flexible, and its tensile deformation can reach up to 27% and 30% in the ZZ and armchair directions, respectively [25].

3.1.2.2 Thermal properties

Thermal conductivity K is the solid ability to conduct heat [37]. This conductivity is related to the Fourier law, given by Eq. (3.2).

$$J = -K.\nabla T \tag{3.2}$$

where J is the thermal flux, and ∇T is the temperature gradient.

Thermal conductivity plays a vital role in thermal management and energy device engineering [38]. It is judged one of the important intrinsic material properties [38]. In fact, materials with higher thermal conductivity are recommended in most advanced applications, such as electronics, storage systems, and energy conversion, to solve overheating problems and improve the heat dissipation ratio [38].

Pure metals, such as gold or silver, exhibit an extremely low level of thermal conductivity because of the limited contribution from phonons. In these materials, heat transfer primarily occurs through the movement of free electrons, which dominate the thermal conductivity behavior. However, when considering 2D metals, this conventional understanding may not hold true. The unique structural and electronic properties of 2D metals introduce complexities that can significantly affect their thermal conductivity characteristics. Therefore the thermal behavior of 2D metals cannot be solely attributed to the presence of free electrons, warranting further investigation and exploration of alternative mechanisms governing thermal transport in these materials [39].

The thermal conductivity of 2D compounds is hugely dependent on their crystalline structure [37]. According to Gu et al., materials with a planar structure have high thermal conductivity, such as graphene and boron nitride. In contrast, low thermal conductivity is typically observed in 2D materials with a two- or three-layer crystalline structure, not including mirror symmetry, for example, 1T-TMDs. Moreover, tri-layer materials with mirror symmetry have medium thermal conductivity, such as 2H-TMDs [37].

Graphene's thermal conductivity is extensively studied in the literature [39]. At room temperature, it is principally because of lattice vibrations, given that electrons contribute around 10%[39]. A single layer of graphene has an exceptional in-plane thermal conductivity that can reach 5000 W m^{-1} K^{-1}, which is 12 times that of silver (406 W m^{-1} K^{-1}) and more than 2 times that of diamond (2310 W m^{-1} K^{-1}), respectively [39,40].

Thermoelectric performance refers to a material's capacity to convert heat into electrical power [41]. This performance is quantitatively evaluated using a dimensionless figure of merit known as "ZT" [42]. The assessment of thermoelectric performance in 2D materials is closely linked to their thermal conductivity [42]. The correlation between thermoelectric performance and thermal conductivity can be described by Eq. (3.3).

$$Z = \frac{\sigma S^2}{K} \tag{3.3}$$

where σ, S, and Z correspond to the electrical conductivity, the Seebeck coefficient, and merit factor, respectively. The comprehension of thermal properties in 2D materials predominantly relies on their specific application, including cooling mechanisms and thermoelectric energy conversion. By investigating and characterizing these properties, scientists can gain valuable insights into the behavior and performance of these materials in various thermal management and energy conversion systems. This knowledge is essential for optimizing the design and efficiency of 2D materials in practical applications, thereby advancing the field of thermal science and engineering [42].

GO exhibits a noticeably reduced thermal conductivity compared to pristine graphene [43]. Chen and Li conducted a comprehensive study on the relationship between the thermal conductivity of GO and its oxidation degree, determining values of approximately 72 and 670 W m^{-1} K^{-1} for oxidation degrees of 0.35 and 0.05, respectively, at room temperature [43]. This disparity in thermal conductivity can be attributed to the average free path of phonons being hindered by the presence of oxygen-containing functional groups within the GO structure, leading to a decrease in thermal conductivity [43].

At room temperature the h-BN's thermal conductivity is estimated to be 400 W m^{-1} K^{-1}[44]. According to Wang et al., its heat conductivity is typically anisotropic; it is estimated at 300 W m^{-1} K^{-1} perpendicularly to the *c*-axis, whereas, in the parallel direction, it is around 30 W m^{-1} K^{-1}[44].

3.1.2.3 Tribological properties

This section will focus on the notable tribological characteristics of 2D materials that have been extensively investigated in the existing body of literature. The term "tribology," derived from Greek, encompasses three fundamental subjects: lubrication, friction, and wear. As a field of study, tribology significantly impacts various aspects of our everyday lives. These three mechanisms are interrelated and interconnected. Friction, as a mechanism, entails the resistance encountered during relative motion, ultimately leading to material degradation, representing the wear mechanism. Lubrication, whether through the use of fluids or solids, serves as a means to mitigate both wear and friction phenomena.

2D-layered materials, such as graphene and h-BN, exhibit remarkable mechanical, chemical, thermal, and electrical properties, which give rise to extraordinary friction behaviors [26]. These materials have garnered significant attention in the field of lubrication because of their exceptional performance as lubricants and nanolubricants [32]. This superior lubricating capability can be attributed to the interlayer interactions characterized by low bonding strength and a slippery nature, significantly reducing friction and wear coefficients [26,32]. In the study conducted by Liu et al., it was observed that the lubrication performance of graphene is influenced by several factors, including the number of layers, morphology, composition, and experimental conditions [45]. It has been found that achieving superlubricity requires the friction coefficients of ultrathin 2D materials to be below 0.01 [28]. Specifically, the mechanical properties of 2D-MoS2, such as superlubricity, exhibit sensitivity to shear strength when subjected to a normal load of 24.6 MPa [46]. This characteristic highlights the significant tribological performance of ultrathin 2D materials in various industrial applications, such as microelectro-mechanical systems and automobiles [3].

Several factors can affect the tribological features of 2D materials. For instance the lubrication performance of graphene is sensitive to the thickness and growth environment [33]. In addition to

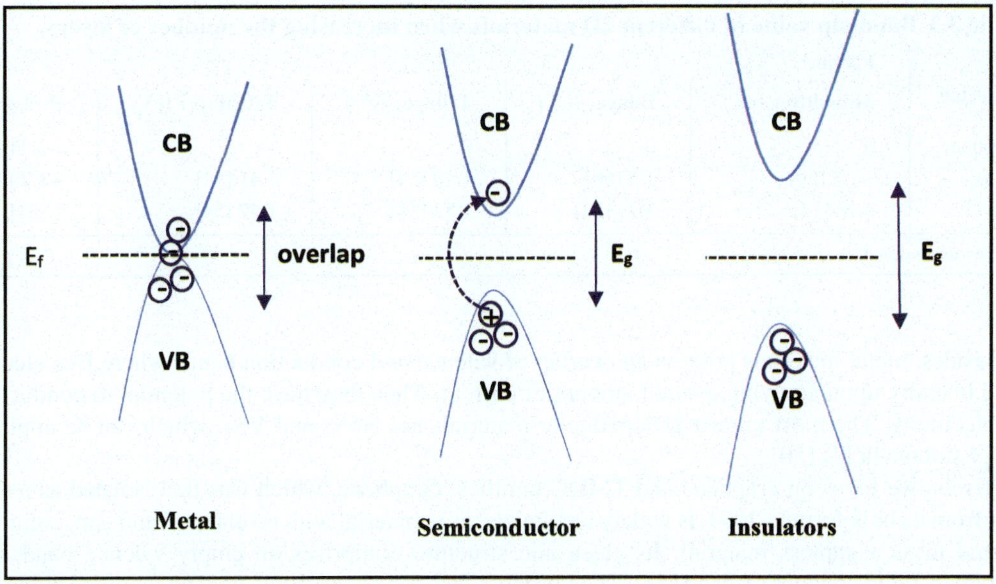

FIGURE 3.6

Classification of materials based on semiconductor theory.

graphene, h-BN can be used as lubricating microparticles integrated into a porous surface, water, or lubricating oil. Therefore lubrication performance depends on concentration, type, and particle size [34]. Furthermore the friction coefficient of h-BN depends on the temperature, according to Podgornik et al.'s investigation [47]. Indeed, this coefficient ranges from 0.2 to 0.26 at room temperature, whereas, in the humid atmosphere, it may be less than 0.1 and reach lower values in vaseline or water [34].

Moreover, the friction behavior of different epitaxial structures on SiC or Si, such as graphene, h-BN, $NbSe_2$, and MoS_2, depends on their thickness [26]. In other words, when the number of layers increases, the friction coefficient decreases, reaching a similar value to that of bulk. For example, the friction coefficient of MoS_2 heterostructure is estimated to be around 10^{-4}[26].

3.1.2.4 Electronic conductivity

Based on the principles of semiconductor theory, materials can be categorized into three distinct groups: insulators, semiconductors, and metals, as illustrated in Fig. 3.6. Specifically an insulator is characterized by a substantial energy bandgap ($\Delta E > 3.2$ eV) existing between its valence and conduction bands. Among the extensively investigated 2D insulator materials are hexagonal boron nitride (h-BN) and mica, as highlighted by Illarionov et al. [48]. These materials are deemed highly promising for the application of 2D field-effect transistors (FETs).

A semiconductor is a material possessing electrical features between those of good insulators and conductors with 1 eV $\leq \Delta E \leq$ 3.2 eV such as molybdenum disulfide (MoS_2), tungsten disulfide (WS_2), and BP. This category is useful for sensing, LED, and FET applications [49].

Table 3.3 Bandgap value of different 2D materials when increasing the number of layers.

Materials	Forms				
	Monolayer (eV)	Bilayer (eV)	Trilayer (eV)	Tetralayer (eV)	Bulk (eV)
Graphene	0	0	0	–	−0.04 [58]
MoS_2	~1.8 [59]	1.5 [60]	1.46 [54]	1.41 [54]	~1.29 [59]
WS_2	1.98 [54]	1.73 [54]	1.53 [54]	1.47 [54]	1 [61]

2D, *two-dimensional.*

Besides, metal materials possess an overlap of valence and conduction bands where free electrons move liberally upon applying a small amount of current. Thus they have the potential to conduct heat and electricity. The most known 2D conductor materials are NbS_2 and VS_2, which can be applied in energy storage fields [50].

Herein, we focus on graphene and TMDC families. Graphene, which was first isolated as a single layer from its bulk form in 2004, is widely recognized as a material with no energy band gap, commonly referred to as a gapless material. Its electronic structure comprises an empty valence band and a completely filled conduction band [51]. The introduction of an electrostatic field or chemical doping can result in the generation of free carriers in graphene [52]. Furthermore, it is noteworthy that altering the number of graphene layers changes its electrical properties. Specifically, when graphene is in a bilayer configuration, it exhibits a semimetal behavior [52].

The graphene family encompasses graphene as well as its derivatives, namely, GO and rGO. The key distinction between rGO and GO lies in the presence of oxygen molecules, which significantly influences their conductivity. Specifically the oxidation process of graphene results in the development of insulator characteristics in GO, whereas the conductive behavior observed in rGO is attributed to the reoxidation process of GO [53]. The electrical properties of GO are highly sensitive to various factors, including the number of layers and the carbon-to-oxygen ratio (C/O). Notably the range of C/O ratios, spanning from 2/1 to values exceeding 6/1, leads to a transition from nonconductive to conductive behavior [53].

Beyond the graphene family, TMDC materials are extensively employed to fabricate potential heterostructure for electronic uses thanks to the indirect-to-direct bandgap crossover [54]. Indeed the 2D materials' bandgap value and nature are susceptible to their thickness and size [55]. In fact a decrease in the number of layers of 2D materials leads to increased carrier mobility and bandgap energy. A direct bandgap is observed in a single layer of 2D TMDCs that turns into an indirect bandgap when the thickness is higher or comparable to two layers. This variation is because of the quantum confinement where the highest bandgap energy is obtained in a single layer (Table 3.3). This outstanding feature conveys an increasing interest worldwide. Furthermore the TMDCs' bandgap is not only sensitive to thickness but also to temperature. Indeed, an increase in the temperatures results in a decrease in the bandgap value [56,57]. Chen and O'Donnell have related temperature and bandgap energy through the following empiric formula presented by Eq. (3.4)[56].

$$E_g(T) = \left\{E_g(0) - < \hbar\omega > \left[\frac{\coth\langle\hbar\omega\rangle}{2kT} - 1\right]\right\} \tag{3.4}$$

where $E_g(0)$ is the indirect/direct bandgap value at 0 K. $\hbar\omega$ is the average phonon energy, and S represents the dimensionless coupling constant [57]. T is the temperature, and K is the Boltzmann constant.

The classification of materials is determined not only by the semiconductor theory but also by the conductivity value, which signifies the material's capacity to conduct electrical current. Conductivity refers to the inherent capability of a substance to transmit current. According to this parameter, metals exhibit high conductivity because of the unrestricted movement of free charge carriers, whereas insulators exhibit low conductivity because these charge carriers are confined and unable to move freely.

The conductivity of rGO is 10^4 S m^{-1}, whereas that of graphene is about 10^6 S m^{-1}. The combination of graphene and its derivatives with other materials can be used to enhance conductivity. In this framework, polymers/graphene significantly improve conductivity in several fields, such as energy storage and sensing materials [62,63]. The conductivity depends on multiple factors. Indeed, temperature and phase type significantly influence, whereas thickness slightly affects conductivity [64]. For instance, these factors reveal different impacts on the conductivity of MoS_2 where a change in variation of one magnitude in thickness remains almost a constant conductivity [65]. Moreover, a decrease in temperature leads to a decrease in conductivity [65]. Furthermore the change from 2H to 1T phase reveals a rise of conductivity of about 10^3, where the conductivity of 1T-MoS_2 varies from 10^3 to 10^4 [62,66]. For graphene the conductivity variation is similar to the TMDCs family, where both the number of layers and temperature affect the carrier mobility. Indeed an increase in the number of layers and temperature leads to a decrease and increase in conductivity, respectively [64].

In addition to graphene and TMDC families, metal carbide and nitride materials abbreviated MXenes belong to 2D materials with $M_{n+1}X_nT_X$ as a general formula. The TM is symbolized as M (Ti, Mo, Nb, V, and Zr); X shows carbon and/or N; and surface termination is symbolized as T (–OH, –F, =O). The index n is equal to 1,2,3. Compared to 2D materials, this family possesses unique electronic properties mainly affected by the M-layer, known as metal layers [67]. Indeed the conductivity of this family depends on the electronic behavior revealed by them (metallic, semiconductor, and insulator) [68]. According to Wang et al., Mo- and Ti-based MXenes have high electrical conductivity [68]. Among this family, Ti_3C_2 is the most studied material where its conductivity is equal to 850 S m^{-1} resulting in great potential for use in fuel cells and Li-ion batteries [69]. Furthermore, this electrical conductivity is manipulated by the surface termination and the M elements [68,70]. As a result, a large electronic transmission is observed in MXenes with –F termination, whereas this characteristic is reduced considerably in O-terminated MXene [68].

3.1.2.5 Magnetic properties

Besides the outstanding properties described above, 2D materials have significant magnetic features, making them useful in the spintronic field [71]. The magnetic features existing in 2D materials provide new nanoscale quantum and spintronic technologies denoting spin FETs, spin filters, and spin valves [72]. Indeed a simple external electric field, also known as spintronics, can tune both charge and spin transport [72].

Magnetism includes a set of physical phenomena that study the magnetic effects of electric current originating from the charge movement and molecule spins constituting the substrate [73]. Thus this phenomenon is purely quantum. According to the orientation of spins, different magnetism states are distinguished. Ferromagnetism occurs when all spins are in the same orientation, unlike an antiferromagnetic state, which is distinguished by spins in an opposite alignment. Furthermore a paramagnetic state is observed where electron spins are oriented randomly [73].

The spintronic field has widened thanks to the abrupt changes in magnetic features from bulk to 2D material form. Among these applications, there are magneto-optic, magnetoelectric, and spin transistors. Moreover, further potentials of 2D magnetic materials are explored, such as high efficiency in quantum computation, data storage, sensors, and biomedical [72].

The first 2D antiferromagnetic and ferromagnetic magnets were reported in 2016 and 2017, respectively. The first type of magnets is made of $TMPS_3$ antiferromagnetic materials where TM can be Mn, Ni, and Fe, whereas the second type is made of $Gr_2Ge_2Te_6$ and Cr_3I [72,74–76]. Several physical effects have shown up in 2D magnetic as well as in their heterostructures because of the interaction between charge carriers, spins, and lattice structure, citing quantum Hall effect and magnon valleytronics [77]. Moreover, the magnetic properties in these materials can be monitored by exterior excitation, citing external perturbations such as doping, electric field, and chemical functionalization. According to Zhang et al., magnetism in 2D intrinsic materials has been foretold by numerous applications citing machine learning [77].

Similar to Fe_2GeT_2 and MnS_2, vanadium diselenide (VSe_2) illustrates a paramagnetic to ferromagnetic crossover from bulk to monolayer form, which can be presevered at ambient conditions [72].

Moreover, Zhang et al. have indicated that the incomplete filling of TM 3d orbitals originates from the magnetic properties in 2D magnets. Based on the symmetry of the TM, four pristine 2D magnets are categorized as honeycomb, square, rectangular, and triangular lattices [77]. The latter lattice comprises $FeGeTe_2$, TMDCs, and TM dihalides that share the same layered structure as well as the common formula MX_2[77]. Furthermore, MXenes are considered intrinsic magnetic 2D materials. Among this family, there are Ti_2C and Cr_2C [77]. A magnetic feature variety is observed in the 2D MXenes family. In fact, 2D Cr_3C_2, TaC_2, Ta_3C_2, and Cr_2C exfoliated from MAX phases reveal a ferromagnetic behavior, whereas an antiferromagnetic behavior is found in Ti_3N_2 and Ti_3C_2[70]. Square and rectangular lattice magnets are also called tetragonal magnets. This class contains FeY, MXY compounds [77].

Regarding the honeycomb lattice, there are layered TM trihalides with a chemical formula MX_3 (M=Cr, V, Ni; X=Br, I). This family and its analogs have gained intense attention since discovering ferromagnetic features in single-layer CrI_3 in 2017 [77]. Besides the $Cr_2Ge_2Te_6$ semiconductor honeycomb lattice reveals FM properties for the FETs application [77]. TM phosphorus trichalcogenides with $M_2P_2Y_6$ stoichiometry where M=Cr, Mn, V and Y=P, Te, Se, are magnetic 2D material with semiconductor behavior that contain a variety of 3d orbitals. The majority of these materials reveal antiferromagnetic behavior [77].

Similar to the electric properties, the magnetic character is sensitive to the number of layers. This change originates from the interaction interlayers [73]. Indeed, these interactions affect the magnetic order strongly. A previous study has shown that an odd number of layers of the 2D material reveals magnetization, whereas an even number of layers leads to a removal of magnetization [73].

3.1.2.6 *Optical properties*

Further to electrical and magnetic properties, 2D materials possess prominent optical features compared with their bulk counterparts, making them applicable in several fields. Materials with luminescent features emit a photon upon an external energy excitation, which can be electrons, photons, or mechanical energy. Upon the excitation nature, luminescence is classified into four major categories based on the different excitation: photoluminescence (PL), cathodoluminescence, electroluminescence, and triboluminescence. This phenomenon is because of the electron transfer from excited to either ground or lower energy states that emerge by an external excitation [58]. Further a luminescent property can be revealed by nonluminescent 2D materials because of a trap level within the bandgap (defect, oxygen vacancies, and poison level) where photons are emitted from this level to a lower state.

Numerous factors affect the luminescence properties of 2D materials, denoting external energy, carrier transport, and optical features of these materials [58]. Inorganic materials have garnered significant attention over organic materials because of several advantageous characteristics. These include the robustness of chemical bonds, which provide enhanced structural integrity and stability. In addition, inorganic materials exhibit highly delocalized electronic wave functions, enabling efficient transport of charge carriers throughout the material. Moreover, these materials display high-efficiency band-to-band electronic transitions, facilitating desirable properties for various applications. The distinct advantages offered by inorganic materials underscore their potential for advancing scientific research and technological development in diverse fields [78,79].

Among 2D materials, BP, TMDC materials, and heterostructure materials are luminescent [58]. However, because of the lack of bandgap, graphene does not illustrate any luminescent properties [80]. Inserting luminescence features into graphene by adding rare earth is a simple route to overcome this limitation without prejudicing graphene's prominent features [80]. Based on Amaro-Gahete et al.'s investigation the obtained luminescence in graphene by adding europium (Eu) is useful in several fields, including tissue cell analysis and biological material [80]. This initial exploration paves the way for future endeavors where graphene will be combined with diverse chemical elements to confer novel features.

Graphene-based materials can be luminescent, such as rGO and carbon quantum dots (QDts), originating from either the presence of the isolated sp^2 or defects, respectively [80]. The exhibited emission from these materials is generally located in the blue and green regions [80]. Some authors have recognized that similar to graphene, rare-earth and lanthanide families have been used to produce or boost luminescence in GO because of their great potential observed in previous investigations [81,82]. According to Esmaeili et al., GO reveals prominent fluorescence emission that can be improved by metal nanoparticles (MNPs), polymers, and molecules [83]. For instance, using MNPs with GO produces a promising nanoplatform for sensing and imaging applications [83].

Both absorption and emission properties of MXene materials depend on several factors, such as the bandgap energy and bandgap type (indirect/direct) [70]. Moreover, the optical properties of MXenes are sensitive to the number of layers, similar to graphene and TMDC materials [70]. The dielectric function ε is written as $\varepsilon(\omega) = \varepsilon_1(\omega) + i\varepsilon_2(\omega)$ where $\varepsilon_1(\omega)$ is the real part and $\varepsilon_2(\omega)$ is the imaginary part [70]. Based on this function the optical properties of MXenes can be analyzed as absorption, PL, and reflection [68].

Once the dielectric function $\varepsilon(\omega)$ has been determined, the refractivity, reflectivity, and absorption or gain coefficients can be estimated from the standard optical relationships.

The PL response of MXene materials is low in an aqueous medium [68]. To bypass this limitation as well as to broaden their biology and optical applications, MXenes QDts are a considerable form because of the edge effect and quantum confinement [68]. These phenomena are observed in QDts because they are atomically thin, resulting in good luminescence, unlike large in-plane nanosheets. Ti_3C_2 is the most studied material among the MXene family [68]. The surface termination affects MXenes' optical features produced during the experimental process [70]. According to Jiang et al., –F and –OH surface termination in Ti_3C decreases its in-plane absorption coefficients compared with its pristine –O terminated surface [70].

In relation to the electrical properties the augmentation of layer count exerts a profound impact on the PL spectrum. Specifically a solitary layer of MoS_2 exhibits robust PL owing to the existence of an excitonic band, which diminishes with the introduction of a second layer of MoS_2[84]. Furthermore the optoelectronic characteristics of graphene are primarily governed by the unhybridized π-orbitals [51].

In summary the remarkable focus on 2D materials stems from their unique atomic structure, which manifests exceptional thermal, mechanical, and tribological properties, thus enhancing their value in industrial applications. In addition, this structure exhibits intriguing attributes encompassing magnetic, electrical, and optical properties. As a result of these remarkable characteristics, 2D materials have garnered significant attention as prospective materials for diverse applications, including spintronics, sensing, and forensic sciences. Various factors, including thickness, temperature, and defects, significantly influence the advantageous properties of 2D materials.

3.1.3 Structure and surface chemistry

3.1.3.1 Classification of defects in two-dimensional materials

During the course of experimentation, it is observed that 2D materials may undergo physical damage, resulting in structural defects that impact their mechanical, electronic, and physicochemical properties. These defects can either be beneficial or detrimental depending on the specific field of application. There are two distinct types of defects, namely, intrinsic and extrinsic, which can be further classified based on their dimensionality as zero-dimensional (0D) and one-dimensional (1D) imperfections [85].

This section will emphasize these classes, focusing on graphene, MXenes, and TMDC families.

3.1.3.1.1 Zero-dimensional impurities

Based on the literature reviews, it has been established that 0D impurities, commonly referred to as point defects, can be readily observed in various 2D materials. The manifestation of these defects varies depending on the lattice structure specific to each 2D material [86]. This particular category of defects encompasses both vacancies, which are essentially vacant sites within the material's lattice, and adatoms, which are foreign atoms that occupy lattice positions.

In the context of graphene, it is important to acknowledge the presence of Stone–Wales (SW) defects and vacancies, which represent significant 0D impurities. SW defects are crystallographic anomalies that arise from a distinct 90 degree-rotation of carbon–carbon (C–C) bonds, leading to a reconfiguration of the connectivity of π bonds. These defects play a notable role in influencing the properties and behavior of graphene structures. Moreover, vacancies, another type of 0D impurity, are also observed in graphene, indicating the absence of carbon atoms within the lattice. Together, SW defects and vacancies contribute to the intricate nature of graphene and warrant further investigation to comprehend its effects on its overall performance and applications (Fig. 3.7) [85]. This type of defect is observed in sp^2 hybridization in graphene, carbon nanotubes, and similar carbon structure [85]. This imperfection is produced either by electron-beam irradiation or by rapid cooling from high temperatures [85]. This impurity cannot be generated in TMDC materials, but they are affluent with vacancies and antisites [87].

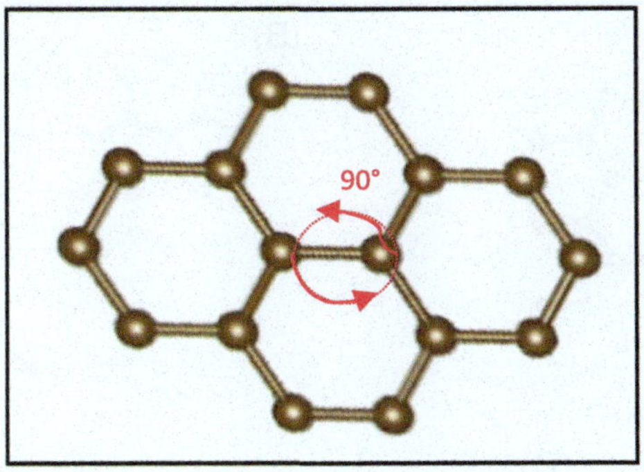

FIGURE 3.7

Stone–Wales defects.

Regarding vacancies, this kind is divided into two groups. The first one is monovacancies (MVs), also known as single vacancies (SVs), that are presented in graphene and TMDC families, whereas the second is double vacancy (DV) imperfection [33,34]. Reconstruction of defects has resulted from the unstable impurities in 2D materials [85]. Graphene and its derivatives reveal both MVs and DVs as impurities [88]. For instance, SVs and DVs are observed in graphene after removing a single carbon (C) atom and 2 adjacent carbon atoms, respectively (Fig. 3.8A and B) [88]. The production process can generate these imperfections, such as ion bombardment (e.g., Ar^+) at different energies. Furthermore, graphene vacancies can be obtained from the synthesis process, such as the reduction of GO [88].

For the TMDCs family, three types of vacancies can be observed. The first one is M vacancy "V_M," whereas the second is X vacancy that can be single or double atoms abbreviated as V_X and V_{X2}, respectively (Fig. 3.8D–F) [33,37]. As reported by the literature, MoS_2 is the most studied 2D TMDC material. In this material, V_{Mo} or V_S can be generated where the V_{Mo} vacancy is presented in these materials as a complex defect "V_{Mo+3S}" or "V_{Mo+6S}" [85]. This imperfection is because of the higher probability of S atoms lost that surround V_{Mo}[85]. Indeed, these defects are observed when the vacancy of Mo is bonded with three S atoms missing in one plane "V_{Mo+3S}" or the loss of all S atoms surrounded with Mo "V_{Mo+6S}" [38].

The concentration of MVs in thermodynamic equilibrium is determined by their energy creation, according to Wu et al. [85]. Furthermore the defect concentration is sensitive exponentially to E_f[86]. Thus the MVs' equilibrium concentration in phosphorene must be larger than the concentration in graphene at similar conditions [86]. Indeed the E_f of MV in graphene is almost five times that of phosphorene and inferior or equal to that of V_S in 1L-MoS_2, ranging from 1.22 to 2.25 eV [86]. Because of the lowest E_f, V_S vacancies in all single layers, TMDC materials are omnipresent, whatever the used synthesis technique of 2D materials in the growth process [38]. In fact, both growth methods and the environment affect V_S formation and their concentrations [38]. The n-type MoS_2 devices can be contributed by V_S defects [38].

The three types of vacancies described above in MoS_2 can also be generated in other single-layer TMDCs in the 1H phase [89]. Both S vacancies and V_{Mo} are observed in the 1H-MoS_2; meanwhile, we notice the absence of the complex impurities "V_{Mo+3S}" or "V_{Mo+6S}" because of the trigonal prism structure [89].

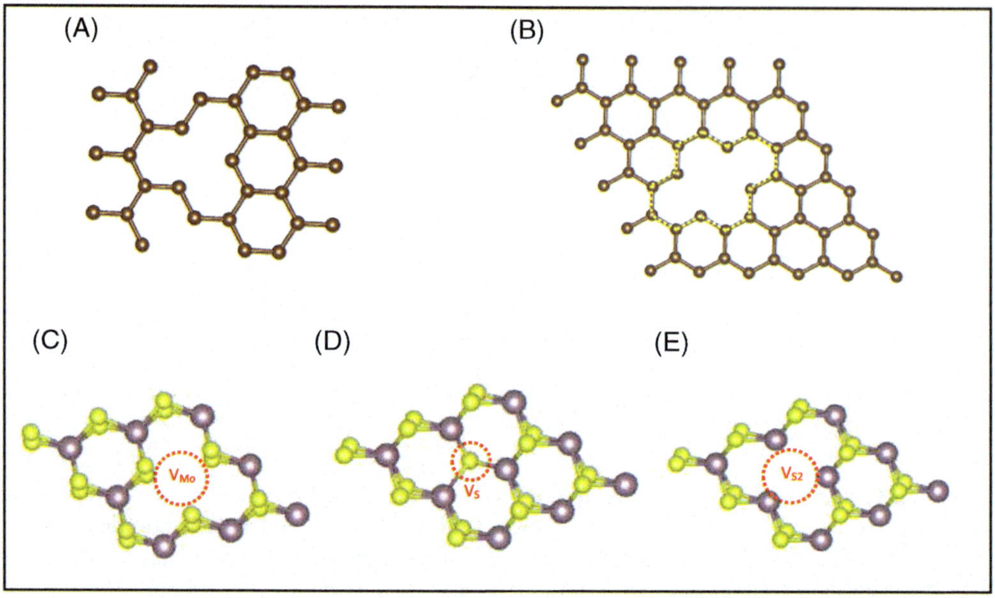

FIGURE 3.8

Vacancy types in graphene: (A) single, (B) double, (C) Mo vacancy (VMo), and (D) single S vacancy (VS) in one plane and (E) double S vacancy (VS_2) in MoS_2.

The chalcogen vacancy is also found in the T phase of $ReSe_2$, ReS_2, and $MoTe_2$[89]. This vacancy is significant in TMDCs materials because of the large structure reconstruction resulting from V_S aggregation. Among the structure reconstruction, there are interlayer fusion and line defects formation. Moreover, a change of the electronic structure is found that has great potential for application as a catalytic.

Besides vacancies, antisites are the most numerous and the simplest impurities in the TMDC class. This type of defect occurs when two different atoms switch their positions. Still, in the case of MoS_2, two types of antisites are distinguished; M_{oS2} and $S2_{Mo}$[89]. The first one occurs when the Mo atom takes the place of the S2 column; otherwise, it is $S2_{Mo}$[87]. This type of defect can often be acquired by the chemical vapor deposition (CVD) technique [87].

Another significant 0D defect in 2D materials is adatoms. This defect occurs when a foreign atom is introduced into the structure, forming either a weak or a strong bond [85]. If the interaction is weak, it is physisorption; otherwise, it is chemisorption between 2D materials and the nearest adatoms [85]. When chemisorption occurs because of the covalent bond between atoms and 2D materials, an sp^3 hybridization is generated; however, this process is complicated in TMDCs [84].

According to a recent study, four positions are distinguished in the 1H phase of TMDCs, where adatoms can be adsorbed [87]. As presented in Fig. 3.8, these positions are above either the chalcogen or the metal atoms abbreviated as T_{Ch} and T_M, respectively. Furthermore the foreign atoms can be adsorbed on the chalcogen-metal bond "$B_{Ch\text{-}M}$," whereas the center of the hexagonal voids is the fourth position where adatoms can take place [87].

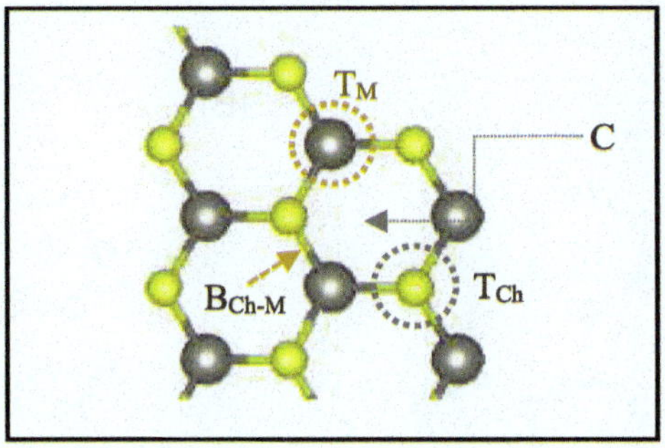

FIGURE 3.9

5|7 Dislocation type in graphene.

According to Kastl et al. the presence of 0D impurities permits the emission of single-photon quantum. Furthermore, these flaws indicate active locations in catalytic processes; otherwise, TMD surfaces are inert [90].

3.1.3.1.2 One-dimensional imperfections

In addition to 0D impurities described in the previous section, 1D defects are also line defects [86]. These defects are unexpected changes along a regular atom order line containing dislocations, grain boundaries (GBs), and edges.

Dislocation defects are generated in all 2D materials that are immobile at normal conditions [86]. Several dislocations are distinguished, such as 5|7, 4|8, 4|4. In fact, 5|7 pair defects are presented in 2D materials such as graphene, TMDCs, and phosphorene (Fig. 3.9) [86]. The atomic structure of TMDCs brings complexity to the dislocation impurities because the TMDCs' single layer is constituted by X–M–X planes [85]. For instance a reaction can occur between 5|7 and 2 S vacancies, forming 4|6 dislocation in a single layer of MoS_2. Moreover, the S and Mo bonding can also create 4|4 and 4|8-fold rings [85].

Besides, GBs are considered a linear arrangement of dislocations. Indeed, they have a significant role in single-layer TMDC materials because of their structure [89]. A monolayer of TMDCs comprises M–X–M layers (three layers of atoms) [89]. Thus they strongly affect the optical and electrical features of these materials generated from either misorientation or misalignment of two crystalline areas during the production process [89]. Moreover, the presence of this defect in the material structure leads to a decrease in its conductivity and thermal features. GB in TMDC materials is complex compared with other single-layer 2D materials [87]. This is because of the TMDCs monolayer composition composed of three atomic layers (X–M–X) [87]. Indeed, mirror twin boundaries are observed in both synthetic MoX_2 (X=Se, S) when the angle between grains is 60 degrees laterally in the ZZ direction [87]. Besides this type of GB, if the angle is smaller (~18 degrees), another twin boundary will be observed in MoS_2, that is, tilt twin boundaries [87]. These boundaries can be observed with transmission electron microscopy (TEM) characterization.

In addition to the described impurities, edge defects arise when an extra half-plane of atoms is introduced into the crystal structure, leading to a distortion of the adjacent atomic planes. Such imperfections are of paramount significance in governing the morphology and characteristic properties of 2D materials [85]. The major defect in TMDC materials is edge imperfections that strongly influence their

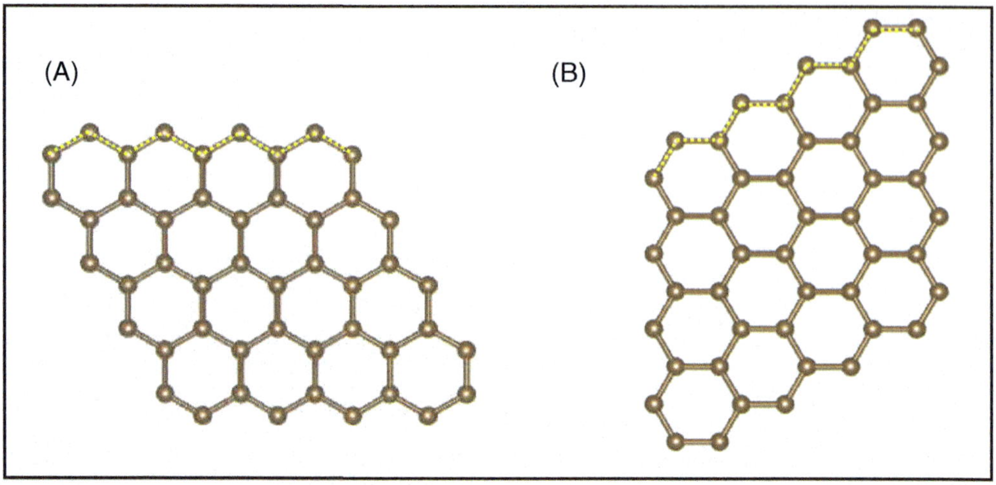

FIGURE 3.10

(A) Zigzag and (B) armchair edges defect in graphene.

electrical, catalytic, magnetic, and optical properties [86]. To investigate this impurity, scientists have used both theoretical and experimental tools, such as DFT and optical characterizations, respectively [87]. The direction of the edge imperfection leads to a couple of edges called "ZZ" and "armchair" edges. Fig. 3.10 shows this type of imperfection in graphene, for instance, that can be distinguished through either Raman or X-ray photoelectron techniques [91]. The different types of edges affect electronic and magnetic features [85]. In fact, ZZ edges illustrate metallic and ferromagnetic, whereas the armchair edges show semiconductor and nonmagnetic behavior [87]. In the WS_2 case, edge imperfections can be surrounded with biexcitons that lead to an intense PL emission originating from a large charge carrier population [87].

In their study, Kastl et al. predicted that 1D impurities behave as spin polarizers hosting charge density waves known as exotic phases [90].

3.1.3.2 *Effect of defects on physical properties*

Based on the literature the defects described in the previous sections reveal a noticeable impact on the features of 2D material, citing both optical and electronic features of layered TMDCs materials [87]. In addition to the presence of 1D impurities such as vacancies and GBs, the electronic characteristics of TM dichalcogenides (TMDCs) are also influenced by 0D defects, specifically foreign atoms [87]. The doping behavior of TMDCs, such as MoS_2, is determined by the type of GBs present in the material, which is highly sensitive to its atomic structure. The introduction of dopants leads to changes in the bandgap and energy shift of the PL peak in TMDCs with semiconductor properties [87].

The presence of impurities in 2D materials, such as GBs, significantly impacts their electrical properties. For example, VS vacancies in MoS_2 introduce donor sites within the material's bandgap, resulting in transport hopping at room temperature. As mentioned previously, MoS_2 exhibits two types of GBs: tilt and twin mirror. Each type of GB affects the conductivity of MoS_2 and consequently influences its potential applications, such as FETs. Tilt GBs in MoS_2 lead to a decrease in conductivity, whereas FETs fabricated with mirror twin MoS_2 demonstrate improved conductance.

In addition to the electrical properties, optical characteristics are sensitive to defects. Although MoS_2 is a nonluminescent material, the PL spectrum of 1LMoS_2, with n-type doping naturally, shows two peaks. Based on various studies the strongest peak is located at 640 nm, generated from direct transition at K′ point in ZB [92]. The weak peak is located at 684 nm, originating from trion (excitons charged negatively) [93]. This peak can be strongly improved owing to the engineering of defects in 1L-MoS_2 by either oxygen plasma or thermal treatment [85].

Furthermore, 2D materials' magnetic features can be influenced by imperfections. For instance a nonmagnetic to ferromagnetic behavior crossover is gained from pristine bulk to a single BP layer when doped with TMs, citing Mn, Fe, Ni, and so on [85]. Moreover, the production of GBs in graphene during the synthesis process is one of the routes to achieve magnetism behavior, according to Yu et al., investigation [94].

The influence of defects on the mechanical properties of materials is well-established [85]. Specifically, in the case of graphene, the in-plane Young's modulus is affected by the density of vacancies. This effect can be observed under two scenarios: low vacancy content and higher vacancy content. The density of vacancies can be accurately determined using Eqs. (3.5) and (3.6) [41].

$$n^2 = \frac{5.9 \times 10^4}{E_L^4}\left[\frac{I(D)}{I(G)}\right] \tag{3.5}$$

$$n_D^2 = \frac{I(D)}{I(G)}\, 7.3 \times 10^{-9}\, E_L^4 \tag{3.6}$$

In this framework an increase in defect density increases in-plane Young's modulus at low defect vacancy [41]. In the context of high vacancy defects the modulus of graphene experiences a decrease. Beyond its mechanical properties the thermal conductivity of graphene is highly influenced by the presence of defects. Even a low concentration of defects can significantly impact the thermal conductivity of graphene, as demonstrated by previous studies [41]. For instance, introducing a mere 0.1% of flaws leads to a remarkable reduction of 83% in heat conductivity [84]. Exploiting the presence of defects in graphene opens up new possibilities for innovative applications, such as developing biocompatible materials [95]. Moreover, graphene-based materials exhibit promising potential for application in spintronic catalyst devices [85].

3.1.3.3 Characterization of defects with optical tools

The different defects described above can be revealed through various characterization tools such as PL and Raman techniques in addition to scanning tunneling microscopy and TEM as well as XPS, which are classified as modern techniques [85]. This section will explain the importance of optical techniques for both the characterization and quantification of defects in 2D layered materials.

Starting with PL spectroscopy, each TMDC material has its unique spectrum that can be affected by several factors, including thickness, defects, and temperature [93]. The presence of impurities in TMDCs affects their bandgap. An intense PL emission is revealed in normal conditions by a single layer of TMDC materials caused by the indirect-to-direct crossover. Two types of emission are observed in the PL spectrum: trions (X_0^-, X_0^+) and exciton (X_0) [93]. Indeed the first one is an exciton emission generated after the occurrence of electron–hole interaction. The second type is a trion emission caused by exciton–hole or exciton–electron interactions [93]. An additional PL peak is observed at low temperature originating from bound exciton labeled "X_b." The energy of this peak is used to identify defects in TMDC materials [93]. Based on the findings of Zheng et al.'s investigations an empirical formula (Eq. 3.7) is used to determine the X_b population as a function of temperature [96–98]:

$$N_{bx}(T) = \frac{N_0}{1 + \left(\frac{\tau}{\tau_0}\right) e^{-E_a/kT}} \tag{3.7}$$

N_{bx} is the X_b population, whereas E_a is the activation energy. τ and τ_0 are the excitonic lifetime and effective scattering time, respectively [93]. K is the Boltzman constant, and T is temperature. As X_b intensity depends on the defect density presented in TMDC materials, the impurity quantification is obtained through X_b/X_0 ratio [93].

Raman spectroscopy, a nondestructive technique employed several decades ago, has proven instrumental in characterizing diverse materials in various states. Through the analysis of Raman spectra, valuable insights can be gleaned regarding vibrational modes, the number of layers, and the quality of produced nanosheets [61]. Remarkably, each distinct 2D material exhibits characteristic vibrational modes that arise from the phonons present in the crystalline network, as manifested in the Raman spectrum. Nevertheless, the presence of imperfections within the material structures can induce shifts in Raman modes or even give rise to the detection of novel vibrational modes.

For instance the prominent peaks of graphene are G and G′ (also known as 2D), corresponding to the high phonon frequency E_{2g} at Γ point in the BZ (Broullin zone) and double resonance from A′1 phonons at K point corner of BZ, respectively [85]. The presence of imperfections in graphene leads to the emergence of distinct peaks, namely, D, D′, and the coupling effect between them (D + D′) caused by structural defects [85]. Specifically the D mode arises from the A′1 phonon near the K point in the BZ [85]. The excitation energy can occur an intense dispersion of the D peak. The D′ induced by imperfection is because of the E_{2g} phonon, whereas the emission of two intervalley phonons generates a D′ + D peak [85]. Based on the intensities of the three Raman vibrational modes, numerous pieces of information can be extracted about the nature as well as the number of defects present in graphene [85]. According to Wu et al. the nature of defects is determined by the ratio $I_D/I_{D'}$, which is around ~1, ~11, ~17 for MVs, DV, and SW defects, respectively [85].

As mentioned previously, defects can be generated through ion bombardment. In this case an increase in induced defects leads to distinguishing two stages between L_D and I_D/I_G ratios [85]. Here, L_D is the average impurity distance, and I_D/I_G is D and G intensities. In the first stage, I_D and $(L_L/L_D)^2$ are proportional to the number of probed defects. The size of the laser spot is L_L[85]. Indeed, $(L_L/L_D)^2$ is the average defect generated in the probed area by the laser. Based on Wu et al.'s investigation the total probed area and G intensity are proportional ($I_G \alpha\ L_L^2$) [85]. Thus D and G intensity ratios (I_D/I_G) are proportional to $1/L_D^2$[85]. In the second stage, I_D decreases when the number of defects increases, and a new relation is obtained: $L_D \propto I_D/I_G$[85].

Similar to graphene the defects presented in TMDC materials can be characterized as well as quantified based on the Raman tool. For instance the MoS_2 Raman spectrum is dominated by two characteristic peaks known as in-plane (E^1_{2g}) and out-of-plane (A_{1g}) modes. These only peaks are observed in a single layer of the MoS_2 spectrum. The presence of defects in MoS_2 reveals three changes in the Raman spectrum [85]. Two changes are observed after electron bombardment; both E^1_{2g} as well as A_{1g} are redshifted and blueshifted with broadenings, respectively [85]. Some peaks observed in the Raman spectrum are activated by defects exciting in MoS_2. Among these peaks the most intense one is designated to the disorder peak "LA," originating from phonons in BZ edges [85].

The tools described above are widely employed to characterize different imperfections in all 2D materials, which affect them either positively or negatively [85]. Although this effect sometimes limits the potential of 2D materials applications, it opens a new path in numerous uses, citing electric and spintronic devices.

3.1.4 Challenges and perspective

Despite the advantages of the nanoparticles resulting from their outstanding features as described above, the environmental benefits and risks will be highlighted in this section.

Nanoparticles offer significant advancements in addressing various daily challenges, such as enhancing energy storage capabilities, facilitating the progress of electronic devices and sensors, and enabling efficient drug delivery systems [99,100]. In the realm of environmental conservation, nanotechnology holds promise in safeguarding the environment through the conservation of essential resources, including raw materials, energy, and water, while also reducing greenhouse gas emissions. Nonetheless, it is crucial to acknowledge the potential hazards associated with certain nanoparticles, which raise concerns among researchers. Specifically, some nanoparticles employed possess detrimental properties that can negatively impact both the environment and human health. The toxicity of such nanoparticles can infiltrate the soil and aquifer systems, posing risks to human well-being [101,102].

From an economic perspective, the cost of nanoparticles emerges as a crucial concern that presents a significant hurdle in their commercialization. The adverse effects associated with these materials necessitate comprehensive evaluations of their potential impacts on human health and the environment. Consequently, researchers are actively engaged in developing models to accurately predict the transport and fate of engineered nanoparticles in the environment. In addition, effective risk assessments are being conducted to ensure the safety of nanoparticle-based products before they are introduced to the market. It is important to note that this process can be financially burdensome, primarily because of the requirement for specialized expertise, as well as the necessary materials and experimental setups.

3.2 Conclusion

The discussion presented in this chapter focuses on the extensive research conducted in the past few decades on 2D materials, which have emerged as a significant area of study. Various material families, including TMDCs, graphene, and Mxenes, have exhibited distinct behaviors such as semiconductor, metal, and insulator properties. These materials possess a remarkable atomic structure that imparts exceptional mechanical, electronic, and optical features. For instance, graphene and h-BN have similar atomic structures and exhibit comparable electronic properties. However, the absence of a bandgap in graphene has limited its applications, particularly in the field of electronics. To address this issue, defect engineering has emerged as a promising approach to enhance the electronic properties of the graphene family. In addition, the introduction of adatoms and dopants has played a crucial role in enhancing the optical characteristics of various 2D materials.

References

[1] J.-W. Jiang, Front. Phys. 10 (2015) 287–302.
[2] G. Yang, L. Li, W.B. Lee, M.C. Ng, Sci. Technol. Adv. Mater. 19 (2018) 613–648.
[3] L. Liu, M. Zhou, L. Jin, L. Li, Y. Mo, G. Su, X. Li, H. Zhu, Y. Tian, Friction 7 (2019) 199–216.
[4] L. Sun, Chin. J. Chem. Eng. 27 (2019) 2251–2260.
[5] R. Tarcan, O. Todor-Boer, I. Petrovai, C. Leordean, S. Astilean, I. Botiz, J. Mater. Chem. C 8 (2020) 1198–1224.
[6] A. Champagne, J.-C. Charlier, J. Phys. Mater 3 (2020) 032006.
[7] R.M. Ronchi, J.T. Arantes, S.F. Santos, Synthesis, structure, properties and applications of MXenes: Current status and perspectives, Ceram. Int. 45 (2019) 18167–18188.
[8] Y. Wang, Y. Xu, M. Hu, H. Ling, X. Zhu, Nanophotonics 9 (2020) 1601–1620.
[9] A.L. Ivanovskii, Graphynes and graphdyines,Progress in, Solid State Chemistry 41 (1–2) (2013) 1–19, doi:10.1016/j.progsolidstchem.2012.12.001.
[10] J. Rosen, M. Dahlqvist, Q. Tao, L. Hultman, 2D Metal Carbides and Nitrides (MXenes): Structure, Properties and Applications, in: B. Anasori, Y. Gogotsi (Eds.), In- and Out-of-Plane Ordered MAX Phases and Their MXene Derivatives, Springer International Publishing, Cham, 2019, pp. 37–52. https://doi.org/10.1007/978-3-030-19026-2_3.
[11] L.W.T. Ng, et al., Printing of Graphene and Related 2D Materials: Technology, Formulation and Applications, in: L.W.T. Ng, G. Hu, R.C.T. Howe, X. Zhu, Z. Yang, C.G. Jones, T. Hasan (Eds.), Printing Technologies, Springer International Publishing, Cham, 2019, pp. 135–178. doi:10.1007/978-3-319-91572-2_5.
[12] T.H. Choudhury, X. Zhang, Z.Y. Al Balushi, M. Chubarov, J.M. Redwing, Annu. Rev. Mater. Res. 50 (2020) 155–177.
[13] K. Khan, A. Khan Tareen, M. Aslam, R. Wang, Y. Zhang, A. Mahmood, Z. Ouyang, H. Zhang, Z. Guo, J. Mater. Chem. C 8 (2020) 387–440.
[14] U. Krishnan, M. Kaur, K. Singh, M. Kumar, A. Kumar, Superlattices Microstruct. 128 (2019) 274–297.
[15] D. Gupta, V. Chauhan, Inorg. Chem. Commun. 121 (2020) 108200.
[16] C.M.O. Bastos, R. Besse, J.L.F. Da Silva, G.M. Sipahi, Phys. Rev. Mater. 3 (2019) 044002.
[17] V. Korsaks, Hexagonal boron nitride luminescence dependent on vacuum level and surrounding gases, Mater. Res. Bull. 70 (2015) 976–979, doi:10.1016/j.materresbull.2015.06.032.
[18] D.J. Late, A. Bhat, C.S. Rout, Fundamentals and Sensing Applications of 2D Materials, in: M. Hywel, C.S. Rout, D.J. Late (Eds.), Chapter 2 - Fundamentals and Properties of 2D Materials in General and Sensing Applications, Woodhead Publishing, 2019, pp. 5–24. https://doi.org/10.1016/B978-0-08-102577-2.00002-6.
[19] X. Mu, J. Wang, M. Sun, Mater. Today Phys. 8 (2019) 92–111.
[20] S. Wu, K.S. Hui, K.N. Hui, Adv. Sci. 5 (2018) 1700491.
[21] A. Chaves, W. Ji, J. Maassen, T. Dumitrică, T. Low, P. Avouris, T. Low, T.F. Heinz, in: 2D Materials: Properties and Devices, Cambridge University Press, Cambridge, 2017, pp. 381–412.
[22] M. Galluzzi, Y. Zhang, X.-F. Yu, J. Appl. Phys. 128 (2020) 230903.
[23] Q. Bao, H.Y. Hoh, 2D Materials for Photonic and Optoelectronic Applications, in: Woodhead Publishing Series in Electronic and Optical Materials, Woodhead Publishing, 2020, pp. 1–35. https://doi.org/10.1016/B978-0-08-102637-3.00001-2.
[24] D. Akinwande, C.J. Brennan, J.S. Bunch, P. Egberts, J.R. Felts, H. Gao, R. Huang, J.-S. Kim, T. Li, Y. Li, K.M. Liechti, N. Lu, H.S. Park, E.J. Reed, P. Wang, B.I. Yakobson, T. Zhang, Y.-W. Zhang, Y. Zhou, Y. Zhu, Extreme Mech. Lett. 13 (2017) 42–77.
[25] J.H. Kim, J.H. Jeong, N. Kim, R. Joshi, G.-H. Lee, J. Phys. D: Appl. Phys. 52 (2018) 083001.

[26] R. Zhang, R. Cheung, Mechanical Properties and Applications of Two-Dimensional Materials', Two-dimensional Materials - Synthesis, Characterization and Potential Applications, InTech, 2016, doi:10.5772/64017.

[27] K. Liu, J. Wu, J. Mater. Res. 31 (2016) 832–844.

[28] S. Zhang, T. Ma, A. Erdemir, Q. Li, Mater. Today 26 (2019) 67–86.

[29] H. Jiang, L. Zheng, Z. Liu, X. Wang, InfoMat 2 (2020) 1077–1094.

[30] A. Yusuf, T. i, S. Umar, U. Salisu & Shasu, H. Alhassan, U. Gupta, U. (2015). Use of polymer matrix composites for conventional steel drivre shafts: A stydy. ELK Asia Pac. j. 1. 179-186.

[31] Y. Liu, B. Zwingmann, M. Schlaich, Polymers 7 (2015) 2078–2099.

[32] K.K. Al-Quraishi, Q. He, W. Kauppila, M. Wang, Y. Yang, Int. J. Smart Nano Mater. 11 (2020) 207–246.

[33] Y. Ibrahim, A. Mohamed, A.M. Abdelgawad, K. Eid, A.M. Abdullah, A. Elzatahry, Nanomaterials 10 (2020) 1916.

[34] A. Lipatov, H. Lu, M. Alhabeb, B. Anasori, A. Gruverman, Y. Gogotsi, A. Sinitskii, Sci. Adv. 4 (2018) eaat0491.

[35] C. Cao, M. Daly, C.V. Singh, Y. Sun, T. Filleter, Carbon 81 (2015) 497–504.

[36] C. Cao, M. Daly, B. Chen, J.Y. Howe, C.V. Singh, T. Filleter, Y. Sun, Nano Lett. 15 (2015) 6528–6534.

[37] X. Gu, Y. Wei, X. Yin, B. Li, R. Yang, et al., Colloquium: Phononic thermal properties of two-dimensional materials, Rev. Mod. Phys. 90 (4) (2018) 041002, doi:10.1103/RevModPhys.90.041002.

[38] B. Mortazavi, E.V. Podryabinkin, I.S. Novikov, S. Roche, T. Rabczuk, X. Zhuang, A.V. Shapeev, J. Phys. Mater. 3 (2020) 02LT02.

[39] T.S. Kasirga, Thermal Conductivity Measurements in Atomically Thin Materials and Devices. Springer Briefs in Applied Sciences and Technology, in: Thermal Conductivity Measurements in 2D Materials, Springer, Singapore, 2020, pp. 11–27. https://doi.org/10.1007/978-981-15-5348-6_2.

[40] A. Kumar, K. Sharma, A.R. Dixit, J. Mater. Sci. 54 (2019) 5992–6026.

[41] Y. Lan, A.J. Minnich, G. Chen, Z. Ren, Adv. Funct. Mater. 20 (2010) 357–376.

[42] G. Zhang, Y.-W. Zhang, Chin. Phys. B 26 (2017) 034401.

[43] J. Chen, L. Li, JETP Lett. 112 (2020) 117–121.

[44] J. Wang, F. Ma, M. Sun, RSC Adv. 7 (2017) 16801–16822.

[45] Y. Liu, Y. Dong, Y. Zhang, S. Liu, Y. Bai, Nanomater. Nanotechnol. 10 (2020) 1847980420946655.

[46] A. Di Bartolomeo, Nanomaterials 10 (2020) 579.

[47] B. Podgornik, T. Kosec, A. Kocijan, Č. Donik, Tribol. Int. 81 (2015) 267–275.

[48] Y.Y. Illarionov, T. Knobloch, M. Jech, M. Lanza, D. Akinwande, M.I. Vexler, T. Mueller, M.C. Lemme, G. Fiori, F. Schwierz, T. Grasser, Nat. Commun. 11 (2020) 3385.

[49] R. Zribi, G. Neri, Sensors 20 (2020) 5404.

[50] L. Li, Z. Li, A. Yoshimura, C. Sun, T. Wang, Y. Chen, Z. Chen, A. Littlejohn, Y. Xiang, P. Hundekar, S.F. Bartolucci, J. Shi, S.-F. Shi, V. Meunier, G.-C. Wang, N. Koratkar, Vanadium disulfide flakes with nanolayered titanium disulfide coating as cathode materials in lithium-ion batteries, Nat Commun. 10 (2019) 1764, doi:10.1038/s41467-019-09400-w.

[51] M K K., M. Jaiswal, Asian J. Phys. 25 (2016) 809–831.

[52] M. Bernardi, C. Ataca, M. Palummo, J.C. Grossman, Nanophotonics 6 (2017) 479–493.

[53] D. Dreyer, S. Park, C. Bielawski, R. Ruoff, Chem. Soc. Rev. 39 (2009).

[54] C. Kumar, S. Das, S. Jit, Micro and Nano Technologies, 2D Nanoscale Heterostructured Materials, in: S. Jit, S. Das (Eds.), 7 - Device physics and device integration of two-dimensional heterostructures, Elsevier, 2020, pp. 195–214. https://doi.org/10.1016/B978-0-12-817678-8.00007-5.

[55] Das, R. Pandey, S.K., Mahadevan, P. (2017). Layer dependent electronic structure changes in transition metal dichalcogenides-the role of geometric confinement. arXiv preprint arXiv:1702.04535.

[56] K.P. O'Donnell, X. Chen, Appl. Phys. Lett. 58 (1991) 2924–2926.

[57] Y.P. Varshni, Physica 34 (1967) 149–154.

[58] W. Jie, Z. Yang, G. Bai, J. Hao, Adv. Opt. Mater. 6 (2018) 1701296.

[59] Z.-Y. Zhao, Q.-L. Liu, Catal. Sci. Technol. 8 (2018) 1867–1879.

[60] N. Lu, H. Guo, L. Wang, X. Wu, X.C. Zeng, Nanoscale 6 (2014) 4566–4571.

[61] K. Abid, N.H. Belkhir, S.B. Jaber, R. Zribi, M.G. Donato, G. Di Marco, P.G. Gucciardi, G. Neri, R. Maâlej, J. Phys. Chem. C. 124 (2020) 20350–20358.

[62] S. Hao, J. Zeng, T. Xu, X. Cong, C. Wang, C. Wu, Y. Wang, X. Liu, T. Cao, G. Su, L. Jia, Z. Wu, Q. Lin, L. Zhang, S. Yan, M. Guo, Z. Wang, P. Tan, L. Sun, Z. Ni, S.-J. Liang, X. Cui, F. Miao, Adv. Funct. Mater. 28 (2018) 1803746.

[63] C. Fang, J. Zhang, X. Chen, G.J. Weng, Nanomaterials 10 (2020) 1129.

[64] X.-Y. Fang, X.-X. Yu, H.-M. Zheng, H.-B. Jin, L. Wang, M.-S. Cao, Phys. Lett. A 379 (2015) 2245–2251.

[65] M.D. Siao, W.C. Shen, R.S. Chen, Z.W. Chang, M.C. Shih, Y.P. Chiu, C.-M. Cheng, Nat. Commun. 9 (2018) 1442.

[66] H. Huang, Y. Cui, Q. Li, C. Dun, W. Zhou, W. Huang, L. Chen, C.A. Hewitt, D.L. Carroll, Nano Energy 26 (2016) 172–179.

[67] M. Xin, J. Li, Z. Ma, L. Pan, Y. Shi, Front. Chem. 8 (2020).

[68] Y. Wang, Y. Xu, M. Hu, H. Ling, X. Zhu, MXenes: focus on optical and electronic properties and corresponding applications, Nanophotonics 9 (2020) 1601–1620.

[69] H. Wang, Y. Wu, J. Zhang, G. Li, H. Huang, X. Zhang, Q. Jiang, Mater. Lett. 160 (2015) ML-BLUED1502621.

[70] X. Jiang, A.V. Kuklin, A. Baev, Y. Ge, H. Ågren, H. Zhang, P.N. Prasad, Phys. Rep. 848 (2020) 1–58.

[71] E.C. Ahn, npj 2D Mater. Appl. 4 (2020) 1–14.

[72] V.P. Ningrum, B. Liu, W. Wang, Y. Yin, Y. Cao, C. Zha, H. Xie, X. Jiang, Y. Sun, S. Qin, X. Chen, T. Qin, C. Zhu, L. Wang, W. Huang, Recent Advances in Two-Dimensional Magnets: Physics and Devices towards Spintronic Applications, Research 2020 (2020), doi:10.34133/2020/1768918.

[73] A.M. Kumar, K. Srinivasa Rao, M.C. Varma, Investigations of surface spin canting in Ni-Zn nanoferrite and its development as magnetic core for microwave applications, J. Magn. Magn. Mater. 471 (2019) 262–266, doi:10.1016/j.jmmm.2018.09.060.

[74] C.-T. Kuo, M. Neumann, K. Balamurugan, H.J. Park, S. Kang, H.W. Shiu, J.H. Kang, B.H. Hong, M. Han, T.W. Noh, J.-G. Park, Sci. Rep. 6 (2016) 20904.

[75] J.-U. Lee, S. Lee, J.H. Ryoo, S. Kang, T.Y. Kim, P. Kim, C.-H. Park, J.-G. Park, H. Cheong, Nano Lett. 16 (2016) 7433–7438.

[76] S. Lee, K.-Y. Choi, S. Lee, B.H. Park, J.-G. Park, APL Mater. 4 (2016) 086108.

[77] S. Zhang, R. Xu, N. Luo, X. Zou, Nanoscale 13 (2021) 1398–1424.

[78] C. Feldmann, T. Jüstel, C.R. Ronda, P.J. Schmidt, Adv. Funct. Mater. 13 (2003) 511–516.

[79] M. Nirmal, L. Brus, Acc. Chem. Res. 32 (1999) 407–414.

[80] J. Amaro-Gahete, A.M. Kaczmarek, D. Esquivel, C. Jiménez-Sanchidrián, P. Van Der Voort, F.J. Romero-Salguero, Chemistry 25 (2019) 6823–6830.

[81] R. Shi, B. Zhu, M. Hu, H. Zhou, J. Fan, Z. Tao, F. Chang, S. Yu, J. Mater. Chem. C 8 (2020) 2336–2342.

[82] M. Gouiaa, I. Bennour, L. Rzouga Haddada, A. Toncelli, J. Xu, A. Mbarek, A. Moscardini, N. Essoukri Ben Amara, R. Maalej, Physica B 582 (2020) 412009.

[83] Y. Esmaeili, E. Bidram, A. Zarrabi, A. Amini, C. Cheng, Sci. Rep. 10 (2020) 18052.

[84] D. Somvanshi, S. Jit, Micro and Nano Technologies, 2D Nanoscale Heterostructured Materials, in: S. Jit, S. Das (Eds.), 5 - Transition metal dichalcogenides based two-dimensional heterostructures for optoelectronic applications, Elsevier, 2020, pp. 125–149. https://doi.org/10.1016/B978-0-12-817678-8.00005-1.

[85] Z. Wu, Z. Ni, Spectroscopic investigation of defects in two-dimensional materials, Nanophotonics 6 (6) (2017) 1219–1237, doi:10.1515/nanoph-2016-0151.

[86] D. Akinwande, C.J. Brennan, J.S. Bunch, P. Egberts, J.R. Felts, H. Gao, R. Huang, J.-S. Kim, T. Li, Y. Li, K.M. Liechti, N. Lu, H.S. Park, E.J. Reed, P. Wang, B.I. Yakobson, T. Zhang, Y.-W. Zhang, Y. Zhou, Y. Zhu, Extreme Mech. Lett. 13 (2017) 42–77.

[87] Z. Lin, B.R. Carvalho, E. Kahn, R. Lv, R. Rao, H. Terrones, M.A. Pimenta, M. Terrones, 2D Mater. 3 (2016) 022002.

[88] A. Farokh Niaei, T. Hussain, M. Hankel, J. Debra, D. Searles, Hydrogenated defective graphene as an anode material for sodium and calcium ion batteries: A density functional theory study, Carbon 136 (2018) 73–84, doi:10.1016/j.carbon.2018.04.034.

[89] J. Lin, W. Zhou, Woodhead Publishing Series in Electronic and Optical Materials, Defects in Advanced Electronic Materials and Novel Low Dimensional Structures, in: J. Stehr, I. Buyanova, W. Chen (Eds.), 6 - Defect in 2D materials beyond graphene, Elsevier, Woodhead Publishing, 2018, pp. 161–187. https://doi.org/10.1016/B978-0-08-102053-1.00006-5.

[90] C. Kastl, R.J. Koch, C.T. Chen, J. Eichhorn, S. Ulstrup, A. Bostwick, C. Jozwiak, T.R. Kuykendall, N.J. Borys, F.M. Toma, S. Aloni, A. Weber-Bargioni, E. Rotenberg, A.M. Schwartzberg, ACS Nano 3 (2019) 1284–1291.

[91] J. Kim, N. Lee, Y.H. Min, S. Noh, N.-K. Kim, S. Jung, M. Joo, Y. Yamada, ACS Omega 3 (2018) 17789–17796.

[92] Y. Kim, Y. Jhon, J. Park, C. Kim, S. Lee, Y.M. Jhon, Sci. Rep. 6 (2016) 21405.

[93] B. Zheng, Y. Chen, IOP Conf. Ser. Mater. Sci. Eng. 274 (2017) 012085.

[94] Q. Yu, L.A. Jauregui, W. Wu, R. Colby, J. Tian, Z. Su, H. Cao, Z. Liu, D. Pandey, D. Wei, T.F. Chung, P. Peng, N.P. Guisinger, E.A. Stach, J. Bao, S.-S. Pei, Y.P. Chen, Nat. Mater. 10 (2011) 443–449.

[95] H. Terrones, R. Lv, M. Terrones, M.S. Dresselhaus, Rep. Prog. Phys. 75 (2012) 062501.

[96] G. Bacher, H. Schweizer, J. Kovac, A. Forchel, H. Nickel, W. Schlapp, R. Lösch, Phys. Rev. B 43 (1991) 9312–9315.

[97] Z. Wu, Z. Luo, Y. Shen, W. Zhao, W. Wang, H. Nan, X. Guo, L. Sun, X. Wang, Y. You, Z. Ni, Nano Res. 9 (2016) 3622–3631.

[98] N. Hamza, A. Toncelli, A. Parchur, E. Alves, R. Maalej, Sens. Actuators B Chem. 248 (2017) 769–776.

[99] ACS Nano. 8 (2014) 5369–5371.

[100] S.T. Manikkoth, D. Panoth, K.M. Thulasi, F. Jahan, A. Paravannoor, B.K. Vijayan, 2021. Chapter 7 - Flexible smart nanosensors, Editor(s): in: S. Thomas, T.A. Nguyen, M. Ahmadi, A. Farmani, G. Yasin (Eds.), In Micro and Nano Technologies, Nanosensors for Smart Manufacturing, Elsevier, 145-182, https://doi.org/10.1016/B978-0-12-823358-0.00009-5.

[101] N. Wilson, Bioscience 68 (2018) 241–246.

[102] P.C. Ray, H. Yu, P.P. Fu, J. Environ. Sci. Health C: Environ. Carcinog. Ecotoxicol. Rev. 27 (2009) 1–35.

CHAPTER

Application of nanotechnology in advancement of heat transfer

4

Laveet Kumar[a], Mamdouh El Haj Assad[b], Yashar Aryanfar[c] and Jorge Luis García Alcaraz[d]

[a]*Department of Mechanical Engineering, Mehran University of Engineering and Technology, Jamshoro, Pakistan,* [b]*Department of Sustainable and Renewable Energy Engineering, University of Sharjah, Sharjah, United Arab Emirates,* [c]*Department of Electric Engineering and Computation, Autonomous University of Ciudad Juárez, Juárez, Chihuahua, México,* [d]*Department of Industrial Engineering and Manufacturing, Autonomous University of Ciudad Juárez, Juárez, Chihuahua, México*

4.1 Introduction

At the nanoscale, heat transport deviates from classical laws [1]. Understanding nanoscale heat transfer is crucial for designing novel materials with tailored thermal transport properties, which have applications in energy conversion, usage, and the thermal management of electronic, optical, and optoelectronic devices. Over the past two decades, significant advancements have been made in studying heat transmission via conduction and convection at the nanoscale, yielding several intriguing observations. Notably, the thermal conductivity of single-walled carbon nanotubes (CNTs), a one-dimensional system, surpasses that of diamond, the best known bulk heat conductor. Computational calculations estimate CNT thermal conductivity at 6600 W/m K at room temperature [2], whereas experimental measurements report a value of 3500 W/m K at room temperature [3]. The one-dimensional nature of heat conduction in CNTs results in exceptionally high and potentially divergent thermal conductivities.

In addition, graphene has been shown to have remarkable thermal conductivity (5800 W/m K) [4]. In the amorphous phase, polymers, such as polyethylene, are typically classified as thermal insulators (0.3 W/m K) [5]. Simulations, however, indicate that thermal conductivity is high and might even be divergent over a polyethylene molecular chain [6]. Thermal conductivities (100 W/m K) of ultradrawn polyethylene nanofibers are discovered to be hundreds of times higher than those of their amorphous counterparts [7]. Thermal conductivity enhancement in fluids dispersed with nanoparticles to improve the convection heat transfer has been observed at amounts far greater than those predicted by conventional effective medium theory [8,9].

The mechanism and potential uses of the increased convection heat transport in nanofluids have been the subject of extensive theoretical and experimental investigation [10–15]. According to theories and a few earlier studies conducted at micrometer or greater separations [16–18], radiative heat transfer between objects separated at tiny distances increases dramatically with decreasing separations.

Recent tests have increased the distances between surfaces to tens of nanometers, and it has been observed that radiative heat transmission is much higher than that expected by Planck's blackbody radiation equation [19].

Graphene and 2D Materials in Heat Transfer: Fundamentals and Applications. DOI: https://doi.org/10.1016/B978-0-12-821964-5.00012-6

Low and reduced thermal conductivities in materials have also been seen and accomplished, in addition to high and raised thermal conductivities. Superlattices exhibit thermal conductivities far lower than those predicted by the Fourier law [20–22], even lower than those of alloys [23,24], and far below those predicted by the lowest thermal conductivity theory. Different nanowires, particularly silicon, have also shown significantly decreased thermal conductivity, some of which defy explanation by existing theories [25–28]. The thermal conductivity of WSe_2 is barely twice that of air, even in a crystal form [29]. These findings have sparked the investigation into thermoelectric energy conversion, and knowledge gained from the research is being used to create bulk nanostructured thermoelectric materials [30].

Understanding the impact of imperfections introduced during production on the operational behavior of single-layer graphene has been a focus of theoretical investigations. Although much attention has been given to atomic flaws, the presence of nanoscale pinholes and bilayer patches bonded to monolayer graphene during synthesis cannot be overlooked. This study employs nonequilibrium molecular dynamics (MD) simulations to explore the influence of these nanoscale flaws on the thermal conductivity of graphene. Both voids and bilayer flaws are considered in modeling the thermal conductivity of single-layer graphene oriented in zigzag and armchair configurations [31].

Despite significant theoretical and experimental research, several questions remain regarding the high thermal conductivities observed in some nanomaterials. This chapter investigates the current understanding of nanoscale thermal transport phenomena using nanoparticles. It explores how computational and experimental methods in heat transfer can be employed to comprehend and explain these observations. Although limitations and unresolved issues persist, the review focuses on key areas of expertise within the field. Readers interested in a broader discussion are encouraged to consult other relevant textbooks [1,32].

4.2 Technological advancements and design improvements of nanoscale heat transfer systems

Many techniques are available to generate surfaces of interest and enhance the thermal performance of heat exchangers [33]. The single convective and phase-change heat transfer is influenced by the energy and topology of the surface, which represent the chemical and geometric nature of the surface. There are a huge number of research articles on these techniques available in the literature, and it has been proved that the dependence of surface morphology on only a few controlling parameters has common trends. For many controlling parameters, surface structure results differ in specific experimental cases [34–36]. Basically, surface modification in micro/nanoscale can be generated by roughening and coating methods. The roughening or subtractive method usually obtains a relatively simple structure, whereas the coating method can provide a more complex structure and variable in material selection. This section briefly presents surface modification methods utilized in heat transfer intensification.

4.2.1 Roughening techniques

In single-phase heat exchangers, roughened surfaces enhance convective heat transfer by promoting turbulence [37]. In phase-change scenarios, these surfaces increase the heat transfer contact area and provide more nucleation sites [38]. Surface roughening can be achieved through mechanical methods such as sandblasting, abrasive blasting, and shot peening or via chemical and electrochemical techniques

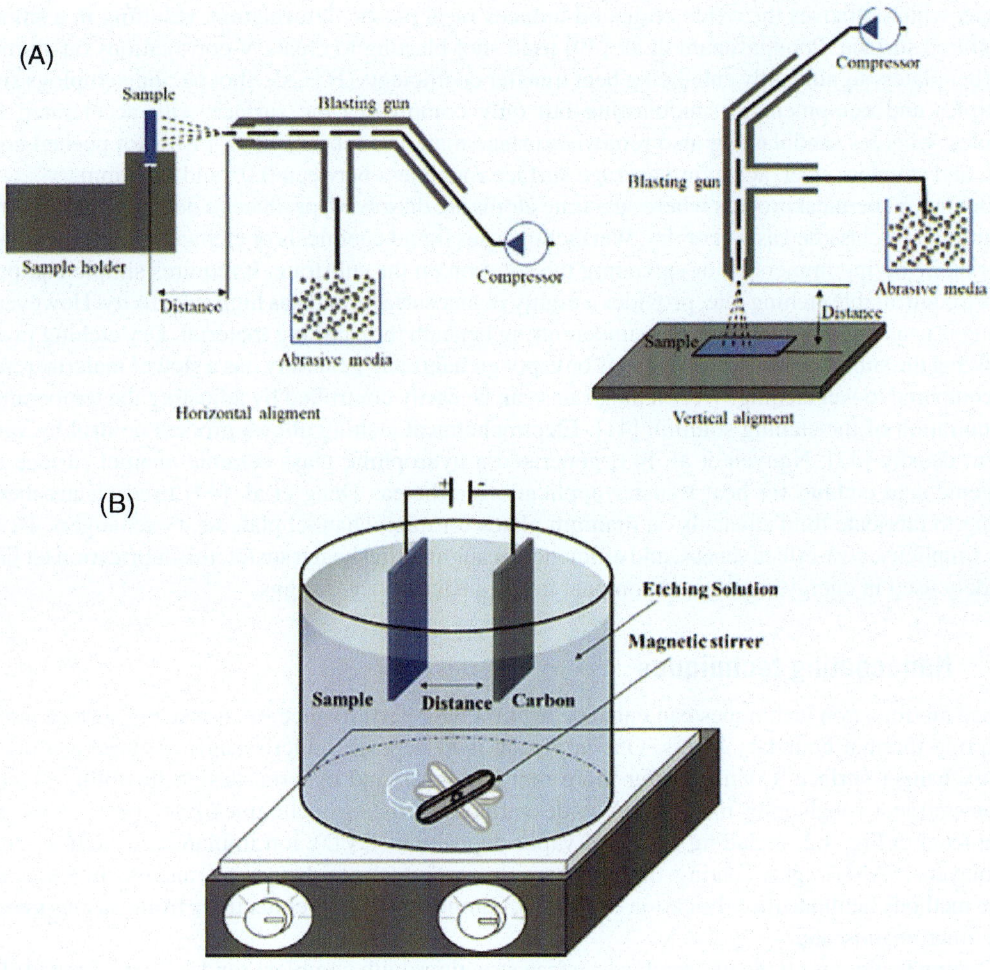

FIGURE 4.1

Roughening surface techniques: (A) sand/abrasive blasting/shot peening and (B) electrochemical etching [33].

such as chemical and electrochemical etching, as depicted in Fig. 4.1. Fig. 4.1A and B illustrates the horizontally and vertically aligned sand or abrasive blasting methods. Sandblasting, also referred to as abrasive blasting, involves propelling a stream of abrasive material through a compressor against a surface at high pressure. This process smooths rough surfaces, shapes surfaces, or removes surface contaminants, thereby enhancing heat transfer efficiency.

Mechanical roughening techniques typically produce simple surface structures characterized by randomly distributed peaks and valleys. In contrast, chemical and electrochemical methods can achieve complex structures with controllable roughness parameters, such as pitch and porosity, by adjusting process parameters. Sand or abrasive blasting involves firing high-speed sand particles at the sample surface, which removes the oxide scale and induces local plastic deformation, resulting in a smoother or textured surface. Poongavanam et al. [39] used shot blasting to create V-corrugations on aluminum absorber plates, significantly enhancing heat transfer coefficients (HTCs). Shot peening employs similar principles and equipment to sandblasting but only compresses the surface with accelerated metal particles, whereas sandblasting also removes surface material. Jabardo et al. [40] shot peened copper, brass, and stainless steel, achieving average surface roughness between 0.03 and 10.5 mm.

Etching, a chemical process where substrate atoms are dissolved, produces a desirable rough surface. Etching can be classified as wet or dry. Wet etching uses liquid chemicals or etchants to remove substrate material, involving immersion or spreading the solution on the substrate. It requires simple equipment, offers a controllable etching rate, provides a highly planar surface, and has high selectivity. However, wet etching is usually isotropic, leading to undercutting beneath the masking material. Dry etching involves dissolving the substrate using reactive ions or vapor etchants and generally has a slower material removal rate compared to wet etching. Wet etching rates can be easily controlled by adjusting the temperature or concentration of the etching solution [41]. Electrochemical etching allows process control by varying current density [42]. Nguyen et al. [43] generated a hydrophilic glass-ceramic enamel surface using halogenic acid etching for heat transfer applications, whereas Deng et al. [44] used electrochemical etching to fabricate fluid channels on titanium plates for microchannel plate heat exchangers. Because of its simplicity, cost-effectiveness, and efficiency in augmenting heat transfer, this fabrication technique is widely used in energy-saving and compact heat transfer device designs.

4.2.2 Nanocoating techniques

Surface modification techniques can enhance heat transfer performance by increasing surface area and disrupting thermal boundary layers. One approach is to deposit a micro/nanostructured coating on a heat exchanger surface. Coatings offer more morphological and material design flexibility compared to conventional roughening methods. A wide variety of coating techniques have been investigated, as depicted in Fig. 4.2, including chemical vapor deposition (CVD), ion implantation, self-assembled monolayers (SAMs), spray coating, dip coating, spin coating, electrodeposition, and electroless coating. These methods facilitate the fabrication of diverse, complex coating architectures from various materials at the micro/nanoscale.

Coating techniques typically produce surface structures with complex, randomly distributed shapes and sizes. The grain size and coating thickness can range from a few nanometers to hundreds of micrometers, depending on the specific technique used. CVD involves forming a film on a heated surface by decomposing gaseous chemical precursors. Advancements in CVD technology have enabled the growth of ultrathin and complex nanostructure coatings, significantly enhancing heat transfer augmentation [45,46]. For example, Preston et al. [47] used both low-pressure CVD and atmospheric pressure CVD to coat single-layer graphene on copper substrates, improving condensation heat transfer. Ion implantation, another coating technique, embeds atoms onto a substrate using a beam of ionized particles, altering the substrate's physical, chemical, or electrical properties. This method is primarily used to create hydrophobic surfaces on heat transfer devices, promoting dropwise condensation and achieving HTC enhancements of more than five times [48,49].

SAMs are single layers of organic molecules that form surface coatings. By selecting appropriate terminal groups on these alkyl molecules the chemical and physical properties of the resulting SAMs can be precisely controlled. SAMs exhibit minimal thermal resistance compared to other hydrophobic coatings [50], making them attractive for enhancing thermal performance on heat transfer surfaces

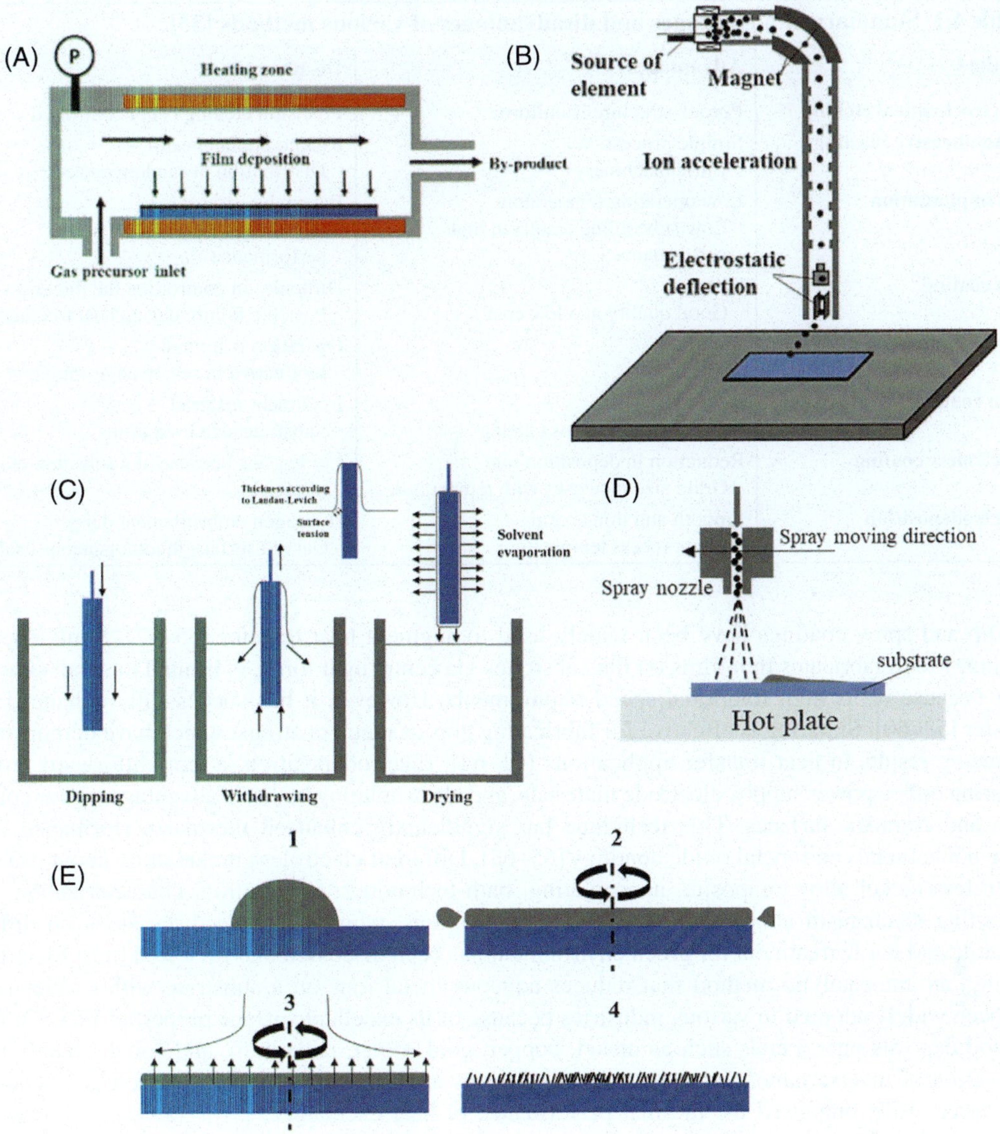

FIGURE 4.2

Nanocoating techniques: (A) chemical vapor deposition, (B) ion implantation, (C) dip coating, (D) spray coating, and (E) spin coating [33].

[51–53]. Simple and cost-effective methods, such as dip and spray coating, can be used to apply thin coatings on various substrates without the need for sophisticated equipment. However, these techniques may result in lower quality coatings than other methods [54] and are best suited for applications involving low shear stress and mild conditions.

Table 4.1 Summary of advantages and disadvantages of various methods [33].

Method	Advantages	Disadvantages
Electrochemical etching	Porous structure is uniform	Precision etching is not controlled
Sand-abrasive blasting	Simple process Little machinery	Risk to human health Distribution of random peaks
Ion implantation	Low operating temperature Low subcooling results in high performance	Expensive High subcooling results in low performance
Dip coating	Simple Good quality at a low cost	Difficulty in controlling the thickness Possible failure during heat treatment
Spray coating	Simple Cheap	Operation is limited Ineffective in severe environments
Spin coating	Low cost Thin homogeneous coating	Inefficient material Substrate size limitations
Electroless coating	Reduction in deposition rate Grain size increases with temperature	Limitations because of a selection of material
Electrodeposition	Smooth and thin coating Low process temperature	Hydrogen embrittlement danger Curved surface inhomogeneous coating

Dip and spray coatings have been widely used to augment heat transfer [55–57]. Similarly, spin coating, which fabricates thin films on flat substrates via centrifugal force, is limited to small substrate sizes because of its high rotational speed requirements. However, it has successfully enhanced heat transfer [58–60]. Sintering is effective for fabricating porous metal coatings, which have demonstrated promising results in heat transfer applications [61–64]. Electrodeposition, a straightforward process requiring only a power supply, electrode materials, and electroplating baths, is advantageous for coating large and complex surfaces. This technique has significantly enhanced thermal performance when using noble metals and metal oxide coatings [65–68]. Different electroless nickel alloy about different electroless nickel alloy, omposite, nano plating, bath techniques, preparation, characterization, new depositing mechanism along with their recent applications, whgich include discussions on difficult substrate and waste treatment for green environment has been discussed and reviewed [69]. Electroless coating, an autocatalytic method that reduces aqueous metal ions on a substrate without electricity, has been widely adopted in various industries because of its excellent surface properties [70–72]. This method deposits pure metals such as nickel, copper, gold, silver, palladium, and cobalt, enabling the fabrication of diverse nano/microstructures for various applications [73–75]. Electroless coating has also successfully enhanced the thermal performance of heat exchangers [76]. Table 4.1 provides an overview of each coating approach, detailing their operating principles, process parameters, effects on surface topology, and their respective advantages and disadvantages.

4.3 Advances in computational and experimental methods

This section overviews various computational and experimental tools used to study nanoscale heat transfer. Significant efforts have been focused on understanding heat conduction in nanostructured materials, particularly crystal-based nanostructures. Heat is primarily carried by vibrations in the crystal lattice, known as phonons, which are quantized lattice waves. Advances in both computational and experimental techniques have enhanced our understanding of phonon transport.

Crystal thermal conductivity is often modeled using approaches developed by researchers such as Klemens [77] and Callaway [78], which are based on the Phonon–Boltzmann equation introduced by Peierls in the 1950s [79]. These models use Fermi's golden rule with fitting parameters, which are typically determined by matching the measured temperature dependence of thermal conductivity to a kinetic expression of the thermal conductivity integral. Although these methods provide a qualitative understanding of phonon transport, the details obtained from such models can be imprecise, especially when applied to explain experimental results in nanostructured materials.

Recent advances in simulation tools have significantly improved our ability to study nanoscale heat transport. Computational techniques such as lattice dynamics based on force constants from empirical force fields or first-principles calculations, MD simulations, Green's functions, Boltzmann transport equations, and Monte Carlo simulations have become more sophisticated. Before the advent of computational methods to determine exact force constants, phonon properties were typically obtained experimentally or through simplified analyses [77,78,80,81].

Experimental studies have also contributed to our understanding of phonon transport. For example, CdTe crystals doped with 1 mol% of Mn, Fe, or ZnTe were formed from the melt [80]. The differences in the covalent, tetrahedral radii of these metal atoms help explain the segregation coefficients of these impurities and of Co. The thermal conductivity of both pure and doped CdTe crystals was tested between 3 and 300 K. In pure CdTe, heat transfer was mainly because of phonons constrained by scattering from crystal borders, isotopes, and other phonons. In the doped crystals, local lattice distortions introduced additional nonmagnetic phonon scattering, similar to Mn, Fe, and Zn. In recent years, atomic models, especially lattice dynamics with input from first-principles density functional theory calculations, have greatly improved the accuracy of predicted phononic properties and thermal conductivity of materials, especially crystals [82–85]. These methods are fitting to experimental data.

Many experimental methods and devices have been optimized and developed to measure heat transport properties and study the fundamental physics of heat transfer. These methods have enabled the observation of many interesting nanoscale heat transport phenomena. In this section, we briefly review these important experimental methods and discuss some perspectives on them.

In many applications, high spatial resolution temperature measurements are essential. Optical methods are constrained by the diffraction limit, leading researchers to explore contact measurements combined with high spatial resolution temperature sensors and scanning probe microscopy. Williams and Wickramasinghe [86] pioneered the first thermal scanning microscope, followed by the development of the atomic force microscope by Binnig et al. Wickramasinghe's scanning thermal microscopy employs thermal conduction between substrates to infer the surface topology of nonconductive samples. This approach was enhanced by replacing the tip of an atomic force microscope with a thermocouple probe for spatially resolved temperature measurements in contact mode [87]. Some commercial scanning thermal microscopes use Wollaston wire resistance thermometers [88]. To further improve spatial resolution, microfabricated cantilevers with thermal sensors at the tip were developed [89–92]. These advanced scanning thermal microscopes are applied in various fields, including surface topology measurements [93], phase-change detection [94], and thermal conductivity measurements.

The authors [95] emphasize research that has significantly advanced the understanding of thermal boundary conductance at the atomic and nanoscale near solid/solid interfaces, which has practical applications. In addition, advances in MD simulation methods have been instrumental in studying thermal boundary conductance and interfacial energy transport. These computational methods inherently account for inelastic effects because of multiple phonon interactions, making them valuable tools for examining thermal boundary conductance across various material systems [96]. MD simulations are particularly effective for analyzing vibrational transport in regimes where the majority of phonon modes participating in energy transport are thermally excited.

4.4 Conclusion and recommendations

This review highlights advancements in nanoscale heat transfer technologies, including roughening and coating techniques, which are critical for various applications, particularly in electronic systems. As electronic devices continue to shrink in size the ability to control, adjust, and utilize nanoscale heat transport will become increasingly important. The review also covers recent progress in both computational and experimental techniques for studying nanoscale heat transport. Lattice mechanics based on first-principles calculations have demonstrated exceptional predictive power in determining the thermal conductivity of perfect crystals. MD simulations using empirical potential models remain vital for examining complex structures. The accuracy of these simulations heavily depends on the precision of the empirical potential models. Reaction potentials, which offer a more realistic depiction of interfacial and surface structures, are particularly promising for studying interfacial heat transport. Multiscale simulations that incorporate results from atomistic simulations are capable of modeling heat transport at larger scales, thus bridging the gap between atomistic simulations and device-level analysis. However, the current frameworks for atomic simulations need improvements in simulating interfacial phonon propagation. Experimental techniques have also seen significant advancements. Microfabricated platforms have been highly effective in measuring the thermal conductivity of nanostructures. Various optical pump-probe methods are widely utilized to measure the interfacial thermal conductivity and overall thermal conductivity of materials. Thermal conductivity spectroscopy, based on the pump-probe method, has been developed to measure the mean free path-dependent thermal conductivity, utilizing the classical size effect. These techniques have proven successful for materials with high thermal conductivity. However, extending their applicability to materials with low thermal conductivity and short phonon mean free paths remains challenging.

References

[1] G. Chen, Nanoscale Energy Transport and Conversion: A Parallel Treatment of Electrons, Molecules, Phonons, and Photons, Oxford University Press, 2005.

[2] S. Berber, Y.-K. Kwon, D. Toma´nek, Unusually high thermal conductivity of carbon nanotubes, Phys. Rev. Lett. 84 (2000) 4613–4616.

[3] E. Pop, D. Mann, Q. Wang, K. Goodson, H. Dai, Thermal conductance of an individual single-wall carbon nanotube above room temperature, Nano Lett. 6 (2005) 96–100.

[4] A.A. Balandin, et al., Superior thermal conductivity of single-layer graphene, Nano Lett 8 (2008) 902–907.

[5] L.H. Sperling, Introduction to Polymer Physical Science, Wiley-Interscience, New Jersey, 2006.
[6] A. Henry, G. Chen, High thermal conductivity of single polyethylene chains using molecular dynamics simulations, Phys. Rev. Lett. 101 (2008) 235502.
[7] S. Shen, A. Henry, J. Tong, R. Zheng, G. Chen, Polyethylene nanofibres with very high thermal conductivities, Nat. Nanotechnol. 5 (2010) 251–255.
[8] H. Masuda, A. Ebata, K. Teramae, N. Hishinuma, Alteration of thermal conductivity and viscosity of liquid by dispersing ultra-fine particles dispersion of Al_2O_3, SiO_2 and TiO_2 ultra-fine particles, Netsu Bussei 7 (1993) 227–233.
[9] S.U.S. Choi, D.A. Siginer, H.P. Wang, in: Developments and Applications of Non-Newtonian Flows, American Society of Mechanical Engineers, New York, 1995, pp. 99–105.
[10] S.K. Das, Nanofluids: Science and Technology, WileyInterscience, New Jersey, 2007.
[11] M. Chandrasekar, S. Suresh, A review on the mechanisms of heat transport in nanofluids, Heat Transfer Eng. 30 (2009) 1136–1150.
[12] S. Ozerinç, S. Kakaç, A. Yazıcıoglu, Enhanced thermal conductivity of nanofluids: a state-of-the-art review, Microfluid. Nanofluid. 8 (2010) 145–170.
[13] J.H.L. Lee, S.H. Lee, C.J. Choi, S.P. Jang, S.U.S. Choi, A review of thermal conductivity data, mechanisms and models for nanofluids, Int. J. Micro-Nano Scale Transp. 1 (2010) 269.
[14] V. Trisaksri, S. Wongwises, Critical review of heat transfer characteristics of nanofluids, Renew Sustain. Energy Rev. 11 (2007) 512–523.
[15] X.-Q. Wang, A.S. Mujumdar, Heat transfer characteristics of nanofluids: a review, Int. J. Therm. Sci. 46 (2007) 1–19.
[16] E.G. Cravalho, C.L. Tien, R.P. Caren, Effect of small spacings on radiative transfer between two dielectrics, J. Heat Transfer 89 (1967) 351–358.
[17] D. Polder, M. Van Hove, Theory of radiative heat transfer between closely spaced bodies, Phys. Rev. B: Solid State 4 (1971) 3303–3314.
[18] C.M. Hargreaves, Anomalous radiative transfer between closely-spaced bodies, Phys. Lett. A 30 (1969) 491–492.
[19] S. Shen, A. Narayanaswamy, G. Chen, Surface phonon polaritons mediated energy transfer between nanoscale gaps, Nano Lett. 9 (2009) 2909–2913.
[20] T. Yao, Thermal properties of AlAs/GaAs superlattices, Appl. Phys. Lett. 51 (1987) 1798–1800.
[21] G. Chen, C.L. Tien, X. Wu, J.S. Smith, Thermal diffusivity measurement of GaAs/AlGaAs thin-film structures, J. Heat Transfer 116 (1994) 325–331.
[22] S.-M. Lee, D.G. Cahill, R. Venkatasubramanian, Thermal conductivity of Si–Ge superlattices, Appl. Phys. Lett. 70 (1997) 2957–2959.
[23] T. Borca-Tasciuc, et al., Thermal conductivity of symmetrically strained Si–Ge superlattices, Superlatt. Microstruct. 28 (2000) 199–206.
[24] D. Li, Y. Wu, R. Fan, P. Yang, A. Majumdar, Thermal conductivity of Si–SiGe superlattice nanowires, Appl. Phys. Lett. 83 (2003) 3186–3188.
[25] G. Chen, Thermal Conductivity of Nanostructures, in: M.T. Terry (Ed.), Semiconductors and Semimetals, Elsevier, California, 2001, pp. 203–259.
[26] D. Li, et al., Thermal conductivity of individual silicon nanowires, Appl. Phys. Lett. 83 (2003) 2934.
[27] A.I. Hochbaum, et al., Enhanced thermoelectric performance of rough silicon nanowires, Nature 451 (2008) 163–167.
[28] J. Lim, K. Hippalgaonkar, S.C. Andrews, A. Majumdar, P. Yang, Quantifying surface roughness effects on phonon transport in silicon nanowires, Nano Lett. 12 (2012) 2482.
[29] C. Chiritescu, et al., Ultralow thermal conductivity in disordered, layered WSe(2) crystals, Science 315 (2007) 351–353.

[30] B. Poudel, et al., High-thermoelectric performance of nanostructured bismuth antimony telluride bulk alloys, Science 320 (2008) 634–638.
[31] M.N. Esfahani, et al., Effect of nanoscale defects on the thermal conductivity of graphene, Mater. Today Commun. 26 (2021) 101856.
[32] Z. Zhang, Nano/Microscale Heat Transfer, McGraw-Hill, Switzerland, 2007.
[33] D.H. Nguyen, H.S. Ahn, A comprehensive review on micro/nanoscale surface modification techniques for heat transfer enhancement in heat exchanger, Int. J. Heat Mass Transfer 178 (2021) 121601.
[34] M. Kul, K.O. Oskay, F. Erden, E. Akca, R. Katirci, E. Koksal, E.J. Akinci, Effect of process parameters on the electrodeposition of zinc on 1010 steel: central composite design optimization, Int. J. Electrochem. Sci. 15 (2020) 9779–9795.
[35] N. Sahu, B. Parija, S.J. Panigrahi, Fundamental understanding and modeling of spin coating process: a review, Indian J. Phys. 83 (4) (2009) 493–502.
[36] D.J. Grosso, How to exploit the full potential of the dip-coating process to better control film formation, J. Mater. Chem. 21 (43) (2011) 17033–17038.
[37] K. Nilpueng, S.J. Wongwises, Experimental study of single-phase heat transfer and pressure drop inside a plate heat exchanger with a rough surface, Exp. Therm. Fluid Sci. 68 (2015) 268–275.
[38] R.L. Webb, The evolution of enhanced surface geometries for nucleate boiling, Heat Transf. Eng. 2 (3–4) (1981) 46–69.
[39] G.K. Poongavanam, K. Panchabikesan, A.J.D. Leo, V.J. Ramalingam, Experimental investigation on heat transfer augmentation of solar air heater using shot blasted V-corrugated absorber plate, Renewable Energy 127 (2018) 213–229.
[40] J.M.S. Jabardo, G. Ribatski, E.J. Stelute, Roughness and surface material effects on nucleate boiling heat transfer from cylindrical surfaces to refrigerants R-134a and R-123, Exp. Therm. Fluid Sci. 33 (4) (2009) 579–590.
[41] J. Heebner, R. Grover, T. Ibrahim, T.A. Ibrahim, Optical Microresonators: Theory, Fabrication, and Applications, Springer Science & Business Media, New York, 2008.
[42] H. Jansen, H. Gardeniers, M. de Boer, M. Elwenspoek, J.J. Fluitman, A survey on the reactive ion etching of silicon in microtechnology, J. Micromech. Microeng. 6 (1) (1996) 14.
[43] H.H. Nguyen, S. Wan, K.A. Tieu, H. Zhu, S.T.J.S. Pham, Rendering hydrophilic glass-ceramic enamel surfaces hydrophobic by acid etching and surface silanization for heat transfer applications, Surf. Coat. Technol. 370 (2019) 82–96.
[44] T. Deng, Z. Zhu, X. Li, T. Ma, Q. Wang, Experimental study on electrochemical etching for titanium printed circuit heat exchanger channels, J. Mater. Process. Technol. 282 (2020) 116669.
[45] A.T. Paxson, J.L. Yague, K.K. Gleason, K.K. Varanasi, Stable dropwise condensation for enhancing heat transfer via the initiated chemical vapor deposition (iCVD) of grafted polymer films, Adv. Mater. 26 (3) (2014) 418–423.
[46] N. Miljkovic, D.J. Preston, R. Enright, E.N. Wang, Electrostatic charging of jumping droplets, Nat. Commun. 4 (1) (2013) 1–9.
[47] D.J. Preston, D.L. Mafra, N. Miljkovic, J. Kong, E.N. Wang, Scalable graphene coatings for enhanced condensation heat transfer, Nano Lett. 15 (5) (2015) 2902–2909.
[48] M. Rausch, A. Leipertz, A.J. Froba, Dropwise condensation of steam on ion implanted titanium surfaces, Int. J. Heat Mass Transf. 53 (1–3) (2010) 423–430.
[49] K. Kim, Y. Lee, J.H. Jeong, Dropwise condensation induced on chromium ion implanted aluminum surface, Nucl. Eng. Technol. 51 (1) (2019) 84–94.
[50] I.O. Ucar, H.Y. Erbi'l, Droplet condensation on polymer surfaces: A review, Turk. J. Chem. 37 (4) (2013) 643–674.

[51] R.W. Bonner III, Dropwise condensation life testing of self assembled monolayers, in: International Heat Transfer Conference, American Society of Mechanical Engineers (ASME), Washington, DC, USA, 2010, pp. 221–226.

[52] B.J. Zhang, C. Kuok, K.J. Kim, T. Hwang, H.J. Yoon, Dropwise steam condensation on various hydrophobic surfaces: Polyphenylene sulfide (PPS), polytetrafluoroethylene (PTFE), and self-assembled micro/nano silver (SAMS), Int. J. Heat Mass Transf. 89 (2015) 353–358.

[53] L. Chen, S. Liang, R. Yan, Y. Cheng, X. Huai, S.J. Chen, *n*-Octadecanethiol self-assembled monolayer coating with microscopic roughness for dropwise condensation of steam, J. Therm. Sci. 18 (2) (2009) 160–165.

[54] V.O. Fasiku, S.J. Owonubi, E. Mukwevho, B. Aderibigbe, E.R. Sadiku, Y. Lemmer, I.D. Ibrahim, J. Mochane, O.O. Daramola, K.J. Selatile, Polymeric materials in coatings for biomedical applications, in: L. Li, Q. Yang (Eds.), Advanced Coating Materials, Wiley, 2018, pp. 481–518.

[55] G. Azimi, R. Dhiman, H.-M. Kwon, A.T. Paxson, K.K. Varanasi, Hydrophobicity of rare-earth oxide ceramics, Nat. Mater. 12 (4) (2013) 315–320.

[56] M.R. Haque, C. Zhu, C. Qu, E.C. Kinzel, A.R. Betz, Experimental investigation of condensation phenomenon on hydrophilic and hydrophobic titanium (Ti) pillared glass surfaces, 2018 Micro Nano Flows Conference, 2018.

[57] X. Ji, D. Zhou, C. Dai, J.J. Xu, Dropwise condensation heat transfer on superhydrophilic-hydrophobic network hybrid surface, Int. J. Heat Mass Transf. 132 (2019) 52–67.

[58] H. Jo, H.S. Ahn, S. Kang, M.H. Kim, A study of nucleate boiling heat transfer on hydrophilic, hydrophobic and heterogeneous wetting surfaces, Int. J. Heat Mass Transf. 54 (25-26) (2011) 5643–5652.

[59] D.-J. Huang, T.-S. Leu, Condensation heat transfer enhancement by surface modification on a monolithic copper heat sink, Appl. Therm. Eng. 75 (2015) 908–917.

[60] R. Parin, A. Martucci, M. Sturaro, S. Bortolin, M. Bersani, F. Carraro, D.J.S. Del Col, Nano-structured aluminum surfaces for dropwise condensation, Surf. Coat. Technol. 348 (2018) 1–12.

[61] M.-S. Kim, K.A.L. Ada, O.K. Kwon, C.W. Park, A comparison of condensation heat transfer performance of MWCNT/Fe composite coatings on steel substrate, J. Mech. Sci. Technol. 28 (4) (2014) 1589–1596.

[62] B. Koro ˘glu, N. Bogan, C.J. Park, Effects of tube row on heat transfer and surface wetting of microscale porous-layer coated, Horizontal-Tube, Falling-Film Evaporator, ASME J. Heat Mass Transf. 135 (4) (2013).

[63] S. Lee, B. Koro ˘glu, C.J. Park, Experimental investigation of capillary-assisted solution wetting and heat transfer using a micro-scale, porous-layer coating on horizontal-tube, falling-film heat exchanger, Int. J. Refrig. 35 (4) (2012) 1176–1187.

[64] H. Xu, Y. Dai, H. Cao, J. Liu, L. Zhang, M. Xu, J. Cao, P. Xu, J.J. Liu, Tubes with coated and sintered porous surface for highly efficient heat exchangers, Front. Chem. Sci. Eng. 12 (3) (2018) 367–375.

[65] A. Hooda, M. Goyat, J.K. Pandey, A. Kumar, R.J. Gupta, A review on fundamentals, constraints and fabrication techniques of superhydrophobic coatings, Prog. Org. Coat. 142 (2020) 105557.

[66] F. Zhou, A. Izgorodin, R.K. Hocking, L. Spiccia, D.R. MacFarlane, Electrodeposited MnO_x films from ionic liquid for electrocatalytic water oxidation, Adv. Energy Mater. 2 (8) (2012) 1013–1021.

[67] M. Ge, S. Wang, J. Zhao, Y. Zhao, L. Liu, Effects of extended surface and surface gold plating on condensation characteristics of steam with large amount of CO_2, Exp. Therm. Fluid Sci. 92 (2018) 13–19.

[68] R. Wen, Q. Li, J. Wu, G. Wu, W. Wang, Y. Chen, X. Ma, D. Zhao, R. Yang, Hydrophobic copper nanowires for enhancing condensation heat transfer, Nano Energy 33 (2017) 177–183.

[69] J. Sudagar, J. Lian, W.J. Sha, Electroless nickel, alloy, composite and nano coatings – A critical review, J. Alloys Compd. 571 (2013) 183–204.

[70] J. Sudagar, G. Bi, Z. Jiang, G. Li, Q. Jiang, J.J. Lian, Electrochemical polarization behaviour of electroless Ni-P deposits with different chromium-free pre-treatment on magnesium alloy, Int. J. Electrochem. Sci. 6 (7) (2011) 2767–2788.

[71] J. Sudagar, K. Venkateswarlu, J. Lian, Dry sliding wear properties of a 7075-T6 aluminum alloy coated with Ni-P (h) in different pretreatment conditions, J. Mater. Eng. Perform. 19 (6) (2010) 810–818.
[72] T. Tamilarasan, R. Rajendran, U. Sanjith, G. Rajagopal, J. Sudagar, Wear and scratch behaviour of electroless Ni-P-nano-TiO_2: effect of surfactants, Wear 346 (2016) 148–157.
[73] D.-W. Ruan, J. Sudagar, Y.-Q. Liang, R. Elansezhian, J.-S. Lian, Effects of surfactants on corrosion behavior of chromium-free electroless nickel deposit on magnesium alloy, J. Jilin Univ. (Eng. Technol. Ed.) 43 (2) (2013) 363–367.
[74] A. Selvakumar, R. Perumalraj, J. Sudagar, S.J. Mohan, Nickel–multiwalled carbon nanotube composite coating on aluminum alloy rotor for textile industries, Proc. Inst. Mech. Eng., L 230 (1) (2016) 319–327.
[75] J. Balaraju, T.S. Narayanan, S. Seshadri, Electroless Ni–P composite coatings, J. Appl. Electrochem. 33 (9) (2003) 807–816.
[76] J. Shim, D. Seo, S. Oh, J. Lee, Y. Nam, Condensation heat-transfer performance of thermally stable superhydrophobic cerium-oxide surfaces, ACS Appl. Mater. Interfaces 10 (37) (2018) 31765–31776.
[77] P.G. Klemens, Thermal conductivity of solids at low temperatures, in: Low Temperature Physics I/Kältephysik I, Springer, Berlin, 1956, pp. 198–281.
[78] J. Callaway, Model for lattice thermal conductivity at low temperatures, Phys. Rev. 113 (4) (1959) 1046.
[79] R. Peierls, Zur kinetischen theorie der wärmeleitung in kristallen, Ann. Phys. 395 (8) (1929) 1055–1101.
[80] P. Klemens, D. Pedraza, Thermal conductivity of graphite in the basal plane, Carbon 32 (4) (1994) 735–741.
[81] M. Holland, Analysis of lattice thermal conductivity, Phys. Rev. 132 (6) (1963) 2461.
[82] G.A. Slack, S. Galginaitis, Thermal conductivity and phonon scattering by magnetic impurities in CdTe, Phys. Rev. 133 (1A) (1964) A253.
[83] A. Ward, D. Broido, D.A. Stewart, G. Deinzer, Ab initio theory of the lattice thermal conductivity in diamond, Phys. Rev. B 80 (12) (2009) 125203.
[84] A. Ward, D. Broido, Intrinsic phonon relaxation times from first-principles studies of the thermal conductivities of Si and Ge, Phys. Rev. B 81 (8) (2010) 085205.
[85] K. Esfarjani, G. Chen, H.T. Stokes, Heat transport in silicon from first-principles calculations, Phys. Rev. B 84 (8) (2011) 085204.
[86] C. Williams, H. Wickramasinghe, Scanning thermal profiler, Microelectron. Eng. 5 (1-4) (1986) 509–513.
[87] M. Fujii, et al., Measuring the thermal conductivity of a single carbon nanotube, Phys. Rev. Lett. 95 (6) (2005) 065502.
[88] R.J. Pylkki, P.J.M.P.J. Moyer, P.E.W.P.E. West, Scanning near-field optical microscopy and scanning thermal microscopy, Japan. J. Appl. Phys. 33 (6S) (1994) 3785.
[89] Y.S.Y. Suzuki, Novel microcantilever for scanning thermal imaging microscopy, Jpn. J. Appl. Phys. 35 (3A) (1996) L352.
[90] K. Luo, Z. Shi, J. Lai, A. Majumdar, Nanofabrication of sensors on cantilever probe tips for scanning multiprobe microscopy, Appl. Phys. Lett. 68 (3) (1996) 325–327.
[91] M. Hinz, O. Marti, B. Gotsmann, M. Lantz, U. Dürig, High resolution vacuum scanning thermal microscopy of HfO_2 and SiO_2,, Appl. Phys. Lett. 92 (4) (2008) 043122.
[92] G. Mills, H. Zhou, A. Midha, L. Donaldson, J. Weaver, Scanning thermal microscopy using batch fabricated thermocouple probes, Appl. Phys. Lett. 72 (22) (1998) 2900–2902.
[93] M.-H. Li, J.J. Wu, Y.B. Gianchandani, Surface micromachined polyimide scanning thermocouple probes, J. Microelectromech. Syst. 10 (1) (2001) 3–9.
[94] A. Hammiche, D. Hourston, H. Pollock, M. Reading, M. Song, Scanning thermal microscopy: subsurface imaging, thermal mapping of polymer blends, and localized calorimetry, J. Vac. Sci. Technol., B 14 (2) (1996) 1486–1491.
[95] A. Giri, P.E. Hopkins, A review of experimental and computational advances in thermal boundary conductance and nanoscale thermal transport across solid interfaces, Adv. Funct. Mater. (2020) 1903857.
[96] M.P. Allen, D.J. Tildesley, Computer Simulation of Liquids, Oxford University Press, New York, 2017.

CHAPTER

5 Two-dimensional material–based nanofluids

Chitra Boobalan[a], Sathishkumar Kannaiyan[a] and Fedal Castro Nagarajan[b]

[a]*Department of Chemical Engineering, Sri Sivasubramaniya Nadar College of Engineering, Kalavakkam, Chennai, Tamil Nadu, India,* [b]*Department of Mechanical Engineering, Aarupadai Veedu Institute of Technology, Paiyanoor, Chennai, Tamil Nadu, India*

5.1 Introduction

Heat dissipation between two fluids is critical in various applications, such as automobile engine cooling, solar thermal systems, and electronic cooling. Conventional fluids, such as water, glycol, and oils, are commonly used for heat transfer but exhibit poor heat dissipation performance because of their low thermal conductivity [1,2]. Nanofluids, which blend nanosized metals with conventional fluids, have become a promising option for enhancing thermal conductivity. Fundamental thermophysical properties, such as viscosity and thermal conductivity, are critical for improving thermal efficiency. Higher viscosity can increase the required pumping power, making it essential to evaluate the rheological properties of nanofluids [3]. The addition of nanomaterials to a base fluid alters its structural properties because of changes in rheological characteristics [4]. Researchers have reported that incorporating nanoparticles in base fluids increases both thermal conductivity and viscosity [5]. However, some studies have documented a reduction in viscosity alongside increased thermal conductivity [6]. Therefore a detailed investigation is needed to identify nanofluids that enhance thermal conductivity while reducing viscosity. Lower viscosity can improve thermal efficiency by decreasing the required pumping power and reducing frictional losses.

Many multifunctional nanofluids are currently in use, with two-dimensional (2D) materials attracting significant attention from researchers because of their remarkable thermal, electrical, optical, and catalytic properties [7]. Owing to their high surface area, 2D materials are excellent candidates for enhancing the thermal conductivity of heat transfer fluids [8]. In 2D materials, only one dimension is nanosized, resembling a large, thin sheet, and the thickness of the material can be reduced to a single atom. The first 2D material to be isolated was graphene. As then, numerous other 2D materials, such as transition metal dichalcogenides (TMDCs), hexagonal boron nitride (hBN), phosphorene, and xenes, have been identified, each exhibiting a wide range of properties, as illustrated in Fig. 5.1.

Graphene is a single layer of carbon atoms arranged in a hexagonal lattice with a thickness of approximately 0.14 nm. The carbon atoms in graphene are covalently bonded, giving it unique properties such as high thermal conductivity, attributed to the rapid movement of electrons at about 1/300 the speed of light. Graphene also exhibits high tensile strength and absorbs 2% of incident light [9]. Another

Graphene and 2D Materials in Heat Transfer: Fundamentals and Applications. DOI: https://doi.org/10.1016/B978-0-12-821964-5.00010-2

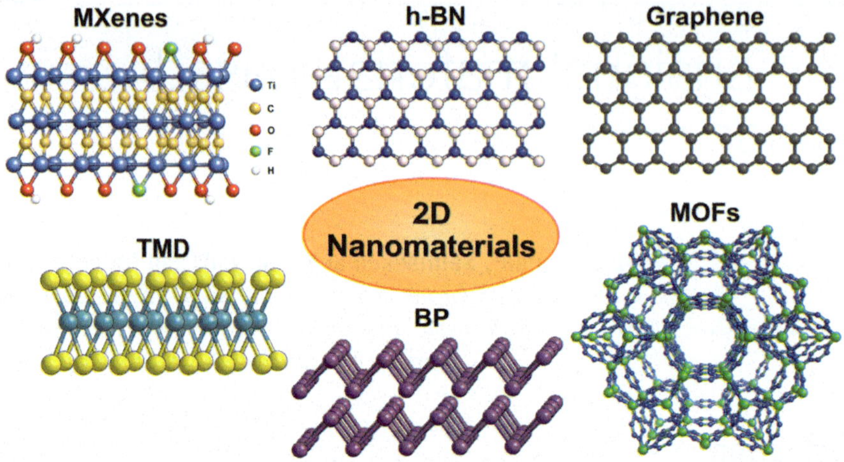

FIGURE 5.1

Schematic representation of 2D nanomaterials. *2D*, Two-dimensional.

notable 2D material is hBN, which is an isomorph of graphene. Unlike graphene, hBN consists of boron and nitrogen atoms and is a wide-bandgap insulator [10]. TMDCs are another class of 2D materials, represented by the chemical formula MX_2, where M stands for a transition metal [e.g., chromium (Cr), molybdenum (Mo), or tungsten (W)] and X represents a chalcogen [e.g., oxygen (O), sulfur (S), selenium (Se), or tellurium (Te)] [11]. TMDC layers are three atoms thick and are van der Waals materials, with a metal layer sandwiched between two chalcogenide layers [12].

Phosphorene, also known as 2D black phosphorus, is a layered allotrope of elemental phosphorus. It is a direct semiconductor with a honeycomb structure, making it suitable for transistors and optoelectronic devices because of its high charge mobility [13]. Xenes are monolayers of materials such as silicene, germanene, and stanene, which have a hexagonal structure similar to graphene. Xenes are used in field-effect transistors and as topological insulators [14].

2D materials differ from bulk materials for three main reasons:

1. **Removal of van der Waals interaction:** Bulk materials are weak because of weak van der Waals interaction, whereas 2D materials are strong since the covalent bonds hold the atoms together tightly [11].
2. **Increase in surface area to volume ratio:** The bulk materials have lower surface area to volume ratio when compared to 2D materials. So, the 2D materials are more reactive than bulk materials [6].
3. **Electron confinement in a plane:** The band structure of a material changes when it moves from bulk to 2D. This change in band structure leads to very high thermal conductivity of graphene [9] and the fluorescence of molybdenum sulfide [12].

This chapter examines the enhanced properties of 2D material–based nanofluids, focusing on thermal conductivity, viscosity, convective heat transfer, and boiling heat transfer coefficients. These characteristics are validated through diverse modeling studies to explore their applications in thermal management.

5.2 Preparation of two-dimensional materials

Although the concept of creating a 2D material might seem straightforward—simply thinning down a bulk material to atomic thickness—it presents significant challenges. Many materials, like diamonds, possess strong chemical bonds that are oriented in three dimensions. Thinning them down disrupts these bonds, leaving dangling bonds on the surface. These dangling bonds are chemically unstable and energetically unfavorable. Consequently the material attempts to rearrange its structure to minimize its surface energy, potentially leading to undesired reconstructions or instabilities [8,15–17].

There are generally two categories of methods used to synthesize 2D materials: top-down and bottom-up approaches.

Top-down approaches involve mechanically or chemically exfoliating a 2D layer from a bulk parent material. Common exfoliation techniques include mechanical cleavage/exfoliation using adhesive tape, liquid exfoliation methods, and intercalation/exfoliation utilizing small molecules or ions inserted between bulk layered planes.

Bottom-up approaches build 2D materials from their constituent atoms or molecules up. Vapor deposition methods like chemical vapor deposition (CVD) and molecular beam epitaxy deposit 2D layers atom-by-atom or molecule-by-molecule on substrates. Wet chemistry techniques such as chemical/hydrothermal synthesis also assemble 2D structures through solution-based reactions.

Each category has advantages—top-down provides an avenue for isolating naturally occurring 2D forms, whereas bottom-up enables precise control over film uniformity, thickness, and material properties. Hybrid methods further expand the toolkit, such as exfoliation followed by CVD overgrowth. The appropriate synthesis pathway is selected based on the targeted material, structure, quality, scale, and application. The preparation methods for 2D nanomaterials are shown in Fig. 5.2.

The mechanical exfoliation method, initially developed for producing monolayer graphene, is now widely used for all van der Waals materials. This technique involves applying sticky tape to the surface of a layered material and peeling it off to collect flakes composed of a small number of layers. These flakes are then transferred to a substrate by pressing the tape onto it, resulting in multilayered flakes of various sizes and shapes. Although the yield of monolayer flakes is low, they can reach sizes of up to 100 microns with very few defects [15]. The transfer of mechanical force is achieved through liquid exfoliation using an organic solvent medium, which results in the layers being suspended in the liquid. Applying tensile stress to these layers, followed by sonication, causes them to separate. However, this method yields a low amount of monolayer flakes, and their size is typically less than 100 nm because of the forces exerted during sonication [8].

CVD is a bottom-up technique that uses one or more precursor gases flowing through a furnace. These gases react either with each other or with the substrate to form a thin layer. Careful regulation of parameters such as gas pressure and composition, reaction time, and temperature is crucial, as they all influence the film's composition, quality, and thickness. Although CVD is scalable, it is more complex and costly compared to top-down methods [16]. For large-scale production the solution-based chemical

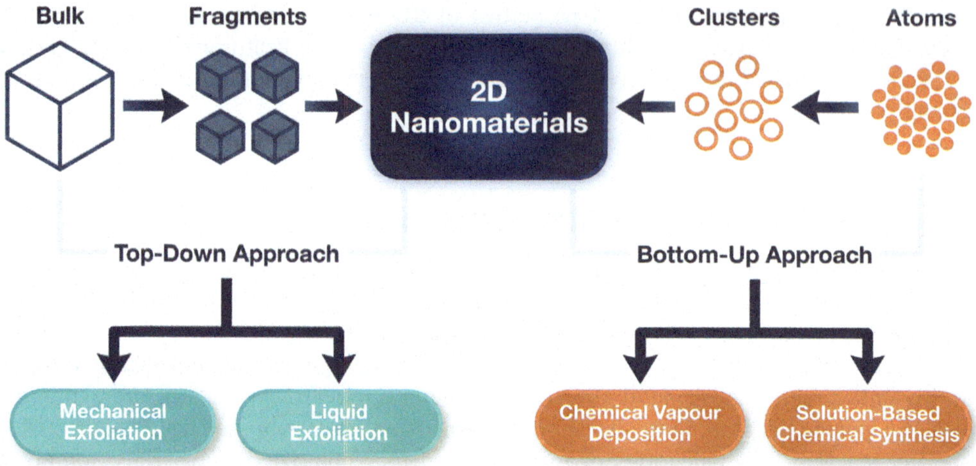

FIGURE 5.2

Preparation of 2D nanostructures. *2D*, Two-dimensional.

synthesis method is preferred. This approach typically yields flakes smaller than 100 nm and utilizes wet chemical techniques to produce 2D materials [17].

5.3 Preparation of nanofluids

Nanofluids can be prepared using two methods: (1) the one-step method and (2) the two-step method. The one-step method simultaneously prepares and disperses the nanomaterials within the base fluid, eliminating the need for drying particles, storage, handling, and dispersion. This minimizes nanoparticle agglomeration and improves stability, resulting in uniformly dispersed and stable nanofluids [18]. The two-step method, the most commonly used approach, involves first producing nanoparticles through mechanical or chemical processes and then dispersing them in base fluids via mechanical stirring or ultrasonication. This method is cost-effective and scalable for large-scale production of nanofluids [19] (Fig. 5.3).

5.4 Thermal conductivity studies

One key factor in enhancing heat transfer is thermal conductivity, which has been extensively studied. Nanofluids exhibit higher thermal conductivity than conventional liquids because of the superior heat conduction properties of solid particles compared to liquids. In addition, nanofluids improve heat transfer because of their large surface area and the Brownian motion of nanoparticles. Numerous experimental and theoretical studies have investigated this phenomenon. It is anticipated that the thermal conductivity of nanofluids will increase with higher concentrations and temperatures because of the random collisions of nanoparticles [1].

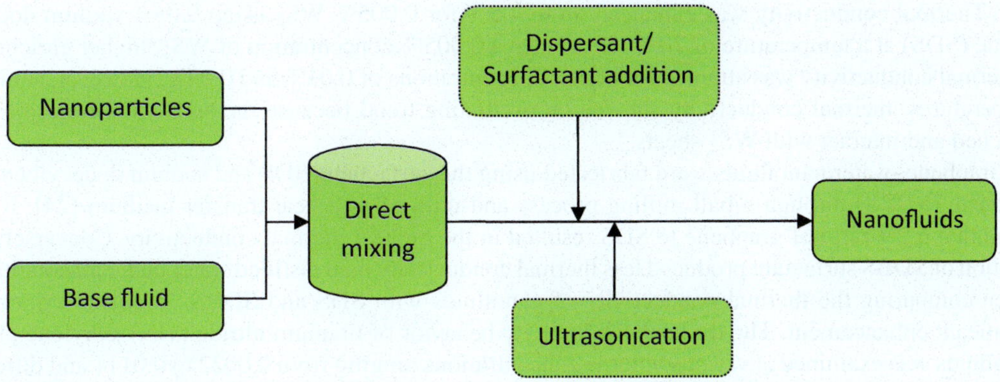

FIGURE 5.3

Two-step method for nanofluid preparation.

The KD2Pro thermal property analyzer is typically used to assess the thermal conductivity of nanofluids at various temperatures. To ensure precise temperature control the nanofluid is maintained within ±0.5°C using a temperature-regulated circulating water bath. Before measurements are taken, the nanofluid is placed in the circulating water bath for 15 minutes to reach thermal equilibrium [20,21]. Studies on the thermal conductivity of 2D materials are highlighted since the higher thermal conductivity leads to enhanced heat transfer. For a single layer of hBN in ethylene glycol–based nanofluids the enhancement of thermal conductivity was studied [22]. The enhancement was found to be about 80% with a 5% volume concentration of boron nitride, which is higher than that of graphene oxide/ethylene glycol nanofluids. It exhibited excellent thermal stability even at a high temperature of 90°C. There was a rapid increase in thermal conductivity for volume concentrations from 0.01% to 0.1%, and above 0.1%, the increase rate became linear up to 5%. For boron nitride nanofluids, the thermal conductivity showed no visible temperature dependence, indicating minimal contribution from temperature-induced Brownian motion and microconvection because of the formation of a 2D sheet network. The experimental and theoretical effective thermal conductivity values of boron nitride nanofluids, with and without surfactant, were studied and compared at room temperature using deionized water as the base fluid [23]. For increased concentration of nanoparticles, it was observed that there is an increase in thermal conductivity. The experimental results of effective thermal conductivity are higher than the theoretically calculated value because of the assumptions made in the calculation, such as neglecting the effect of particle size and particle motion. Boron nitride with no surfactant showed better stability and higher thermal conductivity.

The thermal conductivity studies for plate-like-shaped boron nitride particles in ethylene glycol for various mass concentrations till 20 wt.% and temperature ranging from 293 to 338 K with an increase of 15 K were focused on [20]. The results indicated that there is an increase in thermal conductivity with an increase in volume fraction and temperature of nanoparticles. It was also observed that for the particle volume fraction of 12%, the thermal conductivity increased by more than 260%, which is the largest increase reported for ethylene glycol–based nanofluids. Such a high thermal conductivity increase makes this material suitable for industrial applications. The influence of different surfactants on tungsten disulfide (WS_2)–ethylene glycol nanofluids for various concentrations from 0.05% to 2% was examined

[21]. Thermal conductivity was enhanced up to 4.5% for 0.005% WS_2 using 0.05% sodium dodecyl sulfate (SDS) at a temperature of 70°C. Similar to a 0.005% concentration of WS_2, higher enrichment in thermal conductivity was attained for higher concentrations of 0.01% and 0.02%. However, at higher temperatures, thermal conductivity showed an oscillating trend because the surfactant molecules get detached and interact with WS_2 sheets.

Graphene–water nanofluids were fabricated using the surfactants SDS and sodium dodecylbenzene sulfonate (SDBS) through a ball milling process and utilized as a heat transfer medium [24]. It was found that a 2:1 ratio of graphene to SDS resulted in the highest thermal conductivity. Conversely the addition of SDBS surfactant produced less thermal conductivity than distilled water under all conditions. When comparing the thermal conductivity of nanofluids with SDS and SDBS, SDS showed a more significant enhancement. The thermal conductivity behavior of titanium nitride (TiN)–ethylene glycol nanofluids was examined at various volume concentrations ranging from 0.0022 to 0.0111 and different nanoparticle sizes at a constant temperature. Results indicated that smaller nanoparticles exhibited higher thermal conductivity at the same concentration, demonstrating that nanoparticle size significantly influences thermal conductivity [25]. Predictions for the thermal conductivity improvement of graphene nanofluids suggested substantial enhancement at lower concentrations, with the improvement increasing as concentration rises [26]. The thermal conductivity of these nanofluids lies between that of metal oxide and carbon nanotube nanofluids, suggesting a heat conduction mechanism combining Brownian motion and microconvection effects.

Classical nonequilibrium molecular dynamics simulations calculated the thermal conductivity of phosphorene and graphene [27]. Phosphorene's thermal conductivity is an order of magnitude lower than graphene's because of its limited vibrational frequencies and reduced influence from out-of-plane flexural phonons. Both materials showed a decreasing trend in thermal conductivity with increasing temperature. The thermal conductivity of single-layer and multilayer phosphorene was investigated concerning geometrical and strain factors [28]. Multilayer phosphorene's in-plane thermal conductivity is insensitive to the number of layers, unlike graphene, where interlayer interactions affect conductivity. These findings suggest phosphorene's suitability for thermal management in nanoelectronic devices and thermoelectric applications, with its thermal conductivity influenced by tensile strain and sample length.

The thermal conductivities of hBN, molybdenum disulfide (MoS_2), and hybrid hBN/MoS_2 nanoparticles reinforced with natural and synthetic esters were studied and compared [29]. Enhancements of about 21%–23% for hBN or MoS_2 nanostructures in natural esters and 23%–27% in synthetic esters at 323 K were observed. For hybrid nanofluids, enhancements of 30%–32% in natural esters and 32% in synthetic esters at 323 K were noted, attributed to the synergistic effect of the hybrid nanofluids. Finally, it was shown that the lattice contribution to the thermal conductivity of MoS_2 dominated the carrier contribution over a range from 300 to 800 K [30]. The thermal conductivity of bulk MoS_2 was 2.3 W/mK at 300 K, whereas monolayer MoS_2 exhibited 131 W/mK at 300 K, making it suitable for heat removal in electronic devices.

Nanofluides of hBN using distilled water, ethylene glycol, and an ethylene glycol-distilled water mixture as base fluids and SDS and polyvinylpyrrolidone as surfactants were prepared [31]. Boron nitride was used in this study because of its variable properties such as chemical inertness and high thermal conductivity. The results disclosed that the thermal conductivity is increased nonlinearly with an increase in nanoparticle concentration, which is beyond the prediction of theories for all base fluids. An increased rate of thermal conductivity slowed down after the concentration of 0.5 vol.%. The rise in thermal conductivity at low concentrations is because of the initialization of polydispersed state. The enhancement in thermal conductivity at 3% volume concentration was 16%, 22%, and 26% for ethylene glycol, water/ethylene glycol, and water-based nanofluids, respectively.

The thermal characteristics of highly thermally conductive nanofluids of boron nitride nanosheets were studied [32]. It was observed that the thermal conductivity of boron nitride nanofluids attained 2.39 W/m/K at 24 vol.%, which is 298% more than that of water. The thermal conductivity enrichment was because of the lamellar structures of boron nitride nanosheets and their contacts between the layers. The results explored that the boron nitride can be used as an effective nanofiller to enhance thermal performance.

5.5 Rheological studies

The viscosity of nanofluids is influenced by the morphology of the nanoparticles, an area that has received less attention compared to thermal conductivity. Studies on rheology are limited, but available data show that viscosity decreases as temperature rises. Various methods are used to measure viscosity, including rheometers, capillary viscometers, and Brookfield viscometers. Rheometers assess the shear rate of 2D nanofluids across different temperatures and concentrations. Capillary viscometers measure viscosity based on the time it takes for the nanofluid to pass through a narrow capillary tube, with viscosity being affected by the tube's diameter. Brookfield viscometers determine viscosity by measuring the torque required to rotate a spindle at a constant speed while immersed in the fluid, where the torque correlates directly with the fluid's viscous resistance.

The increase in viscosity associated with the base fluid ethylene glycol for 2D hBN was documented in a study by Lee et al. [22]. They reported shear thinning behavior with increasing shear rates, showing a 710% increase in viscosity at 5 vol.% and a shear rate of 2 s^{-1}. At very high viscosities, above the critical concentration, the nanofluids transitioned to a gel-like state. The viscosity ratio of hBN nanofluids increased while surface tension decreased with higher concentrations because of particle agglomeration in the solution [23]. A 15.82% increase in viscosity was noted at a 0.1% volume concentration of nanoparticles without a surfactant. The transition from viscous to elastic behavior in WS_2 nanofluids using a surfactant was also observed. This transition required more pumping input to initiate fluid flow [21]. At a minimum volume concentration of 0.005% WS_2, an 8.2% reduction in viscosity was observed. Adding 0.05% SDS surfactant to 0.005% WS_2 further reduced viscosity, achieving a total reduction of 10.5%. However, high concentrations of surfactants did not provide a synergistic effect with nanoparticles. At a shear rate of 10 s^{-1}, samples showed a transition from non-Newtonian to Newtonian behavior.

The rheological properties of tungsten oxide nanoparticles in glycerol and ethylene glycol were investigated by Karimi-Nazarabad et al. [33]. They observed that at higher temperatures, the tungsten oxide–glycerol mixture exhibited Newtonian behavior, whereas the tungsten oxide–ethylene glycol mixture demonstrated shear-thinning behavior. The viscosity of these nanofluids decreased exponentially with increasing temperature. Specifically, at 20°C, a viscosity increase of 9% and 14% was observed for 4% tungsten oxide nanoparticles in ethylene glycol and glycerol, respectively. Although the viscosity of tungsten oxide–ethylene glycol nanofluids increased with a higher mass fraction of nanoparticles, the tungsten oxide–glycerol mixture did not exhibit a consistent pattern with respect to mass fraction.

In addition, the rheological behavior and surface tension of titanium nitride–ethylene glycol nanofluids were studied across various concentrations and nanoparticle sizes [25]. It was found that smaller nanoparticles resulted in higher surface tension at the same concentration. Increasing particle concentration led to a transition from Newtonian to shear-thinning yield stress behavior with a viscoelastic structure. Smaller nanoparticles contributed to a more pronounced viscoelastic structure, and the viscosity of 20 nm titanium nitride nanofluids was lower than that of 50 nm nanofluids at lower nanoparticle volume fractions. However, this trend reversed at the highest volume fraction.

The rheological properties of graphene–glycerol nanofluids were measured across mass fractions ranging from 0.0025% to 0.02% and temperatures from 20°C to 60°C [34]. The viscosity of these nanofluids increased with higher mass fractions and decreased with rising temperatures. The results demonstrated the shear-thinning behavior of the graphene nanofluids, with a notable viscosity enhancement of 401.49% observed for a 2% graphene nanofluid at a shear rate of 6.32 s^{-1} and a temperature of 20°C. In addition, the viscosity and thermal conductivity trends of hBN nanofluids were investigated [31]. Within the specified shear rates, hBN nanofluids exhibited Newtonian behavior, except those based on ethylene glycol, which showed non-Newtonian behavior at low shear rates (20–100 s^{-1}) for a 1% concentration. Viscosity increased by 22%, 33%, and 62% for water, ethylene glycol, and ethylene glycol/water-based hBN nanofluids, respectively. It was observed that higher viscosity correlates with improved thermal conductivity, making hBN nanofluids suitable for convective heat transfer applications.

Su et al. [35] studied the surface tension of MoS_2–oil nanofluids and observed that the surface tension of MoS_2 nanofluids decreased with an increase in the mass fraction of nanoparticles because of the increase in the surface layer concentration. Further the surface tension of MoS_2 nanofluids was lower than its base fluid. As the change in pressure and pumping power depends on fluid viscosity for any heat transfer applications, viscosity measurement is very significant. The viscosity measurement of zirconia (ZrO_2)–water nanofluids was studied for different volume concentrations of 0.1%, 0.2%, and 0.3% using a Brookfield viscometer. The obtained viscosity is directly proportional to nanoparticle concentration [36]. In another study the dynamic viscosity of TiO_2–water-based nanofluids was examined across four weight concentrations: 1%, 10%, 20%, and 35%. The temperature range for the experiments was between 293 and 353 K, with increments of 10 K. The results revealed that the nanofluids maintained a Newtonian behavior across all concentrations, and the viscosity enhancement was found to be independent of temperature. At lower concentrations, the nanofluids exhibited viscosity characteristics similar to that of water. However, at higher concentrations, specifically at 35 wt.% and 343 K, the viscosity increased significantly, showing an enhancement of approximately 243% compared to water [37].

In a separate investigation the rheological properties of TiO_2–ethylene glycol nanofluids were assessed up to 8 wt.%. These nanofluids also displayed Newtonian behavior within a shear rate range of 0.5 to 104 s^{-1} over temperatures from 293 to 333 K. It was observed that the viscosity of these nanofluids was dependent on temperature as well as on particle concentration because of aggregation effects. The relative increase in viscosity was found to be independent of temperature but significantly influenced by the concentration of nanoparticles [38]. Furthermore the rheological behavior of TiO_2–water nanofluids with volumetric concentrations ranging from 0.05% to 0.12% was studied under shear rates of 10^1–10^3 s^{-1}. These nanofluids displayed pseudoplastic behavior, indicative of particle aggregation within the fluid. When the concentration exceeded 0.1%, the suspensions exhibited thixotropic characteristics, reflecting changes in viscosity with shear rate [39]. The viscosity dependence of TiO_2 nanofluids with an

80–90 nm hydrodynamic diameter was modeled using the power-law equation for shear rates between 0 and 100 s^{-1}, where the nanofluids demonstrated Newtonian behavior. At higher shear rates, shear-thinning behavior was observed. The addition of polyethylene glycol reduced viscosity, likely because of its adsorption onto the titania surface, which stabilizes the suspension through steric effects and leads to the formation of compact aggregates [40].

5.6 Convective heat transfer studies

To effectively utilize nanofluids, improving thermal conductivity alone is not enough; their performance in convective heat transfer situations is crucial. Therefore the enhancement of the heat transfer coefficient is a more relevant indicator of nanofluid effectiveness than mere thermal conductivity improvements. Besides thermal conductivity, nanofluids exhibit changes in other thermophysical properties, such as density, viscosity, and specific heat, which also influence the heat transfer coefficient. These property changes collectively contribute to the overall effectiveness of nanofluids in practical applications [41]. To determine the heat transfer coefficient in various applications, including internal combustion engines, heat exchangers, computer liquid cooling systems, and solar energy systems such as solar absorbers and collectors, Newton's law of cooling is commonly applied. Measurement of equipment temperatures is typically conducted using thermocouples, and the heat transfer rate is calculated based on specific heat capacity. In addition to heat transfer coefficients, research has explored various dimensionless numbers, shear rates, and pumping power requirements, providing a broader understanding of nanofluids' performance in different contexts.

For instance the heat transfer efficiency of hBN–water nanofluids in computer liquid cooling systems has been thoroughly investigated by Atik and Ertürk [41]. The study examined constant flow rates between 0.3 and 2 L/min and nanoparticle volume fractions ranging from 0.1% to 2%, with and without fins. Results indicated that the presence of hBN nanoparticles enhanced heat transfer performance, particularly at higher volume fractions and lower convection rates. Two experimental setups were employed: one with controlled flow rates through different cold plates and another using a commercial computer liquid cooling system at a specific pumping power for various particle concentrations. The maximum reduction in thermal resistance observed was 11.2% and 7.8% for flat and finned cold plates, respectively, at a 2% volume fraction. In addition, the Stanton number increased by 15.4% and 10% for flat and finned cold plates, respectively, at a flow rate of 0.3 L/min. Conversely, alumina–water nanofluids showed no improvement in performance under restricted flow rate conditions and exhibited performance deterioration at fixed pumping power.

A comparison of the heat transfer properties of MoS_2 and WS_2 nanofluids in concentrated solar power (CSP) systems revealed that MoS_2- and WS_2-based nanofluids exhibited thermal conductivity enhancements of 46% and 35%, respectively, at 90°C [42]. These improvements translated to increases in solar collector efficiency by 21.3% for MoS_2 and 16.8% for WS_2, compared to conventional thermal oils. Despite the higher concentration of WS_2 nanosheets and their smaller agglomerate sizes compared to MoS_2 nanowires, MoS_2 nanofluids demonstrated superior heat transfer performance. The larger agglomerates of MoS_2 facilitated faster and more efficient heat transfer. The findings suggest that metal chalcogenides offer better thermal properties than traditional fluids, highlighting their potential for use in parabolic trough collectors in CSP systems.

Functionalized graphene nanoplatelets in Havoline XLC using SDBS to study the potential for direct solar absorbers were considered [43]. The optical characteristics of the nanofluid change even at very low concentrations of graphene nanoplatelets (0.005% and 0.05% by weight), which produces high sunlight absorption in short layers. The increase in extinction coefficient of nanofluids was 400% for 0.005% concentration when compared with the pure base fluids. The rheology test shows Newtonian behavior of shear rate from 10 to 1000 s^{-1}. It was also found that the viscosity and thermal conductivity do not increase significantly compared to the base fluid, implying that these nanofluids will not need any extra power for pumping to flow.

In a recent study, 2D MXene (Ti_3C_2) nanofluids were synthesized using a heterogeneous solution of ionic liquid and water and tested at three different concentrations to optimize their performance in a hybrid solar photovoltaic/thermal (PV/T) system [44]. Simulation studies compared the thermal efficiencies of the synthesized ionanofluid with olein palm oil/MXene nanofluids and water/alumina nanofluids in the PV/T system. Key thermophysical properties, including thermal conductivity, viscosity, and heat capacity, were assessed across various concentrations and temperatures. The investigation revealed that among the tested concentrations of 0.05%, 0.1%, and 0.2%, the 0.2% concentration provided the highest performance. This concentration led to a 47% enhancement in thermal conductivity, which increased with temperature.

In addition, the introduction of MXene nanosheets reduced the viscosity of the fluid, contributing to its self-lubricating properties. Notably the thermal efficiency of the ionanofluids improved by 81.15%, and the heat transfer coefficient increased by 12.6% compared to water/alumina nanofluids. It was investigated for evacuated tube solar collectors of graphene platelet–water nanofluid effects on the thermal performance [45]. Four different mass fractions of 0.025%, 0.05%, 0.075%, and 0.1% were considered. Thermal efficiency was calculated for the solar collector using different volumetric flow rates of 0.1, 0.5, and 1.5 L/min. Graphene nanofluids showed a high thermal conductivity of 27.6% with Newtonian behavior. The thermal efficiency of the solar collector system was increased to 90.7% for 0.1 wt.% of graphene nanofluids, and it was found to be 35.8% more than that of distilled water at a flow rate of 1.5 L/min. Further, at a low flow rate of 0.5 L/min and a concentration of 0.1 wt.%, the maximum temperature difference of graphene nanofluids was attained. Because of the superior heat transfer property of graphene nanofluids the working fluid efficiency in the solar collector can be enhanced, thus making it suitable for solar energy applications. Bose et al. [46] conducted a study examining the heat transfer characteristics of graphene–water nanofluids in circular stainless steel pipes, focusing on laminar flow under consistent wall heat flux conditions. They varied the nanoparticle concentration between 0.2% and 0.6% by volume and explored Reynolds numbers from 500 to 2000. Their results demonstrated that incorporating graphene nanoparticles enhanced the heat transfer coefficient by up to 56%. However, at the highest concentration of 0.6% by volume, there was a recorded pressure drop of 28.7%. This indicates a substantial improvement in thermal performance because of graphene nanoparticles despite the increased pressure drop in laminar flow systems.

In addition, a mathematical analysis of forced convective heat transfer in circular pipes under laminar flow was performed, comparing several nanofluids, including alumina–water, ZrO_2–water, and TiO_2–water. This study revealed that the thermophysical properties of these nanofluids varied nonlinearly with temperature. For a given Reynolds number the heat transfer coefficient of the nanofluids was consistently higher than that of the base fluid. Furthermore the entropy generation for the nanofluids was lower compared to the base fluid at the same Reynolds number. The Bejan number, indicative of heat transfer irreversibility, approached unity, highlighting minimal irreversibility in these systems. Although particle concentration did increase wall shear stress, the impact was negligible for equal discharge and mass flow rates [47]. Moreover, the friction factor and convective heat transfer characteristics of ZrO_2–water nanofluids in a circular pipe under constant heat flux conditions were examined in laminar flow regimes with volume concentrations of 0.5%, 0.75%, and 1%, utilizing yttrium oxide surfactant. The study revealed that both viscosity and thermal conductivity of the nanofluids increased with higher particle concentrations. The enhancement in heat transfer was significantly influenced by the Reynolds number and particle volume concentration. Specifically, the Nusselt numbers at 0.5%, 0.75%, and 1% particle concentrations were measured at 21.09, 28.05, and 35.73, respectively.

In addition, the investigation found minimal variations in pumping power and friction factor, with nanofluids showing less fluctuation in friction factor compared to the base fluid [48]. Further investigation into the heat transfer characteristics of TiO_2–ethylene glycol nanofluid in a shell-and-tube heat exchanger was performed for volume concentrations of 0.025%, 0.05%, and 0.075%, with volumetric flow rates of 0.2, 0.4, and 0.6 L/min. The results demonstrated that increasing both the nanoparticle concentration and flow rate enhanced the thermophysical properties of the nanofluid. The optimal heat transfer rate of 0.277 J was achieved at a volumetric flow rate of 0.6 L/min and a nanoparticle concentration of 0.075% [49].

5.7 Boiling heat transfer studies

The mechanism of boiling heat transfer is complex, involving the phase change from liquid to vapor. This process is crucial in various industrial applications such as power generation, refrigeration, and cooling, where both pool and flow boiling are highly efficient. A key parameter in boiling heat transfer is the critical heat flux, defined as the point at which bubbles cover the heating surface, significantly affecting the phase change dynamics. A study by Kamel and Lezsovits [50] investigated the heat transfer coefficient in pool boiling behavior using tungsten oxide-based deionized water. The researchers examined the influence of volume concentration and applied heat flux on boiling heat transfer performance in a horizontally heated copper pipe at ambient pressure. They found that the heat transfer coefficient increased with higher applied heat flux, achieving an enhancement ratio of 6.7% for a 0.01% volume concentration at low heat flux. However, the heat transfer coefficient for pool boiling with tungsten oxide nanofluids was reduced compared to the base fluid, with the highest reduction ratio being 15% for a 0.05% volume concentration. This degradation was attributed to the settlement of tungsten oxide flakes on the heating surface during boiling, leading to a reduction in nucleation site density. The study suggested that understanding the stability of nanofluids is crucial in addition to their thermal and flow characteristics.

Another study by Zhang et al. [51] focused on the characteristics of boiling heat transfer using graphene–water nanofluids inside microchannels. The experiments were conducted at different mass flow rates and various concentrations from 0 to 0.05 wt.%. The results indicated that the heat transfer coefficient decreased at higher concentrations because of graphene settling on the heating surface, which covered the nucleation sites. However, the critical heat flux increased compared to water, likely because of enhanced wettability. Further research explored the effect of graphene concentration (from 0 to 0.5 vol.%), boiler temperature, and flow rate (from 2.31 to 7.1 mL/s) on flow boiling in a rectangular channel [52]. The base fluid used was a zinc bromide and acetone salt solution. The study found that increasing nanoparticle concentration led to increased viscosity, density, and thermal conductivity while heat capacity decreased. Thermal conductivity was enhanced by 4.5% at 0.5 vol.% graphene. Across all tested boiler temperatures (411–446K) the salt nanofluid demonstrated nucleate flow boiling behavior with notable enhancement in fluid properties, indicating improved operation of flow boiling at low temperatures.

In addition, experiments with hBN/dichloromethane nanofluids assessed the effects of nanoparticle concentration on thermal performance [53]. Nanofluids prepared at concentrations from 0.5% to 1.5% were tested at power inputs ranging from 50 to 350 W. The addition of nanoparticles altered the base fluid's thermal characteristics, achieving enhancements of 27.6% and 17.65% in boiling and condensation processes, respectively.

5.8 Modeling studies

5.8.1 Mathematical models for thermal conductivity

Various classical models have been used to calculate the viscosity and thermal conductivity of 2D-based nanofluids. Traditional approaches often involved comparing measured thermal conductivity with predicted results because of the absence of an exact theory for calculating the thermal conductivity of nanofluids. Despite the availability of these models, a universally accepted method for calculating thermal conductivity across all expressions has not yet been established. The mathematical equations corresponding to these models are as follows:

1. **Maxwell model:** This model has been used for uniformly sized and randomly dispersed spherical particles, and it is based on effective medium theory [54].

$$\frac{k_{\mathrm{eff}}}{k_{\mathrm{bf}}} = \frac{k_p + 2k_{\mathrm{bf}} + 2(k_p - k_{\mathrm{bf}})\varphi}{k_p + 2k_{\mathrm{bf}} - (k_p - k_{\mathrm{bf}})\varphi} \tag{5.1}$$

2. **Tim model:** The Tim model [55] for spherical-shape particles considering the boundary conditions that resistance is nil for the transfer of heat at the interface of fluid and particle and the heat current is continuous across the interface was given by Eq. (5.2).

$$\frac{k_{\mathrm{nf}}}{k_{\mathrm{bf}}} = 1 + \frac{3(k_p - k_{\mathrm{bf}})\varphi}{(k_p + 2k_{\mathrm{bf}})} \tag{5.2}$$

The Tim model for circular-shaped particles in a 2D nanofluid was given by Eq. (5.3).

$$\frac{k_{\mathrm{nf}}}{k_{\mathrm{bf}}} = 1 + \frac{2(k_p - k_{\mathrm{bf}})\varphi}{(k_p + 2k_{\mathrm{bf}})} \tag{5.3}$$

3. **Hamilton–Crosser model:** This model was developed using shape factor and it has been used for spherical and cylindrical particles [56].

$$\frac{k_{\mathrm{eff}}}{k_{\mathrm{bf}}} = \frac{k_p + (n-1)k_{\mathrm{bf}} - (n-1)(k_p - k_{\mathrm{bf}})\varphi}{k_p + (n-1)k_{\mathrm{bf}} - (k_p - k_{\mathrm{bf}})\varphi} \tag{5.4}$$

For spherical particle, $n = 3$ and for cylindrical particles, $n = 6$.

4. **Wasp model:** For calculating the thermal conductivity, the Wasp model expression was used and given by Eq. (5.5). For spherical particles the Wasp model matched the Hamilton–Crosser model [57].

$$\frac{k_{nf}}{k_{bf}} = \frac{k_p + 2k_{bf} - 2(k_{bf} - k_p)\varphi}{(k_p + 2k_{bf} + 2k_p(k_{bf} - k_p))} \tag{5.5}$$

5. **Bruggeman model:** This model is applicable for homogeneous, randomly dispersed, and spherical nanoparticles (Eq. 5.6) [58].

$$\frac{k_{eff}}{k_{bf}} = \frac{1}{4}\left[(3\varphi - 1)\frac{k_p}{k_{bf}} + (2 - 3\varphi) + \frac{k_{bf}}{4}\sqrt{\Delta}\right] \tag{5.6}$$

where Δ is the thermal conductivity of suspension regardless of particle motion and vibration, k_{eff} is the effective thermal conductivity of nanofluids, k_{bf} is the thermal conductivity of base fluid, k_p is the thermal conductivity of nanoparticle, and Φ is the volume fraction of nanoparticle.

6. **Xue model:** Xue proposed a model for composite materials, including carbon nanotubes [59] and is given in the following expression:

$$\frac{k_{eff}}{k_{bf}} = \frac{1 - \varphi + 2\varphi\frac{k_p}{k_p - k_{bf}}\ln\frac{k_p + k_{bf}}{2k_{bf}}}{1 - \varphi + 2\varphi\frac{k_{bf}}{k_p - k_{bf}}\ln\frac{k_p + k_{bf}}{2k_{bf}}} \tag{5.7}$$

5.8.2 Classical models for viscosity

1. **Einstein model:** Einstein initially derived an expression for calculating the viscosity of a nanofluid under the assumption that the nanofluid contains spherical particles at a very low volume fraction, specifically less than 0.02%[60]. However, this model is not applicable to nanofluids with high particle concentrations. For such cases, the Einstein model is inadequate, and the viscosity must be calculated using alternative models, such as the one given by Eq. (5.8).

$$\frac{\mu_{nf}}{\mu_{bf}} = 1 + 2.5\varphi \tag{5.8}$$

where μ_{nf} is nanofluids viscosity, μ_{bf} is the base fluid viscosity, and ϕ is the volume fraction of the particle in suspension.

2. **Krieger and Dougherty model:** Kreiger and Dougherty [61] proposed a semiempirical model for the behavior of randomly dispersed spherical particles. This model is described by Eq. (5.9) and has been widely applied in the study of particulate systems because of its robust theoretical foundation and practical utility.

$$\frac{\mu_{nf}}{\mu_{bf}} = \left[1 - \frac{\varphi}{\varphi_m}\right]^{-n\varphi_m} \tag{5.9}$$

where ϕ_m is the maximum particle packing fraction, which varies from 0.495 to 0.54 and is approximately 0.605 at higher shear rates, and η is the intrinsic viscosity whose value is 2.5 for monodispersed suspensions of hard spheres.

3. **Batchelor model:** The modified Einstein model considers factors like the Brownian motion effect for spherical nanoparticles, which was given by the Batchelor model [62] in Eq. (5.10).

$$\mu_{\mathrm{nf}} = \left(1 + 2.5\varphi + 6.5\varphi^2\right)\mu_{\mathrm{bf}} \tag{5.10}$$

4. **Brinkman model:** This model has been used for nanoparticles with volume fraction more than 4%[63] and is given by Eq. (5.11).

$$\mu_{\mathrm{nf}} = \mu_{\mathrm{bf}}\left(\frac{1}{(1-\varphi)^{2.5}}\right) \tag{5.11}$$

5. **Mooney model:** Since Einstein's model cannot be used for calculating the viscosity at higher concentrations, Mooney proposed a model [64] given in Eq. (5.12).

$$\frac{\mu_{\mathrm{nf}}}{\mu_{\mathrm{bf}}} = e^{\left(\frac{\varepsilon\varphi}{1-K\varphi}\right)} \tag{5.12}$$

where K is a constant known as the self-crowding factor ($1.35 < k < 1.91$) and ξ is the fitting parameter, whose value is 2.5.

6. **Neilsen model:** A mathematical expression proposed by Neilsen can be used for a particle concentration of more than 0.02%, and it is expressed in Eq. (5.13)[65].

$$\mu_{\mathrm{nf}} = (1 + 1.5\varphi)\, e^{\frac{\varphi}{(1-\varphi_m)\mu_{\mathrm{bf}}}} \tag{5.13}$$

The thermal conductivity of nanofluids using the models given in Eqs. (5.1)–(5.7) can be calculated by substituting the values of thermal conductivity of the base fluid and nanoparticle for various volume concentrations. The viscosity of nanofluids can be found if the viscosity of base fluid and particle concentration are known and can be calculated using Eqs. (5.8)–(5.13).

Many researchers have used different models to calculate the thermal conductivity and compare it with experimental results. For instance, MoS_2 nanofluids were prepared using base fluids such as oil and water for various mass fractions, and their thermal conductivity was measured [35]. Further the Classic Maxwell model (Eq. 5.1) and Tim model (Eq. 5.2) were used to find the theoretical thermal conductivity and compare it with experimental values. It was observed that mass fraction is directly proportional to the thermal conductivity and inversely proportional to surface tension. The predicted thermal conductivity of MoS_2–oil nanofluids was lower than the experimental values.

The effective thermal conductivity of single-walled carbon nanotube (SWCNT)/polyimide and carbon nanofiber/polyimide composition at 24 vol.% was measured [59]. The effect of the addition of hBN to carbon nanofiber was investigated, and the experimental results were compared with the Xue model (Eq. 5.7) and Hamilton–Crosser model (Eq. 5.4). The experimental and predicted thermal conductivity compared using models showed reasonable results. It was observed that the small amount of carbon nanofiber increased the effective thermal conductivity of carbon nanofiber/hBN/polyimide composites. The average absolute deviation of thermal conductivity calculated by Xue and Hamilton–Crosser model for SWCNT/polyimide and carbon nanofiber/polyimide was given in Table 5.1.

The experimental and theoretical thermal conductivity values of alumina, titania, and graphene nanofluids using water as the base fluid have been extensively studied [66]. Theoretical thermal conductivity values were calculated using three different models, and the percentage of error was determined in comparison to experimental results. Notably, the Bruggeman model, as described in

Table 5.1 AAD for SWCNT/PI and CNF/PI for Xue and Hamilton–Crosser model [59].

Nanofluid	Particle thermal conductivity (W/m K)	Xue model (AAD%)	Hamilton–Crosser model (AAD%)
SWCNT/polyimide	3000	11	4.68
Carbon nanofiber (CNF)/polyimide (PI)	2000	33.6	4.41

AAD, *average absolute deviation;* SWCNT, *single-walled carbon nanotube.*

Eq. (5.6), showed a percent error of up to 110% for graphene/water nanofluids when predicting thermal conductivity.

In addition, a new mathematical correlation for the calculation of effective viscosity for 22 different nanofluids, including hBN, graphene, alumina, and silicon carbide, was developed by considering the interfacial layer effect on the nanoparticles [67]. The effective viscosity calculation involves both static and dynamic components. The static part combines the Einstein model (Eq. 5.8) with the effects because of the nanolayer, whereas the dynamic part accounts for the viscosity changes because of the Brownian motion of nanoparticles. Experimental viscosity measurements of hBN–cellulose derivative nanofluids were compared with predicted values using the Krieger–Dougherty model (Eq. 5.9) [68]. Determining viscosity accurately is challenging because of the significant shear thinning behavior of nanofluids. However, the experimental viscosity values, when measured without a dispersant, were found to align with the Krieger–Dougherty model.

5.8.3 Artificial neural network studies

Artificial neural networks (ANNs) are computational models designed to mimic the structure and function of biological neural networks. ANNs are capable of capturing complex, nonlinear relationships between inputs and outputs, making them powerful tools for a variety of tasks, including pattern recognition and prediction. In an ANN, multiple nodes, or neurons, receive input data. Each neuron performs simple computations on the input and transmits the result to other neurons in the network. The output generated by each neuron is referred to as the activation value [69].

ANNs utilize several layers to process and transform data, as illustrated in Fig. 5.4. These layers include the input layer, hidden layers, and the output layer. The input layer, which consists of artificial neurons, receives the initial data and initiates the network's learning process. The output layer evaluates whether the network has successfully learned the task by comparing the final output with expected results. Hidden layers, situated between the input and output layers, process and convert the input data into a form suitable for the output layer. Each hidden layer is fully connected to neurons in both the input and output layers, ensuring comprehensive data integration and transformation [69].

In ANN modeling the feedforward multilayer perceptron (MLP) approach is commonly used. This model comprises multiple layers, each containing neurons interconnected by weight coefficients. The activation function in each neuron computes the weighted sum of inputs plus a bias term to produce the output. The network iteratively adjusts these weights to minimize the prediction error until the output achieves the smallest mean squared error (MSE). Recent studies have applied MLP ANN to

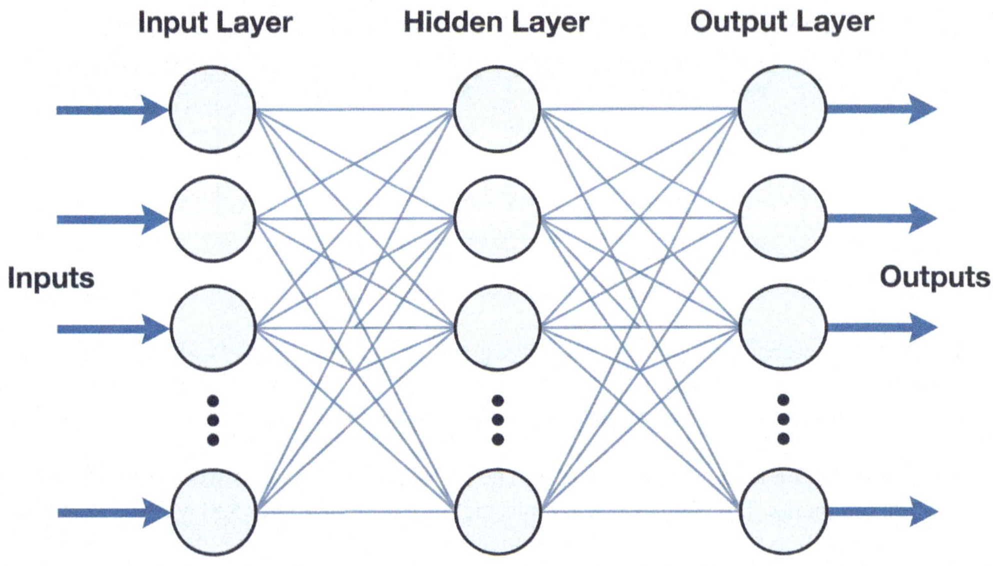

FIGURE 5.4

Structure of artificial neural network.

predict the viscosity of graphene nanofluids, utilizing experimental data collected at varying graphene concentrations (0.025%, 0.075%, and 0.1%) and temperatures (20°C–60°C). The model's performance was evaluated using mean absolute percentage error (MAPE), root mean square error (RMSE), and the coefficient of determination (R^2), yielding values of 0.777, 0.086, and 0.985, respectively. This hybrid ANN-genetic algorithm model demonstrated a high degree of accuracy consistent with experimental observations [70].

Similarly, MLP ANN was employed to predict the thermal conductivity of graphene nanofluids based on experimental data covering concentrations of 0.00025%, 0.0005%, 0.001%, and 0.005% at temperatures ranging from 25°C to 50°C. The model achieved RMSE, MAPE, and R^2 values of 0.04 W/m K, 26%, and 0.99, respectively, reflecting a high level of precision and alignment with experimental results [71]. The viscosity of MXene–palm oil nanofluids was also analyzed using ANN. Experimental measurements of viscosity were taken over a temperature range of 291–373 K with mass concentrations ranging from 0.01% to 0.2%. The data indicated that viscosity decreases with rising temperature and that the impact of MXene concentration is more pronounced at lower temperatures. The ANN model, which used temperature and concentration as inputs to predict dynamic viscosity, achieved MSE, MAPE, and R^2 values of 4.733E − 05, 0.507%, and 0.99975, respectively. These results confirmed the model's strong performance and alignment with experimental data [72]. In addition, an ANN model was developed to estimate the thermal conductivity of ZrO_2 nanofluids, using experimental data across volume percentages of 0.0125%, 0.025%, 0.05%, 0.1%, and 0.2% within a temperature range of 10°C–65°C. This model produced a new correlation with an average error of −0.02% for thermal conductivity predictions, indicating minimal deviation from experimental results [73].

5.9 Challenges in two-dimensional nanofluids

Although 2D nanofluids show promising applications for heat transfer enhancement, several challenges have prevented their widespread industrial adoption. Poor dispersion stability because of particle agglomeration remains an issue, as the strong intersheet attractions can lead to restacking over time. However, recent advances in functionalization techniques are improving stability. There has also been continued work to better understand structure–property linkages and develop standardized characterization protocols. This will help address inaccurate results between early studies. Viscosity penalties also present roadblocks, particularly at higher loading levels critical for thermal performance. Current research explores the use of surfactants, surface modifications, and hybrid nanostructures to minimize the impacts of viscosity. High material costs also remain a barrier to scale-up. However, continuous manufacturing technologies are now enabling lower cost, higher volume production. Computational modeling is additionally providing insights into fluid dynamics and heat transfer mechanisms to optimize nanofluid design. Overall, although challenges persist, the field is making progress toward addressing each issue. Continued advancements in dispersion science, measurement methodologies, material engineering, and modeling offer pathways to realize the full industrial potential of these new heat transfer fluids. With further research, 2D nanofluids show strong promise as economically viable solutions to enhance thermal management applications.

5.10 Conclusion

This chapter provided an overview of the preparation, properties, and applications of 2D nanofluids. Key structure–property relationships were discussed, such as the influence of nanoplate size, shape, concentration, temperature, and surfactant presence on thermal conductivity and viscosity. Although early studies found higher thermal conductivity and Newtonian behavior at low concentrations, more recent work demonstrates non-Newtonian rheology at high loadings. Advances in functionalization are also improving dispersion stability over time. Continued research is expanding the library of 2D nanomaterials investigated as nanofluids. New materials, such as hBN and TMDCs, are showing promise. Hybrid systems and multicomponent mixtures are demonstrating synergistic effects beyond single-component fluids. Computational modeling and machine learning techniques are enhancing predictions of fluid dynamics and heat transfer performance to complement empirical data. Applications in diverse fields, from microelectronics cooling to solar thermal systems, indicate the versatility of tailorable 2D nanofluids. Issues around reproducibility between studies are being addressed through standardized characterization protocols. Overall production costs are decreasing with scaled-up manufacturing techniques. With further progress in resolving remaining challenges, 2D nanofluids have strong potential as next-generation engineered thermal management fluids with applications across many industrial sectors.

Abbreviations

2D	two-dimensional
hBN	hexagonal boron nitride
TMDCs	transition metal dichalcogenides

WS_2	tungsten disulfide
MoS_2	molybdenum disulfide
SDS	sodium dodecyl sulfate
SDBS	sodium dodecylbenzene sulfonate
TiN	titanium nitride
PVP	polyvinyl pyrrolidone
ZrO_2	zirconia
TiO_2	titania
CSP	concentration solar plants
Ti_3C_2	titanium carbide
PV/T	photovoltaic/thermal
k_{eff}	effective thermal conductivity
k_{bf}	thermal conductivity of base fluid
k_p	thermal conductivity of nanoparticle
k_{nf}	thermal conductivity of nanofluid
ϕ	volume concentration of the particle
μ_{nf}	viscosity of nanofluid
μ_{bf}	viscosity of base fluid
ϕ_m	maximum particle packing fraction
ε	fitting parameter
K	self-crowding factor
MLP	multilayer perceptron
ANN	artificial neural network
MAPE	mean absolute percentage error
RMSE	root mean square error
R^2	coefficient of determination

References

[1] S. Aberoumand, A. Jafarimoghaddam, Tungsten(III) oxide (WO_3)—silver/transformer oil hybrid nanofluid: preparation, stability, thermal conductivity and dielectric strength, Alex. Eng. J. 57 (2018) 169–174.

[2] B. Pal, S.S. Mallick, B. Pal, Remarkably improved dispersion stability and thermal conductivity of WO_3–H_2O suspension by SiO_2 coating, J. Nanosci. Nanotechnol. 18 (2018) 3283–3290.

[3] M.T. Naik, G.R. Janardhana, K.V.K. Reddy, B.S. Reddy, Experimental investigation into rheological property of copper oxide nanoparticles suspended in propylene glycol–water based fluids, ARPN J. Eng. Appl. Sci. 5 (2010) 29–34.

[4] S.C. Su, Enhancing thermal conductivity of fluids with nanoparticles, developments and applications of non-Newtonian flows, Proceedings of the 1995 International Mechanical Engineering Congress and Exhibition, ASME, San Francisco, CA, USA, 1995 12–17 November.

[5] M.J. Pastoriza-Gallego, C. Casanova, J.L. Legido, M.M. Piñeiro, CuO in water nanofluid: influence of particle size and polydispersity on volumetric behaviour and viscosity, Fluid Phase Equilib. 300 (2011) 188–196.

[6] L. Chen, H. Xie, Y. Li, W. Yu, Nanofluids containing carbon nanotubes treated by mechanochemical reaction, Thermochim. Acta 477 (2008) 21–24.

[7] S. Yeo, D.K. Nandi, R. Rahul, T.H. Kim, B. Shong, Y. Jang, J.-S. Bae, J.W. Han, S.-H. Kim, H. Kim, Low-temperature direct synthesis of high quality WS_2 thin films by plasma-enhanced atomic layer deposition for energy related applications, Appl. Surf. Sci. 459 (2018) 596–605.

[8] J.N. Colleman, M. Lotya, A. O'Neill, S.D. Bergin, P.J. King, U. Khan, K. Young, A. Gaucher, S. De, R.J. Smith, I.V. Shvet, S.K. Arora, G. Stanton, H. Kim, K. Lee, G.T. Kim, G.S. Duesberg, T. Hallam, J.J. Boland, J.J. Wang, J.F. Donegan, J.C. Grunlan, G. Moriarty, A. Shmeliov, R.J. Nicholls, J.M. Perkins, E.M. Grieveson, K. Theuwissen, D.W. McComb, P.D. Nellist, V. Nicolosi, Two-dimensional nanosheets produced by liquid exfoliation of layered materials, Science 331 (2011) 568–571.

[9] S. Zhang, H. Liu, J. Yu, et al., Multifunctional flexible 2D carbon nanostructured networks, Nat. Commun. 11 (2020) 5134.

[10] N. Li, X. Guo, X. Yang, et al., Direct observation of highly confined phonon polaritons in suspended monolayer hexagonal boron nitride, Nat. Mater. (2020). https://doi.org/10.1038/s41563-020-0763-z.

[11] X. Zheng, Y. Wei, K. Pang, et al., PN/PAs-WSe_2 van der Waals heterostructures for solar cell and photodetector, Sci. Rep. 10 (2020) 17213. https://doi.org/10.1038/s41598-020-73152-7.

[12] J. Pan, Y. Zhang, J. Zhang, et al., Auxetic two-dimensional transition metal selenides and halides, npj Comput. Mater. 6 (2020) 154. https://doi.org/10.1038/s41524-020-00424-1.

[13] S.V. Pedersen, F. Muramutsa, J.D. Wood, et al., Mechanochemical conversion kinetics of red to black phosphorus and scaling parameters for high volume synthesis, npj 2D Mater. Appl. 4 (2020) 36. https://doi.org/10.1038/s41699-020-00170-4.

[14] R. Rodríguez-González, L.M. Gaggero-Sager, I. Rodríguez-Vargas, Self-similar transport, spin polarization and thermoelectricity in complex silicene structures, Sci. Rep. 10 (2020) 14679. https://doi.org/10.1038/s41598-020-71697-1.

[15] S.L EnlaiGao, Z. Qin, M.J. Buehler, X.i-Q. Feng, Z. Xu, Mechanical exfoliation of two-dimensional materials, J. Mech. Phys. Solids 115 (2018) 248–262.

[16] J. Yung xue, J. Li, W. Zhang, H. Chang, Synthesis of high quality two-dimensional materials via chemical vapor deposition, Chem. Sci. 6 (12) (2015) 6705–6716.

[17] D.V. Talapin, J.-S. Lee, M.V. Kovalenko, E.V. Shevchenko, Prospects of colloidal nanocrystals for electronic and optoelectronic applications, Chem. Rev 110 (2009) 389.

[18] Y. Li, J. Zhou, S. Tung, E. Schneider, S. Xi, A review on development of nanofluid preparation and characterization, Powder Technol. 196 (2) (2009) 89–101.

[19] Y. Hwang, J.-K. Lee, J.-K. Lee, Y.-M. Jeong, S. Cheong, Y.-C. Ahn, S.H. Kim, Production anddispersion stability of nanoparticles in nanofluids, Powder Technol. 186 (2) (2008) 145–153.

[20] G. Żyła, J. Fal, J. Traciak, M. Gizowska, K. Perkowski, Huge thermal conductivity enhancement in boron nitride–ethylene glycol nanofluids, Mater. Chem. Phys. 180 (2016) 250–255.

[21] S.N.A. Shah, S. Shahabuddin, M.F.M. Sabri, S.M. Said, K.M. Khedher, N. Sridewi, Two-dimensional tungsten disulfide-based ethylene glycol nanofluids: stability, thermal conductivity, and rheological properties, Nanomaterials 10 (2020) 1340.

[22] D. Lee, J. Park, M. Lee, G. Lee, Aging-resistant nanofluids containing covalent functionalized boron nitride nanosheets, Nanotechnology 28 (2017) 405704–405714.

[23] M. Krishnam, S. Bose, C. Das, Boron nitride (BN) nanofluids as cooling agent in thermal management system (TMS), Appl. Therm. Eng. 106 (2016) 951–958.

[24] H.J. Seong, G.N. Kim, J.H. Jeon, H.M. Jeong, J.P. Noh, Y.J. Kim, H.J. Kim, S.C. Huh, Experimental study on characteristics of grinded graphene nanofluids with surfactants, Materials 11 (2018) 950.

[25] G. Zyła, J. Fal, P. Estellé, Thermophysical and dielectric profiles of ethylene glycol based titanium nitride (TiN–EG) nanofluids with various size of particles, Int. J. Heat Mass Transf. 113 (2017) 1189–1199.

[26] S.S. Gupta, V.M. Siva, S. Krishanan, T.S. Sreeprasad, P.K. Singh, T. Pradeep, S.K. Das, Thermal conductivity enhancement of nanofluids containing graphene nanosheets, J. Appl. Phys. 110 (2011) 084302.

[27] Y. Hong, J. Zhang, X. Huangc, X.C. Zeng, Thermal conductivity of a two-dimensional phosphorene sheet: a comparative study with graphene, Nanoscale 7 (2015) 18716–18724.

[28] Y.-Y. Zhang, Q.-X. Pei, J.-W.u Jiang, N. Wei, Y.-W. Zhang, Thermal conductivities of single- and multi-layer phosphorene: a molecular dynamics study, Nanoscale 8 (2016) 483–491.

[29] J. Taha-Tijerina, H. Ribeiro, K. Aviña, J.M. Martínez, A.P. Godoy, J.M. de Oliveira Cremonezzi, M.A. Luciano, M.A.G. Benega, R.J.E. Andrade, G.J. Macedo Fechine, G. Babu, S. Castro, Thermal conductivity performance of 2D h-BN/MoS_2/-hybrid nanostructures used on natural and synthetic esters, Nanomaterials 10 (2020) 1160.

[30] A.N. Gandi, U. Schwingenschlögl, Thermal conductivity of bulk and monolayer MoS_2, Europhys. Lett. 113 (2016) 36002.

[31] M.K BeybinIlhan, H. Ertürk, Experimental investigation of heat transfer enhancement and viscosity change of hBN nanofluids, Exp. Therm. Fluid Sci. 77 (2016) 272–283.

[32] X. Hou, M. Wang, L.i Fu, Y. Chen, N. Jiang, C.-T.e Lin, Z. Wang, J. Yu, Boron nitride nanosheet nanofluids for enhanced thermal conductivity, Nanoscale 10 (2018) 13004.

[33] M. Karimi-Nazarabad, E.K. Goharshadi, M.H. Entezari, P. Nancarrow, Rheological properties of the nanofluids of tungsten oxide nanoparticles in ethylene glycol and glycerol, Microfluid. Nanofluidics 19 (2015) 1191–1202.

[34] M.B. Moghaddam, E.K. Goharshadi, M.H. Entezari, P. Nancarrow, Preparation, characterization, and rheological properties of graphene–glycerol nanofluids, Chem. Eng. J. 231 (2013) 365–372.

[35] Y. Su, L. Gong, B. Li, D. Chen, An experimental investigation on thermal properties of molybdenum disulfide nanofluids, International Conference on Materials, Environmental and Biological Engineering (MEBE 2015), Atlantis Press, 2015.

[36] S. Kolappan, S. Karthik, K. Logesh, A. Vasudevan, Thermal characterization study of ZrO_2/water nanofluid, Int. J. Ambient Energy 41 (8) (2020) 918–921.

[37] L. Fedele, L. Colla, S. Bobbo, Viscosity and thermal conductivity measurements of water-based nanofluids containing titanium oxide nanoparticles, Int. J. Refrig. 35 (2012) 1359–1366.

[38] H. Chen, Y. Ding, Y. He, C. Tan, Rheological behaviour of ethylene glycol based titania nanofluids, Chem. Phys. Lett. 444 (2007) 333–337.

[39] W.J. Tseng, K.-C. Lin, Rheology and colloidal structure of aqueous TiO_2 nanoparticle suspensions, Mater. Sci. Eng. A355 (2003) 186–192.

[40] P. Alphonse, R. Bleta, R. Soules, Effect of PEG on rheology and stability of nanocrystalline titania hydrosols, J. Colloid Interface Sci. 337 (2009) 81–87.

[41] Ö. Atik, H. Ertürk, Enhancement of central processing unit liquid cooling performance using hexagonal boron nitride nanofluids, J. Therm. Sci. Eng. Appl. 11 (2019) 34501.

[42] P. Martínez-Merino, R. Alcántara, T. Aguilar, J.J. Gallardo, I. Carrillo-Berdugo, R. Gómez-Villarejo, M. Rodríguez-Fernández, J. Navas, Stability and thermal properties study of metal chalcogenide-based nanofluids for concentrating solar power, Energies 12 (2019) 4632.

[43] E. Sani, J.P. Vallejo, D. Cabaleiro, L. Lugo, Functionalized graphene nanoplatelet-nanofluids for solar thermal collectors, Sol. Energy Mater. Sol. Cells 185 (2018) 205–209.

[44] L. Das, K. Habib, R. Saidur, N. Aslfattahi, S.M Yahya, F. Rubbi, Improved thermophysical properties and energy efficiency of aqueous ionic liquid/MXene nanofluid in a hybrid PV/T solar system, Nanomaterials 10 (2020) 1372.

[45] S. Iranmanesh, H.C. Ong, B.C. Ang, E. Sadeghinezhad, A. Esmaeilzadeh, M. Mehrali, Thermal performance enhancement of an evacuated tube solar collector using graphene nanoplatelets nanofluid, J. Clean. Prod. 162 (2017) 121–129.

[46] B.J.R. Bose, L.G. Asirvatham, M.N. Kumar, Experimental convective heat transfer studies on graphene nanofluid for the cooling of next generation electronic components, Int. J. Appl. Eng. Res. 12 (19) (2017) 8534–8539.

[47] N. Purohit, V.A Purohit, K. Purohit, Assessment of nanofluids for laminar convective heat transfer: a numerical study, Eng. Sci. Technol. 19 (2016) 574–586.

[48] S.M Iqbal, C.S Raj, S. Arul, P. Rathnakumar, Heat transfer intensification of zirconia/water nanofluid, J. Adv. Chem. 13 (3) (2017) 1–8.

[49] A.A. Permanasari, B.S. Kuncara, P. Puspitasari, S. Sukarni, T.L. Ginta, W. Irdianto, Convective heat transfer characteristics of TiO_2-EG nanofluid as coolant fluid in heat exchanger, AIP Conf. Proc. 2120 (2019) 050015. https://doi.org/10.1063/1.5115691.

[50] M.S. Kamel, F. Lezsovits, Experimental investigation on pool boiling heat transfer performance using tungsten oxide WO_3 nanomaterial-based water nanofluid, Materials 13 (2020) 1922.

[51] C. Zhang, L. Zhang, H. Xu, D. Wang, B. Ye, Investigation of flow boiling performance and the resulting surface deposition of graphene oxide nanofluid in microchannels, Exp. Therm. Fluid Sci. 86 (2017) 1–10.

[52] M. Hi, D. Giddings, G.S. Walker, Experimental investigation of nanoparticles concentration, boiler temperature and flow rate on flow boiling of zinc bromide and acetone solution in a rectangular duct, Int. J. Heat Mass Transf. 130 (2019) 710–721.

[53] E. ErdemÇiftçi, A. Sözen, Nucleate pool boiling and condensation heat transfer characteristics of hexagonal boron nitride/dichloromethane nanofluid, Heat Transf. Res. 51 (11) (2020) 1043–1059.

[54] J.C. Maxwell, A Treatise on Electricity and Magnetism, Oxford University Press, Cambridge, 1904.

[55] E.V. Timofeeva, A.N. Gavrilov, J.M. McCloskey, et al., Thermal conductivity and particle agglomeration in alumina nanofluids, experiment and theory, Phys. Rev. E 76 (2007) 1–16.

[56] R.L. Hamilton, O.K. Crosser, Thermal conductivity of heterogeneous two-component systems, Ind. Eng. Chem. Fundam. 1 (3) (1962) 187–191.

[57] F.J. Wasp, Solid-Liquid Slurry Pipeline Transportation, Trans Tech Publications, Berlin, 1977.

[58] D.A.G. Bruggeman, Berechnungverschiedenerphysikalischerkkonstanten von heterogenenssubstanzen, Ann. Phys. 24 (1935) 636.

[59] M. Haruki, K. Tanaka, J. Tada, H. Onishi, Y. Tada, Effective thermal conductivity for nanocarbon/polyimide and carbon nanofiber/hexagonal boron nitride/polyimide composites, Polym. Compos. 40 (8) (2019) 3032–3039.

[60] A. Einstein, Eineneuebestimmung der moleküldimensionen, Ann. Phys. 324 (2) (1996) 289–306.

[61] I.M Krieger, J.D Thomas, A mechanism for non-Newtonian flow in suspensions of rigid spheres, Trans. Soc. Rheol. 3 (1) (1957) 137–152.

[62] G.K Batchelor, The effect of Brownian motion on the bulk stress in a suspension of spherical particles, J. Fluid Mech. 83 (1) (1977) 97–117.

[63] H.C. Brinkman, The viscosity of concentrated suspensions and solutions, J. Chem. Phys. 20 (4) (1952) 571 571.

[64] M. Mooney, The viscosity of a concentrated suspension of spherical particles, J. Colloid Sci. 6 (2) (1951) 162–170.

[65] L.E. Nielsen, Generalized equation for the elastic moduli of composite materials, J. Appl. Phys. 41 (11) (1970) 4626–4627.

[66] A. Vakilinejad, M.A. Aroon, M. Al-Abri, H. Bahmanyar, M.T.Z. Myint, G. Reza Vakili-Nezha, Experimental and theoretical investigation of thermal conductivity of some water-based nanofluids, Chem. Eng. Commun. 205 (5) (2018) 610–623.

[67] D.S. Udawattha, M. Narayana, U.P.L. Wijayarathne, Predicting the effective viscosity of nanofluids based on the rheology of suspensions of solid particles, J. King Saud Univ. – Sci. 31 (2019) 412–426.

[68] F. Bouville, S. Deville, Dispersion of boron nitride powders in aqueous suspensions with cellulose, J. Am. Ceram. Soc. 97 (2) (2014) 394–398.

[69] J. Koo, C. Kleinstreuer, A new thermal conductivity model for nanofluids, J. Nanopart. Res. 6 (6) (2004) 577–588.

[70] M. Vakili, S. Khosrojerdi, P. Aghajannezhad, M. Yahyaei, A hybrid artificial neural network-genetic algorithm modeling approach for viscosity estimation of graphene nanoplatelets nanofluid using experimental data, Int. Commun. Heat Mass Transf. 82 (2017) 40–48.

[71] S. khosrojerdi, M. Vakili, M. Yahyaei, K. Kalhor, Thermal conductivity modeling of graphene nanoplatelets/deionized water nanofluid by MLP neural network and theoretical modeling using experimental results, Int. Commun. Heat Mass Transf. 74 (2016) 11–17.
[72] N. Parashar, N. Aslfattahi, S.M. Yahya, R. Saidur, An artificial neural network approach for the prediction of dynamic viscosity of MXene-palm oil nanofluid using experimental data, J. Therm. Anal. Calorim. (2020). https://doi.org/10.1007/s10973-020-09638-3.
[73] A.B. Çolak, Experimental study for thermal conductivity of water-based zirconium oxide nanofluid: developing optimal artificial neural network and proposing new correlation, Int. J. Energy Res. 45 (2020) 1–19.

CHAPTER

Heat conduction in nanofluids

6

Swapnil Narkhede[a] and Anirban Sur[b]

[a]Department of Mechanical Engineering, Dr. D. Y. Patil Institute of Technology, Pune, Maharashtra, India, [b]Symbiosis Institute of Technology (SIT), Symbiosis International (Deemed University), Pune, Maharashtra, India

6.1 Nanofluids preparation

A fluid that contains nanosized solid particles of thermally rich (high thermal conductivity) materials and a homogeneous solution (ideally) is known as a nanofluid. Thermally rich solid particles (copper, gold, silver, copper oxide, aluminum oxide, silica nanoparticles, carbon nanotubes, and zinc oxides) are commonly used for nanofluids. Various base fluids, such as water, acetone, oil, ethylene glycol, and decene, are commonly used to mix with nanoparticles to ensure proper dispersion. It is essential to maintain a suitable concentration of nanoparticles in the base fluid to prevent any adverse chemical reactions while achieving an even, stable, and long-lasting suspension with minimal sedimentation. Thus the concentration of nanoparticles should be carefully controlled to avoid any negative impact on the base fluid's chemical properties.

There are various techniques to prepare nanofluids, such as the single-step direct evaporation approach [1] two-step method [2–4], altering the pH of suspended particles. These techniques often involve ultrasonication, the use of dispersants, and surface activators. Surface activators play a crucial role in modifying the surface properties of solid nanoparticles, such as bonding, hydrophobicity, reactivity, and morphology, and increasing their activation energy. Dispersants are fluids that prevent the agglomeration of nanoparticles in the base fluid. The selection of surface activators and dispersants depends on the size of the solid nanoparticles and the type of base fluid being used.

One frequently used approach to produce nanofluids is the vacuum evaporation onto a running oil substrate technique, as described by A.Hiroshi et al. [5]. In this process a source material is evaporated under a vacuum and deposited onto a substrate, which is a flowing oil film (base oil). The substrate is maintained at a temperature below the boiling point of the oil, and the evaporated material condenses on the substrate as nanoparticles. This technique offers a streamlined approach to nanoparticle production, reducing both processing time and costs while enabling the tunable size of nanoparticles with high yields. However, the process faces several challenges, such as limited material compatibility, potential contamination risks, and difficulty separating solid particles and obtaining dry nanoparticles from the base fluid.

For producing an alumina-based nanofluid, researchers used the single-step laser ablation method [6]. However, subsequent investigations from the same group revealed that a two-step method involving the initial production of nanoparticles followed by their diffusion into the base fluid produced better

Graphene and 2D Materials in Heat Transfer: Fundamentals and Applications. DOI: https://doi.org/10.1016/B978-0-12-821964-5.00011-4

results with oxide-based nanoparticles. In contrast, when working with metallic nanoparticles, this two-step method demonstrated inferior performance compared to the single-step method [7].

6.2 Nanofluid thermophysical properties

The thermophysical properties of nanofluid, namely, specific heat, thermal conductivity, viscosity, enthalpy, and density, play a crucial role in determining its heat transfer performance. These properties are influenced by various factors such as nanoparticle size, shape, mass concentration, and purity level. In addition, the colloidal properties of nanoparticles have a significant impact on the stability of suspended particles, pH values, and zeta potential within the nanofluid. The lubrication properties of nanofluids depend on their viscosity and friction coefficient. Experimental studies have shown that these properties differ significantly from those of the base fluid, resulting in enhanced heat transfer coefficients. Overall a comprehensive understanding of the thermophysical properties of nanofluids is essential to optimize their heat transfer performance.

6.2.1 Thermal conductivity

The heat transfer performance of nanofluid is dependent on the thermal conductivity of nanofluid.

Recent research showed that including solid particles with nanoscale dimensions, possessing high thermal conductivity, into a base fluid could significantly enhance the effective thermal conductivity of the resulting nanofluids. This improvement was observed to be most pronounced when the concentration ratio of the solid particles was carefully controlled [4,8–10].

Researchers have employed a numerical model to determine the effective thermal conductivity of nanofluids. Specifically the Maxwell model was utilized to measure the effective conductivity of a mixture of solid and liquid nanofluids, with the dispersed solid particles being modeled as spheres [11].

Yamada [12] calculated the effective thermal conductivity of nanofluids using Maxwell models and compared it with experimentally investigated results. They observed that the effective thermal conductivity values computed using the numerical model are lower than the experimental values when the solid nanoparticle volume fraction increases.

In a study conducted by Eastman et al. [3] the effective thermal conductivity of ethylene glycol nanofluids containing copper nanoparticles was measured through experimentation. The results showed that a mere 0.3% (by volume) increase in the concentration of copper nanoparticles (mean particle diameter less than 10 nm) led to a substantial 40% increase in the effective thermal conductivity of the nanofluid. In addition, it was observed that reducing the solid particle diameter from 10 μm to 10 nm resulted in a significant increase in surface area to volume ratio by a factor of 1000. Because of this the effective thermal conductivity increased dramatically. The study also concluded that nanofluids prepared by directly dispersing the metal particles in the base fluids exhibit considerably higher thermal conductivity than those prepared by dispersing the metal oxides [13].

The thermal conductivity measured for copper-ethylene glycol nanofluid is abnormally increased and was observed to be twice what the Maxwell-Garnett model predicted [14].

Chopkar et al. [15] were the first to predict the influence of nanoparticle size on the effective thermal conductivity of nanofluids. They measured the thermal conductivity of base fluids and nanofluids using a modified thermal comparator device (metallic temperature probe, heating coil with the controller, microvoltmeter, and voltage stabilizer) and other in-house fabricated equipment.

The widely used method for measuring effective thermal conductivity, known as the transients hot-wire technique, employs a platinum hot wire as a sensing element. However, this technique cannot be utilized for nanofluids that possess adhesive properties, have high Prandtl numbers, or are subjected to high-temperature surroundings, as it may result in damage to the nanofluids. To overcome this limitation, researchers have turned to the use of the transient hot disk method for such nanofluids. This technique offers the additional advantage of measuring both thermal conductivity and thermal diffusivity, which are essential properties for characterizing materials used in time-dependent applications, such as thermal storage [16–18]. The technique involves the use of a sensor in the form of a volume element as a source of heat and temperature-sensing element. The sensor is either fully immersed in the fluid bath or sandwiched between two samples of the solid under study to measure thermal conductivity. The principle underlying this method is that when an external supply is applied to the sensor, it generates heat, causing the temperature of the sensing element to increase. If the material is insulating, the temperature will rise rapidly. On the other hand, if the material is thermally conductive, its temperature will rise slowly, thereby stating its characteristics [19].

Several studies have attempted to reduce the impact of natural convection on thermal conductivity measurements by decreasing sensor input power and measurement times [20,21]. However, Warzoha et al. [19] have found that this approach may lead to inaccuracies in thermal conductivity measurements. Specifically, reducing measurement times can produce misleading results related to interfacial thermal resistance, whereas low power input can result in higher data scattering and overly elevated thermal conductivity values. To address these issues, Warzoha et al. [19] developed a correlation between the sensor input power and key fluid parameters, including expansion coefficient, Prandtl number, and Rayleigh number, to improve the accuracy of thermal conductivity measurements for nanofluids using the transient hot disk technique.

Davis et al. [22] predicted the effective thermal conductivity of two-phase material (nanofluid) with the spherical shape of nanoparticles. The analysis considers the heat transfer rate from a heated body placed in the two-phase material (nanofluid).

Nan et al. [23] proposed a comprehensive model for calculating the effective thermal conductivity of composite materials with interfacial thermal resistance, using multiple scattering theory. Their study showed that neglecting interfacial effects in nanofluids results in a bounded effective thermal conductivity, which is limited by the thermal conductivities of the individual phases. This finding highlights the importance of considering interfacial effects in determining the thermal conductivity of nanofluids.

Yu et al. [24] have proposed a modified Maxwell's model to explain the effective thermal conductivity of nanofluids containing solid nanoparticles. The model considers the presence of interfacial ordered nanolayers that are expected to form around the nanoparticles when their diameter is less than 10 nm. These ordered nanolayers are responsible for improving the thermal conductivity of nanofluids. Unlike traditional nanofluids with larger-sized particles the enhancement in thermal conductivity is achieved with higher particle concentrations (>20% vol) in this case. However, the authors suggest that adding smaller particles is a more promising strategy than increasing the concentration of larger particles.

Literature shows researchers have done several studies for predicting the effective thermal conductivity of nanofluids using numerical and experimental methods. Some of the numerical models for

Table 6.1 Numerical models used for measuring effective thermal conductivity of nanofluids.

Name of the model	Mathematical expression
Maxwell [11]	$\frac{k_{np} + 2k_{bf} + 2(k_{np} - k_{bf})\phi}{k_{np} + 2k_{bf} - (k_{np} - k_{bf})\phi}$
Bruggeman [28]	$\frac{1}{4}\left[(3\phi - 1)\frac{k_{np}}{k_{bf}} + (2 - 3\phi)\right] + \frac{1}{4}\sqrt{\nabla}$
Hamilton and Crosser [25]	$\frac{k_{np} + (n-1)k_{bf} - (n-1)(k_{bf} - k_{np})\phi}{(k_{np} + (n-1)k_{bf} + (k_{bf} - k_{np})\phi)}$
Wasp et al. [27]	$\frac{k_{np} + 2k_{bf} - 2(k_{bf} - k_{np})\phi}{k_{np} - 2k_{bf} + (k_{bf} - k_{np})\phi}$
Davis [22]	$1 + \frac{3\left(\frac{k_{np}}{k_{bf}} - 1\right)[\phi + f\phi^2 + 0(\phi^3)]}{\left(\frac{k_{np}}{k_{bf}} + 2\right) - \left(\frac{k_{np}}{k_{bf}} - 1\right)\phi}$
Nan et al. [23]	$k_{eff} = 1 + \phi\frac{k_{np}}{3k_{bf}}k_{bf}$
Yu and Choi [24]	$\frac{k_{nf}}{k_{bf}} = \frac{k_{pe} + 2k_{bf} + 2(k_{pe} - k_{bf})(1+\beta)^3\phi}{k_{pe} + 2k_{bf} - (k_{pe} - k_b)(1+\beta)^3\phi}$
Jang and Choi [29]	$k_{nf} = k_f(1-\phi) + k_p\phi + 3C_1\frac{d_f}{d_{nano}}k_f Re^2_{nano} P_r\phi$
Kumar et al. [30]	$k_{eff} = k_b + C\frac{2k_bT}{\pi v d_p^2}\frac{\phi r_b}{k_b(1-\phi)r_p}k_b$
Chon et al. [31]	$\frac{k_{nf}}{k_f} = 1 + 64.7\phi^{0.746}\left(\frac{d_f}{d_p}\right)^{0.369}\left(\frac{k_p}{k_f}\right)^{0.7476} Pr^{0.9955}\ Re^{1.2321}$
Li and Peterson [32]	$\frac{k_{eff} - k_b}{k_b} = 0.764\phi + 0.0187(T) - 273.15 - 0.462$
Mintsa et al. [33]	$\frac{k_{nf}}{k_f} = 1.72\phi + 1.0$
Corcione [34]	$\frac{k_{eff}}{k_f} = 1 + 4.4Re^{0.4}Pr^{0.66}\left(\frac{T}{T_{fr}}\right)^{10}\left(\frac{k_p}{k_f}\right)^{0.03}\phi^{0.66}$

determining the effective thermal conductivity of the nanofluids are Maxwell [11], Hamilton and Crosser [25], Jeffery [26], Davis [22], Wasp et al. [27], Bruggeman [28], and Nan [23]. The list of mathematical expressions of these models is given in Table 6.1.

Further the literature suggests that various parameters, including the shape, size, concentration degree of accumulation, surface characteristics, motion, and interaction of solid nanoparticles in the base fluid, influence the effective thermal conductivity of nanofluids. Despite attempts to explain this phenomenon through mathematical expressions, some deficiencies still need to be addressed [35]. Therefore further research is required to enhance our understanding of the fluid flow and heat transfer characteristics of nanofluids and to develop a more effective mathematical model.

6.2.2 Viscosity

Viscosity is a fundamental property that characterizes a fluid's ability to resist flow. As discussed earlier, the effectiveness of nanofluids is influenced by various factors, including their viscosity. Therefore, when designing thermal systems that incorporate nanofluids, it is crucial to consider the fluid's viscosity, as it can impact the pumping power requirements [36]. The first time the effective viscosity of spherical solids suspended in liquid had been estimated by Einstein [37]. Einstein's law to determine the viscosity of nanofluid is given by

$$\frac{\mu}{\mu_{bf}} = 1 + 2.5\varphi \tag{6.1}$$

where μ is viscosity of the nanofluid, μ_{bf} is viscosity of the base fluid, and φ is volume fraction of the nanoparticles.

Yang et al. [38] performed the first work on graphene nanofluid in 2009. They used a two-step method to prepare the silicone oil-based nanofluid with graphene shell (ND-50)-coated nanodiamond particles of 0.35–0.52 volume concentration. They observed ND-50 nanofluid exhibits shear-thinning behavior in the temperature range of 20°C–80°C.

A comparative study on the viscosity of graphene nanosheets, alumina nanoparticles, and CNT was performed by Dhar et al. [39]. They concluded that for enhancing effective thermal conductivity in graphene nanofluid, the sheet size of the graphene played an important role. Heat conduction in polydispersed nanofluids has been explained with the help of sheet percolation and Brownian motion. The critical sheet size required for the phenomena was calculated from the Stokes–Einstein equation. The investigations in this study report that the viscosity of graphene nanofluid follows Einstein's law of viscosity as expressed by Eq. (6.1). Kole and Dey [40] investigated the viscosity and thermal conductivity of functionalized highly exfoliated graphene (HEG) nanofluids with a mixture of ethylene glycol and distilled water as a base fluid. The effect of concentration of graphene and temperature range of 10°C–70°C on the viscosity is investigated. They concluded that the viscosity measurements on the graphene nanofluid with 0.395 volume% concentration display non-Newtonian and shear-thinning behavior along with the base fluid, which is a mixture of ethylene glycol and distilled water. The authors found nearly 100% improved performance of the nanofluid over the base fluid, as shown in Fig. 6.1.

The authors also concluded that the available models for determining the viscosity of the nanofluid failed to explain the dependence of volume concentration of graphene on the viscosity.

Park and Kim prepared and investigated graphene nanofluids M-5 and M-15 [41]. They employed the transient hot-wire method to determine the effective thermal conductivity of the nanofluids and measured their viscosity using a rotational digital viscometer at a temperature of 25°C. Their findings revealed that M-15 nanofluid had a higher rate of viscosity increase compared to M-5 nanofluid. In addition, the authors noted that nanofluids composed of smaller particles with oxidized graphene demonstrated superior heat transfer properties because of improved effective thermal conductivity and viscosity.

Mahbubul et al. [42] have reviewed different viscosity determination methods of nanofluids. The primary focus of their investigation was to examine how the effective viscosity of nanofluids changes with variations in temperature, nanoparticle shape and size, and volume fraction. Brinkman [43] derived an expression for the viscosity of nanofluids having a finite concentration of nanoparticles, which considered the effect of adding a solute molecule to an already existing solution. Frankel and

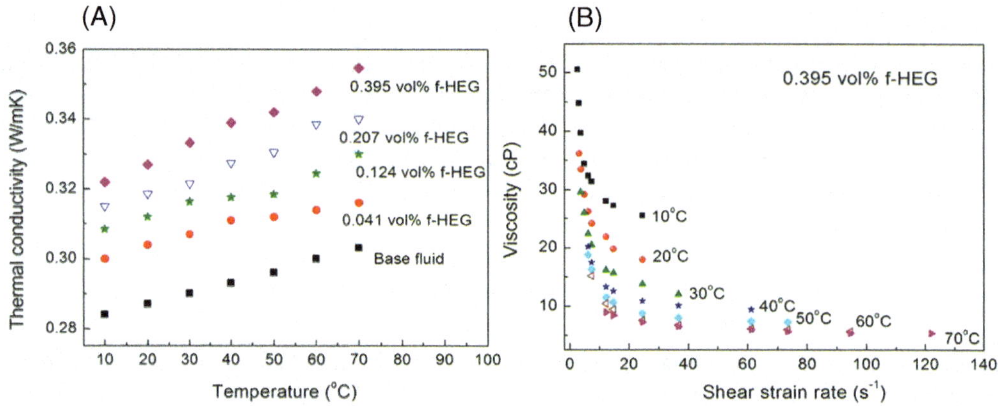

FIGURE 6.1

(A) Influence of nanofluid temperature on thermal conductivity and (B) variation of viscosity with shear strain rate [40].

Acrivos [1] employed an asymptotic approach to elucidate the correlation between the concentration of nanoparticles and the effective viscosity of nanofluids, particularly when the concentration of nanoparticles reaches its maximum value. The purpose of this attempt was to overcome the limitations of previously defined models, which are inadequate for predicting the effective viscosity of nanofluids at higher concentrations of nanoparticles. Lundgren [44] extended Brinkman's work and made it more generalized to be applied to the flow of suspensions where determining effective viscosity was a challenging task. The effective viscosity of Al_2O_3–water nanofluid was measured by Wang et al. [4]. They used different methods such as mechanical blending, polymer coating on particles (Fig. 6.2), and filtration by a precalibrated viscometer. The viscosity of nanofluid prepared by a polymer coating on particles and filtration was lower than that prepared by mechanical blending, depicting that the particles are more uniformly dispersed. They concluded that methods of preparation of nanofluids have a significant influence on the effective viscosity of nanofluids.

For stability and thermal behavior of nanofluid, viscosity plays an important role. So, despite studies on the thermophysical properties of nanofluids, the focus requires the dependence of viscosity on thermal conductivity and vice versa. Researchers have developed several theoretical and empirical models summarized and presented in Table 6.2 to predict viscosity. But they could not accurately predict the viscosity of nanofluids in a wide range of nanoparticle volume fractions [42]. This need demands more sophisticated and rigorous studies to understand the viscosity and its dependence on thermal conductivity from a heat transfer point of view.

6.2.3 Density (ρ)

Heat transfer characteristics of nanofluids depend on the density of the nanofluid. Density is the direct function of nanoparticle volume fraction in base liquid. The density of nanofluids increases with the concentration ratio of solid nanoparticles in base fluids. The density of nanofluids calculated using

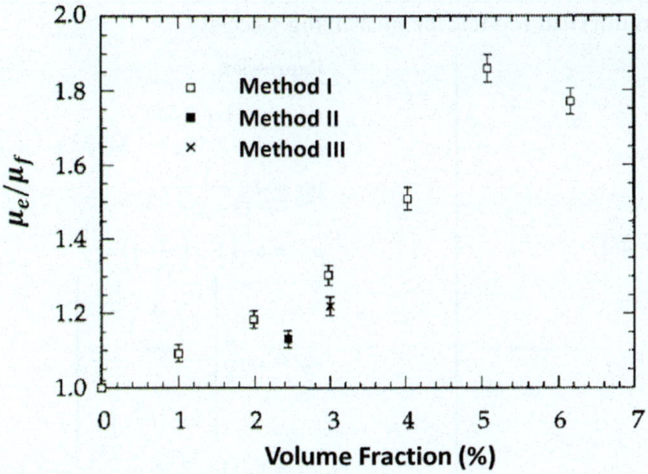

FIGURE 6.2

Variation of relative viscosity of Al_2O_3–water nanofluid with volume fraction variation [4].

mixing theory was found to be valid and can be expressed as [55]

$$\rho_{nf} = (1-\varphi)\rho_{bf} + \varphi\ \rho_s \tag{6.2}$$

Liu et al. [56] observed the variation in the density of graphene nanofluids with temperature variation. The study concluded that the density of graphene nanofluid varies inversely (1.097–1.012 g/cm^3) with a variation temperature range of 30°C–210°C for 0.03% weight of graphene. The authors specifically indicated that the density of graphene nanofluids decreases with increased mass fractions of graphene. Vajjha et al. [57] measured the density of nanofluids having zinc oxide, aluminum oxide, and antimony tin oxide nanoparticles experimentally and compared the experimental result with the numerical model result using the above Eq. 6.2. They found very little variation between the two results.

6.2.4 Specific heat capacity (c_p)

The study of nanofluids and their thermal properties has been a key focus of researchers in various heat transfer applications. It has been observed that the specific heat of a nanofluid plays a significant role in its heat transfer behavior. Mehrali et al. [58] observed that the specific heat capacity of a nanofluid shows a strong dependence on nanoparticle concentration rather than the temperature of the nanofluid. Specific heat can be determined from Eq. (6.3):

$$c_{p,nf} = \varphi\ c_{p,np} + (1-\varphi)\ c_{p,bf} \tag{6.3}$$

and

$$c_{p,nf} = \frac{\varphi\ (\rho c_p)_{,np} + (1-\varphi)\ (\rho c_p)_{,bf}}{\rho_{nf}} \tag{6.4}$$

Table 6.2 Various numerical models for measuring viscosity.

Model	Expression
Einstein [37]	$\mu_{nf} = \mu_{bf}(1 + 2.5\phi)$
Vand [45]	$\mu_{eff} = 1 + 2.5\phi_p + 7.349\phi_p^2 + \ \ldots)\mu_b$
Brinkman [43]	$\mu_{nf} = \mu \dfrac{1}{(1-\phi)^{2.5}}$
Krieger and Dougherty [46]	$\mu_{nf} = \mu_{bf}\left(1 - \left(\dfrac{\phi}{\phi_m}\right)\right)^{-[\eta]\phi_m}$
Frankel and Acrivos [1]	$\mu_{nf} = \mu_{bf}\dfrac{9}{8}\left[\dfrac{\left(\dfrac{\phi}{\phi_m}\right)^{\frac{1}{3}}}{1-\left(\dfrac{\phi}{\phi_m}\right)^{\frac{1}{3}}}\right]$
Lundgren [44]	$\mu_{nf} = \mu_{bf}(1 + 2.5\phi + \dfrac{25}{4}\phi^2 + f\ (\phi^3)$
Batchelor [47]	$\mu_{nf} = \mu_{bf}(1 + 2.5\phi + 6.5\phi^2)$
Kitano et al. [48]	$\mu_{nf} = \mu_{bf}\left[1 - \left(\dfrac{\phi}{\phi_m}\right)\right]^{-2}$
Wang [4]	$\mu_{nf} = \mu_{bf}(1 + 7.3\phi + 123\phi^2)$
Bicerano et al. [49]	$\mu_{nf} = \mu_{bf}(1 + \eta\phi + k_H\phi^2)$
Tseng and Lin [50]	$\mu_{nf} = \mu_{bf} \times 13.47e^{35.98\phi}$
Tseng and Chen [51]	$\mu_{nf} = \mu_{bf} \times 0.4513\ e^{0.6965\phi}$
Chen et al. [52]	$\mu_{nf} = \mu_{bf}(1 + 10.6\phi + (10.6\phi)^2)$
Nguyen et al. [53]	$\mu_{nf} = \mu_{bf} \times 0.904\ e^{0.1482\phi}$ for $d = 47$nm $\mu_{nf} = \mu_{bf}(1 + 0.025\phi + 0.015\phi^2)$ for $d = 36$nm
Masoumi [54]	$\mu_{nf} = \mu_{bf} + \dfrac{\rho P V_B d_p^2}{72\ C\ \delta}$

Specific heat values calculated from Eq. (6.4) can be more reliable, as the specific heat depends on mass. The differential scanning calorimetry method is used to measure the specific heat of nanofluids.

Zhou and Ni [59] used a differential scanning calorimeter to measure the specific heat of water–Al_2O_3 nanofluid. They concluded that the specific heat of the nanofluid is an inverse function of the volume concentration of nanoparticles, which is also evident from the results of Mehrali et al. [58].

Sekhar and Sharma [60] experimentally investigated the specific heat of alumina nanofluids having a mean particle diameter of 47 nm for low volumetric concentration. Their results also indicate that specific heat decreases as the particle concentration increases, similar to that observed by Zhou and Ni [59] and Mehrali et al. [58]. The authors pointed out that this is probably because of the increase in thermal diffusivity.

6.2.5 Heat transfer coefficient

In heat exchangers, improvement in convective heat transfer coefficient can be a superior indicator of overimproved thermal conductivity of a nanofluid. Moreover, the convective heat transfer performance of the heat exchanger is assessed by plotting Nusselt number versus Reynolds number characteristics. When compared on this basis, generally, nanofluids will show improved performance. When different fluids are considered, the pumping power required for pumping both fluids will be different because of the difference in their viscosity. Considering this fact the use of the Reynolds number as a common parameter for comparison is inappropriate, as pointed out by Sadeghinezhad [61]. Considering this fact, many researchers [62–65] choose constant velocity over constant Reynolds number. Moreover, thermal convection is more complex than thermal conduction because of advection. Convective heat transfer performance in the laminar regime is well understood and established, and analytical solutions exist for the same. The most useful and widely used equation for predicting convective heat transfer in laminar flow is given as

$$Nu = \begin{cases} 1.302x_*^{-1/3} - 1 & x_* \leq 0.00005 \\ 1.302x_*^{-1/3} - 0.5 & 0.00005 \leq x_* \leq 0.0015 \\ 4.346 + 8.6810^3 x_*^{-0.506} \exp(-0.41x_*) & x_* \geq 0.001 \end{cases}$$

where $Nu_* = \frac{h(x)d_{\text{tube}}}{k}$, $x_* = \left[\frac{\frac{x}{d_i}}{RePr}\right]$

But for turbulent flow, researchers have to depend on empirical correlations such as Dittus–Boelter, Gnielinski, Petukhov, and Kirillov. But these correlations come with a large error of around 15%[61]. Researchers [62] reported that conventional theory fails to predict the heat transfer behavior of nanofluids.

Sadiq Al-Baghdadi et al. [64] investigated the heat transfer behavior of nanofluids inside a microchannel heat sink using a 3D CFD model. They concluded nanofluids are impractical in microchannel heat sinks, as water is still a safe and cheaper alternative. Purbia et al. [66] studied the heat transfer behavior of graphene–water nanofluid in a shell and tube heat exchanger. They focused on estimating the possible benefits of using graphene nanofluids as heat transfer fluid. The authors found that the operating cost of the overall heat exchanger module reduces dramatically, even though the nanofluid flow brings a higher pressure drop.

Kim and Bang [67] compared the heat transfer performance of water and graphene nanofluid filled in a heat pipe. They found that wall temperatures were low for nanofluid. Moreover, they observed that the thermal resistance of graphene nanofluid is 25% less than that of water.

Alhuyi Nazari et al. [68] presented a review on the use of nanofluids for heat pipe applications. Their literature review concluded that the thermal performance of nanofluid-filled heat pipes depends on the concentration of nanoparticles, their type, and stability. The use of nanofluids in heat pipes increases the number of nucleation sites, which is one of the important reasons behind the enhanced performance of nanofluid-filled heat pipes, alongside the higher thermal conductivity of nanofluids.

Das et al. [69] studied the thermophysical properties of graphene nanofluids inside a thermosyphon at various power inputs. They found that the thermal resistance of nanofluid-filled thermosyphon is less as compared to water-filled thermosyphon, even when the power input is increased.

Case study: Application of nanofluid in heat transfer: The effect of using nanofluid in the enhancement of heat transfer performance is shown here by employing magnesium oxide nanofluid as one of the fluids in the double pipe heat exchanger. A three-dimensional model is constructed and numerically investigated in commercial CFD package ANSYS-Fluent. The properties of magnesium oxide nanofluid are varied considering three different volume fractions of the magnesium oxide

Table 6.3 Models used to calculate thermophysical properties for the case study [69–74].

Property	Equation used
Density	$\rho_{nf} = (1-\varphi)\rho_{bf} + \varphi\,\rho_s$
Specific heat	$c_{p,nf} = \dfrac{\varphi\,(\rho c_p)_{,np} + (1-\varphi)\,(\rho c_p)_{,bf}}{\rho_{nf}}$
Thermal conductivity	$\dfrac{k_{np} + 2k_{bf} + 2(k_{np} - k_{bf})\phi}{k_{np} + 2k_{bf} - (k_{np} - k_{bf})\phi}$
Viscosity	$\mu_{nf} = \mu_{bf}(1 + 2.5\phi)$

nanoparticles in the base fluid. The volume fraction percentages of magnesium oxide nanoparticles considered here are 0.005, 0.01, 0.05, and 0.1, and the thermophysical properties of these nanofluids have been calculated using well-known correlations available in the literature (Table 6.3). The density, specific heat, thermal conductivity, and viscosity of nanofluids have been calculated according to the following equations.

Three-dimensional numerical simulations have been performed, and the results have been compared with water as cold-side fluid. For the sake of comparison, boundary conditions have been kept constant, including the inlet temperature of both the fluids and the pressure outlet boundary condition at the exit of the two channels. Four different coolant flow rates, that is, 6, 7, 8, and 9 LPM, are chosen. The inlet temperature of the coolant is specified as 298 K. The section below summarizes the results of the case study.

6.2.5.1 Results of the case study

The change in the overall heat transfer coefficient of the heat exchanger with the change in volume concentration of the nanoparticles is analyzed. From the analysis, it is observed that with a change in the configuration of the nanofluid, variation in the overall heat transfer coefficient is not much significant and quite difficult to show comparatively in a graphical manner. Moreover, hence, the results of the analysis are presented in tabular form (Table 6.4).

6.3 Conclusion

This chapter provides an overview of the heat transfer characteristics of nanofluids. The impact of several parameters, including nanoparticle size, shape, concentration, and aggregation level, on the thermophysical properties of nanofluids has been examined. In addition, we have reviewed numerical models for forecasting thermophysical property values. Despite the multitude of models proposed by researchers over time the models for predicting effective thermal conductivity and viscosity have not been able to achieve accurate results at higher volume concentrations of nanoparticles as compared to experimental findings. Moreover, very few studies focus on the dependence of viscosity on effective thermal conductivity and vice versa, which are the major properties that affect the thermal performance

Table 6.4 Result of the case study.

Configuration of heat exchanger	V_cold (m/s)	V_hot_nf (m/s)	Qavg (W) (($Q_h + Q_c$)/2)	LMTD (K)	U (W/m^2 K)
Water-nanofluid ($\phi = 0.0012$)	0.1507	0.1	113.925	41.255	332.958
	0.1507	0.5034	221.945	42.562	628.725
	0.1507	1	252.091	42.630	712.985
Water-nanofluid ($\phi = 0.002$)	0.1507	0.1	113.733	41.268	332.292
	0.1507	0.5034	221.974	42.561	628.822
	0.1507	1	252.125	42.629	713.096
Water-nanofluid ($\phi = 0.005$)	0.1507	0.1	114.468	41.205	334.950
	0.1507	0.5034	222.122	42.556	629.326
	0.1507	1	252.618	42.613	714.775
Water–water	0.1507	0.1	113.705	41.275	332.155
	0.1507	0.5034	221.220	42.596	626.171
	0.1507	1	251.955	42.641	712.426

and the pumping power requirements of nanofluids. Hence more sophisticated experiments combined with a better understanding of fluid flow and heat transfer behavior at the nanoscale are required.

Nomenclature

C_p	specific heat capacity
d	diameter
k	thermal conductivity
Nu	Nusselt number
Pr	Prandtl number
r	radius
Re	Reynolds number

Greek letters

μ	viscosity
ϕ	volume fraction
ρ	density
Ψ	sphericity

Subscripts

np	nanoparticle
bf	base fluid
pe	particle equivalent
eff	effective
nano	nanofluids
fr	freezing point
m	maximum

References

[1] N.A. Frankel, A. Acrivos, On the viscosity of a concentrated suspension of solid spheres, Chem. Eng. Sci. 22 (1967) 847–853. https://doi.org/10.1016/0009-2509(67)80149-0.

[2] S. Lee, S. Choi, S. Li, J. Eastman, Measuring thermal conductivity of fluids containing oxide nanoparticles, Heat Transf. 121 (2013) 280–289.

[3] J.A. Eastman, U.S. Choi, S. Li, L.J. Thompson, S. Lee, Enhanced thermal conductivity through the development of nanofluids, Mater. Res. Soc. Symp. – Proc. 457 (1997) 3–11. https://www.osti.gov/servlets/purl/459378.

[4] X. Wang, X. Xu, S.U.S. Choi, Thermal conductivity of nanoparticle–fluid mixture, J. Thermophys. HEAT Transf. 13 (1999) 474–480.

[5] A. Hiroshi, T. Yukihiro, Y. Shigeki, T. Akira, Magnetic properties of ferromagnetic ultrafine particles prepared by vacuum, J. Cryst. Growth 45 (1978) 495–500.

[6] P.X. Tran, Y. Soong, Preparation of nanofluids using laser ablation in liquid technique, in: ASME Applied Mechanics and Materials Conference, American Society of Mechanical Engineers (ASME), Austin, Texas, 2007.

[7] X.Q. Wang, A.S. Mujumdar, Heat transfer characteristics of nanofluids: a review, Int. J. Therm. Sci. 46 (2007) 1–19. https://doi.org/10.1016/j.ijthermalsci.2006.06.010.

[8] A. Abdollahi, H.A. Mohammed, S.M. Vanaki, A. Osia, M.R. Golbahar Haghighi, Fluid flow and heat transfer of nanofluids in microchannel heat sink with V-type inlet/outlet arrangement, Alexandria Eng. J. 56 (2017) 161–170. https://doi.org/10.1016/j.aej.2016.09.019.

[9] S.M.S. Murshed, K.C. Leong, C. Yang, Enhanced thermal conductivity of TiO_2—water based nanofluids, Int. J. Therm. Sci. 44 (2005) 367–373. https://doi.org/10.1016/j.ijthermalsci.2004.12.005.

[10] S.M.S. Murshed, K.C. Leong, C. Yang, Investigations of thermal conductivity and viscosity of nanofluids, Int. J. Therm. Sci. 47 (2008) 560–568. https://doi.org/10.1016/j.ijthermalsci.2007.05.004.

[11] J.C. Maxwell, A Treatise on Electricity and Magnetism, Clarendon Press, Oxford, 1873. https://www.loc.gov/item/03015568/.

[12] E. Yamada, T. Ota, Effective thermal conductivity of dispersed materials, Wärme- Und Stoffübertragung 13 (1980) 27–37. https://doi.org/10.1007/BF00997630.

[13] J.A. Eastman, S.U.S. Choi, S. Li, W. Yu, L.J. Thompson, Anomalously increased effective thermal conductivities of ethylene glycol-based nanofluids containing copper nanoparticles, Appl. Phys. Lett. 78 (2001) 718–720. https://doi.org/10.1063/1.1341218.

[14] J. Garg, B. Poudel, M. Chiesa, J.B. Gordon, J.J. Ma, J.B. Wang, Z.F. Ren, Y.T. Kang, H. Ohtani, J. Nanda, G.H. McKinley, G. Chen, Enhanced thermal conductivity and viscosity of copper nanoparticles in ethylene glycol nanofluid, J. Appl. Phys. 103 (2008). https://doi.org/10.1063/1.2902483.

[15] M. Chopkar, P.K. Das, I. Manna, Synthesis and characterization of nanofluid for advanced heat transfer applications, Scr. Mater. 55 (2006) 549–552. https://doi.org/10.1016/j.scriptamat.2006.05.030.

[16] O. Sanusi, R. Warzoha, A.S. Fleischer, Energy storage and solidification of paraffin phase change material embedded with graphite nanofibers, Int. J. Heat Mass Transf. 54 (2011) 4429–4436. https://doi.org/10.1016/j.ijheatmasstransfer.2011.04.046.

[17] A. Sari, A. Karaipekli, Thermal conductivity and latent heat thermal energy storage characteristics of paraffin/expanded graphite composite as phase change material, Appl. Therm. Eng. 27 (2007) 1271–1277. https://doi.org/10.1016/j.applthermaleng.2006.11.004.

[18] J.M. Khodadadi, S.F. Hosseinizadeh, Nanoparticle-enhanced phase change materials (NEPCM) with great potential for improved thermal energy storage, Int. Commun. Heat Mass Transf. 34 (2007) 534–543. https://doi.org/10.1016/j.icheatmasstransfer.2007.02.005.

[19] R.J. Warzoha, A.S. Fleischer, Determining the thermal conductivity of liquids using the transient hot disk method. Part I: establishing transient thermal-fluid constraints, Int. J. Heat Mass Transf. 71 (2014) 779–789. https://doi.org/10.1016/j.ijheatmasstransfer.2013.10.064.

[20] T. Boumaza, J. Redgrove, Use of the transient plane source technique for rapid multiple thermal property measurements, Int. J. Thermophys. 24 (2003) 501–512. https://doi.org/10.1023/A:1022928206859.

[21] J. Nanda, C. Maranville, S.C. Bollin, D. Sawall, H. Ohtani, J.T. Remillard, J.M. Ginder, Thermal conductivity of single-wall carbon nanotube dispersions: role of interfacial effects, J. Phys. Chem. C 112 (2008) 654–658. https://doi.org/10.1021/jp711164h.

[22] R.H. Davis, The effective thermal conductivity of a composite material with spherical inclusions, Int. J. Thermophys. 7 (1986) 609–620. https://doi.org/10.1007/BF00502394.

[23] C.W. Nan, R. Birringer, D.R. Clarke, H. Gleiter, Effective thermal conductivity of particulate composites with interfacial thermal resistance, J. Appl. Phys. 81 (1997) 6692–6699. https://doi.org/10.1063/1.365209.

[24] W. Yu, S.U.S. Choi, The role of interfacial layers in the enhanced thermal conductivity of nanofluids: a renovated Maxwell model, J. Nanoparticle Res. 5 (2003) 167–171. https://doi.org/10.1023/A:1024438603801.

[25] R.L. Hamilton, O.K. Crosser, Thermal conductivity of heterogeneous two-component systems, Ind. Eng. Chem. Fundam. 1 (1962) 187–191. https://doi.org/10.1021/i160003a005.

[26] D.J. Jefferey, Conduction through a random suspension of spheres, Proc. R. Soc. London. A. Math. Phys. Sci. 335 (1973) 355–367. https://doi.org/10.1098/rspa.1973.0130.
[27] E.J. Wasp, J.P. Kenny, R.L. Gandhi, Solid–liquid flow: slurry pipeline transportation. [Pumps, valves, mechanical equipment, economics], Ser. Bulk Mater. Handl. (United States) 1 (1977) 4.
[28] D.A.G. Bruggeman, Berechnung verschiedener physikalischer Konstanten von heterogenen Substanzen. III. Die elastischen Konstanten der quasiisotropen Mischkörper aus isotropen Substanzen, Ann. Phys. 421 (1937) 160–178. https://doi.org/10.1002/andp.19374210205.
[29] S.P. Jang, S.U.S. Choi, Free convection in a rectangular cavity (Benard convection) with nanofluids, Am. Soc. Mech. Eng. Heat Transf. Div. HTD 375 (2004) 147–153. https://doi.org/10.1115/IMECE2004-61054.
[30] D.H. Kumar, H.E. Patel, V.R.R. Kumar, T. Sundararajan, T. Pradeep, S.K. Das, Model for heat conduction in nanofluids, Phys. Rev. Lett. 93 (2004) 1–4. https://doi.org/10.1103/PhysRevLett.93.144301.
[31] C.H. Chon, K.D. Kihm, S.P. Lee, S.U.S. Choi, Empirical correlation finding the role of temperature and particle size for nanofluid (Al_2O_3) thermal conductivity enhancement, Appl. Phys. Lett. 87 (2005) 1–3. https://doi.org/10.1063/1.2093936.
[32] C.H. Li, G.P. Peterson, Experimental investigation of temperature and volume fraction variations on the effective thermal conductivity of nanoparticle suspensions (nanofluids), J. Appl. Phys. 99 (2006). https://doi.org/10.1063/1.2191571.
[33] H.A. Mintsa, G. Roy, C.T. Nguyen, D. Doucet, New temperature dependent thermal conductivity data for water-based nanofluids, Int. J. Therm. Sci. 48 (2009) 363–371. https://doi.org/10.1016/j.ijthermalsci.2008.03.009.
[34] M. Corcione, Empirical correlating equations for predicting the effective thermal conductivity and dynamic viscosity of nanofluids, Energy Convers. Manag. 52 (2011) 789–793. https://doi.org/10.1016/j.enconman.2010.06.072.
[35] D.K. Devendiran, V.A. Amirtham, A review on preparation, characterization, properties and applications of nanofluids, Renew. Sustain. Energy Rev. 60 (2016) 21–40. https://doi.org/10.1016/j.rser.2016.01.055.
[36] M. Moghimian, B. Rahmanian, M. Safaei, M. Goudarzi, Numerical investigation of heat transfer in circular perforated plates exposed to parallel flow and suction, ADMT J. (2008) 1. http://admt.iaumajlesi.ac.ir/article_534664.html.
[37] A. Einstein, Investigations on the Theory Brownian Movement, Dover Publications, New York, 1956.
[38] Y. Yang, A. Oztekin, S. Neti, S. Mohapatra, Characterization and convective heat transfer with nanofluids, in: ASME/JSME 2011 8th Thermal Engineering Joint Conference, AJTEC 2011, 2011, pp. 1–6. https://doi.org/10.1115/ajtec2011-44448.
[39] P. Dhar, S. Sen Gupta, S. Chakraborty, A. Pattamatta, S.K. Das, The role of percolation and sheet dynamics during heat conduction in poly-dispersed graphene nanofluids, Appl. Phys. Lett. (2013) 102. https://doi.org/10.1063/1.4802998.
[40] M. Kole, T.K. Dey, Investigation of thermal conductivity, viscosity, and electrical conductivity of graphene based nanofluids, J. Appl. Phys. 113 (2013). https://doi.org/10.1063/1.4793581.
[41] S.S. Park, N.J. Kim, Influence of the oxidation treatment and the average particle diameter of graphene for thermal conductivity enhancement, J. Ind. Eng. Chem. 20 (2014) 1911–1915. https://doi.org/10.1016/j.jiec.2013.09.011.
[42] I.M. Mahbubul, R. Saidur, M.A. Amalina, Latest developments on the viscosity of nanofluids, Int. J. Heat Mass Transf. 55 (2012) 874–885. https://doi.org/10.1016/j.ijheatmasstransfer.2011.10.021.
[43] H.C. Brinkman, The viscosity of concentrated suspensions and solutions, J. Chem. Phys. 20 (1952) 571. https://doi.org/10.1063/1.1700493.
[44] T.S. Lundgren, Slow flow through stationary random beds and suspensions of spheres, J. Fluid Mech. 51 (1972) 273–299. https://doi.org/10.1017/S002211207200120X.

[45] V. Vand, Viscosity of solutions and suspensions. I: theory, J. Phys. Colloid Chem. 52 (1948) 277–299. https://doi.org/10.1021/j150458a001.

[46] I.M. Krieger, T.J. Dougherty, A mechanism for non-Newtonian flow in suspensions of rigid spheres, Trans. Soc. Rheol. 3 (1959) 137–152. https://doi.org/10.1122/1.548848.

[47] G.K. Batchelor, The effect of Brownian motion on the bulk stress in a suspension of spherical particles, J. Fluid Mech. 83 (1977) 97–117.

[48] T. Kitano, T. Kataoka, T. Shirota, An empirical equation of the relative viscosity of polymer melts filled with various inorganic fillers, Rheol. Acta. 20 (1981) 207–209. https://doi.org/10.1007/BF01513064.

[49] J. Bicerano, J.F. Douglas, D. a Brune, Model for the viscosity of particle dispersions, J. Macromol. Sci. Part C Polym. Rev. 39 (1999) 561–642.

[50] W.J. Tseng, K.C. Lin, Rheology and colloidal structure of aqueous TiO_2 nanoparticle suspensions, Mater. Sci. Eng. A. 355 (2003) 186–192. https://doi.org/10.1016/S0921-5093(03)00063-7.

[51] W.J. Tseng, C.N. Chen, Effect of polymeric dispersant on rheological behavior of nickel-terpineol suspensions, Mater. Sci. Eng. A. 347 (2003) 145–153. https://doi.org/10.1016/S0921-5093(02)00562-2.

[52] H. Chen, Y. Ding, Y. He, C. Tan, Rheological behaviour of ethylene glycol based titania nanofluids, Chem. Phys. Lett. 444 (2007) 333–337. https://doi.org/10.1016/j.cplett.2007.07.046.

[53] C.T. Nguyen, F. Desgranges, G. Roy, N. Galanis, T. Maré, S. Boucher, H.A. Mintsa, Temperature and particle-size dependent viscosity data for water-based nanofluids—hysteresis phenomenon, Int. J. Heat Fluid Flow. 28 (2007) 1492–1506. https://doi.org/10.1016/j.ijheatfluidflow.2007.02.004.

[54] N. Masoumi, N. Sohrabi, A. Behzadmehr, A new model for calculating the effective viscosity of nanofluids, J. Phys. D. Appl. Phys. 42 (2009). https://doi.org/10.1088/0022-3727/42/5/055501.

[55] B.C. Pak, Y.I. Cho, Hydrodynamic and heat transfer study of dispersed fluids with submicron metallic oxide, Exp. Heat Transf. 11 (2013) 151–170.

[56] J. Liu, F. Wang, L. Zhang, X. Fang, Z. Zhang, Thermodynamic properties and thermal stability of ionic liquid-based nanofluids containing graphene as advanced heat transfer fluids for medium-to-high-temperature applications, Renew. Energy 63 (2014) 519–523. https://doi.org/10.1016/j.renene.2013.10.002.

[57] R.S. Vajjha, D.K. Das, B.M. Mahagaonkar, Density measurement of different nanofluids and their comparison with theory, Pet. Sci. Technol. 27 (2009) 612–624. https://doi.org/10.1080/10916460701857714.

[58] M. Mehrali, E. Sadeghinezhad, A.R. Akhiani, S. Tahan Latibari, S. Talebian, A. Dolatshahi-Pirouz, H.S.C. Metselaar, M. Mehrali, An ecofriendly graphene-based nanofluid for heat transfer applications, J. Clean. Prod. 137 (2016) 555–566. https://doi.org/10.1016/j.jclepro.2016.07.136.

[59] S.Q. Zhou, R. Ni, Measurement of the specific heat capacity of water-based Al_2O_3 nanofluid, Appl. Phys. Lett. 92 (2008) 1–4. https://doi.org/10.1063/1.2890431.

[60] Y.R. Sekhar, K.V. Sharma, Study of viscosity and specific heat capacity characteristics of water-based Al_2O_3 nanofluids at low particle concentrations, J. Exp. Nanosci. 10 (2015) 86–102. https://doi.org/10.1080/17458080.2013.796595.

[61] E. Sadeghinezhad, M. Mehrali, R. Saidur, M. Mehrali, S. Tahan Latibari, A.R. Akhiani, H.S.C. Metselaar, A comprehensive review on graphene nanofluids: recent research, development and applications, Energy Convers. Manag. 111 (2016) 466–487. https://doi.org/10.1016/j.enconman.2016.01.004.

[62] E. Sadeghinezhad, M. Mehrali, S. Tahan Latibari, M. Mehrali, S.N. Kazi, C.S. Oon, H.S.C. Metselaar, Experimental investigation of convective heat transfer using graphene nanoplatelet based nanofluids under turbulent flow conditions, Ind. Eng. Chem. Res. 53 (2014) 12455–12465. https://doi.org/10.1021/ie501947u.

[63] M. Mehrali, E. Sadeghinezhad, M.A. Rosen, A.R. Akhiani, S. Tahan Latibari, M. Mehrali, H.S.C. Metselaar, Heat transfer and entropy generation for laminar forced convection flow of graphene nanoplatelets nanofluids in a horizontal tube, Int. Commun. Heat Mass Transf. 66 (2015) 23–31. https://doi.org/10.1016/j.icheatmasstransfer.2015.05.007.

[64] M.A.R. Sadiq Al-Baghdadi, Z.M.H. Noor, A. Zeiny, A. Burns, D. Wen, CFD analysis of a nanofluid-based microchannel heat sink, Therm. Sci. Eng. Prog. 20 (2020) 100685. https://doi.org/10.1016/j.tsep.2020.100685.

[65] M.N.M. Zubir, A. Badarudin, S.N. Kazi, N.M. Huang, M. Misran, E. Sadeghinezhad, M. Mehrali, N.I. Syuhada, S. Gharehkhani, Experimental investigation on the use of reduced graphene oxide and its hybrid complexes in improving closed conduit turbulent forced convective heat transfer, Exp. Therm. Fluid Sci. 66 (2015) 290–303. https://doi.org/10.1016/j.expthermflusci.2015.03.022.

[66] D. Purbia, A. Khandelwal, A. Kumar, A.K. Sharma, Graphene-water nanofluid in heat exchanger: mathematical modelling, simulation and economic evaluation, Int. Commun. Heat Mass Transf. 108 (2019) 104327. https://doi.org/10.1016/j.icheatmasstransfer.2019.104327.

[67] K.M. Kim, I.C. Bang, Effects of graphene oxide nanofluids on heat pipe performance and capillary limits, Int. J. Therm. Sci. 100 (2016) 346–356. https://doi.org/10.1016/j.ijthermalsci.2015.10.015.

[68] M. Alhuyi Nazari, R. Ghasempour, M.H. Ahmadi, A review on using nanofluids in heat pipes, J. Therm. Anal. Calorim. 137 (2019) 1847–1855. https://doi.org/10.1007/s10973-019-08094-y.

[69] S. Das, A. Giri, S. Samanta, S. Kanagaraj, Role of graphene nanofluids on heat transfer enhancement in thermosyphon, J. Sci. Adv. Mater. Devices 4 (2019) 163–169. https://doi.org/10.1016/j.jsamd.2019.01.005.

[70] G.F. Smaisim, D.B. mohammed, A.M. Abdulhadi, et al., Nanofluids: properties and applications, J. Sol-Gel Sci. Technol. 104 (2022) 1–35. https://doi.org/10.1007/s10971-022-05859-0.

[71] H. Yasmin, S.O. Giwa, S. Noor, M. Sharifpur, Thermal conductivity enhancement of metal oxide nanofluids: a critical review, Nanomaterials 13 (2023) 597. https://doi.org/10.3390/nano13030597.

[72] S. Purree, M. Nadeem, A. Shahzad, H.M Ali, Synthesis, heat transfer properties and stability of nanofluids for commercialization: a review, Chem. Eng. Commun. 210 (5) (2023) 814–836. 10.1080/00986445.2021.1976163.

[73] H. Yasmin, S.O. Giwa, S. Noor, H.Ş. Aybar, Reproduction of nanofluid synthesis, thermal properties and experiments in engineering: a research paradigm shift, Energies 16 (3) (2023) 1145. https://doi.org/10.3390/en16031145.

[74] Y. Sun, D. Yang, L. Shi, H. Wu, Y. Cao, Y. He, T. Xie, Properties of nanofluids and their applications in enhanced oil recovery: a comprehensive review, Energy Fuels 34 (2) (2020) 1202–1218. 10.1021/acs.energyfuels.9b03501.

CHAPTER

Convective heat transfer to nanofluids

7

Debabrata Panda[a], Krunal M. Gangawane[a,b] and Ram Prakash Bharti[c]

[a]*Department of Chemical Engineering, National Institute of Technology Rourkela, Rourkela, Odisha, India,* [b]*Department of Chemical Engineering, Indian Institute of Technology Jodhpur, Karwar, Jodhpur, Rajasthan, India,* [c]*Department of Chemical Engineering, Indian Institute of Technology Roorkee, Roorkee, Uttarakhand, India*

7.1 Introduction

Over the decade, nanofluid technology has evolved as a new interdisciplinary field of great pragmatic relevance. In the recent past, nanofluids have been explored extensively as potential heat enhancers for a range of practical applications, such as electronic devices, heat exchangers, chemical vapor deposition (CVD) reactors, and electronic packaging [1,2][,]. Nanofluids are a class of engineered fluids obtained by suspending nanosized (~1–100 nm) particles and fibers into a base liquid (such as water and oil). A highly stable nanofluid is prepared with proper chemical and thermal treatments. To name a few treatments: control of the pH value of the nanofluid suspensions, use of the ultrasonic bath, overhead magnetic stirring, homogenization, probe sonication, and dispersion by surfactants. These methods intend to alter the suspended nanoparticles' surface properties and subdue the development of groups of particles. However, to tackle the problems of thermal nanofluids (thermal motion, gravitational sedimentation, and aggregation because of static potential), electrostatic and steric stabilization have developed to increase the repulsive interparticle interaction by adsorption, substitution, dissociation, and physical adsorption of ions. Similarly, covalent functionalization is the attachment of chemical moieties to the surface structure via the formation of covalent bonds by sharing at least one pair of electrons between chemical moieties and compound.

Nanofluids are not just a simple suspension of nanoparticles into the base liquids and display exceptional features, unlike the conventional solid–liquid colloidal suspensions of small (i.e., millimeter or micrometer size) particles. For example, nanofluids' potential benefits are higher stability and thermal conductivity, lower pressure drops, and erosion in microchannels compared to millimeter or even micrometer-sized particle suspensions. Because of their excellent thermal characteristics, nanofluids find more significant applications in enhancing heat transfer, even for microscale heat transfer [3,4]. For example, experimental studies [5,6] have indicated about 40% and 150% augmentation in the thermal conductivity by the addition of copper nanoparticles or carbon nanotubes (volume fraction <1%) in ethylene glycol and oil, respectively.

In a similar context of the discussion, graphene-based nanomaterials have received an enormous response from the scientific community owing to their practical significance [7–13]. Graphene is a planar monolayer of sp^2-bonded-carbon (C) atoms firmly crammed into a honeycomb lattice. It has received

Graphene and 2D Materials in Heat Transfer: Fundamentals and Applications. DOI: https://doi.org/10.1016/B978-0-12-821964-5.00001-1

an overwhelming response from the scientific community owing to its 2D structure and exceptional chemical, physical, electrical, mechanical, and thermal characteristics [14]. Primarily, graphene-based nanofluids' enhanced thermal conductivity compared to carbon nanotubes and conventional fluids make it a lucrative choice to enhance heat transfer in various applications. After the remarkable finding of graphite oxide (GtO) [15] the monolayers of graphene oxide (GO) sheets dispersible in the base media were recovered from the shedding of GtO. It is a heterogeneous compound comprising the graphene sheet having covalent bonding to oxygen-bearing groups. The base plane of GO is graced with epoxy and hydroxyl functional groups with carbonyl and carboxylic acid attached to the edges of the plane [16]. Some of the unique features of GO are large surface area, high water dispensability, excellent colloidal stability, ability to easily surface modify, excellent biocompatibility, distinctive amphipathic nature, and superior mechanical properties [17].

Graphene-based nanofluids are used in electronic devices, nuclear energy, solar thermal energy, phase change materials, heat pipes, etc., as an effective way of enhancing convective heat transfer. Different factors affecting the convective heat transfer coefficient of nanofluids are the nature of base fluid, types and shapes of particles, the volume fraction of nanoparticles, temperature, etc. Moreover, the thermal transport properties of nanofluids are influenced by temperature, base fluid, morphology, filling fraction of filler, clustering, acidity, and additives. Encouragingly, graphene has many benefits, such as larger surface area-to-volume ratio, higher thermal conductivity, high dispersion stability, simple preparation, and lower erosion, corrosion, and clogging. Therefore it has become an excellent alternative in nanofluids applications.

This chapter, therefore, addresses the convection heat transfer in both nongraphene and graphene-based nanofluids. The effects of thermophysical properties (namely, effective viscosity and effective thermal conductivity), the volume fraction of nanoparticles in the base fluids, and the applied magnetic field on the convective heat transfer are discussed.

7.1.1 Basic correlations of nanofluids

Before proceeding with the detailed discussion on the empirical correlations of physical properties such as viscosity and thermal conductivity, it would be useful to review some of the basic mathematical expressions of nanofluids correlating the thermophysical properties of nanofluid with those of the base fluid and the solid nanoparticles [18].

- Density:

$$\rho_{nf} = (1-\varphi)\rho_f + \varphi\rho_s \tag{7.1}$$

- Specific heat:

$$(\rho C_p)_{nf} = (1-\varphi)(\rho C_p)_f + \varphi(\rho C_p)_s \tag{7.2}$$

- Coefficient of thermal expansion:

$$(\rho\beta)_{nf} = (1-\varphi)(\rho\beta)_f + \varphi(\rho\beta)_s - \varphi(1-\varphi)(\rho_s-\rho_f)(\beta_s-\beta_f) \tag{7.3}$$

where subscripts *nf*, *f*, and *s* represent the nanofluid suspension, base fluid, and nanoparticles, respectively. Density, the coefficient of thermal expansion, heat capacity, and volume fraction are denoted by ρ, β, C_p, and ϕ, respectively.

7.2 Correlation for the effective viscosity

Viscosity is one of the fluid's critical characteristics, signifying the internal resistance experienced by the fluid to flow. It is also a widely acknowledged parameter of the prime importance of the thermo-fluid behavior of fluids. Some of the essential applications dependent on the viscosity are pumping power, the pressure drop in laminar flow, and heat transfer by convection [62].

For nanofluids the determination of the viscosity is quite different from that of pure fluid. Several theoretical correlations have been proposed to model the nanofluid viscosity. The effective viscosity of nanofluid is estimated as a function of the dynamic viscosity of a base fluid and the volume fraction of nanoparticles. Some of the widely used empirical correlations of viscosity are the Einstein model, Brinkman model, Batchelor model, Graham model, etc. These are discussed in subsequent paragraphs and Table 7.1.

Einstein [19] developed a viscosity model based upon the assumptions of linear variation of viscosity with suspended fine particles with low volume fraction and noninteracting suspensions. The disadvantage of the model was the limitation of low volume fraction (<2%). The subsequent modification of the Einstein model [20] is a popular correlation among researchers. The model is valid for a higher volume fraction of nanoparticles, that is, 4%. Further the effect of the Brownian motion of particles in the suspension on the dynamic viscosity was presented by Batchelor [23] and Masoumi et al. [29].

On the other hand, Masoud Hosseini et al. [30] used dimensionless analysis to develop the viscosity correlation. The effect of temperature on nanofluid viscosity has also been correlated in the literature [30,29] [63]. Recently, Cabaleiro et al. [9] presented the stability and thermophysical properties of nanofluids designed as dispersions of sulfonic acid-functionalized graphene nanoplatelets in the ethylene glycol (10% w) and water (90% w) mixture. They experimentally measured the thermal conductivity, dynamic viscosity, and density for temperatures ranging from 283.15 to 343.15 K and nanoparticle concentration up to 0.5% w. Sarode et al. [31] presented the viscosity correlation for graphene-oxide-copper oxide nanocomposites. Because of the difficulty associated with the viscosity of nanofluid composites, they obtained a nonlinear correlation by curve-fitting the experimental data. The developed correlation was found to predict the GO–CuO nanocomposite-based nanofluid viscosity accurately.

Therefore numerous empirical models are now available for predicting nanofluid viscosity. However, every model has its benefits and limitations. A given model may be suitable for a set of nanoparticles and base fluids. It is not easy to predict all nanofluids' viscosity by using a single model [30], as it is dependent on various parameters, such as volume fraction, size, and shape of nanoparticles and viscosity of the base fluid. The nanofluid viscosity shows a monotonic increase with volume fraction, and for lower volume fraction, the nanofluid behaves like a Newtonian fluid. On the other hand, if the effect of temperature on nanofluid viscosity is considered, a contradiction appears among the nanofluid studies. As reported by Mahbubul et al. [62], few researchers have observed a significant effect of temperature on viscosity, whereas few have reported the negligible influence of temperature on the same.

7.3 Correlation for the effective thermal conductivity

Admittedly, nanofluids have augmented thermal conductivity and a convective heat transfer coefficient equal to the base fluid. Therefore empirical correlations for calculating thermal conductivity have

Table 7.1 Empirical correlations for modeling effective viscosity of nanofluids.

S. no.	Source/model name	Expression	Remarks
1	Einstein model [19]	$\mu_{nf} = \mu_f(1 + 2.5\varphi)$	Simple correlation with assumptions of linear viscous fluid, dilute suspended particles, and low volume fraction (<2%)
2	Brinkman model [20]	$\mu_{nf} = \frac{\mu_f}{(1+2.5\varphi)^{2.5}}$	Modified Einstein formula extended up to higher volume fraction (4%)
3	Krieger and Dougherty [21]	$\mu_{nf} = \mu_f\left(1 - \frac{\varphi_a}{\varphi_m}\right)^{-[\eta]\varphi_m}$ where η is intrinsic viscosity	Semiempirical correlation applicable for wide range of ϕ
4	Franken and Acrivos model [22]	$\mu_{nf} = \mu_f\left(\frac{(\varphi/\varphi_m)^{1/3}}{1-(\varphi/\varphi_m)^{1/3}}\right)$	Effect of volume fraction as a function of maximum ϕ
5	Batchelor model [23]	$\mu_{nf} = \mu_f[1 + 2.5\varphi + 6.5\varphi^2]$	Extension of the Einstein model. Effect of Brownian motion
6	Graham model [24]	$\mu_{nf} = \mu_f\left(1 + 2.5\varphi + 4.5\left[\frac{1}{\left(\frac{h_s}{d_p}\right)\left[2+\left(\frac{h_s}{d_p}\right)\right]\left[2+\left(\frac{h_s}{d_p}\right)^2\right]}\right]\right)$	General form of Franken and Acrivos [22] model
7	Kitano et al. [25]	$\mu_{nf} = \mu_f\left(1 - \frac{\varphi}{\varphi_m}\right)^{-2}$	Viscosity of the two-phase mixture

(*continued on next page*)

8	Wang et al. [26]	$\mu_{nf} = \mu_f[1 + 7.3\varphi + 123\varphi^2]$	Linear model to predict viscosity for $\phi \le 4\%$
9	Kulkarni [63]	$\ln(\mu_{nf}) = [\frac{A}{T}] - B$ $A = 1.8375(\varphi)^2 - 29.643(\varphi) + 165.56$ $B = 4 \times 10^{-6}(\varphi)^2 - 0.001(\varphi) + 0.0186$ For $\varphi \le 6.12$	Not applicable for nanofluids at subzero temperature
10	Avsec and Oblak [27]	$\mu_{nf} = \mu_f\left(\frac{1}{1-2.5\varphi_e}\right)$ where $\varphi_e = \varphi(\frac{h}{r} + 1)^3$	Use of effective volume fraction to predict the nanofluid viscosity
11	Chen et al. [28]	$\mu_{nf} = \mu_f[1 + 10.6\varphi + (10.6\varphi)^2]$	Experimental study: Nanofluid viscosity is only a function of volume fraction but not the temperature
12	Masoumi et al. [29]	$\mu_{nf} = \mu_f\left(1 + \frac{\rho_p V_B d_p^2}{72\,\delta\, C}\right)$ $V_B = \sqrt{\frac{18RT}{\pi N_A \rho_p d_p^3}}$; $\delta = \sqrt[3]{\frac{\pi d_p^3}{6\varphi}}$ $C = \left[(-1.133d_p - 2.771)\varphi + (0.09d_p - 0.393)\right] \times 10^{-6}$	Based upon the tracking of Brownian motion, nanofluid with two different fluid
13	Masoud Hosseini et al. [30]	$\mu_{nf} = \mu_f \exp[m + \alpha(T/T_0) + \beta(\varphi) + \gamma(\frac{d}{1-r})]$ α, β, γ are empirical constants, r is thickness of caping layer, T_0 is reference temperature, T – temperature of nanofluid	Based upon the dependence of dimensionless groups
14	Sarode et al. [31]	$\mu_{nf} = \mu_f(1 + 1135.945\phi + 681571\phi^2)$	GO–CuO nanocomposite-based nanofluid

h—*thickens of nanolayer,* r—*radius of nanoparticle,* ϕ_e—*effective volume fraction,* V_B—*Brownian velocity,* C—*constant,* δ—*distance between nanoparticles,* ρ_p—*particle density,* ϕ_m—*maximum particle packing fraction, and* $[\eta]$—*intrinsic viscosity.* GO - *graphene oxide.*

attracted interest from the research community. Analytical models allow scientists and engineers to predict the thermal conductivity values of nanofluids. The popular empirical models for predicting the thermal conductivity of nanofluid are summarized as follows.

Maxwell's [32] model estimates the effective thermal conductivity of a solid–liquid mixture as the weighted average of both suspended solids and liquid conductivities originating from a point-source method. The model assumes the particle shape to be spherical and expressed as Eq. (7.4):

$$\frac{k_{nf}}{k_f} = 1 + \frac{3\left(\frac{k_s}{k_f} - 1\right)\varphi}{\left(\frac{k_s}{k_f} + 2\right) - \left(\frac{k_s}{k_f} - 1\right)\varphi} \tag{7.4}$$

where k_s and k_f are the thermal conductivities of solid and liquid nanofluids. Corcione [64] pointed out the limitations of the traditional Maxwell theory in accurately estimating the thermal conductivity of nanofluids. According to Corcione, this model tends to underestimate the effective thermal conductivity of nanofluids when the nanoparticle diameter is small and overestimate it when the diameter is large. In contrast the model proposed by Hamilton and Crosser [33] takes into account the shape of suspended particles through a parameter "n" (Eq. 7.5), which provides a more accurate estimation of the effective thermal conductivity of nanofluids.

$$\frac{k_{nf}}{k_f} = 1 + \frac{\frac{k_s}{k_f} + (n-1) - (n-1)\left(1 - \frac{k_s}{k_f}\right)\varphi}{\frac{k_s}{k_f} + (n-1) + \left(1 - \frac{k_s}{k_f}\right)\varphi}, \quad n = \begin{cases} n = 3, & \text{For spheres} \\ n = 6, & \text{For cylinders} \end{cases} \tag{7.5}$$

The Hamilton–Crosser model is suitable for two-phase mixtures with particle diameters on the order of micrometers, providing only a rough estimation of the nanofluid's thermal conductivity. Nevertheless, this model often predicts thermal conductivities that are significantly lower than those measured experimentally.

Jeffery [34] model (Eq. 7.6) included higher expansion terms in empirical correlation, assuring higher accuracy. However, this model was only valid for spherical particles.

$$\frac{k_{nf}}{k_f} = 1 + 3\left(\frac{\frac{k_s}{k_f} - 1}{\frac{k_s}{k_f} + 2}\right)\varphi + \left\{3\left(\frac{\frac{k_s}{k_f} - 1}{\frac{k_s}{k_f} + 2}\right)^2 + 0.75\left(\frac{\frac{k_s}{k_f} - 1}{\frac{k_s}{k_f} + 2}\right)^2 + \frac{9}{16}\left(\frac{\frac{k_s}{k_f} - 1}{\frac{k_s}{k_f} + 2}\right)^3\left(\frac{\frac{k_s}{k_f} + 2}{\frac{2k_s}{k_f} + 3}\right)^3 \cdots\right\}\varphi^2 \tag{7.6}$$

Davis [35] considered higher order terms in their mathematical expression to accurately mimic the pair interaction of randomly dispersed spheres (Eq. 7.7).

$$\frac{k_{nf}}{k_f} = 1 + \frac{3\left(\frac{k_s}{k_f} - 1\right)\varphi}{\left(\frac{k_s}{k_f} + 2\right) - \left(\frac{k_s}{k_f} - 1\right)\varphi}\left(\varphi - f\left(\frac{k_s}{k_f}\right)\varphi^2 + O(\varphi^3)\right) \tag{7.7}$$

One of the early seminal models developed to describe the enhanced effective thermal conductivity of nanofluids was proposed by Yu and Choi [36]. Their theory introduced the concept of a "nanolayer" of immobilized liquid surrounding the nanoparticle surface because of viscosity effects. This model is mathematically expressed in Eq. (7.9).

$$k_{nf} = \frac{k_{pe} + 2k_f + 2(k_{pe} - k_f)(1+\beta)^3 \varphi}{k_{pe} + 2k_f - (k_{pe} - k_f)(1+\beta)^3 \varphi} k_f,$$
$$\text{where } k_{pe} = \frac{\left(2(1-\varsigma) + (1+\beta)^3(1+2\varsigma)\right)\varsigma}{-(1-\varsigma) + (1+\beta)^3(1+2\varsigma)} k_s; \qquad \varsigma = \frac{k_l}{k_s} \text{ and } \beta = \frac{t}{r_s} \tag{7.8}$$

In this model, k_{pe}, k_l, and k_s represent the thermal conductivities of the equivalent nanoparticle, nanolayer, and solid particle, respectively, whereas t and r_s denote the thickness of the nanolayer and the radius of the nanoparticle, respectively. This early work provided an initial framework to incorporate interfacial transport effects from the solid–liquid interaction region. However, it did not consider particle shape anisotropy or Brownian motion impacts, which are limitations that subsequent refined models have aimed to address.

Later et al. [37] proposed a two-component model that considered both static and dynamic contributions to the enhanced thermal conductivity of nanofluids (Eq. 7.9).

$$k_{nf} = k_{static} + k_{Brownian} \tag{7.9}$$

The static part accounted for conductivity increases because of the higher conductivity of nanoparticles via the classical Maxwell effective medium theory. Meanwhile the Brownian motion term accounted for the random impacts of nanoparticles on liquid molecules (Eqs. 7.10 and 7.11). This model improved upon past work by including stochastic motion effects. Later studies found that it overpredicted some experimental data because of limitations that ignored interfacial resistance. Nonetheless, it provided some of the first considerations of particle Brownian motion impacts.

$$k_{Brownian} = 5 \times 10^4 \beta \rho_f \phi c_{p,f} \sqrt{\frac{k_B T}{\rho_s d_p}} f \tag{7.10}$$

$$\text{where } f = (-134.63 + 1722.3\phi) + (0.4705 - 6.04\phi)T \tag{7.11}$$

Xie et al. [65] explored the influence of the interfacial nanolayer on the increment in the thermal conductivity of nanofluids. They modified Yu and Choi [36] model and assumed thermal conductivity changes linearly across the spherical pellets in the radial direction. The expression of the model is given in Eq. (7.12).

$$\frac{k_{nf} - k_f}{k_f} = 3\Theta\varphi_T + \frac{3\Theta^2\varphi_T^2}{1 - \Theta\varphi_T^2}, \quad \text{where} \quad \Theta = \frac{\beta_{lf}\left[(1+\gamma)^3 - \beta_{sf}/\beta_{fl}\right]'}{(1+\gamma)^3 + 2\beta_{lf}\beta_{sl}}$$
$$\beta_{lf} = \frac{k_l + k_f}{k_l + 2k_f}; \quad \beta_{sl} = \frac{k_s + k_l}{k_s + 2k_l}; \quad \beta_{fl} = \frac{k_f + k_l}{k_f + 2k_l}, \quad k_l = \frac{k_f M^2}{(M-\gamma)\ln(1+M) + \gamma M} \tag{7.12}$$

$$M = (k_s/k_f)(1+\gamma) - 1 \quad \text{Total volume fraction, } \varphi_T = \varphi(1+\gamma)^3, \text{ and } \gamma = \frac{t}{r_s} = \frac{\text{thickness of layer}}{\text{radius of particle}}$$

Xue and Xu [66] model assumed that the liquid layer with specific thermal conductivity surrounds the nanoparticle, as shown in Eq. (7.13).

$$\left(1-\frac{\varphi}{\alpha}\right)\frac{k_{nf}-k_f}{2k_{nf}+k_f}+\frac{\varphi(k_{nf}-k_l)(2k_l+k_s)-\alpha(k_s-k_l)(2k_l+k_{nf})}{\alpha(2k_{nf}+k_l)(2k_l+k_s)+2\alpha(k_s-k_l)(k_l-k_{nf})}k_f,$$
$$\text{where } \alpha=\left(\frac{r_s}{r_s+t}\right)^3 \tag{7.13}$$

The effective thermal conductivity (k_{nf}) is determined by Bruggeman's effective media theory.

Empirical correlation (Eq. 7.14) comprising the effect of Brownian motion on nanofluid effective thermal conductivity was presented by Jang and Choi [38].

$$k_{nf}=k_s(1-\varphi)+k_s\varphi+3C_1\frac{d_f}{d_s}k_f\mathrm{Re}_{d_s}\,\mathrm{Pr}\,\varphi \tag{7.14}$$

where d_f is the diameter of a molecule of a base fluid and suspended particle diameter, C_1 is the empirical constant. The Reynolds number (Re_d) based on the particle diameter is expressed as follows:

$$\mathrm{Re}_{d_s}=\frac{\overline{u_s}d_s}{\nu_f},\quad with,\ \overrightarrow{u_s}=\frac{D}{\lambda_f}=\frac{\kappa_{Boltzmann}T}{3\pi\mu_f d_s},\quad \kappa_{Boltzmann}=1.3807\times10^{-23}\ J/K$$

where D is the nanoparticle diffusion coefficient, λ_f is the base fluid molecular mean free path and $\overrightarrow{u_s}$ is the root mean square velocity of suspended particles.

Yang [39] developed an empirical correlation (Eq. 7.15) for thermal conductivity by considering the approximation of kinetic theory under relaxation time along with the Brownian motion.

$$\frac{k_{nf}}{k_f}=\frac{k_s+2k_f+2(k_s-k_f)\varphi}{k_s+2k_f-2(k_s-k_f)\varphi}k_f+157.5\varphi C_{pf}u_p^2\tau,\quad \text{where } u_p=\sqrt{\frac{3k_BT}{m_s}} \tag{7.15}$$

where C_{pf}, τ, u_p, and m_s are the heat capacity of the base fluid, relaxation time, Brownian velocity, and mass of the nanoparticles, respectively.

Sitprasert et al. [67] considered the temperature and nanolayer effects on effective thermal conductivity and expressed the relation as shown in Eq. (7.16).

$$k_{nf}=\frac{(k_s-k_l)\varphi k_l\left[2\beta_1^3-\beta^3+1\right]+(k_s+2k_l)\beta_1^3\left[\varphi\beta^3(k_l-k_f)+k_f\right]}{\beta_1^3(k_s+2k_l)-(k_s-k_l)\varphi\left[\beta_1^3+\beta^3-1\right]},$$
$$\text{where, } \beta=1+\frac{t}{r_s};\quad \beta_1=1+\frac{t}{2r_s},$$
$$\text{thickness of nanolayer, } t=0.01(T-273)r_s^{0.35} \tag{7.16}$$
$$\text{thermal conductivity of nanolayer, } k_l=C(t/r_s)k_f;\quad C=\begin{cases}30 \text{ for } Al_2O_3\\ 110 \text{ for CuO}\end{cases}$$

Corcione [64] presented the relation for the determination of the thermal conductivity of nanofluids (Eq. 7.17).

$$\frac{k_{nf}}{k_f}=1+4.4Re^{0.44}Pr^{0.66}\left(\frac{T}{T_{fp}}\right)^{10}\left(\frac{k_s}{k_f}\right)^{0.33}\phi^{0.66}$$
$$Re=\frac{\rho_f v_{Br} d_s}{\mu_f} \tag{7.17}$$

where T, T_{fp}, k_s, and k_f are the temperature of the nanofluid, freezing point temperature of the base fluid, and thermal conductivities of suspended nanoparticles and base fluid, respectively. This model was obtained by the application of regression analysis. The author reported a standard deviation of about 1.86%.

Sarode et al. [31] experimentally determined the thermal conductivity (k) of GO–CuO nanocomposite-based nanofluid as a function of temperature (T) and volume fraction (ϕ) of nanoparticles using Eq. (7.18).

$$k = 0.056T + 1415.69\phi - 1.252 \tag{7.18}$$

As detailed above, various models have been proposed to estimate the effective thermal conductivity of nanofluid. None of the single models, however, can predict the thermal conductivity for all combinations of base fluid and nanoparticles. These models show the functional dependence of effective thermal conductivity (k_{eff}) on volume fraction, thermal conductivities of base fluid and nanoparticles, nanofluid temperature, Brownian motion, etc. The classical models fail to predict the conductivity of nanofluids (k_{nf}) accurately. The thermal conductivity enhancement is influenced by the temperature of the nanofluid. Empirical models for obtaining k_{eff} of nanofluids are based on Brownian motion and kinetic theory that can effortlessly be affected by variation in temperature. The effective thermal conductivity of nanofluids augments with temperature, therefore making them a promising choice as an efficient heat transfer medium. Both viscosity and thermal conductivity show a strong dependence on the volume fraction of nanoparticles. Hence the significance of the nanoparticle volume fraction has been illustrated in the subsequent section.

7.4 Role of volume fraction of nanoparticles on properties of nanofluids

It is apparent from the ongoing discussion that the essential properties of nanofluids, namely, viscosity and thermal conductivity, show a strong dependence on the volume fraction of nanoparticles. The addition of nanoparticles to the base fluid is found to increase nanofluid viscosity [40,41]. Effective thermal conductivity of nanofluid strongly depends on the nanoparticle volume fraction and, to some extent, on the temperature. Therefore, in this section, the effect of the volume fraction of nanoparticles on the effective thermal conductivity has been highlighted, and the relevant works are tabulated in Table 7.2.

As Al_2O_3 is the most commonly used nanofluid, Ozerinc et al. [49] compared the effect of the volume fraction of Al_2O_3 + water nanofluids on k/k_{eff} (Fig. 7.1) by considering different experimental studies [71–80], and different classical empirical correlations [33,36,37,71–67,75–83] at room temperature. The average particle diameter for experimental and analytical models was taken as 40 nm, as in the experiments, the sizes were close to that value. However, nanoparticle size varies in the study of Wen and Ding[73], with an average value of 41.5 nm. As shown in the figure, there is a significant discrepancy between the experimental data because of the pH of the nanofluid, severity of clustering, and method of production. It should also be noted that all of the experimental results presented here show a nearly linear variation of thermal conductivity with particle volume fraction, and theoretical models are in agreement with experimental results in that aspect.

Table 7.2 Studies of the effect of volume fraction on thermal conductivity of nanofluid.

Sl. no.	Source	Nanofluid	Remarks
1	Eastman et al. [42]	Cu + ethylene glycol	Rise of about 40% in k_{eff} with an increase of 0.3 vol.% of Cu nanoparticles
2	Murshed et al. [43]	TiO_2 + deionized water	Result showed $k_{eff} \propto \phi$ an enhancement of k up to 33% with maximum 5% volume fraction of TiO_2
3	Zhu et al. [44]	Fe_3O_4 + water	Ferrous nanofluids have k_{eff} than other oxide nanofluids for the same volume fraction
4	Liu et al. [45]	Cu + ethylene glycol, water, and synthetic engine oil	A low concentration of Cu in base fluid had higher k_{eff}; 0.3 vol.% of Cu nanoparticles resulted in 48% increase in k_{eff}
5	Jana et al. [46]	CNTs, Cu, Ag, and their hybrids with water	Increase in k_{eff} by 70 vol.% of Cu addition into water
6	Cabaleiro et al. [8]	ZnO–ethylene glycol + water (50%/50% by volume)	k_{eff} enhancements by 8.3% for volume fraction for the 5 wt.%
7	Esfe et al. [47]	CNTs–Al_2O_3 hybrid with water	Particle volume fraction plays a significant role in the high values of thermal conductivity (rather than temperature)
8	Nurdin and Satriananda [48]	Magnetite + water	Enhancement in k_{eff} by 42.5% at a particle volume fraction of 2.5 vol.% and temperature of 60°C
9	Sarode et al. [31]	Graphene oxide– and copper oxide-based nanofluids	Volume fraction: 0.03% of the nanocomposite Estimated thermal conductivity was found to be 1.046 W/m K at 33°C

CNTs, *carbon nanotube.*

On the other hand, Fig. 7.2 depicts the effect of the volume fraction of graphene-based nanofluids (with deionized water and ethylene glycol) on the effective thermal conductivity [50] with different volume fractions ranging from 0.005% to 0.056%. There is an increase in thermal conductivity with increasing volume fraction in a nonlinear manner. The enhancement of thermal conductivity for water-based nanofluid at 25°C for a volume fraction of 0.056% is about 14%, whereas, at 50°C, the enhancement is about 64%. This is a considerably good enhancement for such a small volume fraction, probably because of the extremely high surface area of grapheme sheets. However, there is a small

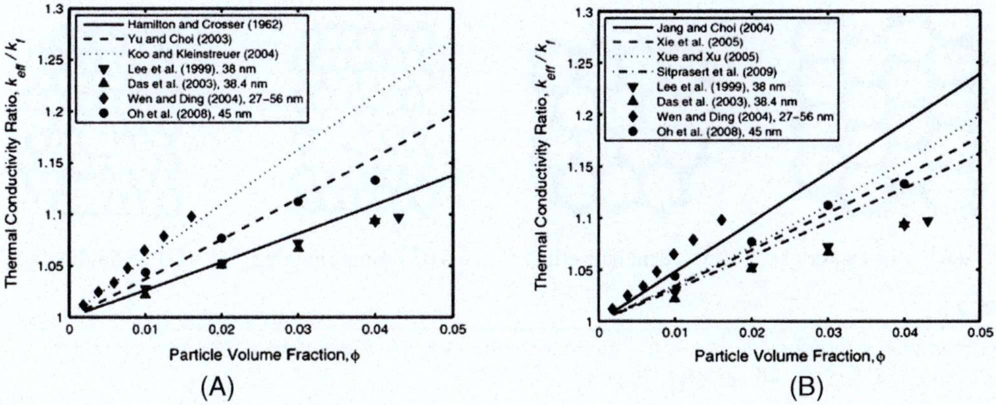

FIGURE 7.1

Comparing the effect of particle volume fraction on k/k_{eff} shown by (A) experimental studies and (B) classical empirical correlations [49].

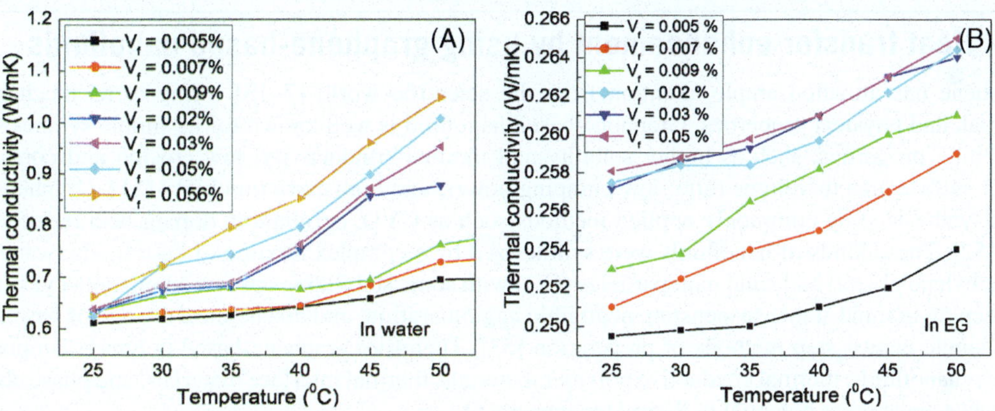

FIGURE 7.2

Effect of temperature and volume fraction on the effective thermal conductivity of graphene-based nanofluid with (A) water and (B) ethylene glycol as a base fluid [50].

enhancement in thermal conductivity of 0.05% volume fraction, which was around 3% at 25°C and 7% at 50°C because of the high viscosity of EG, which in turn affects the thermal conductivity of nanofluids. To check the credibility of the measurements the thermal conductivity of the nanofluids was also measured using another instrument (Lambda), and the variations in the values were less than 2%%[50]. It can also be shown in Fig. 7.2 that the volume fraction has a less significant effect on the thermal conductivity of fluid at a lower temperature. As the temperature increases the effectiveness of volume fraction on thermal conductivity increases.

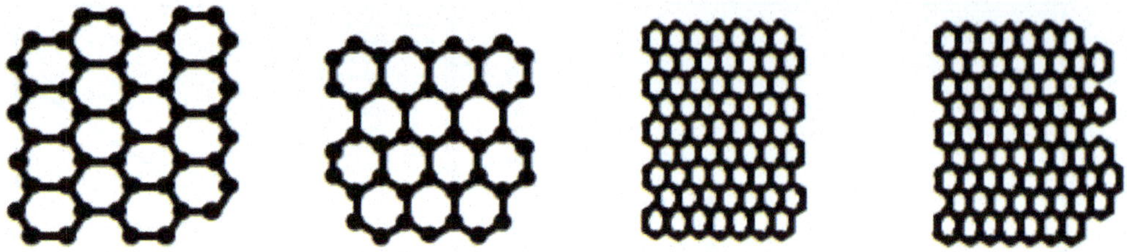

(A) Zig-zag edge (B) Armchair edge (C) Smooth edge (D) Rough edge

FIGURE 7.3

Different graphene morphology affects the thermal conductivity, (A) Zig-zag edge, (B) Armchair edge, (C) Smooth edge, and (D) Rough edge. [54].

Experiments [71–81] have shown an almost linear dependence of thermal conductivity on particle volume fraction and show excellent agreement with those predicted by classical correlations [33,36,37,71–67,76,83].

7.5 Heat transfer enhancement by using graphene-based nanofluids

Graphene has attracted ample attention from the scientific world [7–13] owing to its mechanical, thermal, and physical properties. Graphene-based nanofluid is well known for its simple synthesis and exhibiting properties, such as higher stability and thermal conductivity; low erosion and corrosion; larger surface area to volume ratio; low pumping power; and high carrier mobility [51]. Graphene has been synthesized by commonly applied methods such as CVD, exfoliation, doping, and reduction of GO [52]. These kinds of nanofluids were synthesized by a complex mixture of base fluids (water, oil, and ethylene glycol) and solid nanoparticles (GO) with some stabilizing agent (sodium dodecyl sulfate) by keeping in mind stable suspension, negligible agglomeration, and no chemical change of base fluids with single or two-step methods of preparation [53]. Therefore graphene-based polymer composites, such as nanofluids, thermal conductive polymer materials, thermal interface materials, and phase change materials, have great potential in thermal management.

7.5.1 Thermal conductivity of graphene

Theoretical and experimental predictions have shown a very high thermal conductivity of graphene, ranging from 3000 to 7000 W/m K at 300 K [54]. The thermal conductivity of the graphene is observed to be a function of temperature and morphological behavior (i.e., size, edge, and defects).

Fig. 7.3 shows different morphological patterns of graphene [54]. Significant dependence of edge shapes (Fig. 7.3) on the thermal conductivity of graphene nanoribbons (GNRs) is shown by Guo et al. [55]. However, the edge conditions, defects, and hydrogen functionalization will affect the thermal transport properties of graphene based on theoretical calculations. Moreover, the thermal conductivity of armchair GNRs is monotonously augmented with width. On the other hand the thermal conductivity of zigzag GNRs first enhanced and later declined with the increasing width. By molecular dynamics simulations, researchers found that the highest thermal conductivity occurred in the zigzag direction and the lowest in armchair GNRs. In addition, narrow GNRs with rough edges must show the smallest and most anisotropic thermal conductivity. Guo et al. [55] have analyzed that the effect of edge shapes on the thermal conductivity of GNRs is because of its variety of width, length, and strain.

Similarly the effect of smooth and rough edges of GNRs resulted in the highest thermal conductivity in a smooth edge, irrespective of whether it was a zigzag or armchair edge. The role of functionalized hydrogen on the heat transfer properties of graphene because of nonequilibrium molecular dynamics simulation. The thermal conductivity of the GNRs (zigzag edge) was reduced by about 40% because of the hydrogen coverage of only 2.5% of the total number of carbon atoms. The hydrogen termination can also be attributed to phonon scattering from the edge. The thermal conductivity of graphene is mainly affected by the number of atomic layers, length, temperature, and isotopic modification in experimental results. The determination of the accurate value of the thermal conductivity of graphene is still considered a significant challenge. Conventional thermal conductivity measurement methods are no longer suitable because of scale effect edges. However, some of the novel methods proven to be accurate enough are listed below. Even these methods have their application range [54]; for instance, (1) *Raman optothermal technique* is not suitable under polluted conditions because of the loss of optimal absorption, (2) the *steady-state self-heating method* is useful to evaluate the relationship between thermal conductivity and temperature, and (3) *heat spreader method* is suitable for the estimation of thermal conductivity of graphene attached on the substrate.

7.5.2 Nusselt number correlations for graphene-based nanofluids

Nusselt number (Nu), the dimensionless form of convective heat transfer coefficient (h), is defined as $Nu = hL/k$, where L and K are the characteristics of length and thermal conductivity of the fluid. Various attempts were made to determine the enhancement in the heat transfer by using nanofluids. Recently, Nu correlations for the graphene-based nanofluids were obtained, which are summarized in Table 7.3.

7.5.3 Frictional pressure losses for graphene-based nanofluids

The frictional pressure loss of a nanofluid can be defined as the energy lost in fluid transportation through a pipe because of friction between the pipe and wall. This fictional property of fluid in conduit depends solely upon the type of fluid and flow regime. Different flow equations have been used to calculate pressure losses inside the casing and in the annuli. Based on the Fanning equation, the gradient of frictional pressure drop in a conduit is expressed using Eq. (7.19).

$$\frac{dp_f}{dL} = \frac{f \rho_f \overline{v^2}}{25.8d} \tag{7.19}$$

where ρ_f is frictional pressure in kPa, L is pipe length, f is fanning factor, $\overline{v}$is the average velocity, and d is pipe inner diameter.

For laminar flow,

$$f = \frac{16}{N_{Re}} \tag{7.20}$$

Table 7.3 Nusselt number correlations for graphene-based nanofluids.

Source	Nusselt number (*Nu*)	Remarks
Baby and Ramaprabhu [68]	$Nu = 1.95(\text{Re.Pr D/X})^{0.33}$ $(Re.Pr\ D/X) \geq 33.3$ $4.364 + 0.0722 Re.Pr.\ D/X$ $(Re.Pr\ D/X) < 33.3$	HEG–water, HEG–EG, 16% at 25°C and 75% at 50°C, Re = 10,000–16,000, 69% for 0.005%, and 183% for 0.01%
Sadeghinezhad et al. [56]	$Nu = \frac{\left(\frac{f}{8}\right)(Re-1000)Pr}{1+12.7\left(\frac{f}{8}\right)^{0.5}(Pr^{0.6}-1)}$ $0.5 < Pr < 2000,\ 3000 < Re < 5 \times 10^6$ $Nu = 0.023\ \text{Re}^{0.8} Pr^{0.4}$ $Re > 10^4,\ 0.6 < Pr < 200$ $Nu = \frac{\left(\frac{f}{8}\right)(Re.Pr)}{1.07+12.7\left(\frac{f}{8}\right)^{0.5}(Pr^{0.6}-1)}$ $0.5 < \text{Pr} < 2000,\ 3000 < \text{Re} < 5 \times 10^6$	75%, 79%, and 83% increase of heat flux with velocity of 0.5–1.5 m/s, $\phi = 0.025\%–0.1\%$
Ghozatioo et al. [57]	$Nu = 4.364 + 0.283(x^*/2)^{-0.506}$ $\exp(-41x^*/2);\ x^* = 2(x/D)/(Re.Pr);\ x^* > 0.003$	Water–graphene 0%–0.1%, 25°C–40°C, 25°C increased by 15.0%, 29.2, and 12.6 at 0.05, 0.075, and 0.1 wt.%
Yarmand et al. [58]	$Nu = 0.0017066\ Re^{0.9253} Pr^{-1.29001}$	GNP–Ag/water nanofluids, $Re = 5000–17{,}500$, $\phi = 0\%–0.1\%$; compared with distilled water, a maximum of 32.7% enhancement in Nu for $\phi = 0.1\%$ at $Re = 17{,}500$
Bose et all. [59]	$Nu = 1.953(Re.Pr\ D/X)^{0.33}$ $(Re.Pr\ D/X) \geq 33.3$ $4.364 + 0.0722 Re.Pr.\ D/X$ $(Re.Pr\ D/X) < 33.3$	Graphene-water, $\phi = 0.2\%–0.6\%$, $Re = 500–2000$, 56% enhancement in heat transfer at 0.6%
Barai et al. [60]	$Nu = 5.474\ Re^{0.675} Pr^{-1.216} \phi^{0.524}$	rGO–Fe_3O_4 nanocomposite-based nanofluid
Montazer et al. [61]	$Nu = 1.16\ Re^{0.55} Pr^{-0.312}(1+\phi)^{-0.25}$	F-GNP, SiO_2 and ZnO water-based nanofluids; $Re = 4000–16{,}000$, $\phi = 0\%–0.1\%$; 33.7% and 16.7% heat transfer performance improvement for F-GNP and MWCNT nanofluids, respectively, at $\phi = 0.1$ wt.%$\phi = 0\%–0.03\%$
Sarode et al. [31]	$Nu = 8.1612\ Re^{0.593} Pr^{-0.743} \phi^{0.204}$ $Nu = 5.474\ Re^{0.675} Pr^{-1.216} \phi^{0.524}$	GO–CuO nanocomposite-based nanofluid: $Re = 1000–8000$, 18% enhancement in comparison with water

F-GNP, *functionalized graphene nanoplatelets;* GO, *graphene oxide;* rGO, *redcued graphene oxide;* EG, **ethylene glycol;** ***MWNT, multiwalled carbon nanotubes; HEG, hydrogen exfoliated graphene.***

Table 7.4 Frictional factor for various flows in a cross-section.

Source	Frictional factor (f)	Remarks
Colebrook [69]	$\frac{1}{\sqrt{f}} = -4\log\left(0.269\frac{\delta}{d} + \frac{1.255}{N_{Re}\sqrt{f}}\right)$ $N_{Re} = \frac{928\rho_f \bar{v} d}{\mu}$	Fully developed turbulent flow in a circular pipe
Colebrook [69]	$\frac{1}{\sqrt{f}} = 4\log\left(N_{\mathrm{Re}}\sqrt{f}\right) - 0.395$ $f = \frac{0.0791}{N_{\mathrm{Re}}^{0.25}}$	Turbulent flow in a smooth pipe
Gregory and Fogarsi [70]	$f =$ $\left(-4\log\left\{\frac{\epsilon}{3.7065} - \frac{5.0452}{N_{Re}}\log\left[\frac{\epsilon^{1.1098}}{2.8257} + \left(\frac{7.149}{N_{Re}}\right)^{0.8981}\right]\right\}\right)^{-2}$	Open flow in annulus section

However, for turbulent flow, the friction factor depends on both the Reynolds number and the relative roughness of the conduit wall. Various empirical correlations have been developed to determine the frictional factor for nanofluids, as summarized in Table 7.4.

7.6 Applications of nanofluids

Nanofluids are primarily utilized for their enhanced thermal properties, making them effective coolants in heat transfer equipment such as heat exchangers, electronic cooling systems, and radiators. Numerous studies have analyzed heat transfer over a flat plate using nanofluids. In addition to thermal applications, nanofluids are valuable for their controlled optical properties. For example, graphene-based nanofluids have been shown to improve the efficiency of polymerase chain reactions. Another notable application is in solar collectors, where nanofluids' tunable optical properties enhance performance. Furthermore, nanofluids have been investigated for their potential to improve thermal desalination technologies by altering thermal conductivity and enhancing sunlight absorption, though surface fouling remains a significant challenge. A schematic representation of the various applications of nanofluids is illustrated in Fig. 7.4.

7.7 Conclusion

Nanofluids exhibit significantly enhanced thermophysical properties because of the addition of nanoparticles with high thermal conductivity to base fluids. Key factors influencing their convective heat transfer performance include viscosity, thermal conductivity, and nanoparticle volume fraction. Although several empirical correlations have been developed to model these properties, no single model accurately predicts behavior across all parameter ranges. Recent work has focused on developing multimechanism models incorporating interfacial engineering effects. For viscosity the most widely applied equations are from Brinkman [20] and Batchelor [23]. The Maxwell [32] and Hamilton–Crosser [33] models also remain commonly used for thermal conductivity predictions. However, evolving understandings of Brownian motion, particle shape anisotropy, and interface layer impacts are driving refinement. Graphene nanofluids show particular promise, given graphene's outstanding properties.

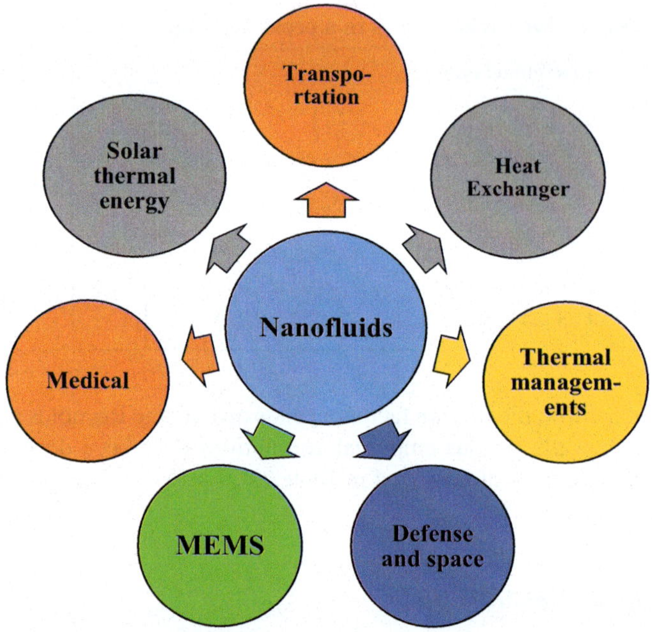

FIGURE 7.4

Applications of nanofluids.

Potential applications in heat transfer equipment, solar technologies, processors, and phase change material composites could benefit from graphene's ability to enhance transport even at low loadings (<1 vol.%). Issues impacting stability under high heat fluxes require further investigation. Overall, ongoing advancements in synthesis methods, modeling techniques, and functionalization strategies hold potential to unlock new capabilities for nanofluid-enabled thermal management.

References

[1] S.K. Das, S.U.S. Choi, W. Yu, T. Pradeep, Nanofluids: Science and Technology, John Wiley & Sons, Inc, 2007.

[2] J. Sarkar, A critical review on convective heat transfer correlations of nanofluids, Renewable Sustainable Energy Rev. 15 (6) (2011) 3271–3277, doi:10.1016/j.rser.2011.04.025.

[3] R. Walvekar, I.A. Faris, M. Khalid, Thermal conductivity of carbon nanotube – experimental and numerical study, Heat Transfer 41 (2) (2012) 145–163.

[4] R. Walvekar, M.K. Siddiqui, S.S. Ong, A.F. Ismail, Application of CNT nanofluids in a turbulent flow heat exchanger, J. Exp. Nanosci. 11 (1) (2016) 1–17.

[5] J.A. Eastman, S.R. Phillpot, S.U.S. Choi., P. Keblinski, Thermal transport in nanofluids, Annu. Rev. Mater. Res. 34 (2004) 219–246.

[6] P. Keblinski, J.A. Eastman, D.G. Cahill, Nanofluids for thermal transport, Mater. Today 8 (6) (2005) 36–44.
[7] M. Azizi-Lalabadi, S.M. Jafari, Bio-nanocomposites of graphene with biopolymers: fabrication, properties, and applications, Adv. Colloid Interface Sci. 292 (2021) 102416.
[8] D. Cabaleiro, L. Colla, F. Agresti, L. Lugo, L. Fedele, Transport properties and heat transfer coefficients of ZnO/(ethylene glycol + water) nanofluids, Int. J. Heat Mass Transfer 89 (2015) 433–443.
[9] D. Cabaleiro, L. Colla, S. Barison, L. Lugo, L. Fedele, S. Bobbo, Heat transfer capability of (ethylene glycol + water)-based nanofluids containing graphene nanoplatelets: design and thermophysical profile, Nanoscale Res. Lett. 12 (2017) 53.
[10] S. Hamze, N. Berrada, D. Cabaleiro, A. Desforges, J. Ghanbaja, J. Gleize, D. Begin, F. Michaux, T. Mare, B. Vigolo, P. Estelle, Few-layer graphene-based nanofluids with enhanced thermal conductivity, Nanomaterials 10 (2020) 1258.
[11] W. Li, M. Mao, N. Hu, J. Wang, J. Huang, W. Zhang, S. Gu, A graphene oxide-cooper nanocomposite for the regeneration of the dentin-pulp complex: an odontogenic and neurovascularization-inducting material, Chem. Eng. Sci. 417 (2021) 129299.
[12] M.-Y. Xia, Y. Xie, C.-H. Yu, G.-Y. Chen, Y.-H. Li, T. Zhang, Q. Peng, Graphene based nanomaterials: the promising active agents for antibiotics-independent antibacterial applications, J. Control. Release 307 (2019) 16–31.
[13] H. Zhao, R. Ding, X. Zhao, Y. Li, L. Qu, H. Pei, L. Yildirimer, Z. Wu, W. Zhang, Graphene-based nanomaterials for drug and/or gene delivery, bioimaging, and tissue engineering, Drug Discov. Today 22 (9) (2017) 1302–1317.
[14] W. Yu, H. Xie, X. Wang, X. Wang, Significant thermal conductivity enhancement for nanofluids containing graphene nanosheets, Phys. Lett. A 375 (2011) 1323–1328.
[15] B. Brodie, Sur le poids atomique du graphite, Ann. Chim. Phys. 59 (1860) 466–472.
[16] M. Hadadian, E.K. Goharshadi, A. Youssefi, Electrical conductivity, thermal conductivity, and rheological properties of graphene oxide-based nanofluids, J. Nanopart. Res. 16 (2788) (2014) 1–17.
[17] T.L. Ba, O. Mahian, S. Wongwises, I.M. Szilágyi, Review on the recent progress in the preparation and stability of graphenebased nanofluids, J. Therm. Anal. Calorim. 142 (2020) 1145–1172.
[18] F. Selimefendigil, H.F. Oztop, Mixed convection of nanofluids in a three dimensional cavity with two adiabatic inner rotating cylinders, Int. J. Heat Mass Transfer 117 (2018) 331–343.
[19] A. Einstein, Eine neue Bestimmung der Molekul-dimensionen, Ann. Phys. 19 (1906) 289–306.
[20] H.C. Brinkman, The viscosity of concentrated suspensions and solutions, J. Chem. Phys. 20 (4) (1952) 571–581.
[21] I.M. Krieger, T.J. Dougherty, A mechanism for non-Newtonian flow insuspensions of rigid spheres, Transactions of Society of Rheology 3 (1) (1959) 137–152.
[22] N. Frankel, A. Acrivos, On the viscosity of a concentrated suspension of solid spheres, Chem. Eng. Sci. 22 (6) (1967) 847–853.
[23] G.K. Batchelor, The effect of Brownian motion on the bulk stress in a suspension of spherical particles, J. Fluid Mech. 83 (1) (1977) 97.
[24] A.L. Graham, On the viscosity of suspensions of solid spheres, Appl. Sci. Res. 37 (3) (1981) 275–286.
[25] T. Kitano, T. Kataoka, T. Shirota, An empirical equation of the relative viscosity of polymer melts filled with various inorganic fillers, Rheol. Acta 20 (2) (1981) 207–209.
[26] X. Wang, X. Xu, S.U.S. Choi, Thermal conductivity of nanoparticle – fluid mixture, J. Thermophys. Heat Transfer 13 (4) (1999) 474–480.
[27] J. Avsec, M. Oblak, The calculation of thermal conductivity, viscosity and thermodynamic properties for nanofluids on the basis of statistical nanomechanics, Int. J. Heat Mass Transf. 50 (21–22) (2007) 4331–4341.
[28] H. Chen, Y. Ding, Y. He, C. Tan, Rheological behaviour of ethylene glycol basedtitania nanofluids, Chem. Phys. Lett. 444 (4–6) (2007) 333–337.

[29] N. Masoumi, N. Sohrabi, A. Behzadmehr, A new model for calculating theeffective viscosity of nanofluids, J. Phys. D 42 (5) (2009) 55501.
[30] S. Masoud Hosseini, A.R. Moghadassi, D.E. Henneke, A new dimensionless group model for determining the viscosity of nanofluids, J. Therm. Anal. Calorim. 100 (3) (2010) 873–877.
[31] H.A. Sarode, D.P. Barai, B.A. Bhanvase, R.P. Ugwekar, V. Saharan, Investigation on preparation of graphene oxide-CuO nanocomposite based nanofluids with the aid of ultrasound assisted method for intensified heat transfer properties, Mater. Chem. Phys. 251 (2020) 123102.
[32] J.C. Maxwell, A Treatise on Electricity and Magnetism, Oxford Clarendon Press, Oxford, 1873.
[33] R.L. Hamilton, O.K. Crosser, Thermal conductivity of heterogeneous two component systems, Ind. Eng. Chem. Fundam. 1 (3) (1962) 187–191.
[34] D.J. Jeffrey, Conduction through a random suspension of spheres, Proc. R. Soc. A 335 (1973) 355–367.
[35] R.H. Davis, The effective thermal conductivity of a composite material with spherical inclusions, Int. J. Thermophys. 7 (1986) 609–620.
[36] W. Yu, S. Choi, The role of interfacial layers in the enhanced thermal conductivity of nanofluids: a renovated Maxwell model, J. Nanopart. Res. 5 (2003) 167–171.
[37] J. Koo, C. Kleinstreuer, A new thermal conductivity model for nanofluids, J. Nanopart. Res. 6 (6) (2004) 577–588.
[38] S.P. Jang, S.U.S. Choi, Effects of various parameters on nanofluid thermal conductivity, J. Heat Transfer 129 (2007) 617–623.
[39] B. Yang, Thermal conductivity equations based on Brownian motion in suspensions of nanoparticles (nanofluids), J. Heat Transfer 130 (2008) 042408.
[40] K. Madhusree, T.K. Dey, Viscosity of alumina nanoparticles dispersed in car engine coolant, Exp. Therm Fluid Sci. 34 (2010) 677–683.
[41] P.C. Mishra, S. Mukherjee, S.K. Nayak, A. Panda, A brief review on viscosity of nanofluids, Int. Nano Lett. 4 (2014) 109–120.
[42] J.A. Eastman, S.U.S. Choi, S. Li, W. Yu, L.J. Thompson, Anomalously increased effective thermal conductivities of ethylene glycol-based nanofluids containing copper nanoparticles, Appl. Phys. Lett. 78 (6) (2001) 718–720.
[43] S.M.S. Murshed, K.C. Leong, C. Yang, Enhanced thermal conductivity of TiO_2–water based nanofluids, Int. J. Therm. Sci. 44 (4) (2005) 367–373.
[44] H.T. Zhu, C.Y. Zhang, S.Q. Liu, Y.M. Tang, Y.S. Yin, Effects of nanoparticle clustering and alignment on thermal conductivities of Fe_3O_4 aqueous nanofluid, Appl. Phys. Lett. 89 (2) (2006) 023123.
[45] M.S. Liu, M.C.C. Lin, C.Y. Tsai, C.C. Wang, Enhancement of thermal conductivity with Cu for nanofluids using chemical reduction method, Int. J. Heat Mass Transfer 49 (17–18) (2006) 3028–3033.
[46] S. Jana, A. Salehi-Khojin, W.H. Zhong, Enhancement of fluid thermal conductivity by the addition of single and hybrid nano-additives, Thermochim. Acta 462 (1–2) (2007) 45–55.
[47] M.H. Esfe, S. Saedodin, M. Biglari, H. Rostamian, Experimental investigation of thermal conductivity of CNTs-Al_2O_3/water: A statistical approach, Int. Commun. Heat Mass Transfer 69 (2015) 29–33.
[48] I. Nurdin, Satriananda, The effect of particle volume fraction and temperature on the enhancement of thermal conductivity of maghemite (γ-Fe_2O_3) water-based nanofluids, AIP Conf. Proc. 1823 (2017) 020011.
[49] S. Ozerinc, S. Kakac, G.A. Yazıcıoglu, Enhanced thermal conductivity of nanofluids: a state-of-the-art review, Microfluid. Nanofluid. 8 (2010) 145–170.
[50] T.T. Baby, S. Ramaprabhu, Investigation of thermal and electrical conductivity of graphene based nanofluids, J. Appl. Phys. 108 (124308) (2010) 1–7.
[51] E. Sadeghinezhad, M. Mehrali, R. Saidur, M. Mehrali, S.T Latibari, A.R Akhiani, H.S. Metselaar, A comprehensive review on graphene nanofluids: recent research, development and applications, Energy Convers. Manage. 111 (2016) 466–487.

[52] C.N Rao, A.K Sood, Graphene: Synthesis, Properties, and Phenomena, John Wiley & Sons, Weinheim, Germany, 2013.

[53] W. Yu, H. Xie, D. Bao, Enhanced thermal conductivities of nanofluids containing graphene oxide nanosheets, Nanotechnology 21 (5) (2009) 055705.

[54] C. Liu, M. Chen, W. Yu., Y. He, Recent advance on graphene in heat transfer enhancement of composites, ES Energy Environ. 2 (2018) 31–42.

[55] Z. Guo, D. Zhang, X.-G. Gong, Thermal conductivity of graphene nanoribbons, Appl. Phys. Lett. 95 (163103) (2009) 1–3.

[56] E. Sadeghinezhad, M. Mehrali, S. Tahan Latibari, M. Mehrali, S.N Kazi, C.S Oon, H.S Metselaar, Experimental investigation of convective heat transfer using graphene nanoplatelet based nanofluids under turbulent flow conditions, Ind. Eng. Chem. Res. 53 (31) (2014) 12455–12465.

[57] A. Ghozatloo, A. Rashidi, M. Shariaty-Niassar, Convective heat transfer enhancement of graphene nanofluids in shell and tube heat exchanger, Exp. Therm. Fluid Sci. 53 (1) (2014) 136–141.

[58] H. Yarmand, S. Gharehkhani, G. Ahmadi, S.F.S. Shirazi, S. Baradaran, E. Montazer, M.N.M. Zubir, M.S. Alehashem, S.N. Kazi, M. Dahari, Graphene nanoplatelets–silver hybrid nanofluids for enhanced heat transfer, Energy Convers. Manage. 100 (2015) 419–428.

[59] B.J Bose, L.G Asirvatham, M.N. Kumar, Experimental convective heat transfer studies on graphene nanofluid for the cooling of next generation electronic components, Int. J. Appl. Eng. Res. 12 (19) (2017) 8534–8539.

[60] D.P. Barai, B.A. Bhanvase, V.K Saharan, Reduced graphene oxide-Fe_3O_4 nanocomposite based nanofluids: study on ultrasonic assisted synthesis, thermal conductivity, rheology, and convective heat transfer, Ind. Eng. Chem. Res. 58 (2019) 8349–8369.

[61] E. Montazer, M.B. Shafii, E. Salami, M.R. Muhamad, H. Yarmand, S. Gharehkhani, Z.Z. Chowdhury, S.N. Kazi, A. Badarudin, Heat transfer in turbulent nanofluids: separation flow studies and development of novel correlation, Adv. Powder Technol. 31 (8) (2020) 3120–3133.

[62] I.M. Mahbubul, R. Saidur, M.A. Amalina, Latest developments on the viscosity of nanofluids, Int. J. Heat Mass Transf 55 (2012) 874–885, doi:10.1016/j.ijheatmasstransfer.2011.10.021.

[63] D.P. Kulkarni, D.K. Das, G.A. Chukwu, Temperature Dependent Rheological Property of Copper Oxide Nanoparticles Suspension (Nanofluid), Journal of Nanoscience and Nanotechnology 6 (4) (2006) 1150–1154, doi:10.1166/jnn.2006.187.

[64] M. Corcione, Empirical correlating equations for predicting the effective thermal conductivity and dynamic viscosity of nanofluids, Energy Conversion and Management 52 (1) (2011) 789–793, doi:10.1016/j.enconman.2010.06.072.

[65] H. Xie, M. Fujii, X. Zhang, Effect of interfacial nanolayer on the effective thermal conductivity of nanoparticle-fluid mixture, International Journal of Heat and Mass Transfer 48 (14) (2005) 2926–2932, doi:10.1016/j.ijheatmasstransfer.2004.10.040.

[66] Q. Xue, W.-M. Xu, A model of thermal conductivity of nanofluids with interfacial shells, Materials Chemistry and Physics 90 (2–3) (2005) 298–301, doi:10.1016/j.matchemphys.2004.05.029.

[67] C. Sitprasert, P. Dechaumphai, V. Juntasaro, A thermal conductivity model for nanofluids including effect of the temperature-dependent interfacial layer, Journal of Nanoparticle Research 11 (2009) 1465–1476, doi:10.1007/s11051-008-9535-4.

[68] T.T. Baby, S. Ramaprabhu, Experimental investigation of the thermal transport properties of a carbon nanohybrid dispersed nanofluid, Nanoscale 3 (2011) 2208–2214, doi:10.1039/C0NR01024C.

[69] C.F. Colebrrok, Turbulent flow in pipes with particular reference to the transition region between the smooth and rough pipe laws, Journal of the Institution of Civil Engineers 11 (4) (1939) 133–156, doi:10.1680/ijoti.1939.13150.

[70] G. A. Gregory, M. Fogarasi, Alternate to standard friction factor equation, Oil Gas Journal 83 (13) (1985) 120–127.

[71] S. Lee, S.U.-S. Choi, J. Eastman, Measuring Thermal Conductivity of Fluids Containing Oxide Nanoparticles, Journal of Heat Transfer 121 (1999) 280–289, doi:10.1115/1.2825978.

[72] S.K. Das, N. Putra, P. Thiesen, W. Roetzel, Temperature Dependence of Thermal Conductivity Enhancement for Nanofluids, Journal of Heat Transfer 125 (2003) 567–574, doi:10.1115/1.1571080.

[73] D. Wen, Y. Ding, Experimental investigation into convective heat transfer of nanofluids at the entrance region under laminar flow conditions, International Journal of Heat and Mass Transfer 47 (24) (2004) 5181–5188, doi:10.1016/j.ijheatmasstransfer.2004.07.01.

[74] D. Kim, Y. Kwon, Y. Cho, C. Li, S. Cheong, Y. Hwang, J. Lee, D. Hong, S. Moon, Convective heat transfer characteristics of nanofluids under laminar and turbulent flow condition, Current Applied Physics 9 (2) (2009) e119–e123, doi:10.1016/j.cap.2008.12.047.

[75] S. Porgar, H.F. Oztop, S. Salehfekr, A comprehensive review on thermal conductivity and viscosity of nanofluids and their application in heat exchangers, Journal of Molecular Liquids 385 (2023) 122213, doi:10.1016/j.molliq.2023.122213.

[76] S.P. Jang, S.U.S. Choi, Role of Brownian motion in the enhanced thermal conductivity of nanofluids, Applied Physics Letters 84 (21) (2004) 4316–4318, doi:10.1063/1.1756684.

[77] G. Coccia, S. Tomassetti, G. Di Nicola, Thermal conductivity of nanofluids: A review of the existing correlations and a scaled semi-empirical equation, Renewable and Sustainable Energy Reviews 151 (2021) 111573, doi:10.1016/j.rser.2021.111573.

CHAPTER

8 Two-dimensional materials in nanolubrication

Abdul Wasy Zia[a], Muhammad Asim[b], Syed Asad Hussain[c], Shahid Rasul[d] and Kashif Ishfaq[e]

[a]*Institute of Mechanical, Process and Energy Engineering (IMPEE), School of Engineering and Physical Sciences, Heriot-Watt University, Edinburgh, United Kingdom,* [b]*School of Professional Education and Executive Development, The Hong Kong Polytechnic University, Hong Kong, China,* [c]*Life Cycle Management Laboratory (LCML), Faculty of Applied Science, School of Engineering, The University of British Columbia, Vancouver, BC, Canada,* [d]*Mechanical and Construction Engineering, Northumbria University, Newcastle upon Tyne, United Kingdom,* [e]*Department of Industrial and Manufacturing Engineering, University of Engineering and Technology, Lahore, Pakistan*

8.1 Introduction

Nanolubricants are emerging for numerous industrial applications, particularly for machining and manufacturing, automotive engines, heavy machinery lubrication, and heating, venting, and air conditioning (HVAC) applications. Nanolubricants are generally made with the addition of nanoadditives, such as nanoparticles, nanosheets, nanoflakes, and nanotubes, to conventional oils, lubricants, and fluids. The superior performance of nanolubricants is governed by nanoadditives whose mechanical, thermal, and fluid properties are attributed to their dimensional aspects, large surface areas, and morphologies such as one-dimensional, two-dimensional (2D), and three-dimensional structures. Usually, these nanoadditives are metals, metal oxides, sulfides, rare-earth elements, polymers, carbon derivatives, and nanocomposites of the aforementioned categories [1]. Table 8.1 summarizes the major types of nanolubricant additives and corresponding elements or compounds used for each category.

The addition of nanoadditives differentiates nanolubricants from conventional lubricants [74] and oils, irrespective of their applications. Referring to a typical example of engine tribology, Fig. 8.1 presents the Stribeck curve, which correlates the frictional behavior of different engine components such as plain bearings, piston rings and skirt, and valve trains as a function of Sommerfeld number, or surface roughness, or film thickness across three different lubrication regimes, that is, boundary lubrication, mixed lubrication, and hydrodynamic/elasto-hydrodynamic lubrication. The friction mapping of engine components suggests that the nanolubricants significantly contribute to the boundary lubrication regime only. Piston rings and valve trains usually have large contacts with liners and piston sleeves; hence the nanoadditives help to improve the tribological performance of contact surfaces, either by rolling and sliding between surfaces [75] to boost the toughness of contact surfaces or by promoting self-repairing [76] surfaces, or through developing protective layers and tribofilms [1].

In the same way, nanolubricants are gaining a reputation in machining and manufacturing industry. The addition of nanoadditives to conventional lubricants fosters heat transfer rates [78] for rapid

Graphene and 2D Materials in Heat Transfer: Fundamentals and Applications. DOI: https://doi.org/10.1016/B978-0-12-821964-5.00014-X

Table 8.1 Common types of nanolubricant additives.

Type of nanolubricant		Elements/compounds	Further reading
Nanoadditives			
Metal	• Pristine	Ag, Au, Bi, Cu, Co, Fe, Ni, Pd, Sn	[2–11]
	• Oxides	Al_2O_3, CuO, Fe_3O_4, TiO_2, ZrO_2, ZnO,	[12–19]
	• Sulfides	WS_2, CuS, CuS_2, MoS_2, $NiMoO_2S_2$, ZnS	[20–26]
	• Hydroxides	$La(OH)_3$, LDHs	[27–30]
Carbon		Fullerenes, diamond, graphite, graphene (G), graphene oxide (GO), carbon nanotubes (SWCNT, MWCNT), carbyne	[31–39]
Polymers		PTFE, PSS, PVP	[40–43]
Rare-earth elements		LaF_3, CeO_2, $Y2O_3$, $CeBO_3$	[44–47]
Salts		$CaCO_3$, LaF_3, ZrP, $Ca_3(BO3)_2$, $Zn_3(PO_4)_2)(H_2O)_4$	[48–50]
Nanocomposites		Cu/Sio_2, Cu/GO, Cu/MoS_2, Al_2O_3/Sio_2, Serpentine/$La(OH)_3$, Al_2O_3/TiO_2, G/MoS_2, G/Ag, Cu/Go, Mn_2O_4, BN. Al_2O_3/MWCNT, FeS_2/G, Fe_2O_3@GO, Zeolite, ZrO_2/SiO_2	[51–59]
Solid nanolubricants		Graphite nanoparticles, hexa-boron-nitride Disulfides (tungsten, molybdenum, calcium) Others: polytetrafluoroethylene, boron-nitride, oxides	[60–72]

MWCNT, *multiwall carbon nanotube.*
Data from [1,73].

cooling of tools and workpieces. Likewise, these nanolubricants improve tool's life and elevate energy efficiency of machining process by reducing frictional torques between the contact surfaces of tools and workpieces. The nanolubricants are more recognized as nanofluids on viscosity differences, particularly when used for HVAC applications where the addition of nanoadditives improves Nusselt number, coefficient of performance (COP), heat transfer coefficient, and thermal conductivity across interfaces. Therefore nanolubricants or nanofluids are widely used in refrigeration systems, heat exchanges, heating pipes, and heat pumps [79].

8.2 Nanolubricant synthesis

Generally, nanolubricants are made with the addition of nanomaterials to conventional oils, aqueous solutions, and typical lubricants. Therefore the performance of nanolubricants highly depends on the physical, chemical, electrical, and thermal characteristics of the nanomaterials. Their performance,

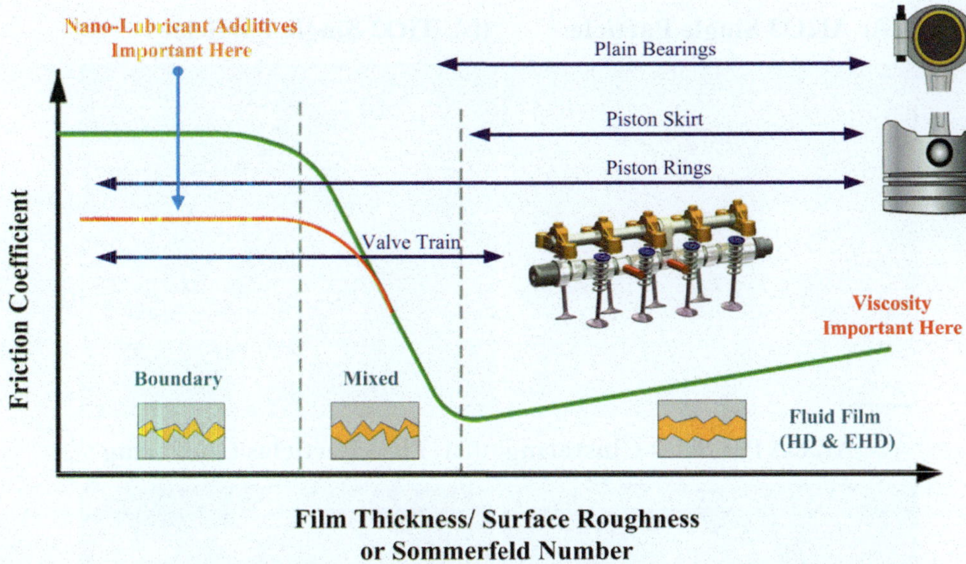

FIGURE 8.1

Mapping friction coefficient of various engine components across Stribeck curve. It can be observed that the nanoadditives' performance is significant in boundary lubrication region only.

Reproduced with permission from [77].

either for tribological setup or for thermal systems, is governed by their morphological and dimensional features and their homogeneous suspension in lubricant. Fig. 8.2 presents the transmission electron micrographs of ~10 nm sized TiO_2 and 8–12 nm sized Al_2O_3 nanoparticles used to prepare TiO_2- and Al_2O_3-based Castrol EDGE A5 5W-30 nanolubricants [80], respectively. Fig. 8.2A and B presents a high-resolution image of single Al_2O_3 and TiO_2 particles, whereas Fig. 8.2C and D presents agglomerations and clustering of 25 to 30 nm sized Al_2O_3 and TiO_2 nanoadditives used to made polyalphaolefin nanolubricants [81]. The nanoparticle morphology is well known for significantly changing the nanolubricants' behavior. The literature suggests that the increasing size of graphite particles is recognized [82] to increase the surface roughness and friction coefficients, and correspondingly, the particle distribution significantly influences tribological performance of nanolubricants [83]. Likewise, the shape and distribution of nanoparticles regulate the thermal properties as well. The literature [84] recommends a brick shape of nanoparticles over a platelet shape to enhance the heat transfer area and performance index of plate heat exchangers. Hence the availability of a large surface area per unit volume of isolated nanoparticles in comparison to clusters and agglomerations can accelerate chemical, electrical, and thermal activities, significantly improving their performance for industrial applications. The nanomaterials are produced using numerous techniques, which vary by type of material and objectives. For example, the nanographite for graphite-based nanolubricants [85] can be produced with several methods, such as exfoliation [86], ball milling [87], physical vapor deposition, chemical vapor deposition [88], reduction [89], emulsion [90], sonication [91], and detonation [92], and the corresponding nanographite produced from each method may vary in crystallography, size, shape, defects, and porosity.

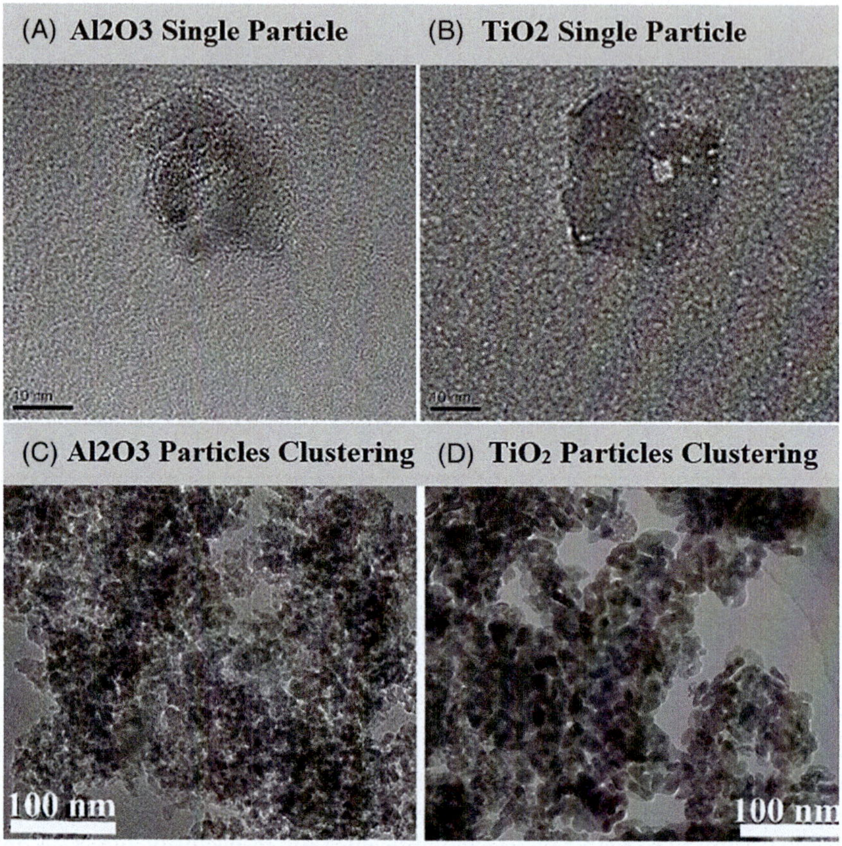

FIGURE 8.2

Transmission electron micrographs of a single (A) 10 nm sized TiO_2 and (B) 8–12 nm sized Al_2O_3 particles and clustering of 25–30 nm sized (C) Al_2O_3 and (D) TiO_2 nanoparticles.

Reproduced with permission from (A and B) [80] and (C and D) [81].

8.3 Nanolubricant properties

Nanolubricants properties depend on the characteristics of nanoadditives and, therefore, are explicitly designed for specific applications. The nanolubricants have demonstrated distinguished performance in reducing friction and wear of materials in numerous tribological systems such as automotive engines, cutting tools, bearings and machinery lubrication, and industrial machining. The nanolubricants assist in the formation of protective layers, tribofilms, transfer layers, or polishing and mending between the contact surfaces to improve the tribology of materials. The nanosized particles efficiently fill the asperities and regulate the surface roughness to govern the friction coefficient. It is observed that graphene/copper nanoadditive-based nanolubricants have significantly reduced friction coefficient and wear rate of automotive engine components by up to 30%[93] while reducing surface roughness up to 47% when compared with conventional 5W-30.

Similarly, graphite-based nanolubricants have increased 20% torque and 7% brake power [94] and have reduced friction coefficient and wear rates by 53% and 33%[95], respectively. In the same way, large surface areas of nanoadditives facilitate thermal and chemical bindings with friction modifiers or host material to improve overall system performance. Besides reducing frictional losses the nanoadditives also reduce energy losses by promoting green emissions as the vehicle efficiency [96] improves with reduced toxic emissions (such as sulfur, nitrogen, zinc, and phosphorus) when compared with a metallic catalyst for ignition engines. In this context the addition of graphite has demonstrated a ~15% decrease in fuel consumption [95] and a 5% improvement in brake thermal efficiency [94]. Nanolubricants for machining applications, typically known as nanofluids for thermal systems, significantly improve energy efficiency by increasing thermal conductivity, heat transfer rate, and COP either for HVAC systems or for industrial drilling and milling processes. Besides the thermal management in machining process the nanolubricants also assist in minimizing the surface residual stresses [97] of machined articles, which improves their fatigue life. These characteristics are attributed to the large surface area of nanoparticles per unit volume, which boost heat transfer at particles and lubricants/fluid interface. The nanosized additives elevate thermal stability because of better dispersion, and their properties can also be regulated by numerous factors such as temperature, viscosity, concentration, surface wettability, and morphological and dimensional characteristics. Particularly for thermal systems the nanoadditives minimize clogging incidents and reduce the pumping power contrary to conventional lubricants and macrofillers. The nanoadditives have demonstrated significant improvements in heat transfer rate, such as the addition of Al_2O_3 and carbon nanotubes up to 0.08% has improved the heat transfer rate of ammonia and water up to 20%[98]. Similarly, the addition of 1.5% CuO to R600a and oil has been reported [99] to increase heat transfer rates up to 80%.

8.3.1 Tribological properties of nanolubricants

8.3.1.1 Role of nanomaterials concentration on tribological performance of nanolubricants

The role of nanolubricants is increasing with industrial transformation. Fig. 8.3 presents the reduction in friction and wear coefficients delivered by 0.1% additions of Sn, Al_2O_3, TiO_2, ZrO_2/SiO_2, Al_2O_3/TiO_2, and $ZnAl_2O_4$ nanoadditives. It can be observed that the addition of 0.1% $ZnAl_2O_4$ and Al_2O_3/TiO_2 has significantly reduced friction coefficients by up to 30% and wear rates by up to 40%, respectively, whereas the addition of $ZnAl_2O_4$ has demonstrated the highest synergetic reduction of ~30% for both friction and wear coefficients. In the same way the literature studies suggest that the addition of 2% silver nanoparticles reduces the friction and wear coefficients by 17%[100]. The Cu-based nanolubricants have demonstrated superior tribological performance, where the addition of 0.15% Cu nanoparticles reduces the friction and wear coefficients by up to 32%[101]. In the same way the addition of 0.5%, 0.1%, and 2.0% of Fe_3O_4 has reduced friction and wear coefficients by 41%, 51%, and 64%[102], respectively. Along with nanoadditive suspensions, nanosized solid lubricants are also recognized to improve tribological performance. Pristine graphite nanoparticles, hexa-boron-nitride, various disulfides (tungsten, molybdenum, calcium, polytetrafluoroethylene, and boron-nitride), and oxides have significantly contributed to reducing cutting forces for drilling, end-milling, and

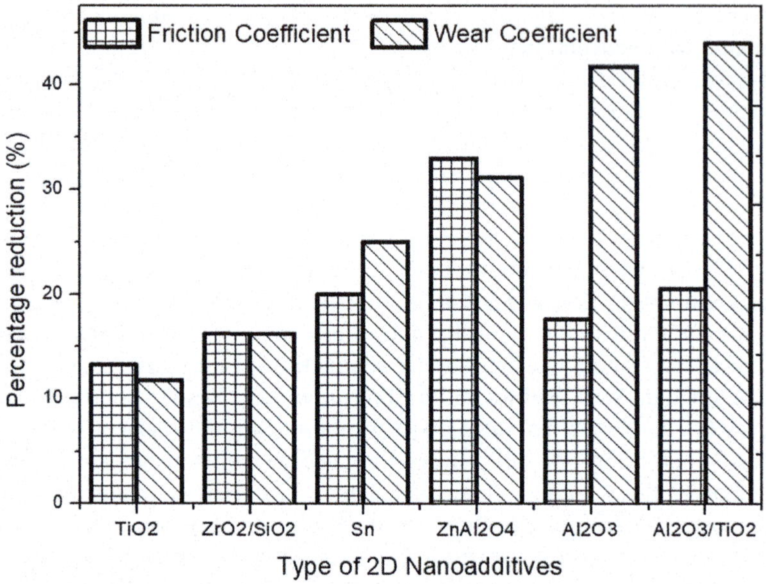

FIGURE 8.3

Percentage decrease in friction and wear coefficients by using nanolubricants having 0.1% TiO_2 [110], ZrO_2/SiO_2 [111], Sn [112], $ZnAl_2O_4$ [113], Al_2O_3 [114], and composite of Al_2O_3 and TiO_2 [115]. *2D*, Two-dimensional.

Data from [116].

turning processes, thus reducing overall mechanical and energy losses in cutting tools and industrial machining procedures [60–72]. The graphitic carbon particles embedded in diamond-like carbon (DLC) coatings have demonstrated superior tribological performance by reducing wear rates up to 60%[103] when compared with conventional DLC coatings. In the same way, MoS_2-based solid lubricants are recognized to improve the tribological performance of steel contacts [104]. Similarly, numerous nanoadditive-based vegetable oils made from coconut, jatropha, soybean, palm, rapeseed, castor, and sesame [105–109] have been reported to boost friction and wear performance of tribological systems.

Fig. 8.4 presents the significance of nanolubricants in reducing the friction coefficient, where fullerene [116] was added by 0.01 and 0.10 vol.% to the mineral oil. It can be observed that the friction coefficient proportionally increases from 0.03 to 0.1 with increasing the normal force from 400 to 1000 N for pristine mineral oil. Meanwhile, the addition of 0.01% fullerene to mineral oil (Nanooil I) significantly reduced the friction coefficient by 50%. Increasing fullerene fraction from 0.01% to 0.1% (Nanooil II) further reduces the friction coefficient, which remains stable around 0.01, that is, a significant decline of 10 times to pristine mineral oil when compared after 2 h of tribological testing. The outstanding decrease in friction coefficient saves energy and mechanical losses and increases product life. Likewise, the addition of 0.06% and 0.1% TiO_2 nanoparticles is reported [117] to save system energy by 21% and 26%, respectively.

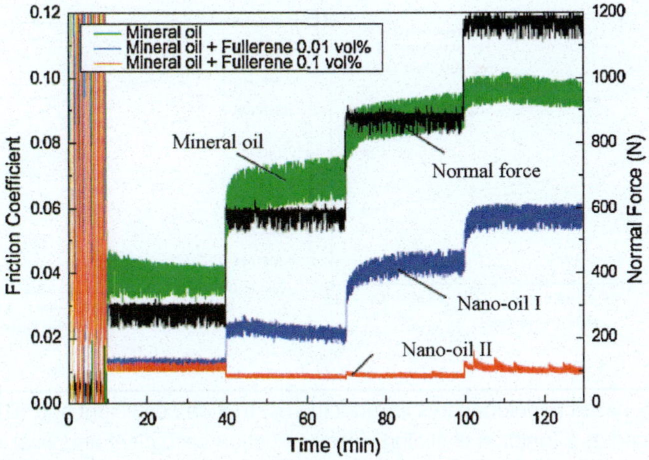

FIGURE 8.4

Decrease in friction coefficient with an addition of 0.01% (Nanooil I) and 0.10% fullerene (Nanooil II) to mineral oil. It can be observed that the friction coefficient of pristine mineral oil reduced from ~0.10 to ~0.05 and ~0.01 with the addition of 0.01% and 0.10% fullerene, respectively.

Reproduced with permission from [117].

8.3.1.2 Role of nanolubricants in tribological systems in automotive applications

Automotive engines usually have 70% losses, which include energy losses [118] and mechanical losses. The energy losses are in the form of unburnt fuel, heat dissipation from the engine block, and exhaust emissions, whereas the mechanical losses include the frictional contacts around the piston cylinder, piston ring, sleeves, valve trains, and crankshaft assembly [119]. The nanolubricants contribute to reducing the friction of the interface surfaces, thus minimizing mechanical losses as well as energy losses by mitigating surface wear that facilitates the escape of unburnt fuel. Fig. 8.5 presents the role of Al_2O_3- and TiO_2-based nanolubricants on friction and wear rate of piston ring as a function of sliding velocity. Generally, the friction coefficient and wear rate reduce with increased velocity. However, as an effect of nanomaterial additives [120], the friction coefficient was reduced by up to 40%, whereas the wear rate was reduced by up to 5% when sliding at a velocity of 1.7 m/s. Similarly, Ali et al. [121] have demonstrated the significance of Al_2O_3/TiO_2-based nanolubricants for automotive engine applications. Their studies suggest that the addition of Al_2O_3/TiO_2 nanoparticles to engine oil generally reduces the friction coefficient and wear rates at 150 N contact load, irrespective of the sliding velocity. At lower velocities of 0.21 m/s the friction coefficient was reduced by 28% (~0.14 to 0.10) and wear rate was reduced by 18% (7.2 to 5.9×10^{-8} mm^3/Nm), whereas the friction coefficient and wear rates were reduced by 50% (0.4 to 0.2) and 5% (5.25 to 5.0×10^{-8} mm^3/Nm) at a higher sliding velocity of 1.75 m/s. Their research outcomes suggest that the addition of Al_2O_3/TiO_2 to engine oil has decreased the friction coefficient and wear rates of the engine ring in the ranges of 39%–53% and 25%–33%, respectively. Similarly, the nanocomposite Al_2O_3/TiO_2-based nanolubricants have demonstrated a 5%–7% decline in frictional power and an improvement in mechanical efficiency by up to 2.5% for gasoline engines [80].

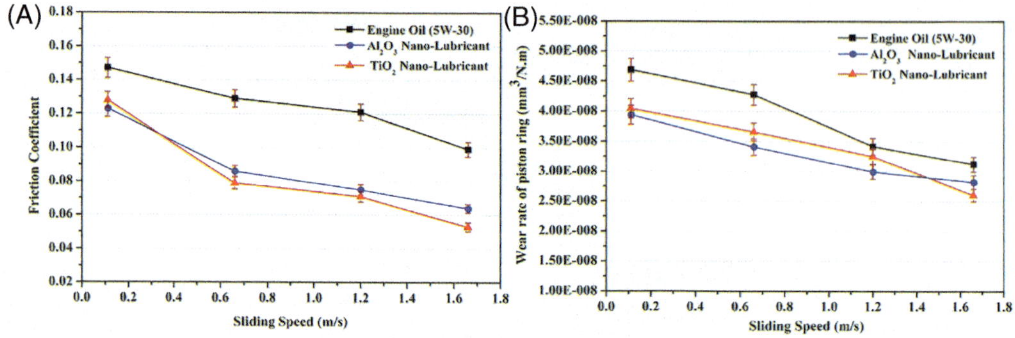

FIGURE 8.5

Role of Al_2O_3- and TiO_2-based nanolubricants in reducing (A) friction coefficient and (B) wear rate of piston ring at 120 N contact load as a function of sliding velocity. Friction coefficient and wear rates were reduced by 40% and 5%, respectively, at a higher sliding velocity of 1.75 m/s.

Reproduced with permission from [120].

8.3.1.3 Role of nanolubricants in tribological performance of industrial machining

Besides saving energy in the automotive industry, nanolubricants have demonstrated superior performance in industrial machining and cutting tool applications. Fig. 8.6A presents the case study of MoS_2 nanolubricants [122] for grinding applications. The nanoadditives penetrate the grinding zone and deform at asperity contact during the machining process; that is, particles are sheared between grinder wheel and workpiece surfaces. The sheared particles form a tribofilm at the interface, which serves as a solid lubricant and reduces the interface friction and wear of grinder wheel. Fig. 8.6B and C presents the SEM micrographs and corresponding EDX spectra of MoS_2 residue on abrasive crystals of grinding wheel and MoS_2 tribofilm at workpiece surfaces, respectively.

8.3.2 Thermal properties of nanolubricants

8.3.2.1 Role of material selection on thermal properties of nanolubricants

The addition of organic, metallic, or metallic oxide nanoparticles significantly improves heat transfer rate, thermal conductivity, and COP. Fig. 8.7 gives an overview of the thermal conductivity of different materials when compared among different material groups and within each subgroup [123]. It can be observed that organic materials and common heat transfer fluids have the lowest thermal conductivity, that is, lower than 1.0 W/mK. Similarly, the thermal conductivity of metal oxides ranges between ~80 and ~100 W/mK and ranges from ~100 to ~500 W/mK for pure metallic nanoparticles. Referring to the group of metal nanoparticles the Cu nanoparticles provide the highest thermal conductivity ~500 W/mK, and Fe particles give the least thermal conductivity of ~100 W/mK, whereas Al, Zn, Ni, and Si remain in between Cu and Fe. Hence pristine or nanomaterial composites are used based on their thermal conductivities specific to application requirements.

8.3.2.2 Role of dimensional features on thermal performance of nanolubricants

The thermal performance of nanolubricants and nanofluids greatly depends on the surface area, thus governed by dimensional features. Fig. 8.8 suggests that the thermal conductivity of conventional

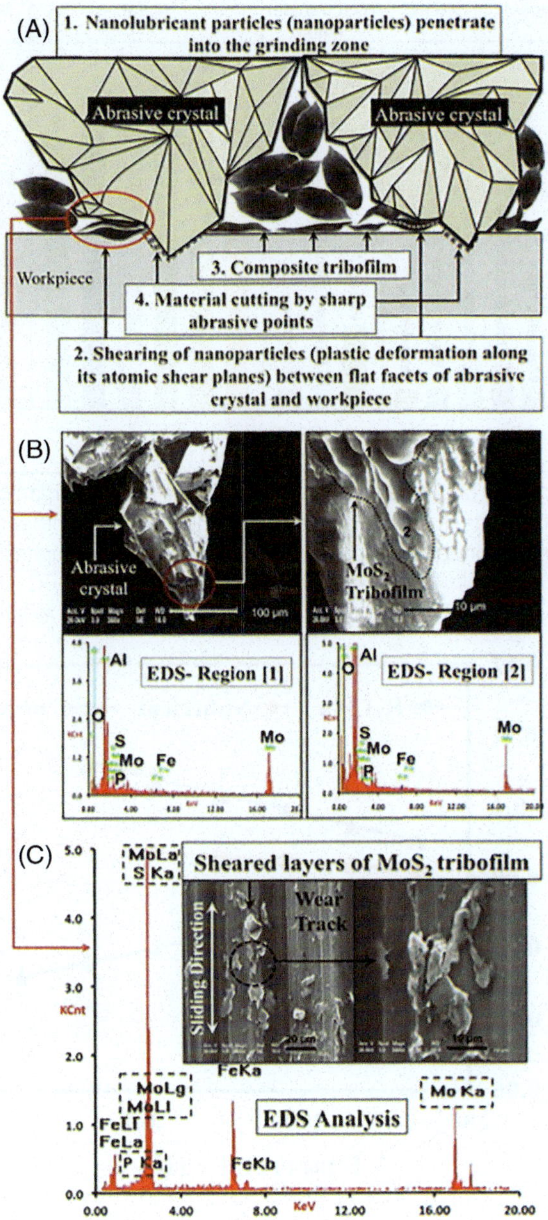

FIGURE 8.6

(A) Tribofilm formation of MoS_2 nanolubricants at the interface of grinder wheel and workpiece. Corresponding scanning electron micrographs and energy-dispersive X-ray spectra identifying MoS_2 nanoadditives on (B) abrasive grain of grinding wheel and (C) workpiece surface.

Reproduced with permission from [122].

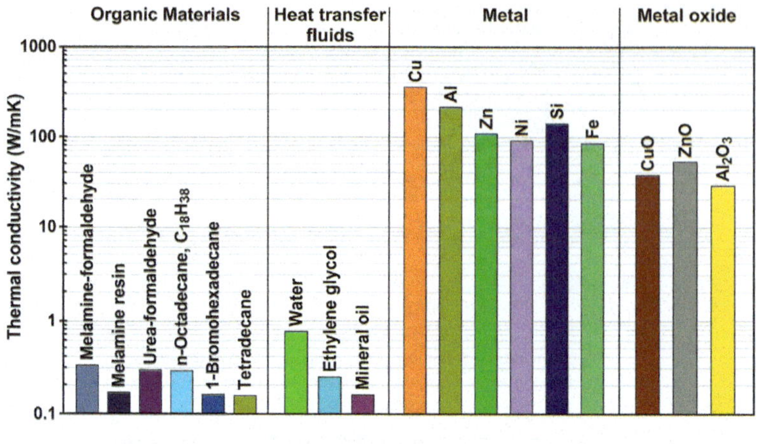

FIGURE 8.7

Mapping of thermal conductivities of different nanoadditive groups such as organic, metals, and metal oxides against common heat transfer fluids. It can be observed that the metals have the highest, whereas organic materials have the least thermal conductivity. The Cu nanoparticles give the highest thermal conductivity (~500 W/mK).

Reproduced with permission from [123].

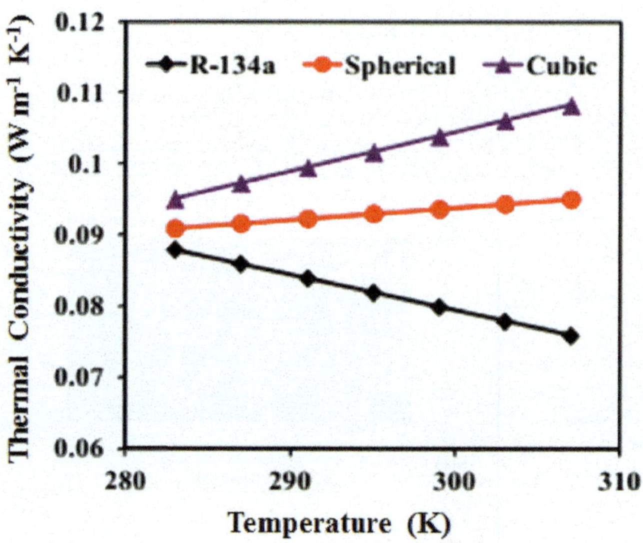

FIGURE 8.8

Addition of ZnO nanoparticles reverses the temperature-dependent thermal conductivity of R134a refrigerants. The thermal conductivity increases up to 25% and 42% for spherical and cubic-shaped ZnO nanoparticles of 25 nm average crystal size.

Reproduced with permission from [124].

R-134a refrigerants reduces by ~14% with increasing temperature from ~283 to ~308 K [79,124], whereas the addition of 10% ZnO nanoparticles by volume reverses the trends, and thermal conductivity improves with increasing temperatures from ~283 to ~308 K. The particles' shape also impacts heat transfer as the surface area changes with geometric shapes. It can be observed that the 25 nm average crystallite-sized ZnO increases thermal conductivity by 25% for spherical shapes and 42% for cubical shapes, respectively. Likewise, the nanoparticle size is also an essential factor in regulating the thermal conductivity of nanolubricants. It is observed that the addition of 13 nm sized Al_2O_3 nanoparticles improves the thermal conductivity of water by 30%[125], whereas further increase in Al_2O_3 nanoparticles size from 13 to 33 nm reversed the trend and reduced the thermal conductivity up to 50%[126]. It is worth noting that the thermal conductivity changes with particle sizes and shapes because of changes in surface area, but it does not always depend on dimensional features because it is a material property thus associated with a type of material, that is, metal, nonmetal, and oxide. Contrary to the aforementioned Al_2O_3 case study where smaller particles improve thermal conductivity, it is observed that the increasing CuO size improves the thermal conductivity of water; that is, a 36 nm particle size corresponds to 12%[126] improvement in thermal conductivity, whereas 50 nm CuO particles have further increased thermal conductivity up to 17%[127].

8.3.2.3 *Improvement in thermal properties of nanolubricants and nanofluids*

8.3.2.3.1 Improvement in thermal conductivity of nanolubricants and nanofluids

Table 8.2 presents the improvement in thermal conductivity of common fluids such as water, ethylene glycol, and R134a refrigerant with the addition of metallic, ceramic, and carbon-based nanoadditives. It can be observed that the utilization of nanoadditives can increase thermal conductivities of conventional lubricants. The confined dataset reflects that the improvement in thermal conductivity remained 30% or lower for Ag, Au, and Fe metals and Al_2O_3, SiO_2, CuO, or TiO_2 oxides. Meanwhile, the addition of Cu has improved thermal conductivity by a range of 40%–80% [123,128]. Carbon nanomaterials, particularly nanotubes, have shown their capabilities in improving the thermal conductivity of nanolubricants and nanofluids. The addition of 1.0% multiwall carbon nanotubes (MWCNTs) by volume has shown improvement in thermal conductivity by 7.0% for water, 13% for ethylene, and 20% for decene [129]. The exceptional improvement of 150% in thermal conductivity with anomalous phenomena has been reported [130] with the addition of MWCNTs to synthetic oil. Similarly, the addition of a large amount of pristine multiwall tubes, such as 30% to water ethylene glycol, has demonstrated [131] ~20% improvement in thermal conductivity and heat transfer coefficient of more than 160%. The remarkable improvements in thermal conductivity and heat transfer coefficient above 150% can significantly improve the overall efficiency of thermal system.

8.3.2.3.2 Improvement in heat transfer coefficient of nanolubricants and nanofluids

The heat transfer coefficient is a measure of heat flux through temperature gradient and is considered an essential design criterion of thermal systems, particularly for heat exchanges. Referring to Table 8.3, it can be observed that the metallic, oxides, and carbon derivates can improve the heat transfer coefficient of various refrigerants from at least 6% to as high as 250%.

Table 8.2 Improvement in thermal conductivity of nanolubricants/fluids with the addition of various nanoadditives.

Sr. no.	Nanolubricants/fluids	Percentage fraction	Percentage improvement (in ascending order)	Further learning
1	SiC/Water	4.200	16.0	[132]
2	Ag/Water	0.001	17.0	[133]
3	CuO/Water	0.400	17.0	[127]
4	Fe/Ethylene glycol	0.550	18.0	[134]
5	MWCNT/Water/Ethylene glycol	0.450	19.0	[131]
6	MWCNT/Decene	1.000	20.0	[129]
7	Au/Water	0.026	21.0	[134]
8	SiO_2/Hydrofluoroether7000	0.020	27.0	[135]
9	Al_2O_3/R134a	5.000	28.0	[136]
10	Al_2O_3/Water	4.300	30.0	[137]
11	TiO_2/Water	5.000	30.0	[138]
12	MWCNT/Water	0.600	38.0	[139]
13	Cu/Ethylene glycol	0.300	40.0	[123]
14	ZnO/R-134a	10.00	42.0	[140]
15	Cu/Water	7.500	78.0	[128]
16	MWCNT/Synthetic Oil	1.000	150.0	[130]

MWCNT, *multiwall carbon nanotube.*
Data partially adopted from [79,116].

Table 8.3 Improvement in heat transfer coefficient of nanolubricants/fluids with the addition of various nanoadditives.

Sr. no.	Nanolubricants/nanofluids	Percentage fraction	Percentage improvement (in ascending order)	Further learning
1	Cu/CuO/Al_2O_3 with R134 and R1234yf	5.000	6.00	[141]
2	MWCNT-COOH/R141b	0.176	23.00	[142]
3	CuO/R113	0.050	29.70	[143]
4	CNT/R123	1.000	36.60	[144]
5	Cu/R141b	0.300	49.00	[98]
6	CNT/R113	0.150	61.00	[145]
7	Diamond/R113	0.050	63.40	[146]
8	Al_2O_3/R141b	0.050	124.00	[147]
9	SiO_2/R134a	0.400	163.00	[148]
10	CuO/R134a	0.500	50–275	[149]

MWCNT, *multiwall carbon nanotube.*
Data adapted from [79,116].

Table 8.4 Improvement in coefficient of performance of nanolubricants/nanofluids with the addition of various nanoadditives.

Sr. no.	Nanolubricants/ nanofluids	Percentage fraction	Percentage improvement (in ascending order)	Further learning
1	Al_2O_3/R134a	5.0	15	[136]
2	Al_2O_3/Water	1.0	17	[150]
3	TiO_2/SDBS	0.5	27	[151]
4	MWCNT-COOH/R141b	0.176	23	[142]
5	TiO_2/R600/Mineral oil	0.2 g/L	62	[152]

Data adopted from [79].

8.3.2.3.3 Improvement in coefficient of performance of nanolubricants and nanofluids

The COP correlates the usable energy inputs either for heating or cooling to the work delivered by the thermal system. Therefore a higher COP is always desirable for an efficient and economical system. Nanolubricants and fluids have shown significant improvements in COP of conventional refrigerants. Referring to Table 8.4, it can be observed that the addition of Al_2O_3 to R134a and water can increase their COP by 15%[136] and 17%[150], respectively. Similarly, TiO_2 has demonstrated 277% and 62% improvements for SDBS [151] and mineral oils [152]. Likewise, the addition of MWCNT-based cysteamine composites has demonstrated a 23% improvement in COP of R141b refrigerants [142].

In addition to the type of nanoadditives, such as organic, metallic, or nonmetallic, the surface functionalization of nanomaterials also improves the thermal performance of nanolubricants. The addition of carbon nanoadditives, particularly the carbon nanotubes, is widely reported [79] for significant improvements in the Nusselt number, thermal conductivity, and heat transfer rates of nanolubricants. Fig. 8.9 gives an overview of pristine-, hydroxyl-, and carboxyl MWCNTs for improving thermal conductivity, heat transfer coefficient, and Nusselt number of R141b refrigerants. It has been observed that pristine MWCNTs when added in fractions up to 0.3 wt.%, can increase the Nusselt number of R141b refrigerant by more than 41%. In addition, incorporating 0.176 vol.% of MWCNTs functionalized with hydroxyl and carboxyl groups can enhance the heat transfer coefficient by 15% and 23%, respectively. In larger fractions, carbon nanotubes can boost the heat transfer coefficient by as much as 160%[131].

8.3.3 Flow properties of nanolubricants/nanofluids

The rheological properties of nanolubricants and nanofluids play a pivotal role in determining their efficacy for lubrication and heat transfer purposes. The concentration of nanoadditives, the size of the dopants, the quality of dispersion, the stability of the suspension, and the operational temperatures all have a profound impact on key characteristics such as viscosity, thermal conductivity, and frictional behavior. It has been observed that an increase in the size and concentration of nanoadditives typically leads to a decrease in the viscosity of nanolubricants and nanofluids. This phenomenon is illustrated in Fig. 8.10, which depicts the relationship between the concentration of nanoparticles and the viscosity of the fluid [153].

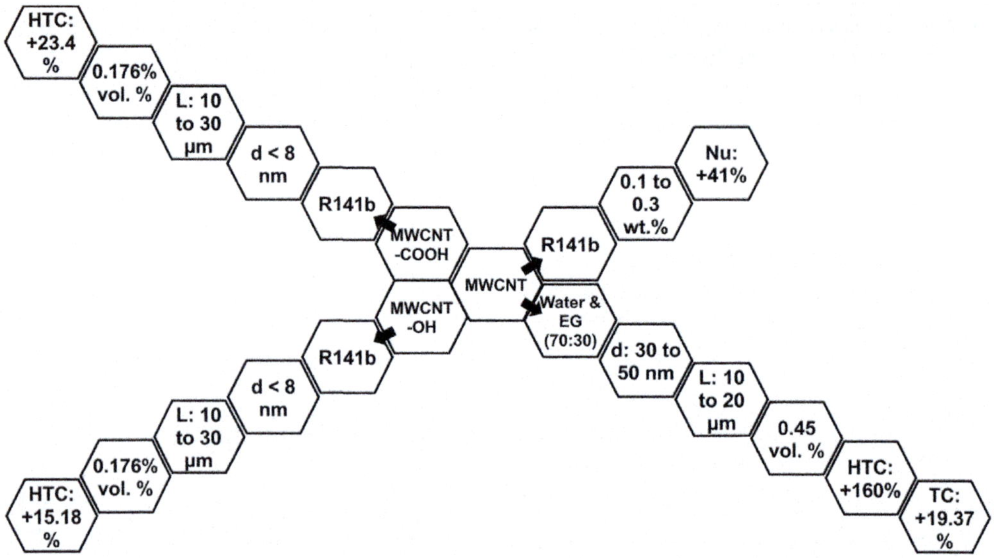

FIGURE 8.9

Overview of pristine and functionalized multiwall carbon nanotubes for improvement in Nusselt number, thermal conductivity, and heat transfer coefficient of nanofluids used for the refrigeration system. *MWCNT*, Multiwall carbon nanotube.

Reproduced with permission from [79].

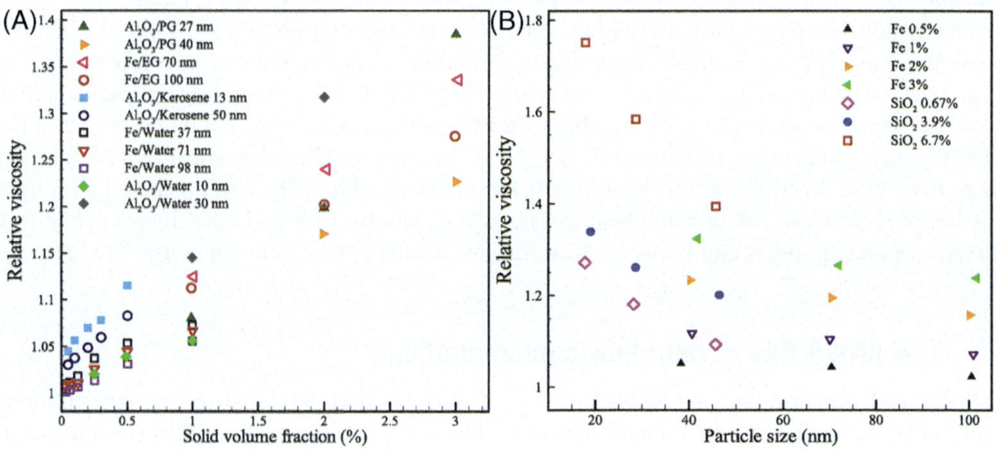

FIGURE 8.10

Positive and inverse relationship of relative viscosity of nanolubricants as function of (A) volumetric fraction and (B) particle size.

Reproduced with permission from [154].

For instance, Fig. 8.10A shows that the viscosity of a fluid tends to increase with the addition of various metallic and nonmetallic nanoparticles into lubricants. Specifically, when the volumetric fraction of iron nanoparticles in water is increased from 0% to 1.0%, the relative viscosity of the nanolubricant increases from approximately 1.01 to about 1.08. This trend highlights the importance of carefully selecting the concentration and type of nanoparticles to achieve the desired performance characteristics in nanolubricants and nanofluids [155].

Similarly, the increasing amount of iron in ethylene glycol from 1.0% to 3.0% increases the relative viscosity from 1.13 to 1.33 [156]. In the same way, higher amounts, such as 2% of TiO_2 nanoparticles, also increase the viscosities of nanolubricants by up to 64%[157]. Referring to Fig. 8.10B the synergistic influence of particle concentration and particle size of nanoadditives on viscosity can be observed. It is observed that the relative velocity of 20 nm sized SiO_2 nanoadditive-based ethylene glycol lubricants increased from ~1.15 to ~1.35 and further increased to 1.77 by increasing SiO_2 amounts from 0.67% to 3.9% and 6.7%[158], respectively. The viscosity trends as a function of particle size and amounts are widely reported for different types of materials. The other research study presents that the relative viscosity of 40 nm sized Fe-based ethylene glycol lubricant increased from ~1.2 to ~1.5 and further from ~1.22 to 1.35 by increasing Fe nanoadditive from 0.5% to 1.0%, 2.0%, and 3.0%, respectively [159]. The analysis of Fe- and SiO_2-based ethylene glycol reveals that the relative viscosity significantly increases with the addition of more nanoadditives. In contrast, the relative viscosity generally decreases as the particle size increases, although this effect is dependent on the type of material used. Data on Fe and SiO_2 indicate that relative viscosity is more sensitive to smaller particle sizes, specifically in the range of 20–50 nm, compared to larger particle sizes of 50–100 nm.

The nanolubricant viscosity is sensitive to working temperatures, thus influencing their performance as well. The literature suggests that Ag-, Ti_2O_3-, SiO_2-, and Al_2O_3-based water lubricants have relative velocities of ~1.04 [160], ~1.07 [161], ~1.43 [162], and ~1.88 [163] at room temperatures (20°C–25°C). However, the viscosity of these nanolubricants generally reduces with increasing temperatures. It is observed that increasing temperature by 100%, that is, from 25°C to 50°C significantly reduces the viscosity of water-based lubricants from 1.58 to 1.18 for Fe_3O_4[164], 1.2 to 0.7 for carbon nanotubes [165], 1.0 to 0.55 for Al_2O_3[166], and 0.95 to 0.45 for TiO_2[167] nanoadditives. The research investigations suggest that the addition of small graphene amounts as low as 0.01 wt.% to engine oils keeps nanolubricant in compliance with Newtonian behavior irrespective of increasing temperatures and shear rates [168]. Referring to the synergistic impact of temperature and amount of nanoadditives the kinematic viscosity of graphene/copper nanolubricants increased from 54.06 to 55.20 mm^2/s with increasing concentrations from 0 to 0.6 wt.% at 40°C, whereas the kinematic viscosity of the same amount (0.6 wt.%) of graphene/copper nanolubricants reduced from 55.20 to 10.00 mm^2/s by increasing temperature from 40°C to 100°C [92]. Fig. 8.11 presents the effect of temperature on the kinematic viscosity of fullerene-based nanolubricants. It can be observed that the kinematic viscosity reduces from ~55 to 20 mm^2/s and further down to 10 mm^2/s by increasing temperature from 40°C to 60°C and 80°C, respectively [169].

8.4 Mechanisms of nanolubrication

Nanolubrication mechanisms are typically governed by physical, chemical, thermal, and electrostatic interactions. Because of varying roughness, material surfaces experience asperity contact, leading to

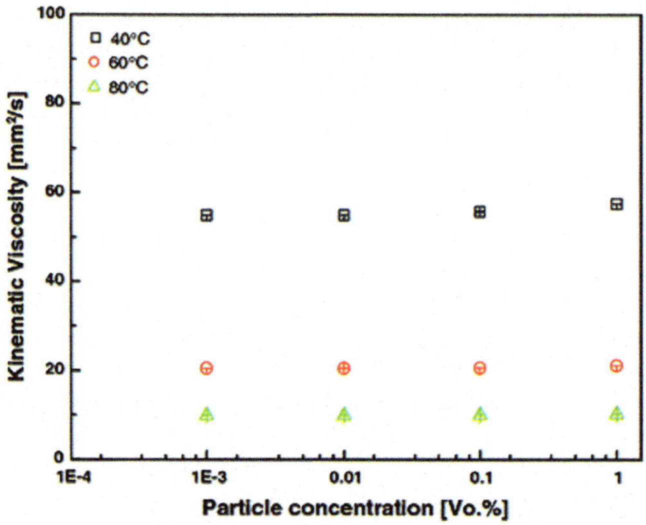

FIGURE 8.11

Kinematic viscosity of fullerene-based nanolubricants as a function of temperature and nanoadditive amounts. The kinematic viscosity reduces with increasing temperatures, irrespective of the amount of nanoadditives.

Reproduced with permission from [169].

wear at frictional interfaces during sliding. The tribological behavior of contact surfaces primarily depends on contact stresses, frictional forces, and the sliding velocity of the interacting surfaces. Nanoadditives enhance tribological performance by rolling between surfaces, shearing under high contact pressures to reduce surface wear, forming transfer layers, protective layers, and tribofilms to improve lubricity, and strengthening contact surfaces to minimize friction and wear. In addition, the quality of nanolubricants, particularly their size, concentration, and dispersion, plays a crucial role in the effectiveness of the lubrication mechanism. It is observed that the addition of nanodiamonds up to 0.2 wt.% to pristine oil reduces the surface roughness from 0.655 μm (without nanodiamonds) to a minimum of 0.218 μm. However, further increasing the concentration of nanodiamonds to 0.5 wt.% increases the surface roughness again to 0.279 μm and wear rate to 0.13×10^6 μm^3[170].

In the same way the addition of graphene nanoplates [171] as additive oil develops a protective film. Therefore the addition of 0.075 wt.% graphene increases the viscosity and film thickness to assist in friction and wear reductions up to 27% and 43%, respectively. Table 8.5 presents the lubrication mechanism of different metallic, nonmetallic, carbon, and rare-earth oxide-based nanolubricants.

8.5 Modeling the nanolubricant behaviors

Nanolubricant behavior is widely being analyzed with numerous analytical and computational tools. The computation studies support identifying nanolubricant behaviors based on material properties, dimensional features, and concentration. Al_2O_3 and CuO nanoadditives are widely investigated with

Table 8.5 Lubrication mechanism of metal, nonmetal, carbon, rare-earth, and nanocomposite-based nanolubricants.

Type of nanoadditive	Element/compounds	Type of mechanism	Lubrication mechanism	Further study
Metal	Sn and Fe	Tribofilm	Sn and Fe reduce friction and wear by forming a soft tribofilm with lower shear stresses to uplift tribological behavior	[172]
	Bi	Protective film	A protective layer produced at contact interfaces and reduced frictional effects	[173]
	Ag	Protective film	Soft protecting film with shearing of Ag particles' contact interface mitigates wear of surfaces	[174]
	Pd	Protective film	A layer with lower shear strength formed to protect contact surfaces	[175]
	Ni	Protective film	Ni particles produce a protective film at the interface	[176]
	Pd and Au	Transfer layer	A transfer layer film formed at contact interface with Au and Pd nanoparticles, which improve tribological performance	[177]
	Fe, Cu, Co	Particles rolling	Low shear Cu film formed at interface, and tribology improves by particles rolling between contact surfaces	[178]
	Cu	Surface repairing	A self-healing Cu film forms at the interface to mitigate wear on contact surfaces	[179]

(*continued on next page*)

Table 8.5 Lubrication mechanism of metal, nonmetal, carbon, rare-earth, and nanocomposite-based nanolubricants—cont'd

Type of nanoadditive	Element/compounds	Type of mechanism	Lubrication mechanism	Further study
Metal oxide	TiO_2	Particles rolling	Particles rolling among sliding surfaces and self-repairing improve tribology	[180]
	CuO and Al_2O_3	Particles sintering	Tribology-induced CuO sintering improves tribological performance	[181]
	La doped TiO_2	Tribofilm	A tribofilm of Fe and Ti oxides to improve tribology	[182]
	ZnO and CuO	Tribofilm	Smooth and compact tribofilm forms to improve tribology	[183]
	ZnO	Protective film	ZnO film form on wear tracks to mitigate wear of materials	[184]
	Al_2O_3	Particles rolling	Rolling nanoparticle mitigates frictional effects	[185]
	Fe_3O_4	Particles rolling	Rolling nanoparticle mitigates frictional effects	[101]
Carbon	Fullerene	Particles rolling	Improves lubrication by rolling between surfaces	[186]
	Nanodiamond	Surface repairing	Improves surface toughness at contact interface to improve tribological performance	[187]
	Graphite	Transfer film	Thin carbon layer forms at the sliding interface to boost lubricity	[188]
	Graphene	Protective film	Interlayer shearing, grain refinement surface mending, reorientations, and the formation of protective film promote lower friction and antiwear mechanisms	[171,189]

(continued on next page)

	Graphene oxide	Tribofilm	Chemically bonded graphene makes a tribofilm of low shear strength at the contact interface	[190]
	Carbon nanotubes	Rolling	Rolling, polishing, and mending at contact surfaces improve tribological performance	[191,192]
Rare-earth compounds	CeBO	Tribofilm	Tribofilm composed of Fe, B, and Ce oxides	[193]
	CeO_2	Tribofilm	Tribofilm contains Ce^{X+}, FeS_X, oxides of Fe and Ti, and organic compounds	[194]
	LaF_3	Tribofilm	Tribofilm of FeF_3, Fe_2O_3, and LaF_3 formed to improve tribological performance. LaF_3 also produced adsorption film in individual cases	[195]
	Al_2O_3/TiO_2	Particles rolling	Friction behavior changes from sliding to rolling movement	[14]
Nanocompsoites	Cu/SiO_2	Tribofilm	Tribofilm formed by FeS, FeSO4, and SiO_2, and self-healing by Cu nanoparticles	[196]
	Cu/GO	Protective film	Combined action of Cu and GO	[197]
	Al_2O_3/SiO_2	Particles rolling	Combined action of Al_2O_3 and SiO_2	[198]
	Serpentine/La(OH)3	Tribofilm	Tribofilm of Fe/Si/O	[199]

Data partially adopted from [1].

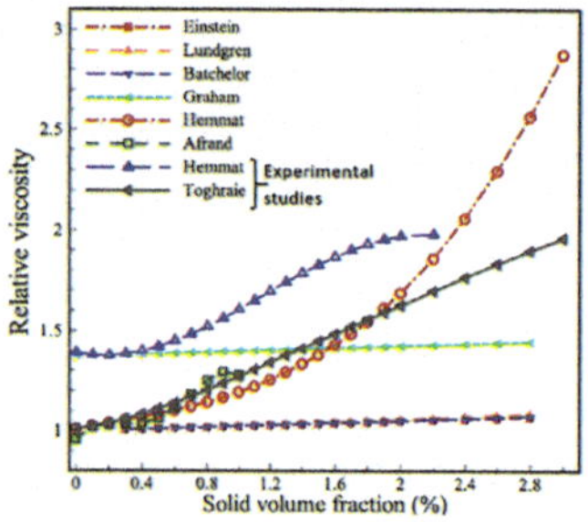

Einstein [201], $\mu_{eff} = (1 + 2.5\varphi_p)$

Brinkman [202], $\mu_{eff} = \frac{1}{(1-\varphi_p)^{2.5}} = (1 + 2.5\varphi_p + 4.375\varphi_p^2 + \ldots)$

Lundgren [203], $\mu_{eff} = \frac{1}{1-2.5\varphi_p} = (1 + 2.5\varphi_p + 6.25\varphi_p^2 + \ldots)$

Batchelor [204], $\mu_{eff} = (1 + \pi\varphi_p + k_H\varphi_p^2) = (1 + 2.5\varphi_p + 6.2\varphi_p^2)$

Frankel and Acrivos [205], $\mu_{eff} = \left(\frac{9}{8}\frac{(\varphi_p/\varphi_{pmax})^{1/3}}{1-(\varphi_p/\varphi_{pmax})^{1/3}}\right)$

Graham [206], $\mu_{eff} = (1 + 2.5\varphi_p) + \left[\frac{4.5}{(h/r_p)\cdot(2+h/r_p)^2}\right]$

Mooney [207], $\mu_{eff} = \exp\left(\frac{2.5\varphi_p}{1-k\varphi_p}\right) = \left\{1 + \frac{2}{5\varphi_p} + \left[\frac{3}{125} + \left(\frac{2}{5k}\right)\right]\varphi_p^2 + \ldots\right\} \rightarrow 1.35 < k < 1.91$

Eilers [208], $\mu_{eff} = \mu_f\left[1 + \frac{1.25\varphi_p}{1-\varphi_p/0.78}\right] = (1 + 2.5\varphi_p + 4.75\varphi_p^2 + \ldots)$

FIGURE 8.12

Common models to estimate relative viscosity of nanolubricants/nanofluids and their comparison with experimental results. It can be observed that the significance of theoretical models is higher for the low amount of nanoadditives when compared with experimental outcomes [201–208].

Reproduced with permission from [154].

the highest proportion of 27% each for their use in nanolubricants for thermal applications, whereas SiO_2 and ZnO had 13% and 11% proportions. However, the TiO_2, Al, Cu, carbon nanotube proportions remained less than 10% [200]. Fig. 8.12 provides an overview of common theoretical models used to estimate the relative viscosity of nanolubricants, comparing these models with experimental results. The comparison shows that the theoretical models align closely with experimental results when the concentration of nanoadditives is low, typically less than 1.0%. However, as the concentration of nanoadditives increases, the discrepancy between theoretical models and experimental results, as well as among the theoretical models themselves, becomes more pronounced. The governing equations of these theoretical models are cited for further study. It is anticipated that advancements in computational tools, driven by Material 4.0 initiatives and the rise of data-driven materials engineering, will help bridge these gaps in the coming years.

Similarly, the frictional behaviors of nanolubricants are widely investigated using different types of nanoadditives and application-specific working conditions. The computational models help to understand the shearing, deformation, elastic-plastic response, and lubrication mechanism of nanoadditives. Fig. 8.13 presents molecular dynamics studies to estimate the friction behavior of carbon onions in contact with DLC. It is deduced that the number of interfacial bonding between carbon onions and DLC is a load-dependent function, and new bonds formed between carbon onions and DLC with increasing contact pressure after a certain threshold. It was found that the carbon onions were free under 2.5 GPa contact pressure, which corresponds to 21.5 nN contact load per carbon onion, whereas interfacial bonding between carbon onion and DLC developed at 5 GPa contact load, which corresponds to 43 nN contact load per carbon onion [209]. The formation of covalent bonds under contact pressure is also reported in other studies [210].

Likewise, the computational studies performed with molecular dynamics simulations help to understand the friction behavior of fullerene and diamond cores. The dangling bonds from fullerene carbon atoms have a higher probability of developing interfacial bonding in diamond cores. The formation of covalent bonds induces strains that promote the formation of additional new bonds. In the same way the addition of catalysts, surfactants, and mineral oils enhances the formation of covalent bonds to induce

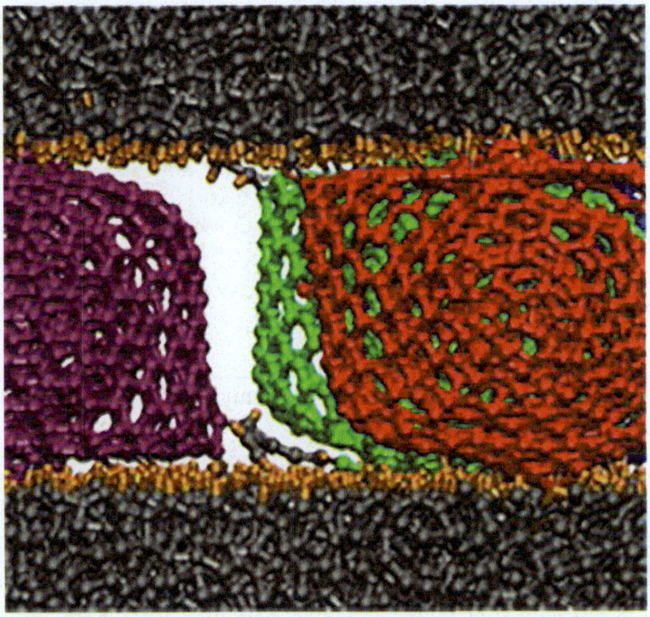

FIGURE 8.13

Computation studies to estimate interfacial bonding among carbon onions and diamond-like carbon (DLC) at contact stress of 5 GPa during sliding wear.

Reproduced with permission from [209].

superlubricity [211–214]. The computational studies develop the understanding of friction behavior of 2D nanomaterials based on their morphologies, such as particles, flakes, pellets, rods, and nanotubes; their sizes; and their concentration.

It is deduced that the flake-like or planner nanomaterials exhibit lower shear resistance [215], thus better reducing friction when compared to other morphologies. Fig. 8.14 presents the lubrication mechanism of single-layer and multilayer graphene to reduce the friction coefficient and advance useable life cycles. Referring to Fig. 8.14A the computation studies suggest that the slipping between graphene sheets and their reorientation significantly promotes lubricity. It can be observed that the graphene sheets are initially aligned and start rotating under the contact loads and sliding motion. In a tribology-induced process the graphene sheets either translate from one place to another having the same orientation or rotate at a certain angle, such as 60 degrees under mechanical and thermal effects. The lubricity also improves with the increasing number of graphene sheets as the probability of slipping and reorientation increases among numerous layers [73]. Referring to Fig. 8.14B, it can be observed that the friction coefficient of bare steel is reduced from ~0.3 to ~0.15 by using multilayer graphene. Similarly, the number of graphene layers also controls the friction coefficient range. It can be observed that the friction coefficient of bare steel remained stable up to ~500 cycles, which increased to 6500 cycles for single-layer graphene and up to 47,000 cycles for multilayer graphene base nanolubricants.

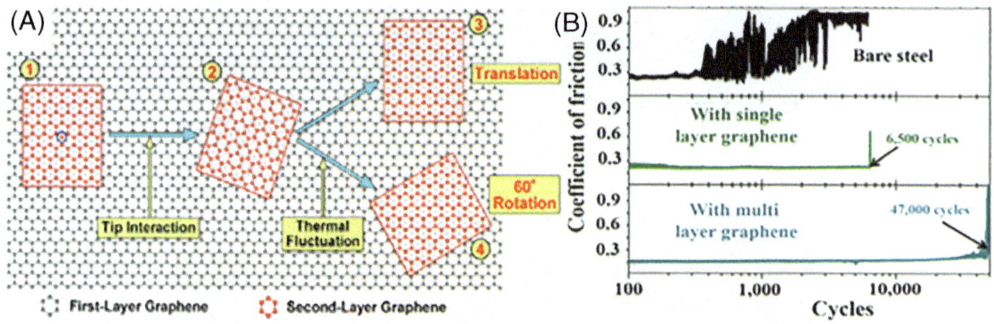

FIGURE 8.14

(A) Computational studies to understand the friction mechanism of graphene layers and (B) experimental investigation of graphene lubrication.

Reproduced with permission from [73].

8.6 Future outlook

Numerous components are expected to be added to nanolubricants with industrial transformation. The nanolubricants may perform multiple actions in numerous fields, such as thickening agents for depressors [216] or antiscoring additives for extreme pressure [217] applications. In the same way the nanolubricants may perform multifunctional actions such as self-regulation [218], self-sensing [219], self-adaption [9,220], self-cleaning, self-healing [221], and self-reporting [222]. Artificial intelligence is emerging in nanolubricants and boosting research and development efficiency by optimizing parameters for desired applications. Various artificial intelligence tools are being used to understand the behaviors of complex nanolubricants to predict their properties and behaviors. Artificial neural networks are used [223] to estimate thermal conductivity ratios of complex nanoadditives, whereas machine learning is used [224] to predict the kinematic viscosity of nanolubricants used in heavy machinery. The data-driven research is using various algorithms, fuzzy logic, and methods such as random forest, decision tree, deviation analysis, and dissimilarity-based boosting technique [225] to optimize large datasets to provide refined parameters. It is anticipated that the artificial intelligence-based computational studies will assist in formulating advanced nanolubricants for complex applications, particularly for renewable energy systems and automotive and aerospace industries.

8.7 Conclusion

This chapter introduces nanolubricants, their properties, and their usability across various industrial applications, such as tribological systems, industrial machining, and thermal systems. It covers nanolubricants based on metallic, nonmetallic, carbon, oxides, polymers, salts, and nanocomposites, along with their corresponding synthesis methods. The findings indicate that nanolubricants significantly reduce friction and wear in various tribological systems, particularly in automotive engines, bearings, cutting tools, and industrial machining. In addition, nanolubricants/fluids markedly improve thermal conductivity, heat transfer coefficient, and COP when applied to various thermal systems. The characteristics of nanolubricants are sensitive to the size, amount, and distribution of their nanoadditives, which, in turn, influence their thermal, fluid, and lubrication behaviors. A broad range of nanoadditives, including metals, nonmetals, oxides, carbon, and composite types, have been tested for nanolubricants. However, further research is needed to enhance nanoadditive distributions, achieve uniformity in size and shape, and improve their dispersion as isolated nanomaterials rather than clusters, thereby increasing the surface area per unit volume.

References

[1] W. Dai, B. Kheireddin, H. Gao, H. Liang, Roles of nanoparticles in oil lubrication, Tribol. Int. 102 (2016) 88–98.

[2] J. Ma, Y. Mo, M. Bai, Effect of Ag nanoparticles additive on the tribological behavior of multialkylated cyclopentanes (MACs), Wear 266 (2009) 627–631.

[3] H. Ghaednia, M.S. Hossain, R.L. Jackson, Tribological performance of silver nanoparticle–enhanced polyethylene glycol lubricants, Tribol. Trans. 59 (2016) 585–592.

[4] H.L. Yu, Y. Xu, P.J. Shi, B.S. Xu, X.L. Wang, Q. Liu, Tribological properties and lubricating mechanisms of Cu nanoparticles in lubricant, Trans. Nonferrous Met. Soc. China 18 (2008) 636–641.

[5] J. Padgurskas, R. Rukuiza, I. Prosyčevas, R. Kreivaitis, Tribological properties of lubricant additives of Fe, Cu and Co nanoparticles, Tribol. Int. 60 (2013) 224–232.

[6] J.C. Sánchez-López, M.D. Abad, L. Kolodziejczyk, E. Guerrero, A. Fernández, Surface-modified Pd and Au nanoparticles for anti-wear applications, Tribol. Int. 44 (6) (2011) 720–726.

[7] M.D. Abad, J.C. Sánchez-López, Tribological properties of surface-modified Pd nanoparticles for electrical contacts, Wear 297 (1–2) (2013) 943–951.

[8] R. Chou, A.H. Battez, J.J. Cabello, J.L. Viesca, A. Osorio, A. Sagastume, Tribological behavior of polyalphaolefin with the addition of nickel nanoparticles, Tribol. Int. 43 (12) (2010) 2327–2332.

[9] X. Liu, Z. Lu, H. Dong, Y. Cao, H. Yang, Study on self-adaptive lubrication mechanism of surface micro-dimple structure filled with multiple lubricants, J. Alloys Compd. 861 (2021) 158479.

[10] M. Flores-Castañeda, E. Camps, M. Camacho-López, S. Muhl, E. García, M. Figueroa, Bismuth nanoparticles synthesized by laser ablation in lubricant oils for tribological tests, J. Alloy Compd. 643 (2015) S67–S70.

[11] J. Padgurskas, R. Rukuiza, I. Prosyčevas, R. Kreivaitis, Tribological properties of lubricant additives of Fe, Cu and Co nanoparticles, Tribol. Int. 60 (2013) 224–232.

[12] H. Ghaednia, R.L. Jackson, J.M. Khodadadi, Experimental analysis of stable CuO nanoparticle enhanced lubricants, J. Exp. Nanosci. 10 (1) (2013) 1–18.

[13] M.V. Thottackkad, R.K. Perikinalil, P.N. Kumarapillai, Experimental evaluation on the tribological properties of coconut oil by the addition of CuO nanoparticles, Int. J. Precis. Eng. Manuf. 13 (1) (2012) 111–116.

[14] R. Melvin Rajkumar, K. Kiruthiga, S. Induja, P.S. Raghavan, M. Shafee, P. Pandiyan, R. Sathyamurthy, S. Chandra Sekhar, A.E. Kabeel, Studies on tribological behaviour of ZnO nanorods suspended in SAE 20w 40 engine oil, Solid State Commun. 328 (2021) 114235.

[15] E. Prakash, K.K. Siva, K.K. Muthu, Experimental studies on vibration characteristics on ball bearing operated with copper oxide nano particle mixed lubricant, Int. J. En. Technol. 5 (5) (2013) 4127–4130.

[16] L. Peña-Parás, J. Taha-Tijerina, L. Garza, D. Maldonado-Cortés, R. Michalczewski, C. Lapray, Effect of CuO and Al_2O_3 nanoparticle additives on the tribological behavior of fully formulated oils, Wear 332–333 (2015) 1256–1261.

[17] W.G. Xue, Z.H. Zhao, P. Wang, Z.L. Jin, X.H. Xu, X.G. Zhou, Performance study of zinc oxide nanoparticles for lubricant oil, Adv. Mater. Res. 1118 (2015) 195–204.
[18] D. Jiao, S. Zheng, Y. Wang, R. Guan, B. Cao, The tribology properties of alumina/silica composite nanoparticles as lubricant additives, Appl. Surf. Sci. 257 (2011) 5720–5725.
[19] I.C. Harta, K. Owens, S.D.J. Santiago, D. Schall, S. Thrush, G. Barber, Q. Zou, Tribological performance of ZnO–oil nanofluids at elevated temperatures, SAE Int. J. Fuels Lubr. 6 (1) (2013) 126–131.
[20] S. Rengifo, C. Zhang, S. Harimkar, B. Boesl, A. Agarwal, Effect of WS2 addition on tribological behavior of aluminum at room and elevated temperatures, Tribol. Lett. 65 (3) (2017) 24.
[21] X.H. Zhang, H.X. Xu, J.T. Wang, X. Ye, W.N. Lei, M.Q. Xue, H. Tang, C. Li, Synthesis of ultrathin WS_2 nanosheets and their tribological properties as lubricant additives, Nanoscale Res. Lett. 11 (1) (2016) 442.
[22] M. Kalin, J. Kogovšek, M. Remškar, Nanoparticles as novel lubricating additives in a green, physically based lubrication technology for DLC coatings, Wear 303 (2013) 480–485.
[23] Y. Xu, E. Hu, K. Hu, Y. Xu, X. Hu, Formation of an adsorption film of MoS2 nanoparticles and dioctyl sebacate on a steel surface for alleviating friction and wear, Tribol. Int. 92 (2015) 172–183.
[24] P. Nallasamy, N. Saravanakumar, S. Nagendran, E.M. Suriya, D. Yashwant, Tribological investigations on MoS2-based nanolubricant for machine tool slideways, Proc. Inst. Mech. Eng. J. J. Eng. Tribol. 229 (5) (2014) 559–567.
[25] M.R. Yi, C.H. Zhang, The synthesis of two-dimensional MoS2 nanosheets with enhanced tribological properties as oil additives, RSC Adv. 8 (17) (2018) 9564–9573.
[26] L.J. Chen, D.Y Zhu, Preparation and tribological properties of unmodified and oleic acid-modified CuS nanorods as lubricating oil additives, Ceram. Int. 43 (5) (2017) 4246–4251.
[27] F.Y Zhao, Z.M. Bai, Y. Fu, D. Zhao, C.M. Yan, Tribological properties of serpentine, $La(OH)_3$ and their composite particles as lubricant additives, Wear 288 (2012) 72–77.
[28] S. Li, H. Qin, R. Zuo, Z. Bai, Tribological performance of Mg/Al/Ce layered double hydroxides nanoparticles and intercalated products as lubricant additives, Appl. Surf. Sci. 353 (2015) 643–650.
[29] H.D. Wang, Y.H. Liu, Z. Chen, B.B Wu, S.L. Xu, J.B. Luo, Layered double hydroxide nanoplatelets with excellent tribological properties under high contact pressure as water-based lubricant additives, Sci. Rep. 6 (2016) 22748.
[30] H.D. Wang, Y.H. Liu, F.M. Guo, H.P. Sheng, K.L. Xia, W.R. Liu, J.G. Wen, Y.J. Shi, A. Erdemir, J.B. Luo, Catalytically active oil-based lubricant additives enabled by calcining Ni–Al layered double hydroxides, J. Phys. Chem. Lett. 11 (1) (2020) 113–120.
[31] L. Joly-Pottuz, N. Ohmae, Carbon-based nanolubricants, in: J.M. Martin, N. Ohmae (Eds.), Nanolubricants, Wiley, England, 2008.
[32] B.C. Ku, Y.C. Han, J.E. Lee, J.K. Lee, S.H. Park, Y.J. Hwang, Tribological effects of fullerene (C60) nanoparticles added in mineral lubricants according to its viscosity, Int. J. Precis. Eng. Manuf. 11 (4) (2010) 607–611.
[33] D.X. Peng, Y. Kang, R.M. Hwang, S.S. Shyr, Y.P. Chang, Tribological properties of diamond and SiO_2 nanoparticles added in paraffin, Tribol. Int. 42 (6) (2009) 911–917.
[34] H.T. Liu, H.M. Ji, H.P. Hong, Y. Hammad, Tribological properties of carbon nanotube grease, Ind. Lubr. Tribol. 66 (5) (2014) 579–583.
[35] S. Bobbo, L. Fedele, M. Fabrizio, S. Barison, S. Battiston, C. Pagura, Influence of nanoparticles dispersion in POE oils on lubricity and R134a solubility, Int. J. Refrig. 33 (2010) 1180–1186.
[36] S. Bhaumik, S. Prabhu, K.J. Singh, Analysis of tribological behavior of carbon nanotube based industrial mineral gear oil 250 cSt viscosity, Adv. Tribol. 2014 (2014) 341365.
[37] H.M. Xie, B. Jiang, J.H. Dai, C. Peng, C.X. Li, Q. Li, F.S. Pan, Tribological behaviors of graphene and graphene oxide as water-based lubricant additives for magnesium alloy/steel contacts, Materials 11 (2) (2018) 206.

[38] J. Yang, Y.F. Xia, H.J. Song, B.B. Chen, Z.Z. Zhang, Synthesis of the liquid-like graphene with excellent tribological properties, Tribol. Int. 105 (2017) 118–124.
[39] N. Nunn, Z. Mahbooba, M.G. Ivanov, D.M. Ivanov, D.W. Brenner, O. Shenderova, Tribological properties of polyalphaolefin oil modified with nanocarbon additives, Diam. Relat. Mater. 54 (2015) 97–102.
[40] D.M. Kumar, J. Bijwe, S.S.V. Ramakumar, PTFE based nano-lubricants, Wear 306 (1–2) (2013) 80–88.
[41] N. Kumar, V. Saini, J. Bijwe, Performance properties of lithium greases with PTFE particles as additive: controlling parameter-size or shape? Tribol. Int. 148 (2020) 106302.
[42] J.M. Wang, J.H. Wang, C.S. Li, G.Q. Zhao, X.B. Wang, Tribological performance of poly(sodium 4-styrenesulphonate) as additive in water-glycol hydraulic fluid, Lubricat. Sci. 24 (3) (2012) 140–151.
[43] T.T. Tu, W.X. Wang, Y. Wan, J.B. Pu, Tribology properties of the aqueous solution of polyvinylpyrrolidone, Mater. Sci. 4 (3) (2014) 103–110.
[44] M. Zhang, X.B. Wang, W.M Liu, Tribological behavior of LaF_3 nanoparticles as additives in poly-alpha-olefin, Ind. Lubr. Tribol. 65 (4) (2013) 226–235.
[45] G.Y. Bai, J.Q. Wang, Z.G. Yang, H.G. Wang, Z.F. Wang, S.R. Yang, Preparation of a highly effective lubricating oil additive–Ceria/graphene composite, RSC Adv. 4 (87) (2014) 47096–47105.
[46] B.S. Chen, K.C. Gu, J.H. Fang, J. Wu, J. Wang, N. Zhang, Tribological characteristics of monodispersed cerium borate nanospheres in biodegradable rapeseed oil lubricant, Appl. Surf. Sci. 353 (2015) 326–332.
[47] T. Shen, D. Wang, J. Yun, Q. Liu, X. Liu, Z. Peng, Tribological properties and tribochemical analysis of nano-cerium oxide and sulfurized isobutene in titanium complex grease, Tribol. Int. 93 (2016) 332–346.
[48] M. Zhang, X.B. Wang, X.S. Fu, Y.Q. Xia, Performance and anti-wear mechanism of $CaCO_3$ nanoparticles as a green additive in poly-alpha-olefin, Tribol. Int. 42 (7) (2009) 1029–1039.
[49] X. Hou, J. He, L. Yu, Z. Li, Z. Zhang, P. Zhang, Preparation and tribological properties of fluorosilane surface-modified lanthanum trifluoride nanoparticles as additive of fluoro silicone oil, Appl. Surf. Sci. 316 (2014) 515–523.
[50] X. He, H. Xiao, H. Choi, A. Díaz, B. Mosby, A. Clearfield, et al., α-Zirconium phosphate nanoplatelets as lubricant additives, Colloids Surf. A Physicochem. Eng. Asp. 452 (2014) 32–38.
[51] N.N.M. Zawawi, W.H. Azmi, M.F. Ghazali, Tribological performance of Al_2O_3–SiO_2/PAG composite nanolubricants for application in air-conditioning compressor, Wear 492 (2022) 204238.
[52] X.H. Jia, J. Huang, Y. Li, J. Yang, H.J. Song, Monodisperse Cu nanoparticles @MoS_2 nanosheets as a lubricant additive for improved tribological properties, Appl. Surf. Sci. 494 (2019) 430–439.
[53] W. Song, J.C. Yan, H.B. Ji, Fabrication of GNS/MoS2 composite with different morphology and its tribological performance as a lubricant additive, Appl. Surf. Sci. 469 (2019) 226–235.
[54] H.J. Song, X.H. Jia, N. Li, X.F. Yang, H. Tang, Synthesis of α-Fe_2O_3 nanorod/graphene oxide composites and their tribological properties, J. Mater. Chem. 22 (3) (2012) 895–902.
[55] M.S. Zhang, B.B. Chen, H. Tang, G.G. Tang, C.S. Li, L. Chen, H.M. Zhang, Q. Zhang, Hydrothermal synthesis and tribological properties of FeS2 (pyrite)/reduced graphene oxide heterojunction, RSC Adv. 5 (2) (2015) 1417–1423.
[56] Y. Meng, F.H. Su, Y.Z. Chen, Supercritical fluid synthesis and tribological applications of silver nanoparticle-decorated graphene in engine oil nanofluid, Sci. Rep. 6 (2016) 31246.
[57] L. Wang, P.W. Gong, W. Li, T. Luo, B.Q. Cao, Mono-dispersed Ag/Graphene nanocomposite as lubricant additive to reduce friction and wear, Tribol. Int. 146 (2020) 106228.
[58] Y. Meng, F.H. Su, Y.Z. Chen, Synthesis of nano-Cu/graphene oxide composites by supercritical CO_2-assisted deposition as a novel material for reducing friction and wear, Chem. Eng. J. 281 (2015) 11–19.
[59] J. Zhao, Y.R. Li, Y.Y. He, J.B. Luo, In situ green synthesis of the new sandwichlike nanostructure of Mn_3O_4/graphene as lubricant additives, ACS Appl. Mater. Interfaces 11 (40) (2019) 36931–36938.
[60] T. Sugihara, Y. Nishimoto, T. Enomoto, On-machine tool resharpening process for dry machining of aluminum alloys employing LME phenomenon, Precis. Eng. 40 (2015) 241–248.

[61] S.J. Raykar, D.M. D'Addona, D. Kramar, Analysis of surface topology in dry machining of EN-8 steel, Proc. Mater. Sci. 6 (2014) 931–938.

[62] A. Thakur, S. Gangopadhyay, Dry machining of nickel-based super alloy as a sustainable alternative using TiN/TiAlN coated tool, J. Clean. Prod. 129 (2016) 256–268.

[63] V.S. Sharma, G. Singh, K. Sørby, A review on minimum quantity lubrication for machining processes, Mater. Manuf. Process. 30 (8) (2015) 935–953.

[64] V. Rodriguez, J. Sukumaran, A.K. Schlarb, P. De Baets, Influence of solid lubricants on tribological properties of polyetheretherketone (PEEK), Tribol. Int. 103 (2016) 45–57.

[65] J. Bijwe, K. Kumar, J.N. Panda, T. Parida, P. Trivedi, Design and development of high performance tribocomposites based on synergism in two solid lubricants, Compos. B Eng. 94 (2016) 399–410.

[66] R. Padmini, P. Vamsi Krishna, G. Krishna Mohana Rao, Experimental evaluation of nano-molybdenum disulphide and nano-boric acid suspensions in vegetable oils as prospective cutting fluids during turning of AISI 1040 steel, Proc. Inst. Mech. Eng., Part J: J. Eng. Tribol. 230 (2015) 1350650115601694.

[67] J. Deng, T. Can, J. Sun, Microstructure and mechanical properties of hot-pressed Al_2O_3/TiC ceramic composites with the additions of solid lubricants, Ceram. Int. 31 (2) (2005) 249–256.

[68] M. Matsui, H. Kakishima, Improvement of tribological performance of steel by solid lubricant shot-peening in dry rolling/sliding contact wear tests, Wear 260 (6) (2006) 669–673.

[69] N.S.K. Reddy, P.V. Rao, Experimental investigation to study the effect of solid lubricants on cutting forces and surface quality in end milling, Int. J. Mach. Tools Manuf. 46 (2) (2006) 189–198.

[70] M. Zalaznik, M. Kalin, S. Novak, G. Jakša, Effect of the type, size and concentration of solid lubricants on the tribological properties of the polymer PEEK, Wear 364 (2016) 31–39.

[71] P. Vamsi Krishna, D. Nageswara Rao, Performance evaluation of solid lubricants in terms of machining parameters in turning, Int. J. Mach. Tools Manuf. 48 (10) (2008) 1131–1137.

[72] K. Kiliçay, M. Ulutan, Investigation of the solid lubrication effect of commercial boron-based compounds in end milling, Int. J. Precis. Eng. Manuf. 17 (4) (2016) 517–524.

[73] J. Zhao, Y. Huang, Y. He, et al., Nanolubricant additives: A review, Friction 9 (2021) 891–917.

[74] N.F. Azman, S.A. Samion, M.N.H.M. Sot, Investigation of tribological properties of CuO/palm oil nanolubricant using pin-on-disc tribotester, Green Mater. 6 (1) (2018) 30–37.

[75] H. Li, R. Papadakis, S. Jafri, et al., Superior adhesion of graphene nanoscrolls, Commun. Phys. 1 (2018) 44.

[76] M. Asnida, S. Hisham, N.W. Awang, A.K. Amirruddin, M.M. Noor, K. Kadirgama, D. Ramasamy, G. Najafi, F. Tarlochan, Copper (II) oxide nanoparticles as additve in engine oil to increase the durability of piston-liner contact, Fuel 212 (2018) 656–667.

[77] K. Chowdary, A. Kotia, V. Lakshmanan, A.H. Elsheikh, M.K.A. Ali, A review of the tribological and thermophysical mechanisms of bio-lubricants based nanomaterials in automotive applications, J. Mol. Liq. 339 (2021) 116717.

[78] I.O. Sadiq, S. Sharif, M.A. Suhaimi, N.M. Yusof, D.W. Kim, K.H. Park, Enhancement of thermo-physical and lubricating properties of SiC nanolubricants for machining operation, Procedia Manuf. 17 (2018) 166–173.

[79] L. Yang, W. Jiang, W. Ji, O. Mahian, S. Bazri, R. Sadri, I.A. Badruddin, S. Wongwises, A review of heating/cooling processes using nanomaterials suspended in refrigerants and lubricants, Int. J. Heat Mass Transfer 153 (2020) 119611.

[80] M.K.A. Ali, P. Fuming, H.A. Younus, M.A.A. Abdelkareem, F.A. Essa, A. Elagouz, H. Xianjun, Fuel economy in gasoline engines using Al_2O_3/TiO_2 nanomaterials as nanolubricant additives, Appl. Energy 211 (2018) 461–478.

[81] M.K.A. Ali, H. Xianjun, Role of bis(2-ethylhexyl) phosphate and Al_2O_3/TiO_2 hybrid nanomaterials in improving the dispersion stability of nanolubricants, Tribol. Int. 155 (2021) 106767.

[82] J.K. Katiyar, S.K. Sinha, A. Kumar, Friction and wear durability study of epoxy-based polymer (SU-8) composite coatings with talc and graphite as fillers, Wear 362 (2016) 199–208.

[83] H. Ghaednia, R.L. Jackson, The effect of nanoparticles on the real area of contact, friction, and wear, ASME. J. Tribol. 135 (4) (2013) 041603.

[84] A. Bhattad, J. Sarkar, Effects of nanoparticle shape and size on the thermohydraulic performance of plate evaporator using hybrid nanofluids, J. Therm. Anal. Calorim. 143 (2021) 767–779.

[85] J.P. Singh, S. Singh, T. Nandi, S.K. Ghosh, Development of graphitic lubricant nanoparticles based nanolubricant for automotive applications: thermophysical and tribological properties followed by IC engine performance, Powder Technol. 387 (2021) 31–47.

[86] S. Perry, W. Tysoe, Frontiers of fundamental tribological research, Tribol. Lett. 19 (3) (2005) 151–161.

[87] H. Shinde, et al., Mechanically stable nano-graphite polyaniline composites: synthesis and characterization, Intern. J. Innov. Res. Sci. Eng. Technol. 3 (6) (2014) 14164–14171.

[88] Y.M. Zhao, et al., Preparation of nano-graphite films and field emission properties, Chin. Phys. Lett. 21 (5) (2004) 904–906.

[89] Y. Zhang, et al., Preparation of nanographite sheets supported Si nanoparticles by in situ reduction of fumed SiO_2 with magnesium for lithium ion battery, J. Power Sources 281 (2015) 425–431.

[90] Q. Chen, et al., Preparation of water-soluble nanographite and its application in water-based cutting fluid, Nanoscale Res. Lett. 8 (1) (2013) 52.

[91] Y.V. Ioni, et al., Preparation of finely dispersed nanographite, Inorg. Mater. 47 (6) (2011) 597–602.

[92] Z.-H. Jin, et al., Nano-graphite synthesized by explosive detonation and its application in preparing diamond under high-pressure and high-temperature, Acta Phys. Sin. 53 (4) (2004) 1260–1264.

[93] M.K.A. Ali, X. Hou, M.A.A. Abdelkareem, Anti-wear properties evaluation of frictional sliding interfaces in automobile engines lubricated by copper/graphene nanolubricants, Friction 8 (2020) 905–916.

[94] A. Ahmed, A.N. Shah, A. Azam, G.M. Uddin, M.S. Ali, S. Hassan, H. Ahmed, T. Aslam, Environment-friendly novel fuel additives: Investigation of the effects of graphite nanoparticles on performance and regulated gaseous emissions of CI engine, Energy Convers. Manage. 211 (2020) 112748.

[95] J.P. Singh, S. Singh, T. Nandi, S.K. Ghosh, Development of graphitic lubricant nanoparticles based nanolubricant for automotive applications: Thermophysical and tribological properties followed by IC engine performance, Powder Technol. 387 (2021) 31–47.

[96] M.H. Cho, J. Ju, S.J. Kim, H. Jang, Tribological properties of solid lubricants (graphite, Sb_2S_3, MoS_2) for automotive brake friction materials, Wear 260 (2006) 855–860.

[97] O.R. Raftar, M. Kaveh, M. Khajehzadeh, A. Rahimi, M.R. Razfar, Nano-lubricant influence on surface residual stresses in hard milling, Proc. Inst. Mech. Eng., Part E (2021). https://doi.org/10.1177/09544089211004793.

[98] K.L. Jin, K. Junemo, H. Hiki, T.K. Yong, The effects of nanoparticles on absorption heat and mass transfer performance in NH_3/H_2O binary nanofluids, Int. J. Refrig. 33 (2010) 269–275.

[99] M.A. Akhavan-Behabadi, M.K. Sadoughi, M. Darzi, M. Fakoor-Pakdaman, Experimental study on heat transfer characteristics of R600a/POE/CuO nano-refrigerant flow condensation, Exp. Therm. Fluid Sci. 66 (2015) 46–52.

[100] W. Zhang, D. Demydov, M.P. Jahan, K. Mistry, A. Erdemir, A.P. Malshe, Fundamental understanding of the tribological and thermal behavior of Ag–MoS2 nanoparticlebased multi-component lubricating system, Wear 288 (2012) 9–16.

[101] X.L. Wang, Y.L. Yin, G.N. Zhang, W.Y. Wang, K.K. Zhao, Study on antiwear and repairing performances about mass of nano-copper lubricating additives to 45 steel, Phys. Procedia 50 (2013) 466–472.

[102] G. Zhou, Y. Zhu, X. Wang, M. Xia, Y. Zhang, H. Ding, Sliding tribological properties of 0.45% carbon steel lubricated with Fe_3O_4 magnetic nano-particle additives in base oil, Wear 301 (2013) 753–757.

[103] A.W. Zia, New generation carbon particles embedded diamond-like carbon coatings for transportation industry, in: A.S.H. Makhlouf, N.Y. Abu-Thabit (Eds.), Advances in Smart Coatings and Thin Films for Future Industrial and Biomedical Engineering Applications, Elsevier, India, 2020, pp. 307–332.
[104] M. Matsui, H. Kakishima, Improvement of tribological performance of steel by solid lubricant shot-peening in dry rolling/sliding contact wear tests, Wear 260 (2006) 669–673.
[105] M. Anthony Xavior, M. Adithan, Determining the influence of cutting fluids on tool wear and surface roughness during turning of AISI 304 austenitic stainless steel, J. Mater. Process. Technol. 209 (2009) 900–909.
[106] N. Talib, E.A. Rahim, The effect of tribology behavior on machining performances when using bio-based lubricant as a sustainable metalworking fluid, Proc. CIRP 40 (2016) 504–508.
[107] Y. Zhang, C. Li, D. Jia, D. Zhang, X. Zhang, Experimental evaluation of MoS2 nanoparticles in jet MQL grinding with different types of vegetable oil as base oil, J. Clean. Prod. 87 (2015) 930–940.
[108] Y. Wang, C. Li, Y. Zhang, M. Yang, B. Li, D. Jia, Y. Hou, C. Mao, Experimental evaluation of the lubrication properties of the wheel/workpiece interface in minimum quantity lubrication (MQL) grinding using different types of vegetable oils, J. Clean. Prod. 127 (2016) 487–499.
[109] R.R. Srikant, V.S.N.V. Ramana, Performance evaluation of vegetable emulsifier based green cutting fluid in turning of American Iron and Steel Institute (AISI) 1040 steel–an initiative towards sustainable manufacturing, J. Clean. Prod. 108 (2015) 104–109.
[110] T. Luo, X. Wei, X. Huang, L. Huang, F. Yang, Tribological properties of Al_2O_3 nanoparticles as lubricating oil additives, Ceram. Int. 40 (2014) 7143–7149.
[111] W. Li, S. Zheng, B. Cao, S. Ma, Friction and wear properties of ZrO_2/SiO_2 composite nanoparticles, J. Nanopart. Res. 13 (2011) 2129–2137.
[112] S. Zhang, L. Hu, D. Feng, H. Wang, Anti-wear and friction-reduction mechanism of Sn and Fe nanoparticles as additives of multialkylated cyclopentanes under vacuum condition, Vacuum 87 (2013) 75–80.
[113] X. Song, S. Zheng, J. Zhang, W. Li, Q. Chen, B. Cao, Synthesis of monodispersed $ZnAl_2O_4$ nanoparticles and their tribology properties as lubricant additives, Mater. Res. Bull. 47 (2012) 4305–4310.
[114] J. Deng, T. Can, J. Sun, Microstructure and mechanical properties of hot-pressed Al_2O_3/TiC ceramic composites with the additions of solid lubricants, Ceram. Int. 31 (2) (2005) 249–256.
[115] T. Luo, X. Wei, H. Zhao, G. Cai, X. Zheng, Tribology properties of Al_2O_3/TiO_2 nanocomposites as lubricant additives, Ceram. Int. 40 (2014) 10103–10109.
[116] I. Ali, A. Basheer, A. Kucherova, N. Memetov, T. Pasko, K. Ovchinnikov, V. Pershin, D. Kuznetsov, E. Galunin, V. Grachev, A. Tkachev, Advances in carbon nanomaterials as lubricants modifiers, J. Mol. Liq. 279 (2019) 251–266.
[117] R. Saidur, S.N. Kazi, M.S. Hossain, M.M. Rahman, H.A. Mohammed, A review on the performance of nanoparticles suspended with refrigerants and lubricating oils in refrigeration systems, Renew. Sustain. Energy Rev. 15 (2011) 310–323.
[118] K. Holmberg, P. Andersson, A. Erdemir, Global energy consumption due to friction in passenger cars, Tribol. Int. 47 (2012) 221–234.
[119] R.I. Taylor, R.C. Coy, Improved fuel efficiency by lubricant design: a review, Proc. Inst. Mech. Eng. 214 (1999) 1–15.
[120] M.K.A. Ali, X.J. Hou, L.Q. Mai, Q.P. Cai, R.F. Turkson, B.C. Chen, Improving the tribological characteristics of piston ring assembly in automotive engines using Al_2O_3 and TiO_2 nanomaterials as nano-lubricant additives, Tribol. Int. 103 (2016) 540–554.
[121] M.K.A. Ali, H. Xianjun, F.A. Essa, M.A.A. Abdelkareem, A. Elagouz, S.W. Sharshir, Friction and wear reduction mechanisms of the reciprocating contact interfaces using nanolubricant under different loads and speeds, ASME. J. Tribol. 140 (5) (2018) 051606.

[122] P. Kalita, A.P. Malshe, S.A. Kumar, V.G. Yoganath, T. Gurumurthy, Study of specific energy and friction coefficient in minimum quantity lubrication grinding using oil-based nanolubricants, J. Manuf. Process. 14 (2012) 160–166.

[123] D. Wen, G. Lin, S. Vafaei, K. Zhang, Review of nanofluids for heat transfer applications, Particuology 7 (2009) 141–150.

[124] P.B. Maheshwary, C.C. Handa, K.R. Nemade, Effect of shape on thermophysical and heat transfer properties of ZnO/R-134a nanorefrigerant, Mater. Today 5 (2018) 1635–1639.

[125] H. Masuda, A. Ebata, K. Teramae, N. Hishinuma, Alteration of thermal conductivity and viscosity of liquid by dispersing ultra-fine particles (dispersion of y-$A1_2O_3$, SiO_2, and TiO_2 ultrafine particles), Netsu Bussei 4 (1993) 227–233.

[126] S. Lee, Measuring thermal conductivity of fluids containing oxide nanoparticles, ASME J. Heat Transfer 121 (2) (1999) 280–289.

[127] L.P. Zhou, B.X. Wang, Experimental research on the thermophysical properties of nanoparticle suspensions using the quasi-steady method, Ann. Proc. Chin. Eng. Thermophys. (2002) 889–892.

[128] Y. Xuan, Q. Li, Heat transfer enhancement of nanofluids, Int. J. Heat Fluid Flow 21 (2000) 58–64.

[129] H. Xie, Nanofluids containing multiwalled carbon nanotubes and their enhanced thermal properties, J. Appl. Phys. 94 (8) (2003) 4967–4971.

[130] S.U.S. Choi, Z.G. Zhang, W. Yu, F.E. Lockwood, E.A. Grulke, Anomalous thermal conductivity enhancement in nanotube suspensions, Appl. Phys. Lett. 79 (14) (2001) 2252–2254.

[131] V. Kumaresan, R. Velraj, K. Das Sarit, Convective heat transfer characteristics of secondary refrigerant based CNT nanofluids in a tubular heat exchanger, Int. J. Refrig. 35 (2012) 2287–2296.

[132] H. Xie, Thermal conductivity enhancement of suspensions containing nanosized alumna particles, J. Appl. Phys. 91 (2002) 4568–4572.

[133] H.E. Patel, S.K. Das, T. Sundararajan, Thermal conductivities of naked and monolayer protected metal nanoparticle based nanofluids: manifestation of anomalous enhancement and chemical effects, Appl. Phys. Lett. 83 (2003) 2931–2933.

[134] T.K. Hong, H.S. Yang, C.J. Choi, Study of the enhanced thermal conductivity of Fe nanofluids, J. Appl. Phys. 97 (2005) 064311.

[135] M.R.M. Nawi, M.Z.A. Rehim, W.H. Azmi, S.A. Razak, The characterization and thermo-physical property investigations of SiO_2/HFE7000 nanorefrigerants, Int. J. Refrig. 88 (2018) 275–283.

[136] I.M. Mahbubul, A. Saadah, R. Saidur, M.A. Khairul, A. Kamyar, Thermal performance analysis of Al_2O_3/R-134a nanorefrigerant, Int. J. Heat Mass Transf. 85 (2015) 1034–1040.

[137] L. Yang, K. Du, X.S. Zhang, B. Cheng, Preparation and stability of Al_2O_3 nano-particle suspension of ammonia–water solution, Appl. Therm. Eng. 31 (17–18) (2011) 3643–3647.

[138] S.M.S. Murshed, K.C. Leong, C. Yang, Enhanced thermal conductivity of TiO_2–water based nanofluids, Int. J. Therm. Sci. 44 (4) (2005) 367–373.

[139] M.J. Assael, Thermal conductivity of suspensions of carbon nanotubes in water, Int. J. Thermophys. 25 (4) (2004) 971–985.

[140] P.B. Maheshwary, C.C. Handa, K.R. Nemade, Effect of shape on thermophysical and heat transfer properties of ZnO/R-134a nanorefrigerant, Mater. Today 5 (2018) 1635–1639.

[141] B. Sun, D. Yang, Flow boiling heat transfer characteristics of nano-refrigerants in a horizontal tube, Int. J. Refrig. 38 (2014) 206–214.

[142] H. Peng, G. Ding, H. Hu, W. Jiang, Influence of carbon nanotubes on nucleate pool boiling heat transfer characteristics of refrigerant–oil mixture, Int. J. Therm. Sci. 49 (2010) 2428–2438.

[143] P. Hao, D. Guoliang, J. Weiting, H. Haitao, G. Yifeng, Heat transfer characteristics of refrigerant-based nanofluid flow boiling inside a horizontal smooth tube, Int. J. Refrig. 32 (2009) 1259–1270.

[144] K.J. Park, D.S. Jung, Boiling heat transfer enhancement with carbon nanotubes for refrigerants used in building air conditioning, Energy Build. 39 (9) (2007) 1061–1064.

[145] Z. Liu, R.H.S. Winterton, A general correlation for saturated and subcooled flow boiling in tubes and annuli, based on a nucleate pool boiling equation, Int. J. Heat Mass Transf. 34 (1991) 2759.

[146] P. Hao, D. Guoliang, H. Haitao, J. Weiting, Z. Dawei, W. Kaijiang, Nucleate pool boiling heat transfer characteristics of refrigerant/oil mixture with diamond nanoparticles, Int. J. Refrig. 33 (2010) 347–358.

[147] Y.A. Cengel, A.J. Ghajar, Heat and Mass Transfer, Fundamentals and Applications, 5th Ed., McGraw-Hill Education, New York, USA, 2014.

[148] S.S. Sanukrishna, M. Shafi, M. Murukan, M.Jose Prakash, Effect of SiO_2 nanoparticles on the heat transfer characteristics of refrigerant and tribological behaviour of lubricant, Powder Technol. 356 (2019) 39–49.

[149] M.A. Kedzierski, M. Gong, Effect of CuO Nanolubricant on R134a Pool Boiling Heat Transfer With Extensive Measurement and Analysis Details, USA: NISTIR 7336, National Institute of Standards and Technology, Gaithersburg, USA, 2007.

[150] E.I. Eid, R.A. Khalaf-Allah, M. Tolan, Enhancement of pool boiling characteristics by an addition of nano Aluminum oxide to R-141b over a rough horizontal steel circular heater, Int. J. Refrig. 98 (2019) 311–322.

[151] M.R.A. Hady, M.S. Ahmed, G. Abdallah, Experimental investigation on the performance of chilled-water air conditioning unit using alumina nanofluids, Therm. Sci. Eng. Prog. 5 (2018) 589–596.

[152] G. Jatinder, O.S. Ohunakin, D.S. Adelekan, O.E. Atiba, A.B. Daniel, J. Singh, A.A. Atayero, Performance of a domestic refrigerator using selected hydrocar- bon working fluids and TiO_2–Mo nanolubricant, Appl. Therm. Eng. 160 (2019) 114004.

[153] V. Kumaresan, R. Velraj, K. Das Sarit, Convective heat transfer characteristics of secondary refrigerant based CNT nanofluids in a tubular heat exchanger, Int. J. Refrig. 35 (2012) 2287–2296.

[154] H. Khodadadi, S. Aghakhani, H. Majd, R. Kalbasi, S. Wongwises, M. Afrand, A comprehensive review on rheological behavior of mono and hybrid nanofluids: effective parameters and predictive correlations, Int. J. Heat Mass Transfer 127 (2018) 997–1012.

[155] M.H. Esfe, S. Saedodin, S. Wongwises, D. Toghraie, An experimental study on the effect of diameter on thermal conductivity and dynamic viscosity of Fe/water nanofluids, J. Therm. Anal. Calorim. 119 (2015) 1817–1824.

[156] R. Prasher, D. Song, J. Wang, P. Phelan, Measurements of nanofluid viscosity and its implications for thermal applications, Appl. Phys. Lett. 89 (2006) 133108.

[157] A. Turgut, I. Tavman, M. Chirtoc, H.P. Schuchmann, C. Sauter, S. Tavman, Thermal conductivity and viscosity measurements of water-based TiO_2 nanofluids, Int. J. Thermophys. 30 (2009) 1213–1226.

[158] V.Y. Rudyak, A.V. Minakov, S.L. Krasnolutskii, Physics and mechanics of heat exchange processes in nanofluid flows, Phys. Mesomech. 19 (3) (2016) 298–306.

[159] M.H. Esfe, S. Saedodin, O. Mahian, S. Wongwises, Efficiency of ferromagnetic nanoparticles suspended in ethylene glycol for applications in energy devices:effects of particle size, temperature, and concentration, Int. Commun. Heat Mass Transf. 58 (2014) 138–146.

[160] G.A. Oliveira, E.P. Bandarra Filho, D. Wen, Synthesis and characterization of silver/water nanofluids, High Temp.—High Press. 43 (2014) 69–83.

[161] Y. He, Y. Jin, H. Chen, Y. Ding, D. Cang, H. Lu, Heat transfer and flow behaviour of aqueous suspensions of TiO_2 nanoparticles (nanofluids) flowing upward through a vertical pipe, Int. J. Heat Mass Transf. 50 (11–12) (2007) 2272–2281.

[162] A.V. Minakov, D.V. Guzei, M.I. Pryazhnikov, V.A. Zhigarev, V.Y. Rudyak, Study of turbulent heat transfer of the nanofluids in a cylindrical channel, Int. J. Heat Mass. Transf. 102 (2016) 745–755.

[163] D. Kwek, A. Crivoi, F. Duan, Effects of temperature and particle size on the thermal property measurements of Al_2O_3—water nanofluids, J. Chem. Eng. Data 55 (12) (2010) 5690–5695.

[164] L.S. Sundar, M.K. Singh, A.C. Sousa, Investigation of thermal conductivity and viscosity of Fe_3O_4 nanofluid for heat transfer applications, Int. Commun. Heat Mass Transfer 44 (2013) 7–14.

[165] L. Chen, H. Xie, W. Yu, Y. Li, Rheological behaviors of nanofluids containing multiwalled carbon nanotube, J. Dispers. Sci. Technol. 32 (2011) 550–554.

[166] K.B. Anoop, T. Sundararajan, S.K. Das, Effect of particle size on the convective heat transfer in nanofluid in the developing region, Int. J. Heat Mass Transf. 52 (2009) 2189–2195.

[167] S. Bobbo, L. Fedele, A. Benetti, L. Colla, M. Fabrizio, C. Pagura, S. Barison, Viscosity of water based SWCNH and TiO_2 nanofluids, Exp. Therm. Fluid Sci. 36 (2012) 65–71.

[168] A.K. Rasheed, Heat Transfer, Tribology and Performance of Graphene Nanolubricants in an IC Engine (PhD thesis), University of Nottingham, United Kingdom, 2017.

[169] L. Jaekeun, C. Sangwon, H. Yujin, L. Changgun, H.K. Soo, Enhancement of lubrication properties of nano-oil by controlling the amount of fullerene nanoparticle additives, Tribolo. Lett. 28 (2007) 203–208.

[170] W. Zhai, W. Lu, X. Liu, L. Zhou, Nanodiamond as an effective additive in oil to dramatically reduce friction and wear for fretting steel/copper interfaces, Tribol. Int. 129 (2019) 75–81.

[171] L. Wu, L. Gu, R. Jian, Lubrication mechanism of graphene nanoplates as oil additives for ceramics/steel sliding components, Cera. Int. 47 (2021) 16935–16942.

[172] Y.J.J. Jason, H.G. How, Y.H. Teoh, H.G. Chuah, A study on the tribological performance of nanolubricant, Processes 8 (2020) 1372.

[173] M. Flores-Castañeda, E. Camps, M. Camacho López, S. Muhl, E. García, M. Figueroa, Bismuth nanoparticles synthesized by laser ablation in lubricant oils for tribological tests, J. Alloy Compd. 643 (2015) S67–S70.

[174] J. Ma, Y. Mo, M. Bai, Effect of Ag nanoparticles additive on the tribological behavior of multialkylated cyclopentanes (MACs), Wear 266 (2009) 627–631.

[175] M.D. Abad, J.C. Sánchez-López, Tribological properties of surface-modified Pd nanoparticles for electrical contacts, Wear 297 (2013) 943–951.

[176] R. Chou, A.H. Battez, J.J. Cabello, J.L. Viesca, A. Osorio, A. Sagastume, Tribological behavior of polyalphaolefin with the addition of nickel nanoparticles, Tribol. Int. 43 (2010) 2327–2332.

[177] J.C. Sánchez-López, M.D. Abad, L. Kolodziejczyk, E. Guerrero, A. Fernández, Surface-modified Pd and Au nanoparticles for anti-wear applications, Tribol. Int. 44 (2011) 720–726.

[178] J. Padgurskas, R. Rukuiza, I. Prosyčevas, R. Kreivaitis, Tribological properties of lubricant additives of Fe, Cu and Co nanoparticles, Tribol. Int. 60 (2013) 224–232.

[179] X.L. Wang, Y.L. Yin, G.N. Zhang, W.Y. Wang, K.K. Zhao, Study on antiwear and repairing performances about mass of nano-copper lubricating additives to 45 steel, Phys. Procedia 50 (2013) 466–472.

[180] M.-J. Kao, C.-R. Lin, Evaluating the role of spherical titanium oxide nanoparticles in reducing friction between two pieces of cast iron, J. Alloy Compd. 483 (2009) 456–459.

[181] L. Peña-Parás, J. Taha-Tijerina, L. Garza, D. Maldonado-Cortés, R. Michalczewski, C. Lapray, Effect of CuO and Al_2O_3 nanoparticle additives on the tribological behavior of fully formulated oils, Wear 332–333 (2015) 1256–1261.

[182] K. Gu, B. Chen, Y. Chen, Preparation and tribological properties of lanthanum-doped TiO_2 nanoparticles in rapeseed oil, J. Rare Earths 31 (2013) 589–594.

[183] S.M. Alves, B.S. Barros, M.F. Trajano, K.S.B. Ribeiro, E. Moura, Tribological behavior of vegetable oil-based lubricants with nanoparticles of oxides in boundary lubrication conditions, Tribol. Int. 65 (2013) 28–36.

[184] A. Hernandez Battez, J.E. Fernandez Rico, A. Navas Arias, J.L. Viesca Rodriguez, R. Chou Rodriguez, J.M. Diaz Fernandez, The tribological behaviour of ZnO nanoparticles as an additive to PAO6, Wear 261 (2006) 256–263.

[185] Yosef Jazaa, in: Effect of Nanoparticle Additives on the Tribological Behavior of Oil Under Boundary Lubrication (Graduate theses and dissertations), Lowa State University, USA, 2018, p. 16773.
[186] B. Bhushan, B.K. Gupta, G.W.V. Cleef, C. Capp, J.V. Coe, Fullerene (C60) films for solid lubrication, Tribol. Transac. 36 (1993) 573–580.
[187] C. Chou, S. Lee, Tribological behavior of nanodiamond-dispersed lubricants on carbon steels and aluminum alloy, Wear 269 (2010) 757–762.
[188] C.E. Morstein, M. Dienwiebel, Graphite lubrication mechanisms under high mechanical load, Wear 477 (2021) 203794.
[189] W. Zhai, X. Shi, M. Wang, Z. Xu, J. Yao, S. Song, Y. Wang, Grain refinement: a mechanism for graphene nanoplatelets to reduce friction and wear of Ni3Al matrix self-lubricating composites, Wear 310 (2014) 33–40.
[190] B. Gupta, N. Kumar, K. Panda, et al., Energy efficient reduced graphene oxide additives: mechanism of effective lubrication and antiwear properties, Sci. Rep. 6 (2016) 18372.
[191] B. Wang, W. Tang, X. Liu, et al., Synthesis of ionic liquid decorated muti-walled carbon nanotubes as the favorable water-based lubricant additives, Appl. Phys. A 123 (2017) 680.
[192] B. Yu, Z. Liu, C. Ma, J. Sun, W. Liu, F. Zhou, Ionic liquid modified multi-walled carbon nanotubes as lubricant additive, Tribol. Int. 81 (2015) 38–42.
[193] Z. Li, X. Hou, L. Yu, Z. Zhang, P. Zhang, Preparation of lanthanum trifluoride nanoparticles surface-capped by tributyl phosphate and evaluation of their tribological properties as lubricant additive in liquid paraffin, Appl. Surf. Sci. 292 (2014) 971–977.
[194] T. Shen, D. Wang, J. Yun, Q. Liu, X. Liu, Z. Peng, Tribological properties and tribochemical analysis of nano-cerium oxide and sulfurized isobutene in titanium complex grease, Tribol. Int. 93 (2016) 332–346.
[195] X. Hou, J. He, L. Yu, Z. Li, Z. Zhang, P. Zhang, Preparation and tribological properties of fluorosilane surface-modified lanthanum trifluoride nanoparticles as additive of fluoro silicone oil, Appl. Surf. Sci. 316 (2014) 515–523.
[196] C. Zhang, S. Zhang, L. Yu, Z. Zhang, Z. Wu, P. Zhang, Preparation and tribological properties of water-soluble copper/silica nanocomposite as a water-based lubricant additive, Appl. Surf. Sci. 259 (2012) 824–830.
[197] Y. Meng, F. Su, Y. Chen, Synthesis of nano-Cu/graphene oxide composites by supercritical CO_2-assisted deposition as a novel material for reducing friction and wear, Chem. Eng. J. 281 (2015) 11–19.
[198] D. Jiao, S. Zheng, Y. Wang, R. Guan, B. Cao, The tribology properties of alumina/silica composite nanoparticles as lubricant additives, Appl. Surf. Sci. 257 (2011) 5720–5725.
[199] F. Zhao, Z. Bai, Y. Fu, D. Zhao, C. Yan, Tribological properties of serpentine, $La(OH)_3$ and their composite particles as lubricant additives, Wear 288 (2012) 72–77.
[200] M. Molana, H. Wang, A critical review on numerical study of nanorefrigerant heat transfer enhancement, Powder Technol. 368 (2020) 18–31.
[201] A. Einstein, Eine neue bestimmung der molek dimensionen, Ann. Phys. 324 (2) (1906) 289–306.
[202] H.C. Brinkman, The viscosity of concentrated suspensions and solutions, J. Chem. Phys. 20 (1952) 571.
[203] T. Lundgren, Slow flow through stationary random beds and suspensions of spheres, J. Fluid Mech. 51 (1972) 273–299.
[204] G. Batchelor, The effect of Brownian motion on the bulk stress in a suspension of spherical particles, J. Fluid Mech. 83 (1977) 97–117.
[205] N.A. Frankel, A. Acrivos, On the viscosity of a concentrate suspension of solid spheres, Chem. Eng. Sci. 22 (1967) 847–853.
[206] A. Graham, On the viscosity of suspensions of solid spheres, Appl. Sci. Res. 37 (1981) 275–286.
[207] M. Mooney, A shearing disk plastometer for unvulcanized rubber, Ind. Eng. Chem. Anal. Ed. 6 (1934) 147–151.

[208] V.H. Eilers, Die viskocitat von emulsionen hochviskoser stoffe als funktion der konzentration, Kolloid Z. 97 (1941) 313–321.

[209] E.W. Bucholz, S.R. Phillpot, S.B. Sinnott, Molecular dynamics investigation of the lubrication mechanism of carbon nano-onions, Comput. Mater. Sci. 54 (2012) 91–96.

[210] K. Hayashi, K. Tezuka, N. Ozawa, T. Shimazaki, K. Adachi, M. Kubo, Tribochemical reaction dynamics simulation of hydrogen on a diamond-like carbon surface based on tight-binding quantum chemical molecular dynamics, J. Phys. Chem. C 115 (46) (2011) 22981–22986.

[211] A. Erdemir, O. Eryilmaz, Ramirez Gonzalez, Superlubricious carbon films derived from natural gas (U.S. Patent No. 10,702,862). U.S, Patent and Trademark Office (2020).

[212] A. Erdemir, A.U. Mane, J.W. Elam, G. Ramirez, O. Eryilmaz, Producing Carbon-Based Boundary Films From Catalytically Active Lubricant Additives (US Pat. US9951291B2), 2018.

[213] D. Berman, B. Narayanan, M.J. Cherukara, S.K.R.S. Sankaranarayanan, A. Erdemir, A. Zinovec, A.V. Sumant, Operando tribochemical formation of onion-like-carbon leads to macroscale superlubricity, Nat. Commun. 9 (2018) 1164.

[214] N. Argibay, T.F. Babuska, J.F. Curry, et al., In-situ tribochemical formation of self-lubricating diamond-like carbon films, Carbon 138 (2018) 61–68.

[215] J. Zhang, Q. Xu, L. Gao, T. Ma, M. Qiu, Y. Hu, H. Wang, J. Luo, A molecular dynamics study of lubricating mechanism of graphene nanoflakes embedded in Cu-based nanocomposite, Appl. Surf. Sci. 511 (2020) 145620.

[216] P. Ghosh, M. Das, Study of the influence of some polymeric additives as viscosity index improvers and pour point depressants e synthesis and characterization, J. Pet. Sci. Eng. 119 (2014) 79–84.

[217] E.S. Forbes, Antiwear and extreme pressure additives for lubricants, Tribology 3 (1970) 145–152.

[218] J. Cui, D. Daniel, A. Grinthal, K. Lin, J. Aizenberg, Dynamic polymer systems with self-regulated secretion for the control of surface properties and material healing, Nat. Mater. 14 (8) (2015) 790–795.

[219] Q. Han, Z. Ding, Z. Qin, T. Wang, X. Xu, F. Chu, A triboelectric rolling ball bearing with self-powering and self-sensing capabilities, Nano Energy 67 (2020) 104277.

[220] H. Cao, F. Wen, J.T.M. De Hosson, Y.T. Pei, Instant WS2 platelets reorientation of self-adaptive WS_2/a-C tribocoating, Mater. Lett. 229 (2018) 64–67.

[221] O. Rifaie-Graham, E.A. Apebende, L.K. Bast, N. Bruns, Self-reporting fiber-reinforced composites that mimic the ability of biological materials to sense and report damage, Adv. Mater. 30 (19) (2018) 16.

[222] H. Gong, C. Yu, L. Zhang, G. Xie, D. Guo, J. Luo, Intelligent lubricating materials: a review, Compos. Part B Eng. 202 (2020) 108450.

[223] M. Vafaei, M. Afrand, N. Sina, R. Kalbasi, F. Sourani, H. Teimouri, Evaluation of thermal conductivity of MgO-MWCNTs/EG hybrid nanofluids based on experimental data by selecting optimal artificial neural networks, Physica E Low Dimens. Syst. Nanostruct. 85 (2017) 90–96.

[224] G. Sharma, A. Kotia, S.K. Ghosh, et al., Kinematic viscosity prediction of nanolubricants employed in heavy earth moving machinery using machine learning techniques, Int. J. Precis. Eng. Manuf. 21 (2020) 1921–1932.

[225] S.A. Hussain, E.W.M Lee, R.K.K. Yuen, W. Wang, L. Wang, Dissimilarity-based boosting technique for the modelling of complex HVAC systems, Energy Build. 247 (2021) 111151.

CHAPTER

Studies of nanorefrigerants—Properties, characterization, and applications

9

O.A. Alawi[a], H.M. Kamar[a], A.R. Mallah[b] and Kazi Salim Newaz[b]

[a]*Department of Thermofluids, School of Mechanical Engineering, Universiti Teknologi Malaysia, Johor Bahru, Malaysia,* [b]*Department of Mechanical Engineering, Faculty of Engineering, Universiti Malaya, Kuala Lumpur, Malaysia*

9.1 Introduction

A nanorefrigerant refers to a novel working fluid containing suspended nanoparticles in a pure refrigerant. Maxwell [1] originally developed the dissolving approach of solid nanoparticles. Solid microparticles were dispersed in the base fluid to enhance the working fluid's thermophysical properties. However, the produced nanorefrigerant suffered significant difficulties, including long-term stability, corrosion, and erosion. Choi [2] introduced the first principle of nanofluids by mixing solid nanoparticles in base fluids. The reason behind nanofluids development was usually to enhance the effectiveness of heat transfer investigation of various heat transfer working fluids, which has recently been much extended to refrigerants.

Many previous studies showed that the base fluids' effective thermal conductivity was enhanced by adding solid nanoparticles [3,4]. The effective thermal conductivity was increased by about 104% by adding carbon nanotubes (CNTs) to pure R-113 [5]. Nanofluid investigations are not limited to thermal conductivity; there are different thermophysical properties, including dynamic viscosity, density, specific heat capacity, and surface tension.

Boiling heat transfer is among the most significant issues that need to be examined for refrigerants. Once nanoparticles were added to the pure refrigerants, the boiling heat transfer became a complex phenomenon. Investigators noticed different conclusions in evaluating the heat transfer rates of refrigerant-based nanofluids [6,7]. Previously, results stated that the heat transfer efficiency of nanoparticles was greater than that of nanoparticle-based refrigerants. Pool boiling and flow boiling were discussed, but several pieces of research centered on nanorefrigerant pool boiling. The pool boiling and flow boiling of nanorefrigerants are discussed in this chapter.

The research area of nanorefrigerants can be divided into two categories such as (1) pure nanorefrigerants and (2) nanolubricants [8]. Nanolubricants are the new trend among investigators as efficient chilling units [9–12]. The cooling unit performance was enhanced by introducing nanoparticles to the coolant lubricants. Lee et al. [13] found that using nanooil decreased the friction coefficient by about 90% relative to raw oil. Fullerene C60 with nanooil demonstrated an improvement of 5.26% and 5.3% in two different compressors [14]. They stated a reduction of 3°C–5°C in the shell compressor temperature, where reduced shell compressor temperature verified the improved lubricant stability.

Graphene and 2D Materials in Heat Transfer: Fundamentals and Applications. DOI: https://doi.org/10.1016/B978-0-12-821964-5.00006-0

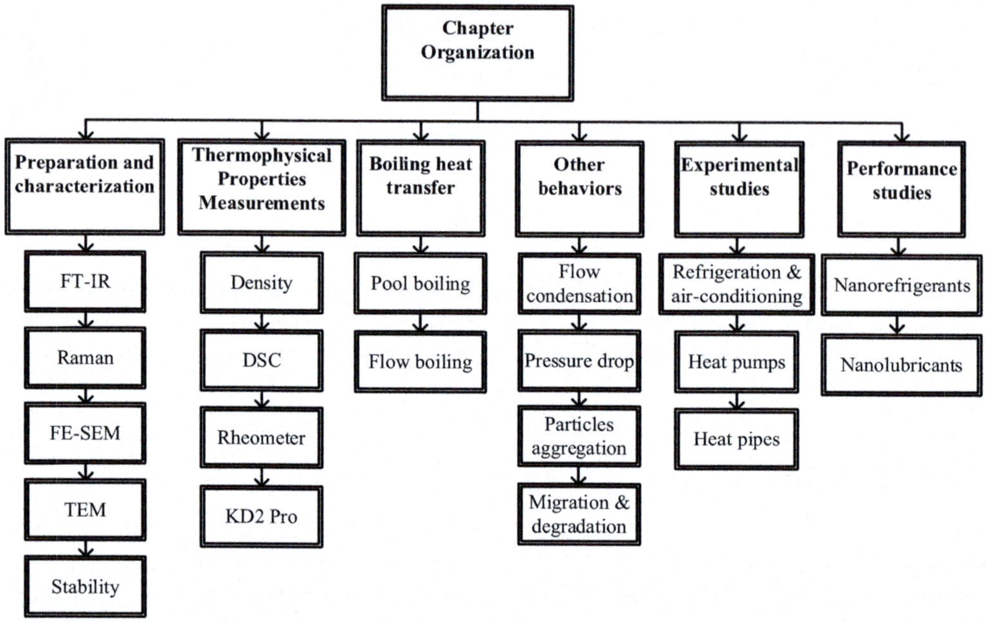

FIGURE 9.1

Organization of the chapter. *FT-IR*, Fourier-transform infrared spectroscopy; *TEM*, transmission electron microscope.

The investigation of cooling unit efficiency is still in its early stages. Every analysis, research endeavor, and experimental study, regardless of scale, is critical for enhancing the thermal properties of nanofluids and nanorefrigerants. This chapter provides a comprehensive overview of nanorefrigerants, covering key areas such as preparation methods, thermal and physical properties, pressure loss, pool boiling, and flow boiling, both with and without surfactants. In addition, this chapter includes various experimental studies on heating, ventilation, and air conditioning (HVAC) systems (Fig. 9.1).

9.2 Preparation and characterization

Synthesis and preparation of nanofluids are considered the most significant steps in examining this innovative category of heat transfer working fluids. Nanofluids' production requires several remarkable techniques for long-term stability and dealing with agglomeration, sedimentation, and settlement problems. Various solid nanoparticles can be employed in the secondary fluid cooling unit, including metal dispersion, metal oxides, and nanosized nonmetals in base fluids [15–20]. Nanofluids can be prepared using one- and two-step methods (Fig. 9.2). In a single-step process, solid nanoparticles are produced and added to a liquid simultaneously. Meanwhile, in a two-step method, nanoparticles are made individually in two stages and then dispersed in a base liquid to ensure that the nanofluid is always stable. Generally the two-stage technique is preferred over the one-step process. The nanooil preparation

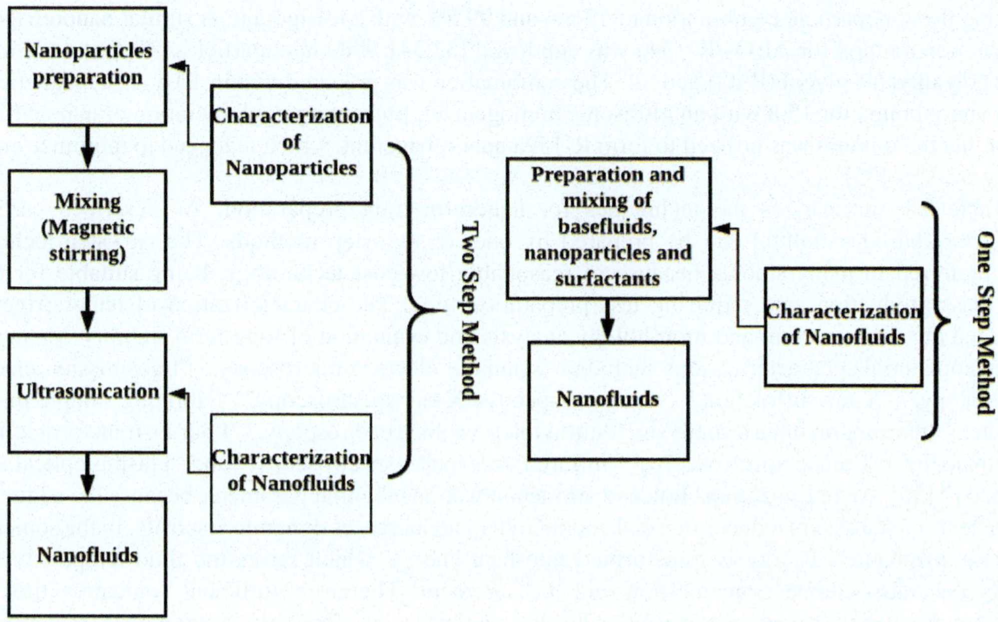

FIGURE 9.2

One- and two-step methods of nanofluids preparation.

that might enter the refrigerator compressor is also well described in many analyses [15–20]. Several authors also summarized the strategies utilized to prepare hybrid nanofluids [15,21]. Surfactants or additives are added to produce stable nanofluids. The stability time depends on combining a surfactant and a nanofluid and various other conditions, including the production technique, ultrasound control, temperature, and others.

Similarly, one- or two-step processes can be used to make refrigerant-based nanofluids [6,22]. However, as the boiling point of most base refrigerants is lower than the air temperature, they are not ideal for preparing nanorefrigerants in the gas phase under these conditions. Boiling points of R-113 and R-141b are 47.6°C and 32.1°C, respectively; therefore the two refrigerants are mainly utilized to produce nanorefrigerants at normal environmental temperatures [23]. Peng et al. [24] developed a nanorefrigerant by combining R-113 refrigerant with CuO nanoparticles. To determine the precise amount of CuO nanoparticles needed, a digital electronic balance was employed. The mixture required 30 min of ultrasound vibration to achieve proper dispersion. This process resulted in a well-dispersed nanorefrigerant that remained stable for 12 h without sedimentation.

Similarly, Trisaksri and Wongwises [25] incorporated low concentrations of TiO_2 (0.01–0.05 vol.%) into R-141b refrigerant. They used ultrasound vibration to maintain nanoparticle dispersion, which was stable for 6 h. Two methods are recommended for producing nanorefrigerants with a lower normal boiling point [5,26–30]: (1) one- or two-step preparation of refrigerant-based nanofluids at different temperatures and pressure parameters [31,32], (2) preparation of refrigerant-based nanolubricants. Al_2O_3/R-141b was prepared in an orbital incubator shaker at 15°C and 240 rpm for 24 h by constantly shaking the nanoparticle combination of 13 nm and 99.5% with a refrigerant. A similar nanorefrigerant preparation method for Al_2O_3/R-134a was employed [33,34]. ZnO nanoparticles (40 nm) were mixed with polyalkylene glycol (PAG) and oil. The combination was prepared within 10 h via a magnetic agitator and vibrated for 15 h with an ultrasonic homogenizer, preventing agglomeration of nanoparticles. After that the mixture was utilized to form R-152a nanorefrigerant, which managed to minimize energy usage by 0.5%[35].

Table 9.1 summarizes the techniques for nanorefrigerant preparation. As described earlier a refrigerant-based nanofluid can be prepared by one- or two-step methods. The two-step technique was preferred in many studies because of reasonably low-cost technology, being suitable for basic and mass production, and requiring less preparation time. The characterization of nanorefrigerants included the nanostructure and morphology analysis and evaluation of long-term stability. Meanwhile the nanomaterials characterizations included "scanning electron microscopy," "transmission electron microscopy," "X-ray diffraction," "energy-dispersive X-ray spectroscopy," "vibrating sample magnetometer," "thermogravimetric analysis," "ultraviolet–visible spectroscopy," "Fourier-transform infrared spectroscopy," "Raman spectroscopy," "infrared absorption spectroscopy," and "plasma-optical spectrometry" [36]. An optimization choice of surfactant was an essential parameter because the addition of surfactants or stabilizers reduces thermal conductivity and increases dynamic viscosity. In the sonication process, mechanical energy is transformed into heat energy, which raises the fluid temperature and varies the mass/volume concentration and measurement. Therefore sufficient sonication time and vibration frequencies supply the energy to disperse nanoparticles efficiently. A nanorefrigerant might be chemically stable for about 3–4 weeks if the surfactant and volume concentrations were suitable [20]. TiO_2 with three different sizes of 25, 40, 60, and 100 nm was added to R-141b with the presence of three surfactant types (CTAB, SDBS, and NP-10) at 25–500 mg/L [33]. They claimed that as nanoparticle size and concentration increased, the stable hydrodynamic diameter improved. In addition, as the surfactant concentration was increased, the size of the nanoparticles decreased. Moreover, Lin et al. [37] investigated the long-term stability of MWCNTs/R-141b in the presence of SDBS.

9.3 Thermophysical properties of refrigerant-based nanofluids

Thermophysical properties are critical for estimating nanorefrigerant efficiency. The most discussed properties to calculate heat transfer efficiency are density, specific heat capacity, active thermal conductivity, dynamic viscosity, latent heat, and surface tension. The thermal conductivity of refrigerant-based nanofluid had previously been the subject of several studies. Viscosity-based analyses have recently become popular. However, as the study provides a more critical concept of nanorefrigerant thermal transfer efficiency, it should be expanded to include other thermophysical properties.

9.3.1 Thermal conductivity

Thermal conductivity plays a significant role in heat transfer enhancements. Nanofluids' thermal conductivity changes according to different variables such as weight/volume concentrations of nanoparticles; temperature; shape and size of the nanoparticle; pH value; adding surfactant or stabilizer; and preparation method [38–41]. Moreover, the boiling and condensation of nanorefrigerants influenced thermal conductivity measurements. Jiang et al. [42] discussed the

Table 9.1 Previous research on the synthesis of nanorefrigerants [36].

References	Nanorefrigerants	Parameters	Method of preparation	Observations
[24]	CuO/R-113	Mass flux = 100, 150, and 200 kg/m^2 s; $\varphi = 0.1$, 0.2, and 0.5 wt.%; heat flux = 3.08, 4.62, and 6.16 kW/m^2	30 min of ultrasonication	12 h long-term stability
[5]	CNTs/R-113	$\varphi = 0.2$–1.0 vol.%, $D = 15$ and 80 nm, $L = 1.5$ and 10 μm, $T = 303$ K	30 min of ultrasonication	Thermal conductivity was influenced by CNTs' diameter and aspect ratio
[27]	Diamond/VG68 oil/R-113	$\varphi = 5$, 10, and 15 wt.%, $D = 10$ nm, heat flux = 10–80 kW/m^2	The mixture of nanolubricants was weighed, mixed, and vibrated for 1 h	12 h long-term stability
[28]	Cu/R-113/SDS, CTAB, and Span-80	$\varphi = 0.1$, 0.5, and 1.0 wt.%, $D = 20$ nm, heat flux = 10–80 kW/m^2	Cu, surfactants, and pure R113 were weighed, mixed, and vibrated for 1 h	Stability for 24 h
[31]	Al_2O_3/R-141b	$T = 15°C$, $\varphi = 0.5$ to 2 vol.%	An orbital incubator shaker was used with a two-step method	No stability tool was used
[35]	ZnO/R-152a	$\varphi = 0.1$, 0.3, and 0.5 ol.%, $D = 50$ nm	ZnO was suspended with PAG, stirred via a magnetic stirrer for 10 h, and vibrated for 15 h	The purpose of the surfactant was not to avoid reducing system performance
[66]	Cu/R-141b/SDBS	$\varphi = 0.008$, 0.015, and 0.05 vol.%, $D = 30$ nm	Cu powder was suspended in R-141b and SDBS with ultrasonication for 8 h	The critical heat flow was reduced because of SDBS
[37]	MWCNT/R-141b	/	Two-step method	Long-term stability was tested
[33]	TiO_2/R-141b	$D = 25$, 40, 60, and 100 nm	Ultrasonication for 20 min using two-step method	Aggregation behavior was checked
[54]	MWCNT/R-141b and Span-80	$D = 20$ nm, $\varphi = 0.1$, 0.2, and 0.3 vol. %	Ultrasonication for 30 min using two-step method	Spectrophotometer controlled stability
[118]	TiO_2/R-141b/CTAB	$D = 30$, $\varphi = 0.01$ and 0.03 vol. %	TiO_2 suspended in R-141b and CTAB, stirred, and ultrasonicated for 90 min	Stable for 7 days

CNTs, *carbon nanotubes;* MWCNTs, *multiwalled carbon nanotubes.*

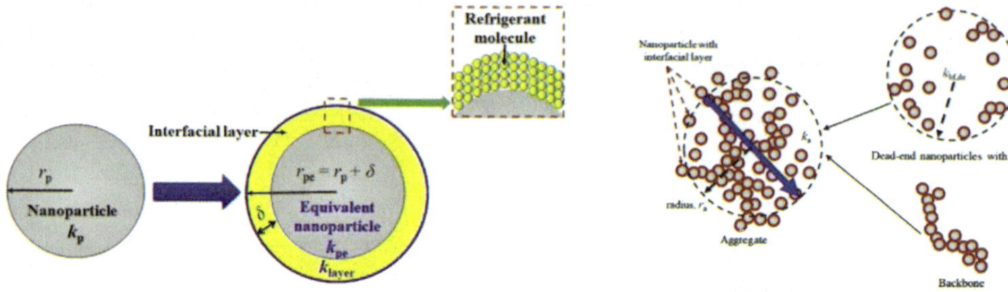

FIGURE 9.3

Schematic diagram of single nanoparticle aggregate and interfacial layer [48].

impacts of aggregation/agglomeration for refrigerant-based nanofluids. Moreover, they introduced a model to approximate the thermal conductivity of nanorefrigerants. CNTs nanorefrigerants presented a higher thermal conductivity than CNT-based H_2O or nanosphere-based nanofluids [5].

Table 9.2 summarizes the available findings on the thermal conductivity of nanorefrigerants. Different nanomaterials, such as Al_2O_3, CuO, Al, Cu, and Ni, were added to different pure refrigerants such as (R-113, R-134a, and R-141b) [31,34,43–47]. They came to the following conclusions: (1) the thermal conductivity of refrigerant-based nanofluids improved with the increase of nanoparticle volume/mass concentrations, (2) the thermal conductivity of refrigerant-based nanofluids increased with the decrease of particles nanosize, and (3) the thermal conductivity of refrigerant-based nanofluids improved with the increase of preparation temperature. Thermal conductivity of R-141b-based Al_2O_3, TiO_2, and SiO_2 with low concentration was measured in the range of 278–298 K [48]. A mathematical model for thermal conductivity of refrigerant-based nanofluids with low concentration was developed, including Brownian motion, the interfacial layer, and nanoparticle aggregation (Fig. 9.3). They showed the highest rise in thermal conductivity of 16.87% using SiO_2-R141b-based SiO_2 with 0.1 vol.% at 298 K.

9.3.2 Dynamic viscosity

The measurements of heat transfer efficiency and pressure loss are functions of dynamic viscosity. Sharp increases in dynamic viscosity are undesirable because they cause the heat transfer convective coefficient to decrease, the friction pressure drop to increase, and the pumping power to increase. Most of the published studies (Table 9.3) lack an objective discussion of the effect of nanolubricants viscosity [41], which explains why this section focuses on the analysis of nanorefrigerant viscosity.

Mahbubul et al. [31,43,44,49] in different studies, discussed the effects of temperature and nanoparticle weight/volume concentrations on the value of viscosity in the presence of R-134a, R-141b, and R-123. They observed that the dynamic viscosity of refrigerant-based nanofluids increased as the nanoparticle concentration increased and decreased if the temperature was raised. Compared to pure refrigerant at similar temperatures the viscosity of refrigerant-based nanofluids improved by approximately 13.68%. Alawi et al. [45,46] came to the same conclusions as mentioned above after testing the effects of volume fraction and temperature on the thermophysical properties of R-134a-based CuO and R-134a-based Al_2O_3. Furthermore, Alawi et al. [47] confirmed that the dynamic viscosity of R-134a-based single-walled carbon nanotubes (SWCNTs) was improved by using higher concentrations

Table 9.2 Previous research on thermal conductivity of nanorefrigerants [36].

References	Nanorefrigerants	Parameters	Observations
[42]	Cu/R-113, Al/R-113, Ni/R-113, CuO/R-113, and Al_2O_3/R-113	At room temperature and pressure	Thermal conductivity increased by up to 20% with 1 vol.%
[5]	CNTs/R-113	$\varphi = 0$–1 vol.%, CNTs dia. = 15 and 80 nm, L= 1.5 and 10 μm	The thermal conductivity improvement was higher at smaller CNTs' diameter and larger CNTs' aspect ratio
[31,43]	Al_2O_3/R-141b	T= 5°C–30°C, $\varphi = 0.1$–0.4 vol. % and 0.5–2.0 vol. %	Thermal conductivity increased by 1.6 times with 2 vol.% and 20°C
[34,44]	Al_2O_3/R-134a	T= 283–308 K, u = 5 m/s, $\varphi = 1$–5 vol.%, mass flux = 100 kg/m^2	Thermal conductivity improved by 28.58%
[45]	CuO/R-134a	$\varphi = 1$–5 vol. %, $d_p = 20$ nm, $T = 300$–325 K	Thermal conductivity improved by a maximum improvement of 43% at 325 K and 5 vol.%
[46]	Al_2O_3/R-134a	$\varphi = 1$–5 vol.%, $d_p = 20$ nm, $T = 300$–325 K	Higher concentration and temperature with small nanoparticles size improved the thermal conductivity
[47]	SWCNT/R-134a	$\varphi = 1$–5 vol. %, $d_p = 20$ nm, $T = 300$–325 K	Higher concentration and temperature increased the value of thermal conductivity

CNTs, *carbon nanotubes*; SWCNTs, *single-walled carbon nanotubes.*

Table 9.3 Previous research on the viscosity of nanorefrigerants [36].

References	Nanorefrigerants	Parameters	Observations
[49]	TiO_2/R-123	$\varphi = 0.5$–2.0 vol.%, $T = 5°C$–20°C	Higher concentration increased the dynamic viscosity, and vice versa with higher temperature
[31,43]	Al_2O_3/R-141b	$\varphi = 0.1$–0.4 and 0.5–2.0 vol.%, $T = 5°C$–30°C	The dynamic viscosity increased 179 times at 5°C and 2 vol.%
[34,44]	Al_2O_3/R-134a	$\varphi = 1$–5 vol.%, $T =$ 283–308 K, $u = 5$ m/s, mass flux = 100 kg/m^2	Dynamic viscosity increased by 13.68%
[45]	CuO/R-134a	$T =$ 300–325 K, $\varphi = 1$–5 vol. %, $d_p = 20$ nm	Higher mass/volume fraction with lower temperature increased the dynamic viscosity
[46]	Al_2O_3/R-134a	$T =$ 300–325 K, $\varphi = 1$–5 vol.%, $d_p = 20$ nm	A higher vol.% fraction increased the value of dynamic viscosity
[47]	SWCNT/R-134a	$T =$ 300–325 K. $\varphi = 1$–5 vol.%, $d_p = 20$ nm	Higher mass/volume fraction with lower temperature increased the dynamic viscosity
[50]	TiO_2/R-123	$T =$ 300–325 K, $\varphi = 1$–5 vol.%, $d_p = 20$ nm	Higher mass/volume fraction with lower temperature increased the dynamic viscosity

SWCNTs, *single-walled carbon nanotubes.*

and lower temperatures. Moreover, the nanorefrigerant viscosity of refrigerant-based nanofluids varied based on different nanoparticles' volume percentages and temperatures [50].

9.3.3 Density and specific heat capacity

The specific heat capacity and density of refrigerant-based nanofluids flow have a significant impact on heat transfer enhancements [20,51,52]. The density of R-141b-based Al_2O_3 was increased with the increase of Al_2O_3 nanomaterials concentration and decreased with the increase of preparation temperature [49]. Mahbubul et al. [34] showed that the density of R-134a-based Al_2O_3 increased by nearly 11% at a similar temperature relative to pure R-134a. Moreover, Alawi et al. [45,47] evaluated the thermophysical properties with different volume percentages and temperatures using R-134a-based CuO and R-134a-based SWCNTs. The specific heat capacity was increased with the increasing of temperature and volume fraction. Meanwhile the density decreased with higher temperatures and increased with higher concentration. The previous studies on specific heat capacity and density of different refrigerant-based nanofluids are listed in Table 9.4.

9.3.4 Surface tension

Surface tension is the desirable force applied upon the surface molecules of a liquid by the molecules underneath that tend to attract the surface molecules in the liquid bulk, besides making the liquid presume the shape of the minor surface area. Formation of bubbles, dry-out, and heat transfer are significant parameters in discussing the nanorefrigerant flow. This phenomenon was taken into consideration in different investigations. Experimentations in a smooth horizontal tube examined the heat transfer properties of CuO/polyolester (POE)/R-600a nanolubricants during condensation at several nanoparticle concentrations [53]. Nanoparticles were added to increase the surface tension of the refrigerant-based nanofluids and improve the heat transfer.

9.4 Boiling heat transfer of nanorefrigerants

Two configurations of heat transfer—pool boiling (natural convection) and flow boiling (forced convection)—are utilized in various applications, including heating units, ventilation systems, air conditioning, heat exchangers, thermal–chemical operations, and nuclear reactors. An innovative method to enhance heat transfer efficiency involves incorporating nanoparticles into phase-change fluids. The subsequent subsections will review experiments involving pool and flow boiling with nanorefrigerants, highlighting recent advancements and findings in this field [54–56].

9.4.1 Pool boiling of nanorefrigerants

There has been limited research published on the boiling characteristics of nanorefrigerants. However, the incorporation of nanoparticles significantly enhances the efficiency of heat transfer processes. Table 9.5 provides a summary of recent studies on nanorefrigerant boiling, both with and without the presence of lubricants.

Table 9.4 Previous research on specific heat and density of nanorefrigerants [36].

References	Nanorefrigerants	Parameters	Observations
[49]	Al_2O_3/R-141b	$\varphi = 0$–0.4 vol.%, $T = 5°C$–$30°C$	Higher mass/volume fraction with lower temperature increased the density
[45]	CuO/R-134a	$T = 300$–325 K, $\varphi = 1$–5 vol. %, $d_p = 20$ nm	Higher mass/volume fraction with lower temperature increased the density
[34]	Al_2O_3/R-134a	$\varphi = 5$ vol.%, $T = 283$–308 K	Density and specific heat capacity increased by 3.2% and 2.6% relative to pure R-134a
[47]	SWCNTs/R-134a	$T = 300$–325 K, $\varphi = 1$–5 vol.%, $d_p = 20$ nm	Specific heat capacity increased with increased temperature and nanoparticle concentration, whereas higher mass/volume fraction with lower temperature increased the density

SWCNTs, *single-walled carbon nanotubes.*

Table 9.5 Applications of nanorefrigerants in pool boiling heat transfer.

References	Base refrigerant	NPs	Surfactant	Oil	Test section	Observations
[57]	R-123	CNTs	/	/	Horizontal circular tube	Results showed a 36.6% improvement in nucleate boiling heat transfer
[59]	R-22	CNTs	/	/	Horizontal tube	Results showed a 24.7% improvement in nucleate boiling heat transfer
[58]	R-141b	Au	/	/	Horizontal tube	Boiling heat transfer was more than twice at nanoparticle concentration of 1%
[24]	R-113	CuO	/	/	Horizontal smooth tube	Mass fraction increased significantly with a drop in frictional pressure
[55]	R-134a	CuO		POE	Straight surface	Boiling heat transfer enhancements were between 50% and 275%
[25]	R-141b	TiO_2	/	/	Boiling tube	Deterioration in heat transfer was based on the nanoparticle's addition
[27]	R-113	Diamond	/	VG68	Horizontal flat surface	Nucleate pool boiling heat transfer enhancement was lower than 63.4%
[60]	R-113	CNTs	/	VG68	Boiling vessel	Nucleate pool boiling heat transfer increased by 61%
[119]	R-141b	Au	/	/	Plain copper tube	At 1% nanoparticle concentration, the pool boiling heat transfer was twice relative to pure R-141b

(*continued on next page*)

[56]	R-113	Cu	SDS, CTAB, and Span-80	/	Boiling vessel	The nucleate pool boiling heat transfer enhancement was because of the addition of surfactants
[28]	R-113	Cu	/	VG68 Ester Oil	Boiling vessel	The percentage of 49% improved pool boiling heat transfer
[62]	R-134a	Al_2O_3	/	POE	Horizontal flat surface	Heat transfer was increased by boiling
[63]	R-113	Cu	/	VG 68	Boiling surface	Boiling heat transfer enhancement was because of the addition of surfactants
[64]	R-141b	Al_2O_3	SDBS	/	Boiling vessel	δ-Al_2O_3/R-141b enhanced pool boiling heat transfer using SDBS surfactant
[65]	R-134a	Al_2O_3		POE	Turbo-BII-HP boiling surface	The average heat transfer increased by about 10% with low weight concentration
[66]	R-141b	Cu	SDBS	/	Boiling vessel	Cu particles improved the efficiency of R-141b heat transfer
[54]	R-141b	MWCNTs	Span-80	/	Corrugated tube	The best enhancement in heat transfer was by using 0.3 wt.%

CNTs, *carbon nanotubes;* MWCNTs, *multiwalled carbon nanotubes;* POE, *polyolester.*

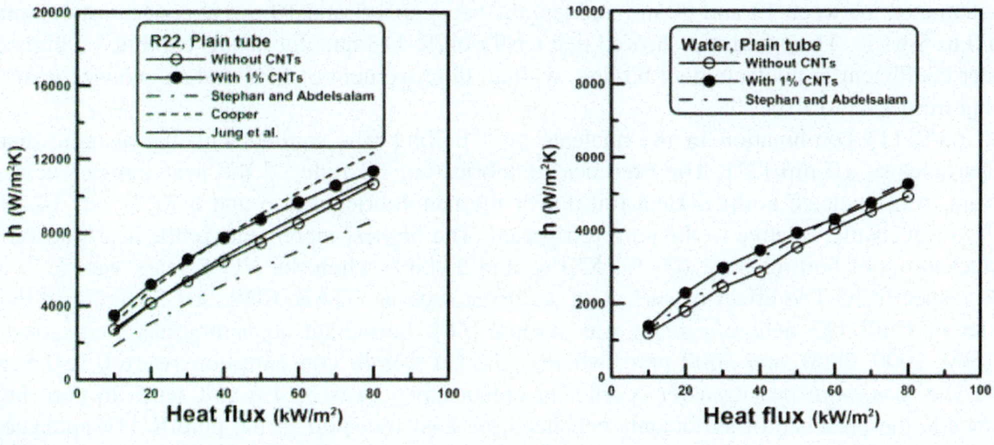

FIGURE 9.4

Nucleate boiling heat transfer coefficients for base R22/DW with 1.0 vol.% CNTs [59]. *CNTs*, Carbon nanotubes.

The impact of boiling heat transfer on CNTs/R-22 nanorefrigerant using 1 vol.% was examined by Park and Jung [57]. The diameter and length of CNTs were 20 nm and 1 μm, respectively. The boiling was conducted in a test rig using a visible glass. The heat flux was arranged in the range of decreasing order from 80 to 10 kW/m^2. The temperature of the pool was established at 7°C. The test results demonstrated that the heat transfer coefficient of the nanorefrigerant was increased. The highest heat transfer coefficient was 24.7%, whereas the low heat flux was attained. Moreover, they formulated a relationship between heat flux and heat transfer coefficient; as the heat flux improved, heat transfer efficiency was reduced because of CNTs. The availability of low heat flux bubbles was ineffective, with excessive CNTs' thermal conductivity potentially enhancing the surface area of bubbles and influencing the thermal boundary layer.

The boiling properties were examined by Liu and Yang [58] using gold nanoadditives and R-141b with 0.09, 0.4, and 1.0 vol.%. Horizontal flat tubes (100 mm in length and 18 mm in diameter) were used in the study, where the heat flux was increased from 3.5 to 100 kW/m^2. The results revealed that the gold particles improved the heat transfer through the pool. Park and Jung [59] examined the impact of CNTs on R-123 and R-134a refrigerants. In particular the heat transfer was increased to 36.6% and exhibited poor heat flow (see Fig. 9.4), but the increase of heat transfer at a more significant heat flux was insignificant. Kedzierski [55] examined the boiling characteristics for combining CuO/POE/R-134a. The heat flow ranged from 10 to 120 kW/m^2. When the mass percentage increased from 0.5% to 1%, the heat transfer was improved from 50% to 275% compared to the pure POE/R-134a. They reported that the heat transfer using a mass fraction of 2% deteriorated by an average of 12%.

Trisaksri and Wongwises [25] investigated the heat transfer coefficients of pure refrigerant 141b and R-141b-based TiO_2 nanofluids at concentrations of 0.01%, 0.03%, and 0.05%. Their experiments were conducted within a pressure range of 200–500 kPa. The results confirmed that TiO_2 nanoparticle concentration significantly influences thermal transport properties. In addition, Peng et al. [60] evaluated the boiling performance of oil/R-113 nanolubricants with varying nanoparticle sizes. They used CNTs with diameters between 15 and 80 nm and lengths between 1.5 and 10 μm at concentrations ranging from 0 to 5 wt.%. Their findings showed that CNTs/oil/R-113 nanolubricants exhibited a higher heat transfer coefficient in nucleate pool boiling, with an improvement of 61% and an enhancement factor ranging from 1.23 to 1.61.

VG68/R-113 combination in the nucleate pool boiling was studied with the average diamond nanoparticles of 10 nm [27]. The presence of lubricating oil reduced the heat transfer in nanorefrigerant from nucleate boiling. Heat transfer of the nanolubricant at 3 and 5 wt.% was 14.4% and 19.8%, respectively, relative to the pure refrigerant. The highest increments in the heat transfer from the nucleate pool boiling were 63.4%, 57.1%, and 53.04% when the VG68 mass was 1, 3, and 5 wt.%, respectively. The effect of surfactant additives such as CTAB, SDS, and Span-80 on the heat transfer of Cu/R-113 nanorefrigerant was studied [56]. Surfactant concentrations were considered 200, 500, 1000, 2000, and 5000 ppm, whereas the Cu weight concentrations were 0.1, 0.5, and 1 wt.%. The maximum heat transfer coefficient enhancement was 55.4% at 1 wt.% in pool boiling. Meanwhile the presence of surfactants enhanced the heat transport in the pure R-113 nucleate pool boiling.

Peng et al. [61] used oil/R-113 to examine the pool boiling's heat transfer coefficient with different nanoparticles. The average diameters of Cu nanoparticles were 20, 50, and 80 nm. A variety of heat fluxes were investigated with the mass percentage of 0–5 wt.%. The pool heat transfer coefficient was enhanced by approximately 49% using the 20 nm nanoparticles. Moreover, using nanoparticles >20 nm in size decreased the heat transfer coefficient.

The boiling effectiveness of Al_2O_3/POE/R-134a was discussed by Kedzierski [62] using a geometrically flat surface at several heat fluxes. They demonstrated that Al_2O_3 with nanolubricant increased the heat transfer relative to POE/R-134a. The typical heat transfer improvement uses 0.5, 1, and 2 wt.% at the heat fluxes lower than 40 kW/m^2 were 105%, 149%, and 155%, respectively. In another study, R-113 refrigerant was used, and the surfactants' effect was examined [63]. A unique terminology was introduced as a surfactant effect factor (SEF) $\left(\text{SEF} = \frac{\text{Heat transfer (nanorefrigerant+surfactant)}}{\text{Heat transfer (nanorefrigerant)}}\right)$. The SEF improved at higher surfactant concentrations and decreased at lower molecular surfactant weight, heat flux, and nanolubricant concentration.

The application of nucleate pool boiling heat transfer using R-141b-based Al_2O_3 nanofluid was studied within a flat copper surface at the operating condition of 10–200 kW/m^2[64]. Nanoparticle concentrations with and without a surfactant (SDBS) addition varied from 0.01 to 0.1 vol.%. Moreover, the roughness of the boiling surface was examined. The experimental outcomes confirmed that the mixture of solid nanoparticles and surfactants enhanced the heat transfer features of nanorefrigerant pool boiling. Meanwhile the heat transfer coefficient was reduced using 0.1 vol.% without a surfactant.

Kedzierski [65] tested the pool boiling efficiency of Al_2O_3/R-134a nanorefrigerant. Polyester lubricant was applied as a nanolubricant and Al_2O_3 (10 nm). Nanolubricants were POE lubricants mixed with R134a at various volume fractions in the range of (1.6, 2.3, and 5.1 vol.%). The volume fractions of 1.6% and 2.3% revealed an average increment of about 13% and 10%, respectively. The typical heat transfer degradation of the nanolubricant was 14% at 5 vol.%. Furthermore, under atmospheric pressure, the critical heat flux and boiling characteristics of different nanoparticle concentrations (0.008, 0.015, and 0.05 vol.%) were evaluated [66]. Cu nanoparticles with a 30-nm average diameter and the surfactant SDBS were used. According to their findings the nanorefrigerant heat transfer coefficient was superior to base R-141b.

9.4.2 Flow boiling of nanorefrigerants

In general, boiling occurs in HVAC units. The issue of boiling for a nanorefrigerant must also be experimentally highlighted. Furthermore the flow scheme and the modest layer development of nanoparticles should be seriously analyzed. Not much research can be found in the literature that reviews the boiling of nanorefrigerant. Table 9.6 summarizes the previous studies on boiling heat transfer in refrigerant-based nanofluids.

A horizontal tube was tested for boiling flow using CuO/POE/R-134a [67]. They demonstrated that the nanolubricant mass fraction of 5% showed a minor impact on the two-phase heat transfer coefficient. On the other hand the heat transfer coefficient for nanolubricant improved from 42% to 82% at mass flows from 100 to 400 kg/m^2 s. Moreover, the heat transfer coefficient improved by more than 100% at 2 wt.%. The presence of nanoparticles presented a minimal impact on the unit pressure. Peng et al. [68] tested a smooth horizontal tube to show the effects of CuO/R-113 nanorefrigerant flow. The heat transfer coefficient was measured numerically and experimentally, and the results showed a 93% agreement with a 20% variance. The heat transfer coefficient of the refrigerant-based nanofluid was higher than that of the pure refrigerant. The heat transfer coefficient increased by approximately 29.7% in the most extreme case.

Henderson et al. [69] examined the effect of flow boiling of SiO_2/POE/R-134a in 100–400 kg/m^2 s. The addition of SiO_2 nanoparticles decreased the heat transfer coefficient relative to the base refrigerant. The degradation of the heat transfer coefficient was attributable to the steady dispersion difficulties. No noticeable impact was observed at the concentration of 0.02 vol.% on the heat transfer coefficient for SiO_2/POE/R-134a. An increase of 42% and 101% in the heat transfer coefficient was noticed at 0.04 and 0.08 vol.%, respectively.

In experimental studies, Sun and Yang [70] tested the flow boiling characteristics of R-141b-based Cu, Al, Al_2O_3, and CuO nanofluids in the horizontal tubes. The mass percentage, vapor quality, and mass flow varied from 0.1 to 0.3 wt.%, 0.3 to 0.8, and 120 to 330 kg/m^2 s, respectively. Span-80 surfactant was utilized to enhance the quality of dispersion. They observed that, because of the nanoparticles' existence, the heat transfer of the boiling flow improved. They reported that different nanoparticles could impact the nanorefrigerant heat transfer coefficient. For the Cu/R141b mixture the highest heat transfer coefficient was attained using 0.3 wt.% and vapor quality of 0.5%–49%.

Sun and Yang [30] tested copper tubes with the dimensions of $L = 1400$ mm, $D_o = 9.52$ mm, and $D_i = 8.22$ mm using R-141b-based Cu, Al, Al_2O_3, and CuO nanofluids. To determine nanorefrigerant efficiency, several inlet vapor qualities and mass fluxes were examined with 0.1, 0.2, and 0.3 wt.%. The vapor quality was in the range of 0.3–0.8, whereas the mass flow varied from 120, 210, and 330 kg/m^2 s. Span-80 was added for stability purposes. Their experimental data showed that (Fig. 9.5) the heat transfer coefficient was improved noticeably at a constant mass flow rate compared to a nearly linear mass flow. Meanwhile the concentration of 0.3 wt.%, Cu/R-141b provided the highest average thermal transfer factor of 25%.

A corrugated tube was used as a test section for the boiling effectiveness of MWCNTs/R-141b nanorefrigerant [54]. The nanorefrigerant was in different weight concentrations such as 0.1, 0.2, and 0.3 wt.%. The maximum heat transfer improvement was achieved using 0.3 wt.% of MWCNTs/R-141b relative to pure refrigerants (R-141b). The rise in the Nusselt number (Nu) was 41%, and the pressure loss of nanorefrigerants increased with the Reynolds number (Re). The maximum Nu_{nf}/Nu_f value was improved by using a higher mass flow rate. The highest heat transfer coefficient, with the developing

Table 9.6 Impact of nanorefrigerants on flow boiling heat transfer.

References	Base refrigerant	NPs	Surfactant	Oil	Test section	Observations
[67]	R-134a	CuO	/	POE	Horizontal smooth tube	Boiling heat transfer was enhanced by 42%
[68]	R-113	CuO	/	/	Horizontal smooth tube	Heat transfer was enhanced by 29.7%
[69]	R-134a	CuO	/	Polyolester	Horizontal smooth tube	Flow boiling heat transfer enhanced by 101%
	R-134a	SiO_2	/	/	Horizontal smooth tube	The heat transfer using convective boiling was decreased
	R-134a	SiO_2	/	POE (RL68H)	Horizontal tube	Heat transfer decreased because of increasing nanoparticle concentration
[70]	R-141b	Cu, Al, Al_2O_3, CuO	Span-80		Horizontal tube	Flow boiling heat transfer enhanced by 49% using 0.3 wt.% and a quality of 0.5
[30]	R-141b	CuO	/	/	Horizontal pipe	Boiling heat transfer increased by 1.14 times
[72]	R-600	CuO	/	POE	Horizontal smooth tube	Boiling heat transfer forced convection improved by 42.2% at 1.5 wt.%
[34]	R-134a	Al_2O_3	/	/	Horizontal smooth tube	There was a substantial rise in heat transfer amount
[54]	R-141b	MWCNTs	Span-80	/	Corrugated tube	The best enhancement in heat transfer was observed at 0.3 wt.%

MWCNTs, *multiwalled carbon nanotubes;* POE, *polyolester.*

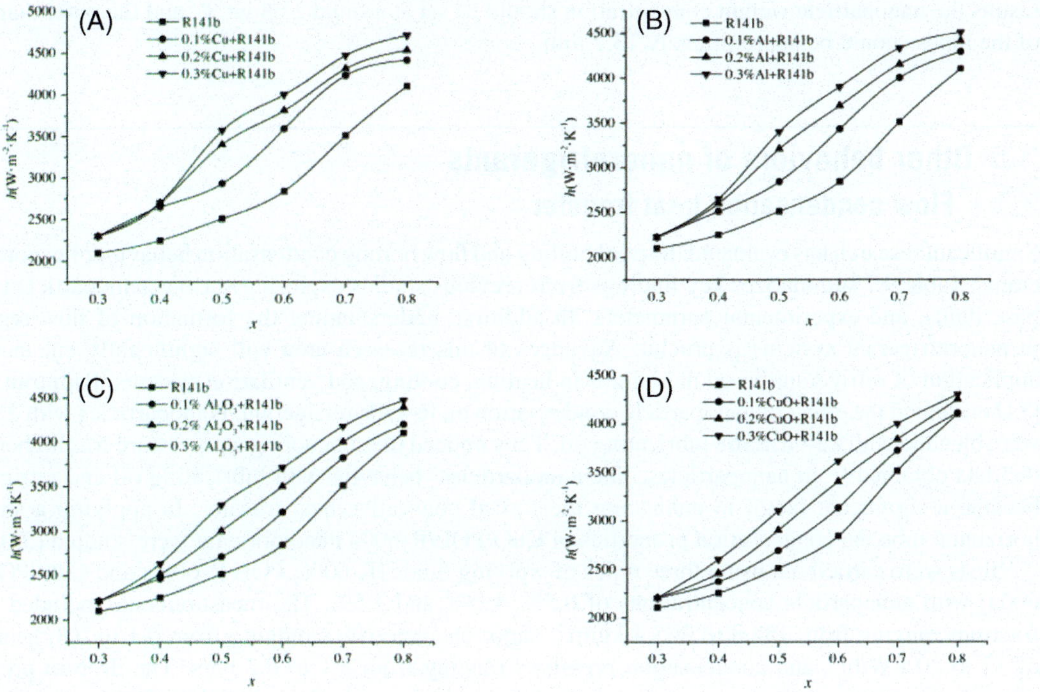

FIGURE 9.5

Heat transfer coefficients of R-141b-based: (A) Cu, (B) Al, (C) Al_2O_3, and (D) CuO [30].

effect of 1.21 times, was recorded using 1 wt.% Span-80. On the other hand the highest pressure drop was observed using 0.3 wt.% of MWCNTs/R-141b.

A smooth horizontal tube was tested for forced heat transfer using CuO/oil/R-600a [71]. A wide range of nanoparticle concentrations was prepared as 0.5, 1, 1.5, 2, and 5 wt.% in the presence of RL68H POE. Moreover, the experimental conditions included heat flux at 3–6 kW/m^2 and inlet vapor of <0.25. The best reading of heat transfer was noticed at the concentration of 2%. However, the heat transfer coefficient increased to 4.56%, 18.25%, and 32.5% using 0.5, 1, and 1.5 wt.%, respectively. On the other hand the heat transfer coefficient decreased by 7.94% at the higher concentration (5 wt.%). Moreover, a smooth tube was evaluated for CuO/oil/R-600a flow [72]. The examination variables covered nanoparticle mass fraction from 0 to 1.5 wt.%, mass flow from 50 to 400 kg/m^2 s, inlet vapor quality from 0 to 0.9, and heat flux from 3 to 8 kW/m^2. Finally the highest average heat transfer was recorded using 1.5 wt.%, up to 42.2%, 37%, 30%, and 28.6%, at the conditions of 77, 154, 232, and 348 kg/m^2 s, respectively.

In a shell-and-tube evaporator the efficiency of R-600a-based Al_2O_3 nanofluid was evaluated mathematically [73]. For any tube diameter in the range, evaporator cooling capacity using 5 wt.%-Al_2O_3/R-600a increased more than five times relative to pure R-600a. Heat transfer coefficient, pressure drop, and pumping power increased as nanoparticle weight fraction increased. According to the optimization results the nanoparticle weight concentration should be set at around 2.68 wt.%, and the outer diameter of the tubes should be approximately 13.7 mm.

9.5 Other behaviors of nanorefrigerants

9.5.1 Flow condensation heat transfer

Significant research has evaluated the pool boiling and flow boiling condensation behavior of nanorefrigerants. Table 9.7 summarizes key findings from previous studies exploring various nanoparticle types, base fluids, and experimental parameters. In addition, understanding the formation of flow regimes in nanorefrigerant systems is crucial. Advances in this research area will significantly enhance the application of refrigerant-based nanofluids in heating, cooling, and ventilation systems. Pamitran et al. [74] evaluated the effect of nanoparticle condensation on R-410a refrigerant. Nanoparticles with 1 wt.% were blended with Ze-GLES68 lubricating oil. They noticed the subcooling temperature was higher than the data obtained from nanoparticles. The nanoparticles' behavior with lubricating oil and refrigerant became a significant factor in enhancing the forced convective heat transfer. In applying a smooth horizontal tube the condensation properties of CuO/POE/R-600a nanolubricant were studied [53].

Tests were carried out using three types of working fluids (R-600a, POE/R-600a, and CuO/POE/R-600a) with nanoparticle concentrations of 0.5%, 1.0%, and 1.5%. The measurements included mass fractions ranging from 154.8 to 265.4 kg/m^2, vapor characteristics ranging from 0.1 to 0.8, heat flux of 17 to 20 kW/m^2, and condensation pressures ranging from 5.1 to 6.2 bars. The drop-in pressure of nanorefrigerants was experimentally confirmed in a horizontal tube using CuO/POE/isobutene with nanoparticle percentages of 0.5%–1.5%[75]. The impact of mass flow, vapor quality, and concentration on reducing condensation pressure was assessed. A new parameter for the frictional penalty factor (FPR) was introduced as $\text{FPR} = \frac{\text{frictional pressure drop}_{nr}}{\text{frictional pressure drop}_{r}}$. The results showed that the nanorefrigerant enhanced the FPR with a smaller quantity of vapor. The effects of CuO/R-141b on heat transfer flow condensation were evaluated for various mass fractions [76]. To determine their influence, mass flow and inlet vapor characteristics were analyzed. The analysis revealed that increasing the nanoparticle mass percentages improved the heat transfer coefficient, with the most significant improvement observed at lower mass flows. The highest recorded improvement in heat transfer was 39.68%.

9.5.2 Pressure drop properties

In the initial stages, research on the pressure drop of phase-change materials characteristics using nanorefrigerants remains. The high-pressure drop needs to be subsided to enhance the thorough pumping capability. Currently, there is no summary of the impact of the nanoparticles on nanorefrigerant pressure drop characteristics. In designing and optimizing air-conditioning units with nanorefrigerants, considerable studies with pressure drop properties are significant. Several investigations on the features of nanorefrigerant pressure drop properties are classified in Table 9.8.

Peng et al. [68] tested a smooth horizontal tube for frictional pressure drop using CuO/R-113 nanorefrigerant. They suggested a mathematical correlation to approximate the drop-in pressure (Fig. 9.6). A flow of 100–200 kg/m^2 s, heat fluxes of 3.08–6.16 kW/m^2, vapor attributes of 0.2–0.7, and mass of nanoparticles at 0 to 0.5 wt.% were used as operating conditions. At the mass flow of 100 kg/m^2 s

Table 9.7 Impact of nanorefrigerants on flow condensation heat transfer.

References	Base refrigerant	NPs	Oil	Test section	Observations
[74]	R-410A	Ni	Ze-GLES68	Condenser tube	Fluid subcooling was achieved to a high degree
[72]	R-600a	CuO	POE	Horizontal smooth tube	The condensation heat transfer using refrigerant-based nanofluid was significantly higher than pure R-600a
[76]	R-141b	CuO	/	Vertical circular tube	The condensation heat transfer increased by 39.68% at 1wt.%
[75]	R-600a	CuO	POE	Horizontal tube	The frictional penalty factor was independent of higher vapor quality during condensation

POE, *polyolester.*

Table 9.8 Impact of nanorefrigerants on pressure drop properties.

References	Base refrigerant	NPs	Test section	Observations
[68]	R-113	CuO	Horizontal smooth tube	Frictional pressure loss increased with increasing nanoparticle weight concentration
[77]	R-141b	Al_2O_3	Horizontal tube	The pressure drop increased up to 181%
[50]	R-123	TiO_2	Horizontal tube	The pressure loss increased to 42.5% at a concentration of 0.5 vol.%

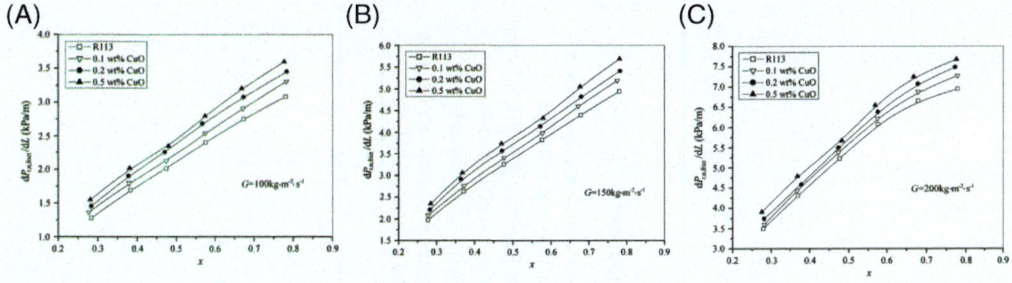

FIGURE 9.6

Frictional pressure drop of R-113-based CuO nanofluid against local vapor quality with various mass fluxes: (A) $G = 100$ kg/m^{2}s; (B) $G = 150$ kg/m^{2}s; and (C) $G = 200$ kg/m^{2}s [68].

and 0.5 wt. % the frictional pressure drop was increased by about 20.8%. Furthermore, as the vapor quality improved, the flow regime developed an annular flow, and it caused the pressure to drop. Mahbubul et al. [77] studied the pressure drop using Al_2O_3/R-141b nanorefrigerant. Even small volume percentages of nanoparticles reduced the nanorefrigerant pressure. Their investigation demonstrated that the pressure loss increased significantly when the concentration was >3 vol.%, regardless of the vapor characteristics. This deviation was because of more significant collisions and subsequent interactions with the molecular wall. Alawi et al. [50] recommended the correlations for calculating the frictional pressure of TiO_2/R-123 nanorefrigerant. The experiments were performed at temperatures ranging from 27°C to 52°C, with nanoparticle percentages ranging from 0.5% to 2%, mass fluxes ranging from 150 to 200 kg/m^2 s, and inlet vapor characteristics ranging from 0.2 to 0.7. Their analyses were aligned with a former study in which the concentration of nanoparticles had a vital part in minimizing frictional pressure drop.

Numerical study in a 3-mm horizontal circular tube subjected to uniform heat flux and slip boundary condition using SiO_2 and Al_2O_3 suspended in HFE7000 [78]. Results indicate that the mean pressure drop reduces by ~25% using the wall's slip boundary condition. In addition, although using a slipping wall pipe with a slip length of 100 μm and a 6% volume concentration of Al_2O_3 dispersed in HFE7000, total entropy generation reduces by ~20% compared to pure HFE7000 flowing in a no-slip pipe.

9.5.3 Studies on particle aggregation

Van der Waals forces, Brownian motion, and excessive free interface energy cause nanoparticles to cluster and aggregate in pure fluids. This clustering primarily leads to settlement and sedimentation, negatively impacting the colloidal stability of nanoparticles in concentrated solutions and reducing potential heat transfer enhancements. Understanding the aggregation behavior of nanoparticles in base fluids is crucial, prompting the need for further research in this area. Numerous studies have investigated the general behavior of nanoparticles in refrigerants, shedding light on these dynamics [79–83]. The summary of the available studies is described in Table 9.9.

TiO_2 nanoparticles were examined for stability and sedimentation in R-113, R-123, and R-141b refrigerants [9]. The ratio of light transmission and the nanoparticle size distribution were identified in the analysis. TiO_2 nanoparticles in the refrigerants demonstrated long-term stability over other working

Table 9.9 Impact of nanorefrigerants on particle aggregation behavior.

References	Base Refrigerant	NPs	Oil	Surfactant	Observations
[12]	R-113, R-123, R-141b	TiO_2	/	Span-80	Long-term stability was shown with TiO_2 nanoparticles
[33]	R-141b	MWCNTs	/	Anionic, cationic, nonionic	The aggregation of MWCNTs was prevented because of the existence of surfactants
[84]	R-141b	TiO_2	ATMOS, NM56	/	No aggregation behavior because of the presence of oil

MWCNTs, *multiwalled carbon nanotubes.*

fluids because of pure refrigerants' polarity and dielectric constant. In addition, temperature was a crucial factor for stability measurements. Peng et al. [33] examined the performance of TiO_2/R-141b on the primary nanoparticle size and nanorefrigerant surfactant aggregation qualities. TiO_2 nanoparticles in the size of 20, 40, 60, and 100 nm were dispersed in R-141b in 25–500 mg/L. The experimental results suggested an increment of 127.6%. Nonetheless, the investigators did not concentrate on the general behavior of the lubricating oil. The significance of lubricants on the behavior of nanoparticle aggregation was determined in a previous study [84]. The effect of nanoparticle diameter, nanoparticle concentration, and temperature on aggregation efficiency was investigated using TiO_2, R-141b, ATMOS, and NM56. To research the action of particle aggregation the time for the creation of hydraulic diameter was quantified.

9.5.4 Migration and degradation properties

Information on migration and degradation properties is needed when discussing the performance of nanorefrigerants in various thermal applications. The migration of nanoparticles throughout the phase change procedure degrades the nanoparticle mass/volume concentration, so this problem needs to be addressed further. The effects of nanoparticle diameter, nanoparticle percentage, nanoparticle type, and the presence of surfactant on migration qualities need to be investigated further [85]. With evaporation and condensation the mass of nanoparticles typically tends to decrease. Meanwhile, nanorefrigerant thermal conductivity is linked to suspended nanoparticles and nanorefrigerant degradation. Table 9.10 summarizes the features of migration and deterioration. The information about migration and degradation properties is necessary to discuss nanorefrigerants' performance for different thermal applications. The migration of nanoparticles in the phase change procedure deteriorates the nanoparticle concentration; therefore this issue needs to be further considered. Further studies need to be extended about the influences of nanoparticle size, nanoparticle percentage, nanoparticle type, and presence of surfactant on migration qualities [85]. Usually the nanoparticle mass showed an apparent reduction with evaporation and condensation. Meanwhile, the thermal conductivity of nanorefrigerant is related to the suspended nanoparticles and degradation of nanorefrigerant. Table 9.10 presents the information about migration and degradation characteristics.

Experimentation and numerical study were conducted using CuO/R-113 on the migration characteristics [26]. RB68EP was perfectly soluble in R-113 as lubricating oil. A radiation heater was applied to boil the working fluid (refrigerant). The migrated weight varied as the nanoparticle/refrigerant combination was vaporized. The migration percentages were 5.84% and 1.14% at nanoparticle concentrations of 0.0912% and 1.536%, respectively, and the values of 5.25% and 0.93% were observed for the nanolubricant. Experimental discussion on the refrigerant formation and heating variables throughout the nanoparticles' migration during boiling was conducted by Peng et al. [56]. Various nanoparticles were utilized in the study, including Cu (20, 50, and 80 nm), Al (20 nm), $Al2O_3$ (20 nm), and CuO (40 nm). The base refrigerants used were R-113, R-141b, and *n*-pentane, mixed with RB68EP lubricant (0–10 wt.%). The study identified that the most critical factors influencing the results were the size of the nanoparticles, the dynamic viscosity of the refrigerant, and the mass percentage of the lubricating oil. Notably, the migration of Cu was observed to be 315.6% for 20 nm particles and 448% for 50 nm particles, as shown in Fig. 9.7. Peng et al. [86] discussed the influence of migration behavior using CNTs nanoparticles with (R-113 and R-141b). Different parameters were considered, such as oil concentration, heat flux, and the initial liquid rate. Their results revealed that the migration ratio

Table 9.10 Impact of nanorefrigerants on nanoparticle migration/degradation properties.

References	Base refrigerant	NPs	Oil	Vessel type	Observations
[26]	R-113	CuO	RB68EP	Boiling	The migrated mass increased as the nanoparticle mass concentration increased
[56]	R-113, R-141b, *n*-pentane	Cu, Al, Al_2O_3, CuO	RB68EP	Boiling	The migration ratio increased as the heat flux increased
[86]	R-113, R-141b, *n*-pentane	CNTs	RB68EP	Boiling	Thermophysical properties of CNTs, oil concentration, and liquid level showed impacts on the migration properties
[29]	R-141b	TiO_2	MO	Boiling	The migrated mass was influenced by heat flow and initial mass
[87]	R-141b	Al_2O_3, TiO_2	–	Boiling	The migrated mass increased with bigger nanoparticle size and decreased with density
[85]	R-141b	TiO_2	NM56	Boiling and condensation	The range of degradation ratio was from 28% to 73%

CNTs, *carbon nanotubes.*

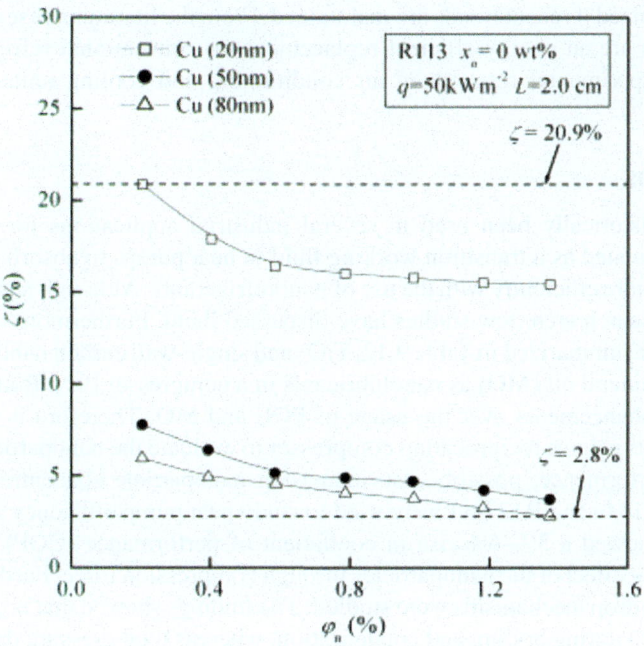

FIGURE 9.7

Effect of nanoparticle size (20, 50, and 80 nm) on the migration ratio of Cu nanoparticles [56].

was increased with different CNTs lengths and diameters. Furthermore the increase in the initial liquid concentration enhanced the migration percentage. Two sizes of boiling vessels were used to analyze the effect of the boiling vessel size on migration features using TiO_2/R-141b [29]. The tested vessels were subjected to heating systems with/without isolation to study the influence of isolation on migration. TiO_2 (21 and 40 nm) and Al_2O_3 (13 and 50 nm) nanoparticles were used with R-141b [87]. The variance in density showed a significant effect on the behavior of nanoparticle migration. Their study assumed that the higher density led to quicker sedimentation. TiO_2/NM56/R-141b was tested with different parameters such as the lubricant mass fraction in the range of 5%–20%, nanoparticle weight concentration in the range of 0.2%–1.0%, the heating temperature in the range of 50°C–80°C, and cooling temperature in the range of 5°C–15°C [85]. The nanolubricant/refrigerant degraded by 28%–77% after 20 alternating times. In the long run the degradation was dropped by using lower heating and cooling temperatures, higher oil mass fraction, and lower nanoparticle mass fraction.

9.6 Experimental studies on nanorefrigerants

9.6.1 Refrigeration and air-conditioning systems

Modern advances make it possible to manage household conditions, the most significant energy consumers in domestic and industrial areas. However, it could also result in considerable savings, which can always be maximized if recent trends are maintained. From the literature, research has demonstrated that nanorefrigerants are an energy-efficient replacement for conventional refrigerants [88–90]. This section discusses experimental research in air conditioning and cooling units for nanocooling use (Table 9.11).

9.6.2 Heat pumps

Heat pumps have historically been used in several industrial applications for heating and cooling. Refrigerants are circulated as a transition working fluid in heat pumps to absorb heat energy. The heat pumps are working more efficiently with the use of nanorefrigerants. Although nanorefrigerants are very applicable in this research area, few studies have discussed them. Furthermore the conclusions of the available analyses are summarized in Table 9.12. TiO_2 and single-wall carbon nanohorns (SWCHs) were added as POE and mineral oil (MO) as nanolubricants in a compressor [91]. Both TiO_2 and SWCNHs did not show any enhancements over the usage of POE and MO. Therefore it was recommended to repeat the current tests using a reciprocating compressor to evaluate the nanoparticles' possible positive effect on system performance, possibly also increasing nanoparticle concentration. Cu, Al_2O_3, and CuO suspended in R134a and R1234yf were used to study heat pump efficiency and cost-effectiveness [92]. Both modes showed a 5%–6% rise in coefficient of performance (COP), which is principally because of the cooling effect of the nanoparticles through compression force. Furthermore, both the heat transfer and pressure drop mechanisms were studied. The findings showed that heat transfer coefficients increased by up to 6% during boiling and condensation, whereas total pressure drop increased by up to 3.5%.

9.6.3 Heat pipes

Minimization is a fundamental reason for nanotechnology. Heat diffusion is a frequent complication for advanced electronics industries. Because of their excellent heat transfer capability, heat pipes are traditionally used in a number of electronics applications, particularly in high-speed electronics systems that need high heat diffusion rates. The stability and sturdiness of these devices depend mainly on the thermal transfer capability of the circulated fluids. The efficiency of thermal piping is limited by the low thermal transport properties of standard heat transfer fluids. Scientific tests revealed that the fluid nanoadditives improved heat transfer capability and heat pipe characteristics [93].

A heat pipe was tested using TiO_2 nanoparticles (21 nm) and R-11 refrigerant [94]. The experimental conditions were the heat pipe inclination angles from 0 to 90 degrees, nanoparticle concentrations of 0.01%–1.0%, and various heat fluxes. Their results demonstrated that the efficiency of heat pipes was improved as the heat flow increased. Eventually the performance of the heat pump was improved using nanorefrigerant. The maximum efficiency of 83% was attained with the best possible inclination angle of 60 degrees.

9.7 Performance studies on nanorefrigerants and nanolubricants

9.7.1 Performance studies on nanorefrigerants

The heat transfer coefficient of boiling and condensation were enhanced as a result of using nanorefrigerants [53,57,66,95,96], resulting in more compact and light cooling and refrigeration units using an

Table 9.11 Applications of nanorefrigerants in refrigeration systems.

References	Base refrigerant	NPs	Lubricant	Operating system	Observations
[9]	R-134a	TiO_2, Al_2O_3	POE	Refrigeration	Power consumption decreased by 26.1% with 0.1wt.% TiO_2
[120]	R-134a	Al_2O_3	/	Refrigerator	(1) COP increased by 4.4%; (2) energy consumption decreased by 2.4%; (3) refrigerant load was reduced by 40%
[106]	R-600a	TiO_2	/	Refrigeration	Energy consumption was reduced by 5.94% and 9.60% at Al_2O_3-0.1 and Al_2O_3-0.5 vol.%, respectively
[107]	R-134a	Al_2O_3	POE	Refrigeration	Energy savings were about 25%
[121]	R-134a	CuO	/	Vapor compression	The optimum nanoparticles diameter and nanoparticles concentrations were 25 nm and 0.55%, respectively
[122]	R-12	TiO_2	MO	Refrigeration	COP increased by 17%, and the average compressor work was reduced by 11%
[123]	R-134a	Al_2O_3	PAG oil	Refrigeration	Energy consumption was reduced by 10.32% at Al_2O_3-0.2 vol.%
[124]	R-134a	CuO	MO	Refrigeration	The overall refrigeration system performance increased
[114]	R-134a, R-600a	$MoFe_2O_4$-$NiFe_2O_4$, $MoFe_2O_4$-$NiFe_2O_4$-Fullerene (C_{60}–C_{70})	KFR32	Domestic refrigerator	The coefficient of performance increased by 5.33%

(*continued on next page*)

Table 9.11 Applications of nanorefrigerants in refrigeration systems—cont'd

References	Base refrigerant	NPs	Lubricant	Operating system	Observations
[125]	R-600a	Graphite	MO	Domestic refrigerator	Power consumption and pull-down time were reduced by 4.55% and 15.22%, respectively
[126]	Ammonia/Water	Al_2O_3	/	Diffusion absorption refrigeration	System performance improved by 51%
[123]	R-290/R-600a (50/50)	ZnO	MO	Refrigeration	Adding ZnO (0.2–1 wt.%) reduced energy consumption by 7.48% and increased COP by 45%
[127]	LPG	TiO_2	MO	Domestic refrigerator	Energy consumption and cooling capacity were reduced and increased, respectively.
[105]	R-410a	$NiFe_2O_4$	VG32	Air-conditioning	EER improved by nearly 6%
[112]	R-1342	SiO_2	PAG	Automotive air conditioner	COP increased by 24%

Table 9.12 Applications of nanorefrigerants in refrigeration system performance.

References	Nanorefrigerant	Parameters	Observations
[12]	TiO_2, Al_2O_3/R-134a	$d_p = 50$ nm, $\varphi = 0.06$, and 1.0 wt.%	Energy usage decreased by 26.1% using TiO_2-0.1wt.%
[105]	$NiFe_2O_4$/R-410a	$d_p = 35$ nm	The energy efficiency ratio increased by nearly 6%
[106]	TiO_2/R-600a	$\varphi = 0$, 0.1, and 0.5 g/L, $d_p = 50$ nm	Energy consumption reduced by 5.94%, and the freezing rate increased by 9.60%
[107]	Al_2O_3/R-134a	Size < 50 nm, $\varphi = 0.06$ wt.%	Energy consumption was reduced by 25%, and COP increased by 33%
[108]	Al_2O_3/PAG/R-134a	$d_p = 40$–50 nm, $\varphi = 0.2$ vol.%	Energy consumption decreased by 10.32%, and COP increased by 3.5% using 0.2 vol.% nanorefrigerant
[109]	TiO_2/R-134a, TiO_2/R-436A, TiO_2/R-436B	$\varphi = 0.1$–0.5 g/L	Irreversibility was reduced significantly for the compressor
[35]	ZnO/R-152a	$D = 40$ nm, $\varphi = 0.1$–0.5 vol.%	COP increased by 3.56%, and power consumption was reduced by 21% using 0.5vol.%
[34]	Al_2O_3/R-134a	$T = 283$–308 K, $\varphi = 5$ vol.%	COP increased by 15%

energy-efficient compressor. The solid nanoparticles can also grow the solubility of additives between the pure refrigerant and the oil lubricant, increasing the compressor's life cycle. The evaluation of refrigerant-based nanofluids consists of studies regarding the condensation and boiling heat transfer processes resulting from the phase change approach in cooling units, which usually rely on multiple thermophysical characteristics.

Table 9.13 summarizes the previous investigations on the thermal efficiency of nanorefrigerants. In the literature, researchers studied topics related to nanorefrigerants, including basic principles, preparation techniques, thermophysical properties, complications/difficulties, and applications [6,22,97–101]. Ajayi et al. [102] performed numerical studies to replace base refrigerants with nanorefrigerants because of the latter's properties, including improved heat transfer coefficient and reduced pressure drop. In the long-term enhancement of evaporation and condensation for the nanolubricant/refrigerant the degradation recognized because of nanoparticle suspension might be minimized by smaller nanoparticle concentration, higher lubricant mass fraction, and lower cooling/heating temperatures. The whole refrigerant dry-out process is shown in Fig. 9.8[103]. Only a small part of the nanoparticles was produced in the cooling unit throughout the nanolubricants flow. This complexity could be resolved by calculating the migration ratio, which may be optimized by minimizing the concentration of nanoparticles, increasing the oil fraction, and limiting the heat flow [85]. Yang et al. [104] applied nanorefrigerants to compare a smooth tube's thermal transfer efficiency and an internally threaded tube. The heat transfer capabilities inside the threaded tube were higher than in the smooth tube at 0.2 wt.%. The performance coefficient could be enhanced by 26.1% using TiO_2/R-134a at a mass fraction of 0.1%[12]. The energy effectiveness improved by 6% with the application of nanorefrigerant MNRO/R-410a [105]. TiO_2/R-600a nanorefrigerant with 0.5 g/L minimized energy consumption by about 9.6%[106]. Subramani and Prakash [107] used nanorefrigerants to show improved freezing capacity, reduction in compressor electric power (25%), and enhancement of COP (33%). Kumar et al. [108] experimented using Al_2O_3/R-134a nanorefrigerant with PAG oil and noticed a reduction of 10.3% in the nanoparticle concentration and 0.2% in energy utilization. Padmanabhan and Palanisamy [109] examined the use of TiO_2 nanorefrigerant and lubricants (MO and POE) to enhance the COP and vapor compression of cooling units. The efficiency of vapor compression cooling units was examined using ZnO/R-152a nanorefrigerant [35]. They noticed that suction temperature, suction pressure, and discharge pressure were reduced with the increase of nanoparticle concentration, and an optimum COP was acquired with 0.5 vol.%. The enhancement of COP improved the thermal conductivity, density, and specific heat by 15%, 3.2%, and 2.6%, respectively [34]. An experimental investigation [110] showed improved heat absorption and mass transfer efficiency with the nanoparticles in the NH_3/H_2O binary system.

9.7.2 Performance studies on nanolubricants

Using lubricants is essential inside a refrigeration unit's compressor to reduce losses produced by friction and, therefore, reduce pressure drop and power usage. A nanolubricant is made by adding lubricating oil to nanoparticles/nanofluids to enhance the heat transfer efficiency [111]. Experimental findings indicated that the addition of nanoparticles enhanced the lubricant's solubility with the pure refrigerant and improved the heat transfer efficiency [61,63,84,112,113]. Increased lubricant solubility minimized the condenser's oil deposition or other components that could affect the working conditions. Table 9.13 describes the performance of refrigeration applications using nanolubricants. Bi et al.

Table 9.13 Applications of nanolubricants in refrigeration system performance.

References	Nanolubricant	Parameters	Observations
[12]	TiO_2/MO	$D = 50$ nm, $\varphi = 0.06$, and 1.0 wt.%	Increased solubility of R-134a increased and energy consumption reduced
[13]	Fullerene/MO	$T = 20–80°C$, $\varphi = 0$, 0.01, and 0.1 vol.%	The friction coefficient was reduced by 90% relative to the oil
[11]	TiO_2/MO	$d_p = 30–40$ nm, $T = 40°C–90°C$, $\varphi = 0$, 0.005, 0.01, and 0.015 vol.%	Friction coefficient was reduced by 11%
[14]	Fullerene C60 nanooil	$\varphi = 1–3$ g/L	The friction coefficient decreased by 19.6%, and input power decreased by 5.6%
[114]	$MoFe_2O_4$-$NiFe_2O_4$/Fullerene	Span-80, $\varphi = 0.25$ g/L	COP of the refrigerator increased by 5.33%

COP, *coefficient of performance.*

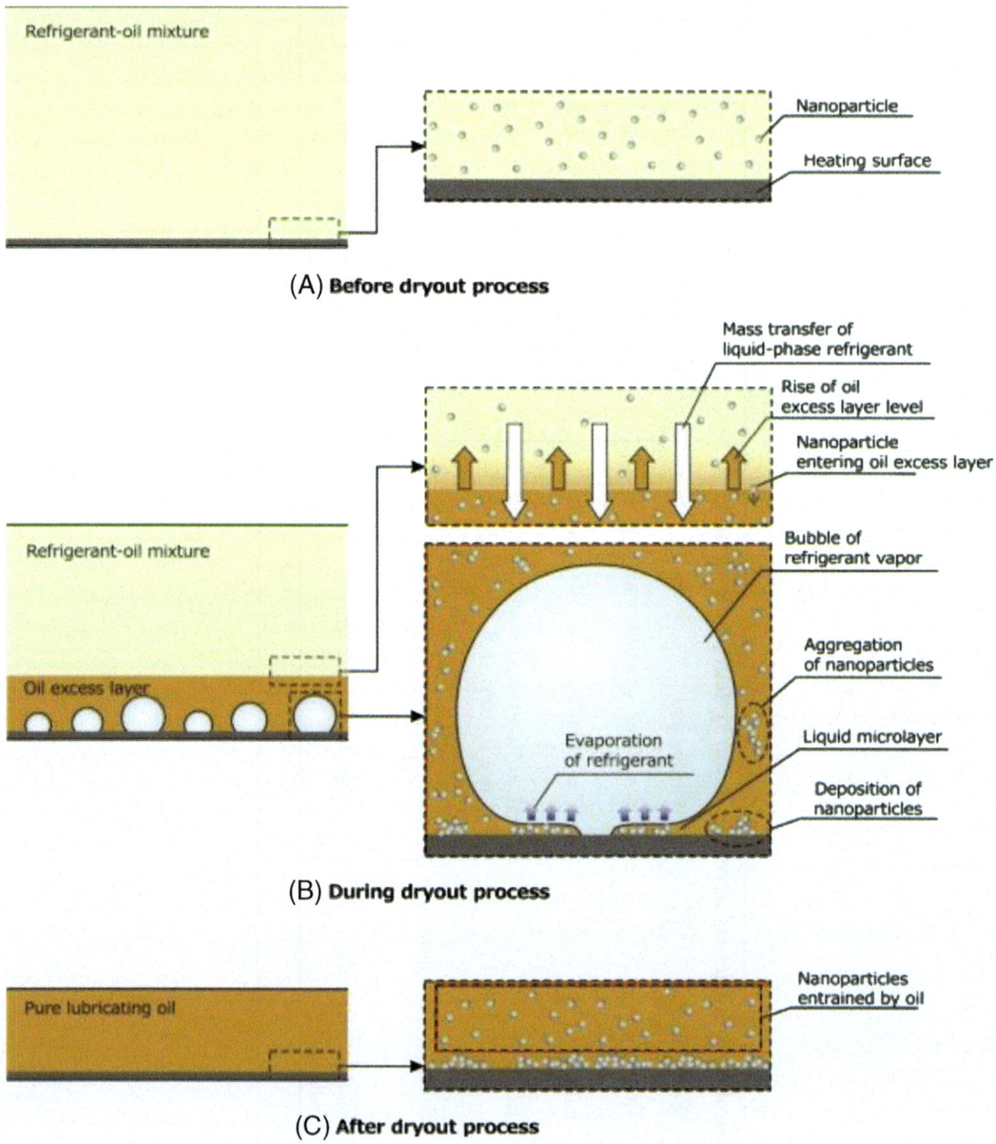

FIGURE 9.8

Diagram of the mixture-to-oil migration of nanoparticles in three stages: (A) before dryout process, (B) during dryout process, and (C) after dryout process [103].

[12] confirmed that the mixture of nanoparticles and lubricants improved the pure refrigerant solubility and the oil return rate [13]. Moreover, minimizing the friction factor reduced the compressor work. Sabareesh et al. [11] reduced compressor work by 11% using a TiO_2-based MO compressor [14]. Fullerene C60 suspended with nanooil reduced the surface temperature of the compressor shell. Moreover, compressor work decreased because of the enhancement of the COP [114]. The COP of an automotive air conditioning unit was improved after using SiO_2-based PAG nanolubricant in the compressor [115]. They observed an increase in the COP by 24% at the concentration of 0.05 vol.%.

The exergy efficiency of vapor compression refrigeration system was tested with CNTs, gold, and $HAuCl_4$ nanoparticles suspended with a conventional lubricant, PAG [116]. The lowest cooling time of about 1600 s was recorded using 0.2% gold with 0.005% CNTs, and the COP increased by 31.7% using 0.1% gold with 0.005% CNTs. A study [117] examined graphene-based nanolubricants (0, 0.2, 0.4, and 0.6 g/L) and selected mass charges (40, 50, 60, and 70 g) of pure R-600a refrigerant. When infused with either 40 g and 0 g/L or 60 g and 0.2 g/L nanofluid mixtures, the lowest compressor power consumption and cabinet temperature observed inside the device were 65 W and −12°C, respectively. Moreover, the highest PPTR value of 5.22 was obtained using 60 g and 0.2 g/L mixtures, whereas the highest COP value of 0.76 was obtained using 70 g and 0 g/L mixtures, respectively.

9.8 Challenges and future prospects

Because of the drawbacks of base refrigerants, such as toxicity and global warming potential, nanorefrigerants have been used in several cases, such as cooling, air-conditioning units, and heating purposes. Nanorefrigerants increase the thermal and heat transport properties of a system. A nanorefrigerant is employed as a heat absorber in a cooling unit. The literature found that the application of refrigerant-based nanofluids instead of pure refrigerants would be possible to minimize energy consumption by reducing the operation of pumps and compressors. Moreover, nanorefrigerants improved unit efficiency by increasing the COP, heat transfer coefficient, and freezing capacity and lowering discharge pressure, energy usage, and irreversibility. Nanorefrigerants improved the evaporator's effectiveness by raising the critical heat flow throughout the flow and pool boiling. In this chapter, many interesting properties of nanofluids are discussed. Thermal conductivity has gotten the most attention in previous studies, but many researchers have recently started looking into other thermo-physical properties. Nanofluids seem to be possible in an extensive choice of applications. However, the advance of the field is hindered by (1) an absence of reliability in the results found by different academics, (2) poor characterization of suspensions, and (3) an absence of theoretical understanding of the procedures that source property variations. Consequently, this chapter focused on a few key subjects that should be given more consideration shortly:

1. Researchers will continue to investigate the extensive research avenue of improving heat transfer at the atomic level. To improve chemical interactions at the atomic level, further quantum molecular mechanics research must be made.
2. Although nanomaterials are normally used with HCFCs, CFCs, and HFCs, which are mandatory to be phased out under the international Kyoto and Montreal procedures, there are few studies on the use of nanomaterials with pure refrigerants such as CO_2 and NH_3.
3. As a result of the inconsistent performance of nanofluids, forthcoming studies will determine the ideal mass fraction of any nanoparticle with the essential pure base fluid to reach the maximum

efficiency of heat transfer in refrigeration or air conditioning systems at an insignificant pumping power rise. Nanorefrigerant optimization in shape, scale, density, and other variables will track this pattern.

4. There is no sufficient comparative study on metal and metal oxide nanomaterials.
5. Because of the pressure drop produced by higher viscosity of refrigerant-based nanofluids, additional, comprehensive studies should study the optimal concentration of various nanoparticle suspensions in the base fluid.
6. Whereas the role of phase change enthalpy level in determining the use of any refrigerant type is one of the influential factors, there are inadequate experimental correlations.
7. Insufficient studies have been directed on the condensation process or the entire refrigeration cycle, which includes the boiling and condensation stages, and more research is essential.
8. The proposed relationships between the thermophysical properties and heat transfer mechanisms should be investigated further.

9.9 Conclusion

Regardless of different questionable findings, nanorefrigerants could be regarded as an energy-efficient replacement for base refrigerants. Various complications in the innovative analysis must be performed before these refrigerants can be employed in various uses. This investigation was designed to cover the experimental research of nanorefrigerants, that is, from preparation, characterization, thermophysical properties, and heat transfer features to actual utilization in cooling equipment, heat pumps, and air conditioning units. The findings of the present study are stated as follows:

1. Nanoparticles have a tremendous effect on the thermophysical properties and heat transfer enhancements of pure refrigerants. The thermal conductivity of nanorefrigerants improved with mass percentage and temperature but decreased with nanoparticle size. The dynamic viscosity of nanorefrigerants improved by increasing the nanoparticle volume concentration. However, it declined as temperature increased. Consequently, friction factor and pressure drop also reduced with increasing temperature. Specific heat capacity increased with mass concentration and temperature. Density improved with the mass/volume concentration of nanoparticles suspended in the refrigerant/lubricant and decreased with temperature.
2. Nanomaterials in refrigerants and lubricants typically increase heat transfer rates significantly. Although the increase in pressure drop caused by dynamic viscosity growth is a negative result, the improvement in heat transfer is much greater. A rise in the mass fraction of nanoparticles and a decrease in their nanosized, assuming both boiling and condensation procedures, increases both the intensive parameters of dynamic viscosity and thermal conductivity.
3. In the refrigeration and air conditioning industries, refrigerant-based nanofluids and nanolubricants are exemplary passive methods for reducing energy consumption. The use of refrigerant-based nanofluid and based nanolubricants increased the overall COP and reduced freezing capacity.
4. According to most available papers, heat flux, mass velocity or vapor quality, and nanomaterial dosing level are the parameters that influence flow boiling and condensation heat transfer coefficients.

Acknowledgment

This chapter was made possible with funding from Universiti Technology Malaysia (04E76) and Universiti Malaya (RMF0400-2021, ST049-2022).

Nomenclatures

Al	aluminum
Al_2O_3	aluminum oxide
CFC	chlorofluorocarbon
CNTs	carbon nanotubes
COP	coefficient of performance
CTAB	hexadecyl trimethyl ammonium bromide
Cu	copper
CuO	copper (II) oxide
EDX	energy-dispersive X-ray spectroscopy
EER	energy efficiency ratio
FPF	frictional penalty factor
FT-IR	Fourier-transform infrared spectroscopy
GNS	graphene nanosheet
HCFC	hydrochlorofluorocarbon
HFC	hydrofluorocarbon
HTC	heat transfer coefficient
HVAC	heating, ventilation, and air conditioning
MO	mineral oil
MWCNTs	multiwalled carbon nanotubes
Ni	nickel
Nu_f	Nusselt number of base fluids
Nu_{nf}	Nusselt number of nanofluid
PAG	polyalkylene glycol
POE	polyolester
R-113	1,1,2-trichloro-1,2,2-trifluoroethane
R-134a	1,1,1,2-tetrafluoroethane
R-141b	1,1-dichloro-1-fluoroethane
R-152a	1,1-difluoroethane
R-22	chlorodifluoromethane
SDBS	sodium dodecylbenzene sulfonate surfactant
SDBS	sodium dodecylbenzene sulfonate
SDS	sodium dodecyl sulfate
SEF	surfactant effect factor
SEM	scanning electron microscopy
SiO_2	silicon dioxide
SPM	scanning probe microscope
SWCHs	single-wall carbon nanohorns

SWCNTs	single-walled carbon nanotubes
TEM	transmission electron microscope
TEM	transmission electron microscopy
TGA	thermogravimetric analysis
TiO_2	titanium dioxide
TPS	transient plane source
UV–Vis	ultraviolet–visible spectroscopy
VCRS	vapor compression refrigeration system
VSM	vibrating sample magnetometer
XRD	X-ray diffraction
ZnO	zinc oxide

References

[1] J.C. Maxwell, A Treatise on Electricity and Magnetism, 1, Clarendon Press, 9 High St, Wotton-under-Edge, GL12 7DE, UK, 1881.

[2] In ASME international mechanical engineering congress and exposition, vol. 17421, pp. 99–105. American Society of Mechanical Engineers, (1995).

[3] X. Wang, X. Xu, S.U.S. Choi, Thermal conductivity of nanoparticle—fluid mixture, J. Thermophys. Heat Transf. 13 (4) (1999) 474–480, doi:10.2514/2.6486.

[4] D.-H. Yoo, K.S. Hong, H.-S. Yang, Study of thermal conductivity of nanofluids for the application of heat transfer fluids, Thermochim. Acta 455 (1–2) (2007) 66–69.

[5] W. Jiang, G. Ding, H. Peng, Measurement and model on thermal conductivities of carbon nanotube nanorefrigerants, Int. J. Therm. Sci. 48 (6) (2009) 1108–1115, doi:10.1016/j.ijthermalsci.2008.11.012.

[6] V. Nair, P.R. Tailor, A.D. Parekh, Nanorefrigerants: a comprehensive review on its past, present and future, Int. J. Refrig. 67 (2016) 290–307, doi:10.1016/j.ijrefrig.2016.01.011.

[7] J.E.S.P. Carvalho, P.O. Sotomayor, J.A.R. Parise, F. Pradelle, Numerical assessment of critical properties of nanofluids: applications to nanorefrigerants and nanolubricants, J. Mol. Liq. 318 (2020) 113938.

[8] L. Yang, et al., A review of heating/cooling processes using nanomaterials suspended in refrigerants and lubricants, Int. J. Heat Mass Transf. 153 (2020) 119611.

[9] S.-S. Bi, L. Shi, L.-L. Zhang, Application of nanoparticles in domestic refrigerators, Appl. Therm. Eng. (2008), doi:10.1016/j.applthermaleng.2007.11.018.

[10] L. Fu, R. Wang, W. Cong, Q. Li, Y. Wu, Experiment study on performance of refrigerator using nano-particle additive, J. Xi'an Jiaotong Univ. 42 (2008) 852–854.

[11] R.K. Sabareesh, N. Gobinath, V. Sajith, S. Das, C.B. Sobhan, Application of TiO_2 nanoparticles as a lubricant-additive for vapor compression refrigeration systems–an experimental investigation, Int. J. Refrig. 35 (7) (2012) 1989–1996.

[12] S.-s. Bi, L. Shi, L.-l. Zhang, Application of nanoparticles in domestic refrigerators, Appl. Therm. Eng. (2008), doi:10.1016/j.applthermaleng.2007.11.018.

[13] K. Lee, Y. Hwang, S. Cheong, L. Kwon, S. Kim, J. Lee, Performance evaluation of nano-lubricants of fullerene nanoparticles in refrigeration mineral oil, Curr. Appl. Phys. 9 (2) (2009) e128–e131.

[14] M. Xing, R. Wang, J. Yu, Application of fullerene C60 nano-oil for performance enhancement of domestic refrigerator compressors, Int. J. Refrig. 40 (2014) 398–403.

[15] J. Sarkar, P. Ghosh, A. Adil, A review on hybrid nanofluids: recent research, development and applications, Renew. Sustain. Energy Rev. 43 (2015) 164–177.

[16] Y. Li, J. Zhou, S. Tung, E. Schneider, S. Xi, A review on development of nanofluid preparation and characterization, Powder Technol. 196 (2) (2009) 89–101, doi:10.1016/j.powtec.2009.07.025.
[17] W. Yu, H. Xie, A review on nanofluids: preparation, stability mechanisms, and applications, J. Nanomater. 2012 (2012), doi:10.1155/2012/435873.
[18] Z. Haddad, C. Abid, H.F. Oztop, A. Mataoui, A review on how the researchers prepare their nanofluids, Int. J. Therm. Sci. 76 (2014) 168–189, doi:10.1016/j.ijthermalsci.2013.08.010.
[19] N.A.C. Sidik, H.A. Mohammed, O.A. Alawi, S. Samion, A review on preparation methods and challenges of nanofluids, Int. Commun. Heat Mass Transfer 54 (2014) 115–125, doi:10.1016/j.icheatmasstransfer.2014.03.002.
[20] D.K. Devendiran, V.A. Amirtham, A review on preparation, characterization, properties and applications of nanofluids, Renew. Sustain. Energy Rev. 60 (2016) 21–40, doi:10.1016/j.rser.2016.01.055.
[21] S.U. Ilyas, R. Pendyala, N. Marneni, Preparation, sedimentation, and agglomeration of nanofluids, Chem. Eng. Technol. (2014), doi:10.1002/ceat.201400268.
[22] O.A. Alawi, N.A.C. Sidik, H.A. Mohammed, A comprehensive review of fundamentals, preparation and applications of nanorefrigerants, Int. Commun. Heat Mass Transf. 54 (2014) 81–95.
[23] A.A.M. Redhwan, W.H. Azmi, M.Z. Sharif, R. Mamat, Development of nanorefrigerants for various types of refrigerant based: a comprehensive review on performance, Int. Commun. Heat Mass Transf. (2016), doi:10.1016/j.icheatmasstransfer.2016.06.007.
[24] H. Peng, G. Ding, W. Jiang, H. Hu, Y. Gao, Heat transfer characteristics of refrigerant-based nanofluid flow boiling inside a horizontal smooth tube, Int. J. Refrig. 32 (6) (2009) 1259–1270.
[25] V. Trisaksri, S. Wongwises, Nucleate pool boiling heat transfer of TiO_2-R141b nanofluids, Int. J. Heat Mass Transf. (2009), doi:10.1016/j.ijheatmasstransfer.2008.07.041.
[26] G. Ding, H. Peng, W. Jiang, Y. Gao, The migration characteristics of nanoparticles in the pool boiling process of nanorefrigerant and nanorefrigerant–oil mixture, Int. J. Refrig. 32 (1) (2009) 114–123.
[27] H. Peng, G. Ding, H. Hu, W. Jiang, D. Zhuang, K. Wang, Nucleate pool boiling heat transfer characteristics of refrigerant/oil mixture with diamond nanoparticles, Int. J. Refrig. (2010), doi:10.1016/j.ijrefrig.2009.11.007.
[28] H. Peng, G. Ding, H. Hu, Effect of surfactant additives on nucleate pool boiling heat transfer of refrigerant-based nanofluid, Exp. Therm. Fluid Sci. (2011), doi:10.1016/j.expthermflusci.2011.01.016.
[29] I.M. Mahbubul, A. Kamyar, R. Saidur, M.A. Amalina, Migration properties of TiO_2 nanoparticles during the pool boiling of nanorefrigerants, Ind. Eng. Chem. Res. (2013), doi:10.1021/ie302006n.
[30] B. Sun, D. Yang, Experimental study on the heat transfer characteristics of nanorefrigerants in an internal thread copper tube, Int. J. Heat Mass Transf. (2013), doi:10.1016/j.ijheatmasstransfer.2013.04.031.
[31] I.M. Mahbubul, R. Saidur, M.A. Amalina, Influence of particle concentration and temperature on thermal conductivity and viscosity of Al_2O_3/R141b nanorefrigerant, Int. Commun. Heat Mass Transf. 43 (2013) 100–104, doi:10.1016/j.icheatmasstransfer.2013.02.004.
[32] I.M. Mahbubul, S.S. Khaleduzzaman, R. Saidur, M.A. Amalina, Rheological behavior of Al_2O_3/R141b nanorefrigerant, Int. J. Heat Mass Transf. (2014), doi:10.1016/j.ijheatmasstransfer.2014.01.073.
[33] H. Peng, L. Lin, G. Ding, Influences of primary particle parameters and surfactant on aggregation behavior of nanoparticles in nanorefrigerant, Energy (2015), doi:10.1016/j.energy.2015.05.116.
[34] I.M. Mahbubul, A. Saadah, R. Saidur, M.A. Khairul, A. Kamyar, Thermal performance analysis of Al_2O_3/R-134a nanorefrigerant, Int. J. Heat Mass Transf. 85 (2015) 1034–1040.
[35] D. Sendil Kumar, R. Elansezhian, ZnO nanorefrigerant in R152a refrigeration system for energy conservation and green environment, Front. Mech. Eng. (2014), doi:10.1007/s11465-014-0285-y.
[36] A. Bhattad, J. Sarkar, P. Ghosh, Improving the performance of refrigeration systems by using nanofluids: a comprehensive review, Renew. Sustain. Energy Rev. (2018), doi:10.1016/j.rser.2017.10.097.
[37] L. Lin, H. Peng, G. Ding, Dispersion stability of multi-walled carbon nanotubes in refrigerant with addition of surfactant, Appl. Therm. Eng. (2015), doi:10.1016/j.applthermaleng.2015.08.011.

[38] S.M.S. Murshed, K.C. Leong, C. Yang, Investigations of thermal conductivity and viscosity of nanofluids, Int. J. Therm. Sci. (2008), doi:10.1016/j.ijthermalsci.2007.05.004.
[39] K.H. Solangi, et al., A comprehensive review of thermo-physical properties and convective heat transfer to nanofluids, Energy 89 (2015) 1065–1086, doi:10.1016/j.energy.2015.06.105.
[40] S.A. Angayarkanni, J. Philip, Review on thermal properties of nanofluids: recent developments, Adv. Colloid Interface Sci. 225 (2015) 146–176, doi:10.1016/j.cis.2015.08.014.
[41] W.H. Azmi, K.V. Sharma, R. Mamat, G. Najafi, M.S. Mohamad, The enhancement of effective thermal conductivity and effective dynamic viscosity of nanofluids—a review, Renew. Sustain. Energy Rev. (2016), doi:10.1016/j.rser.2015.09.081.
[42] W. Jiang, G. Ding, H. Peng, Y. Gao, K. Wang, Experimental and model research on nanorefrigerant thermal conductivity, HVAC&R Res. (2009), doi:10.1080/10789669.2009.10390855.
[43] I.M. Mahbubul, R. Saidur, M.A. Amalina, Thermal conductivity, viscosity and density of R141b refrigerant based nanofluid, Procedia Eng. 56 (2013) 310–315, doi:10.1016/j.proeng.2013.03.124.
[44] I.M. Mahbubul, S.A. Fadhilah, R. Saidur, K.Y. Leong, M.A. Amalina, Thermophysical properties and heat transfer performance of Al_2O_3/R-134a nanorefrigerants, Int. J. Heat Mass Transf. 57 (1) (2013) 100–108, doi:10.1016/j.ijheatmasstransfer.2012.10.007.
[45] O.A. Alawi, N.A.C. Sidik, Influence of particle concentration and temperature on the thermophysical properties of CuO/R134a nanorefrigerant, Int. Commun. Heat Mass Transf. 58 (2014) 79–84, doi:10.1016/j.icheatmasstransfer.2014.08.038.
[46] O.A. Alawi, N.A.C. Sidik, Mathematical correlations on factors affecting the thermal conductivity and dynamic viscosity of nanorefrigerants, Int. Commun. Heat Mass Transf. 58 (2014) 125–131, doi:10.1016/j.icheatmasstransfer.2014.08.033.
[47] O.A. Alawi, N.A.C. Sidik, The effect of temperature and particles concentration on the determination of thermo and physical properties of SWCNT-nanorefrigerant, Int. Commun. Heat Mass Transf. 67 (2015) 8–13, doi:10.1016/j.icheatmasstransfer.2015.06.014.
[48] S. Zhang, et al., Measurement and modeling of the thermal conductivity of nanorefrigerants with low volume concentrations, Thermochim. Acta 688 (2020) 178603.
[49] I.M. Mahbubul, A. Saadah, R. Saidur, M.A. Khairul, A. Kamyar, Thermal performance analysis of Al_2O_3/R-134a nanorefrigerant, Int. J. Heat Mass Transf. 85 (2015) 1034–1040, doi:10.1016/j.ijheatmasstransfer.2015.02.038.
[50] O.A. Alawi, N.A.C. Sidik, A.S. Kherbeet, Measurements and correlations of frictional pressure drop of TiO_2/R123 flow boiling inside a horizontal smooth tube, Int. Commun. Heat Mass Transf. 61 (2015) 42–48, doi:10.1016/j.icheatmasstransfer.2014.12.006.
[51] I.M. Shahrul, I.M. Mahbubul, S.S. Khaleduzzaman, R. Saidur, M.F.M. Sabri, A comparative review on the specific heat of nanofluids for energy perspective, Renew. Sustain. Energy Rev. (2014), doi:10.1016/j.rser.2014.05.081.
[52] H. Riazi, T. Murphy, G.B. Webber, R. Atkin, S.S.M. Tehrani, R.A. Taylor, Specific heat control of nanofluids: a critical review, Int. J. Therm. Sci. (2016), doi:10.1016/j.ijthermalsci.2016.03.024.
[53] M.A. Akhavan-Behabadi, M.K. Sadoughi, M. Darzi, M. Fakoor-Pakdaman, Experimental study on heat transfer characteristics of R600a/POE/CuO nano-refrigerant flow condensation, Exp. Therm. Fluid Sci. (2015), doi:10.1016/j.expthermflusci.2015.02.027.
[54] D. Yang, B. Sun, H. Li, X. Fan, Experimental study on the heat transfer and flow characteristics of nanorefrigerants inside a corrugated tube, Int. J. Refrig. (2015), doi:10.1016/j.ijrefrig.2015.04.011.
[55] M.A. Kedzierski, Effect of CuO nanoparticle concentration on R134a/lubricant pool-boiling heat transfer, J. Heat Transfer (2009), doi:10.1115/1.3072926.
[56] H. Peng, G. Ding, H. Hu, Influences of refrigerant-based nanofluid composition and heating condition on the migration of nanoparticles during pool boiling. Part I: experimental measurement, Int. J. Refrig. (2011), doi:10.1016/j.ijrefrig.2011.07.009.

[57] K.J. Park, D. Jung, Boiling heat transfer enhancement with carbon nanotubes for refrigerants used in building air-conditioning, Energy Build. (2007), doi:10.1016/j.enbuild.2006.12.001.

[58] D.-W. Liu, C.-Y. Yang, Effect of nano-particles on pool boiling heat transfer of refrigerant 141b, International Conference on Nanochannels, Microchannels, ASME, 2007, doi:10.1115/ICNMM2007-30221.

[59] K.J. Park, D. Jung, Enhancement of nucleate boiling heat transfer using carbon nanotubes, Int. J. Heat Mass Transf. (2007), doi:10.1016/j.ijheatmasstransfer.2007.03.012.

[60] H. Peng, G. Ding, H. Hu, W. Jiang, Influence of carbon nanotubes on nucleate pool boiling heat transfer characteristics of refrigerant-oil mixture, Int. J. Therm. Sci. (2010), doi:10.1016/j.ijthermalsci.2010.06.025.

[61] H. Peng, G. Ding, H. Hu, W. Jiang, Effect of nanoparticle size on nucleate pool boiling heat transfer of refrigerant/oil mixture with nanoparticles, Int. J. Heat Mass Transf. (2011), doi:10.1016/j.ijheatmasstransfer.2010.12.035.

[62] M.A. Kedzierski, Effect of Al_2O_3 nanolubricant on R134a pool boiling heat transfer, Int. J. Refrig. (2011), doi:10.1016/j.ijrefrig.2010.10.007.

[63] H. Hu, H. Peng, G. Ding, Nucleate pool boiling heat transfer characteristics of refrigerant/nanolubricant mixture with surfactant, Int. J. Refrig. (2013), doi:10.1016/j.ijrefrig.2012.12.015.

[64] X. Tang, Y.-H. Zhao, Y. Diao, Experimental investigation of the nucleate pool boiling heat transfer characteristics of δ-Al_2O_3-R141b nanofluids on a horizontal plate, Exp. Therm. Fluid Sci. (2014), doi:10.1016/j.expthermflusci.2013.08.025.

[65] M.A. Kedzierski, Effect of concentration on R134a/Al_2O_3 nanolubricant mixture boiling on a reentrant cavity surface, Int. J. Refrig. (2015), doi:10.1016/j.ijrefrig.2014.09.012.

[66] Y.H. Diao, C.Z. Li, Y.H. Zhao, Y. Liu, S. Wang, Experimental investigation on the pool boiling characteristics and critical heat flux of Cu-R141b nanorefrigerant under atmospheric pressure, Int. J. Heat Mass Transf. 89 (2015) 110–115, doi:10.1016/j.ijheatmasstransfer.2015.05.043.

[67] K. Bartelt, Y. Park, A. Jacobi, Flow-Boiling of R-134a/POE/CuO nanofluids in a horizontal tube, International Refrigeration and Air Conditioning Conference, West Lafayette, Indiana, USA, 2008.

[68] H. Peng, G. Ding, W. Jiang, H. Hu, Y. Gao, Measurement and correlation of frictional pressure drop of refrigerant-based nanofluid flow boiling inside a horizontal smooth tube, Int. J. Refrig. (2009), doi:10.1016/j.ijrefrig.2009.06.005.

[69] K. Henderson, Y.G. Park, L. Liu, A.M. Jacobi, Flow-boiling heat transfer of R-134a-based nanofluids in a horizontal tube, Int. J. Heat Mass Transf. (2010), doi:10.1016/j.ijheatmasstransfer.2009.11.026.

[70] B. Sun, D. Yang, Flow boiling heat transfer characteristics of nano-refrigerants in a horizontal tube, Int. J. Refrig. 38 (1) (2014) 206–214, doi:10.1016/j.ijrefrig.2013.08.020.

[71] S. Baqeri, M.A. Akhavan-Behabadi, B. Ghadimi, Experimental investigation of the forced convective boiling heat transfer of R-600a/oil/nanoparticle, Int. Commun. Heat Mass Transf. (2014), doi:10.1016/j.icheatmasstransfer.2014.04.005.

[72] M.A. Akhavan-Behabadi, M. Nasr, S. Baqeri, Experimental investigation of flow boiling heat transfer of R-600a/oil/CuO in a plain horizontal tube, Exp. Therm. Fluid Sci. (2014), doi:10.1016/j.expthermflusci.2014.06.013.

[73] P. Dey, B.K. Mandal, Performance enhancement of a shell-and-tube evaporator using Al_2O_3/R600a nanorefrigerant, Int. J. Heat Mass Transf. 170 (2021) 121015.

[74] A.S. Pamitran, K.I. Choi, J.T. Oh, H.K. Oh, Forced convective boiling heat transfer of R-410A in horizontal minichannels, Int. J. Refrig. (2007), doi:10.1016/j.ijrefrig.2006.06.005.

[75] M. Darzi, M.K. Sadoughi, M. Sheikholeslami, Condensation of nano-refrigerant inside a horizontal tube, Phys. B Condens. Matter 537 (2018) 33–39, doi:10.1016/j.physb.2018.02.002.

[76] Q. Peng, L. Jia, C. Dang, X. Zhang, Q. Huang, Experimental investigation on flow condensation of R141b with CuO nanoparticles in a vertical circular tube, Appl. Therm. Eng. (2018), doi:10.1016/j.applthermaleng.2017.10.108.

[77] I.M. Mahbubul, R. Saidur, M.A. Amalina, Heat transfer and pressure drop characteristics of Al_2O_3-R141b nanorefrigerant in horizontal smooth circular tube, Procedia Eng 56 (2013) 323–329.
[78] S. Mohamadi, M.H. Yazdi, E. Solomin, A. Fudholi, K. Sopian, P.L. Chong, Heat transfer and entropy generation analysis of internal flow of nanorefrigerant with slip condition at wall, Therm. Sci. Eng. Prog. 22 (2021) 100829.
[79] R. Azizian, E. Doroodchi, T. McKrell, J. Buongiorno, L.W. Hu, B. Moghtaderi, Effect of magnetic field on laminar convective heat transfer of magnetite nanofluids, Int. J. Heat Mass Transf. 68 (2014) 94–109, doi:10.1016/j.ijheatmasstransfer.2013.09.011.
[80] M.J.L. Ismay, E. Doroodchi, B. Moghtaderi, Effects of colloidal properties on sensible heat transfer in water-based titania nanofluids, Chem. Eng. Res. Des. (2013), doi:10.1016/j.cherd.2012.10.005.
[81] J. Jiang, G. Oberdörster, P. Biswas, Characterization of size, surface charge, and agglomeration state of nanoparticle dispersions for toxicological studies, J. Nanoparticle Res. (2009), doi:10.1007/s11051-008-9446-4.
[82] X. Li, D. Zhu, X. Wang, Evaluation on dispersion behavior of the aqueous copper nano-suspensions, J. Colloid Interface Sci. 310 (2) (2007) 456–463, doi:10.1016/j.jcis.2007.02.067.
[83] C. Pang, J.Y. Jung, Y.T. Kang, Aggregation based model for heat conduction mechanism in nanofluids, Int. J. Heat Mass Transf. (2014), doi:10.1016/j.ijheatmasstransfer.2013.12.055.
[84] L. Lin, H. Peng, G. Ding, Experimental research on particle aggregation behavior in nanorefrigerant-oil mixture, Appl. Therm. Eng. (2016), doi:10.1016/j.applthermaleng.2015.12.052.
[85] L. Lin, H. Peng, Z. Chang, G. Ding, Experimental research on degradation of nanolubricant–refrigerant mixture during continuous alternation processes of condensation and evaporation, Int. J. Refrig. 76 (2017) 97–108, doi:10.1016/j.ijrefrig.2016.12.021.
[86] H. Peng, G. Ding, H. Hu, Migration of carbon nanotubes from liquid phase to vapor phase in the refrigerant-based nanofluid pool boiling, Nanoscale Res. Lett. (2011), doi:10.1186/1556-276X-6-219.
[87] A. Kamyar, I.M. Mahbubul, R. Saidur, A.M. Afifi, Influence of nanoparticle type, size and weight on migration properties of nanorefrigerant, Adv. Mater. Res. (2013), doi:10.4028/www.scientific.net/AMR.832.45.
[88] W.H. Azmi, M.Z. Sharif, T.M. Yusof, R. Mamat, A.A.M. Redhwan, Potential of nanorefrigerant and nanolubricant on energy saving in refrigeration system—a review, Renew. Sustain. Energy Rev. (2017), doi:10.1016/j.rser.2016.11.207.
[89] M.R. Salem, Performance enhancement of a vapor compression refrigeration system using R134a/MWCNT-oil mixture and liquid-suction heat exchanger equipped with twisted tape turbulator, Int. J. Refrig. 120 (2020) 357–369.
[90] S.S. Chauhan, Performance evaluation of ice plant operating on R134a blended with varied concentration of $Al_2O_3SiO_2$/PAG composite nanolubricant by experimental approach, Int. J. Refrig. 113 (2020) 196–205.
[91] L. Fedele, et al., Nanofluids characterization and application as nanolubricants in heat pump systems, Sci. Technol. Built Environ. (2015), doi:10.1080/23744731.2015.1023164.
[92] G. Kosmadakis, P. Neofytou, Investigating the effect of nanorefrigerants on a heat pump performance and cost-effectiveness, Therm. Sci. Eng. Prog. 13 (2019) 100371.
[93] S.S. Sanukrishna, M. Murukan, P.M. Jose, État des lieux des études expérimentales sur les nanofrigorigènes: Recherches récentes, développement et utilisations, Int. J. Refrig. 88 (2018) 552–577, doi:10.1016/j.ijrefrig.2018.03.013.
[94] P. Naphon, D. Thongkum, P. Assadamongkol, Heat pipe efficiency enhancement with refrigerant-nanoparticles mixtures, Energy Convers. Manag. (2009), doi:10.1016/j.enconman.2008.09.045.
[95] P. Naphon, C. Thongjing, Pool boiling heat transfer characteristics of refrigerant-nanoparticle mixtures, Int. Commun. Heat Mass Transf. (2014), doi:10.1016/j.icheatmasstransfer.2014.01.014.

[96] L. Cheng, L. Liu, Boiling and two-phase flow phenomena of refrigerant-based nanofluids: fundamentals, applications and challenges, Int. J. Refrig. 36 (2) (2013) 421–446, doi:10.1016/j.ijrefrig.2012.11.010.

[97] A.K. Tiwari, P. Ghosh, J. Sarkar, Particle concentration levels of various nanofluids in plate heat exchanger for best performance, Int. J. Heat Mass Transf. (2015), doi:10.1016/j.ijheatmasstransfer.2015.05.118.

[98] R. Saidur, S.N. Kazi, M.S. Hossain, M.M. Rahman, H.A. Mohammed, A review on the performance of nanoparticles suspended with refrigerants and lubricating oils in refrigeration systems, Renew. Sustain. Energy Rev. (2011), doi:10.1016/j.rser.2010.08.018.

[99] A. Celen, A. Çebi, M. Aktas, O. Mahian, A.S. Dalkilic, S. Wongwises, A review of nanorefrigerants: flow characteristics and applications, Int. J. Refrig. 44 (2014) 125–140.

[100] O.A. Alawi, N.A.C. Sidik, M. Beriache, Applications of nanorefrigerant and nanolubricants in refrigeration, air-conditioning and heat pump systems: a review, Int. Commun. Heat Mass Transf. 68 (2015) 91–97, doi:10.1016/j.icheatmasstransfer.2015.08.014.

[101] M.S. Patil, S.C. Kim, J.H. Seo, M.Y. Lee, Review of the thermo-physical properties and performance characteristics of a refrigeration system using refrigerant-based nanofluids, Energies (2016), doi:10.3390/en9010022.

[102] O.O. Ajayi, D.E. Ibia, M. Ogbonnaya, A. Attabo, A. Michael, CFD analysis of nanorefrigerant through adiabatic capillary tube of vapour compression refrigeration system, Procedia Manuf. (2017), doi:10.1016/j.promfg.2016.12.102.

[103] L. Lin, H. Peng, Z. Chang, G. Ding, Experimental investigation on TiO_2 nanoparticle migration from refrigerant–oil mixture to lubricating oil during refrigerant dryout, Int. J. Refrig. (2017), doi:10.1016/j.ijrefrig.2017.02.026.

[104] D. Yang, B. Sun, H. Li, C. Zhang, Y. Liu, Comparative study on the heat transfer characteristics of nano-refrigerants inside a smooth tube and internal thread tube, Int. J. Heat Mass Transf. 113 (2017) 538–543, doi:10.1016/j.ijheatmasstransfer.2017.05.069.

[105] R. Wang, Q. Wu, Y. Wu, Use of nanoparticles to make mineral oil lubricants feasible for use in a residential air conditioner employing hydro-fluorocarbons refrigerants, Energy Build. (2010), doi:10.1016/j.enbuild.2010.06.023.

[106] S. Bi, K. Guo, Z. Liu, J. Wu, Performance of a domestic refrigerator using TiO_2-R600a nano-refrigerant as working fluid, Energy Convers. Manag. 52 (1) (2011) 733–737, doi:10.1016/j.enconman.2010.07.052.

[107] N. Subramani, M. Prakash, Experimental studies on a vapour compression system using nanorefrigerants, Int. J. Eng. Sci. Technol. (2012), doi:10.4314/ijest.v3i9.8.

[108] D.S. Kumar, R. Elansezhian, Experimental study on Al_2O_3-R134a nano refrigerant in refrigeration system, Int. J. Mod. Eng. Res. 2 (2012) 3927–3929.

[109] V.M.V. Padmanabhan, S. Palanisamy, The use of TiO_2 nanoparticles to reduce refrigerator ir-reversibility, Energy Convers. Manag. (2012), doi:10.1016/j.enconman.2012.03.002.

[110] J.K. Lee, J. Koo, H. Hong, Y.T. Kang, The effects of nanoparticles on absorption heat and mass transfer performance in NH_3/H_2O binary nanofluids, Int. J. Refrig. (2010), doi:10.1016/j.ijrefrig.2009.10.004.

[111] M. Youbi-Idrissi, J. Bonjour, The effect of oil in refrigeration: current research issues and critical review of thermodynamic aspects, Int. J. Refrig. (2008), doi:10.1016/j.ijrefrig.2007.09.006.

[112] M.Z. Sharif, W.H. Azmi, A.A.M. Redhwan, R. Mamat, Investigation of thermal conductivity and viscosity of Al_2O_3/PAG nanolubricant for application in automotive air conditioning system, Int. J. Refrig. (2016), doi:10.1016/j.ijrefrig.2016.06.025.

[113] S. Bobbo, L. Fedele, M. Fabrizio, S. Barison, S. Battiston, C. Pagura, Influence of nanoparticles dispersion in POE oils on lubricity and R134a solubility, Int. J. Refrig. (2010), doi:10.1016/j.ijrefrig.2010.04.009.

[114] T. Jia, R. Wang, R. Xu, Performance of $MoFe_2O_4$-$NiFe_2O_4$/Fullerene-added nano-oil applied in the domestic refrigerator compressors, Energy Econ. (2014), doi:10.1016/j.ijrefrig.2014.06.001.

[115] M.Z. Sharif, W.H. Azmi, A.A.M. Redhwan, R. Mamat, T.M. Yusof, Performance analysis of SiO_2/PAG nanolubricant in automotive air conditioning system, Int. J. Refrig. (2017), doi:10.1016/j.ijrefrig.2017.01.004.

[116] K. Mohan, S. Sundararaj, K.G. Kannan, A. Kannan, Experimental analysis on refrigeration system using CNT, gold & HAUCL4 nano fluids, Mater. Today Proc. 33 (2020) 360–366.

[117] D.S. Adelekan, O.S. Ohunakin, J. Gill, I.P. Okokpujie, O.E. Atiba, Performance of an Iso-Butane driven domestic refrigerator infused with various concentrations of graphene based nanolubricants, Procedia Manuf. 35 (2019) 1146–1151.

[118] S. Tazarv, M. Saffar-Avval, F. Khalvati, E. Mirzaee, Z. Mansoori, Experimental investigation of saturated flow boiling heat transfer to TiO_2/R141b nanorefrigerant, Exp. Heat Transf. (2016), doi:10.1080/08916152.2014.973976.

[119] X.F. Yang, Z.H. Liu, Pool boiling heat transfer of functionalized nanofluid under sub-atmospheric pressures, Int. J. Therm. Sci. (2011), doi:10.1016/j.ijthermalsci.2011.07.009.

[120] C.-S. Jwo, L.-Y. Jeng, T.-P. Teng, H. Chang, Effects of nanolubricant on performance of hydrocarbon refrigerant system, J. Vac. Sci. Technol. B Microelectron. Nanom. Struct. (2009), doi:10.1116/1.3089373.

[121] E.A. Abdel-hadi, S.H. Taher, Heat transfer Analysis of vapor compression system using nano, Int. Conf. Adv. Mater. Eng. Vol. 15, pp. 80—84, Publisher: IACSIT Press, Singapore (2011).

[122] R.K. Sabareesh, N. Gobinath, V. Sajith, S. Das, C.B. Sobhan, Application of TiO_2 nanoparticles as a lubricant-additive for vapor compression refrigeration systems—an experimental investigation, Int. J. Refrig. (2012), doi:10.1016/j.ijrefrig.2012.07.002.

[123] R. Kumar, J. Singh, Effect of ZnO nanoparticles in R290/R600a (50/50) based vapour compression refrigeration system added via lubricant oil on compressor suction and discharge characteristics, Heat Mass Transf. (2017), doi:10.1007/s00231-016-1921-3.

[124] T. Coumaressin, K. Palaniradja, Performance analysis of a refrigeration system using nano fluid, Int. J. Adv. Mech. Eng. 4 (4) (2014) 459–470.

[125] J.F. Lou, H. Zhang, R. Wang, Experimental investigation of graphite nanolubricant used in a domestic refrigerator, Adv. Mech. Eng. (2015), doi:10.1177/1687814015571011.

[126] A. Sözen, E. Özbaş, T. Menlik, M.T. Çakir, M. Gürü, K. Boran, Improving the thermal performance of diffusion absorption refrigeration system with alumina nanofluids: an experimental study, Int. J. Refrig. (2014), doi:10.1016/j.ijrefrig.2014.04.018.

[127] D.S. Adelekan, et al., Experimental performance of LPG refrigerant charges with varied concentration of TiO_2 nano-lubricants in a domestic refrigerator, Case Stud. Therm. Eng. (2017), doi:10.1016/j.csite.2016.12.002.

CHAPTER 10

Two-dimensional nanomaterials for bioheat transfer: Applications in hyperthermia, drug delivery, biosensing, and nucleic acid amplification

Faaiza Naaz[a], Shobana Gabriel[b], Samson Eugin Simon[c] and Abdul Khaliq Rasheed[d]

[a]*Department of Biochemistry, Auxilium College, Vellore, Tamil Nadu, India,* [b]*Department of Biomedical Sciences, Faculty of Medical and Life Sciences, Sir Jeffrey Cheah Sunway Medical School, Sunway University, Bandar Sunway, Selangor, Malaysia,* [c]*Department of Hematology and Oncology, College of Medicine, Cancer Research Center, University of Tennessee Health Science Center, Memphis, TN, United States,* [d]*Department of New Energy Science and Engineering, School of Energy and Chemical Engineering, Xiamen University Malaysia, Sepang, Selangor, Malaysia*

10.1 Introduction

Bioheat transfer refers to the transport of thermal energy within biological tissues, encompassing heat generation, conduction, convection, and radiation processes that are critical for maintaining physiological homeostasis and enabling therapeutic interventions. [1]. Two-dimensional (2D) nanostructures, such as graphene nanoparticles (GNP), hexagonal boron nitride (hBN), black phosphorus (BP), molybdenum disulfide (MoS_2), and transition metal carbides and nitrides (MXenes), have atomically thin and wide surface areas, allowing them to absorb and store electrical or thermal energy in response to various spectral emissions [2,3]. Today the integration of 2D nanomaterials in bioheat transfer is widely studied for treating physiological disorders, injuries, cancer, and microbial infections [4]. This chapter discusses the recent applications of 2D nanoparticles (NPs) and their complexes in cancer hyperthermia, biophotonics, and biosensing. Nanoscale heat transfer in biological systems enhances hyperthermia, diathermy, and ablation therapies for cancer treatment. NP-induced hyperthermia is achieved with precise manipulation of radio waves, microwaves, ultrasound, or magnetic fields, contributing nanoheat transport into solid tumors [5]. Likewise, carbon-based GNPs have free atomic groups to develop excessive heat and stable bonds with photosensitizers such as metallic or magnetic conjugates; they potentially elevate hyperthermia in solid tumors in response to spectral radiation [6]. Although 2D nanomaterials in bioheat transfer are significant, several challenges remain. The biological tissue exhibits complex thermal properties influenced by blood perfusion, metabolic heat, and thermal conductivity. The Pennes' bioheat transfer equation is widely used to model temperature distribution in soft tissues such as the retina and brain during laser irradiation-based therapies. [7]. The equation illustrates the energy balance between the conductive heat transfer, metabolic processes generated by endogenous heat, and the effects because

Graphene and 2D Materials in Heat Transfer: Fundamentals and Applications. DOI: https://doi.org/10.1016/B978-0-12-821964-5.00007-2

of the blood vessels acting as thermal sinks or sources [8,9].

$$\rho C \frac{\partial T}{\partial t} = \nabla(k\nabla T) + W_b C_{pb}(T_a - T) + Q_m + H$$

where ρ is the density, C is the specific heat capacity of the tissue, W_b represents the volumetric perfusion rate, C_{pb} is the specific heat capacity of the blood, Q_m is the volumetric heat generated by specific tissue, and H is the heat deposited in the tissue because of some external source, which would also be any heat source in the case of other therapies.

Moreover, the bloodstream administration of NPs is debated, considering the viscosity of nanofluid accumulation, porosity, thermophoresis, Brownian relaxation, and magnetic field [9,10]. The transport behavior of NPs in blood flow requires computational modeling using constant-viscosity, Reynolds', and Vogel's models to determine effective nanofluid viscosity and homotopy perturbation method for magnetic stimulation in blood flow [11]. The biocompatibility of 2D nanomaterials remains a concern due to potential toxicity in off-target environments. This can be addressed through surface functionalization, PEGylation, and careful material design. Additionally, the Fermi energy (E_F) of these materials, a quantum mechanical property governing their electronic behavior, can be tuned to optimize their photothermal conversion efficiency and surface plasmon resonance properties for specific therapeutic applications. Fermi energy (E_F) is a concept in quantum mechanics, especially the NPs' electronic properties in the conversion of thermal, optical, and electromagnetic energy [12]. In materials like graphene the Fermi energy can be calculated using

$$E_F = \hbar v_F \sqrt{\pi n_s}$$

where v_F is the Fermi velocity, and n_s is the surface carrier density (in m^{-2}).

The position of the Fermi energy level affects the surface plasmon resonance (SPR) when conduction electrons oscillate collectively in response to incident light, raising the surrounding surface temperature [13]. Optimal SPR frequency is also crucial for NP-mediated drug delivery and biosensing. The size, shape, and composition of NPs influence surface plasmon resonance, which can be harnessed to enhance targeted drug release through photothermal effects and enable precise detection of tumor markers [14]. In biophotonics and biosensing, manipulating the Fermi energy and SPR frequency of 2D nanomaterials enhances their optoelectronic properties, allowing precise control over light absorption and heat generation [15]. This strategy improves the sensitivity and efficiency of biosensing applications by tuning the interaction of nanomaterials with biological targets at specific wavelengths. In addition to cellular activity the applications of 2D NPs are expanding, particularly in the development of improved biosensing and rapid diagnostic tools, including nanoparticle-assisted polymerase chain reaction (nanoPCR) for genomic sequencing and high-throughput pathogen detection [16]. 2D nanomaterials such as graphene oxide (GO), hBN, carbon nanotubes (CNTs), and nanometallic composites such as copper oxide (CuO) and alumina (Al_2O_3) have strong thermal conductivity, utilized in nanoparticle-assisted polymerase chain reaction [17]. NanoPCR facilitates uniform heat distribution, improving the efficiency of the denaturation, annealing, and extension phases, resulting in faster and more efficient DNA amplification. Overall the use of 2D nanomaterials in bioheat transfer has a wide application. Here, we illustrate the recent findings and the mechanisms behind the applications of 2D materials in bioheat transfer.

10.2 Applications of two-dimensional nanomaterial in hyperthermia

10.2.1 Hyperthermia and photothermal therapy in cancer

Eliminating cancer remains a significant clinical challenge due to limitations in diagnostics and treatment modalities. However, over the past decade, NP-mediated hyperthermia treatments have demonstrated improved outcomes in cancer patients. A decade of studies has shown that NP-mediated HTs can improve positive outcomes in cancer patients. 2D nanomaterials, such as graphene with its sp^2-hybridized carbon structure, possess extraordinary electrical and thermal conductivity [6]. These 2D NPs have distinct thin nanostructures, potentially converting photonic energy to thermal energy, typically known as photothermal therapy (PTT). PTT eliminates malignant solid tumors by increasing cellular level temperature, which is recognized by the American Cancer Society (ACS) as a potential treatment in human cancer patients [18]. Elevated temperatures disrupt cellular protein and enzymes, leading to cell death, making HT an effective approach in cancer therapy [19]. Temperatures above 43°C–44°C are clinically defined as mild hyperthermia (MHT), while temperatures reaching 45°C–50°C constitute high hyperthermia (HHT). Temperatures exceeding 50°C induce rapid cell death primarily through necrosis[20,21]. Studies have demonstrated that applying MHT for 30 minutes to various cancer cells or solid tumors promotes programmed cell death rather than necrosis, offering an advantage in anticancer therapies [22]. Moreover, hyperthermia increases reactive oxygen species (ROS) activity, leading to cytotoxicity in colon, leukemia, prostate, pancreatic, and breast cancer cells [23,24]. The activation of ROS can influence mitochondrial function in blood platelets and adipose tissue, induce autophagic vacuolation in human colon cancer cells, and lead to heat shock protein 27 elimination in leukemia cells [25,26]. Furthermore, studies have provided evidence that pre-, post-, or cotreatment of HT has accelerated mitochondrial apoptosis, DNA damage, and cell cycle arrest in cancer cells [27,28].

The concept of enhancing treatments with 2D nanomaterials, such as GNP, emerged from promising findings on radiation therapy enhancement with short exposure durations and hyperthermic sensitization in solid tumors [29,30]. NPs can be administered in different ways to target tumors (Fig. 10.1). Two major methods are local subcutaneous administration (LSA) and intravenous administration (IVA). Cancers like melanoma are suitable for LSA, as spectra like near-infrared (NIR) light penetrate a few centimeters into the skin. Internal tumors, such as adenocarcinomas, require IVA of metallic NPs combined with radioactive emissions to achieve atomic-level magnetic vibrations, producing magnetic hyperthermia therapy [31,32]. The discovery of GNPs, carbon-based with a unique 2D structure and photoactive sites, offers significant potential in cancer therapy [33]. Under spectral radiation, GNP and GO were known to induce apoptosis in numerous cancers. In a recent study, GNP demonstrated a reduction in functional groups or a shift to reduced GO (rGO) under UV-B radiation, resulting in an active structural remodeling that enhanced cellular cytotoxicity by increasing ROS activity and promoting mitochondrial apoptosis in ER+ breast cancer cells [34]. GO structural modification under UV-B irradiation is believed to be involved in the exfoliation of positively charged atoms from the surface, accumulating negative free radicals that attack cancer cells, leading to increased production of ROS [35]. Similarly, 2D MXene NPs with metallic layering increase surface and edge functional atoms under optimal UV-B irradiation at 5 mJ/cm^2, inducing synergistic apoptosis in breast cancer cells by disrupting nuclear and cellular-bound microtubules [36]. A study combined low-intensity LED

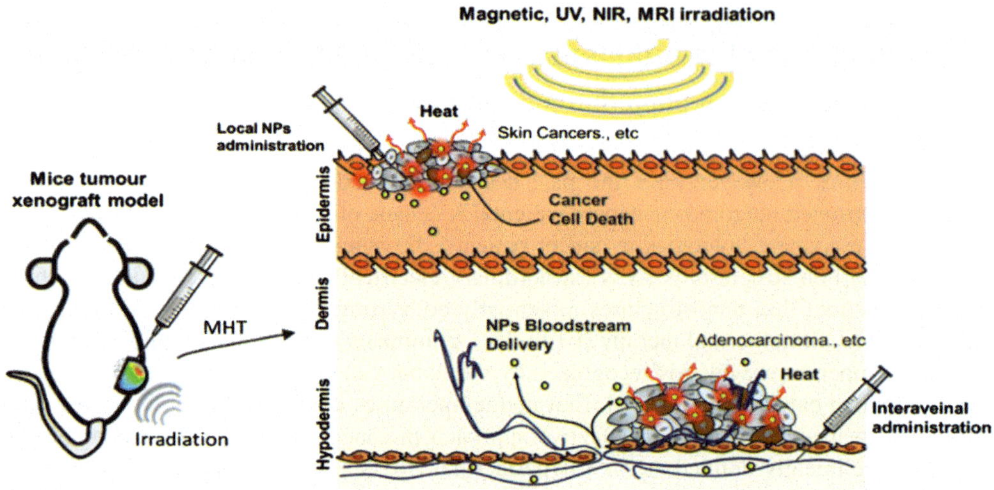

FIGURE 10.1

Schematic representation of the nanoparticle's administration methods: tumor-site accumulation, interveinal administration, moderate hyperthermia (MHT) induction in response to spectral radiation; and photothermal therapy in vivo mice model. *NIR*, Near-infrared; *NPs*, nanoparticles.

phototherapy with rGO treatment to ablate melanoma tumors in mice. The combination promoted inflammation-enhanced necrosis, immune response, and tumor mass reduction [37]. A similar 2D material, BP, demonstrated a synergistic effect of PTT and immunotherapy. BP was administered near the tumor via subcutaneous injections, whereas PD-L1 was given intravenously, leading to a reduction in 4T1 tumor mass under 980 nm laser irradiation in mice [38]. Overall the use of 2D nanomaterials in PTT shows promising tools for improving cancer treatment outcomes by promoting tumor ablation and apoptosis, reducing recurrence, and minimizing toxicity.

10.2.2 Biophotonics in drug delivery using two-dimensional nanomaterial

Biophotonics is multidisciplinary and deals with the interaction between photons (light) and biological matter. In imaging and diagnostics, it involves fluorescence microscopy, optical coherence tomography, and Raman spectroscopy, and in therapeutics, it covers photodynamic therapy and PTT as described in the earlier topics. A recent growing area of interest in biophotonics is NP-assisted, light-triggered drug release; this allows NPs to release their therapeutic load when exposed to specific wavelengths of light, reducing systemic side effects [39]. GO, MXenes, BP, and MoS_2 are some of the most convenient 2D nanomaterials for light-assisted drug release because of their strong NIR absorption, high drug-loading capabilities, and biocompatibility [40,41]. The remarkable properties of MXenes, which are metal carbides or nitride layers, make them suitable for dual PTT and drug delivery. Under NIR light, MXene drug conjugates exfoliate at specific sites through a temperature-induced mechanism [40,41]. The components of MXenes are 2D carbides, nitrides, and carbonitrides of transition metals, forming an effective layered structure that can be modified with various conjugates. These modifications improve

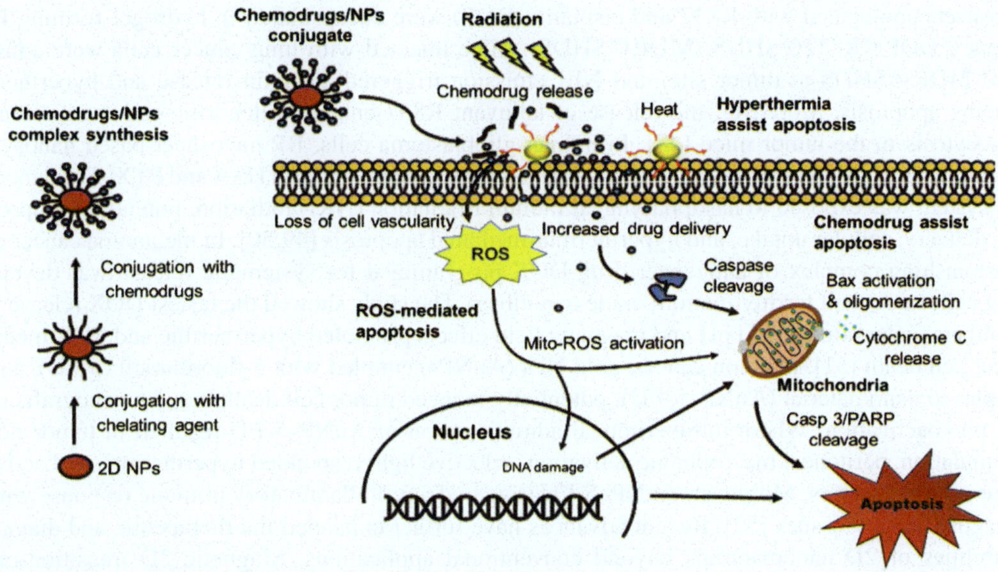

FIGURE 10.2

Schematic representation of the nanoparticles' complex system functions on a cancer cell: left, indicating nanocomplex system synthesis. Right, controlled chemotherapeutic drug release facilitated by radiation, nanoparticle enhancement of hyperthermia, loss of cellular vascularity because of heat, and ROS-mediated apoptotic signaling in cancer cells. *2D*, Two-dimensional; *NPs*, nanoparticles; *ROS*, reactive oxygen species.

photothermal activity or enhance drug delivery mechanisms in the treatment of cancer and other diseases [42,40]. Mostly, nanomolecules (2D structure) administered along with cancer drugs alone or with mediator complex (magnetic nanocomposites) promoted controlled drug delivery and hyperthermia enhancement (Fig. 10.2). In a recent study, doxorubicin hydrochloride (DOX) and methoxy poly(ethylene glycol) benzaldehyde (mPEG-CHO) were complexed with $Ti_3C_2T_x$ MXene nanosheets to form a 2D nano-prodrug (TDP). Following intravenous administration and NIR irradiation at the tumor site, TDP induced localized hyperthermia and sustained DOX release, resulting in 4T1 tumor regression in mice [43].

Recently, in liver cancer cells using the anticancer drug sorafenib complexed with indocyanine green nanodrug (SNIP) developed by the self-assembled nanoprecipitation method, exposure under NIR laser irradiation produced hyperthermia, and mechanical functions showed ROS-mediated cell death [44]. In pancreatic cancer cells, an iron oxide core coated with dimercaptosuccinic acid (MF66) magnetic NP, a 2D metal complex carrying gemcitabine, and the pseudo-peptide NucAnt promoted sustained drug release, magnetic hyperthermia-mediated cell death, ROS activation, and mitochondrial and ERK1/2-JNK pathways of apoptosis in a mice xenograft model using 4.2×10^{-5} times lower gemcitabine doses [45,46]. In breast cancer cells, superparamagnetic iron oxide NPs (MF66) functionalized with doxorubicin (DOX) and pseudopeptide NuCant (N6L) made into a nanocomplex material (MF66-DOX-N6L). Administered to the intratumor region reduced tumor volume in the mouse xenograft model. DOX-N6L release is mediated by magnetically induced hyperthermia [47]. In lung cancer, MXene nanosheets conjugated with R837 and cisplatin (DDP) were encapsulated in hydrogel-forming PEG-MXene@DDP@R837@SHDS (MDR@SHDS). Mice injected with lung cancer cells were administrated MDR@SHDS at tumor site, and NIR emission triggered cisplatin release and hyperthermia, inducing apoptosis. Moreover, the release of adjuvant R837 enhances dendritic cell maturation and phagocytosis in the tumor mice [48]. In human glioblastoma cells, BP nanosheet-based nanosystem [BP–human serum albumin (HSA)–paclitaxel (PTX)] functionalized with HSA and PTX. The nanocomplex system was exposed to nasopharyngeal radium irradiation (NRI) radiation, potentially improving drug delivery, cellular uptake, and hyperthermia-mediated apoptosis [49,50]. In melanoma cancer cells, an astonishing complex of anticancer drug DOX programmed for systematic release was developed using chitosan/cobalt ferrite/titanium oxide nanofibers. The study showed the fastest DOX release from nanofibers under altered acid pH and magnetic field effects promoted hyperthermic and ROS-mediated cancer cell death [51]. In colon cancer, gold NPs (AuNPs) coupled with 5-fluorouracil (5-FU), a dual-complexed nanomaterial (AuNP–5-FU), potentially induced tumor cell death in mice xenografts using NIR photoactivation. Where intraperitoneal administration of AuNP–5-FU resulted in tumor nodule accumulation, peritoneal macrophagic activation, selective light responded hyperthermia, and activated tumor necrosis factors. Moreover, AuNP–5-FU triggered a proinflammatory immune response without damaging healthy tissues [52]. Recent advances have further expanded the therapeutic and diagnostic capabilities of 2D nanomaterials beyond conventional applications. Magnetic 2D transition-metal-based nanocomposites now integrate hyperthermia induction, real-time imaging, and photothermal or photodynamic therapy within a single nanosystem [53], enabling magnetic field-triggered drug release and NIR-activated hyperthermia. Simultaneously, stimuli-responsive MXene-based hydrogels have enabled spatiotemporally controlled drug release and deep-tissue photothermal ablation, with "spiky" biodegradable MXene nanosheets demonstrating enhanced cellular internalization and improved drug-loading capacity [54]. In biosensing, MXene-based electrochemical biosensors have achieved ultra-sensitive detection of multiple cancer biomarkers, including CEA, CA-125, HER2, and miRNAs, with exceptional specificity and sensitivity. Furthermore, terahertz surface plasmon resonance sensors integrating MXene heterostructures have enabled dual-modal detection of both cancer biomarkers and malaria antigens, showcasing the remarkable versatility of 2D material-based sensing platforms. These emerging developments underscore the rapidly evolving landscape of 2D nanomaterials in bioheat transfer, bridging fundamental research and clinical translation.

10.3 Two-dimensional materials in biosensing

Biosensors are modern analytical devices used to transfer the biochemical signals generated from the interaction of the targets (enzyme, substrate, and antigen) with the biorecognition elements (enzyme, antibody, and DNA) into another signal with the help of optical and electrical techniques for the detection of chemical compounds [55]. Similar to other applications, there are various types of 2D nanomaterials used in biosensing. Primarily, 2D nanomaterials are used for the detection of several cancer biomarkers in electrochemical sensors, graphene, and graphene-related 2D nanomaterials [50]. The emerging 2D nanomaterial MXenes also serves as the best candidate in biosensors for sensing applications [40]. Recently the combination of 2D nanomaterials with graphene-like structures has gained widespread interest in the fields of catalysis, clean energy, electronics, fuel cells, and supercapacitors [44]. Besides, carbon nanomaterials also offer better possibilities for enhancing the performance of biosensors because of their extraordinary distinctive features such as good biocompatibility, high electrical conductivity, and catalytic activity [56]. Comparatively the unique essence of molybdenum disulfide (MoS2) can hold vast potential by promoting the large surface area and appreciable band gaps, and simultaneously, it lowers the current leakage in the fields of electronic, optical, mechanical, and electrochemical [55]. Hence a tremendous and broad application in the field of biosensing could be inevitably provided by 2D molybdenum disulfide nanomaterials in recent times, and they are also expected to surpass other 2D nanomaterials [55]. Preceding, looking onto other newly launched 2D nanomaterials, graphdiyne, composed of benzene rings and alkyne units, has also proved to be a powerful tool in the field of the biosensor. Moreover, graphdiyne provides an immense application in the platform of a biosensor for biomedical detection and environmental analyses by immobilizing biomolecules [55]. SPR technology has gained a growing interest in recent decades because of its humongous applications. Hence, recently, researchers have combined 2D materials such as graphene, MoS2, and WS2 onto the traditional SPR biosensors, which have helped to improve the applications of diagnostics and biomedical detection [57]. Among the family of 2D layered materials, transition metal oxides and dichalcogenides are also used for various applications of biosensors. In addition to this, for various applications of biosensors, Group 5 elements, also known as pnictogens, have also gained interest among researchers because of their attractive physical, electronic, and chemical properties [58].

10.3.1 Two-dimensional nanomaterial-assisted polymerase chain reaction

Polymerase chain reaction (PCR) is a method used to amplify DNA molecules, which is crucial for clinical diagnosis and research purposes. It has been a cornerstone in science since its invention in 1985, a conventional PCR amplifies copies of DNA using thermal cycling, which involves rapid heating and cooling processes [59]. Reverse transcriptase PCR is a similar process that involves high temperatures for converting RNA to cDNA and making copies of DNA molecules. Low thermal transmission and sensitivity are major issues in the PCR process. Using 2D NPs such as GO and metal oxide NPs (CuO, Al_2O_3, and TiO_2) improves thermal conductivity and allows for the detection of low-level DNA concentrations for amplification [60]. The addition of NPs into PCR reactions is referred to as nanoPCR, as described in Fig. 10.3. It not only improves the quality of the process but also shortens the reaction duration and increases accuracy [61,62]. Below, we have elaborated on the efficiency of NanoPCR, specifically how 2D nanostructures such as GO, hBN, metallic oxides, and CNTs improve PCR reactions.

10.3.1.1 Graphene oxide

GO is a known 2D nanosheet composed of carbon atoms in a hexagonal (honeycomb) configuration. Besides, GO has outstanding characteristics, such as high current density, chemical inertness, high surface area, and high thermal conductivity properties, which make various biological applications [63]. GO is extensively used in PCR for its heat transfer and conduction properties and mixture with single-stranded DNA or protein [17,64,65]. Many studies have reported the use of GO in the enhancement of PCR yield products, mainly in eliminating the nonspecific amplification of the PCR reaction. According to Jia et al., an increasing concentration of GO ranging from 12 to 60 mg/mL further enhanced the specificity of PCR, whereas no effects in the production of PCR yield were observed with less than 12 mg/mL and it showed inhibition when GO concentration was greater than 70 mg/mL. Nevertheless, rGO improved the efficiency of PCR at an optimal concentration of 0.8 mg/mL and had an inhibitory

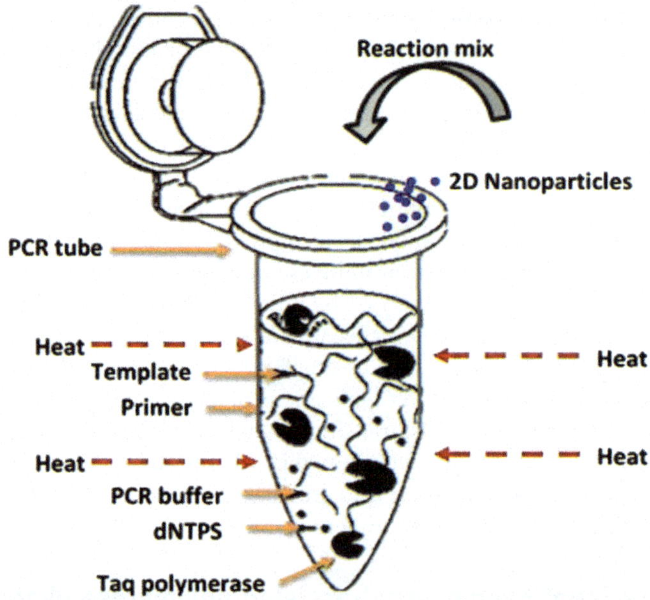

FIGURE 10.3

Illustration of the addition of 2D nanoparticles in a PCR reaction. *2D*, Two-dimensional.

effect of more than 12 mg/mL. Reduced graphene oxide (rGO) improved PCR efficiency at an optimal concentration of 0.8 mg/mL, while concentrations above 12 mg/mL showed inhibitory effects. Notably, rGO maintained enhanced PCR amplification through eight successive cycles at 12 mg/mL, suggesting superior performance compared to GO, Fig. 10.4 [64].

In addition, graphene nanoflakes (GNFs) enhanced the thermal conductivity of PCR in a dose-dependent manner, whereas the addition of excessive GNF results in the inhibition of PCR productivity. Simultaneously the size of the NP was said to influence the surface area. When there is an increase in surface area, there is an increase in the PCR yield, suggesting that the outstanding thermal conductivity property of GNF was the key to the enhancement of PCR efficiency, as shown in Fig. 10.5 [65].

In addition, Gabriel et al. suggested that the efficiency of PCR was further enhanced in the presence of GO at a concentration range of 40–0.004 μg/mL. A reduction in PCR efficiency was seen when GO concentration was below 0.04 μg/mL, and an inhibitory effect was observed at more than 4 μg/mL, suggesting that the optimum concentration of GO required to enhance the efficiency of PCR is 0.04 μg/mL, as shown in Fig. 10.5A. The existence of GO led to a 1.5-fold increase in PCR products [17].

10.3.1.2 Hexagonal boron nitride

Synthesized hBN NP, an isoelectric analog of graphene, is a white crystalline compound well-arranged in a 2D plane of boron and nitrogen atoms, making it a very stable hexagonal honeycomb layered structure. Its unique properties, such as excellent thermal conductivity, chemical inertness, high resistance to oxidation, and nontoxicity, then its similar structural and physical properties to its analog make it an appropriate 2D NP application in the PCR amplification system [66,50]. From a recent study done by

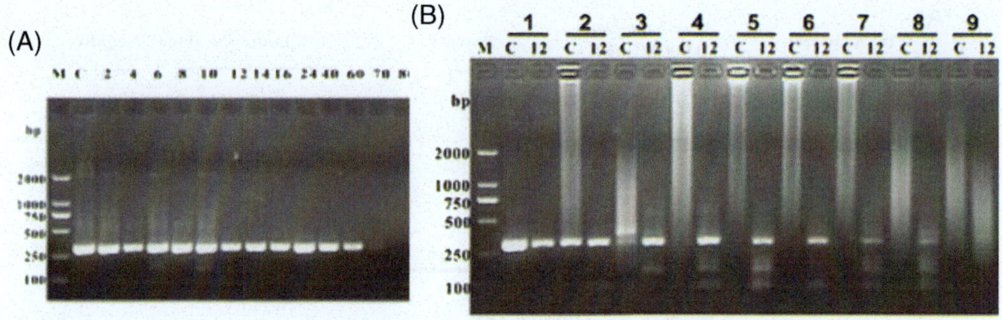

FIGURE 10.4

(A) The effect of GO in the amplification of PCR. M: DNA marker; C: control, no GO; the numbers marked above the lanes are the concentrations of GO in μg/mL. (B) The effect of rGO in nine rounds of repeated PCR. The template is a 300 bp pET-32a plasmid DNA. M: DNA marker; C: control, no rGO; the concentration of rGO is 12 μg/mL [64]. *GO*, Graphene oxide; *rGO*, reduced graphene oxide.

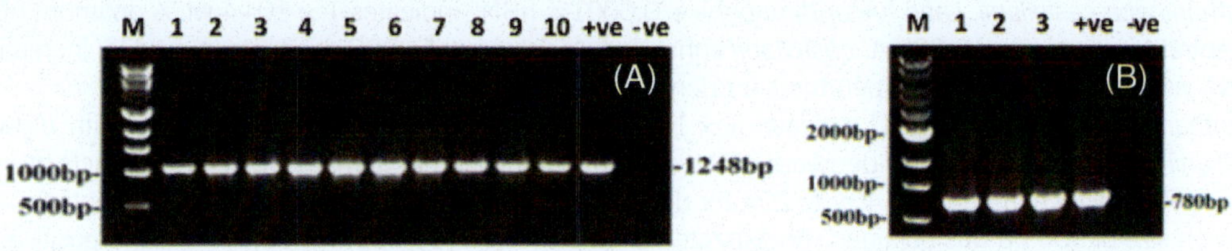

FIGURE 10.5

(A) Agarose gel electrophoresis of PCR products showing the concentration-dependent effect of PCR amplification using a region (1248 bp) of EG-1. Lanes 1–9 represent 0.001 w/w%–84.864 AU, 0.002 w/w%–103.13 AU, 0.004 w/w%–115.813 AU, 0.006 w/w%–128.881 AU, 0.008 w/w%–143.143 AU, 0.01 w/w%–154.194 AU, 0.02 w/w%–122.168 AU, 0.04 w/w%–108.839 AU, 0.08 w/w%–99.199 AU, 0.1 w/w%–88.272 AU, and 1.0 w/w%–89.343 AU (GNF concentration and band intensities, respectively; AU, arbitrary unit). (B) Agarose gel electrophoresis of PCR products showing the effect of the GNFs' size on PCR using 1248 bp of EG-1. Lanes 1, 2, and 3 represent the product obtained using EG-1 with 0.01 w/w% GNFs of 60, 12, and 8 nm, respectively. The lane labeled as "+ve" indicates the PCR products obtained under identical conditions without the addition of the 2D nanoparticles, and the lane labeled as "−ve" indicates the negative control reaction in which water was added instead of template DNA. The lane labeled as "M" shows the positions of the DNA molecular weight markers [65]. *2D*, Two-dimensional; *GNFs*, graphene nanoflakes.

Rasheed et al. the optimum concentration of hBN at 1 × 10−4 wt.% enhanced the PCR DNA yield by 16%. A twofold increase in PCR yield was observed at 1 × 10−4 wt.% when there was a temperature reduction in PCR steps, which includes denaturation at 91.5°C, annealing at 46.2°C, and extension at 68°C as shown in Fig. 10.6A. Besides, there was an enhancement in the amplification of PCR product at 1 × 10−4 wt.% when the number of PCR cycles was reduced from 40 to 29 cycles (Fig. 10.6B), suggesting a reduction in the overall number of PCR cycles by 11× with a shortened PCR reaction by 32 minutes [67].

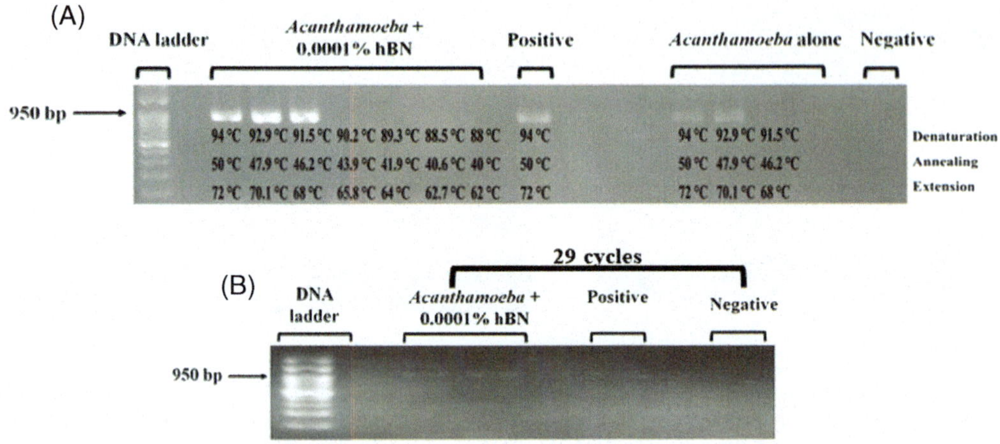

FIGURE 10.6

(A) Agarose gel image showing a range of reduced temperature-dependent effects on PCR amplification of *Acanthamoeba castellanii.* Lane "*Acanthamoeba* + 0.0001% hBN" indicates 1 × 10–4 wt.% optimum hBN concentration along with 10,000 *A. castellanii* enhanced *Acanthamoeba* PCR products in gradual increments of DNA yield at reduced PCR temperatures for denaturation, annealing, and extension. (B) Lane "*Acanthamoeba* + 0.0001% hBN" indicates 1×10^{-4} wt.% optimum hBN concentration along with 10,000 *A. castellanii* cells did not cause any decrease in the amount of amplified *Acanthamoeba* PCR products at reduced PCR cycle number from 40x to 29x for denaturation, annealing, and extension. An expected amplicon size of 950 bp was obtained. An expected amplicon size of 950 bp was obtained. Lane positive control and "*Acanthamoeba* alone" indicate PCR reaction with DNA template but without hBN nanoparticles. Lane negative control indicates PCR reaction with distilled water but without added DNA template and hBN nanoparticles [67]. *hBN*, hexagonal boron nitride.

10.3.1.3 Two-dimensional metal oxide nanoparticles (CuO, Al_2O_3, and TiO_2)

Copper oxide (CuO), alumina (Al_2O_3), and other metal oxides similarly influence the PCR amplification system. Gabriel et al. found that the addition of CuO and Al_2O_3 improved the PCR productivity in the range of 40–0.004 μg/mL. When high concentrations of metal oxide NPs were added into the PCR reaction, there were no changes in the production of PCR yield, but there was a reduction in PCR efficiency with a lower concentration of metal oxide. The optimal concentration of CuO and Al_2O_3 at 4 and 0.04 μg/mL, respectively, was required to enhance the efficiency of PCR (Fig. 10.7A). The presence of CuO and Al_2O_3 resulted in a 1.4- and 1.5-fold increase in PCR productivity. Inspiringly the combination of CuO and Al_2O_3 in a PCR reaction resulted in a clear enrichment effect on PCR amplification, as shown in Fig. 10.7B. The addition of these metal oxide NPs in the PCR reaction leads to an excellent surface effect, whereas the thermal conductivity property enhances PCR productivity [17].

Besides, titanium dioxide (TiO_2) enhances the thermal conductivity of the PCR reaction buffer, facilitating more efficient denaturation of the DNA template [68]. According to Khaliq et al. the 25-nm diameter of TiO_2 NP improved PCR efficiency by 2.9–6.9-fold, with 0.4 nM being the optimal concentration of TiO_2 as shown in Fig. 10.8A. Moreover, maximal amplification of PCR was observed

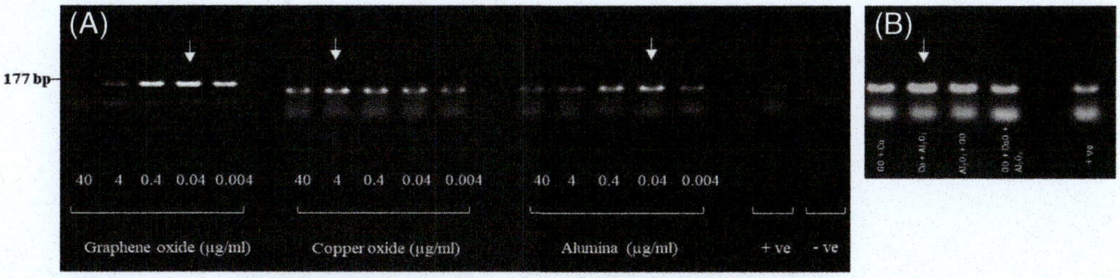

FIGURE 10.7

(A) Representative effects of 2D nanoparticles, GO, CuO, and Al_2O_3 at 40–0.004 μg/mL on PCR amplification of a 177 bp region of *Naegleria* DNA. The lane labeled as +ve indicates PCR products with the DNA template but without the addition of 2D nanoparticles. 2D nanoparticles improved PCR efficiency, yield, and fidelity by >50% at their maximal optimal concentrations. (B) Agarose gel electrophoresis of PCR products showing the effect of hybrid 2D nanoparticles on PCR reactions. PCR amplification was carried out in the presence of optimum concentrations of nanoparticles. Note that hybrid nanoparticles further improved PCR productivity. The lane labeled as "+ve" indicates the PCR products obtained under identical conditions without the addition of the 2D nanoparticles, and the lane labeled as "−ve" indicates the negative control reaction in which water was added instead of template DNA [17]. *2D*, Two-dimensional; *GO*, graphene oxide.

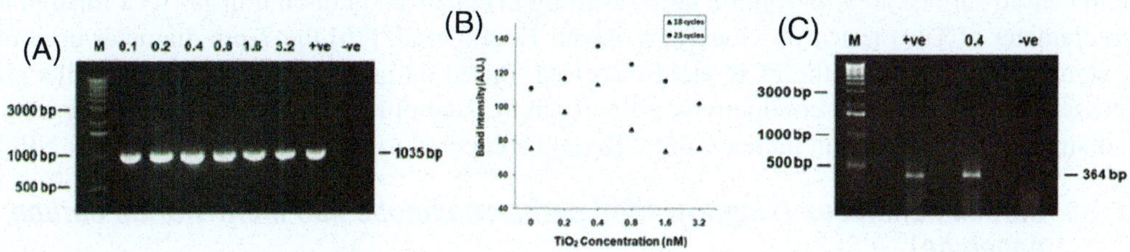

FIGURE 10.8

(A) Agarose gel electrophoresis of PCR products showing concentration-dependent effect of TiO_2 nanoparticles. PCR amplification of a 1035 bp region of human HSPA1A promoter was carried out in the presence of 0.1–3.2 nM TiO_2 nanoparticles. (B) Amplification of a 394 bp human MAO-C promoter DNA segment was carried out in the presence and absence of 0.2–3.2 nM TiO_2 nanoparticles with the regular number of cycles (23 cycles) and a reduced number of PCR cycles (18 cycles). (C) PCR amplification of a 364 bp mouse HMG-CoA reductase gene fragment after 50% time reduction in each step (denaturation, annealing, and extension) of PCR cycles was carried out in the presence or absence of 0.4 nM TiO_2 nanoparticles. The lane labeled as "+ve" indicates the PCR products obtained under identical conditions without the addition of the nanoparticles, and the lane labeled as "−ve" indicates the negative control reaction in which water was added instead of template DNA. The lane labeled as "M" shows positions of the DNA molecular weight markers and AU: arbitrary unit [61].

at 0.4 nM when the number of PCR cycles was reduced from 23 to 18 cycles (Fig. 10.8B). Although there was a 50% reduction in time for the denaturation, annealing, and extension steps, TiO_2 enhanced the PCR reaction without compromising the PCR yield (Fig. 10.8C). This study suggests that the excellent thermal conductivity property of TiO_2 quickens the denaturation of the DNA template, resulting in an overall reduction in PCR reaction time [61].

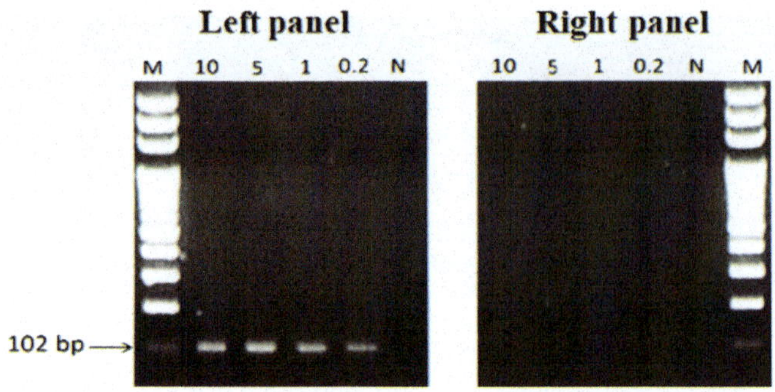

FIGURE 10.9

Template-dependent PCR amplification showing the increased yield of 102 bp-sized products. The left panel represents the PCR products in the presence of 0.2 nM of TiO_2 and the right panel for PCR products without TiO_2 nanosuspension. In both the panels, 10, 5, 1, and 0.2 indicate the concentration of cDNA template in nanograms (ng) used in the PCR. *N* indicates for negative control in which no DNA was added [69].

In addition, the smaller size (7-nm diameter) of TiO_2 at its optimal concentration, 0.2 nM, enhanced the efficiency of PCR even at low initial DNA template concentrations (Fig. 10.9). It could be because of the increased surface area-to-volume ratio resulting in enhanced phonon transport. Compared to the 25-nm diameter of TiO_2 tested previously by Abdul Khaliq et al. [76] the 7-nm diameter used in this study significantly increased the PCR yield threefold. Based on previous research the smaller size of NPs possesses higher thermal conductivity property in PCR amplification system, which suggests why 7-nm diameter TiO_2 showed an increase in PCR yield compared to the 25-nm diameter TiO_2 NPs [69].

10.3.1.4 Carbon nanotubes (single-walled carbon nanotube and multi-walled carbon nanotube)

Single-walled CNTs (SWCNTs) are cylindrical fullerenes with single-carbon atom thickness that can interact with DNA and protein with excellent biocompatibility [70]. Cui et al. reported that the addition of SWCNTs up to 3 μg/μL into PCR reactions in the presence of Mg^{2+} further enhanced the amplification of PCR yield but led to an inhibitory effect at a larger dose of SWCNTs (Fig. 10.10A). These could be because of the accumulation of reaction components, such as DNA templates, PCR primers, and Taq enzymes, triggered by van der Waals force with SWCNTs added in the PCR reaction [71,72]. However, the addition of SWCNTs into PCR reactions in the absence of Mg^{2+} produces a similar effect up to 3 μg/μL, whereas it has an inhibitory effect at higher dosages (Fig. 10.10B). A small amount of SWCNTs added into PCR reactions gives rise to a better enhancement in the PCR yield [73]. These results suggest that SWCNT NPs function similarly to Mg^{2+} ions and also ensure the Taq enzyme activity in the PCR system [73].

As mentioned earlier, SWCNTs can play the role of Mg^{2+} in preserving the activity of Taq DNA polymerase. Nonetheless, the possible activity of other types of CNTs, such as multiwalled CNTs (MWCNTs) in long PCR amplification systems, remains unknown [74]. Long PCR around 100 kb DNA sequences usually produces nonspecific PCR products even after the addition of other PCR additives. Thus the addition of CNT into PCR reactions may help to gradually reduce the smearing of bands and eventually improve long-term PCR efficiency [74]. According to Du et al. the addition of SWCNTs and

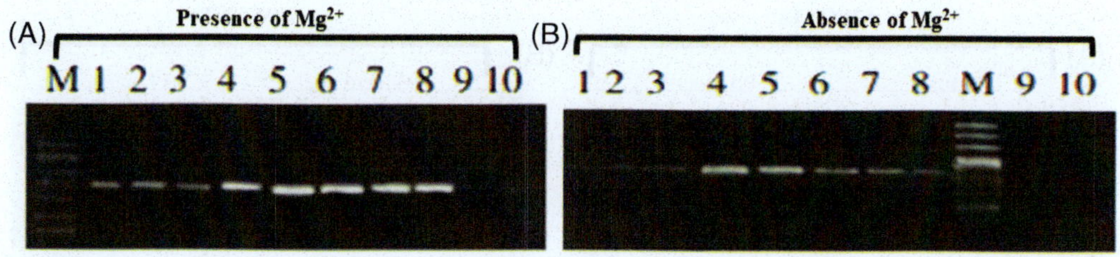

FIGURE 10.10

(A) Electrophoresis result of PCR products in the presence of Mg^{2+} and increasing SWCNT amounts. *M* is molecular marker. (B) Electrophoresis result of PCR products in the absence of Mg^{2+} ions and with increasing SWCNTs. 1–10 are the PCR products with 0.33, 0.66, 0.99, 1.32, 1.65, 1.98, 2.31, 2.64, 2.97, and 3.3 μg/μL SWCNTs in reaction liquid. M is the molecular marker [73]. *SWCNTs*, Single-walled carbon nanotubes.

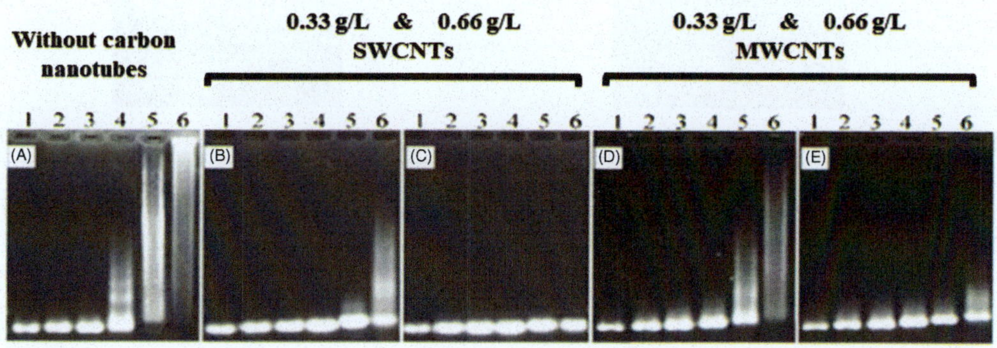

FIGURE 10.11

CNT impact on the repeated PCR of 549 bp DNA. Lanes 1, 2, 3, 4, 5, and 6 are reactions of the first, second, third, fourth, fifth, and sixth rounds, respectively. (A) Repeated PCR without CNTs. (B) and (C) Six rounds of repeated PCR with 0.33 and 0.66 g/L SWCNTs, respectively. (D and E) Six rounds of repeated PCR with 0.33 and 0.66 g/L MWCNTs, respectively [75]. *CNTs*, Carbon nanotubes; *MWCNTs*, multi-walled carbon nanotubes; *SWCNTs*, single-walled carbon nanotubes.

MWCNTs enhanced the specificity of the repeated PCR and prevented long DNA thermal degradation. With the addition of 0.66 g/L SWCNTs, there was a complete absence of nonspecific amplification PCR products, and desirable PCR specificity and yield were produced. Moreover, enhanced PCR specificity was obtained when 0.66 g/L MWCNTs were included for repeated PCR, resulting in a high PCR yield (Fig. 10.11). In contrast, SWCNTs displayed a greater specificity rate compared to MWCNTs, as MWCNTs' binding affinity to DNA was more toward protecting long DNA from thermal deprivation [75].

Besides, heat transfer can contribute to DNA amplification with the presence of CNT in repeated PCR. The addition of SWCNTs and MWCNTs in a dose-dependent manner inhibited DNA thermal degradation. Total DNA degradation for SWCNTs was inhibited at 80°C and 85°C at 1.67 and 3.33 g/L, respectively. However, the addition of MWCNTs showed more inhibition of DNA degradation than

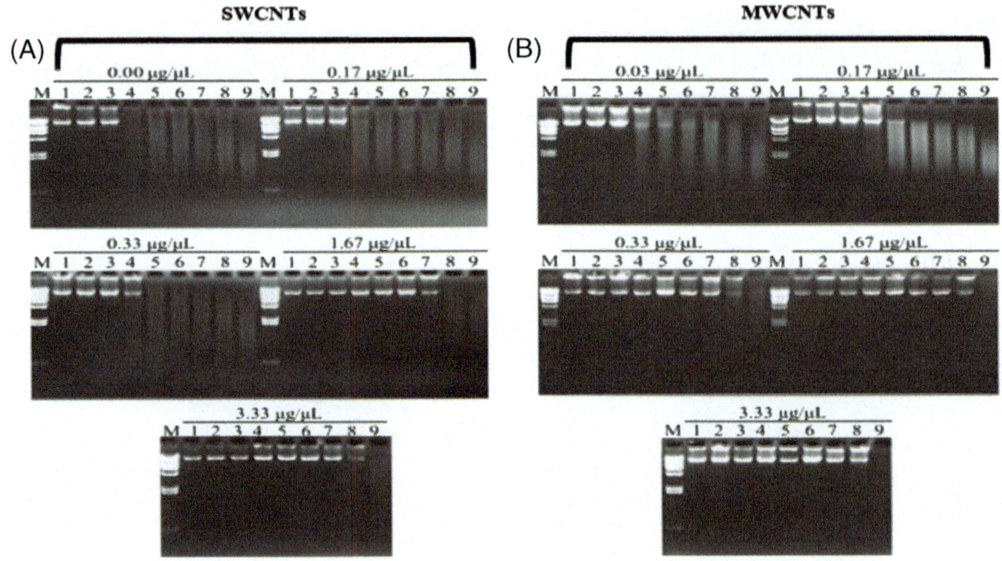

FIGURE 10.12

(A) Effect of temperature on lambda DNA integrity in the presence of SWCNTs. (B) Effect of temperature on lambda DNA integrity dependent on MWCNT dose. Temperatures at lanes 1, 2, 3, 4, 5, 6, 7, 8, and 9 were 25°C, 37°C, 50°C, 60°C, 70°C, 75°C, 80°C, 85°C, and 90°C, respectively. Lanes M for makers (lambda DNA/HindIII with the following digest fragment in base pairs: 23,130, 9416, 6557, 4361, 2322, 2027, 564, and 125) [75]. *MWCNTs*, Multi-walled carbon nanotubes; *SWCNTs*, single-walled carbon nanotubes.

SWCNTs when only a small amount of MWCNTs was required to inhibit DNA thermal degradation. From the study, only 0.33 g/L of MWCNTs inhibited DNA degradation at 85°C but required 3.33 g/L of SWCNTs to inhibit DNA degradation, as shown in Fig. 10.12 [75]. Despite the remarkable progress discussed above, several critical challenges must be addressed to facilitate the clinical translation of 2D nanomaterials in bioheat transfer. Scalable and reproducible manufacturing of high-quality 2D nanomaterials with consistent physicochemical properties remains a significant bottleneck. The long-term biocompatibility, biodistribution, and biodegradation mechanisms of these materials in vivo require comprehensive systematic investigation, particularly for non-titanium MXenes and emerging materials such as black phosphorus. Furthermore, the complex interplay between NP properties and biological systems—including blood perfusion, metabolic heat generation, and tissue heterogeneity—necessitates sophisticated computational modeling approaches that integrate the Pennes bioheat equation with NP distribution dynamics. Looking forward, several promising directions are poised to accelerate the field. First, machine learning and artificial intelligence (AI) offer powerful tools for predicting optimal nanomaterial compositions, structures, and surface functionalization strategies for specific therapeutic applications. Second, the convergence of photothermal and plasmonic properties with microfluidic technologies is enabling ultrafast photonic PCR and point-of-care diagnostic devices that leverage the exceptional thermal conductivity of 2D nanomaterials. Third, the development of 2D heterostructure-based wearable sensors for noninvasive sweat and heat monitoring represents a transformative approach to personalized health management. Finally, bioengineered 2D nanomaterials (Bio-2DNMs) designed with biomimetic surface coatings and targeting ligands are expected to enhance tumor specificity while minimizing off-target toxicity. Addressing these challenges and pursuing these opportunities will be essential for realizing the full potential of 2D nanomaterials in next-generation bioheat transfer applications.

10.4 Conclusion

This chapter has comprehensively reviewed the applications of 2D nanomaterials in bioheat transfer, encompassing hyperthermia-based cancer therapy, biophotonic drug delivery, biosensing, and nanoPCR. The exceptional photothermal conversion efficiency, high surface area, and tunable electronic properties of materials such as graphene, MXenes, and black phosphorus have enabled significant advances in photothermal therapy and controlled drug release. In diagnostics, 2D nanoPCR has demonstrated enhanced sensitivity, specificity, and reduced reaction times through improved thermal conductivity, while electrochemical biosensors incorporating graphene and MXenes have achieved ultrasensitive detection of cancer biomarkers. Despite these promising developments, several challenges remain. The clinical translation of 2D nanomaterials requires comprehensive assessment of long-term biocompatibility, biodistribution, and potential off-target toxicity. Scalable and reproducible manufacturing processes are needed to meet regulatory standards. Furthermore, the complex interplay between NP properties and biological systems necessitates sophisticated computational modeling approaches.Looking forward, emerging technologies are poised to accelerate the field. Machine learning and AI offer powerful tools for predicting optimal nanomaterial compositions and structures for specific therapeutic applications. The convergence of photothermal and plasmonic properties with microfluidics is enabling ultrafast photonic PCR for point-of-care diagnostics. Meanwhile, wearable biosensors incorporating 2D nanomaterials are bringing continuous health monitoring closer to clinical reality. The continued integration of 2D nanomaterials in bioheat transfer applications promises to transform cancer therapy, infectious disease diagnostics, and personalized medicine in the coming decade.

References

[1] N.A. Shah, B. Almutairi, D. Vieru, B. Lee, J.D. Chung, Bioheat transfer with thermal memory and moving thermal shocks, Fractal Fract. 7 (8) (2023) 629.

[2] V. Brusko, A. Khannanov, A. Rakhmatullin, A.M. Dimiev, Unraveling the infrared spectrum of graphene oxide, Carbon 229 (2024) 119507.

[3] A. Elbanna, H. Jiang, Q. Fu, J.F. Zhu, Y. Liu, M. Zhao, … J. Teng, 2D material infrared photonics and plasmonics, ACS Nano 17 (5) (2023) 4134–4179.

[4] Q. Liu, J. Zou, Z. Chen, W. He, W. Wu, Current research trends of nanomedicines, Acta Pharm. Sin. B 13 (11) (2023) 4391–4416.

[5] M. Chen, H. Tang, S. Chen, M. Lyu, H. Quan, Two-dimensional multifunctional nanosheets as radiosensitizers for chemodynamic/radio-therapy, Colloids Surf. B Biointerfaces 234 (2024) 113699.

[6] J. Zhou, T. Yang, J. Chen, C. Wang, H. Zhang, Y. Shao, Two-dimensional nanomaterial-based plasmonic sensing applications: advances and challenges, Coord. Chem. Rev. 410 (2020) 213218.

[7] T. Lillicrap, M. Tahtalı, A. Neely, X. Wang, A. Bivard, C. Lueck, A model based on the Pennes bioheat transfer equation is valid in normal brain tissue but not brain tissue suffering focal ischaemia, Australas. Phys. Eng. Sci. Med. 40 (4) (2017) 841–850.
[8] S.F. Megahid, A.E. Abouelregal, H.M. Sedighi, Modified Moore–Gibson–Thompson Pennes' bioheat transfer model for a finite biological tissue subjected to harmonic thermal loading, Mech. Time-Depend. Mat. 28 (3) (2023) 1441–1463.
[9] M. Singh, Modified Pennes bioheat equation with heterogeneous blood perfusion: a newer perspective, Int. J. Heat Mass Transfer 218 (2024) 124698.
[10] M. Singh, Biological heat and mass transport mechanisms behind nanoparticles migration revealed under microCT image guidance, Int. J. Therm. Sci. 184 (2023) 107996.
[11] R. Satish, B.T. Raju, P.D. Prasad, S.S.R. Raju, C.S.K. Raju, M.D. Kumar, Dynamics of comparative analysis of Reynold's and Vogel's models (variable viscosity) in a wire coating process filled with magnetized porous, Int. J. Mod. Phys. B 38 (2024) 2450417.
[12] S.A. Ghopry, B. Liu, A. Shultz, J.Z. Wu, Tuning Fermi energy of graphene using platinum nanoparticles and ultraviolet irradiation to increase charge transfer for surface-enhanced Raman spectroscopy, ACS Appl. Nano Mater. 6 (23) (2023) 21626–21633.
[13] W. Yanan, Z. Fei, Z. Meng, C. Gang, W. Xiao, M. Yan, Z. Houyu, … Z. Bing, Dependence relationship between doping modes and surface plasmon resonance properties of doped ZnO nanocrystals, Colloids Surf. A Physicochem. Eng. Asp. 676 (2023) 132103.
[14] S. Das, R. Devireddy, M.R. Gartia, Surface plasmon resonance (SPR) sensor for cancer biomarker detection, Biosensors 13 (3) (2023) 396.
[15] L. Yuan, Q. Chen, J.E. Riviere, Z. Lin, Pharmacokinetics and tumor delivery of nanoparticles, J. Drug Delivery Sci. Technol. 83 (2023) 104404.
[16] R. Jiao, Nanomaterial-based clinical testing and diagnostic technologies, Highl. Sci. Eng. Technol. 40 (2023) 132–141.
[17] S. Gabriel, A.K. Rasheed, R. Siddiqui, J.N. Appaturi, L.B. Fen, N.A. Khan, Development of nanoparticle-assisted PCR assay in the rapid detection of brain-eating amoebae, Parasitol. Res. 117 (6) (2018) 1801–1811.
[18] C. He, L. Yu, L. Ding, H. Yao, Y. Chen, Y. Hao, Lysine demethylase KDM3A regulates nanophotonic hyperthermia resistance generated by 2D silicene in breast cancer, Biomaterials 255 (2020) 120181.
[19] M. Overchuk, R.A. Weersink, B.C. Wilson, G. Zheng, Photodynamic and photothermal therapies: synergy opportunities for nanomedicine, ACS Nano 17 (9) (2023) 7979–8003.
[20] A. Samali, C.I. Holmberg, L. Sistonen, S. Orrenius, Thermotolerance and cell death are distinct cellular responses to stress: dependence on heat shock proteins, FEBS Lett. 461 (3) (1999) 306–310.
[21] Y. Zhang, Z. Li, Y. Huang, B. Zou, Y. Xu, Amplifying cancer treatment: advances in tumor immunotherapy and nanoparticle-based hyperthermia, Front. Immunol. 14 (2023) 1258786.
[22] Y. Gu, R. Piñol, R. Moreno-Loshuertos, C.D.S. Brites, J. Zeler, A. Martínez, … A. Millán, Local temperature increments and induced cell death in intracellular magnetic hyperthermia, ACS Nano 17 (7) (2023) 6822–6832.
[23] E. Obrador, A. Jihad-Jebbar, R. Salvador-Palmer, R. López-Blanch, M. Oriol-Caballo, M.P. Moreno-Murciano, … J.M. Estrela, Externally applied electromagnetic fields and hyperthermia irreversibly damage cancer cells, Cancers 15 (13) (2023) 3413.
[24] Y. Yang, L. Huangfu, H. Li, D. Yang, Research progress of hyperthermia in tumor therapy by influencing metabolic reprogramming of tumor cells, Int. J. Hyperthermia 40 (1) (2023) 2270654.
[25] H. Kurokawa, H. Ito, M. Terasaki, H. Matsui, Hyperthermia enhances photodynamic therapy by regulation of HCP1 and ABCG2 expressions via high level ROS generation, Sci. Rep. 9 (1) (2019) 1–8.
[26] P. Viana, P. Hamar, Targeting the heat shock response induced by modulated electro-hyperthermia (mEHT) in cancer, Biochim. Biophys. Acta Rev. Cancer 1879 (2024) 189069.

[27] M.B. Ahmed, A.A. Alghamdi, S.U. Islam, H. Ahsan, Y.S. Lee, The complex roles of DNA repair pathways, inhibitors, hyperthermia, and contact inhibition in cell cycle halts, Mini Rev. Med. Chem. 23 (5) (2023) 514–529.

[28] A. Loboda, I. Smolanka Sr, V.E. Orel, L. Syvak, T. Golovko, I. Dosenko, … O. Mokhonko, Efficacy of combination neoadjuvant chemotherapy and regional inductive moderate hyperthermia in the treatment of patients with locally advanced breast cancer, Technol. Cancer Res. Treat. 19 (2020) 1533033820963599.

[29] M. Haque, M.S. Shakil, K.M. Mahmud, The promise of nanoparticles-based radiotherapy in cancer treatment, Cancers 15 (6) (2023) 1892.

[30] B. Tian, C. Wang, S. Zhang, L. Feng, Z. Liu, Photothermally enhanced photodynamic therapy delivered by nano-graphene oxide, ACS Nano 5 (9) (2011) 7000–7009.

[31] I. Pereira, C. Monteiro, M. Pereira-Silva, D. Peixoto, C. Nunes, S. Reis, … A.C. Paiva-Santos, Nanodelivery systems for cutaneous melanoma treatment, Eur. J. Pharm. Biopharm. 184 (2023) 214–247.

[32] Y. Yuan, X. Peng, X. Weng, J. He, C. Liao, Y. Wang, … J. Qu, Two-dimensional nanomaterials as enhanced surface plasmon resonance sensing platforms: Design perspectives and illustrative applications, Biosens. Bioelectron. 241 (2023) 115672.

[33] D.F. Báez, Graphene-based nanomaterials for photothermal therapy in cancer treatment, Pharmaceutics 15 (9) (2023) 2286.

[34] S.E. Simon, A. Numan, F.A. Jayakumar, M. Khalid, K.O. Tan, Investigating the impact of UV irradiated graphene oxide on human breast cancer cells, Mater. Today Commun. 29 (2021) 102830.

[35] S. Tiwari, B.A. Rudani, P. Tiwari, P. Bahadur, S.J. Flora, Photodynamic therapy of cancer using graphene nanomaterials, Expert Opin. Drug Deliv. 21 (9) (2024) 1331–1348.

[36] E.W. Tan, S.E. Simon, A. Numan, M. Khalid, K.O. Tan, Impact of UV radiation on Mxene-mediated tubulin dissociation and mitochondrial apoptosis in breast cancer cells, Colloids Surf. B Biointerfaces 235 (2024) 113793.

[37] R.F. de Paula, I.A. Rosa, I.F. Gafanhão, J.L. Fachi, A.M.G. Melero, A.O. Roque, … E.C. de Oliveira, Reduced graphene oxide, but not carbon nanotubes, slows murine melanoma after thermal ablation using LED light in B16F10 lineage cells, Nanomedicine 28 (2020) 102231.

[38] D. Xu, J. Liu, Y. Wang, Y. Jian, W. Wu, R. Lv, Black phosphorus nanosheet with high thermal conversion efficiency for photodynamic/photothermal/immunotherapy, ACS Biomater. Sci. Eng. 6 (9) (2020) 4940–4948.

[39] A. Nomani, A. Kianmehr, S. Abdoli, S. Javani, Metal-based nanomaterials photodynamic action with a focus on Au and Ag nanomaterials, Drug Formulation Design, IntechOpen, United Kingdom, 2023.

[40] A. Szuplewska, D. Kulpińska, A. Dybko, M. Chudy, A.M. Jastrzębska, A. Olszyna, Z. Brzózka, Future applications of MXenes in biotechnology, nanomedicine, and sensors, Trends Biotechnol. 38 (3) (2020) 264–279.

[41] X. Wang, X. Yu, J. Song, W. Huang, Y. Xiang, X. Dai, H. Zhang, Two-dimensional semiconducting antimonene in nanophotonic applications–a review, Chem. Eng. J. 406 (2020) 126876.

[42] K. Elumalai, S. Srinivasan, A. Shanmugam, Review of the efficacy of nanoparticle-based drug delivery systems for cancer treatment, Biomed. Technol. 5 (2024) 109–122.

[43] J. Wang, Q. Qiao, Y. Feng, Y. Guo, T. Liao, L. Li, … C. Li, Two-dimensional MXene-based nano-prodrug for synergistic chemo-photothermal therapy on cancer treatment, Mater. Today Chem. 38 (2024) 102048.

[44] H. Wu, C. Wang, J. Sun, L. Sun, J. Wan, S. Wang, … H. Jin, Self-assembled and self-monitored sorafenib/indocyanine green nanodrug with synergistic antitumor activity mediated by hyperthermia and reactive oxygen species-induced apoptosis, ACS Appl. Mater. Interfaces 11 (47) (2019) 43996–44006.

[45] M. Gibertini, M. Koperski, A.F. Morpurgo, K.S. Novoselov, Magnetic 2D materials and heterostructures, Nat. Nanotechnol. 14 (5) (2019) 408–419.

[46] M. Sanhaji, J. Göring, P. Couleaud, A. Aires, A.L. Cortajarena, J. Courty, … I. Hilger, The phenotype of target pancreatic cancer cells influences cell death by magnetic hyperthermia with nanoparticles carrying gemicitabine and the pseudo-peptide NucAnt, Nanomedicine 20 (2019) 101983.

[47] S. Piehler, H. Dähring, J. Grandke, J. Göring, P. Couleaud, A. Aires, … I. Hilger, Iron oxide nanoparticles as carriers for DOX and magnetic hyperthermia after intratumoral application into breast cancer in mice: impact and future perspectives, Nanomaterials 10 (6) (2020) 1016.

[48] Y. Ma, T. Jiang, R. Zhang, F. Liu, S. Song, H. Zhang, … Z. He, The application of 2D Mxene nanosheet-based thermosensitive gel delivery system loaded with cisplatin and imiquimod for lung cancer, Int. J. Nanomed. 19 (2024) 4719–4733.

[49] X. Wang, T. Tang, Q. Miao, S. Xie, X. Chen, J. Tang, … J. Xu, Detection of transgenic rice line TT51-1 in processed foods using conventional PCR, real-time PCR, and droplet digital PCR, Food Control 98 (2019) 380–388.

[50] Y. Wang, F. Wang, H. Wang, M. Song, Graphene oxide enhances the specificity of the polymerase chain reaction by modifying primer-template matching, Sci. Rep. 7 (1) (2017) 1–10.

[51] M. Radmansouri, E. Bahmani, E. Sarikhani, K. Rahmani, F. Sharifianjazi, M. Irani, Doxorubicin hydrochloride-loaded electrospun chitosan/cobalt ferrite/titanium oxide nanofibers for hyperthermic tumor cell treatment and controlled drug release, Int. J. Biol. Macromol. 116 (2018) 378–384.

[52] V. Mulens-Arias, A. Nicolás-Boluda, A. Pinto, A. Balfourier, F. Carn, A.K. Silva, … F. Gazeau, Tumor-selective immune-active mild hyperthermia associated with chemotherapy in colon peritoneal metastasis by photoactivation of fluorouracil–gold nanoparticle complexes, ACS Nano 15 (2) (2021) 3330–3348.

[53] Ying Yuan, Bo Chen, Luping Song, Xingxing An, Qinrui Zhang, Hao Lu, Chang Ming Li, Chunxian Guo;, Magnetic two-dimensional nanocomposites for multimodal antitumor therapy: a recent review, J. Mater. Chem. 12 (6) (2024) 1404–1428, doi:10.1039/d3tb02333h.

[54] Tiankun Hui, Wang C, Fu C, Song Z, Zhang Q, Zheng B, Fu J, Qiu M, Yue B., Spikey biodegradable MXene nanosheets with enhanced cellular internalization and stimuli-responsive drug delivery for synergistic cancer therapy, J Nanobiotechnology 23 (1) (2025) 560, doi:10.1186/s12951-025-03628-z.

[55] L. Wu, J. Gao, X. Lu, C. Huang, J. Chen, Graphdiyne: a new promising member of 2D all-carbon nanomaterial as robust electrochemical enzyme biosensor platform, Carbon 156 (2020) 568–575.

[56] F. Zhao, J. Wu, Y. Ying, Y. She, J. Wang, J. Ping, Carbon nanomaterial-enabled pesticide biosensors: design strategy, biosensing mechanism, and practical application, TrAC Trends Anal. Chem. 106 (2018) 62–83.

[57] M.M. Rahman, M.M. Rana, M.S. Rahman, M.S. Anower, M.A. Mollah, A.K. Paul, Sensitivity enhancement of SPR biosensors employing heterostructure of PtSe2 and 2D materials, Opt. Mater. 107 (2020) 110123.

[58] M.Z.M. Nasir, M. Pumera, Emerging mono-elemental 2D nanomaterials for electrochemical sensing applications: from borophene to bismuthene, TrAC Trends Anal. Chem. 121 (2019) 115696.

[59] H. Zhu, H. Zhang, Y. Xu, S. Laššáková, M. Korabečná, P. Neužil, PCR past, present and future, BioTechniques 69 (4) (2020) 317–325.

[60] Z. Xue, M. You, P. Peng, H. Tong, W. He, A. Li, … C. Yao, Taqman-MGB nanoPCR for highly specific detection of single-base mutations, Int. J. Nanomed. 16 (2021) 3695–3705.

[61] A. Khaliq, P.J. Sonawane, B.K. Sasi, B.S. Sahu, T. Pradeep, S.K. Das, N.R. Mahapatra, Enhancement in the efficiency of polymerase chain reaction by TiO_2 nanoparticles: crucial role of enhanced thermal conductivity, Nanotechnology 21 (25) (2010) 255704.

[62] Y. Wang, Y. Fei, T. Yang, Z. Luo, Y. Xu, B. Su, X. Lin, Nanotechnology for ultrafast nucleic acid amplification, Nano Today 48 (2023) 101749.

[63] J. Hu, J. Wu, W. Liu, D. Jin, H.E. Dirani, S. Kerdiles, … D.J. Moss, 2D graphene oxide: a versatile thermo-optic material, Adv. Funct. Mater. 34 (46) (2024) 2406799.

[64] J. Jia, L. Sun, N. Hu, G. Huang, J. Weng, Graphene enhances the specificity of the polymerase chain reaction, Small 8 (13) (2012) 2011–2015.

[65] A. Khaliq, R. Kafafy, H.M. Salleh, W.F. Faris, Enhancing the efficiency of polymerase chain reaction using graphene nanoflakes, Nanotechnology 23 (45) (2012) 455106.

[66] Z. Rafiei-Sarmazdeh, S.M. Zahedi-Dizaji, A.K. Kang, Two-dimensional nanomaterials, Nanostructures, IntechOpen, United Kingdom, 2019.

[67] A.K. Rasheed, R. Siddiqui, S.M.K. Ahmed, S. Gabriel, M.Z. Jalal, A. John, N.A. Khan, hBN nanoparticle-assisted rapid thermal cycling for the detection of *Acanthamoeba*, Pathogens 9 (10) (2020) 824.

[68] X. Zhou, X. Yi, Z. Liang, J. Zhang, G. Li, Y. Tang, … S. Zhang, Rapid quantitative detection system constructed via photonic PCR based on the photothermal effect of NH2-MWCNTs/TiO_2, Vacuum 229 (2024) 113614.

[69] G. Lenka, W.H. Weng, Nanosized particles of titanium dioxide specifically increase the efficiency of conventional polymerase chain reaction, Digest J. Nanomater. Biostruct. 8 (4) (2013) 1435–1445.

[70] F. Ke, J. Chen, R. Wu, Y. Chen, Dispersion quality of single-walled carbon nanotubes reveals the recognition sequence of DNA, Nanotechnology 31 (25) (2020) 255708.

[71] T.K. Mandal, M.S. Fleming, D.R. Walt, Preparation of polymer coated gold nanoparticles by surface-confined living radical polymerization at ambient temperature, Nano Lett. 2 (1) (2002) 3–7.

[72] C.V. Nguyen, L. Delzeit, A.M. Cassell, J. Li, J. Han, M. Meyyappan, Preparation of nucleic acid functionalized carbon nanotube arrays, Nano Lett. 2 (10) (2002) 1079–1081.

[73] D. Cui, F. Tian, Y. Kong, C.S. Ozkan, I. Titushikin, H. Gao, Effects of single-walled carbon nanotube on polymerase chain reaction, MRS Online Proc. Library 773 (1) (2003) 231–235.

[74] Z. Zhang, C. Shen, M. Wang, H. Han, X. Cao, Aqueous suspension of carbon nanotubes enhances the specificity of long PCR, BioTechniques 44 (4) (2008) 537–545.

[75] X. Du, H. An, B. Jin, L. Meng, Q. Liu, Carbon nanotubes altering specificity of repeated PCR and DNA integrity properties, J. Nanosci. Nanotechnol. 14 (7) (2014) 5547–5551.

[76] R. Abdul Khaliq, P.J. Sonawane, B.K. Sasi, B.S. Sahu, T. Pradeep, S.K. Das, N.R. Mahapatra, Enhancement in the efficiency of polymerase chain reaction by TiO_2 nanoparticles: crucial role of enhanced thermal conductivity, Nanotechnol. 21(25) (2010). doi:10.1088/0957-4484/21/25/255704.

CHAPTER

11 Graphene- and boron nitride–based phase-change materials for heat transfer and energy storage applications

Pooja Singh[a], Alok Kumar Ansu[b] and Ravi Kumar Sharma[a,b]

[a]*Department of Mechanical Engineering, Manipal University Jaipur, Jaipur, Rajasthan, India,*
[b]*Integrated Centre for Energy and Sustainable Technologies, Manipal University Jaipur, Jaipur, Rajasthan, India*

11.1 Introduction

Other than water and air, energy is the most important need of human beings, and owing to the increasing human population, this need and its demand gap are increasing continuously. To bridge this gap, new and renewable energy sources are required. This gap can also be effectively bridged by harvesting the excessively available natural energy sources, such as solar energy. The Sun is an infinite source of energy that provides energy for free and in an immense amount. If this energy can be stored using an appropriate medium, it can be stored in later hours when needed. Thermal energy storage (TES) plays a crucial role in storing large amounts of energy in suitable materials.

In developed nations, buildings play a vital role and are the leading sector in energy consumption. According to the European Union report by Asadi et al. [1] and Diakaki et al. [2], nearly 40% of the entire energy is utilized by the building sector, and it also produces the same amount of CO_2 emissions, which is a matter of concern. This increase in energy demand is primarily because of the high standard of living and comfort zones of people for heating and cooling purposes. To achieve proper balance, a TES system is convenient and effective.

The sun, which is the principal source of solar energy, emits more energy during the daytime on Earth than is required. If this excess energy can be stored, it can be used later for various applications. Energy storage can be performed in the form of sensible and latent heat.

Sensible heat is the amount of heat required to change the temperature of a body, whereas in the case of latent heat, the heat exchange occurs at an almost constant temperature while changing the phase of the substance. The total heat contained was calculated using the following equation:

$$Q = \int_{T_i}^{T_m} mC_p dT + ma_m \Delta h_m + \int_{T_m}^{T_f} mC_p dT \tag{11.1}$$

Graphene and 2D Materials in Heat Transfer: Fundamentals and Applications. DOI: https://doi.org/10.1016/B978-0-12-821964-5.00009-6

Latent heat thermal energy storage systems (LHTES) are generally preferred over sensible heat TES systems because LHTES have a greater capacity to store heat and can store and release energy by melting and solidification.

However, these phase-change materials (PCMs) possess excellent thermophysical properties but suffer from low thermal conductivity. Numerous investigations have been conducted to enhance the thermal conductivity of organic PCMs, and the incorporation of two-dimensional (2D) materials, such as graphene and boron nitride, has been the most favorable supporting materials for increasing thermal conductivity [3]. They possess exceptional physical and thermal transport properties, particularly in 2D space. An early breakthrough work on this was done by Novoselov et al. [4]. Xie et al. [5] use hexagonal boron nitride (BN) as a supporting material to prepare a stearic-based composite PCM. The thermal conductivity of the composite was increased by 73.36% compared to that of pure stearic acid. The latent heat value recorded was 136.20 J/g, and the melting point was 51.81°C. Similarly, Tafrishi et al. [6] examined the effect of adding BN to octadecane, which is paraffin. Their results showed that the thermal conductivity of the composite has increased by 23.0%, 26.5%, and 27.7% at 280, 300, and 320 K, respectively. The heat capacity of octadecane was also enhanced by the addition of BN.

They mentioned that a monocrystalline graphitic film contains a thickness of a few atoms but is very stable under atmospheric conditions. This film appears to have a 2D semimetal with slight overlapping of valence and conductance bonds. Experimental observations have shown that it has an electron mobility of 1.5×10^5 $cm^2/V/s$ at room temperature and Young's modulus of 1 TPa. They also stated that graphene can be an excellent material for metallic transistor applications. Graphene sheets also possess high thermal conductivity and are 2D materials with low-density carbon. They also possess good electrical properties and a high specific area. The adoption of graphene as a porous material for PCM impregnation has substantially increased owing to the excellent properties of the developed composites with good compatibility with the base PCM [7]. As its inception, graphene has been developed using various methodologies, such as mechanical stretching [8] and chemical development [9].

Graphene-in-plane possesses the maximum known thermal conductivity of 2000–4000 W/m K [10]. Materials with smaller grain sizes and isotopic impurities have a thermal conductivity of ~2000 W/m K, whereas those with larger grain sizes have higher thermal conductivity 4000 W/m K [11]. The extraordinary properties of graphene and its derived materials, such as graphene oxide (GO), graphene nanoplatelets (GNPs), and carbon nanotubes (CNTs), have attracted researchers worldwide to use them as energy storage materials.

This review provides insights into the PCMs and 2D energy storage materials used for this purpose. The use of graphene for energy harvesting has been presented and discussed in detail. In addition to graphene the use of other 2D materials, such as BN, a carbon-based material, has also been discussed. The notable limitations and merits of these materials are highlighted. The findings of this work will be very useful for future researchers and application engineers working in this area.

11.2 Thermal energy storage system

The available thermal energy can be stored in a TES system. The energy demand varies from time to time; for example, it may be on a daily, monthly, or seasonal basis. The TES system helps meet this demand during peak hours. The TES system has the potential to bridge the gap between energy demand and supply and smoothen the operation of a well-designed system effectively. TES systems can store

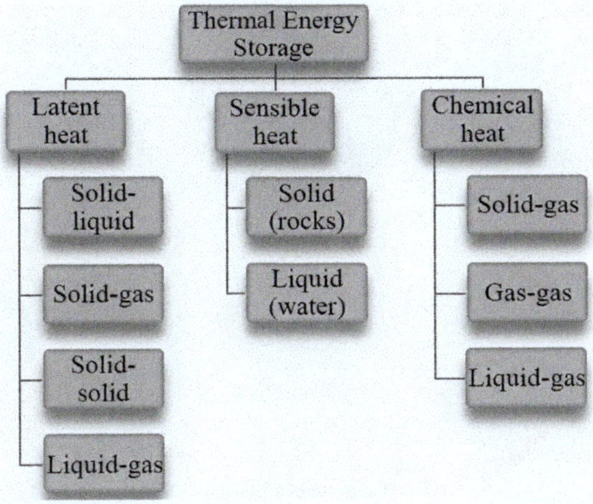

FIGURE 11.1

Classification of thermal energy storage [12].

ample amounts of thermal heat energy and control the emission of the greenhouse effect depending on the capacity of the system and can release heat depending on the requirements or applications. This storage can be performed in two major ways: sensible and latent heat storage.

Sensible energy storage is preferred in the industry because of its cost-effectiveness; however, latent heat storage offers more benefits in terms of high energy storage capacity. The classification of these systems is presented in Fig. 11.1.

11.2.1 Types of phase-change materials

PCMs can store large amounts of energy and release it whenever required. During the daytime, as the temperature increases, these PCMs melt and store the energy, and when the sun sets or the temperature decreases, these materials start to solidify owing to the temperature difference and release the stored energy as per the requirements. The temperature difference and storage and release of these materials work in a thermal cycle. A pictorial depiction of this process is shown in Fig. 11.2.

PCMs have a higher storage density in a smaller volume than other materials and require a smaller temperature change between storing and releasing energy. Phase can be classified into three categories. A brief classification of PCM is presented in Fig. 11.3.

11.2.2 Selection criteria of phase-change materials

The selection criteria of PCMs for TES applications depend on many factors, such as thermal, chemical, and economic properties (Table 11.1). The durability of PCM is also an important criterion for latent heat TES applications [12,14].

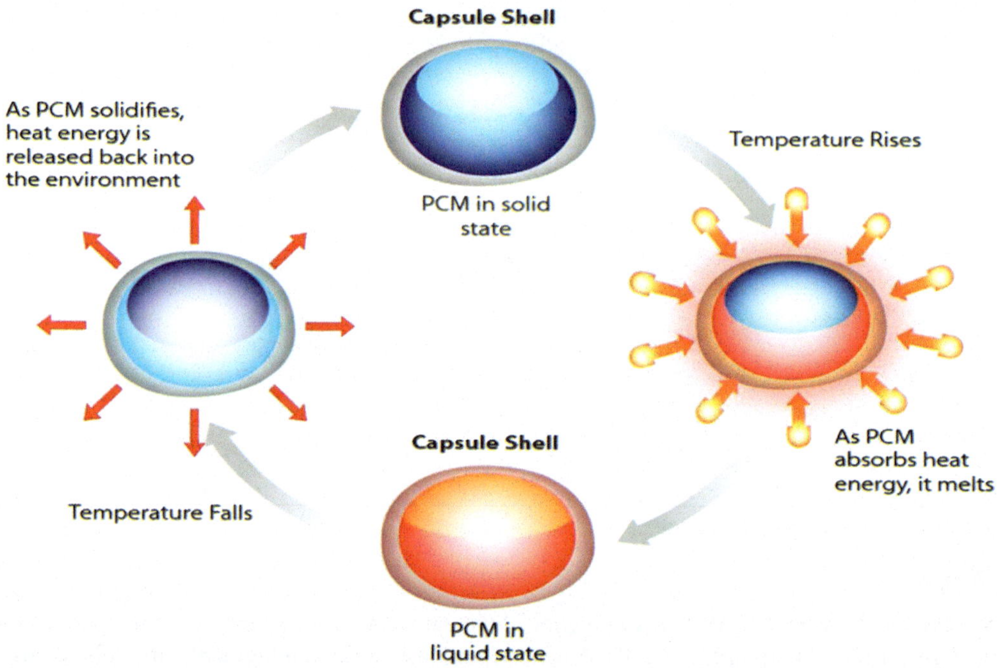

FIGURE 11.2

Phase-change process. *PCM*, Phase-change material.

Table 11.1 Thermo-physical properties of PCMs.

Thermal properties	Physical properties	Kinetic properties	Chemical properties	Economic properties
Phase change temperature in desired range	High density and high specific heat	No supercooling	Chemically stable	Less cost
High latent heat	Congruent melting	High rate of growth in crystal	Noncorrosive	Easily available
High thermal conductivity	Low vapor pressure	–	Inflammable	–
–	–	–	Nontoxic	–

PCM, *phase-change material.*

11.3 Types of phase-change materials

11.3.1 Organic phase-change materials

Organic PCMs mainly consist of paraffins and nonparaffins. Paraffins are hydrocarbon compounds that consist of straight-chain alkanes represented by C_nH_{2n+2}. With an increase in the number of

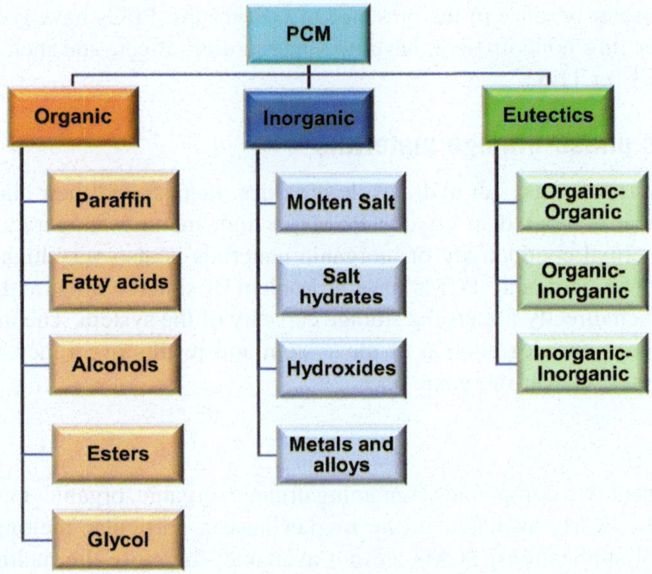

FIGURE 11.3

Classification of phase-change materials [13]. *PCM*, Phase-change material.

alkanes in the hydrocarbon the melting temperature and the latent heat of fusion also increase. Paraffins are the most widely used materials owing to their excellent energy storage capacity, which is widely used in buildings with a heat capacity ranging between 2.14 and 2.4 kJ/kg k. Paraffins exhibit unique characteristics as PCMs over other materials, such as being noncorrosive, colorless, nontoxic, chemically inert, highly durable, and abundantly available. Moreover, paraffin also shows no phase segregation over a number of repeated thermal phase-change cycles [15–18]. However, paraffins possess low thermal conductivity and exhibit slow oxidation when exposed to oxygen. The thermal conductivity can be enhanced by inserting or adding filler materials to the base materials.

11.3.2 Fatty acids

Vegetable and animal sources form fatty acids, which are nonpolluting and represented by the formula $CH_3(CH_2)_{2n}COOH$. Examples of fatty acids include lauric, myristic, palmitic, stearic, capric, and caprylic acids. The single material in this class has its own characteristics, unlike paraffins, which have very similar characteristics, as noted by Sharma et al. [19]. Compared to paraffins, these materials exhibit outstanding solid–liquid phase change properties; however, they are more expensive than paraffins.

11.3.3 Polyethylene glycol

Polyethylene glycol (PEG) is derived from petroleum, which belongs to dimethyl ether chains and is expressed as the chemical formula $HO–CH_2–(CH_2–O–CH_2–)_n–CH_2–OH$. PEG is a polymer with molecular weights ranging from 400 to 100,000 and melting temperatures varying from 3.2°C to 68.7°C.

PEGs are soluble in water because of the presence of OH groups. PEGs have good latent heat storage capacity, are nontoxic, are noncorrosive, have no supercooling effect, and melt congruently, making them promising PCMs for TES.

11.3.4 Inorganic phase-change materials

Inorganic PCMs consist of salts, salt hydrates, hydroxides, metals, and their alloys. Inorganic PCMs have a high latent heat compared to organic PCMs, a high melting temperature, and are available at a low cost. The thermal conductivity of inorganic materials is also very high compared to that of organic PCMs. However, inorganic PCMs have a problem of supercooling, and the materials do not melt congruently, which directly affects the storage capacity of the system. The materials are corrosive in nature, which leads to the degradation of the system and results in an increase in cost; therefore organic PCMs are preferred over inorganic ones.

11.3.5 Eutectics

Eutectics are mixtures of two compounds containing organic–organic, organic–inorganic, or inorganic–inorganic compounds. PCMs available in the market have a particular melting point. For specific melting-point thermal applications, PCMs are not available; therefore the melting temperature must be amended by preparing eutectics. This is the reason eutectic PCMs have gained significant attention from researchers. Thermal properties, such as latent heat and specific heat, are altered or tailored by preparing eutectics. Ansu et al. [20] prepared a binary eutectic mixture comprising PEG 2000 and PEG 10,000. The mixture was prepared using varying weight percentages of both PEGs. The results of the analysis revealed that the prepared mixture is a potential material that can be used for solar-based applications. In the above context, Sari [21] prepared an eutectic mixture comprising lauric and stearic acids in a ratio of 75.5:24.5. The results show a fair amount of latent heat with 182.7 J/g and a melting temperature of 37°C. Similarly, Sarı [22] in one of his experiments prepared eutectic mixture PCMs of some fatty acids, such as lauric–myristic acid (LA–MA), lauric–palmitic acid, and myristic–stearic acid (MA–SA), in different mass fraction ratios. A thermal cycle (melt/freeze) test was also performed to evaluate the thermal stability and reliability of the prepared eutectic mixture. Differential Scanning Calorimeter (DSC) analysis shows that LA–MA at a ratio of 66:34 has a latent heat of 166.8 J/g and melting temperature of 34.2°C. LA–MA has a latent heat of 166.3 J/g with a melting temperature of 35.2°C in the ratio of 69:31, whereas in the case of MA–SA, it is 182.4 J/g and 44.1°C when the ratio is 64:36. An accelerated thermal cycle test for 1460 cycles was performed to determine the thermal stability, and it was found that the variation in the melting temperature and latent heat was within 5%, which means that the eutectic mixture was very stable after the cycle test. Similarly, several experiments have been conducted in the past to determine the melting temperature and latent heat by preparing eutectic mixture PCMs for real-time applications.

11.4 Two-dimensional materials for energy storage

The primary attractions of carbon materials in energy storage are their outstanding thermal properties, excellent chemical stability, and stable and fine heat resistance. Graphene is a carbon-based 2D material with a low density and excellent thermal and mechanical properties. Compared to other carbon-based materials, graphene has high thermal conductivity and specific heat properties, which have attracted considerable interest among researchers. Graphene was obtained by oxidizing graphite using $KMnO_4$ and $NaNO_3$ as oxidizing agents in H_2SO_4/H_3PO_4. Graphene has also been developed using various techniques, such as mechanical stretching [8], epitaxial growth [23], chemical vapor deposition growth [24], and chemical development [9]. Compared to CNTs, graphene, being a porous material, is a better filler for preparing composites at a relatively lower cost.

11.4.1 Graphene oxide

The chemically developed operationalized graphenes prepared from GO are remarkably more productive and lower in cost, therefore it's been considerably interesting. The characterization of GO contributes significantly to the research on graphene and implies evaluations based on numerous microscopic and spectroscopic methods. Qi et al. [25] utilized the sheets of GO for the preparation of shape-stabilized PEG of molecular weight 6000. GO sheets were produced from natural graphite flakes and mixed with PEG 6000 by rigorous magnetic stirring for 12 h at 35°C. The maximum weightage of PEG was kept at 96%, which resulted in a very high latent heat of 142 J/g. The developed composite also maintained its thermophysical properties for 200 thermal cycles. Chen et al. [26] developed *n*-octadecane and GO-based microencapsulated PCMs, which show a high latent heat of approximately 200 J/g with high thermal conductivity. Wang et al. [27] developed GO-modified microcapsules of capric acid as core material with urea-formaldehyde resin. The inflection point was achieved at 0.6% addition of GO at a very high thermal conductivity of the order of above 70%. The developed composites also exhibited significant stability for 100 thermal cycles. He et al. [28] developed D-mannitol-based microencapsulated materials in which GO was used as a thermal conductivity enhancer. The SEM micrographs show that the size of the spherical capsules was between 100 and 400 nm without any visible breakage. The DSC thermograms show that the latent heat of these capsules is 216.7 J/kg during melting and 174.7 J/g during solidification. An increase of 128% in the thermal conductivity was also observed in the developed capsules.

11.4.2 Graphene nanoplatelets

GNP are also 2D materials with a thickness ranging between 5 and 10 nm and a surface area of 50–750 m^2/g. These materials were produced in various sizes, approximately between 1 and 50 μm. Researchers have taken a lot of interest in GNPs because of their very high thermal conductivity, which is as large as 3000–5000 W/m K [29]. GNPs possess outstanding thermal and mechanical characteristics, which significantly increase the heat transfer rate of composite PCMs. GNPs can also be used in a granular form, which can easily be dispersed in solvents and water. The properties of GNPs, such as electric field-assisted band gaps, linear monochromatic optical contrasts, and saturable absorption, depend on the number of layers.

11.4.3 Enhancement of thermal properties of phase-change materials using graphene

Organic phases have very low thermal conductivity, which directly affects the heat transfer rate. To increase the conductivity, several methods have been adopted, such as the dispersion of nanoparticles in various mass fractions in the base PCMs. The addition of carbon-based materials, such as CNTs, carbon nanofibres, and expanded graphite, has also produced significant results. The addition of graphene is also a method by which the thermal conductivity has increased exponentially. Several studies have been conducted, and the results revealed that the conductivity was significantly enhanced by huge amount as compared to other additions/dispersions of materials. In this context, Amin et al. [30] used beeswax as a PCM, which has a very high thermal capacity, and added graphene at different concentrations ranging from 0.05 to 0.3 wt.% to form the composite. The results show that the composite thermal conductivity increases by 1056% with 0.3 wt.% of graphene in the composite. The latent energy of the composite also increased from 141.49 to 186.75 kJ/kg with the heat. The increase in latent heat can be explained by the Brownian motion and particle clustering mechanisms [31]. The specific heat of the material increased by 12%, which may be because of the Brownian motion of the nanoplatelets.

Similarly, Yang et al. [32] prepared polyethylene glycol/boron nitride composite PCMs, in which a very low content of GNPs was introduced. The weight fraction of BN ranges from 5% to 30% with 1% of GNPs. The results show an exponential increase in thermal conductivity, which increases by 158% when the weight % of BN is 30. The melting point temperature and the latent heat measured by DSC analysis were 62.9°C and 123.8 J/g, respectively. The conductivity further increased by 336% when GNPs were added to the composite PEG/30BN/1GNP. According to Balandin et al. [33], graphene exhibits various interesting properties, such as excellent thermal conductivity, which ranges from 4840 to 5300 W/Mk for single-layer graphene, which has been claimed to be higher than that of copper by and order of magnitude. Hence graphene can be considered a favorable thermally conductive filler because of its outstanding thermal conductivity and low density. Malekpour et al. [34] investigated the thermal conductivity of laminate graphene films set on polyethylene terephthalate membrane. The study of two types of laminate graphene, as deposited or uncompressed and compressed, was conducted to determine the physical specifications that most significantly affect heat transfer by conduction. At room temperature, the thermal conductivity of graphene was in the range of 40–90 W/m K. Compressed graphene exhibits better thermal conductivity with better alignment for the same flake size. Wu et al. [35] reported the synthesis of composites consisting of polystyrene/reduced graphene oxide (PS/rGO) using a two-step in situ reduction method. A PS/rGO composite with 5 wt% rGO was prepared by hot press molding, yielding a conductivity as high as 22.68 S/m. With a high degree of GO reduction the conductivity was found to be superior. These composites are potential candidates for multiple electric devices.

11.5 Boron nitride as two-dimensional energy storage materials

Another class of 2D material for TES is carbon-based BN, which is gaining continuous interest from researchers as an alternative to graphene with similar properties. This is also a great thermal conductivity enhancer of organic PCMs as nanofillers. Boron, which is adjacent to carbon and lies in the second row of the periodic table, can form a stable covalent bond network in a manner similar to graphene [36]. They form this network by a simple substitution of carbon atoms with alternate B and N atoms, whereas the bonding of boron atoms is more complex [37]. The structure of BN is similar to that of honeycomb and hexagonal layers, with a bond length of 1.466 Å, which is slightly higher than that of graphene (1.421 Å) [38]. Compared to graphite, BN is flame-resistant and can withstand higher temperatures, whereas graphite is prone to ignition in the presence of air [39].

The thermal conductivity of BN at room temperature is also significantly higher (~550 W/m K), which has attracted considerable interest from researchers [40]. Fang et al. [41] developed a composite

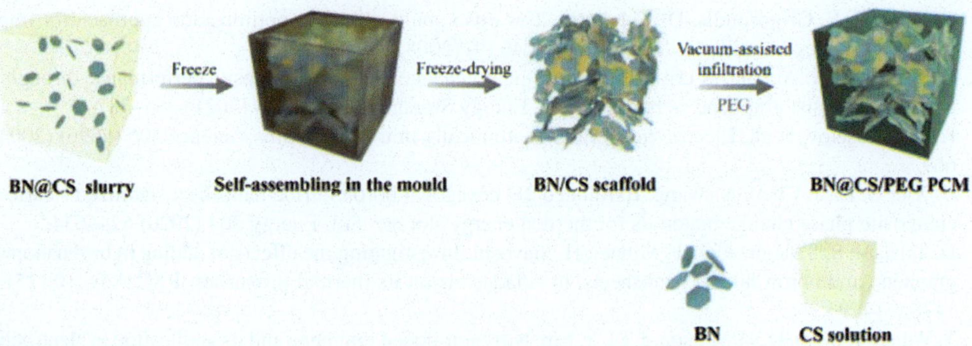

FIGURE 11.4

Steps for preparation of shape-stabilized composites [42]. *BN*, Boron nitride; *PCM*, phase-change material; *PEG*, polyethylene glycol.

of paraffin and hexagonal BN (in different wt.%). The results indicated that with an increase in the BN fraction, the thermal conductivity of paraffin increased. Overall, 10% of h-BN results in a 60% increase in the thermal conductivity of paraffin. Jia et al. [42] developed a shape-stabilized form of stable PCM (PEG 6000) of BN@chitosan using a process depicted in Fig. 11.4 and investigated its thermal performance. The results show that the addition of 27.01% BN increases thermal conductivity by 804%.

11.6 Conclusion

This chapter demonstrates that base organic PCMs inherently exhibit low thermal conductivity, which limits their heat transfer capability and, consequently, overall system performance. Therefore thermal conductivity enhancement, typically achieved through the incorporation of conductive fillers, is essential for practical thermal management applications. However, the filler loading must be carefully optimized, as increasing the filler content generally improves the thermal conductivity while simultaneously reducing the latent heat storage capacity. Achieving an effective balance between these competing properties is critical for preserving both the energy storage density and heat transfer efficiency. The literature also reveals substantial variability in reported thermal property data, arising from differences in material preparation, measurement techniques, and testing conditions. This lack of methodological consistency complicates direct comparisons across studies and hinders reliable performance assessment. The establishment of standardized procedures for thermal characterization would improve reproducibility, enable meaningful benchmarking, and accelerate the development and deployment of high-performance PCM composites.

References

[1] E. Asadi, M.G. Da Silva, C.H. Antunes, L. Dias, Multi-objective optimization for building retrofit strategies: a model and an application, Energy Build. 44 (2012) 81–87.

[2] C. Diakaki, E. Grigoroudis, D. Kolokotsa, Towards a multi-objective optimization approach for improving energy efficiency in buildings, Energy Build. 40 (9) (2008) 1747–1754.

[3] U.B. Shahid, A. Abdala, A critical review of phase change material composite performance through figure-of-merit analysis: graphene vs boron nitride, Energy Storage Materials 34 (2021) 365–387.

[4] K.S. Novoselov, et al., Electric field effect in atomically thin carbon films, Science 306 (5696) (2004) 666–669.

[5] B. Xie, C. Li, J. Chen, N. Wang, Exfoliated 2D hexagonal boron nitride nanosheet stabilized stearic acid as composite phase change materials for thermal energy storage, Sol. Energy 204 (2020) 624–634.

[6] H. Tafrishi, S. Sadeghzadeh, F. Molaei, H. Siavoshi, Investigating the effects of adding hybrid nanoparticles, graphene and boron nitride nanosheets, to octadecane on its thermal properties, RSC Adv. 10 (25) (2020) 14785–14793.

[7] Y. Wang, Y. Shao, D.W. Matson, J. Li, Y. Lin, Nitrogen-doped graphene and its application in electrochemical biosensing, ACS nano 4 (4) (2010) 1790–1798.

[8] H.A. Becerril, J. Mao, Z. Liu, R.M. Stoltenberg, Z. Bao, Y. Chen, Evaluation of solution-processed reduced graphene oxide films as transparent conductors, ACS nano 2 (3) (2008) 463–470.

[9] T. Szabó, et al., Evolution of surface functional groups in a series of progressively oxidized graphite oxides, Chem. Mater. 18 (11) (2006) 2740–2749.

[10] A.A. Balandin, Thermal properties of graphene and nanostructured carbon materials, Nat. Mater. 10 (8) (2011) 569–581.

[11] S. Chen, et al., Thermal conductivity of isotopically modified graphene, Nat. Mater. 11 (3) (2012) 203–207.

[12] R. Sharma, P. Ganesan, V. Tyagi, H. Metselaar, S. Sandaran, Developments in organic solid–liquid phase change materials and their applications in thermal energy storage, Energy Convers. Manage. 95 (2015) 193–228.

[13] P. Singh, R. Sharma, A. Ansu, R. Goyal, A. Sarı, V. Tyagi, A comprehensive review on development of eutectic organic phase change materials and their composites for low and medium range thermal energy storage applications, Sol. Energy Mater. Sol. Cells 223 (2021) 110955.

[14] P. Singh, R.K. Sharma, A.K. Ansu, R. Goyal, Study on thermal properties of organic phase change materials for energy storage, Mater. Today: Proc. 28 (2020) 2353–2357.

[15] A. Abhat, Low temperature latent heat thermal energy storage: heat storage materials, Sol. Energy 30 (4) (1983) 313–332.

[16] G.A. Lane, Solar Heat Storage: Latent Heat Materials, CRC Press, Boca Raton, FL, 1983.

[17] S. Hasnain, Review on sustainable thermal energy storage technologies, Part I: heat storage materials and techniques, Energy Convers. Manage. 39 (11) (1998) 1127–1138.

[18] S. M. Hasnain, Latent Heat Thermal Energy Storage for Solar Heating Applications (Ph.D. thesis), University of Leeds, Leeds, 1990.

[19] A. Sharma, V.V. Tyagi, C. Chen, D. Buddhi, Review on thermal energy storage with phase change materials and applications, Renew. Sustain. Energy Rev. 13 (2) (2009) 318–345.

[20] A.K. Ansu, R.K. Sharma, V.V. Tyagi, D. Tripathi, Prediction of thermal properties and reliability testing of binary eutectic mixture of polyethylene glycol 2000 and 10000 as phase change materials, ChemistrySelect 5 (31) (2020) 9745–9757.

[21] A. Sari, K. Kaygusuz, Thermal performance of a eutectic mixture of lauric and stearic acids as PCM encapsulated in the annulus of two concentric pipes, Sol. Energy 72 (6) (2002) 493–504.

[22] A. Sarı, Eutectic mixtures of some fatty acids for low temperature solar heating applications: thermal properties and thermal reliability, Appl. Therm. Eng. 25 (14–15) (2005) 2100–2107.

[23] M. Inagaki, New Carbons-Control of Structure and Functions, Elsevier, Amsterdam, 2000.

[24] F. Bundy, W. Bassett, M. Weathers, R. Hemley, H. Mao, A. Goncharov, The pressure-temperature phase and transformation diagram for carbon; updated through 1994, Carbon 34 (2) (1996) 141–153.

[25] G.-Q. Qi, et al., Polyethylene glycol based shape-stabilized phase change material for thermal energy storage with ultra-low content of graphene oxide, Sol. Energy Mater. Sol. Cells 123 (2014) 171–177.

[26] D.-Z. Chen, et al., Fabrication, morphology and thermal properties of octadecylamine-grafted graphene oxide-modified phase-change microcapsules for thermal energy storage, Compos. B. Eng. 157 (2019) 239–247.

[27] X. Wang, Z. Chen, W. Xu, X. Wang, Capric acid phase change microcapsules modified with graphene oxide for energy storage, J. Mater. Sci. 54 (24) (2019) 14834–14844.

[28] L. He, S. Mo, P. Lin, L. Jia, Y. Chen, Z. Cheng, D-mannitol@silica/graphene oxide nanoencapsulated phase change material with high phase change properties and thermal reliability, Appl. Energy 268 (2020) 115020.

[29] Y. Kim, et al., A composite of a graphene oxide derivative as a novel sensing layer in an organic field-effect transistor, J. Mater. Chem. C 2 (23) (2014) 4539–4544.

[30] M. Amin, N. Putra, E.A. Kosasih, E. Prawiro, R.A. Luanto, T. Mahlia, Thermal properties of beeswax/graphene phase change material as energy storage for building applications, Appl. Therm. Eng. 112 (2017) 273–280.

[31] A. Zabalegui, D. Lokapur, H. Lee, Nanofluid PCMs for thermal energy storage: latent heat reduction mechanisms and a numerical study of effective thermal storage performance, Int. J. Heat Mass Transf. 78 (2014) 1145–1154.

[32] J. Yang, et al., Largely enhanced thermal conductivity of poly (ethylene glycol)/boron nitride composite phase change materials for solar-thermal-electric energy conversion and storage with very low content of graphene nanoplatelets, Chem. Eng. J. 315 (2017) 481–490.

[33] A.A. Balandin, et al., Superior thermal conductivity of single-layer graphene, Nano Lett. 8 (3) (2008) 902–907.

[34] H. Malekpour, et al., Thermal conductivity of graphene laminate, Nano Lett. 14 (9) (2014) 5155–5161.

[35] N. Wu, X. She, D. Yang, X. Wu, F. Su, Y. Chen, Synthesis of network reduced graphene oxide in polystyrene matrix by a two-step reduction method for superior conductivity of the composite, J. Mater. Chem. 22 (33) (2012) 17254–17261.

[36] T. Ishii, T. Sato, Y. Sekikawa, M. Iwata, Growth of whiskers of hexagonal boron nitride, J. Cryst. Growth 52 (1981) 285–289.

[37] Z. Zhang, E.S. Penev, B.I. Yakobson, Two-dimensional boron: structures, properties and applications, Chem. Soc. Rev. 46 (22) (2017) 6746–6763.

[38] L. Pauling, The structure and properties of graphite and boron nitride, Proc. Nat. Acad. Sci. U.S.A. 56 (6) (1966) 1646.

[39] R.S. Ambekar, et al., 2D hexagonal boron nitride-coated cotton fabric with self-extinguishing property, ACS Appl. Mater. Interfaces 12 (40) (2020) 45274–45280.

[40] P. Jiang, X. Qian, R. Yang, L. Lindsay, Anisotropic thermal transport in bulk hexagonal boron nitride, Phys. Rev. Mater. 2 (6) (2018) 064005.

[41] X. Fang, et al., Thermal energy storage performance of paraffin-based composite phase change materials filled with hexagonal boron nitride nanosheets, Energy Convers. Manage. 80 (2014) 103–109.

[42] X. Jia, et al., High thermal conductive shape-stabilized phase change materials of polyethylene glycol/boron nitride@chitosan composites for thermal energy storage, Compos. Part A Appl. Sci. Manuf. 129 (2020) 105710.

CHAPTER 12

2D materials in solar heating and cooling

L. Harish Kumar[a], C. Bhinitha[b], Kazi Salim Newaz[b], H.H. Masjuki[c] and M.N.M. Zubir[b]

[a]*Department of Mechanical Engineering, NITTE Meenakshi Institute of Technology, NITTE (Deemed to be University), Bengaluru, Karnataka, India,* [b]*Department of Mechanical Engineering, Faculty of Engineering, Universiti Malaya,, Kuala Lumpur, Malaysia,* [c]*Department of Mechanical Engineering, Faculty of Engineering, International Islamic University Malaysia, Kuala Lumpur, Malaysia*

12.1 Introduction

Most industries mainly depend on fossil fuels to meet the increasing energy demand. However, fossil fuel sources are bound to deplete with time because of their very limited reserve. Moreover, there is enhanced concern about rising global temperature because of the emissions of harmful gases and radiation by using fossil fuel, which has become a very vital socioeconomic issue [1]. It was estimated that the surge in global energy demand was about 2.3% in the year 2018, with 1.7% related to CO_2 emission [2]. A continued increase in global temperatures is leading to more extreme climatic conditions. Current projections estimate that global temperatures will rise by 1.5°C–1.7°C above preindustrial levels between 2030 and 2052 [3]. Thus several regulations, such as the Paris Climate Change Agreement [4] and the United Nations Environment Programme [5], have been set up to limit global emissions and temperature. Therefore continuous research is being carried out intensively, looking for affordable and clean energy to resolve the global warming issue. Using renewable resources, mainly solar energy, eradicates the need for fossil fuels and also can solve the challenges associated with the enhanced global temperature [6,7]. Solar energy is easily available, safe, and in abundance. Devices such as solar collectors give an advantage to directly use solar energy by transforming it into other types of useful energy [8]. Direct utilization of solar energy by converting it to heat is one of the keys to the solar harnessing technique and is readily available for a wide range of applications. Moreover, solar thermal technology is one of the potential ways of meeting the energy demand of many countries. This technology collects the thermal energy from the sun. It converts it into useful work for various applications such as water heating, space heating, space cooling, and refrigeration for residential, commercial, and industrial applications. Enhancing thermal performance is very critical to developing higher efficient solar thermal systems toward meeting the demands of sustainability regulations [9, 67,68].

Solar thermal energy has been rapidly developing and used widely to provide domestic and industrial heating and cooling [10,69]. The role of heat transfer and energy management is highly crucial and

Graphene and 2D Materials in Heat Transfer: Fundamentals and Applications. DOI: https://doi.org/10.1016/B978-0-12-821964-5.00008-4

challenging in achieving energy and cost savings along with improved systems. Thus developing heat transfer fluids with superior heat transfer capabilities is always vital in such systems. Conventional heat transfer fluids, such as water and oil, provide limited heat exchange capabilities because of their poor thermal conductivity values [11,12]. Dispersal of solid particles (milli and micro) in these conventional heat transfer fluids enhances their thermal conductivity [13,14,70,71]]. Besides the course, milli- and microparticles in heat transfer fluids resulted in instability, fouling, thermal degradation, and blocking of tubes in the systems. Nanoparticles developed by Choi [15] provided a new outlook in different heat transfer systems as they provided superior qualities. Notable augmentation in thermal conductivity and heat transfer coefficient of heat transfer fluids has garnered significant attention among scientific researchers in the past decade. Several nanoparticles, such as metals [16,17], metal oxides [18,19], metal carbides and nitrides [20–22], carbon-based [23,24], and hybrid [25,26], have been utilized in solar applications. Among them, single-layer nanomaterials or 2D materials (2DMs) have gained significant attention and become the apparent choice for future solar technologies because of their robust physical and chemical properties. These materials have immense potential for enhancing performance in solar applications. Graphene is the most used 2DM for solar photovoltaic (PV) and thermal technology because of its high transparency and better capability of heat transport, as it can provide higher electrical and thermal conductivity [6,27–29]. There is no notable substantial review on only 2DMs in solar heating and cooling, respectively. Therefore this chapter summarizes the effect of 2DMs in different solar heating, cooling, and storage systems. Further the use of 2DMs in different concentrating and nonconcentrating solar collectors has been discussed.

In the present chapter a brief introduction to solar energy and its thermal conversion is provided in Section 12.2, along with the classification of solar collectors. Section 12.3 deals with the effect of 2DMs on the thermal performance of solar collectors for heating purposes. Furthermore the effect of 2DMs on solar cooling and thermal storage is also discussed. Section 12.4 summarizes the vital findings concerning the 2DMs in solar thermal applications.

12.2 Solar energy

The sun is the primary source of solar energy, which reaches the Earth's surface in an attenuated form. Solar energy, being inexhaustible, has the potential to meet all current and future energy demands. This makes it the ultimate source among all renewable energy options, providing a clean and sustainable form of energy. Solar energy can be harnessed through two primary methods, as illustrated in Fig. 12.1. Among various techniques, solar thermal energy is particularly prominent because of its efficiency and convenience. It is extensively used in applications such as water and space heating, industrial heat generation, and space cooling.

Solar energy is increasingly in demand across many countries because of growing concerns about the environmental impact of fossil fuels used for heating and cooling applications. Water and space heating constitute a significant portion of energy consumption. Utilizing converted heat from solar radiation is the most straightforward method of harnessing solar energy. Consequently, solar water heating and cooling systems are highly effective in utilizing solar thermal energy [30,31]. Solar thermal systems offer sustainable solutions for water and development resources, particularly in arid regions. However, substantial political and financial supports from stakeholders, especially during the initial implementation phase, are essential to address and overcome various barriers.

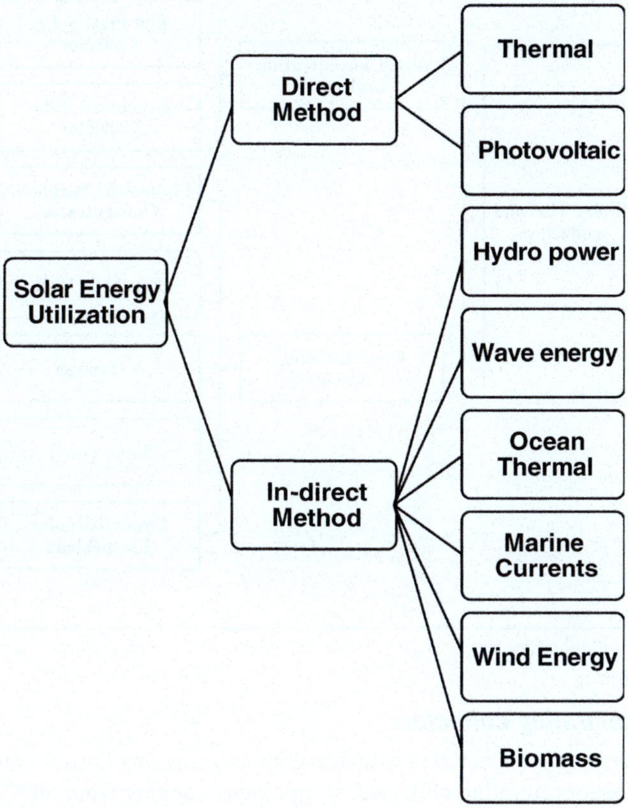

FIGURE 12.1

Classification of solar energy utilization.

12.2.1 Solar thermal conversion

The solar thermal collection works on a basic principle that results in a surge in temperature of the surface, which absorbs radiation partially when the solar radiation is incident on it. An equilibrium state is achieved when the heat gain rate by the absorber and heat loss rate to surroundings are alike. Solar collectors are used to harness the incident solar radiation into thermal energy. Different types of solar collectors are available and are used for a wide range of applications.

12.2.2 Solar thermal collectors

Solar thermal collectors are conservatively signified in two systems, as shown in Fig. 12.2:

1. Nonconcentrating collectors
2. Concentrating collectors

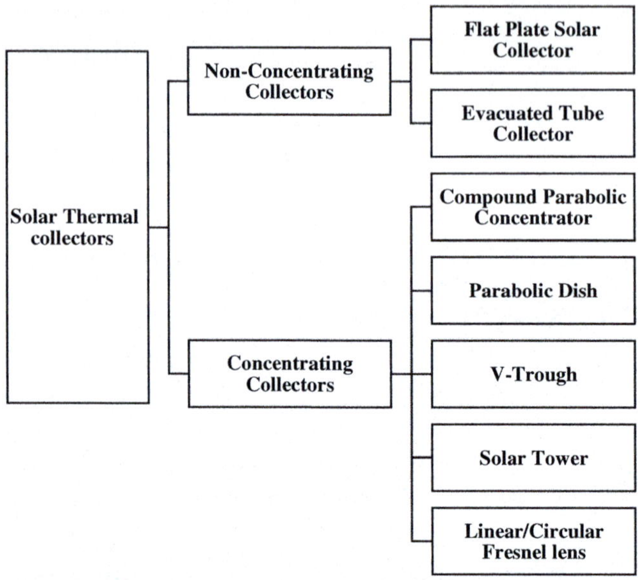

FIGURE 12.2

Classification of solar thermal collectors.

12.2.2.1 Nonconcentrating collectors

The incident solar radiation is collected in a device without using any optical concentrating parts. Such a device is called a nonconcentrating collector. Temperature ranging from 50°C to 90°C can be easily achieved. However, solar radiation level and collector type highly influence the output. The construction and maintenance are simpler for nonconcentrating collectors, as there is no presence of any moving parts. These types of collectors are used mainly for lower temperature applications. Flat-plate solar collectors (FPSCs) and evacuated tube solar collectors (ETSCs) are the most used ones.

12.2.2.1.1 Flat-plate solar collector

An FPSC is a distinct kind of heat exchanger that converts incident solar energy into heat with no optical concentrating part. In an FPSC, heat exchange takes place between the two media through conduction and convection at the interface of the absorber plate and tubes through which the working fluid is flowing. It is the most used solar collector because of its design simplicity, no parts under motion, and maintenance ease. The operating temperature of FPSC is in the range from 30°C to about 110°C. It is used in various applications such as water and building heating, air drying, and conditioning in both domestic and industries. Fig. 12.3 shows the cross-section of a basic FPSC.

The total solar spectrum falling on the absorber plate through the covers is absorbed and converted to useful energy [32]. At the interface the heat energy partly gets transferred to the working liquid flowing through the tube. The absorbed heat is carried away for different applications [12–33].

The thermal performance of FPSC is considerably treated in detail to understand the technical aspects of the components. Thus FPSC is given much importance in thermal processes. The working fluid is highly crucial in achieving better thermal performance. The design and thermophysical properties of

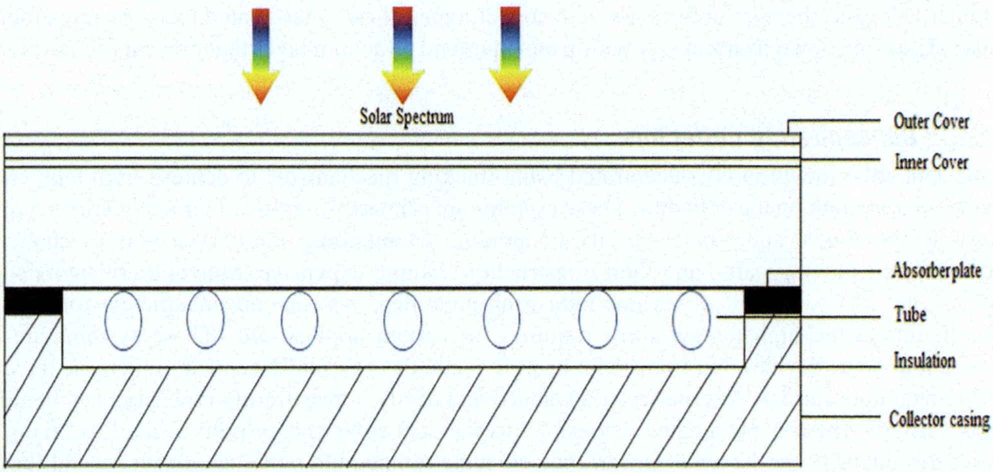

FIGURE 12.3

Schematic diagram of a flat-plate solar collector.

working fluids are the other significant factors that influence the thermal performance of FPSC. With less viscosity and high specific heat properties a heat transfer fluid can be pumped with ease in a system. This enhances the gain in the final available energy. Moreover, the stability and corrosiveness characteristics of working fluids are considered vital in achieving a better system [34].

12.2.2.1.2 Evacuated tube collector

ETSCs consist of parallel glass tubes arranged in a series. The design of an evacuated tube collector (ETC) can be likened to a flat-plate collector placed inside an evacuated cylinder. When the receiver end is metal, a glass-to-metal seal is essential to maintain the vacuum within the cylinder. This vacuum condition inside each tube eliminates conduction and convection losses. The evacuated glass tubes are filled with water and exposed to sunlight, which heats the water inside the tubes. Because of the higher density of cold water compared to hot water, the heated water in the glass tubes rises into the insulated water tank, whereas the cold water from the tank descends into the glass tubes. This continuous process of heating is driven by natural convection, known as the thermosiphon effect. The tank is insulated to keep the water hot for use later in the day or even the next day. Two key features that ensure the high efficiency of this solar heater are the vacuum layer, which prevents both convective and conductive losses, and the one-way heat transfer design of the heat pipes, which prevents the absorber from losing heat back to the water.

The inner shield of the evacuated tubes is protected by a specialized coating that maximizes energy absorption while minimizing heat radiation. Inside the glass heat pipes a liquid is heated and vaporizes, increasing the temperature of the pipes. When heat is exchanged between the condenser end of the heated pipe and the cold water in the tank, the vapor condenses back into a liquid and collects at the bottom of the heat pipe. This cycle continues as long as there is sunlight. A syringe mechanism is used to draw hot water from the top by injecting cold water into the bottom of the tank. In countries with four seasons, this ETC acts as a quasitracking system. The cylindrical shape of the tubes ensures that a similar surface area is always oriented toward the sun as it moves across the sky. This system remains functional as long as there is sunlight. Even in cold climates, these solar-heated tubes operate efficiently because of unidirectional heat transfer within the pipes and a vacuum layer that prevents convective heat loss.

12.2.2.2 *Concentrating collectors*

The incident solar radiation is concentrated using tracking mechanisms to achieve high temperatures from the solar medium in the collector. These systems are primarily employed for applications requiring elevated temperatures, such as electricity generation. Maintaining these collectors is challenging because of their moving parts, and their construction is quite expensive. Standard collectors such as flat plates and ETCs, as well as various industrial-grade heat systems, are insufficient for supplying carrier liquids at the high temperatures required for certain applications. These systems are often used as initial-stage devices, but traditional heating methods are usually needed to increase the carrier fluid's temperature further. Alternatively, advanced and costly concentrating collectors can be utilized for this purpose. These sophisticated devices focus incident solar energy onto a small receiving area, significantly intensifying the solar power and achieving temperatures that can reach several thousand degrees Celsius. Concentrating collectors operate efficiently by continuously adjusting their position to track the sun. They capture direct solar radiation over a large area and concentrate it onto a smaller absorber area. This design allows them to achieve higher temperatures more effectively than flat-plate collectors. However, a significant drawback is the necessity for additional mechanical support to track the sun, as diffused sky radiation cannot be easily directed onto the absorber.

Most concentrating collector designs incorporate mechanical systems that continuously orient the collectors toward the sun and focus the absorber on the radiation point. Common types of concentrating collectors include heliostat field collectors, parabolic collectors, linear Fresnel reflectors, and parabolic trough collectors (PTCs). Although these four are the main types, there are numerous other designs tailored to specific situations. The basic principle of a solar cell (PV) is to produce electricity using light energy; a concentrating solar system, however, generates electricity by utilizing heat from the sun's rays. The absorber of a concentrating collector comprises mirrors and lenses that centralize sunlight onto a thermal receiver, functioning similarly to a boiler. The conversion of light to heat occurs in a thermal receiver, and this heat is then transferred to a steam turbine or engine to be converted into electricity.

Single- and dual-purpose plants of multiple types have been researched and analyzed in the field. Installations began in the early 1980s to gather user experience and have been continually upgraded for enhanced efficiency of concentrating solar power (CSP). It is believed that solar thermal power plants will significantly contribute to an economical, efficient, and environmentally sustainable energy supply, whether in remote distributed markets or large-scale grid-connected markets. For solar-only power plants and fossil/solar hybrid plants, possible installations include parabolic dishes, central receivers, and parabolic and Fresnel troughs. Correspondingly, for industrial applications, decentralized process heat will be provided by a concentrating collector that is cost-effective.

12.3 2D materials in solar applications

Solar energy is proposed as an alternative to meet the rising concerns about environmental protection. Solar thermal conversion is the most simple and direct application of solar energy, having tremendous potential over other resources. Hot water heating consumes a significant amount of energy. The most

Table 12.1 Operating range of solar collectors in solar water heating systems.

Collector category	Temperature	Output (%)
Flat-plate solar collector	Operating temperature: 28°C–62°C Output water temperature: 40°C–65°C	50–80
Evacuated tube solar collector	Operating temperature: 25°C–60°C Output water temperature: 40°C–80°C	55–80
Concentrating collectors	Operating temperature: 65°C–80°C Output water temperature: 35°C–65°C	60–80

widely used water heating systems at present are solar-based water heating systems. With a low cost and maintenance, solar water heaters can effectively and efficiently produce concentrated thermal energy from collected solar energy. Solar collectors significantly affect the performance of solar water heating systems. FPSC evacuated tubes are the most commonly used ones in a solar heating system [10,35,36]. The selection of collectors mainly depends on the technical aspects of the collectors, location of installation, and energy demand. The effectiveness of the corresponding collectors in a water heating system is concluded in Table 12.1.

The thermal performance of solar collectors is significantly influenced by the choice of working fluids used for heat transport. Fig. 12.4 illustrates the general applications of nanofluids. A variety of working fluids are employed in solar collectors, including air, water, glycol, refrigerants, silicones, oils, and fine solid particles ranging in size from milli- to nanoscales. Each of these fluids has its own set of disadvantages, such as leakage, cost, corrosiveness, flammability, low thermal conductivity, instability, and thermal degradation. However, nanofluids have demonstrated superior stability, higher thermophysical properties, and enhanced heat exchange capabilities. This makes them highly effective for achieving better performance in water heating systems. One of the key methods to improve the efficiency of solar heating systems is by enhancing the properties of the heat transfer fluids used. Incorporating 2DMs into solar heating systems has been shown to provide superior performance compared to other materials.

Research on 2DMs expanded rapidly after 2004 with the first isolation of single-layer graphene, launching the new field of two-dimensional nanosheets [38–40]. As 2DMs are inherently hydrophobic, surface functionalization techniques are commonly used to modify wettability and enhance dispersion stability in solvents and polymers [41–44]. Graphene nanoplatelets (GNPs), consisting of stacked graphene sheets in a 2D arrangement, show particular promise for heat transfer applications because of their planar geometry. GNPs are readily incorporated into base fluids to form nanofluids with tunable thermophysical properties. Fig. 12.5 shows various synthesized volumetric concentrations of GNPs dispersed in deionized water, with particle thicknesses ranging from 1 to 100 nm and surface areas of 10–750 m^2/g [45].The strong optical absorption properties of graphene, even at low temperatures, make GNP nanofluids well-suited as an absorptive medium in solar collectors and thermal storage units. Continued work optimizes dispersions, functionalization, and characterization methods to realize performance enhancement. However, addressing long-term stability under repeated thermal cycling also remains critical for practical solar thermal applications [6,27,45].

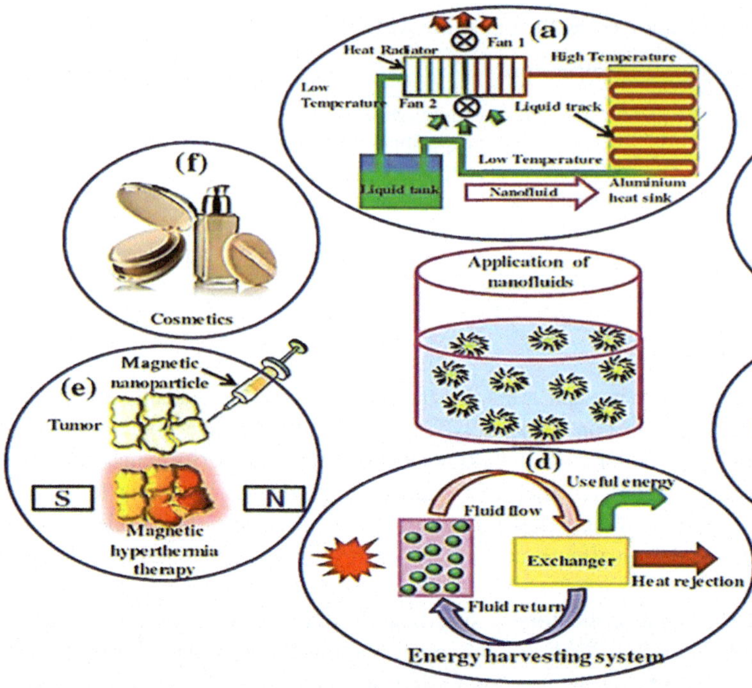

FIGURE 12.4

General applications of graphene nanofluids in heating and cooling systems [37]

(permission from Elsevier) [37].

FIGURE 12.5

Different volume concentrations of GNP nanofluids prepared and compared with deionized water for solar application [27]. *GNP*, Graphene nanoplatelet.

Reproduced with permission from Elsevier [27].

12.3.1 2DMs in flat-plate solar collector

Kumar et al. [6] examined gallic acid-doped GNPs in the FPSC under forced circulation conditions. They reported that even a lower concentration of GNPs in aqueous resulted in enhanced performance of FPSC. The experimental investigation reported a 23% increase in collector efficiency. However, there was a rise in pumping power with the increase in the concentration of GNPs. Functionalized TEA-GNPs incorporated in an aqueous medium were examined in FPSC under forced circulation [23]. About 10.5% augmentation in thermal efficiency was reported. Alawi et al. [46] have considered the effect of using pentaethylene glycol treated GNP dispersed in water experimentally for calculating the thermal performance of FPSC. It was observed that thermophysical properties of working fluid increased except for specific heat. Particularly the thermal conductivity of the nanofluid had an enhancement in its value up to 32.4% in comparison with water. The efficiency measured experimentally increased by 13.3% for polyethylene glycol (PEG)-GNP of 0.1 wt.%. Moreover, the rise in flow rate and heat flux showed significant augmentation in thermal performance. However, it decreased subsequently with the increase in the input temperature of the working fluid.

Graphene oxide (GO) was incorporated in an aqueous medium and tested in FPSC [29]. Augmentation of 7% was reported for 0.02% concentration. Verma et al. [47] reported about 17% increment in the thermal efficiency of FPSC using GNP/water nanofluids. Vakili et al. investigated GNP/deionized water nanofluids in the solar collector for domestic hot water systems [27]. For different mass flow rates, augmentation inefficiency was obtained with a maximum efficiency of 93.24% for GNP nanofluid than using base fluid. Hence they concluded that the addition of GNP in a base fluid such as water increases the absorption capacity and also the efficiency of solar thermal systems for different residential applications.

GO with Multi-walled Carbon Nano Tubes (MWCNTs) hybrid nanofluids were utilized in a direct absorber collector operated at a lower temperature [48]. Their investigation revealed that water-based nanofluid had improved optical absorption than ethylene glycol-based nanofluid. Moreover, they are very effective in enhancing the efficiency of Direct Absoprtion Solar Collectors (DASC). Nanofluid with water as base fluid showed enhancement of up to 17.7% to that of 15.1% with ethylene glycol as base fluid. Thus the addition of even a small fraction of 2DM enhances the thermal performance of solar collectors.

Ahmadi et al. investigated the hot water production capability and thermal performance of an FPSC closed system, as depicted in Fig. 12.6 [45]. Their main objective was to develop a high-quality solar water heating system for household applications. They prepared graphene GNP–water nanofluids with concentrations of 0.01% and 0.02% by weight using the two-step method. The study demonstrated an enhancement in thermal efficiency of approximately 12.20% and 18.90%, as shown in Fig. 12.7, with water temperatures reaching up to 68°C and 71°C for the 0.01% and 0.02% GNP nanofluids, respectively. The comparison of theoretical and experimental efficiencies revealed that incorporating 2DMs like graphene into the working fluids significantly improved the thermal performance of the solar water heating systems.

Further, GNP nanofluids showed greater improvement in heat transfer even at lower fractions of concentrations. Moreover, the enhancement in the thermal conductivity up to 13.6% and 7% was observed in the GNP nanofluid with water and ethylene glycol as base fluids [49]. The experimental results conducted on graphene-based nanofluids displayed much enhanced thermal conductivity than

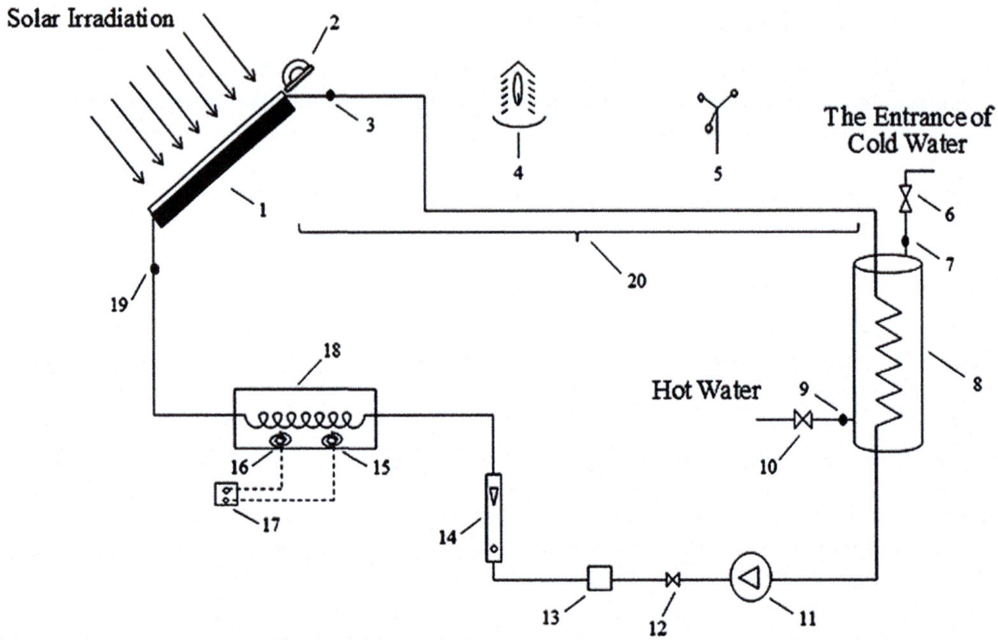

FIGURE 12.6

Experimental setup [(1) FPSC, (2) pyranometer, (3, 7, 9, 19) thermometer, (4) ambient temperature sensor, (5) anemometer, (6, 10, 12) valve, (8) water heater, (11) pump, (13) ventilation valve, (14) flow meter, (15, 16) heater, (17) controller, (18) tank, and (20) thermal insulator] used for the investigation of the thermal performance of FPSC using GNP nanofluids [45]. *FPSC*, Flat-plate solar collector; *GNP*, graphene nanoplatelet.

Reproduced with permission from Elsevier [45].

other nanofluids [50,51]. Thus the enhanced performance of FPSC was reported. Besides the increase in heat flux and flow rate the thermal efficiency was also increased. However, all the above reviews reported an increase in pressure drop and pumping power with the increase in the concentration of GNPs in the different base fluids.

12.3.2 2D materials in an evacuated tube collector

In the area of solar thermal systems the ETC has been a subject of significant research and development, particularly in the context of enhancing its efficiency using advanced working fluids. As noted by Iranmanesh et al. [28], there has been a concerted effort to replace conventional fluids with those exhibiting higher thermal conductivity, aiming to improve the overall efficiency of these collectors. The ETC, known for its application in low-temperature systems, offers the advantage of simplicity and cost-effectiveness, making it an attractive candidate for such enhancements.

The integration of nanofluids into solar collectors has been a focus of study, with graphene emerging as a particularly promising material. Graphene, a single-layer carbon atom structure, boasts exceptional thermal conductivity, which can significantly enhance heat transfer properties. The interest in GNPs

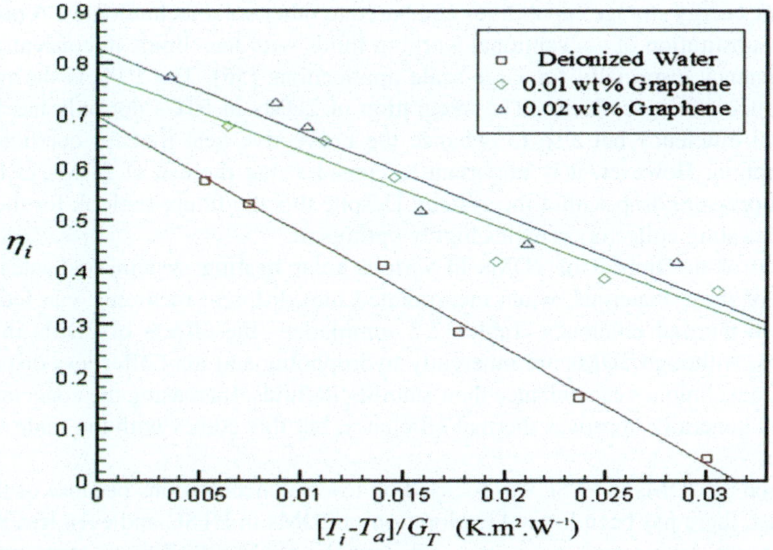

FIGURE 12.7

Thermal efficiency obtained for different mass fractions of GNP dispersed in water in comparison with deionized water [45]. *GNP*, Graphene nanoplatelet.

Reproduced with permission from Elsevier [45].

among researchers has been piqued because of their superior electrical properties, thermal conductivity, stability, and thermophysical characteristics [52].

Iranmanesh et al. [28] conducted an experimental study on the use of GNPs in an ETSC. The GNPs were characterized using transmission electron microscopy and field emission scanning electron microscopy. Their findings revealed that the thermal efficiency of the ETSC could be enhanced by up to 91% with the use of GNPs. In addition, an increase in nanoparticle concentration was found to augment the thermal energy gain, resulting in higher fluid outlet temperatures. Another study reported an impressive 95% enhancement in thermal efficiency when GNPs were used in methanol-based fluids within an ETSC [53]. Further enhancements to the ETSC have been explored by integrating phase change materials (PCMs) within the tubes, as suggested by Papadimitratos et al. [54]. This approach allows for the storage of energy in the form of latent heat, which can be utilized at a later, more convenient time, such as late evening, to extend the cooling period for a solar water heater system. The exploration of material modifications in ETCs to improve the overall performance of solar systems has been the subject of extensive research. The incorporation of 2DMs, particularly graphene, has demonstrated the potential to elevate the performance of ETCs, emphasizing the promise of advanced materials in enhancing solar thermal technologies.

12.3.3 2D materials in concentrating collectors

2DMs have garnered significant interest because of their exceptional material properties, making them particularly suitable for high-temperature applications. After years of research and development, CSP systems are increasingly being adopted for both industrial and residential uses. These systems are noted for their superior energy storage capabilities compared to other solar technologies. A promising area of research is the substitution of conventional working fluids with nanofluids in concentrating collectors to enhance efficiency, especially for large-scale applications [55]. The PTC is the most commonly used type of concentrating collector. The integration of 2DMs in PTCs not only has the potential to improve thermal efficiency but also to enhance the convective heat transfer coefficient and overall energy consumption. However, it is important to consider that the use of 2DMs in PTCs may lead to an increased pressure drop within the system. Despite this the future outlook for the application of 2DMs in concentrating collectors remains highly optimistic.

An overview of the impact of 2DMs in various solar heating systems indicates that even low concentrations of these materials, when incorporated into different mediums, can lead to significant improvements in thermal efficiency. Table 12.2 summarizes the effects of 2DMs in different solar thermal systems. Although 2DMs are inherently hydrophobic and may offer less dispersion stability, functionalization techniques can enhance their stability in fluids. Increasing the concentration of 2DMs in the base fluid generally improves thermal efficiency, but this comes with the trade-off of increased pumping power.

Most research on 2DMs in FPSC has focused on lower concentrations because of these considerations. In contrast, there has been limited exploration of 2DMs in ETSC and even less in concentrating collectors. This gap in research presents a substantial opportunity for scientists and engineers to investigate the potential of various 2DMs in ETSC and concentrating collectors, opening new avenues for innovation and efficiency improvements in solar thermal technologies.

12.3.4 Solar cooling system

The extensive use of fossil fuels over the years has led to their depletion and has also resulted in significant environmental damage because of the combustion of these fuels. The consequences of burning fossil fuels have been a cause for concern, leading to a rapid increase in the Earth's temperature and contributing to global warming. As a result, there has been a growing demand for air conditioning worldwide. Grid-powered air conditioners have become dominant in the global market, but they put a strain on the electrical grid, especially during peak usage times. To address the imbalance between energy supply and demand, there has been an increase in the number of thermal power stations, which in turn has led to the burning of more fossil fuels. This has created a vicious cycle between power generation and the emission of greenhouse gases. In this context, refrigeration systems powered by solar energy have emerged as a promising alternative to conventional systems, offering a way to reduce or even replace traditional refrigeration under the pressure of environmental protection.

Solar cooling systems operate in two main modes: solar thermal refrigeration and solar PV refrigeration. Recently, solar PV systems have become a strong contender because of the decline in the cost of PV cells, making them competitive with solar thermal cooling systems. Future developments in solar cooling systems are likely to focus on enhancing the absorption machines within these systems, aiming to optimize the balance between thermal input and cooling output. According to Suresh and Salyan [57], GNPs are being explored as energy storage materials because of their importance in solar cooling and heating applications. These materials could help bridge the gap between energy supply and demand, offering new possibilities for the advancement of solar-powered refrigeration systems.

Table 12.2 The effect of 2DMs in different solar thermal systems.

References	Type of study	Application	2DM	Surfactant	Characterization	Important remarks
[6]	Experimental	FPSC	GGNP	No surfactant	Energy Dispersive X-ray Spectroscopy(EDX), Field Emission Scanning Electron Microscope(FESEM)	Highest enhancement of about 23% in collector efficiency
[23]	Experimental and theoretical	FPSC	TEA-GNP	No surfactant	Fourier Transform Infrared Spectrocoty(FTIR), Raman spectra, EDX, UV–Vis, zeta potential, TEM	Highest augmentation of about 10.53% in collector efficiency
[26]	Experimental	FPSC	GNP hybrid	Tw-80	FTIR, XRD, FESEM, High Resolution Transmission Electron Microscope(HRTEM), UV–Vis, EDX	Highest enhancement of 21.9% in collector efficiency
[46]	Experimental and theoretical	FPSC	Pentaethylene glycol-GNP	No surfactant	FESEM, TEM	Highest augmentation of about 13.13% in collector efficiency
[56]	Experimental	FPSC	Graphene	No surfactant	TEM	About 6% augmentation in collector efficiency
[47]	Experimental	FPSC	GNP	Triton-X 100	TEM, zeta potential	Highest augmentation of about 17% in collector efficiency
[29]	Experimental	FPSC	GO	No surfactant	XRD, UV–Vis, SEM	Highest augmentation of about 7.3% in collector efficiency
[27]	Experimental	FPSC	GNP	No surfactant	TEM, XRD, zeta potential	Highest augmentation of 33% in collector efficiency

(*continued on next page*)

Table 12.2 The effect of 2DMs in different solar thermal systems—cont'd

References	Type of study	Application	2DM	Surfactant	Characterization	Important remarks
[28]	Experimental	ETSC	GNP	No surfactant	FESEM, TEM	About 91% collector efficiency was reported
[53]	Experimental	ETSC	GNP	No surfactant	XRD, SEM	About 95% collector efficiency was reported

2DM, *2D material;* ETSC, *evacuated tube solar collector;* FESEM, *field emission scanning electron microscopy;* FPSC, *flat-plate solar collector;* GNP, *graphene nanoplatelet;* GO, *graphene oxide;* TEM, *transmission electron microscopy.*

Otanicar et al. [58] have noted that solar PV systems are projected to emit less carbon dioxide and require a larger collector area than thermal systems, thereby absorbing more radiation per use. Economic analysis is also crucial for the adoption of solar cooling systems, with low initial and maintenance costs being key factors in making them attractive [10]. The integration of solar cooling with solar heating could further enhance benefits, although this presents complex engineering challenges. However, the need for peak electricity for cooling in summer highlights the potential advantages of solar cooling technology. Furthermore, Dutil and Rousse [59] suggest that if solar cooling technologies were to replace conventional methods, there could be a significant reduction in total energy consumption, with estimates of savings of up to 40%–50% in the Mediterranean and some European countries. The incorporation of 2DMs in solar systems is not only beneficial for enhancing heating applications but also offers a broader scope for cooling. It is important to note that in colder regions or during winter, the heat or electricity generated by solar collectors or PV systems can be used for heating water or powering devices in addition to air conditioning.

12.3.5 Energy storage

Energy storage plays a crucial role in modern technologies, providing essential flexibility to energy systems. Its quick and modular flexibilities enhance its attractiveness, allowing deployment according to demand. Energy storage involves accumulating energy produced for later use. This energy, in various forms, is typically stored in a device known as an accumulator. Accumulators are utilized across a range of applications involving electrical, thermal, chemical, and mechanical technologies. Forms of energy that are difficult to store are often converted and stored in more convenient and economical ways. Ensuring continuous availability and managing the variability of energy are major concerns in energy utilization across different applications, making energy storage essential. The primary goal of energy storage is to store energy when it is available and deliver it efficiently when needed.

The variability of solar energy presents a significant challenge for its utilization. Hence an energy storage system is crucial for many solar applications. These systems store excess energy when solar generation exceeds demand and provide sufficient energy when solar energy is unavailable. Energy storage plays a key role in balancing supply and demand, contributing to a more flexible and reliable energy system. Energy can be stored in various forms, including thermal, chemical, electrical, and mechanical. Technologies for storing energy include thermal storage, compressed air, hydrogen, pumped hydroelectric storage, flywheels, and batteries, each suited to specific energy conversion applications.

Solar heating and cooling applications primarily require thermal energy for direct and indirect use. Thermal energy can be stored and utilized in two main forms: sensible heat and latent heat storage. Sensible heat storage involves using a properly insulated tank containing either a working fluid or permeable solid, operating over a range of temperatures. Latent heat storage, on the other hand, employs PCMs. As PCMs melt, they store heat, which is then extracted when they solidify.

PCMs are substances that melt or solidify at specific temperature ranges, enabling them to store or release significant amounts of energy. They undergo phase transitions to provide useful heating or cooling, with energy stored for later use. PCMs are extensively used for latent heat storage across various applications. Available in temperature ranges from 5°C to 200°C, PCMs absorb heat as they melt and release it as they solidify. This property allows for effective energy storage and utilization. For example, ice melting stores heat when heat is added and releases heat when the ice solidifies, providing an example of phase change-based storage systems operating fundamentally at phase change temperatures.

The carbon-based group of nanomaterials is much preferred to be embedded in PCMs to enhance their thermal conductivity and shape stability [37,60]. Importantly, 2DMs like graphene are the most interesting material in the present scenario of research because of their exclusive properties. Graphene-based nanostructures have unique characteristics such as low clogging and erosion, higher heat transport capability, and higher thermal conductivity. Moreover, graphene is a highly dense carbon single atom in sheet form with a honeycomb pattern and is hydrophobic. Good characteristics and chemical composition have made graphene a promising working nanoparticle for extensive engineering applications [61]. In contrast, GO, another 2D nanomaterial, is a unique chemically modified single monolayer graphene containing oxygen-influenced functionalized groups such as carboxyl, carbonyl, epoxides, and hydroxyl. Their chemical investigation reveals that GO's oxygen-to-carbon ratio is about 1:3, respectively.

Energy storage nanocomposites based on 2DMs are prepared by direct blending of 2D nanomaterials in the form of GNP, GO, or functionalized GNPs with PCMs. Functionalized GNP and GNP provide much higher thermal conductivities in the PCMs than GO nanosheets. The thermal properties of PCMs are significantly enhanced using graphene or GO. However, the overall cost of the system must be considered to achieve effective and economical results. It is important to note that the presence of impurities or defects in the functional groups of graphene or GO can lead to a reduction in thermal conductivity values. Two-dimensional carbon nanomaterials, whether single- or multilayer, improve the thermal conductivity of composite PCMs. This improvement is primarily due to the lower thermal resistance at the interface with the matrix PCMs. Graphene's high thermal conductivity, ranging from 3000 to 5000 W/m K, and its large surface area make it an ideal material for this application [62]. It also helps to improve the thermal transport properties of ceramic materials and working fluids as well. Thermal transport properties depend on the quality and crystallinity of the graphene. With just an atom thickness of 2.3% graphene, white light can be absorbed [63].

The incorporation of graphene/GO in the system results in a significant effect in PCM-based solar-to-thermal energy storage. They act as effective photothermal materials for solar thermal energy harvesting systems and enhance the thermal conductivity of PCMs. Moreover, they act as good energy transfer materials from solar receivers to the storage system in the case of solar thermal systems. It has high light absorption and an efficiency rate of light-to-thermal conversion [64]. Comparatively, PCMs with

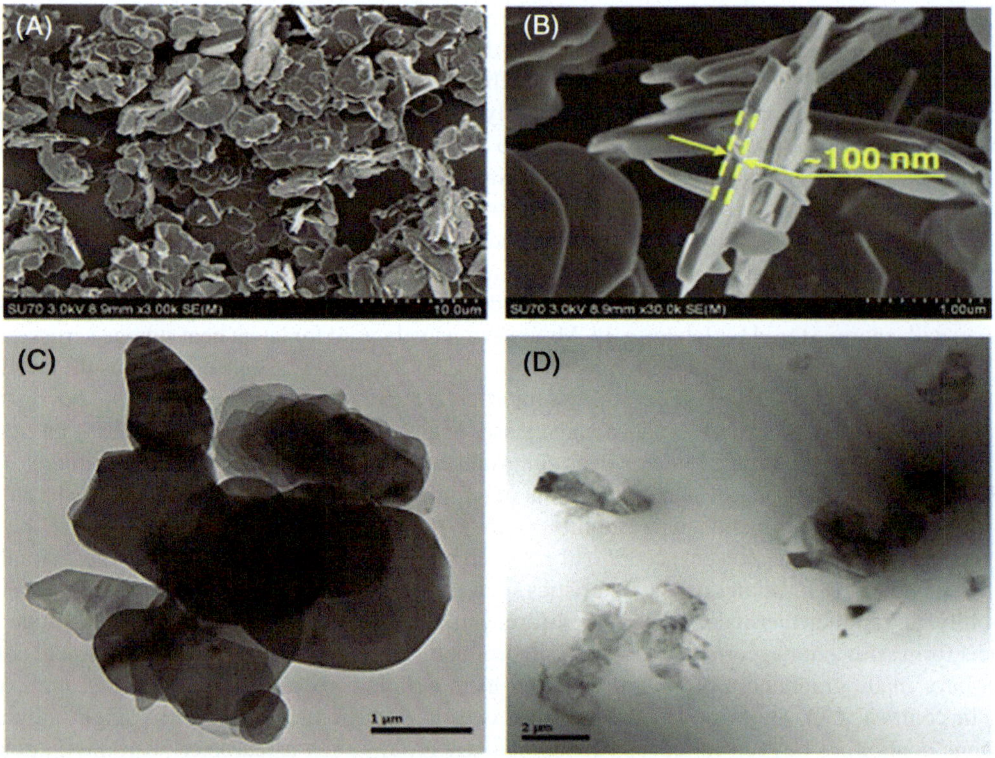

FIGURE 12.8

Morphology of the h-BN nanosheets and its dispersion in PCMs: (A and B) SEM images and (C and D) TEM images of h-BN [65]. *h-BN*, Hexagonal boron nitride; *PCMs*, phase change materials; *TEM*, transmission electron microscopy.

Reproduced with permission from Elsevier [65].

the incorporation of graphene-related nanoparticles result in higher storage densities and conversion rates than conventional solar water systems. Consequently, research on graphene-based PCM storage systems has grown significantly. Recent developments in 2D nanomaterials and their composites toward performance augmentation of PCM systems and their role in energy harvesting and storage systems are presented.

GO nanoparticles blended in different molecular weights of polymeric PCMs such as PEG can enhance the performance of solar thermal energy systems. GO of 4 wt.% was incorporated in PEG-4000 using direct mixing through the microwave method. Intermolecular reactions between PEG and GO cause negligible leakage of the composites during the phase transition. The thermal conductivity of the composites is enhanced even with the addition of less amount of GO. Thus the conversion rate of incident solar radiation to thermal energy and energy storage is increased. Meanwhile, the melting temperature and solidifying temperatures of PEG-4000 are decreased with the addition of GO. An energy storage capacity of about 95% was achieved for the composites in comparison with pure PEG-4000. The addition of more amounts of GO until the optimum amount in PEG results in better storage and release performance of the system [64].

Another 2DM that holds significant promise in enhancing the thermal conductivity of PCMs is boron nitride nanosheets. Hexagonal boron nitride nanosheets (h-BN) share a similar crystal structure with graphene but offer higher thermal conductivity. Because of their economic advantage, h-BN nanosheets are considered potential fillers for the preparation of composites with enhanced thermal conductivity, as they are more cost-effective compared to graphene. A notable investigation into the preparation of composite PCMs using h-BN nanosheets for thermal energy storage has been conducted. As depicted in Fig. 12.8, h-BN nanosheets were added to paraffin to create a composite PCM for thermal energy storage applications. The nanosheets were effectively dispersed within the paraffin PCM. At a concentration of 10 wt.%, h-BN nanosheets with sizes ranging from 0.5 to 2 μm were found to increase the thermal conductivity of the composite by up to 60%. The addition of h-BN nanosheets resulted in only a minor alteration of the temperature range of the PCMs, which was considered negligible. The use of the PCM composite with 2DMs significantly improved the thermal storage rate, highlighting the potential of h-BN nanosheets in enhancing the performance of thermal energy storage systems [65].

Boron nitride (thermal conductivity ranging from 260 to 300 W/m K) can be added to PEG to enhance the thermal conductivity of the system. Applications with BN are limited as they require large amounts of BN to be incorporated with PCMs for overall thermal conductivity to increase, and they are not cost-effective. However, graphene/BN-based hybrid nanoparticles can be embedded with PEG for better results. With just 1 wt.% of graphene incorporated with PEG in the presence of BN, thermal conductivity is enhanced up to 330%. The composite graphene/BN/PEG converts solar radiation to thermal energy at a higher rate. The efficiency of thermal energy conversion was found to be around 72%, mainly because of the presence of graphene. Energy density storage decreased from 180 to 122 J/g in comparison with pure PEG. This reduction is mainly caused because of the presence of BN [66].

12.4 Conclusion

2D nanomaterials have shown great potential for enhancing solar thermal technologies. Graphene and other 2D nanocarbons have been demonstrated to improve the performance of key solar collecting and harvesting devices like flat plates and ETCs when added in small amounts. However, further exploration of 2DM applications in concentrating solar technologies is still needed. PCMs incorporated with graphene and other 2D nanocarbons show promise for energy harvesting and thermal storage. The ability of these 2D fillers to absorb solar radiation and conduct heat allows for enhanced thermal conductivity and energy density in composite PCM systems. Graphene additions also improve the structural stability of PCM composites during phase transition cycles. Optimization of 2D nanomaterial type, concentration, and dispersion achieves tunable melting and crystallization properties. Looking ahead, continued research on 2DM-based solar thermal systems offers opportunities to advance device efficiency, lower costs, and bring new technologies to market. Multiscale theoretical and experimental studies are still required to fully understand structure-property relationships and long-term material stability. Life cycle and techno-economic analyses can guide scalable manufacturing and deployment. With further development, 2D nanomaterials may find widespread roles in next-generation solar energy solutions, including water heating, space conditioning, and industrial processes.

Acknowledgment

The authors gratefully acknowledge the Fundamental Research Grant Scheme awarded by the Ministry of Higher Education Malaysia, grant number FRGS/1/2019/TK03/UM/02/12 (FP143-2019A). The authors are also grateful to RMF0400-2021, ST049-2022, AMMP Centre, CES, Department of Mechanical Engineering, and University of Malaya for support in conducting the research works and dissemination of research output.

References

[1] S. Evans, Coronavirus set to cause largest ever annual fall in CO_2 emissions, Carbon Brief (2020).
[2] H.E. Murdock, D. Gibb, T. André, Renewables 2019 Global Status Report. Vol. 8, Resources, UNEP - UN, Environment Programme, 2019.
[3] IPCC, in: Summary for Policymakers—Global warming of 1.5°C, an IPCC Special Report, IPCC, 2018, p. 32. https://www.researchgate.net/publication/329866816_Summary_for_Policymakers_In_Global_warming_of_15C_An_IPCC_Special_Report%0Ahttps://www.ipcc.ch/sr15/chapter/spm/.
[4] F. Johnsson, J. Kjärstad, J. Rootzén, The threat to climate change mitigation posed by the abundance of fossil fuels, Clim. Policy 19 (2) (2019) 258–274.
[5] United Nations Environment Programme, Towards a zero-emissions, efficient and resilient buildings and construction sector, in: 2019 Global Status Report, UNEP - UN, Environment Programme, 2019, pp. 1–48.
[6] L.H. Kumar, S.N. Kazi, H.H. Masjuki, M.N.M. Zubir, A. Jahan, O.C. Sean, Experimental study on the effect of bio-functionalized graphene nanoplatelets on the thermal performance of liquid flat plate solar collector, J. Therm. Anal. Calorim. [Internet] (2021). https://doi.org/10.1007/s10973-020-10527-y.
[7] R. Pathak, K.C. Pramod, K.H.L. Madhusudhan, N. Giri, S. Uprety, Power generation using parabolic dish CSP technology, AIP Conf. Proc. 2204 (2020) 40023.
[8] A. Alarcón Villamil, J.E. Hortúa, A. López, Comparison of thermal solar collector technologies and their applications, Tecciencia 8 (15) (2013) 27–35.
[9] R. Blaga, A. Sabadus, N. Stefu, C. Dughir, M. Paulescu, V. Badescu, A current perspective on the accuracy of incoming solar energy forecasting, Prog. Energy Combust. Sci. 70 (2019) 119–144.
[10] T.S. Ge, R.Z. Wang, Z.Y. Xu, Q.W. Pan, S. Du, X.M. Chen, et al., Solar heating and cooling: present and future development, Renew. Energy 126 (2018) 1126–1140.
[11] P. Raj, S. Subudhi, A review of studies using nanofluids in flat-plate and direct absorption solar collectors, Renew. Sustain. Energy Rev. 84 (2018) 54–74.
[12] L. Anto Joseph Deeyoko, K. Balaji, S. Iniyan, C. Sharmeela, Exergy, economics and pumping power analyses of flat plate solar water heater using thermal performance enhancer in absorber tube, Appl. Therm. Eng. 154 (2019) 726–737.
[13] A.S. Ahuja, Augmentation of heat transport in laminar flow of polystyrene suspensions. I. Experiments and results, J. Appl. Phys. 46 (8) (1975) 3408–3416.
[14] H. Masuda, A. Ebata, K. Teramae, N. Hishinuma, Alteration of thermal conductivity and viscosity of liquid by dispersing ultra-fine particles. dispersion of Al_2O_3, SiO_2 and TiO_2 ultra-fine particles, Netsu Bussei 7 (4) (1993) 227–233.

[15] S.U.S. Choi, Enhancing thermal conductivity of fluids with nanoparticles, in: American Society of Mechanical Engineers, Fluids Engineering Division (Publication) FED, 1995, pp. 99–105.
[16] M. Sheikholeslami, D.D. Ganji, Heat transfer of Cu-water nanofluid flow between parallel plates, Powder Technol. 235 (2013) 873–879.
[17] N.R. Chowdhury, M. MacGregor-Ramiasa, P. Zilm, P. Majewski, K. Vasilev, 'Chocolate' silver nanoparticles: synthesis, antibacterial activity and cytotoxicity, J. Colloid Interface Sci. 482 (2016) 151–158.
[18] N. Pauzi, N.M. Zain, N.A.A. Yusof, Gum arabic as natural stabilizing agent in green synthesis of ZnO nanofluids for antibacterial application, J. Environ. Chem. Eng. 8 (3) (2020) 103331.
[19] S. Choudhary, A. Sachdeva, P. Kumar, Time-based analysis of stability and thermal efficiency of flat plate solar collector using iron oxide nanofluid, Appl. Therm. Eng. [Internet] 183 (2021) 115931. http://www.sciencedirect.com/science/article/pii/S135943112033413X.
[20] M. Krishnam, S. Bose, C. Das, Boron nitride (BN) nanofluids as cooling agent in thermal management system (TMS), Appl. Therm. Eng. 106 (2016) 951–958.
[21] N.L. Cadena-de la Peña, C.I. Rivera-Solorio, L.A. Payán-Rodríguez, A.J. García-Cuéllar, J.L. López-Salinas, Experimental analysis of natural convection in vertical annuli filled with AlN and TiO_2/mineral oil-based nanofluids, Int. J. Therm. Sci. 111 (2017) 138–145.
[22] X. Li, C. Zou, L. Zhou, A. Qi, Experimental study on the thermo-physical properties of diathermic oil based SiC nanofluids for high temperature applications, Int. J. Heat Mass Transf. 97 (2016) 631–637.
[23] W.S. Sarsam, S.N. Kazi, A. Badarudin, Thermal performance of a flat-plate solar collector using aqueous colloidal dispersions of graphene nanoplatelets with different specific surface areas, Appl. Therm. Eng. 172 (2020) 115142.
[24] R. Sadri, M. Hosseini, S.N. Kazi, S. Bagheri, A.H. Abdelrazek, G. Ahmadi, et al., A facile, bio-based, novel approach for synthesis of covalently functionalized graphene nanoplatelet nano-coolants toward improved thermo-physical and heat transfer properties, J. Colloid Interface Sci. 509 (2018) 140–152.
[25] M.E. Zayed, J. Zhao, Y. Du, A.E. Kabeel, S.M. Shalaby, Factors affecting the thermal performance of the flat plate solar collector using nanofluids: a review, Sol. Energy 182 (2019) 382–396.
[26] O.A. Hussein, K. Habib, A.S. Muhsan, R. Saidur, O.A. Alawi, T.K. Ibrahim, Thermal performance enhancement of a flat plate solar collector using hybrid nanofluid, Sol. Energy 204 (2020) 208–222.
[27] M. Vakili, S.M. Hosseinalipour, S. Delfani, S. Khosrojerdi, M. Karami, Experimental investigation of graphene nanoplatelets nanofluid-based volumetric solar collector for domestic hot water systems, Sol. Energy 131 (2016) 119–130.
[28] S. Iranmanesh, H.C. Ong, B.C. Ang, E. Sadeghinezhad, A. Esmaeilzadeh, M. Mehrali, Thermal performance enhancement of an evacuated tube solar collector using graphene nanoplatelets nanofluid, J. Clean. Prod. 162 (2017) 121–129.
[29] D. Anin Vincely, E. Natarajan, Experimental investigation of the solar FPC performance using graphene oxide nanofluid under forced circulation, Energy Convers. Manag. 117 (2016) 1–11.
[30] J.A. Duffie, W.A. Beckman, Flat-plate collectors, in: Solar Engineering of Thermal Processes, Wiley, 2013, pp. 236–321.
[31] S.P.P. Sukhatme, J.K. Nayak, J.K. Naik, in: Solar Energy Principles of Thermal Collection and Storage, fourth ed, Tata McGraw-Hill Education, New Delhi, 2014, pp. 1–431.
[32] H. Pal Singh, A. Jain, A. Singh, S. Arora, Influence of absorber plate shape factor and mass flow rate on the performance of the PVT system, Appl. Therm. Eng. 156 (2019) 692–701.
[33] A. Shafieian, M. Khiadani, A. Nosrati, Strategies to improve the thermal performance of heat pipe solar collectors in solar systems: a review, Energy Convers. Manag. 183 (2019) 307–331.
[34] H.K. Gupta, A.G. Das, J. Mathur, Investigations for effect of Al_2O_3-H_2O nanofluid flow rate on the efficiency of direct absorption solar collector, Case Stud. Therm. Eng. [Internet]. 5 (2015) 70–78. http://www.sciencedirect.com/science/article/pii/S2214157X15000039.

[35] S. Jaisankar, J. Ananth, S. Thulasi, S.T. Jayasuthakar, K.N. Sheeba, A comprehensive review on solar water heaters, Renew. Sustain. Energy Rev. [Internet] 15 (6) (2011) 3045–3050. http://www.sciencedirect.com/science/article/pii/S1364032111001092.
[36] R. Shukla, K. Sumathy, P. Erickson, J. Gong, Recent advances in the solar water heating systems: a review, Renew. Sustain. Energy Rev. [Internet] 19 (2013) 173–190. http://www.sciencedirect.com/science/article/pii/S1364032112006089.
[37] S.A. Angayarkanni, J. Philip, Review on thermal properties of nanofluids: recent developments, Adv. Colloid Interface Sci. [Internet] 225 (2015) 146–176. http://www.sciencedirect.com/science/article/pii/S0001868615001633.
[38] C. Alvial-Palavicino, K. Konrad, The rise of graphene expectations: anticipatory practices in emergent nanotechnologies, Futures [Internet] 109 (2019) 192–202. http://www.sciencedirect.com/science/article/pii/S0016328718301253.
[39] A.K. Geim, K.S. Novoselov, The rise of graphene, Nat. Mater. 6 (3) (2007) 183–191.
[40] J. Liu, F. Wang, L. Zhang, X. Fang, Z. Zhang, Thermodynamic properties and thermal stability of ionic liquid-based nanofluids containing graphene as advanced heat transfer fluids for medium-to-high-temperature applications, Renew. Energy [Internet] 63 (2014) 519–523. http://www.sciencedirect.com/science/article/pii/S0960148113005326.
[41] V. Ravichandran, S. Vasanthi, S. Shalini, S.A.A. Shah, M. Tripathy, N. Paliwal, Green synthesis, characterization, antibacterial, antioxidant and photocatalytic activity of *Parkia speciosa* leaves extract mediated silver nanoparticles, Results Phys. 15 (2019) 102565.
[42] Y. Si, E.T. Samulski, Synthesis of water soluble graphene, Nano Lett. 8 (6) (2008) 1679–1682.
[43] R. Sadri, M. Hosseini, S.N. Kazi, S. Bagheri, S.M. Ahmed, G. Ahmadi, et al., Study of environmentally friendly and facile functionalization of graphene nanoplatelet and its application in convective heat transfer, Energy Convers. Manag. 150 (2017) 26–36.
[44] W.S. Sarsam, A. Amiri, M.N.M. Zubir, H. Yarmand, S.N. Kazi, A. Badarudin, Stability and thermophysical properties of water-based nanofluids containing triethanolamine-treated graphene nanoplatelets with different specific surface areas, Colloids Surf. A Physicochem. Eng. Asp. 500 (2016) 17–31.
[45] A. Ahmadi, D.D. Ganji, F. Jafarkazemi, Analysis of utilizing graphene nanoplatelets to enhance thermal performance of flat plate solar collectors, Energy Convers. Manag. 126 (2016) 1–11.
[46] O.A. Alawi, H. Mohamed Kamar, A.R. Mallah, S.N. Kazi, N.A.C. Sidik, Thermal efficiency of a flat-plate solar collector filled with pentaethylene glycol-treated graphene nanoplatelets: an experimental analysis, Sol. Energy 191 (2019) 360–370.
[47] S.K. Verma, A.K. Tiwari, D.S. Chauhan, Experimental evaluation of flat plate solar collector using nanofluids, Energy Convers. Manag. 134 (2017) 103–115.
[48] R. Shende, R. Sundara, Nitrogen doped hybrid carbon based composite dispersed nanofluids as working fluid for low-temperature direct absorption solar collectors, Sol. Energy Mater. Sol. Cells [Internet] 140 (2015) 9–16. http://www.sciencedirect.com/science/article/pii/S0927024815001233.
[49] S. Sen Gupta, V. Manoj Siva, S. Krishnan, T.S. Sreeprasad, P.K. Singh, T. Pradeep, et al., Thermal conductivity enhancement of nanofluids containing graphene nanosheets, J. Appl. Phys. 110 (8) (2011) 84302.
[50] A. Amiri, M. Shanbedi, G. Ahmadi, H. Eshghi, B.T. Chew, S.N. Kazi, Microwave-assisted direct coupling of graphene nanoplatelets with poly ethylene glycol and 4-phenylazophenol molecules for preparing stable-colloidal system, Colloids Surf. A Physicochem. Eng. Asp. [Internet] 487 (2015) 131–141. http://www.sciencedirect.com/science/article/pii/S092777571530220X.
[51] M. Mehrali, E. Sadeghinezhad, S.T. Latibari, S.N. Kazi, M. Mehrali, M. Zubir, et al., Investigation of thermal conductivity and rheological properties of nanofluids containing graphene nanoplatelets, Nanoscale Res. Lett. 9 (1) (2014) 15.

[52] M. Mehrali, S. Tahan Latibari, M.A. Rosen, A.R. Akhiani, M.S. Naghavi, E. Sadeghinezhad, et al., From rice husk to high performance shape stabilized phase change materials for thermal energy storage, RSC Adv. 6 (51) (2016) 45595–45604.

[53] M.M. Sarafraz, M.R. Safaei, Diurnal thermal evaluation of an evacuated tube solar collector (ETSC) charged with graphene nanoplatelets-methanol nano-suspension, Renew. Energy [Internet] 142 (2019) 364–372. http://www.sciencedirect.com/science/article/pii/S0960148119305750.

[54] A. Papadimitratos, S. Sobhansarbandi, V. Pozdin, A. Zakhidov, F. Hassanipour, Evacuated Tube Solar Collectors Integrated With Phase Change Materials and Silicone Oil, 2, American Society of Mechanical Engineers, Power Division (Publication) POWER, 2017 V002T09A014.

[55] P. Martínez-Merino, R. Alcántara, T. Aguilar, J.J. Gallardo, I. Carrillo-Berdugo, R. Gómez-Villarejo, et al., Stability and thermal properties study of metal chalcogenide-based nanofluids for concentrating solar power, Energies 12 (24) (2019) 4632.

[56] F.E.B. Bioucas, S.I.C. Vieira, M.J.V. Lourenço, F.J.V. Santos, C.A. Nieto de Castro, Performance of heat transfer fluids with nanographene in a pilot solar collector, Sol. Energy 172 (2018) 171–176.

[57] S. Suresh, S. Salyan, Solar thermal energy storage using graphene nanoplatelets-added phase change materials, in: H. Tyagi, A. Agarwal, P. Chakraborty, S. Powar (Eds.), Applications of Solar Energy. Energy, Environment, and Sustainability, Springer, Singapore, ,2018, pp. 187–205.

[58] T. Otanicar, R.A. Taylor, P.E. Phelan, Prospects for solar cooling—an economic and environmental assessment, Sol. Energy [Internet] 86 (5) (2012) 1287–1299. http://www.sciencedirect.com/science/article/pii/S0038092X12000369.

[59] Y. Dutil, D. Rousse, A review of active solar cooling technologies, Industrial Research Chair in Technologies of Energy and Energy Efficiency, 9, 2010. http://www.t3e.info/pdf/Publications/2010_PALENC_Solar_cooling.pdf.

[60] L. Feng, C. Wang, P. Song, H. Wang, X. Zhang, The form-stable phase change materials based on polyethylene glycol and functionalized carbon nanotubes for heat storage, Appl. Therm. Eng. 90 (2015) 952–956.

[61] C. Amaral, R. Vicente, P. Marques, A. Barros-Timmons, Phase change materials and carbon nanostructures for thermal energy storage: a literature review, Renew. Sustain. Energy Rev. 79 (2017) 1212–1228.

[62] A.A. Balandin, S. Ghosh, W. Bao, I. Calizo, D. Teweldebrhan, F. Miao, et al., Superior thermal conductivity of single-layer graphene, Nano Lett. 8 (3) (2008) 902–907.

[63] A.R. Bin Mohd Yusoff, D. Kim, F.K. Schneider, W.J. Da Silva, J. Jang, Au-doped single layer graphene nanoribbons for a record-high efficiency ITO-free tandem polymer solar cell, Energy Environ. Sci. 8 (5) (2015) 1523–1537.

[64] X. Wei, F. Xue, X.-d. Qi, J.-h. Yang, Z.-w. Zhou, Y.-p. Yuan, et al., Photo- and electro-responsive phase change materials based on highly anisotropic microcrystalline cellulose/graphene nanoplatelet structure, Appl. Energy 236 (2019) 70–80.

[65] X. Fang, L.W. Fan, Q. Ding, X.L. Yao, Y.Y. Wu, J.F. Hou, et al., Thermal energy storage performance of paraffin-based composite phase change materials filled with hexagonal boron nitride nanosheets, Energy Convers. Manag. 80 (2014) 103–109.

[66] J. Yang, L.S. Tang, R.Y. Bao, L. Bai, Z.Y. Liu, W. Yang, et al., Largely enhanced thermal conductivity of poly (ethylene glycol)/boron nitride composite phase change materials for solar-thermal-electric energy conversion and storage with very low content of graphene nanoplatelets, Chem. Eng. J. 315 (2017) 481–490.

[67] H. Kumar, K.S., Aryamol, Green synthesis of carbon-based nanomaterials from plant extracts and their applications in nanofluids, Nanostructured Carbon Materials from Plant Extracts, Elsevier, 2025, pp. 601–634, Submitted for publication.

[68] L.H., Kumar, S.N., Kazi, H.H., Masjuki, M.N.M., Zubir, A review of recent advances in green nanofluids and their application in thermal systems. Chemical Engineering Journal 429, 132321.

[69] L.H., Kumar, S.N., Kazi, H.H., Masjuki, M.N.M., Zubir, A. Jahan, C. Bhinitha. Energy, exergy and economic analysis of liquid flat-plate solar collector using green covalent functionalized graphene nanoplatelets. Applied thermal engineering 192, 116916.

[70] L.H., Kumar, S.N., Kazi, H.H., Masjuki, M.N.M., Zubir. 2D materials in phase change materials. Solar Energy Harvesting, Conversion, and Storage, 143-170.

[71] K.S., Aryamol, L.H., Kumar, S., Thomas, S., Gopi, M.S., Sreekala. Synthesis methods of nanostructured carbon materials. Nanostructured Carbon Materials from Plant Extracts, 27-52

CHAPTER

Heat transfer application of graphene and other 2D material in thermal coating

13

Afrin Jahan[a], Bilal Akbar[b], Naveed Akram[c,d] and Kazi Salim Newaz[d]

[a]*Department of Mechanical Engineering, Faculty of Engineering, Nanotechnology & Catalysis Research Centre (NANOCAT), Institute for Advanced Studies, Universiti Malaya, Kuala Lumpur, Malaysia,* [b]*Department of Mechanical Engineering, Brunel University of London, Uxbridge, London, United Kingdom,* [c]*Department of Mechanical Engineering, Mirpur University of Science and Technology (MUST), Mirpur, Azad Jammu and Kashmir, Pakistan,* [d]*Department of Mechanical Engineering, Faculty of Engineering, Universiti Malaya, Kuala Lumpur, Malaysia*

13.1 Introduction

The field of thermal science and engineering has witnessed significant advancements in heat exchangers, which are essential components in various industrial, commercial, and residential applications. Consistent research has led to the development of different methods for enhancing the heat transfer rate in heat exchange processes. New heat transfer designs are emerging in heating, ventilation, air conditioning, automobiles, aviation, electrical energy storage, power plants, and many other engineering sectors. Hence improving the thermal-hydraulic efficiency of heat exchangers is critical. Active and passive techniques, or a combination of both, can be used to increase efficiency.

The implementation of active methods requires external power to increase the heat transfer. Enhancement techniques, such as mechanical aid, use external power for forceful stirring and an electromagnetic field to apply high voltage and current to the fluid. Other purposes of this external power are surface vibration to increase the flow of fluids to the system, injection or suction of fluid, vibration of fluid, and magnetic field disruption of nanoparticles. Active methods often require mechanical aids to increase their surface area. Hence active approaches have not been widely industrialized because of their high financial cost for pumping and the energy loss associated with vibrations or acoustic disturbances [1]

However, passive heat transfer enhancement methods operate without external power input and rely on structural or material modifications to improve thermal performance. These methods typically include surface modifications—such as extended, roughened, or treated surfaces—as well as the incorporation of nanoadditives into working fluids and geometric alterations of heat pipes. Together, these approaches increase the effective heat transfer area or alter fluid dynamics to improve thermal exchange efficiency. Recently, hybrid systems that integrate both passive and active techniques have gained prominence for achieving optimized performance under varying operational conditions. The design of heat exchangers remains inherently complex because it requires precise evaluation of the heat transfer rate and flow resistance. Performance assessment is often expressed through the overall enhancement ratio, which represents the balance between heat transfer improvement and the associated

Graphene and 2D Materials in Heat Transfer: Fundamentals and Applications. DOI: https://doi.org/10.1016/B978-0-12-821964-5.00013-8

pressure drop. Quantitatively, this is defined as the ratio of the Nusselt number enhancement to the corresponding friction factor increase [2].

$$\text{Overall enhancement ratio} = \frac{\frac{N_u}{N_{u0}}}{\left(\frac{f}{f_0}\right)^{\frac{1}{3}}} \tag{13.1}$$

Hence a higher friction factor ratio causes less heat to be transferred owing to the pressure drop.

Modern nanotechnology in surface modification has generated remarkable results and has great potential for improving the heat transmission in heat exchangers. In single-phase heat exchangers, research has demonstrated that a micro/nanosurface increases heat transmission by generating turbulence and fluid mixing [2]. Coating and roughening of the surface structure on a fine scale, from micro to nanoscale, are examples of surface modification. The roughening or subtractive process usually yields a simple structure, whereas the coating method might yield a more complex structure with more material-selection variables. Although still in its early stages, research indicates that increasing the contact angle between a surface and fluid can significantly reduce frictional losses [3]. This finding highlights the potential of surface coatings as a powerful strategy for application in process engineering. Among the various coating approaches, nanostructured or nanocoatings have emerged as the most effective for enhancing the performance.

Metals, polymers, ceramics, and composites have long been employed as micro- and nanoscale coatings to improve surface performance. Advanced techniques, such as self-assembled monolayers, chemical vapor deposition (CVD), ion implantation, electrodeposition, and electroless plating, enable the fabrication of intricate nanostructures, including porous layers and pyramid-like pillars on heat exchanger surfaces. These nanocoatings have been widely applied to single-phase heat exchangers to enhance their heat transfer efficiency. For instance, Jing et al. demonstrated that a composite coating comprising SiO_2, $BaSO_4$, TiO_2, AlN, and BN significantly improved the overall heat transfer coefficient. In their study, silicon, modified polyester, and amino resins were deposited onto copper heat exchangers, and the superior thermal conductivity of the coating directly contributed to the observed increase in heat transfer performance.

One of the most significant contributions of surface nanocoatings is their ability to protect heat exchangers from corrosion. Corrosion, a redox reaction, leads to the gradual degradation of materials via an electrochemical process. This natural phenomenon incurs a global cost of approximately $2.5 trillion annually, as reported by the National Association of Corrosion Engineers. When combined with the expenses of conventional prevention measures the total cost is approximately 4% of a developed nation's GDP. Therefore advancements in corrosion mitigation can significantly impact the economy. This has attracted considerable attention from researchers, making heat transfer enhancement through nanocoating a prominent focus of research [4].

Over the past few decades a wide range of protective strategies, including polymer coatings, metal multilayers, organic films, and nitride-based coatings, have been investigated; however, none provide a universal solution. The discovery of graphene in 2004 marked a turning point, as this two-dimensional (2D) material rapidly drew attention because of its extraordinary physicochemical properties. The honeycomb lattice of sp^2-hybridized carbon atoms enables diverse coating applications, including single-layer films, multilayer assemblies, functionalized graphene oxide (GO), and composite systems. However, the limitations of graphene, most notably its lack of an intrinsic bandgap, have driven the exploration of alternative 2D materials designed to overcome these drawbacks and expand the scope

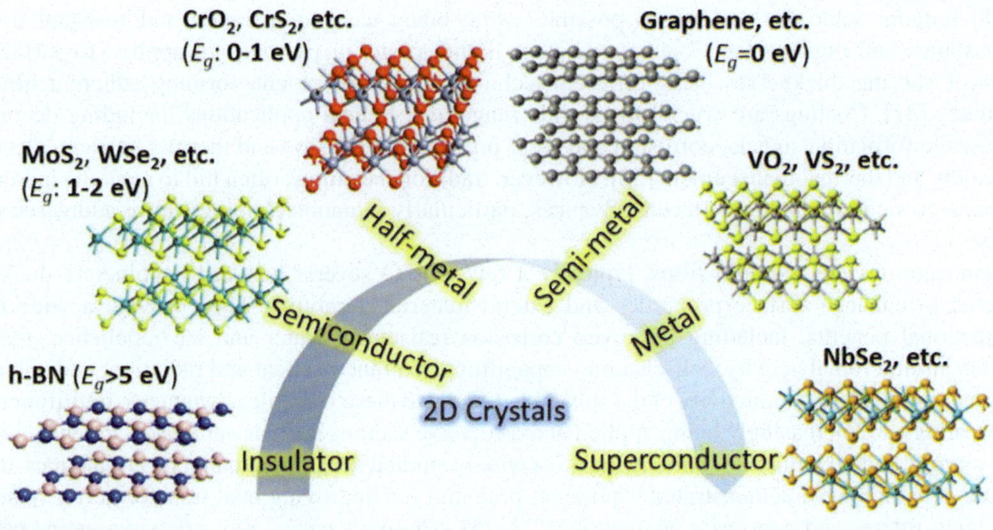

FIGURE 13.1

Atomic configuration of different 2D materials with disparate electrical properties. *2D*, Two-dimensional.

of surface engineering [5–7]. Transition metal dichalcogenides (TMDCs), for example, offer tunable bandgaps, although their charge-carrier mobility remains limited [8]. Other synthetic 2D materials, such as silicane, germanene [9,10], and stanine [11], have further expanded the family of 2D materials [12]. However, unlike graphene and TMDCs, these materials cannot be easily isolated from their bulk forms and must be placed on dielectric substrates for their use in devices.

Ongoing research has also led to the development of additional 2D materials, as shown in Fig. 13.1[13], such as hexagonal boron nitride (h-BN), TMDCs, and MXenes, which have been made feasible through advances in synthesis techniques. These materials exhibit a range of advantageous properties, including a large surface area [14], impenetrability [15], tunable bandgap [16], thermal stability [17], and unique wetting translucency [16,18,19]. Notably, 2D materials are impermeable to even the smallest atoms, such as helium, a trait not observed in traditional materials, making them highly advantageous for corrosion mitigation and passivation. In addition, their ultrathin nature renders them practically invisible, further enhancing their potential for use as nanoscale anticorrosive coatings.

Compared to roughening procedures, coating technology allows for a greater variety of complex surface structures and material selection. Coating processes include CVD, ion implantation, self-assembly, spray coating, dip coating, spin coating, electrodeposition, and electroless coating. Coating processes produce complex surface structures with random distributions of shapes and sizes. The grain size and coating thickness can range from a few nanometers to hundreds of micrometers, depending on the coating process used [20].

13.2 Concept of nanocoating

A surface serves not only as a physical barrier between an object and its environment but also performs essential functions that are critical to product performance. For optimal functionality, surfaces should remain stable for as long as possible while being easy to maintain and resistant to dirt, water stains, and fingerprints. Coatings, whether liquid, paste, or powder, are applied to surfaces in layers of varying thicknesses using different techniques and equipment, forming adherent films on substrates [21]. Coatings are crucial for a wide range of technical applications, including decorative purposes, information signals, corrosion and wear protection, diffusion and thermal barriers, electrical insulation, and thermal management [22]. However, traditional coatings often fail to meet the increasing demands of modern materials. Recent advances, particularly in nanotechnology, have addressed these issues.

Nanocoatings are ultrathin films, typically a few tens to several hundred nanometers thick, engineered to enhance surface properties and extend material durability. They provide a wide range of functional benefits, including improved corrosion resistance, water and ice repellence, reduced friction, antimicrobial activity, self-cleaning capabilities, tolerance to heat and radiation, and enhanced electrical and thermal conductivity of the substrate. Owing to these versatile advantages, multifunctional nanocoatings are increasingly being applied across diverse sectors, such as aerospace, defense, healthcare, marine, and oil industries. Among the most widely studied designs are nanoporous and nanofinned structures, which have demonstrated significant potential for improving heat transfer performance by leveraging micro- and nanoscale engineering [23–25]. These advances offer efficiency gains beyond those of traditional coating technologies, positioning nanostructured coatings as key enablers of next-generation thermal management solutions.

13.3 Benefits of nanocoating

Nanostructured coatings offer distinct advantages over conventional microstructured coatings, primarily owing to their unique structural characteristics and interactions at reduced length scales. One of the primary benefits arises from the surface area effect; as the particle size decreases, the surface area-to-volume ratio increases, enabling nanocoatings to offer significantly larger active surfaces for the same volume. This expanded surface availability enhances heat transfer rates far more effectively than microstructural alternatives. Another critical advantage is the reduction in the pressure drop. When microstructured coatings are applied to microchannels, they often obstruct the flow paths, leading to significant pressure drops. In contrast, nanostructured coatings form ultrathin layers that minimally alter the channel dimensions, thereby keeping the pressure losses negligible.

Nanocoatings also promote capillary rise through their micro- and nanoporous architectures, which function as capillary tubes. This effect induces localized turbulence and supports more efficient fluid film formation, thereby improving the heat transfer performance. Furthermore, the enlarged surface area of the nanostructured layers ensured an extended contact time between the fluid and substrate. Their atomic-scale thickness and strong adhesion to the base material further ensure long-term stability without impairing heat exchanger operation. Finally, the nanostructured surfaces contributed to a reduced contact angle, facilitating enhanced wettability and improved film dynamics. Optimizing the spacing of fluid streams on such surfaces not only elevates the heat transfer efficiency but also mitigates the frictional losses.

These synergistic effects enhance heat transfer, stabilize fluid dynamics, and minimize flow resistance, rendering nanostructured coatings highly advantageous for high-performance thermal management applications.

13.4 Coating materials

The term coating is often used interchangeably with paint; however, in practice, coatings represent a much broader category of materials and their applications. A coating is a protective or decorative layer applied to a surface or substrate, which may be formulated in liquid, paste, or powder form. Paints are one of the most common types of coating; however, the term also encompasses floor coatings, emulsions, fillers, and highly specialized systems designed for particular functions.

Coatings are generally composed of film-forming binders, pigments, solvents, and a range of additives that are used to tailor their performance. Once applied and cured, these components form a thin, continuous film that provides both aesthetic and functional benefits. The selection of coating materials is influenced by multiple factors, including performance requirements, application conditions, film formation mechanisms, design considerations, and specific protective or decorative properties.

Despite their diversity, all coating materials share the following fundamental characteristics: They wet the surface, form an adhesive film, flow to create uniform coverage, and harden to provide mechanical strength and chemical resistance. These features enable coatings to serve as indispensable materials for extending durability, enhancing appearance, and protecting substrates in a wide range of applications [22,26].

The selection of raw materials for pigments, film-forming binders, extenders, solid additives, and plasticizers has traditionally relied on their chemical and physicochemical properties. However, in recent years, increasing attention has been directed toward the role of particle size, shape, and surface interactions in determining material performance. Nanoparticles with structures engineered at the nanometer scale exhibit properties distinct from those of their bulk counterparts and molecular precursors [27]. These nanoscale attributes introduce new dimensions to material design and performance, particularly in the formulation and structural optimization of coatings. Consequently, an emerging segment of the coating industry is being defined by nanospecific innovations and applications.

Nanocoatings, typically comprising a single molecule–thick layer on a substrate, form atomic-scale films that are invisible to the naked eye. Their durability varies with application; for instance, they are stable for months in electronics and years in automotive waxes [22]. As the particle size decreases the surface and interfacial effects become increasingly dominant, imparting size-dependent functionalities. Consequently, nanoparticles can enhance the coating performance by improving the hardness, scratch resistance, antimicrobial properties, ease of cleaning, and thermal conductivity of the coating.

2D materials have attracted considerable interest as emerging nanostructures. Characterized by their atomic-scale thickness and large lateral dimensions, 2D oxides possess exceptionally high aspect ratios and expose a substantial fraction of their surface atoms. Consequently, they exhibit physicochemical properties that differ markedly from those of their bulk counterparts [28]. These materials are commonly obtained from layered precursors, in which strong in-plane covalent bonding and weak interlayer van der Waals interactions enable the exfoliation into nanosheets. Although the ideal structure is a single monolayer, most experimentally accessible 2D oxides consist of a few stacked layers, typically fewer than 10 layers.

Current research has focused on 2D nanomaterials, such as graphene, h-BN, and TMDCs, owing to their exceptional properties and broad applications in electronics, optoelectronics, catalysis, energy storage, sensors, photovoltaics, lithium batteries, and composites. Graphene is an atomically thin sheet of sp^2-hybridized carbon atoms arranged in a honeycomb lattice. It combines an ultrahigh specific surface area, impermeability to gases and water, outstanding electrical and thermal conductivity, high thermal stability, and remarkable mechanical strength. These features have established graphene as a cornerstone material in nanoscience and nanotechnology, spurring intensive exploration across scientific and engineering disciplines.

Similarly, h-BN often referred to as "white graphene" due to its structural similarity to graphene, consists of alternating boron and nitrogen atoms arranged in a honeycomb lattice, forming boron nitride nanosheets [9]. Owing to its exceptionally high in-plane thermal conductivity, h-BN is widely employed in thermal management systems to facilitate efficient heat dissipation. In addition, h-BN coatings act as robust fire-resistant barriers, providing surface protection against environmental degradation [29].

Following the discovery of graphene, TMDs emerged as another prominent class of 2D materials. Similar to graphene, TMDs are atomically thin and highly versatile; however, they exhibit distinct electronic and optical characteristics arising from quantum confinement and strong-surface interactions. Notably, when reduced from bulk to monolayer, many TMDs transition from an indirect to a direct bandgap, thereby enabling tunable electronic properties of the material. This bandgap engineering imparts strong photoluminescence (PL) and large exciton binding energies, positioning TMDs such as MoS_2 and WS_2 as exceptionally promising candidates for the development of a new generation of optoelectronic devices, photonic components, and energy-harvesting systems.

Beyond their intrinsic electronic properties the potential applications of TMDs have been further broadened by their robust environmental stability. For instance, research has demonstrated the efficacy of TMD-based coatings in maintaining performance integrity across a range of humidity conditions, highlighting their suitability for practical and real-world applications [30].

Beyond TMDs the rapidly evolving domain of **nanocoatings** exploits a broad spectrum of nanoscale materials to engineer surfaces with highly specific functions. These include **antimicrobial efficiency, optical transparency, mechanical robustness, and environmental durability**, all of which arise from precise control over nanoscale architecture and interfacial chemistry. The classification of nanocoatings is primarily dictated by their dominant material constituents, with each class exhibiting a distinctive combination of structure-dependent functionalities and application-specific advantages.

Metal-based nanocoatings exhibit superior catalytic, optical, and antimicrobial properties owing to their surface plasmon resonance and redox-active properties. For example, titanium dioxide (TiO_2) exhibits pronounced photocatalytic and self-cleaning performance under UV illumination, enabling its integration into antifogging and UV-protective surfaces. Silver nanoparticles are distinguished by their broad-spectrum antimicrobial action, attributed to ion release and membrane disruption, making them indispensable in bioprotective coatings. Zinc oxide (ZnO) combines UV-blocking efficiency with antibacterial capability, whereas gold nanoparticles contribute exceptional biocompatibility and optical tunability, driving innovations in biosensing and biomedical coatings.

Ceramic-based nanocoatings provide a balance between mechanical resilience and thermal stability. Silicon dioxide (SiO_2) coatings enhance optical devices through antireflective and hydrophobic effects, whereas aluminum oxide (Al_2O_3) provides wear resistance and high-temperature endurance, which are critical for industrial and aerospace protection layers. Polymer-based nanocoatings leverage chain flexibility and chemical inertness to achieve adaptive surface behavior. Polytetrafluoroethylene (PTFE) yields low-friction, nonstick, and hydrophobic coatings, whereas polyurethane offers mechanical elasticity and scratch resistance, facilitating its use in automotive, marine, and consumer applications. Epoxy-based nanocoatings are distinguished by their strong adhesion and corrosion resistance. Their densely crosslinked structures confer chemical resistance and structural integrity in harsh industrial conditions.

Organic and composite nanocoatings, such as organosilanes and sol–gel hybrids, epitomize interfacial engineering at the molecular scale. Organosilane films exhibit hydrophobic, antifouling, and anticorrosive properties, whereas sol–gel coatings integrate ceramic-like hardness with optical transparency, expanding their role in protective and functional thin film applications. Emerging nanomaterials, such as graphene, carbon nanotubes (CNTs), and quantum dots (QDs), represent the frontier of multifunctional coatings. Graphene exhibits ultrahigh tensile strength and electrical conductivity, along with near-impermeable barrier performance. CNT-based coatings reinforce mechanical durability and electron mobility, whereas QDs introduce tunable PL, unlocking pathways for smart coatings, sensors, and optoelectronic interfaces.

13.5 Different coating techniques

Although nanocoating research is advancing rapidly, it remains in its early developmental stages. Owing to their nanoscale dimensions, nanocoatings are invisible to the naked eye and barely perceptible to touch. Despite this, they offer numerous benefits postapplication, such as enhanced scratch resistance, increased hardness, resistance to mold and bacteria, and improved heat transfer properties of the substrate. In addition, nanocoatings address many of the limitations associated with traditional coatings, including poor adhesion to substrates, reduced flexibility, diminished strength, low abrasion resistance, and low durability [28]. Various nanocoating techniques were employed, as described in the following sections.

13.5.1 Sol–gel method

The sol–gel process has emerged as one of the most versatile and effective methods for producing chemically homogeneous coatings and powders, offering broad applications across diverse fields such as optics, electronics, energy, environmental science, biology, and medicine. It is particularly useful for fabricating membranes, separation devices, fuel and solar cells, catalysts, and sensors. Recently, hybrid sol–gel coatings have attracted significant attention because of their multifunctional properties, including self-cleaning, depollution, scratch resistance, antiicing and antifogging, antibacterial activity, UV protection, corrosion resistance, and waterproofing [31].

A bottom-up strategy for the sol–gel technique is illustrated in Fig. 13.2 [32]. This process involves two interconnected steps: sol formation and gelation. The term "sol" refers to a colloidal suspension of hydrolyzed nanoparticles or molecular precursors, whereas the "gel" represents a three-dimensional, cross-linked network formed during polymerization and hydrolysis. Typically, nanoparticles are dispersed in a monomer solution to form a sol, which subsequently undergoes condensation reactions to generate a gel network. Polymers within the system can act as nucleating agents, facilitating crystal growth and infiltrating the crystal layers to produce nanocomposites.

To further enhance material performance, wear-resistant inhibitors, such as inorganic compounds and nanoparticles, have been incorporated into sol–gel systems, significantly improving the corrosion resistance of the resulting films [34]. Thin sol–gel coatings can be deposited onto substrates using techniques such as spin or dip coating, with gelation achieved either by condensation or by the addition of polymers. This flexibility enables precise control over the phase, morphology, and size of the final material, making the sol–gel method a powerful tool for advanced material design [35].

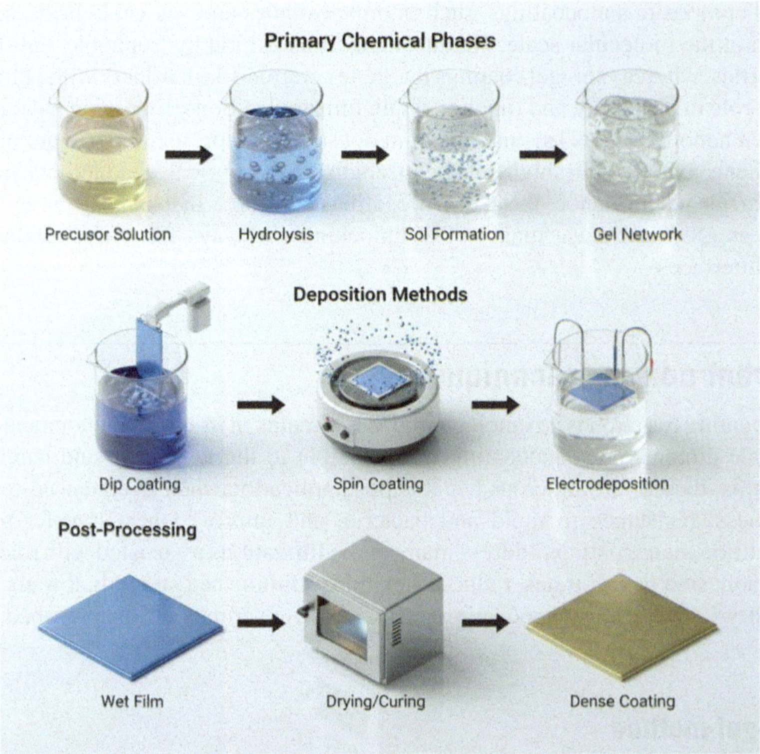

FIGURE 13.2

Processing routes for obtaining sol–gel coatings.

Modified from [33].

13.5.2 Dip coating

Dip coating is a widely used technique for depositing thin liquid films on solid substrates. In this process the substrate is immersed in a colloidal coating medium and withdrawn at a controlled rate, as illustrated in Fig. 13.3 [36]. Upon withdrawal a uniform liquid layer adheres to the surface via a combination of physical and chemical interactions. The wet film subsequently dries, accompanied by solvent evaporation and chemical transformations that promote its solidification. In many cases an additional curing or sintering step is required to ensure complete consolidation and obtain the final functional coating [37].

13.5.3 Spin coating

Spin coating is a widely employed thin-film deposition technique used in semiconductor, nanotechnology, and materials research applications. This method is versatile, enabling the coating of substrates ranging from a few square millimeters to large flat-panel displays with diameters exceeding 1 m. It has been extensively applied to fabricate photoresists, barrier layers, natural and synthetic semiconductors, nanoparticles, metal and metal oxide precursors, and transparent conductive oxides.

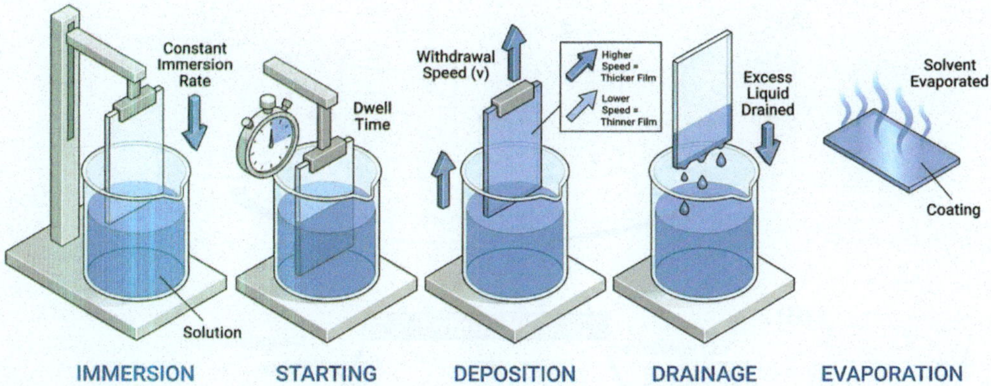

FIGURE 13.3

Dip-coating method with illustration of every step.

Adapted from [36].

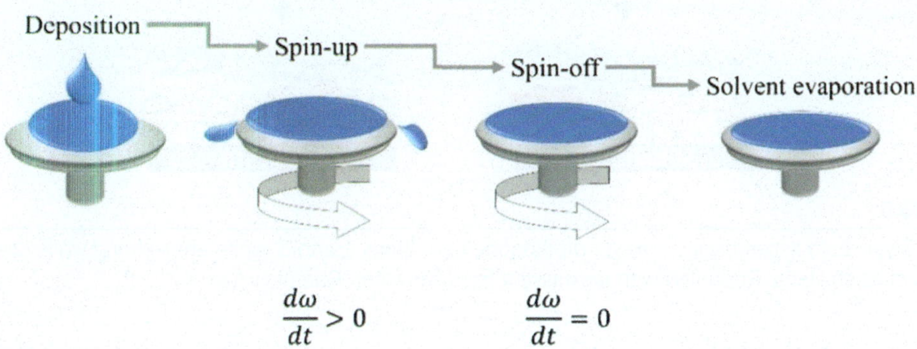

FIGURE 13.4

Spin coating [38].

As illustrated in Fig. 13.4, this process relies on centrifugal force to form a uniform film on a solid substrate. A liquid precursor solution was dispensed at the center of a circular substrate, which was subsequently rotated at high speed. The spinning action drives the liquid outwards, spreading it evenly over the surface. The film thickness, typically ranging from 1 to 10 μm, is governed by the dynamic interplay of centrifugal forces, surface tension, and solvent evaporation at the liquid–vapor interface of the droplet. This balance enables the production of highly uniform coatings, which are essential for fabricating advanced devices.

13.5.4 Layer-by-layer method

The layer-by-layer (LbL) approach is a technique for fabricating thin films. This is a promising method for producing layers of thin films with precisely regulated composition, thickness, and architecture on a

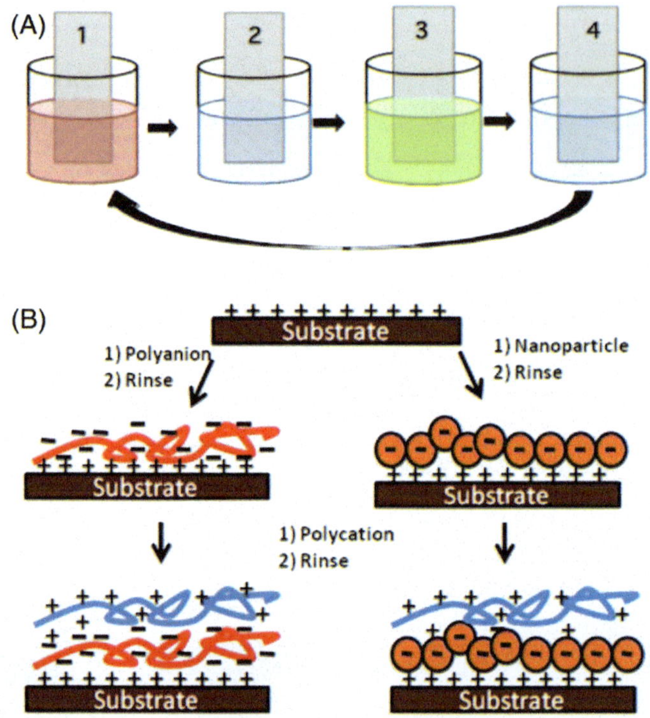

FIGURE 13.5

Layer by layer (LbL) assembly scheme for dipping method. Steps 1 and 3 represent layering while steps 2 and 4 represent wash steps. Reprinted with permission from American Chemical Society [39].

nanoscale, which is an LbL process. Films are created by depositing alternating layers of oppositely charged elements, as shown in Fig. 13.5 [39]. These layers can be applied using various methods, including dipping, spinning, and spray coating. Polyions, metals, ceramics, graphene, nanoparticles, and biological molecules can be deposited using the LbL assembly technique.

13.5.5 Self-assembly coating

Self-assembly is a spontaneous process in which disordered building units organize into larger, well-defined structures through interactions that minimize the free energy of the system. Self-assembly can occur at multiple length scales and manifest in both static and dynamic forms, governed by fundamental physical principles [40]. Early studies have demonstrated that these assemblies produce highly ordered patterns that may be permanent or metastable [41]. For example, when thiols and dithiols are adsorbed onto Au(1[redacted]1[redacted]1), they form highly organized structures. Research indicates that during gas-phase deposition, the growth rate exhibits a complex relationship with pressure, characterized by distinct phases: linear, quadratic, and saturated growth. The most straightforward model of this process

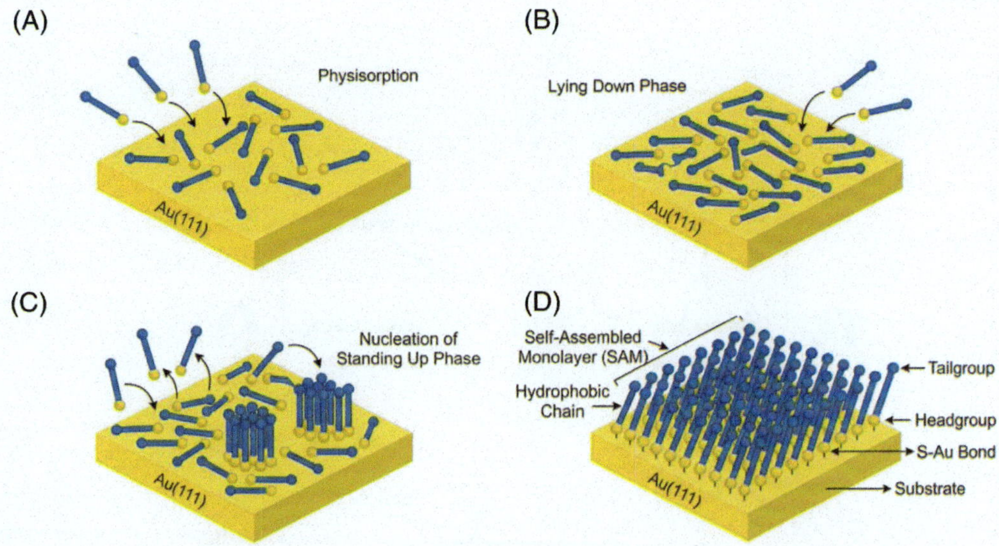

FIGURE 13.6

Scheme of different steps occurring during self-assembly of alkanethiols on Au(1[redacted]1[redacted]1): (A) physisorption, (B) lying downphase formation, (C) nucleation of the standing-up phase, and (D) completion of the standing-up phase.

Adapted from [42].

(Fig. 13.6) suggests an initial step of physisorption, followed by the chemisorption of molecules, culminating in the development of crystalline, ordered domains where the molecules are arranged in a densely packed manner [42].

The controlled self-assembly of micro- and nanoparticles has enabled advances in diverse fields, including nanophotonics, solar cells, catalysis, data storage, and thermal management. Compared with conventional top-down nanofabrication methods, which typically require expensive cleanroom facilities, self-assembly offers a cost-effective alternative. Static self-assembly is generally slow, arising naturally from aging processes that drive systems toward stable configurations via free energy minimization. In contrast, dynamic self-assembly produces patterns on demand, governed by dipole–dipole magnetic interactions between the building blocks and hydrodynamic drag forces [43]. External magnetic fields are frequently employed to enhance control, offering long-range, precise actuation and tunability [44].

13.5.6 Plasma or ion-beam-assisted techniques

Ion-beam-assisted deposition (IBAD) is a sophisticated vacuum-based process for thin-film growth that employs a remote broad-beam ion or plasma source to deliver energetic and reactive ions (typically 100–2000 eV) directly onto the substrate during deposition (Fig. 13.7). This concurrent bombardment alters the fundamental dynamics of film nucleation and growth when combined with conventional

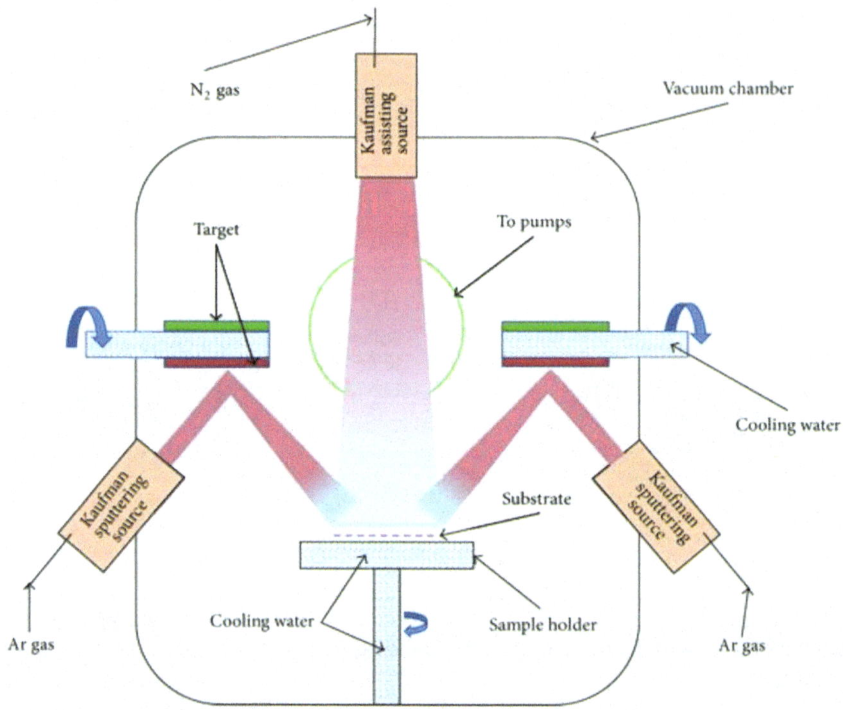

FIGURE 13.7

Schematic of Ion beam assisted deposition system [45].

techniques, such as evaporation or ion-beam sputtering, allowing precise tailoring of microstructural features and material properties. For optical thin films, where precise control of absorption and refraction is essential, IBAD yields significant advantages, including enhanced mechanical durability, reduced internal stresses, and greater resistance to environmental degradation, thereby safeguarding the inherent optical functionality of the deposited materials.

IBAD operates by independently controlling the ion energy, flux, and species while depositing the material via physical vapor deposition (PVD). The ions promote adatom mobility, densification, and intermixing at the film-substrate interface, leading to more uniform and robust layers than those obtained using unassisted methods. Reactive variants, which use gases such as oxygen or nitrogen, further enable the formation of compound films at lower temperatures.

13.5.7 Vapor deposition

Vapor deposition refers to a diverse range of processes, including PVD (thermal and electron beam evaporation, sputtering, laser ablation, IBAD, oblique-angle deposition, and conformal evaporated film by rotation), CVD, atomic layer deposition (ALD), and molecular beam epitaxy.

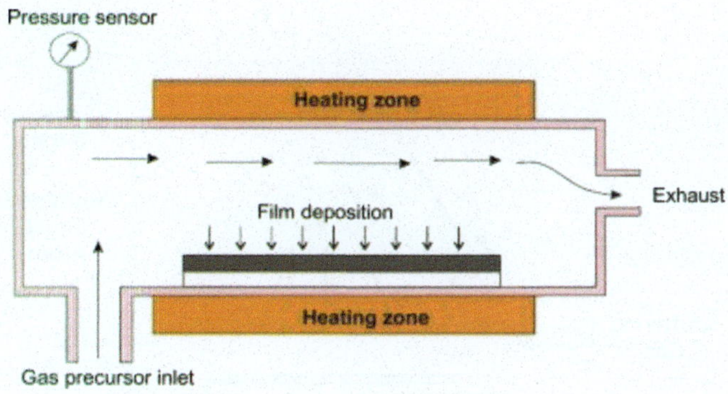

FIGURE 13.8

Illustration of chemical vapor deposition.

Adopted from [46].

13.5.7.1 Chemical vapor deposition

CVD is a process in which gaseous or vapor-phase chemicals react at the gas–solid interface to form solid deposits [46]. CVD can be broadly classified based on the source material and energy used to drive deposition. When heating methods, such as resistance or electromagnetic induction, are employed, the process is referred to as thermal CVD. CVD can be further categorized into thermally activated, laser-assisted, and plasma-assisted techniques [47]. The CVD process occurs in a vacuum chamber under pressures ranging from atmospheric (atmospheric-pressure CVD) to less than 10^{-8} Torr (ultrahigh-vacuum CVD).

The conventional CVD process consists of several key stages, as depicted in Fig. 13.8: (1) transport of reactive gases to the substrate surface, (2) adsorption of gas species onto the surface, (3) surface-catalyzed heterogeneous reactions, (4) diffusion of the adsorbed species to active growth sites, (5) nucleation and film growth on the substrate, and (6) desorption and removal of gaseous byproducts from the surface. Critical parameters, such as the temperature, pressure, reactant gas concentration, and total gas flow, must be carefully controlled and monitored. The chemical reactions catalyzed by the substrate include pyrolysis, oxidation, reduction, hydrolysis, or a combination thereof. The operating temperature is primarily dictated by the nature of the chemical reactions [48,49].

13.5.7.2 Physical vapor deposition

PVD involves the production of a vapor flux, followed by vapor condensation on a substrate in a vacuum chamber to create a thin layer (Fig. 13.9). PVD is an umbrella term that refers to various processes, including thermal and electron-beam evaporation, sputtering, and laser ablation. The primary distinction between these PVD methods is the manner in which vapor flows from a target composed of a particular material [50]. It employs physical processes such as evaporation or sputtering to extract metals from a target source in terms of atoms or molecules and then deposits these energy-carrying vapor particles on the substrate or component through a vacuum or semivacuum space to create a film layer [51].

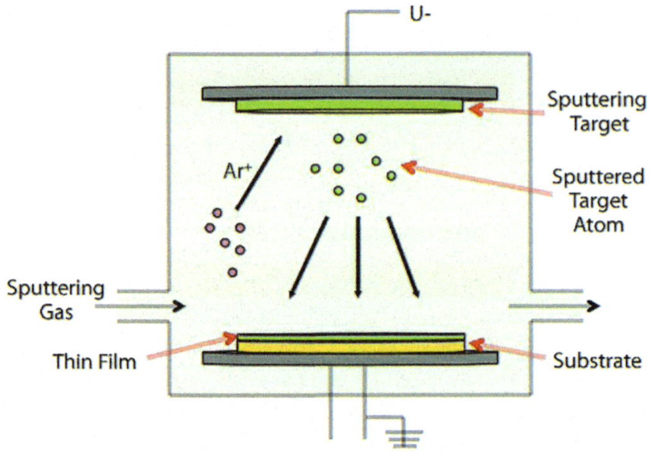

FIGURE 13.9

Schematic view of physical vapour deposition (PVD) [50].

13.5.7.3 Sputtering

Magnetron sputtering is a deposition method (Fig. 13.10) that involves directing intense ions generated in a glow-discharge plasma toward the target, which may consist of an element, alloy, compound, or combination of these [52,53]. Energetic ions, often argon, transfer momentum to target atoms, ejecting them in a process known as sputtering. The ejected atoms form a vapor that condenses on the substrate, thereby creating a thin layer. This process is typically conducted in a vacuum chamber with low-pressure plasma (~5×10^{-3} Torr) or higher pressure plasma (5×10^{-3} to 30×10^{-3} Torr). In addition, ion bombardment leads to the emission of secondary electrons from the target, which play a crucial role in sustaining plasma. In addition, ion bombardment leads to the emission of secondary electrons from the target, which play a crucial role in sustaining plasma.

13.5.8 Laser ablation

Laser ablation, also known as pulsed laser deposition, is the process of irradiating a target with an intense pulsed laser beam, as shown in Fig. 13.11 [54]. When a laser pulse is absorbed by a target, its energy is first transformed into electronic excitation and then into thermal, chemical, and mechanical energy, resulting in evaporation, ablation, plasma production, and exfoliation. The ejected material spreads into the surrounding vacuum as a plume comprising various energetic species, such as atoms, molecules, electrons, ions, clusters, particles, and molten globules. Finally, these species condensed into a thin film on the substrate. Laser ablation is often performed in chambers with high or ultrahigh vacuum. Reactive gaseous species, such as oxygen, can be added for the reactive deposition of oxides or other compounds [55].

13.5.9 Thermal spraying

The term thermal spray refers to a group of coating techniques in which metallic, ceramic, and certain polymeric materials, typically provided in the form of powders, wires, or rods, are thermally energized

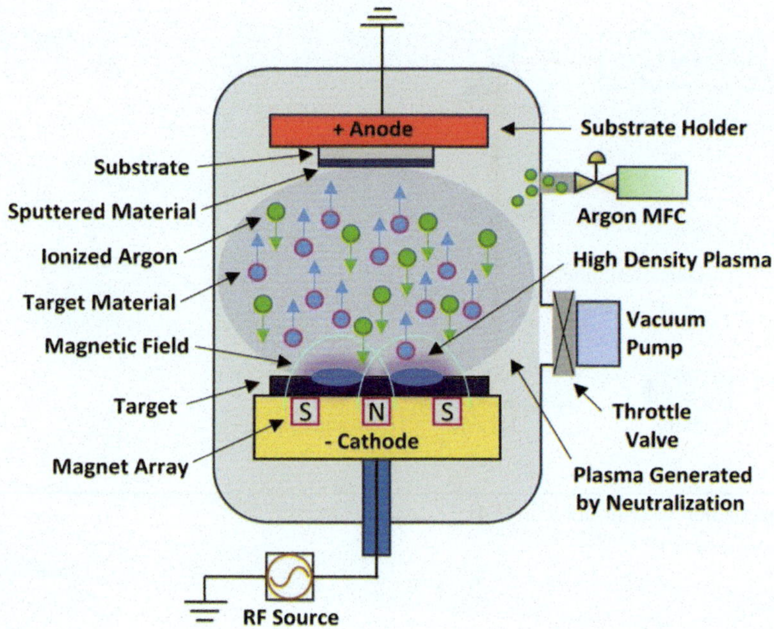

FIGURE 13.10

Illustration of magnetron sputtering [53].

and directed toward a substrate to create a protective or functional layer [56]. In these processes the feedstock is heated, often to temperatures exceeding its melting point, using either electrical sources (such as plasma or arc discharges) or chemical combustion flames [57]. As illustrated in Fig. 13.12, these heating mechanisms differentiate the major subclasses of thermal spray.

Thermal spraying differs from conventional coating technologies, such as electroplating, PVD, and CVD, because of its capacity to produce substantially thicker coatings, ranging from approximately 20 μm to several millimeters, at significantly higher deposition rates [58]. This scalability renders it particularly advantageous for large-area applications that require enhanced wear, corrosion, or thermal resistance. Furthermore, because the substrate experiences minimal thermal loading, the process is compatible with heat-sensitive or flammable materials, which broadens its industrial applicability.

The coating is generated through the successive deposition and solidification of individual molten or semimolten particles on a substrate surface. The resulting microstructure and performance characteristics, such as hardness, adhesion strength, and surface roughness, are strongly influenced by particle velocity and thermal energy during deposition. Empirically, higher particle velocities correlate with denser, more adherent coatings and reduced porosity. The primary categories of thermal spray processes include arc, flame, plasma, detonation gun, and high-velocity oxy-fuel spraying. Each variant offers distinct thermal and kinetic profiles that govern microstructural evolution and ultimately determine coating performance in specific operational environments.

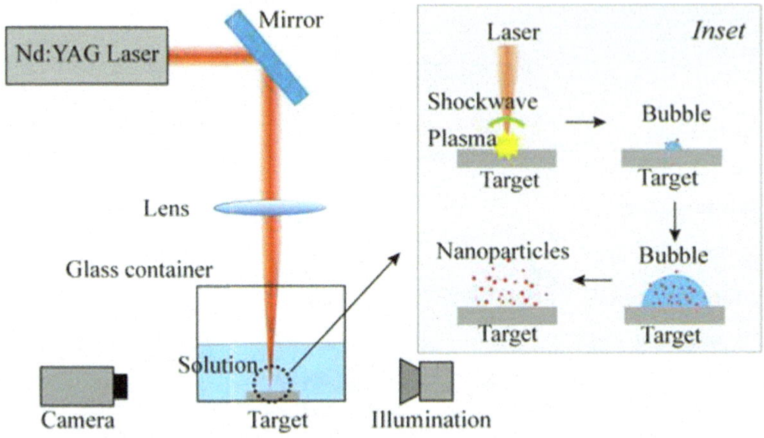

FIGURE 13.11

Schematic representation of laser ablation [54].

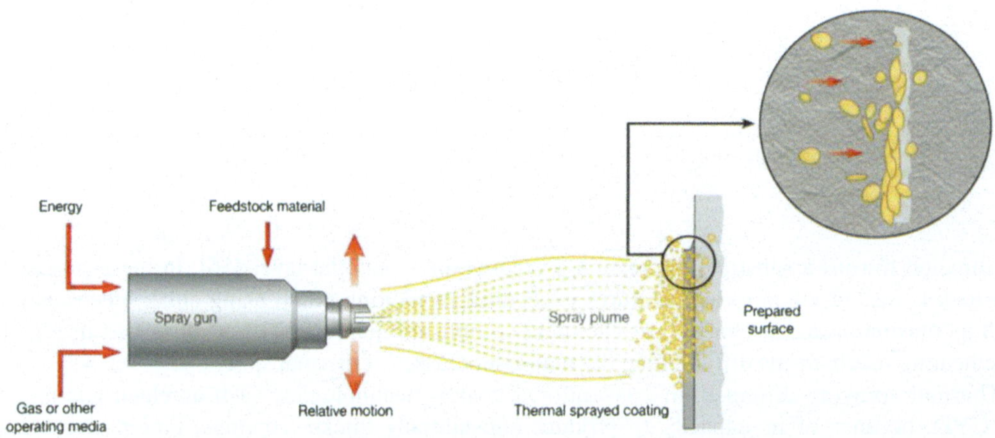

FIGURE 13.12

Principles of thermal spraying.

Adapted from [57].

13.6 Mechanical properties of nanocoating

Industrial protective coatings are systematically designed to satisfy performance benchmarks dictated by their intrinsic chemical composition and microstructural characteristics. These properties critically influence the coating behavior under specific service conditions, determining factors such as resistance to degradation, mechanical integrity, and long-term stability. As illustrated in Fig. 13.13 the coating process followed a controlled sequence to optimize film formation and interfacial bonding. Prior to industrial deployment, coatings undergo panel-based evaluations to quantify their key physical

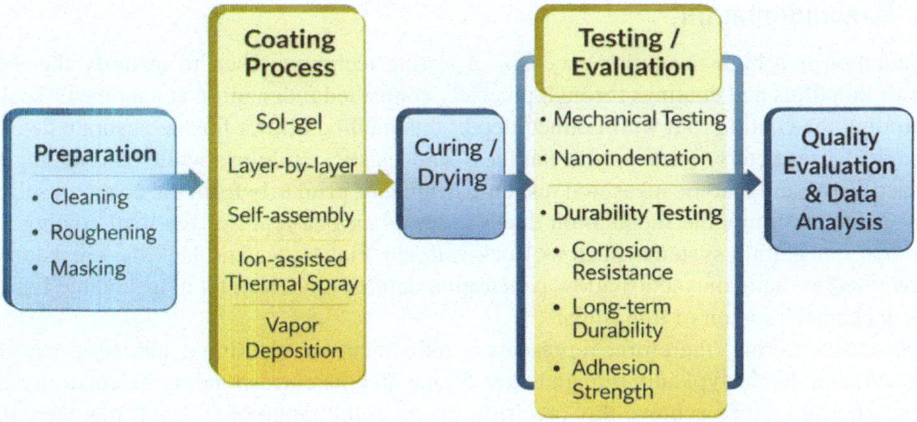

FIGURE 13.13

Process flow chart of coating.

attributes, including indentation resistance, adhesion strength, hardness, and corrosion resistance. Such empirical assessments form the basis for comparative performance analyses, enabling data-driven refinement of formulation parameters. Through iterative testing and performance correlation, engineers have advanced coating technologies to enhance their reliability, efficiency, and life cycle resilience in industrial environments.

13.6.1 Hardness of coating

In various industrial applications the concept of coating hardness is interpreted differently depending on functional requirements and material performance expectations. Hardness testing serves a dual purpose: It not only provides a measure of the mechanical robustness of coatings but also acts as an indirect indicator of the degree of cure or crosslinking in polymeric systems. The experimental evaluation of coating hardness typically involves measuring the indentation resistance across test specimens to determine both the surface and through-thickness hardness profiles of the coating.

Traditional indentation methods, such as Knoop, Vickers, Superficial, and Rockwell tests, share a common principle but differ in terms of load application and geometry of the indenter. Despite their widespread use, these techniques often suffer from operator subjectivity and variability in visual interpretation of indentation dimensions, leading to inconsistencies in reported hardness values. Such limitations can propagate through manufacturing processes, resulting in nonuniform product quality, compromised coating adhesion, and increased failure rates or warranty claims.

In contrast, **instrumented nanoindentation** represents a significant advancement in coating characterization techniques. This technique provides quantitative, high-resolution measurements of the mechanical response under precisely controlled load and displacement conditions. Its automated and repeatable nature minimizes human bias, yielding reproducible data that can be directly correlated with the microstructural and compositional parameters of coatings. Consequently, nanoindentation is a robust analytical tool for elucidating the mechanical integrity and performance of advanced coatings at the nanoscale level.

13.6.2 Nanoindentation

Nanoindentation is a high-resolution mechanical testing technique used to quantify the mechanical behavior of thin films and coatings through precisely controlled indentation at nanometer-scale depths. By penetrating the coating to a well-defined depth, this method allows for the accurate determination of mechanical parameters, such as the hardness and elastic modulus, while minimizing substrate interference. To ensure that the measured response originates primarily from the coating itself a general empirical guideline limits the indentation depth to less than one-tenth of the total coating thickness. Conventional indentation systems, such as Rockwell and Vickers testers, lack the precision and load control required to maintain such shallow penetration depths, rendering them unsuitable for nanoscale mechanical characterization of thin films.

Nanohardness testing, therefore, represents a refinement of traditional hardness measurements, where indentation depths typically remain below 50 nm and the corresponding indentation dimensions do not exceed 150 nm. To achieve this precision, loads in the range of 0.1–100 mN were applied to the indenter using instrumentation capable of load and displacement resolutions of approximately 1 N and 0.2 nm, respectively. The Berkovich diamond indenter is commonly employed because of its well-defined geometry and its mechanical stability. Prior to testing the exact shape of the indenter tip was characterized to enable an accurate calculation of the contact area between the indenter and specimen surface [59].

Beyond conventional hardness determination, nanoindentation provides comprehensive insights into the elastic-plastic behavior of coating materials. Parameters, such as the elastic modulus, fracture toughness, creep resistance, and time-dependent deformation, can be extracted from load–displacement data, offering a multidimensional understanding of coating performance under service conditions. The integration of these parameters is essential for assessing functional performance, particularly for coatings exposed to cyclic or high-stress environments.

Recent studies have highlighted the exceptional toughness and mechanical resilience of graphene-based and other 2D nanocomposite coatings. This improvement is primarily attributed to the uniform dispersion of the nanoparticles within the matrix and the strong interfacial bonding between the reinforcement and host material. Conversely, coatings with higher porosities exhibited reduced hardness owing to stress concentration and void propagation. Therefore optimizing the nanoparticle concentration is critical. Moderate addition of nanoparticles enhances the hardness through load transfer and grain refinement mechanisms, whereas excessive incorporation leads to agglomeration, which disrupts the microstructure and diminishes the overall mechanical integrity of the material [60].

13.6.2.1 Adhesion test

The durability and performance of a coating are profoundly influenced by its adhesion to the substrate, an aspect that is not adequately captured by conventional mechanical parameters such as hardness and elasticity. Therefore it is imperative to employ testing methodologies that accurately replicate real-world service conditions. Among these, **scratch testing** has emerged as a robust instrumented technique capable of evaluating not only the adhesive and cohesive strengths but also the tribological behavior of thin films.

In a typical scratch test a **diamond indenter** traverses the coated surface at a controlled velocity, whereas a normal load is applied either incrementally or continuously, most commonly following a linear increase. This procedure is often referred to as a ramping scratch test. This progressive loading allowed for the precise identification of the critical loads corresponding to the onset of film delamination, cracking, or cohesive failure. The applied forces generally span several tens to hundreds of millinewtons, whereas the resulting displacements typically range from a few nanometers to several hundred nanometers. The Rockwell-type diamond conical indenter remains the most widely utilized geometry because of its reproducibility and well-characterized contact mechanics. However, alternative indenter types, such as Berkovich or cube-corner tips, are occasionally employed to enhance the stress concentration and facilitate the evaluation of brittle or ultrathin coatings [61,62].

13.6.3 Corrosion resistance

Nanocoating technology has emerged as a pivotal strategy for enhancing the corrosion resistance of materials and structural components. These coatings act as protective barriers against aggressive environmental agents, thereby prolonging their service lives and reducing maintenance costs. The corrosion resistance of such coatings is typically evaluated through accelerated testing, most commonly in a controlled salt spray (fog) environment. During this assessment, samples are exposed to a highly corrosive atmosphere for a prescribed duration, after which surface degradation, such as pitting, delamination, or discoloration, is examined to quantify the coating integrity and performance.

A significant advancement in this field is the integration of nanoindentation as a diagnostic tool for the mechanical and microstructural evaluation of protective coatings. The advent of contemporary automated nanoindenters has minimized human error and improved experimental reproducibility by precisely controlling the applied forces and indentation depths. These instruments employ electromagnetic actuation and capacitive depth sensing to characterize both the elastic and plastic responses of materials at the nanoscale, enabling the detailed quantification of hardness, modulus mapping, depth profiling, and time-dependent deformation (creep). Depending on the testing configuration, dual load heads can be employed: a low load (up to 500 mN) for thin coatings and a high load (up to 30 N) for thicker or more robust films.

The nanoindentation technique determines the mechanical behavior of a coating by inducing controlled penetration into a defined fraction of the film's total thickness, which is commonly limited to 1/10 of the coating thickness to prevent the interference of the substrate. Unlike traditional indentation methods, such as Rockwell or Vickers testing, which lack nanoscale precision, nanoindentation derives the hardness and elastic modulus from the slope of the unloading curve (Fig. 13.14).

Modern nanoindentation systems, such as the MML NanoTest, provide exceptional measurement precision, achieving force sensitivities as low as 3 nN and displacement resolutions of 0.001 nm over a broad operational range (0–500 mN load and 0–50 μm displacement). This high level of accuracy enables the comprehensive characterization of a wide variety of material systems, allowing the reliable determination of key mechanical parameters, including hardness, elastic modulus, creep behavior, plastic and contact depths, elastic recovery, and plasticity indices.

Nanoindentation provides a distinct analytical advantage over conventional mechanical testing by simultaneously delivering localized hardness and modulus data from minimal sample volumes. The automation of instrumented indentation testing further enables high-throughput measurements and facilitates statistical robustness. Moreover, as materials transition from bulk to micro- and nanoscales, their mechanical behavior becomes scale-dependent, a phenomenon that is uniquely suited for probing using nanoindentation. Its adaptability to various sample geometries further extends its applicability to advanced composites, thin films, and nanostructured devices.

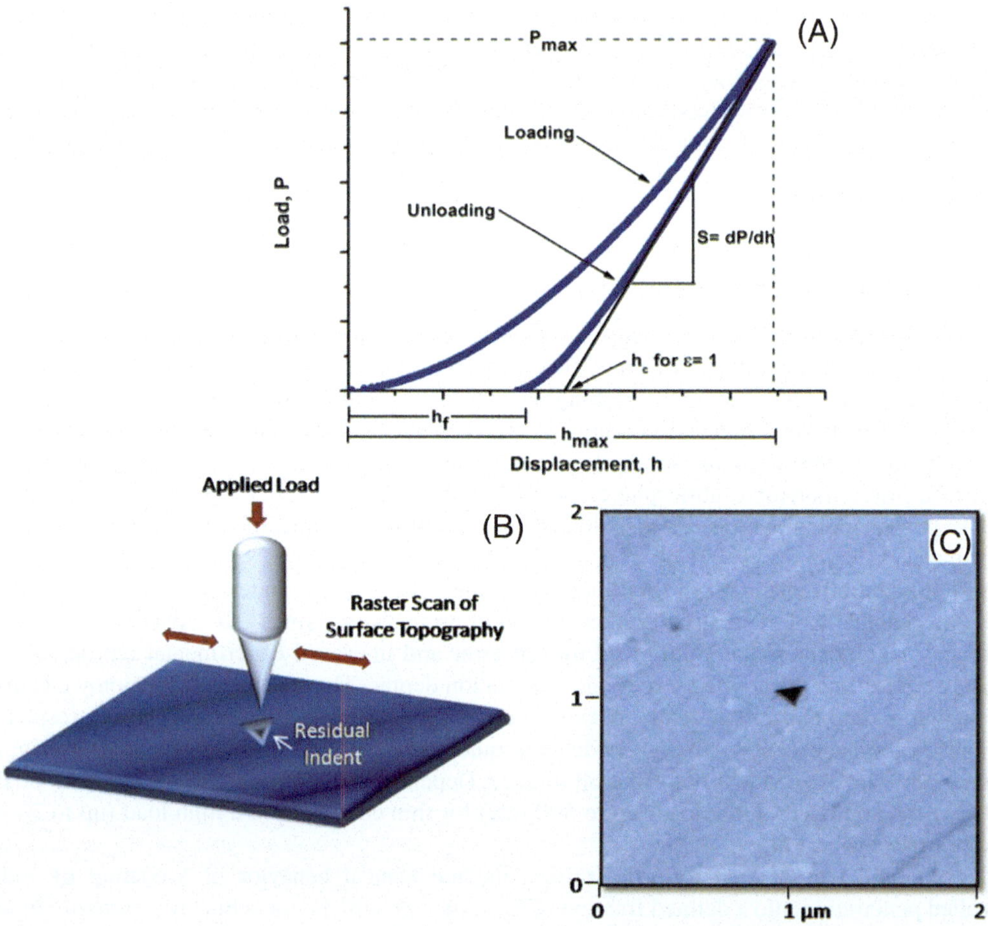

FIGURE 13.14

(A) Schematic of the nanoindenter capable of performing indentation and raster scanning to produce a topographical image. (B) Example of a topographic image created using in situ SPM imaging, showing the sample surface after the nanoindentation test with a visible residual indent impression. (C) Example load-displacement curve from the nanoindentation test showing the measurements required for the calculation of mechanical properties [63].

Despite these advancements the micromechanical characterization of coating systems for predictive-design applications remains a challenge. Parameters such as elastic modulus, coefficient of thermal expansion, inter- and intragranular strength, and fracture toughness are vital for the rational design of **thermal barrier coatings (TBCs)** and **functionally graded materials**. However, their measurement at micrometric scales is constrained by instrumental limitations. The continuing refinement of nanoindentation and complementary microscale test methods is progressively addressing this gap, enabling deeper integration of experimental mechanics with computational material design.

13.7 Application of nanocoating

13.7.1 Anticorrosive coating

Metallic corrosion remains an intrinsic challenge in modern engineering systems, even when protective layers are applied. Environmental factors, such as salt, pH variations, microbial activity, and moisture, accelerate corrosion through thermodynamically driven reactions that drive metals toward their lowest energy states. In corrosive environments rich in water, oxygen, and salts, metals often undergo accelerated degradation [64]. The electrochemical nature of corrosion, which is driven by redox interactions with oxidizing species, leads to progressive structural degradation of the material. Given the ubiquity of metals in critical sectors such as transportation, maritime infrastructure, and public utilities, corrosion represents not only a materials engineering challenge but also a major societal risk, as demonstrated by failures such as the Flint water crisis [65].

Corrosion kinetics depend intricately on both the intrinsic properties of the metal and corrosivity of the environment. Higher corrosion rates correspond to lower corrosion resistance and more aggressive external conditions. Widely adopted strategies to mitigate corrosion include cathodic protection, alloying, and surface coatings. Sacrificial anodes provide galvanic protection by the preferential oxidation of a more active metal; however, their use is often constrained by installation complexities [49]. Conversely, impressed current cathodic protection eliminates the need for sacrificial materials but introduces challenges such as hydrogen embrittlement, current nonuniformity, and electrical hazards.

Surface coatings remain one of the most versatile approaches for isolating metals from corrosive environments. Organic and polymer coatings, including multilayered systems, provide effective physical barriers but often suffer from limitations such as high thickness, thermal/electrical insulation effects, microbial degradation, and microstructural porosity [66], as well as multilayered polymer coatings [67,68] and nitride-based films [69,70]. Each class offers distinct trade-offs. Although polymeric coatings can achieve substantial thicknesses, their high dimensional contribution may alter the thermal and electrical characteristics of the coated components [71]. Furthermore, conventional polymer coatings suffer from porosity, microbial degradation, and thermal instability, all of which compromise their long-term performance [67].

In contrast, nanoscale 2D materials, such as graphene, h-BN, TMDCs, and MXenes, have opened new pathways for nanoscale corrosion protection. These atomically thin layers exhibit exceptional impermeability, mechanical strength, and chemical stability, making them promising candidates for ultrathin coating applications [14] (Kang et al., 2014). The impermeability of graphene to helium atoms (Bunch et al., 2008) represents a paradigm shift in barrier protection. Supplementary properties include tunable electronic bandgaps [16], high thermal conductivity [17], and wetting translucency [18,19]. These characteristics enable the formation of nearly invisible barrier layers capable of obstructing molecular and ionic diffusion to protect against corrosion.

Graphene was initially explored as a direct passivation coating grown on Cu or Ni using CVD. Although effective in short-term protection, long-term exposure often leads to accelerated corrosion owing to galvanic coupling effects and defect-mediated permeation of corrosive species. Consequently, research has shifted toward graphene- or GO-based additives incorporated into polymeric and composite coatings. Functionalized graphene enhances barrier properties by increasing the tortuosity of ionic diffusion pathways and improving coating adhesion. LbL graphene architectures and laser-assisted in situ growth on steel, such as the Ni/Fe catalyst-enabled approach by Ye et al. [72], offer scalable, localized, and efficient routes for integrating graphene into metal protection systems.

The electrochemical fabrication of metal–graphene composite coatings has gained attention because of its simplicity and adaptability. Incorporating GO into commonly used metallic coatings (Ni, Zn, Cu, and their alloys) increases hardness, wear resistance, and corrosion resistance by modifying microstructures and improving passive film stability. Similarly, plasma electrolytic oxidation coatings reinforced with graphene or GO exhibit superior corrosion resistance, particularly in lightweight metals, such as magnesium. In addition, h-BN, a structural analogue of graphene, provides a robust 2D barrier owing to its chemical inertness, high thermal stability, and anisotropic thermal conductivity Monolayer h-BN grown by CVD reduces copper corrosion by an order of magnitude and offers applications in heat management and as gas barriers [55].

13.7.2 Thermal conduction and heat transport through a coated surface

Thermal conduction constitutes the diffusive transport of thermal energy through a material driven by a temperature gradient, fundamentally governed by the material's intrinsic thermal conductivity, *k*. At the atomic scale, heat propagation in nonmetallic solids occurs predominantly via lattice vibrations (phonons), which transport energy as atoms oscillate about their equilibrium positions [73]. The efficiency of phonon-mediated transport is critically dependent on structural order and interatomic bonding characteristics. In highly ordered crystalline solids with strong covalent bonds, such as diamonds, the phonon mean free paths are maximized due to the minimization of Umklapp and boundary scattering, yielding thermal conductivities exceeding 2000 W m^{-1} K^{-1}[74]. Conversely, amorphous materials and liquids lack long-range periodicity, resulting in diffusive, wave-like phonon transport with substantially reduced thermal conductivities (typically <1 W m^{-1} K^{-1}) owing to strong phonon localization and scattering [75].

In coated systems, heat transfer becomes more complex because of the introduction of additional layers and interfaces. These interfaces act as phonon scattering sites and are often associated with mismatches in the thermal diffusivity and vibrational spectra between adjacent materials. Heat transfer proceeds spontaneously from regions of higher temperature to regions of lower temperature until thermal equilibrium is achieved. Under steady-state, one-dimensional conditions, the conductive heat transfer is described by Fourier's law (Eq. 13.2).

$$q = -k\frac{dT}{dx} \tag{13.2}$$

where q is the heat flux (W m^{-2}), k is the thermal conductivity of the material (W m^{-1} K^{-1}), and dT/dx is the temperature gradient in the direction of heat flow. For multilayer-coated architectures the effective thermal resistance R_{tot} accumulates contributions from bulk conduction and interfacial resistance (Eq. 13.3).

$$R_{tot} = \sum_i \frac{L_i}{k_i A} + \sum_j R_{K,j} \tag{13.3}$$

where L_i and k_i denote the thickness and conductivity of layer i, A is the cross-sectional area, and $R_{K,j}$ represents the Kapitza resistance at interface j.

Polymers exhibit intrinsically low thermal conductivities (0.1–0.5 W m^{-1} K^{-1}) due to their disordered chain morphology and weak intermolecular forces. A prevalent strategy to enhance thermal transport in polymeric coatings involves the incorporation of high-conductivity nanofillers, such as graphene, which possesses an exceptional in-plane thermal conductivity of ~3000–5000 W m^{-1} K^{-1} [76,77]. However, the effective thermal conductivity of graphene–polymer nanocomposites often remains substantially below theoretical predictions owing to interfacial thermal resistance (Kapitza resistance), which impedes phonon transmission across the graphene-matrix boundary [78]. This resistance arises from the acoustic mismatch and weak van der Waals coupling, which disrupts the phonon continuity at the interface. Increasing the filler loading can mitigate this limitation by promoting percolating networks that establish filler–filler thermal pathways, thereby reducing the dominance of interfacial resistance in the overall thermal transport [79]. Recent theoretical frameworks incorporating percolation thresholds, Kapitza resistance, and graphene–graphene contact resistance have successfully modeled the nonlinear enhancement of composite thermal conductivity with filler fraction [80].

GO exhibits significantly reduced thermal conductivity (2–10 W m^{-1} K^{-1}) relative to pristine graphene owing to the disruption of the sp^2-hybridized carbon network by oxygen functional groups (epoxy, hydroxyl, and carboxyl) that introduce phonon scattering centers [81]. Thermal reduction of GO to reduced GO (rGO) partially restores the conjugated sp^2 network, yielding conductivity enhancements of up to 30-fold, depending on the reduction temperature and method. High-temperature thermal reduction (>2000°C) has been shown to maximize restoration of the graphitic lattice, achieving rGO thermal conductivities approaching 300 W m^{-1} K^{-1}. Nevertheless, the residual defects and functional groups in rGO maintain substantial phonon scattering, limiting thermal transport relative to that of pristine graphene [82,83].

For coated surfaces employed in phase-change heat transfer applications (e.g., condensation), surface wettability critically influences the thermal performance through its effect on condensate morphology. Nonwetting (hydrophobic) coatings promote dropwise condensation, wherein discrete droplets form and shed rapidly under gravity, minimizing the thermal resistance of the liquid film compared to filmwise condensation on wetting surfaces [84]. The equilibrium contact angle θ governing this behavior is described by Young's equation:

$$\gamma_{SV} - \gamma_{SL} = \gamma_{LV} \cos\theta \tag{13.4}$$

where γ_{SV}, γ_{SL}, and γ_{LV} represent solid–vapor, solid–liquid, and liquid–vapor interfacial tensions, respectively [85]. The reversible work of adhesion follows the Young–Dupré relation (Eq. 13.5).

$$W_A = \gamma_{LV} (1 - \cos\theta) \tag{13.5}$$

Recent experimental studies have demonstrated that strategically patterned nonwetting coatings (e.g., PTFE strips) on condenser tubes can enhance heat transfer coefficients by 35%–50% by directing condensate drainage pathways and promoting early droplet shedding [86]. Advanced nonwetting architectures, including solid-infused slippery surfaces, have achieved a fourfold increase in condensation heat transfer coefficients by sustaining dropwise condensation under high subcooling conditions [87]. These advances highlight the dual role of coatings in thermal management: (1) modifying bulk thermal transport through filler incorporation and (2) engineering interfacial phenomena to optimize phase-change heat transfer.

13.7.3 Heat transfer performance of graphene coating in an antideicing composite

Graphene is being utilized in the area of antiicing and deicing due to its high conductivity. Graphene conducts electricity, and when a voltage is applied to the poles of the graphene coating, the current generates Joule heat, which melts the ice. As a result, electrical energy is transformed into Joule heat and used to melt ice from a stationary helicopter rotor blade [88,89]. Subsequently, by applying the lubricant, a graphene cover with antiicing and deicing properties was created (Clarke & Levi, 2003). Chen et al. (2017) observed that the increasing spraying pressure progressively enhanced the heat conductivity of the graphene covering. Under the same heating conditions as the anti-/deicing component the graphene coating produced with a high spraying pressure had a much higher surface temperature than that produced with a low spraying pressure. The spraying pressure affected the heat conductivity of the graphene covering. The graphene concentration in the unit spraying area increased when the spraying pressure was increased. Thus the heat conductivity of the graphene coating was boosted, whereas the anti-/deicing capacity was enhanced. By increasing the spray pressure a thinner graphene layer can be applied, resulting in increased heat conductivity. Consequently, increasing the spray pressure improves the adhesion of the graphene coating [88,89]. Subsequently, a coating with antiicing and deicing abilities was developed by applying a lubricant on graphene [90].

13.7.4 Heat dissipation coating

High-thermal-conductivity heat dissipation coatings play a pivotal role in thermal management across a broad range of industrial applications. Despite their importance the development of coatings that simultaneously exhibit high thermal conductivity and effective heat dissipation remains a considerable technological challenge [91]. In practical systems, heat is transferred from the heat source [e.g., integrated circuits (ICs), LEDs, or power devices] to the surrounding environment primarily through conduction and convection. Natural convection, although ubiquitous in both natural and engineered systems, generally contributes only modestly to heat removal, particularly in air, where its effectiveness is limited. Under these conditions, thermal radiation becomes increasingly significant and can be comparable to or even exceed natural convective heat transfer, particularly at elevated surface temperatures. When the surface temperature exceeds that of the ambient environment, radiative heat transfer provides an additional and often substantial pathway for energy dissipation. Consequently, the accurate evaluation and optimization of heat-dissipation coatings must consider radiative heat transfer alongside conduction and convection, rather than treating natural convection as the sole cooling mechanism.

A study showed that an aluminum plate coated with graphene nanoparticles and epoxy-polyester blended powder exhibited better heat dissipation. For comparison, the thermal conductivity of aluminum plates was improved from 5 to 6 and 33.3 W m^{-1} K^{-1} for the boron nitride-loaded coating plates and graphene-loaded coating, respectively. The heat dissipation performance of the resulting plates was further investigated under forced and natural convection. Under forced convection the radiative heat transfer coefficient (h_r) of the bare Al plate was approximately 1.8% of the total heat transfer coefficient (h_T), whereas for the graphene-loaded coating, h_r was approximately 16% of h_T. Therefore radiative heat transfer cannot be neglected in heat dissipation through forced convection. Under natural convection the h_r of the bare Al plate was approximately 4% of h_T, whereas the h_r/h_T of the graphene-loaded coating was approximately 33%, indicating that thermal radiation cannot be ignored in the dissipation through natural convection [92].

13.7.5 Heat transfer/removal for electronic devices

The continuous increase in the power density of modern electronic systems has made efficient heat transfer and removal critical design requirements for devices used in energy storage, communication, and information technologies. Effective thermal management is directly linked to device performance, as higher cooling rates enable microprocessors to operate at higher speeds and with greater reliability than conventional cooling methods. In particular, electronic components operating at ultrafast and high frequencies generate substantial amounts of heat within highly confined regions, leading to the formation of localized hotspots where the power density can exceed the average value by more than an order of magnitude [93]. Under such conditions, materials and coatings with insufficient thermal conductivity can cause performance degradation or premature device failure, even when the overall operating temperature of the system remains relatively moderate [94].

Thermal management in electronics is the attempt to keep the temperature of devices, such as ICs, under control through temperature monitoring and hardware cooling. A thermal management setup generally consists of a temperature controller, liquid/air cooling system, heat sink, and thermal interface materials (TIMs). The selection of materials with high thermal conductivity, as well as the design of heat sinks and other thermal management methodologies for hotspot removal, are all part of the thermal management methodologies for hotspot removal. The advancement of these techniques is particularly relevant to the field of materials science. 2D materials open up new possibilities for thermal management [95–98]. Because of their high surface area, good mechanical strength, and tunable physical properties, 2D materials, such as graphene, h-BN, TMDCs, and black phosphorous, have piqued the interest of researchers and engineers since their inception in 2004 [99]. Graphene has been demonstrated to show an ultrahigh thermal conductivity of 4800–5300 W m^{-1} K^{-1}, exceeding that of its bulk counterpart graphite, which already has a high thermal conductivity of 2000 W m^{-1} K^{-1}[100]. Modern electronics are made up of a flat copper plate coated with graphene that is connected to a microprocessor and runs on passive cooling rather than a fan. This new design resulted in a significant reduction in the microprocessor temperature. The new design reduces the temperature by increasing both the surface area of the heat transfer and the thermal conductivity of graphene. Compared with existing active cooling solutions the suggested passive cooling solution has the benefits of noise-free operation, lower energy consumption, and higher reliability.

13.7.6 Thermal barrier coating

Thermodynamic factors impose well-understood, temperature-dependent limitations on the combustion engine efficiency. As a result of the drive to increase the efficiency of aircraft and power plants, engine operation is pushed to higher temperatures. TBCs are critical for the development of this area because they improve operating temperatures by lowering the temperatures encountered by essential engine components, such as turbine blades and rotors, while remaining within material limits [56].

TBCs are thin coatings of a highly insulating material, such as ceramics, that are bonded to a substrate, which is often a metal structure, to protect it against temperature excursions caused by operating circumstances or accidents. TBCs may be used to substantially increase working temperatures for a variety of practical applications, as well as enhance the efficiency and longevity of structural components and machine parts that use thermal energy. TBCs have benefited a variety of sectors, including the aeronautical, aerospace, automotive, and nuclear industries, as well as heavy-duty utilities such as diesel vehicles [101].

13.7.7 Thermal barrier coating in jet engine

Enhancing the operating temperature of gas turbines in jet engines has always been a top priority for industrial manufacturers because it is directly related to the efficiency and performance of the engine. The growth of TBCs is the most recent achievement in the race to increase the turbine entry temperature.

Most gas turbine systems are used in power plants or for aerospace propulsion, such as jet or helicopter engines. The most recent advancements in turbine stage components are increasingly centered on complex coating designs to provide environmental and thermal protection. Materials for high-temperature components are typically selected based on their high-temperature mechanical properties and manufacturability. Therefore surface optimization is required. Environmental damage shield coatings have been developed to extend the life of high-temperature components. Such sealers do not act as inert barriers; instead, they protect by reacting with oxygen to produce a dense oxide scale that prevents damaging species from infiltrating the substrate, 2. The principle of an environmentally protective layer is relatively simple: When a protective material is deposited onto a part using various methods, it must lead to the production of a steadier oxide than the substrate when the component is in contact with its working environment. Chromia and/or silica formers, as well as alumina formers at higher temperatures, are the most commonly used systems by manufacturers. This coating acts as a surface coating, oxidizing instead of the substrate and preserving the integrity of the actual part over time.

13.7.7.1 Thermal barrier coating in gas turbine

During their service life, modern gas turbine blades are exposed to severe thermal, mechanical, and environmental loads. Operating gas temperatures commonly exceed 1000°C, and the continuous drive for higher thermal efficiency has led to a steady increase in the turbine inlet temperatures. Efficiency is primarily improved by increasing the working temperature while minimizing the use of cooling air. To withstand these extreme conditions a combination of advanced strategies has been developed, including the use of high-temperature superalloys, film cooling techniques, and TBC systems [101].

TBCs are applied to turbine blades to provide thermal insulation and protect the underlying substrate from oxidation, hot corrosion, and particulate or debris impact, thereby significantly reducing the risk of component failure [102,103]. In addition to thermal exposure, turbine blades experience substantial centrifugal stress and transient shock loading, which further challenge coating durability. Numerous experimental and field studies have demonstrated that the temperature drop across a TBC system strongly depends on the material characteristics and geometric parameters of the ceramic topcoat, particularly its thickness and thermal conductivity [101]. Ceramic materials have proven to offer superior resistance to high-temperature environments, although their inherent brittleness renders them susceptible to cracking and other microstructural discontinuities under service conditions .

Recently, the partial replacement of metallic components with ceramic materials and ceramic-based composites has represented a major advancement in gas turbine hot-section technology. These material systems offer superior thermal stability and oxidation resistance at high temperatures. In parallel, polymer-based nanocomposites, particularly those reinforced with graphene, have attracted attention because of their enhanced mechanical, thermal, electrical, and gas barrier properties [104,105]. Polyamides are important engineering polymers, and polyamide-based nanocomposites exhibit high stiffness, improved barrier performance, low heat distortion temperatures, and inherent flame resistance [106,107].

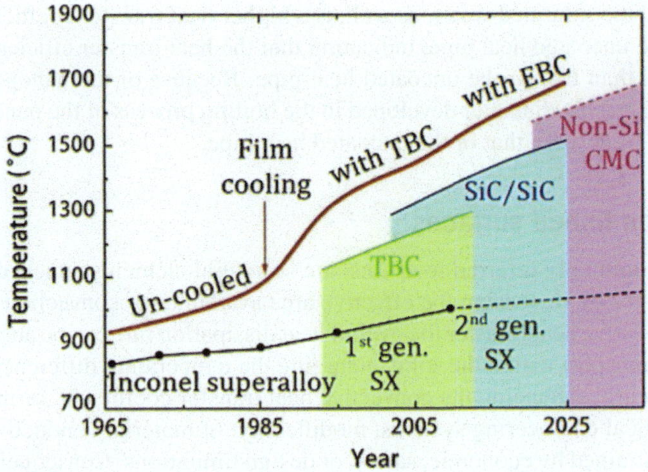

FIGURE 13.15

Evolution of turbine inlet gas temperature with the development of high-temperature technologies, including single-crystal superalloys, ceramic matrix composites, film cooling, thermal barrier coatings, and environmental barrier coatings [108,109]. *CMCs*, Ceramic matrix composites; *EBCs*, environmental barrier coatings; *TBCs*, thermal barrier coatings.

To further reduce metal temperatures and extend component lifetimes, advanced cooling strategies that combine internal heat transfer enhancement, external film cooling, and TBCs are widely employed in turbine hot components. A comprehensive understanding of the interactions among these thermal protection techniques and the resulting conjugate heat transfer behavior is essential for effective turbine design, as illustrated in Fig. 13.15. The evolution of turbine inlet gas temperature with successive generations of high-temperature materials and coatings highlights the importance of integrated approaches [108,109].

The failure of TBC systems is often associated with the formation and growth of thermally grown oxide layers at the interface between the substrate and coating. Conjugate heat transfer studies have shown that the application of TBCs to film-cooled surfaces significantly reduces the substrate temperature, although lateral heat conduction may lead to localized hot spots [110]. Overall, the addition of TBCs markedly improved the cooling effectiveness, particularly when combined with optimized coolant mass flow rates. Although TBCs and modified cooling hole geometries can introduce aerodynamic penalties and additional energy losses, these effects diminish as the coolant flow rate increases [111].

The incorporation of graphene as a nanofiller in TBC systems has been shown to substantially enhance their morphological stability, mechanical strength, and thermal insulation performance. Graphene-reinforced nanocomposites also exhibit improved thermal resistance, reduced flammability, and superior barrier properties, underscoring their potential for next-generation TBC applications [107].

13.7.8 The nanocoated heat pipe

Nanocoated heat pipes exhibit better thermal performance than uncoated heat pipes. The boiling curve for the nanocoated heat pipe showed a higher heat flux at a lower temperature difference between the heat pipe surface and the saturated water, as well as a higher heat transfer coefficient of approximately 215% than that of the uncoated heat pipe, indicating that the heat transfer efficiency of the nanocoated heat pipe was higher than that of the uncoated heat pipe. Because of the nanoporous structure of the nanocoating layer the bubble diameter developed in the boiling process of the nanocoated heat pipe was smaller and released faster than that of the uncoated heat pipe.

13.7.9 Coating on finned surfaces

Extended surfaces, commonly referred to as fins, are structural elements attached to a primary surface to enhance heat transfer by increasing the effective area available for convection with the surrounding environment. In convective heat transfer the overall heat dissipation rate can be augmented by increasing the conductive heat transport within the solid, enlarging the temperature difference between the surface and ambient temperature, enhancing the convective heat transfer coefficient, or increasing the exposed surface area. In practical engineering systems, modification of material conductivity or operating temperature is often constrained by economic, safety, or design limitations. Consequently, the incorporation of fins is a cost-effective and widely adopted approach for improving thermal performance.

Many engineering systems inherently generate significant amounts of heat during operation, which, if not adequately dissipated, can lead to thermal degradation, reduced efficiency, or catastrophic failure. Finned surfaces are extensively employed in applications such as internal combustion engines, heat exchangers, and electronic cooling systems to facilitate effective heat removal. With the increasing demand for compact, lightweight, and high-performance thermal systems, particularly in modern electronics with high packaging densities, considerable research has focused on optimizing fin geometry and surface characteristics to maximize heat transfer while minimizing size and weight.

Recent investigations have demonstrated that surface modification and the use of internal or external fins can substantially enhance the thermal performance. In particular, coating fins with high-thermal-conductivity nanomaterials has emerged as a promising strategy for further improvements. Nanomaterials such as graphene exhibit exceptional thermal conductivity and can reduce thermal resistance at the fin–fluid interface. Accordingly, graphene-based coatings deposited using techniques such as PVD have been explored to fabricate composite fin surfaces with superior heat dissipation characteristics, enabling efficient steady-state heat transfer with minimal additional material or structural complexity.

13.7.10 Thermal barrier coating on cutting tools

During the metal cutting process, approximately 90% of the energy is converted to thermal heat flux, resulting in a severe increase in the cutting temperature in the cutting zone [112]. An increased cutting temperature causes excessive wear on the tool rake and flank faces, reducing the tool life [113,114]. Furthermore, induced thermal softening of the tool and workpiece impairs chemical element diffusion and affects surface durability, dimensional precision, and features of the machined part [91]. Consequently, the cutting temperature is an important metric for predicting tool wear and cutting efficiency [115]. Compared to uncoated tools the coated tools exhibited higher hardness, lower surface roughness, lower thermal conductivity, and lower affinity when in contact with the workpiece.

Knowledge of the coating effects on cutting temperature is important for machining different workpieces by proper selection of tool coating and design of conventional and microcoated tools. Fig. 13.16 shows the current tool coating design [91].

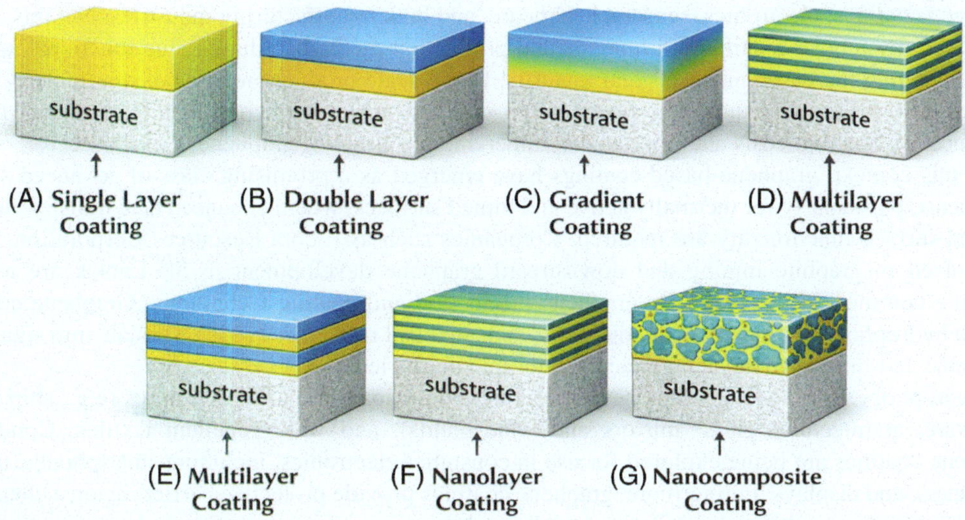

FIGURE 13.16

Current tool coating designs on a substrate: (A) Single layer coating (B) Double layer coating (C) Gradient coating (D) & (E) Multilayer coating (F) Nanolayer coating and (G) Nanocomposite coating [91].

If graphene is used for the tool coating, undesirable hazards such as unexpected heat production due to friction and vibration, wear, shear, tear, surface texture (rough surface), and build-up edge will not occur. In addition, heat generation was significantly reduced because of the lower coefficient of friction. Consequently, the above research shows that graphene is superior to steel and the traditionally used TiN in many ways [116]. It was shown that graphene coating on cutting tools can remove the requirement of coolant by comparing the cutting force, frictional force, and total work done of graphene, steel, and titanium nitride.

13.8 Economic effect of two-dimensional material coating

Graphene, a 2D carbon material, exists in several forms ranging from single atomic layers to multilayer nanoplatelets with diverse morphologies. Its distinctive honeycomb lattice structure imparts exceptional mechanical strength, high thermal and electrical conductivity, optical transparency, and unique electronic and optical properties to the material. Since its experimental isolation in 2004, graphene has attracted sustained scientific and industrial attention, resulting in major advances in scalable and economically viable production techniques as well as a rapid expansion of application-oriented research.

Recent industrial assessments indicate a strong growth trajectory for graphene-based technologies. According to updates from the Graphene Council, more than 2300 graphene-related patents have been granted worldwide, predominantly to industrial organizations and academic institutions. Graphene is projected to impact over 45 industrial sectors, including energy storage, semiconductors, composites, advanced coatings, electronics, sensors, healthcare, and bulk manufacturing industries. Notably, recent patent activity reflects a shift from fundamental production methods to high-value functional applications, signaling increased technological maturity. Large-scale production methods, particularly CVD, have enabled the manufacture of graphene in ton-scale quantities with controlled quality, contributing to reduced production costs and improved commercial feasibility of graphene.

In this context, graphene-based coatings have emerged as a promising class of advanced surface treatments, particularly for thermally active and finned surfaces, where enhanced heat transfer, durability, and surface functionality are required. Companies such as Elcora Resources Corporation, which is involved in graphite mining and downstream graphene development in Sri Lanka, are actively pursuing commercial pathways for graphene production and coating technology. Graphene coatings exhibit hydrophobicity, electrical conductivity, and chemical resistance, which translate into significant economic and functional advantages across multiple industries.

Hydrophobic graphene coatings have potential applications in marine structures (e.g., ship hulls), cookware, architectural glass (mirrors and windshields), and water-repellent textiles. Conductive graphene coatings are being explored for use in consumer electronics, including smartphones, tablets, computers, and displays. Furthermore, graphene coatings provide protective barriers against chemicals, moisture, corrosion, ultraviolet radiation, and fire, while offering biocompatible and corrosion-resistant surfaces that are suitable for medical devices. These multifunctional properties make graphene coatings attractive for aerospace, energy, electronics, and personal care products.

Beyond coatings, graphene applications extend to high-efficiency and flexible solar cells, nanoelectronic devices, supercapacitors, high-sensitivity gas sensors, molecular separation membranes, and medical implants, among others. Although graphene-enhanced products are not yet ubiquitous in the consumer market, continued advances in manufacturing, cost reduction, and application-specific design suggest that graphene-based coatings, particularly on complex geometries such as finned surfaces, are likely to play an increasingly important role in next-generation thermal, structural, and protective systems.

13.9 Current and future potential applications

The industrial coatings market has experienced steady growth in recent years, reflecting broader recovery in global manufacturing. For example, European coating demand increased by approximately 5% in Q1 2021, following a rebound in industrial activity that began in Q3 2020 [117]. Growth was most pronounced in emerging markets, such as China and India, whereas Europe saw moderate expansion and North America exhibited slower growth trends. Graphene and related 2D materials have demonstrated promising, though still limited, applications in thermal management for industrial use. In electronics, graphene coatings have been employed on LED filaments, enhancing heat dissipation and consequently extending their operational lifetime and efficiency. Graphene-based heat spreaders are commercially available with thicknesses ranging from 20 to 100 μm, achieving in-plane thermal conductivities of up to 3200 W m^{-1} K^{-1}[95]. Vertically aligned graphene TIMs have also been developed, providing cross-plane thermal conductivities of up to 1000 W m^{-1} K^{-1}, significantly improving heat transfer in stacked electronic architectures [118]. Graphene-enhanced thermal greases have been reported to achieve thermal conductivities exceeding 13.4 W m^{-1} K^{-1}, offering potential for broader thermal management applications [119].

Consumer electronics have begun integrating graphene films; for instance, the Huawei Mate 20 smartphone was among the first devices to utilize a graphene-based thermal film to reduce the surface temperature perception. However, to achieve meaningful temperature reductions, further improvements in film quality and layer uniformity are necessary. Beyond electronics, graphene and other 2D materials are increasingly being explored for industrial coatings on cutting tools and wear-resistant surfaces, leveraging their mechanical strength and thermal conductivity [120].

Market projections indicate robust growth in graphene-based coatings. According to market research the global graphene coatings market is expected to reach $133.21 million by 2028, growing at a compound annual growth rate of 6.8% from 2021 to 2028 [121]. Future expansion is expected to be driven by applications in electric vehicles, power electronics, 5G/6G infrastructure, and energy storage, where efficient thermal management is a critical enabler. However, widespread adoption depends on overcoming persistent challenges related to scalable production, material standardization, long-term reliability, and integration into existing industrial coating processes.

13.10 Conclusion

Efficient regulation of heat generation, dissipation, and transfer is pivotal to the design and performance of next-generation engineering systems. Contemporary research continues to advance passive thermal management strategies, focusing on material innovations that enhance energy efficiency and the reliability of the system. Among these the optimization of heat transfer fluids using nanoparticle additives and the surface modification of critical components via thin films and coatings have demonstrated measurable improvements in thermal performance. Thin-film and coating technologies are now utilized for both structural protection and enhancing functionality in increasingly challenging thermal conditions. Structural coatings extend the component lifespan by mitigating abrasion, oxidation, and corrosion, whereas functional coatings fine-tune the thermal conductivity, optical response, and electrical or magnetic behavior. Beyond conventional metallic coatings, recent breakthroughs have highlighted 2D materials, such as graphene, h-BN, and TMDs, as transformative candidates for thermal interface engineering. These materials exhibit exceptional in-plane thermal conductivities (often exceeding 2000 W m^{-1} K^{-1} for graphene) and superior chemical stability, making them ideal for use in heat exchangers, energy storage systems, aerospace components, and microelectronic devices. This chapter examines fabrication techniques, including CVD, ALD, and solution-based methods, along with thermomechanical characterization protocols that guide the selection of appropriate coating systems. Emerging evidence suggests that 2D coatings not only enhance thermal transport but also impart notable anticorrosive and antioxidative properties, thereby improving component reliability in aggressive environments. Ultimately, these developments in nanostructured coating technologies present a pathway toward more sustainable, energy-efficient, and cost-effective heat-transfer systems. The widespread adoption of graphene and other low-dimensional materials could yield substantial industrial and environmental benefits, such as reducing material waste, lowering energy demand, and supporting the broader global transition to clean and sustainable manufacturing paradigms.

Acknowledgments

The authors are grateful to the funding authorities of HORIZON-MSCA-2022-SE-01-01, awarded by the European Research Executive Agency with project number 101130406 (ice-Link) (IF049-2024);

RMF 0400-2021; NANOCAT Centre, AMMP, and Energy Centre; and the Department of Mechanical Engineering, University of Malaya, for supporting this research. The authors are also grateful to Brunel University, England, Mirpur University of Science and Technology, and the University of Malaya for their support in this study.

References

[1] M. Jadhav, Awari R., D. Bibe, A. Bramhane, et al., Review on Enhancement of Heat Transfer by Active Method, International Journal of Current Engineering and Technology 6 (2016) 221–225.

[2] S. Liu, M. Sakr, A comprehensive review on passive heat transfer enhancements in pipe exchangers, Renew. Sustain. Energy Rev. 19 (May) (2013) 64–81, doi:10.1016/j.rser.2012.11.021.

[3] D.H. Nguyen,, H.S. Ahn, A comprehensive review on micro/nanoscale surface modification techniques for heat transfer enhancement in heat exchanger, International Journal of Heat and Mass Transfer 178 (2021) 121601, doi:10.1016/j.ijheatmasstransfer.2021.121601.

[4] R.K. Singh Raman, A. Tiwari, Graphene: the thinnest known coating for corrosion protection, JOM 66 (2014) 637–642, doi:10.1007/s11837-014-0921-3.

[5] T. Niu, J. Zhang, W. Chen, Surface engineering of two-dimensional materials, ChemNanoMat 5 (1) (2019) 6–23, doi:10.1002/cnma.201800181.

[6] D. Zhang, F. Peng, X. Liu, Protection of magnesium alloys: from physical barrier coating to smart self-healing coating, J. Alloys Compd. 853 (2021) 157010, doi:10.1016/j.jallcom.2020.157010.

[7] G.R. Bhimanapati, Z. Lin, V. Meunier, Y. Jung, J. Cha, S. Das, D. Xiao, Y. Son, M.S. Strano, V.R. Cooper, L. Liang, S.G. Louie, E. Ringe, W. Zhou, S.S. Kim, R.R. Naik, B.G. Sumpter, H. Terrones, F. Xia, … J.A. Robinson, Recent advances in two-dimensional materials beyond graphene, ACS Nano 9 (12) (2015) 11509–11539, doi:10.1021/acsnano.5b05556.

[8] M. Chhowalla, Z. Liu, H. Zhang, Two-dimensional transition metal dichalcogenide (TMD) nanosheets, Chem. Soc. Rev. 44 (9) (2015) 2584–2586, doi:10.1039/c5cs90037a.

[9] R. Zhang, X. Yu, Q. Yang, G. Cui, Z. Li, The role of graphene in anti-corrosion coatings: a review, Constr. Build. Mater. 294 (66) (2021) 123613, doi:10.1016/j.conbuildmat.2021.123613.

[10] L. Li, S.Z. Lu, J. Pan, Z. Qin, Y.Q. Wang, Y. Wang, G.Y. Cao, S. Du, H.J. Gao, Buckled germanene formation on Pt(111), Adv. Mater. 26 (28) (2014), doi:10.1002/adma.201400909.

[11] F. Zhu, W. Chen, Y. Xu, C. Gao, D. Guan, C. Liu, D. Qian, S.-C. Zhang, J.-f. Jia, Epitaxial growth of two-dimensional stanene, Nat. Mater. 14 (10) (2015) 1020–1025.

[12] M. Ezawa, Monolayer topological insulators: silicene, germanene, and stanene, J. Phys. Soc. Jpn. 84 (12) (2015) 1–11, doi:10.7566/JPSJ.84.121003.

[13] M.R.E. Tanjil, Y. Jeong, Z. Yin, W. Panaccione, M.C. Wang, Ångström-scale, atomically thin 2D materials for corrosion mitigation and passivation, Coatings 9 (2) (2019), doi:10.3390/coatings9020133.

[14] H.K. Chae, D.Y. Siberio-Pérez, J. Kim, Y.B. Go, M. Eddaoudi, A.J. Matzger, M. O'Keeffe, O.M. Yaghi, A route to high surface area, porosity and inclusion of large molecules in crystals, Nature 427 (6974) (2004) 523–527, doi:10.1038/nature02311.

[15] J. Scott Bunch, S. S Verbridge, J. S Alden, A. M van der Zande, J. M Parpia, H. G Craighead, P. L McEuen, Impermeable atomic membranes from graphene sheets, Nano Lett. 8 (8) (2008) 2458–2462, doi:10.1021/nl801457b.

[16] T. Chu, H. Ilatikhameneh, G. Klimeck, R. Rahman, Z. Chen, Electrically tunable bandgaps in bilayer MoS_2, Nano Lett. 15 (12) (2015) 8000–8007, doi:10.1021/acs.nanolett.5b03218.

[17] H.Y. Nan, Z.H. Ni, J. Wang, Z. Zafar, Z.X. Shi, Y.Y. Wang, The thermal stability of graphene in air investigated by Raman spectroscopy, J. Raman Spectrosc. 44 (7) (2013) 1018–1021, doi:10.1002/jrs.4312.

[18] C.J. Shih, M.S. Strano, D. Blankschtein, Wetting translucency of graphene, Nat. Mater. 12 (10) (2013) 866–869, doi:10.1038/nmat3760.

[19] J. Rafiee, X. Mi, H. Gullapalli, A.V. Thomas, F. Yavari, Y. Shi, P.M. Ajayan, N.A. Koratkar, Wetting transparency of graphene, Nat. Mater. 11 (3) (2012) 217–222, doi:10.1038/nmat3228.

[20] A. Ashraf, Y. Wu, M. Cai Wang, K. Yong, T. Sun, Y., T. Jing, R.R. Haasch, N. Aluru, S Nam, Doping-induced tunable wettability and adhesion of graphene, Nano Lett. 16 (7) (2016) 4708–4712, doi:10.1021/acs.nanolett.6b02228.

[21] K.J. Karger, in: Paints, coatings and solvents, 51, Wiley -VCH, 1994, pp. 613–614.

[22] H.-J. Streitberger, A. Goldschmidt, BASF Handbook Basics of Coating Technology, 2019, doi:10.1515/9783748600251.

[23] S. Kalaiselvam, M.S. Gugan, E. Kuraloviyan, R. Meganathan, A. Niruthiya Priyan, M.R. Swaminathan, Experimental investigation of anodized/spray pyrolysed nanoporous structure on heat transfer augmentation, J. Therm. Sci. 18 (4) (2009) 358–363, doi:10.1007/s11630-009-0358-0.

[24] W. Lin, R. Zhang, K.S. Moon, C.P. Wong, Synthesis of high-quality vertically aligned carbon nanotubes on bulk copper substrate for thermal management, IEEE Trans. Adv. Packag. 33 (2) (2010) 370–376, doi:10.1109/TADVP.2009.2034335.

[25] M.S. Sarwar, Y.H. Jeong, S.H. Chang, Subcooled flow boiling CHF enhancement with porous surface coatings, Int. J. Heat Mass Transfer 50 (17–18) (2007) 3649–3657, doi:10.1016/J.IJHEATMASSTRANSFER.2006.09.011.

[26] P. Mischke, Film Formation, 2019, doi:10.1515/9783748602262.

[27] B. Nadeem, K. Irshad, F. Wail, Nanomaterials: a review of synthesis methods, properties, recent progress, and challenges, Mater. Adv. 2 (2024) 1821–1871, doi:10.1039/D0MA00807A.

[28] M.D. Faysal Hossain, N. Akther, Y. Zhou, Recent advancements in graphene adsorbents for wastewater treatment: current status and challenges, Chin. Chem. Lett. 31 (10) (2020) 2525–2538, doi:10.1016/j.cclet.2020.05.011.

[29] A.L. Davesne, S. Lazar, S. Bellayer, S. Qin, J.C. Grunlan, S. Bourbigot, M. Jimenez, Hexagonal boron nitride platelet-based nanocoating for fire protection, ACS Appl. Nano Mater. 2 (9) (2019) 5450–5459, doi:10.1021/acsanm.9b01055.

[30] M. Evaristo, A.E. Tomas, P. Ae, A. Cavaleiro, Tribological behaviour of C-alloyed transition metal dichalcogenides (TMD) coatings in different environments, Int. J. Mech. Mater. Des. 4 (2008) 137–143, doi:10.1007/s10999-007-9034-2.

[31] R. Zandi-Zand, A. Ershad-langroudi, A Rahimi., J. Non-Cry Sol. 351 (2005) 1307–1311, doi:10.1016/j.jnoncrysol.2005.02.022.

[32] V. Madhavi, A.V.B. Reddy, B.K. Chennuri, Insights into nanocomposite materials for gas sorption applications, Nano-Structures and Nano-Objects, 38, 2024, doi:10.1016/j.nanoso.2024.101127.

[33] G.J. Owens, R.K. Singh, F. Foroutan, M. Alqaysi, et al., Sol-gel based materials for biomedical applications, Progress in Materials Science 77 (2016) 1–79, doi:10.1016/j.pmatsci.2015.12.001.

[34] B.A., Rasool Hassan, Z.B., Mohd Yusoff, M., Azmi, & S., Bin, (2013). Supportive and Palliative Care in Solid Cancer Patients. In Cancer Treatment - Conventional and Innovative Approaches. InTech. https://doi.org/10.5772/55358.

[35] M. Joshi, B. Adak, Advances in nanotechnology based functional, smart and intelligent textiles: a review, Compr. Nanosci. Nanotechnol. 1–5 (2019) 253–290, doi:10.1016/B978-0-12-803581-8.10471-0.

[36] I.A. Neacşu, A.I. Nicoară, O.R. Vasile, B.Ş. Vasile, Inorganic micro- and nanostructured implants for tissue engineering, in: Nanobiomaterials in Hard Tissue Engineering: Applications of Nanobiomaterials, 2016, pp. 271–295, doi:10.1016/B978-0-323-42862-0.00009-2.

[37] P. Adámek, Construction of dip-coater, Edukacja – Technika – Informatyka 16 (2) (2016) 152–156, doi:10.15584/eti.2016.2.20.

[38] S. Jamileh, S.B. Dilpreet, M.S. Nicole, M.B. Teresa, L.K. Autumn, The spin-coating method improved the performance characteristics of the films obtained from poly(lactic acid) and cellulose nanocrystals, Sustain. Mater. Technol. (2020) e00212, doi:10.1016/j.susmat.2020.e00212.

[39] S. Srivastava, N.A. Kotov, Composite Layer-by-Layer (LBL) assembly with inorganic nanoparticles and nanowires. Acc. Chem. Res., 41 (12) (2008) 1831–1841.

[40] B. Wang, Y. Zhang, L. Zhang, Self-assembly of nanoparticles: Static and dynamic, Materials Today 25 (2019) 112–113, doi:10.1016/J.MATTOD.2019.04.001.

[41] G.M. Whitesides, B. Grzybowski, Self-assembly at all scales, Science 295 (2002) 2418–2421, doi:10.1126/science.1070821.

[42] C. Vericat, M.E. Vela, G. Benitez, P. Carro, R.C. Salvarezza, Chem. Soc. Rev. 39 (2010) 1805–1834, doi:10.1039/B907301A.

[43] Z. Nie, A. Petukhova, E. Kumacheva, Properties and emerging applications of self-assembled structures made from inorganic nanoparticles, Nat. Nanotechnol. 5 (1) (2010) 15–25, doi:10.1038/nnano.2009.453.

[44] M. Kuang, J. Wang, L. Jiang, Bio-inspired photonic crystals with superwettability, Chem. Soc. Rev. 45 (24) (2016) 6833–6854, doi:10.1039/c6cs00562d.

[45] H. Shang, J. Li, T. Shao, Mechanical properties and thermal stability of TiN/Ta multilayer film deposited by ion beam assisted deposition. Adv. Mat. Sci. Eng. (2014). 1–8, doi:10.1155/2014/639461.

[46] Q. Zhang, D. Sando, V. Nagarajan, Chemical route derived bismuth ferrite thin films and nanomaterials, J. Mater. Chem. C 4 (2016) 4092–4124.

[47] Y. Kajikawa, Roughness evolution during chemical vapor deposition, Mater. Chem. Phys. 112 (2) (2008) 311–318, doi:10.1016/J.MATCHEMPHYS.2008.06.008.

[48] M. Tortello, I. Pasternak, K. Zeranska-Chudek, W. Strupinski, R.S. Gonnelli, A. Fina, Chemical-vapor-deposited graphene as a thermally conducting coating, ACS Appl. Nano Mater. 2 (5) (2019) 2621–2633, doi:10.1021/acsanm.8b02243.

[49] M. Mohamed, M.N. Omar, M.S.A. Ishak, R. Rahman, N.Z. Yahaya, M.K.A. Abdul Razab, M.Z. Ahmad Thirmizir, J. Karger-Kocsis, C.S. Oon, S.N. Kazi, M.A. Hakimin, A.H. Abdelrazek, A.R. Mallah, F.W. Low, S.K. Tiong, I.A. Badruddin, S. Kamanger, J.J.W. Wu, B. Ao, … R. Li, Synergically effective thermal conductivity of PU coating using different combinations of additives, Sol. Energy 10 (5) (2021) 129938, doi:10.1016/j.carbon.2012.06.043.

[50] https://www.sigmaaldrich.com/content/dam/sigma-aldrich/product9/163/pvd_schematic_l.eps/_jcr_content/renditions/pvd_schematic_l-large.jpg.

[51] A. Baptista, F.J.G. Silva, J. Porteiro, J.L. Míguez, G. Pinto, L. Fernandes, On the physical vapour deposition (PVD): evolution of magnetron sputtering processes for industrial applications, Procedia Manuf. 17 (2018) 746–757, doi:10.1016/J.PROMFG.2018.10.125.

[52] I.V. Tudose, F. Comanescu, P. Pascariu, S. Bucur, L. Rusen, F. Iacomi, E. Koudoumas, M.P. Suchea, Chemical and physical methods for multifunctional nanostructured interface fabrication, in: Functional Nanostructured Interfaces for Environmental and Biomedical Applications, 2019, pp. 15–26, doi:10.1016/B978-0-12-814401-5.00002-5.

[53] O. Öztürk, E. Gür, Layered transition metal sulfides for supercapacitor applications, ChemElectroChem (11) (2024) e202300575, doi:10.1002/celc.202300575.

[54] H Huang, J Lai, J Lu, Z Li, Pulsed laser ablation of bulk target and particle products in liquid for nanomaterial fabrication, AIP Adv. 9 (2019) 015307, doi:10.1063/1.5082695.

[55] M. Kim, S. Osone, T. Kim, H. Higashi, T. Seto, Synthesis of nanoparticles by laser ablation: a review, KONA Powder Part. J. 2017 (34) (2017) 80–90, doi:10.14356/kona.2017009.

[56] D. Josell, J.E. Bonevich, T.M. Nguyen, R.N. Johnson, Heat transfer through nanoscale multilayered thermal barrier coatings at elevated temperatures, Surf. Coat. Technol. 275 (2015) 75–83, doi:10.1016/j.surfcoat.2015.05.036.

[57] M.F. Smith, Comparing cold spray with thermal spray coating technologies, in: V.K. Champagne (Ed.), *The Cold Spray Materials Deposition Process, Woodhead Publishing Series in Metals and Surface Engineering*, Woodhead Publishing, 2007, pp. 43–61, doi:10.1533/9781845693787.1.43.

[58] S. Pungaiya, C. Kailasanathan, A Review of Surface Coating Technology to Increase the Heat Transfer, International Journal of Mechanical Engineering and Robotics Research 7 (5) (2018) 458–465, doi:10.18178/ijmerr.7.5.458-465.

[59] J.N. Hasnidawani, H.N. Azlina, H. Norita, N.N. Bonnia, Hardness and adhesion performances of nanocoating on carbon steel, IOP Conf. Ser.: Mater. Sci. Eng. 290 (1) (2018), doi:10.1088/1757-899X/290/1/012078.

[60] J. Chapman, F. Regan, Nanofunctionalized superhydrophobic antifouling coatings for environmental sensor applications-advancing deployment with answers from nature, Advanced Engineering Materials 14 (4) (2012) 174–184, doi:10.1002/adem.201180037.

[61] V. Jardret, H. Zahouani, J.L. Loubet, T.G. Mathia, Understanding and quantification of elastic and plastic deformation during a scratch test, Wear 218 (1) (1998) 8–14, doi:10.1016/S0043-1648(98)00200-2.

[62] C.T. Pan, T.T. Wu, C.F. Liu, C.Y. Su, W.J. Wang, J.C. Huang, Study of scratching Mg-based BMG using nanoindenter with Berkovich probe, Mater. Sci. Eng. A 527 (9) (2010) 2342–2349, doi:10.1016/J.MSEA.2009.11.070.

[63] ME Dickinson, JP. Schirer, Probing more than the surface, Mater. Today 12 (7–8) (2009) 46–50, doi:10.1016/S1369-7021(09)70202-0.

[64] M. Mohamed, M.N. Omar, M.S.A. Ishak, R. Rahman, et al., Synergically effective thermal conductivity of PU coating using different combinations of additives, Solar Energy 10 (5) (2021).

[65] S. Böhm, Graphene against corrosion, Nat. Nanotechnol. 9 (2014) 741–742, doi:10.1038/nnano.2014.220.

[66] M. Stratmann, R. Feser, A. Leng, Corrosion protection by organic films, Electrochim. Acta 39 (8–9) (1994) 1207–1214, doi:10.1016/0013-4686(94)E0038-2.

[67] D.P. Le, Y.H. Yoo, J.G. Kim, S.M. Cho, Y.K. Son, Corrosion characteristics of polyaniline-coated 316L stainless steel in sulphuric acid containing fluoride, Corros. Sci. 51 (2) (2009) 330–338, doi:10.1016/J.CORSCI.2008.10.028.

[68] T. Tüken, G. Arslan, B. Yazici, M. Erbil, The corrosion protection of mild steel by polypyrrole/polyphenol multilayer coating, Corros. Sci. 46 (11) (2004) 2743–2754, doi:10.1016/J.CORSCI.2004.03.003.

[69] A.Y. Adesina, Z.M. Gasem, A. Madhan Kumar, Corrosion resistance behavior of single-layer cathodic arc PVD nitride-base coatings in 1M HCl and 3.5 pct NaCl solutions, Metall. Mater. Trans. B 48 (2) (2017) 1321–1332, doi:10.1007/s11663-016-0891-7.

[70] P.M. Perillo, Corrosion behavior of coatings of titanium nitride and titanium-titanium nitride on steel substrates, Corrosion 62 (02) (2006) 182–185.

[71] D. Prasai, J.C. Tuberquia, R.R. Harl, G.K. Jennings, K.I. Bolotin, Graphene: corrosion-inhibiting coating, ACS Nano 6 (2) (2012) 1102–1108, doi:10.1021/nn203507y.

[72] X. Ye, Z. Lin, H. Zhang, H. Zhu, Z. Liu, M. Zhong, Protecting carbon steel from corrosion by laser in situ grown graphene films, Carbon 94 (2015) 326–334, doi:10.1016/j.carbon.2015.06.080.

[73] N. Tambo, Y. Liao, C. Zhou, E.M. Ashley, K. Takahashi, P.F. Nealey, Y. Naito, J. Shiomi, Ultimate suppression of thermal transport in amorphous silicon nitride by phononic nanostructure, Sci. Adv. 6 (2020) eabc0075.

[74] X. Shen, N. Ouyang, Y. Huang, Y.-H. Tung, C.-C. Yang, M. Faizan, N. Perez, R. He, A. Sotnikov, K. Willa, C. Wang, Y. Chen, E. Guilmeau, Amorphous-like ultralow thermal transport in crystalline argyrodite Cu7PS6, Adv. Sci. 11 (2024) 2400258, doi:10.1002/advs.202400258.

[75] R. Cravero, A. Tlili, J. Paterson, M. Tomelleri, P. Marcello, R. Debord, S. Pailhès, O. Bourgeois, F. Hippert, D. Le Qui, J. Raty, P. Noe, V.M. Giordano, Glass-like phonon dynamics and thermal transport in a GeTe nano-composite at low temperature, Small 20 (2024) 2310209, doi:10.1002/smll.202310209.

[76] A.Y. Serov, Z. Ong, E. Pop, Effect of grain boundaries on thermal transport in graphene, Appl. Phys. Lett. 102 (2013) 033104.

[77] K. Ruan, Y. Guo, C. Lu, X. Shi, T. Ma, Y. Zhang, et al., Significant reduction of interfacial thermal resistance and phonon scattering in graphene/polyimide thermally conductive composite films for thermal management, Research 2021 (2021) 8438614, doi:10.34133/2021/8438614.

[78] M.W. Boomstra, B.J. Geurts, A.V. Lyulin, Reducing Kapitza resistance of graphene–paraffin interfaces by alkyl functionalisation, Chem. Phys. Lett. 846 (2024) 141341, doi:10.1016/j.cplett.2024.141341.

[79] S Chen, N Xu, L Gorbatikh, D. Seveno, Find the maximum thermal conductivity of graphene reinforced polymer composite: a molecular dynamics approach, Polym. Compos. 45 (1) (2024) 302–314, doi:10.1002/pc.27777.

[80] Y Su, J.J. Li, G.J. Weng, Theory of thermal conductivity of graphene-polymer nanocomposites with interfacial Kapitza resistance and graphene-graphene contact resistance, Carbon 137 (2018) 222–233, doi:10.1016/j.carbon.2018.05.033.

[81] X. Mu, X. Wu, T.Go Zhang, B D., T Luo, Thermal transport in graphene oxide – from ballistic extreme to amorphous limit, Sci. Rep. 4 (2014) 3909, doi:10.1038/srep03909.

[82] Y.S. Lee, et al., Effects of high-temperature thermal reduction on thermal conductivity of reduced graphene oxide polymer composites, Appl. Surf. Sci. 650 (2024) 159140, doi:10.1016/j.apsusc.2023.159140.

[83] L. Hu, A. Wang, W Wang, Structural evolution of graphene oxide and its thermal stability during high temperature sintering, J. Wuhan Univ. Technol.—Mat. Sci. Edit. 37 (2022) 342–349, doi:10.1007/s11595-022-2537-8.

[84] J. Ma, S. Sett, Cha H, X. Yan, N. Miljkovic, Recent developments, challenges, and pathways to stable dropwise condensation: a perspective, Appl. Phys. Lett. 29 (26) (2020) 260501, doi:10.1063/5.0011642.

[85] S. Hatte, R. Pitchumani, Limits of dropwise condensation heat transfer on dry nonwetting surfaces, iScience 27 (11) (2024) 111059, doi:10.1016/j.isci.2024.111059.

[86] S. Shahlaei, E. Aminian, M. Kamali, H. Saffari, Experimental study of condensation heat transfer on thermally improved hybrid teflon-copper vertically patterned surfaces, Heat Transfer Eng. 47 (3) (2026) 228–246, doi:10.1080/01457632.2024.2445914.

[87] S. Hatte, R. Stoddard, R. Pitchumani, Novel solid-infused durable nonwetting surfaces for sustained condensation heat transfer enhancement, Appl. Therm. Eng. 219 (2023) 119458, doi:10.1016/j.applthermaleng.2022.119458.

[88] V. Volman, Y. Zhu, A.-R.O. Raji, B. Genorio, W. Lu, C. Xiang, C. Kittrell, J.M. Tour, Radio-frequency-transparent, electrically conductive graphene nanoribbon thin films as deicing heating layers, ACS Appl. Mater. Interfaces 6 (1) (2014) 298–304, doi:10.1021/am404203y.

[89] A.-R.O. Raji, T. Varadhachary, K. Nan, T. Wang, J. Lin, Y. Ji, B. Genorio, Y. Zhu, C. Kittrell, J.M. Tour, Composites of graphene nanoribbon stacks and epoxy for joule heating and deicing of surfaces, ACS Appl. Mater. Interfaces 8 (5) (2016) 3551–3556, doi:10.1021/acsami.5b11131.

[90] N. Wang, S. Chen, A. Nkansah, Q. Wang, X. Wang, M. Chen, L. Ye, J. Liu, Vertically aligned graphene-based thermal interface material with high thermal conductivity, in: THERMINIC 2018 – 24th International Workshop on Thermal Investigations of ICs and Systems, Proceedings, September 1–4, 2018, 2018, doi:10.1109/THERMINIC.2018.8593303.

[91] J. Zhao, Z. Liu, B. Wang, J. Hu, Y. Wan, Tool coating effects on cutting temperature during metal cutting processes: comprehensive review and future research directions, Mech. Syst. Sig. Process. 150 (2021) 107302, doi:10.1016/j.ymssp.2020.107302.

[92] F. Kung, M.C. Yang, Improvement of the heat-dissipating performance of powder coating with graphene, Polymers 12 (6) (2020), doi:10.3390/polym12061321.

[93] H.F. Hamann, A. Weger, J.A. Lacey, Z. Hu, P. Bose, E. Cohen, J. Wakil, Hotspot-limited microprocessors: direct temperature and power distribution measurements, IEEE J. Solid-State Circuits 42 (1) (2007) 56–64, doi:10.1109/JSSC.2006.885064.

[94] E. Pop, Energy dissipation and transport in nanoscale devices, Nano Res. 3 (3) (2010) 147–169, doi:10.1007/s12274-010-1019-z.

[95] A.A. Balandin, Thermal properties of graphene and nanostructured carbon materials, Nat. Mater. 10 (8) (2011) 569–581, doi:10.1038/nmat3064.

[96] T. Mueller, F. Xia, P. Avouris, Graphene photodetectors for high-speed optical communications, Nat. Photonics 4 (5) (2010) 297–301, doi:10.1038/NPHOTON.2010.40.

[97] A.H. Zehri, A. Nylander, L. Ye, J. Liu, Graphene-coated copper nanoparticles for thermal conductivity enhancement in water-based nanofluid, 2019 22nd European Microelectronics and Packaging Conference and Exhibition, EMPC 2019, 2019, doi:10.23919/EMPC44848.2019.8951883.

[98] B. Liu, L. Chen, G. Liu, A.N. Abbas, M. Fathi, C. Zhou, High-performance chemical sensing using Schottky-contacted chemical vapor deposition grown monolayer MoS_2 transistors, ACS Nano 8 (5) (2014) 5304–5314, doi:10.1021/nn5015215.

[99] D. Golberg, Y. Bando, Y. Huang, T. Terao, M. Mitome, C. Tang, C. Zhi, Boron nitride nanotubes and nanosheets, ACS Nano 4 (6) (2010) 2979–2993, doi:10.1021/nn1006495.

[100] H. Malekpour, K.H. Chang, J.C. Chen, C.Y. Lu, D.L. Nika, K.S. Novoselov, A.A. Balandin, Thermal conductivity of graphene laminate, Nano Lett. 14 (9) (2014) 5155–5161, doi:10.1021/nl501996v.

[101] N.P. Padture, M. Gell, E.H. Jordan, Thermal barrier coatings for gas-turbine engine applications, Science 296 (5566) (2002) 280–284, doi:10.1126/science.1068609.

[102] X. Fan, W. Jiang, J. Li, T. Suo, T.J. Wang, R. Xu, Numerical study on interfacial delamination of thermal barrier coatings with multiple separations, Surf. Coat. Technol. 244 (2014) 117–122, doi:10.1016/j.surfcoat.2014.01.045.

[103] R. Ghasemi, R. Razavi, R. Mozafarinia, H. Jamali, Comparison of microstructure and mechanical properties of plasma-sprayed nanostructured and conventional yttria stabilized zirconia thermal barrier coatings, Ceramurgia Int. 39 (2013) 8805–8813.

[104] X. Chen, Z. Yi, J. Lei, H. Yi, W. Yao, W. Zhu, T. Duan, Preparation and perfomance of an aging-resistant nanocomposite film of binary natural polymer–graphene oxide, ACS Omega 1 (6) (2021) 1173–1181, doi:10.1021/acsomega.6b00291.

[105] S. Güryel, M. Alonso, B. Hajgató, Y. Dauphin, G. Van Lier, P. Geerlings, F. De Proft, A computational study on the role of noncovalent interactions in the stability of polymer/graphene nanocomposites, J. Mol. Model. 23 (2) (2017) 43, doi:10.1007/s00894-017-3214-2.

[106] W. Ali, A. Kausar, T. Iqbal, Reinforcement of high performance polystyrene/polyamide/polythiophene with multi-walled carbon nanotube obtained through various routes, Compos. Interfaces (2015) 22, doi:10.1080/09276440.2015.1076251.

[107] A. Kausar, Composite coatings of polyamide/graphene: microstructure, mechanical, thermal, and barrier properties, Compos. Interfaces 25 (2) (2018) 109–125, doi:10.1080/09276440.2017.1340020.

[108] B. Lv, X. Jin, J. Cao, B. Xu, Y. Wang, D. Fang, Advances in numerical modeling of environmental barrier coating systems for gas turbines, J. Eur. Ceram. Soc. 40 (9) (2020) 3363–3379, doi:10.1016/J.JEURCERAMSOC.2020.03.036.

[109] F.Y. Lv, P. Zhang, Drag reduction and heat transfer characteristics of water flow through the tubes with superhydrophobic surfaces, Energy Convers. Manage. (2016), doi:10.1016/j.enconman.2016.01.034.

[110] S. Na, B. Williams, R.A. Dennis, K.M. Bryden, T.I. Shih, GT-2007-27599 Internal and Film Cooling of a Flat Plate, American Society of Mechanical Engineers (ASME), New York, NY, USA, 2015, pp. 1–10.

[111] X. Yang, Z. Feng, T.W. Simon, Conjugate heat transfer modeling of a turbine vane endwall with thermal barrier coatings, Aeronaut. J. 123 (1270) (2019) 1959–1981, doi:10.1017/aer.2019.56.

[112] K. Bobzin, High-performance coatings for cutting tools, CIRP J. Manuf. Sci. Technol. 18 (2017) 1–9, doi:10.1016/J.CIRPJ.2016.11.004.

[113] M. Nouari, G. List, F. Girot, D. Coupard, Experimental analysis and optimisation of tool wear in dry machining of aluminium alloys, Wear 255 (2003) 1359–1368, doi:10.1016/S0043-1648(03)00105-4.

[114] G.K. Dosbaeva, M.A. El Hakim, M.A. Shalaby, J.E. Krzanowski, S.C. Veldhuis, Cutting temperature effect on PCBN and CVD coated carbide tools in hard turning of D2 tool steel, Int. J. Refract. Met. Hard Mater 50 (2015) 1–8, doi:10.1016/J.IJRMHM.2014.11.001.

[115] V.P. Astakhov, J. Outeiro, Importance of Temperature in Metal Cutting and Its Proper Measurement/Modeling, 2019, doi:10.1007/978-3-030-03822-9_1.

[116] D.K. Das, S. Santra, S. Sahoo, Graphene coating on cutting tools can remove the use of coolants, J. Nanosci. Nanoeng. Appl. 6 (2016) 1–5. www.stmjournals.com.

[117] European CoatingsIndustrial Coatings Market Trends Q1, 2021. https://www.european-coatings.com/news/markets-companies/strong-increase-in-global-demand-for-paints-and-coatings/ Accessed 31 December 2025.

[118] C Tian, H Fu, Z Wang, Z Zhang, W Qian, H Zhang, et al., Deformable surface design of vertical graphene thermal interface materials for efficient heat dissipation, Cell Rep. Phys. Sci. 5 (6) (2024) 101978, doi:10.1016/j.xcrp.2024.101978.

[119] C.-N. Hsu, K.-W. Lee, C.-C. Chen, Using graphene-based grease as a heat conduction material for hectowatt-level LEDs: a natural convection experiment, Processes 9 (5) (2021) 847, doi:10.3390/pr9050847.

[120] K.S. Novoselov, A. Mishchenko, A. Carvalho, H A., Castro Neto. 2D materials and van der Waals heterostructures, Science 353 (2016) aac9439, doi:10.1126/science.aac9439.

[121] MarketsandMarkets. (2023). Graphene Coating Market by Product Type (Solvent-Based and Water-Based), Application (Corrosion-Resistant Coating, Scratch-Resistant Coating, Antifouling Coating, Flame-Retardant Coating), End-Use Industry, and Region—Global Forecast to 2028 (Report Code: CH 8862). https://www.marketsandmarkets.com/Market-Reports/graphene-coating-market-69854439.html

CHAPTER

Graphene and 2DM-based thermal interface materials for thermal management

14

Kamaljit Singh[a], Priyanka Mahajan[b] and Hasan Koten[c]

[a]*Department of Mechanical Engineering, School of Engineering and Technology, CGC University, Mohali, Punjab, India,* [b]*University Centre for Research & Development, Chandigarh University, Mohali, Punjab, India,* [c]*Faculty of Engineering and Pure Sciences, Istanbul Medeniyet University, Istanbul, Turkey*

14.1 Introduction

The continued miniaturization and functional integration of electronic and optoelectronic systems have driven device power densities to unprecedented levels, such that thermal management has emerged as a primary constraint on the performance, operational stability, and service lifetime of these devices. Advanced devices, including 5G transceivers, high-brightness light-emitting diodes (LEDs), and next-generation artificial intelligence (AI) accelerators such as NVIDIA's GB200 Graphic Processing Unit (GPU), routinely generate local heat fluxes exceeding 100 W cm^{-2}, thereby testing the limits of current thermal interface materials (TIMs) and substrate-level cooling strategies [1,2]. The emergence of localized "hot spots" in processors, power modules, and radio-frequency (RF) hardware often surpasses the dissipation capacity of conventional packaging, rendering the thermal interface between heat-generating chips and their heat sinks a critical bottleneck in thermal management. TIMs are specifically engineered to conform to surface roughness and fill interstitial air gaps, thereby minimizing interfacial thermal resistance (R_{th}) and facilitating efficient heat transfer from the device to a heat spreader or sink [3,4]. However, traditional particle-filled polymer composites, typically loaded with high volume fractions of ceramic fillers such as Al_2O_3, AlN, or BN, rarely achieve thermal conductivities exceeding 1–5 W $m^{-1}K^{-1}$ and are increasingly inadequate for managing rising heat fluxes [4].

Therefore the search for materials capable of transcending these limits has intensified interest in atomically thin solids with fundamentally different heat-transport physics. The isolation of graphene established a new paradigm in which phonon-dominated thermal conduction can approach or exceed that of the best bulk crystalline conductors. High-quality graphene exhibits in-plane thermal conductivities on the order of 2000–5000 W m^{-1} K^{-1}, arising from strong sp^2 covalent bonding, low phonon scattering, and long mean free paths (MFPs) within its two-dimensional lattice. The pronounced anisotropy between in-plane and cross-plane transport, governed by weak van der Waals coupling, introduces both challenges and opportunities in interface engineering, alignment control, and composite design [5]. Beyond graphene, layered materials such as hexagonal boron nitride (h-BN) offer high thermal conductivity combined with electrical insulation, chemical inertness, and compatibility with diverse device environments. Selected transition-metal dichalcogenides further expand the design space by

Graphene and 2D Materials in Heat Transfer: Fundamentals and Applications. DOI: https://doi.org/10.1016/B978-0-12-821964-5.00005-9

coupling thermal functionality with tunable electronic and optical behaviors [6]. These attributes have rapidly positioned two-dimensional materials (2DMs) as leading candidates for thermal interface layers, heat spreaders, and integrated thermal management components in next-generation electronics [7].

Early studies have demonstrated that incorporating liquid-phase exfoliated graphene and few-layer graphene into polymer matrices can produce record thermal conductivity enhancements at relatively low filler loadings ($\leq$10 vol.%), outperforming composites based on Carbon nanotubes (CNTs)or metallic particles [3]. Subsequent studies have generalized this approach to a wide range of matrices and architectures, reinforcing the concept that graphene's geometry and interfacial phonon coupling enable efficient thermal percolation networks [3]. However, it became evident that simply adding two more two-dimensional (2D) fillers is insufficient, as R_{th}, filler orientation, and microstructure control the effective $\lambda_{\perp}$ and contact resistance, which are most relevant for real devices [4,7].

This realization has driven a shift from random dispersions toward architected graphene and 2DM frameworks designed explicitly for heat transport across interfaces. Vertically aligned graphene (VAG) monoliths, foams, and fiber arrays have achieved metal-level $\lambda_{\perp}$ while maintaining a low compressive modulus, directly addressing the long-standing trade-off between stiffness and thermal performance in TIMs [8]. For example, mechanically engineered graphene monoliths with predominantly vertical alignment have achieved $\lambda_{\perp}$ above 140 W m^{-1} K$^{-1,}$ enabling system-level cooling efficiencies several times higher than those of state-of-the-art commercial TIMs. Complementary strategies exploit VAG foams or graphene fibers embedded in soft matrices, achieving $\lambda_{\perp}$ of tens to > 80 W m^{-1} K^{-1} at modest filler contents, along with low contact thermal resistance and excellent compressibility [8]. Metal–graphene hybrid architectures, such as laminated graphene–Cu composites with strain-tuned electron–phonon coupling, further illustrate how interfacial engineering can push thermal conductivities of structural components beyond those of pure copper while preserving process compatibility [9].

Simultaneously, advanced 2DM TIMs have evolved beyond graphene-centric architectures. h-BN, distinguished by its high in-plane thermal conductivity and intrinsic electrical insulation, functions as an efficient heat spreader and TIM filler in power electronics and RF platforms where dielectric integrity is mandatory [10,11]. Beyond single-component systems, diverse 2DMs and their van der Waals heterostructures provide a versatile framework for the engineering of anisotropic heat transport. For example, measurements of interfacial thermal conductance across graphene/h-BN and MoS_2/h-BN junctions demonstrated that phonon transmission spectra govern interfacial heat leakage in layered device stacks [6]. These atomic-scale insights into phonon-mediated transport establish quantitative design principles for heterostructure TIMs and encapsulation layers that regulate lateral heat spreading within 2D channels while facilitating efficient cross-plane dissipation into underlying heat sinks [6].

Despite rapid progress, substantial challenges and opportunities remain. Achieving low and reproducible contact thermal resistance under realistic packaging pressures over large areas and through repeated thermal cycling is nontrivial, particularly when relying on anisotropic 2D fillers whose basal planes are naturally parallel to substrates [4]. Although several grapheme- and h-BN-based products have entered the market, issues such as scalable synthesis, defect control, and reliable integration with conventional packaging processes continue to limit their widespread commercialization [7]. Simultaneously, emerging applications, ranging from flexible and wearable electronics to mid-infrared (IR) radiative thermal management, phase-change composites, and smart, thermally switchable interfaces, are broadening the functional requirements imposed on TIMs, including mechanical compliance, environmental stability, spectral selectivity, and multifunctionality, such as electromagnetic interference (EMI) shielding or Joule heating [10,12].

Against this backdrop the present chapter surveys graphene and advanced 2DM-based TIMs from fundamentals to application-oriented design. It first outlines the essential physics of heat transport in 2DMs and across their interfaces, emphasizing their anisotropy, phonon scattering mechanisms, and R_{th}. It then reviews the main classes of graphene and 2DM TIM architectures, dispersed composites, three-dimensional (3D) frameworks, vertically aligned monoliths and foams, fiber-based systems, and metal–2D hybrids, highlighting the structure–property–performance relationships in terms of $\lambda_{\perp}$, mechanical compliance, and contact resistance [9,8]. The subsequent sections discuss device-level demonstrations in high-power electronics and explore the remaining barriers to large-scale adoption, including manufacturing, reliability, and cost considerations [7]. Finally, this chapter outlines future directions in interfacial engineering, multiscale modeling, and multifunctional design, with a view toward enabling practical, industry-ready 2DM TIMs that can meet the escalating demands of thermal management in emerging electronic technologies.

14.2 Fundamentals of heat transport in two-dimensional materials

The thermal transport properties of 2DMs have attracted increasing attention owing to their critical role in nanoscale energy management and device performance. Unlike their bulk counterparts, 2D systems exhibit distinct phonon transport characteristics governed by reduced dimensionality, strong anisotropy, and interfacial effects. Understanding these mechanisms is essential for optimizing heat dissipation in emerging nanoelectronic and optoelectronic technologies. Thermal conductivity is a key parameter that determines the efficiency and reliability of miniaturized devices. Advances in nanotechnology have demonstrated that controlling heat flow at the atomic level enables superior thermal management in high-power electronic components, flexible circuits, and next-generation computing systems. The ability to engineer phonon scattering pathways through defects, strain, or heterostructures further extends the potential of 2DMs for tunable thermal performance.

14.2.1 Phonon physics in layered solids

In layered solids, such as graphite, graphene, and transition metal dichalcogenides (TMDs), the thermal transport properties are governed by the physics of phonons, which are quantized lattice vibrations that carry heat. Although in-plane transport is dominated by sp^2 bonding, out-of-plane transport is governed by weak van der Waals interactions. The strong bonding and high stiffness because of the unique structure facilitate in-plane phonon transport, leading to exceptional thermal conductivity values. For instance the thermal conductivity of silicone rubber with vertical carbon fibers in the in-plane analysis was 4.72 $W\,m^{-1}\,K^{-1}$, with a 2045.5% thermal conductivity enhancement [13]. The molecular dynamics simulation showed that the out-of-plane thermal conductivity could be reduced significantly compared to the in-plane thermal conductivity with an increase in the interlayer coupling.

Heat transport in 2DMs exhibits distinct in-plane and out-of-plane behaviors. The high thermal conductivity of 2DMs, such as graphene, is the result of longitudinal and transverse acoustic phonons, which primarily influence the in-plane thermal conductance. The efficiency of the flexural acoustic mode is reduced by factors such as strain and defects, even though it also aids in in-plane heat transfer. Out-of-plane heat transport is limited by weak layer contacts and a slower phonon rate, which are significantly influenced by the support material when stacked. Compared to unsupported materials, this reduces the thermal conductivity. Although out-of-plane transport is dependent on external factors and structural information, models and simulations have shown that phonon scattering affects in-plane transport [14].

14.2.2 Thermal anisotropy

Thermal transport in TIMs is inherently direction-dependent and is commonly resolved into through-plane ($\lambda_\perp$) and in-plane ($\lambda_\parallel$) components according to the Fourier heat conduction formalism. $\lambda_\perp$ governs the heat flow across the bond-line thickness from the chip to the heat spreader or heat sink, whereas $\lambda_\parallel$ describes the lateral heat spread within the TIM layer. In electronic packaging a high $\lambda_\perp$ is essential to minimize the interfacial thermal resistance ($R_{th} = t/\lambda_\perp$, where t is bond line thickness) and to enable rapid evacuation of waste heat from high-power-density devices. Conversely an adequate $\lambda_\parallel$ mitigates lateral temperature gradients and suppresses hot-spot formation arising from nonuniform heat flux or imperfect interfacial conformity [15].

Anisotropy frequently emerges from the geometry and orientation of the filler. One-dimensional (1D) and 2D fillers, such as carbon fibers and graphene nanoplatelets (GNPs), exhibit intrinsically anisotropic phonon transport owing to their crystallographic symmetries and bonding topologies [16]. For example, in graphene, the in-plane thermal conductivity can exceed ~2000–3000 W m^{-1} K^{-1} at room temperature in high-quality samples, whereas the cross-plane values are typically two orders of magnitude lower (~5–20 W m^{-1} K^{-1}), reflecting the strong sp^2 covalent bonding within the basal plane and weak interlayer van der Waals coupling. Direction-dependent phonon dispersion and scattering along armchair and zigzag directions further modulate the thermal conductance spectrum across temperature regimes, as illustrated in Fig. 14.1A and B [17]. The strategic alignment of anisotropic fillers via shear-assisted processing, magnetic/electric field alignment, freeze casting, or templated assembly can significantly enhance λ along the alignment axis by promoting continuous phonon transport pathways and reducing interfacial phonon scattering. Indeed, architected composites with vertically aligned high-aspect-ratio fillers have been reported to have through-plane conductivities exceeding 300 W m^{-1} K^{-1} under optimized conditions. However, such configurations often exhibit comparatively modest $\lambda_\parallel$ because heat spreading orthogonal to the alignment direction remains constrained by matrix-dominated transport and filler–matrix interfacial resistance [15]. Although the construction of continuous, percolated thermal networks suppresses phonon–phonon and phonon–interface scattering relative to randomly dispersed systems, the resulting anisotropy complicates device-level thermal management. Excessively high $\lambda_\perp$ coupled with insufficient $\lambda_\parallel$ can lead to efficient vertical heat extraction but inadequate lateral homogenization of temperature, particularly in heterogeneous integration platforms. Therefore rational microstructural design must target a balanced anisotropy by optimizing both $\lambda_\perp$ and $\lambda_\parallel$ to minimize the overall thermal resistance, ensure a uniform temperature distribution, and enhance the long-term reliability of advanced electronic packaging architecture.

14.2.3 Scattering mechanisms

The MFP of phonons is the basis for determining the thermal conductivity of 2DMs. For instance, in ideal graphene, phonons can travel long distances, leading to high thermal conductivity measurements. However, experimental samples face several challenges, such as boundaries and defects that reduce thermal transfer and MFP in 2DMs. Grain boundaries, size, and defects are a few reasons for scattering

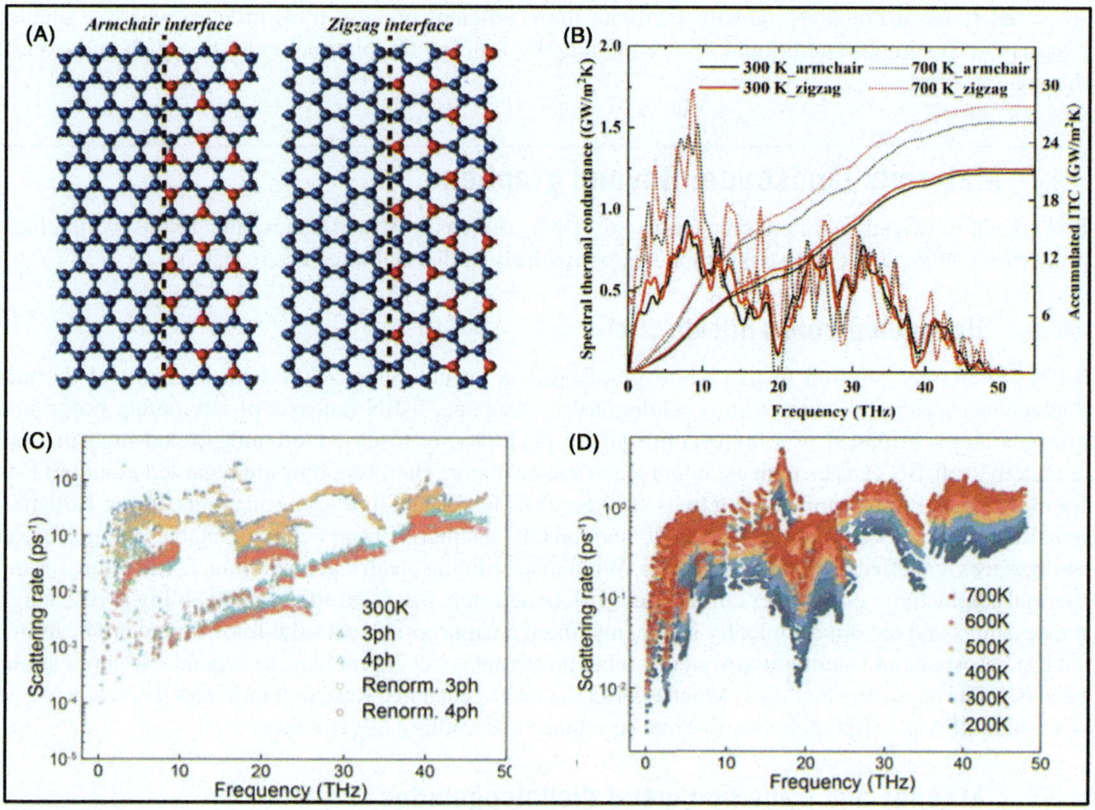

FIGURE 14.1

(A) Armchair and zigzag directions of graphene/C_3B heterointerfaces, (B) thermal conductance spectrum as a function of frequency (THz) under varying scattering rates for different temperature ranges [17]. Phonon scattering rate with respect to frequency (C) comparison of the 3ph and 4ph processes at 300 K with or without renormalization and (D) temperature-dependent scattering mechanism [18].

that have a significant influence on heat conduction in these materials. Phonon mobility is significantly influenced by defects such as vacancies; structural issues; variations in conditions such as temperature; and grain boundaries (Fig. 14.1C and D) [19,18].

Phonons are dispersed by natural defects or foreign materials in the graphene lattice, which lowers heat conductivity. Edge scattering is crucial in pure graphene because phonons interact with its edges. Both Umklapp and edge scattering are significant, with their effects varying according to the sample size and shape. Yang et al. [20] demonstrated that Umklapp phonon scattering is negligible in graphene nanoribbons shorter than ~775 nm, where thermal transport is dominated by edge scattering. In these narrow ribbons, edge scattering strongly suppresses the thermal conductivity. Conversely, in wider ribbons, the increased phonon population and reduced energy separation between modes enhance Umklapp scattering, whereas the relative impact of edge scattering diminishes. Moreover, larger grain sizes and reduced boundary density facilitate more efficient phonon transmission, whereas smaller graphene flakes impose additional scattering, thereby limiting the phonon mean-free path and overall thermal transport.

14.3 Materials landscape: Beyond graphene

It is crucial to investigate the larger family of 2DMs, such as h-BN, MXenes, and TMDs, which have properties comparable to those of graphene, particularly in thermal interface research.

14.3.1 Hexagonal boron nitride

h-BN, commonly referred to as "white graphene," is an emerging 2DM with exceptional thermal management properties. Structurally analogous to graphene, h-BN consists of alternating boron and nitrogen atoms arranged in a honeycomb lattice [21]. Owing to its wide bandgap and high thermal conductivity, h-BN can act as an excellent electrical insulator. Therefore they are regarded as ideal TIMs for electronics. Furthermore, owing to its stronger B–N bonds and lighter boron atoms, heat conduction in h-BN primarily occurs through acoustic phonons. Its distinct thermal behavior enables the insulation of interfaces and effective heat distribution. Without sacrificing electrical insulation, h-BN improves the thermal conductivity of polymer composites [18]. In addition, h-BN exhibits a great ability to resist high temperatures and remain chemically stable, making it an appropriate material for different applications, such as aerospace and automotive systems. It has an advantage over graphene in dependable applications because of its insulating qualities, which restrict the loss of currents. Research indicates that when paired with other 2DMs, h-BN increases thermal resistance and cooling effectiveness.

14.3.2 MXenes and transition metal dichalcogenides

2D MXenes are transition metal carbides, nitrides, and/or carbonitrides with the general formula $M_{n+1}X_nT_x$, where M is an early transition metal (e.g., Ti, Zr, V, Nb, Ta, or Mo), X is carbon and/or nitrogen, and Tx represents the functional groups on the surface of MXene [22]. TMDs are regarded as inorganic materials with the chemical formula MX_2 (M = transition metal = Mo, W; X = chalcogen = S, Se, Te) [23]. Highly efficient shielding materials must either reflect or absorb electromagnetic waves to protect electronic gadgets against EMI. Highly conductive metal foils have been effective, although reflections can generate secondary waves of interference. Therefore modern devices require materials that can absorb electromagnetic radiation.

MXenes are ideal for reducing and absorbing Electromagnetic (EM) waves owing to their superior electrical and thermal conductivities. Their layered structure facilitates the creation of supplementary intersurfaces that help reflect EM waves. This increases the likelihood of interaction with the shield and promotes internal wave movement, which enhances EM wave absorption [24]. Absorption is increased by internal scattering, whereas the overall effectiveness of shielding is reduced by surface reflection [25]. TMDs can be split into single-layer structures owing to weak van der Waals interactions. MoS_2, a class of TMDs, has been widely explored for protection against EMI. MoS_2 is a naturally available anisotropic semiconductor that is accessible in various shapes and sizes. It has multiple polytypes, including the 3R phase with synthesized MoS_2, and is typically found in the 2H phase. TMDs are ideal for EM wave suppression because of their reduced conductivity and dielectric constant compared to graphene [25].

Table 14.1 Comparative properties of 2D materials versus traditional TIM fillers.

2D material	Thermal conductivity	Young's modulus	References
MoS_2	62.2 W m^{-1} K^{-1} @ 300 K		[32]
MoS_2	131 W m^{-1} K^{-1} @ 300 K	0.33 ± 0.07 TPa	[31]
Monolayer garphene	600 W m^{-1} K^{-1} @ 300 K		[33]
Five-layer h-BN	250 W m^{-1} K^{-1} @ 300 K		[34]
Eleven-layer h-BN	360 W m^{-1} K^{-1} @ 300 K		[34]
h-BN	~600 W m^{-1} K^{-1} @300 K	~0.8 TPa	[26]
$TiS_2(HS)_x$	0.43 W m^{-1} K^{-1} @ 300 K		[35]
WSe_2	0.05 W m^{-1} K^{-1}		[36]
MXene (Ti_2C)	19.57 W m^{-1} K^{-1} @ room temperature		[37]

2D, *two-dimensional;* TIM, *thermal interface material.*

14.3.3 Comparison of intrinsic properties: Thermal conductivity, Young's modulus, and dielectric constants

In addition to graphene, MXenes, TMDs, and h-BN have shown great promise as TIMs. They exhibit diverse characteristics and play a critical role in thermal management devices. h-BN exhibits a bandgap ranging between 3.6 and 7.1 eV, which depends on various aspects, such as doping, layer number, defects, and purity. The monolayer h-BN has a high in-plane thermal conductivity of ~600 W m^{-1} K^{-1} and Young's modulus of ~0.8 TPa. It can be an effective solution for insulating heat spreaders because of its low dielectric constant of approximately 3–4 [26].

For oxygen-terminated MXene, Young's modulus drops dramatically from 0 to 1000 K. $Ti_3C_2O_2$ shows a decrease from 482 to 442 GPa, whereas Ti_2CO_2 shows a more pronounced loss from 565 to 469 GPa. According to this data, Ti_2CO_2 is more susceptible to thermal softening than $Ti_3C_2O_2$, although both materials maintain significant mechanical stiffness at high temperatures [27]. Similarly the thermal conductivity examined using experimental methods of $Ti_3C_2T_x$ may rise to ~55 W m^{-1} K^{-1}[28]. In high-performance dielectric applications, $Ti_3C_2T_x$ MXene/polymer nanocomposites incorporating boron nitride exhibited a dielectric constant of ~31.8 with an exceptionally low dielectric loss [29]. The bandgap of monolayer MoS_2 (TMD) is 1.8 eV, whereas that of the bulk material is nearly 1.2 eV (Table 14.1). Although MoS_2 exhibits high strength and Young's modulus (0.33 ± 0.07 TPa), its properties are still incomparable to those of graphene [30]. The thermal conductivity of MoS_2 at 300 K was determined to be 131 W m^{-1} K^{-1}[31].

14.4 Fabrication strategies for high-performance thermal interface materials

The development of high-performance TIMs based on 2DMs critically depends on precise fabrication strategies. The integration approach directly influences thermal transport efficiency, mechanical integrity, and large-scale manufacturability. Advanced fabrication paradigms, such as interface optimization, structural engineering, and scalable synthesis, are therefore central to realizing the full potential of 2DM-based TIMs.

14.4.1 Vertically aligned architectures

Among the various structural configurations, VG architectures represent a significant advancement in thermal management design. In these systems, graphene nanosheets are deliberately oriented perpendicular to the substrate, creating a highly anisotropic network that facilitates efficient through-plane heat conduction. This vertically aligned arrangement effectively overcomes the intrinsic in-plane heat transport dominance of 2DMs, thereby minimizing interfacial thermal resistance and enhancing thermal conductivity across device layers. Such architectures can be achieved through methods, including plasma-enhanced chemical vapor deposition (CVD), template-assisted assembly, or directional freeze-casting. Beyond graphene, similar vertically oriented frameworks have been explored with other 2DMs, such as h-BN and MXenes, demonstrating the general applicability of this strategy. Notably, these aligned structures offer not only superior thermal transport but also tunable mechanical compliance, making them highly suitable for next-generation, high-power-density electronic systems.

14.4.2 Mechanical compression/folding

Reduced graphene oxide (rGO) or stacked graphene oxide (GO) films can be modified by compression, cutting, or rolling. The rolling-cutting approach, which involves coiling GO films, cutting them, and then lowering them at high temperatures, results in vertically aligned rGO films with higher conductivity. By applying mechanical forces to multilayer graphene sheets, shrinking-compressing processes create tiny vertical structures that enhance phonon pathways. These methods work well with polymers, are easily scalable, and utilize solutions, making them suitable for commercial production of TIMs [38].

14.4.3 Magnetic and electric field alignment

Magnetic and electric fields can significantly and efficiently orient 2DMs, such as graphene, in an organized manner to improve thermal transport. Aligned flakes assist in reducing scattering and increasing conductivity by ensuring that phonons have a continuous path to flow. Utilizing the inherent properties of graphene, these techniques preserve the behavior of graphene by reducing energy through the rotation and lining of flakes under external fields. Research has shown that graphene flakes have excellent diamagnetic properties that allow them to arrange themselves under weak magnetic fields. Investigations have also demonstrated that flakes can be aligned using commercial magnets. Depending on the type of field used, the orientation can be altered and reversed after the field is removed [39]. Flakes, especially those with surface charges, can also be aligned using electric fields. Regulated electric fields have been used to achieve vertical orientation in materials such as GO and MXene, which enhances their electrical properties and increases their thermal conductivity, thereby creating multifunctional composites with improved performances.

14.4.4 Ice-templating/freeze-drying

The freeze-drying method has found widespread application in several fields, including biologics, pharmaceutical industries, and food and beverage industries, particularly when using temperature-sensitive compounds/materials [40]. This technique is helpful in attaining anisotropic cellular monoliths within the material along with the required porosity and structural morphology. This process involves the uniform dispersion of colloidal particles within a solvent, typically water, followed by complete freezing. Because of their low solubility, dispersed particles are expelled from ice crystals, resulting in the formation of either a well-connected lamellar structure or dendritic structures [40].

Interestingly the structured porous architecture is preserved, which is extremely crucial, by additional sublimation of the solid solvent under vacuum. This enables the production of monolithic scaffolds that resemble the morphology of ice crystals. Factors such as the size of the particles, type of solvent, and freezing conditions can have a significant effect on the final structure attained, which may range between <2 and >50 nm. Interestingly the anisotropy in cellular monoliths is a result of directional freezing, in which the growth of the ice crystal is typically governed by changes in temperature. This facilitates the formation of aligned channels and ordered pores, which improves the electrochemical performance by providing more active sites and areas for absorbing thermal stress during chemical reactions [41].

14.4.5 Three-dimensional continuous frameworks

14.4.5.1 *Chemical vapor deposition–grown foams*

CVD is regarded as the most promising method for form development. Techniques such as template-assisted CVD have been widely adopted to develop linked 3D graphene networks [42]. Notably, this process helps in designing lightweight, extremely porous materials with good mechanical and transport properties by incorporating the traits of nanomaterials, such as graphene. A 3D graphene interconnected network is formed after the decomposition of the carbon precursor on a metallic or nonmetallic substrate using a mixture of hydrogen and argon gas [43]. The source of carbon can be in gaseous, liquid, or solid form. The process may include cleaning of the template, hydrogenation of metals to remove the oxide layer, oxidation of seashells to form a porous structure, CVD growth, and etching of the template to free the 3D graphene form [42].

14.4.5.2 *Polymer-derived carbons: Hybridizing two-dimensionals with carbon scaffolds*

To effectively use carbon and its composite materials, cost-effective and efficient synthesis techniques must be explored. Therefore it is also important to select an inexpensive, abundantly available carbon precursor that can exhibit diverse structural forms. Based on these necessities, carbon-rich polymers have emerged as an ideal choice for the role of precursors.

To form polymer-derived carbon materials, solid organic materials are decomposed at elevated temperatures. The size, structure, and defects within the polymer-based carbon materials can be tailored using optimal precursor types and synthesis conditions. Polymer-based carbon materials are more controllable than biomass-based carbon materials. Several carbonization techniques, including pyrolytic and hydrothermal carbonization, are typically used to synthesize carbon-based substances for diverse applications [44].

14.4.5.3 *Graphene fiber/Film vertically arranged arrays*

VAG nanosheet arrays, another class of 3D graphene family, have garnered more attention recently because of their better mass and heat transport and reaction kinetics than graphene [45]. A comparison of the microstructures using Scanning electron microscopy (SEM) images of random and vertical

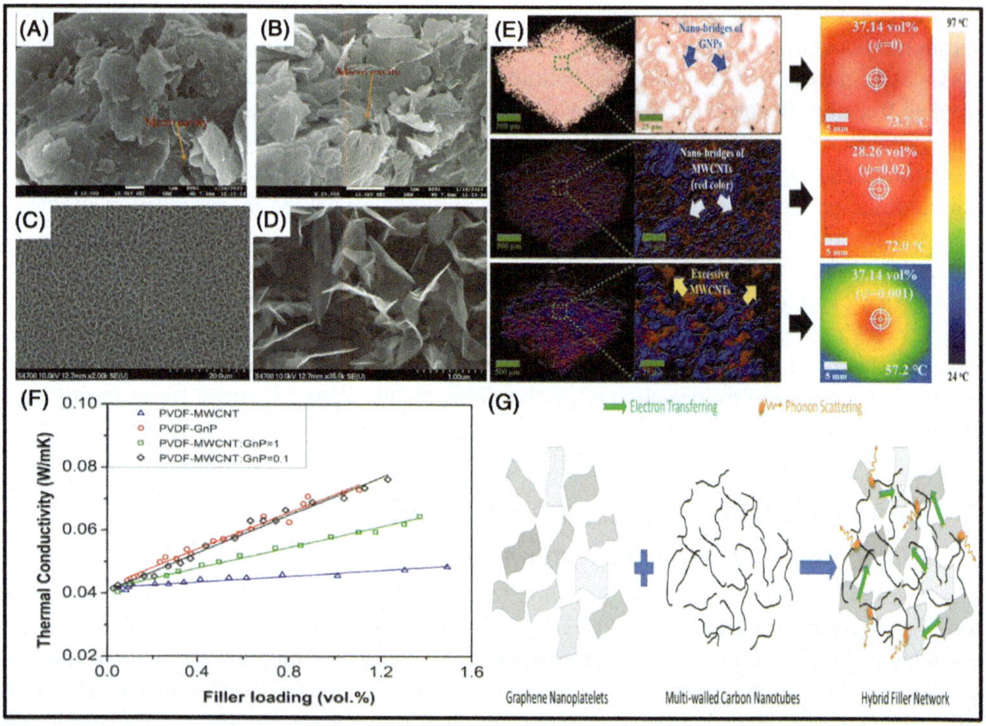

FIGURE 14.2

(A and B) SEM micrographs of random graphene at low and high resolutions [46]; (C and D) vertically aligned graphene at low and high resolutions [47]; (E) micro-CT images representing the presence of GNPs and CNTs in the hybrid composites [48]; (F) plot of thermal conductivity against filler loading; (G) schematic illustration of phonon bridging via 1D/2D hybrid fillers [49]. *1D*, One-dimensional; 2D, two-dimensional; *GNPs*, graphene nanoplatelets.

graphene at different magnifications is shown in Fig. 14.2A–D. These arrays, often developed through plasma-enhanced CVD or deposition assisted by electric fields, can produce freestanding films or fibers featuring open channels for swift ion movement and numerous exposed activated edge sites. In contrast to randomly stacked graphene, their vertical alignment ensures that they remain in close contact with the substrate and reduces the complexity of diffusion paths [45]. In addition, vertically arranged arrays show great efficiency when applied as electrodes for energy storage and conversion, supporting improved batteries, water-splitting reactions, and high-frequency capacitors. They are excellent prospects for next-generation electrochemical devices because of their structural plasticity and scalability.

14.4.6 Hierarchical and hybrid fillers

One of the most effective methods for improving the performance of 2DMs for distinct applications involving mass and heat transport, such as electrochemical energy conversion and storage, is the formation and utilization of hybrid and hierarchical fillers. These fillers enable synergistic increases in conductivity, mechanical stability, and ion transport by combining structural complexity with multifunctionality. The multiscale architecture of hierarchical fillers integrates micro-, meso-, and microporous sizes into a single structure. In addition, hierarchical structures with a greater number of active sites for redox reactions facilitate interconnected pathways for the rapid transport of electrons and ions. For instance, it has been observed that hierarchical metal–organic frameworks as well as their derivatives can control electrochemical characteristics at various scales, increasing the catalytic activity and diffusion rate through compositional variations [50]. These fillers are highly suitable for high-performance batteries and supercapacitors owing to their hierarchical structure. This leads to enhanced structural integrity even during repeated cycles.

In a composite, two or more hybrid fillers, such as carbon, polymers, and inorganic nanostructures, are added. Their incomparable characteristics were utilized in this hybridization. For example, polymer-based matrices are typically reinforced with flame-retardant additives or carbon nanostructures (CNTs) to produce multipurpose composites with enhanced mechanical strength, safety, and conductivity [51].

14.4.6.1 0D–two-dimensional and one-dimensional–two-dimensional hybrids

Graphene is the most explored material for diverse applications owing to its unmatched properties. However, graphene sheets can restack, which limits their effectiveness. To address this issue, hybrid structures that include nanoparticles or nanostructures have been developed with graphene. 0D materials, such as fullerenes, quantum dots, or organic molecules, can be decorated on the surface and edges of graphene. Different interaction mechanisms, such as π–π interactions and electrostatic attractions, stabilize the hybrids. For example, C60 fullerenes form homogeneous layers when deposited on graphene via thermal evaporation or chemical reactions [52]. This restricts graphene restacking and improves transport. In addition, strong covalent or dipole–dipole interactions can be generated by anchoring nanoparticles to the hydroxyl and carboxyl groups of GO.

Carbon nanotubes (CNTs), which bridge graphene sheets, are extensively explored 1D materials. By forming a conductive channel the tubular configuration of CNTs not only improves electron transport but also increases the strength and flexibility of the material. Incorporating CNTs into graphene results in a hybrid material that forms highly conductive 3D structures with an increased surface area. This synergistic blend enhances the performance of energy storage devices by preventing the restacking of graphene and allowing rapid cycling. In addition to anchoring defects and decorating the edges of CNTs, this process stabilizes CNTs within graphene frameworks, leading to a robust hybrid architecture [53].

14.4.6.2 Metallic hybrids

Graphene–copper (Gr–Cu) hybrids are promising metallic composites for improved energy and electrical applications. Although copper is extensively used owing to its excellent electrical and thermal conductivity, ductility, and affordability, its long-term stability, oxidation resistance, and mechanical strength are limited, restricting its applications [54]. In copper/graphene composites, graphene acts as a reinforcing filler that helps decrease electron scattering and enhances grain boundary consistency. Effective mass and heat transport were attained owing to the strong interface bond between the graphene and copper matrix. Besides, adding graphene to copper composites improves their hardness and strength. For example, well-developed graphene/copper composites have shown better oxidation resistance and mechanical strength than pure copper, with conductivity levels comparable to bulk copper [55].

In addition to graphene–Cu composites, graphene-indium laminates represent another category of emerging metallic hybrids being investigated to address the increasing demand for advanced thermal management and high-performance electronic materials. The synergistic bonding within these laminates enables the composite to exhibit balanced and multifunctional properties required for various applications, including energy devices and electronic gadgets. The laminated graphene-indium hybrid structure facilitates efficient mass transport and reduces interfacial resistance. A recent study reported that a hybrid laminated composite of a graphene film integrated with indium and aluminum demonstrated superior thermal conductivity and mechanical characteristics. This improvement was attributed to the presence of an indium layer that bridged the graphene film in the composite matrix [56].

14.5 Structure–property relationships

The performance of TIMs is intrinsically governed by their interfacial architecture and the resulting physicochemical interactions. Therefore the rational design of TIMs requires a fundamental understanding of how structural features, such as filler morphology, surface chemistry, and dispersion, translate into macroscopic thermal and mechanical behaviors. Optimizing these relationships is essential for achieving high interfacial thermal conductance, mechanical compliance, and long-term stability under operational stress. Tailoring interfacial bonding, minimizing phonon scattering at heterogeneous boundaries, and ensuring mechanical integrity are central strategies for realizing next-generation high-performance TIMs.

14.5.1 Percolation thresholds

The percolation threshold represents the minimum concentration of filler materials required to form a continuous pathway to increase the conductivity of the composites. A study revealed that incorporating GNPs into the polymer matrix initially increases the thermal conductivity at a gradual rate owing to limited phonon transfer because of interfacial resistance. However, once the critical fraction limit for the filler material was reached, the conductivity improvement was rapid, highlighting the onset and importance of percolation behavior. Hence the threshold is a critical point for filler materials, which facilitates the efficient transport of phonons via interlinked networks to achieve high thermal conductivity in composite materials [57]. Interestingly, composites with larger GNPs exhibited better conductivity and lower percolation thresholds than those with smaller fillers. This is mainly attributed to the capacity of larger fillers to maintain a larger contact area among the neighboring filler materials, forming a continuous pathway for heat transport [57].

In another study, epoxy composites containing either graphene or boron nitride particles were investigated. Both filler materials demonstrated a significant increase in thermal conductivity after the filler level exceeded 20%. However, graphene outperformed boron nitride in terms of increasing heat transport and required less material to become electrically conductive. Heat is primarily transported through interconnected filler networks at high loadings, and the ability of the composites to control heat is significantly influenced by the cross-plane conductivity of the 2D fillers [58]. These studies indicate the effectiveness of 2D geometry in altering the limited R_{th} into an efficient percolated conductive path for thermal transport.

14.5.2 Thermal conductivity versus mechanical compliance

Graphitic nanofillers, including GNPs and CNTs, are widely used to enhance phonon-mediated heat transport in polymer-based TIMs. Their intrinsically high in-plane thermal conductivities, typically on the order of ~2000–5000 W m^{-1} K^{-1} for graphene and up to ~3000 W m^{-1} K^{-1} for defect-free CNTs, facilitate efficient phonon propagation along extended sp^2-bonded networks. When incorporated at sufficient loading levels, these fillers form continuous percolated pathways that substantially reduce interfacial phonon scattering. Consequently, polymer composites containing large lateral-dimension GNPs have been reported to exhibit bulk thermal conductivities exceeding 8 W m^{-1} K^{-1} at filler fractions of approximately 30 wt.%, highlighting the effect of enhanced filler connectivity and interface optimization on heat transport performance [57]. However, such high filler contents introduce a critical trade-off between thermal conductivity and mechanical compliance. Increasing the volume fraction and lateral size of rigid graphitic platelets substantially increases the composite modulus, suppresses conformability, and compromises flexibility. For TIM applications, mechanical compliance is not a secondary parameter but a functional requirement: effective thermal transport depends not only on intrinsic conductivity but also on minimizing R_{th} at the solid–solid contact. A mechanically stiff TIM may exhibit high bulk conductivity but perform poorly in device operation if it cannot conform to surface asperities and generate a sufficiently "wet" interface under moderate contact pressure.

Insufficient compliance leads to incomplete interfacial contact, entrapped air pockets, and microscopic voids, each acting as a low-conductivity thermal barrier (~0.026 W m^{-1} K^{-1} for air). These defects dramatically increase contact resistance, often dominating the overall thermal impedance of the interface. Accordingly the effective thermal performance of graphene-based TIMs is governed by the coupled optimization of filler-induced network conductivity and interfacial conformability [57]. Achieving this balance through controlled filler morphology, hybrid filler architectures, or elastomeric matrices is essential for translating the high intrinsic conductivity of GNPs and CNTs into reliable, low-resistance thermal interfaces under realistic device conditions.

14.5.3 Pressure dependence

Packaging pressure plays an indispensable role in measuring the thermal conductivity and resistance at the interface of TIMs. During device packaging, TIMs are compressed owing to the applied pressure, thereby reducing the voids and defects at the interface. Owing to this, both the surface wetting and the conformity are enhanced, lowering the contact resistance and improving the heat transport efficiency. Studies have shown that at high pressure, λ increases because of the improved contact between the TIM and the mating device surface. Nonetheless, excessive pressure can severely affect the effectiveness of the filler material and substrate by damaging their softness or squeezing out the TIMs in extreme cases. For instance, it has been depicted in a study that both the roughness and imperfect contacts are the major reasons for thermal contact resistance, and to address this issue, packaging pressure should be kept in controlled limits by forcing the TIMs into the irregularities, which can enhance the contact further [59].

14.6 Characterization techniques for nano–thermal interface materials

Characterization techniques are essential for elucidating the thermal, structural, and interfacial behavior of nanoengineered TIMs (nano-TIMs). Advanced microscopy, spectroscopy, and thermal analysis methods enable precise evaluation of microstructural features, composition, and heat transport pathways. Such comprehensive characterization provides fundamental insights into the mechanisms governing thermal conductivity, interfacial resistance, and material stability, thereby guiding the rational design of high-performance nano-TIMs for next-generation thermal management applications.

14.6.1 Laser flash analysis

Laser flash analysis (LFA) is one of the most widely explored characterization techniques for analyzing the thermal diffusivity of bulk materials, including nanoengineered TIMs. This method has been significantly refined in terms of instrumentation and data processing since its first publication by Parker et al., establishing it as a remarkable technique for exploring thermophysical properties. Parker et al. [60] showed that applying a brief, intense light wave onto a single side of the object being studied may cause the temperature on the other side to rise noticeably, allowing for the very accurate calculation of thermal diffusivity. This method is simple yet effective and has been standardized according to the American Society for Testing and Materials (ASTM) standard to ensure reproducibility and comparability across testing laboratories worldwide.

A powerful laser beam is focused on the sample, producing heat waves that are detected by a thermocouple and analyzed to determine the thermal diffusivity (α). Laser flash techniques can operate in two modes: by adjusting the distance between the source and sensor or by altering the intensity of the heat source. The frequency modulation method might underestimate α because of phase lag. With a 5% uncertainty for pyrolytic graphite, this method accurately determined the thermal diffusivity of metals such as stainless steel and copper. However, this system is complex and requires additional equipment, making it costly [61]. Using the sample thickness (d) and the recorded half-rise time ($t_{1/2}$), the thermal diffusivity ($\alpha = 0.1388 \times \frac{d^2}{t_{1/2}}$) can be computed. The thermal conductivities of the samples were determined using $\lambda = \alpha \times \rho \times C_p$, where C_p and ρ are the specific heat capacity and density of the sample, respectively [62]. When assessing the thermal attributes of nano-TIMs, the incorporation of mathematical modeling and computations is advantageous because it increases accuracy.

The transient performance of the LFA reduces the faults related to complex boundary conditions and heat losses. It requires only small sample sizes and simple geometries, which makes it more useful and convenient than steady-state approaches. The LFA integrated with the material design of electronic device applications in nano-TIMs significantly improves the realization of heat dissipation using graphene and CNTs [62]. The interfacial resistance and bulk diffusivity of a material are the primary determinants of LFA performance. By measuring diffusivity, LFA assists researchers in interlinking deviations at the nanoscale to thermal behavior.

LFA is highly utilized for TIM characterization because it is an approach used in commercial programs that automatically gather data and fit curves using advanced approaches. These devices have enabled TIMs to be explored under varying temperature conditions, making them ideal for real-world situations. The LFA can also be used in conjunction with density and specific heat measurements to determine the thermal conductivity, providing a comprehensive analysis of the thermal properties of materials. As nano-TIMs are being developed for next-generation devices, LFA is crucial for measuring bulk diffusivity [63].

14.6.2 Time-domain thermoreflectance

Time-domain thermoreflectance (TDTR) has gained substantial interest as a versatile and prevailing technique for investigating R_{th} at the nanoscale level. It utilizes femtosecond or picosecond laser pulses to determine the heat propagation through thin films and at interfaces with exceptional temporal precision. TDTR is a technique that monitors thermal properties by observing thermoreflectance, that is, by measuring the changes in reflectance with variations in temperature. Samples with a thin metal layer usually show a linear change in their reflectance when the temperature is slightly increased. This method, developed between the 1970s and the 1980s, initially employed continuous light sources for both heating and sensing. Subsequently the development of laser technology became a key factor in measuring various thermal characteristics. This technique was primarily used by Paddock and Eesley to examine the thermal diffusivity within metallic sheets. Later, Maris's group focused on this technique using picosecond light flashes for ultrasonic research. Over the past few decades, TDTR has been used to measure the anisotropic thermal conductivity and assess phonon transport in materials [64].

Typically, in a TDTR experiment, a thin metal layer is coated on the testing sample to absorb the laser energy and transform it into heat. The variations in the reflectivity of the materials were measured using a time-delayed probe beam [64]. Finally, one or two important thermal parameters were governed until the model precisely reflected the observed behavior after comparing the experimental results to the predictions of a thermal model. In polymer-based nano-TIMs, a thermal transfer bottleneck arises between the matrix and reinforcement interface. This interfacial bottleneck between the surfaces is measured directly by TDTR, which traces the rate at which heat escapes from the heated surface into the underlying layers of material. For instance, at heterogeneous interfaces, TDTR can resolve interfacial conductance values as low as a few MW m^{-2} K, offering new insights into phonon scattering and mismatch processes [65]. This aspect is highly essential for forming TIMs with improved heat dissipation rates, particularly for high-power devices. In addition, the optical nature of TDTR facilitates a noncontact measurement system, limiting the requirement for direct temperature sensors or electrodes, which could impact fragile nanoscale structures.

Furthermore, TDTR is renowned for its accuracy and versatility in experimental design. Researchers can regulate the probe delay or modulation frequency of the pump beam to alter the depth of thermal penetration and focus on specific layers at the interfaces. This approach can be extremely useful for investigating 2DMs, superlattices, and multilayered structures, particularly if the thermal transport varies. Machine learning (ML) for data processing and multidelay probe configurations are recent developments that have increased measurement speed and accuracy [66]. These advancements position TDTR as a valuable research tool and potentially a standard for nanoscale thermal measurements, in addition to more traditional methods such as the LFA.

14.6.3 Steady-state heat flow methods for assessing thermal interface materials

Steady-state heat flow techniques remain among the most widely recognized approaches for evaluating the effective thermal performance of TIMs under realistic operating pressure and boundary conditions. Unlike transient methods, such as LFA and TDTR, steady-state techniques require the establishment of a uniform and constant temperature gradient across the specimen while accurately determining the resulting heat flux. These methods rely on direct application of Fourier's law of heat conduction, with variations depending on the specific measurement configuration. Common steady-state techniques include the absolute method, comparative cut-bar method, radial heat flow method, and parallel thermal conductance technique. Although these approaches are valued for their physical transparency and simplicity of theoretical interpretation, their implementation is experimentally demanding, often involving long equilibration times and stringent control of thermal losses, which limits their practicality compared to transient methods.

Among these the ASTM D5470 standard remains the benchmark protocol for industrial assessment of TIMs. This method employs a 1D heat flow configuration, typically realized with a guarded hot plate (GHP), in which the TIM is sandwiched between two highly conductive metal blocks, often made of copper, and subjected to controlled mechanical pressure. Heat is applied to one side of the assembly, and the resulting temperature gradient across the specimen is precisely measured. From this temperature drop, both the thermal resistance and effective thermal conductivity of the TIM can be quantified. The ASTM D5470 approach has been extensively validated and remains the benchmark technique for industrial TIM characterization because of its reproducibility, accuracy, and relevance to practical device conditions [67].

The absolute GHP method is one of the primary techniques used to determine the thermal conductivity of materials, particularly low-to-medium conductive materials. It operates by generating a consistent heat flow around a flat specimen using a central heater surrounded by a guard heater to prevent heat loss [68]. This technique enables precise measurements of thermal conductivity based on the size of the specimen and temperature gradient without the use of any reference standard. The GHP method is highly recognized for its accuracy and capacity to calibrate various other measuring instruments. However, it requires a large sample and can take a long time to achieve the desired temperature balance. Despite these limitations, it is still regarded as the optimal approach for insulating materials and thermal substrates. In addition, the GHP technique efficiently decreases errors caused by contact resistance and heat loss [68].

Notably the radial heat flow approach minimizes the errors caused by the edge effect by allowing the heat to diffuse outward from the center of the cylindrical specimen. Temperatures were measured at various distances from the core center, where the heater was located [69]. This design is suitable for powders that cannot be molded into solid shapes and for high-temperature testing, as it decreases end losses. This method is particularly useful in high-power thermal management applications, such as nuclear and aircraft systems. This creates a consistent environment for studying the effects of temperature on material changes.

14.6.4 Infrared thermography

IR imaging has become one of the most critical techniques for the characterization and evaluation of TIMs for real-world applications, particularly for electronic devices, where hot spots can have detrimental effects. Hot spots can substantially reduce the reliability and efficiency of such devices. Unlike LFA and TDTR, this technique can generate a direct and spatial visualization of the temperature distribution across the operating specimen. High-resolution IR cameras capture the emitted IR radiation from the component surface, generating thermal maps that indicate regions with higher temperatures. This allows researchers to measure the effectiveness of TIMs in reducing these hotspots. IR is particularly useful for nano-TIMs, which are designed to improve heat dispersion and reduce local thermal resistance. IR imaging provides a platform for nondestructive and real-time monitoring, making it an essential characterization method for evaluating materials in realistic operating environments [70].

IR thermography has shown great promise for evaluating hotspot reduction techniques, especially for nano-TIMs. The CT scan graph and IR image for 2DMs, along with the thermal behavior and scattering mechanism in 2DMs, are shown in Fig. 14.2E–G. Nano-TIMs are intended not only to reduce the total thermal resistance but also to distribute the heat more homogeneously, as well as to decrease localized overheating. IR imaging directly measures this effect by comparing the thermal maps of devices with and without the improved TIMs. According to studies, devices using nano-TIMs have much lower hot spot intensity and exhibit improved thermal uniformity, which can be easily recorded via IR visualization. Better IR thermography methods, for example, have been used to measure hotspot temperatures in power electronics, demonstrating how sophisticated TIMs can reduce peak operating temperatures and increase device lifespans [71]. These findings highlight the value of IR thermography as a diagnostic tool, allowing researchers to link nanoscale material design to macroscopic thermal performance in real-world applications.

IR thermography has emerged as a powerful diagnostic tool for evaluating TIMs under realistic operating conditions. Unlike steady-state or quasi-static laboratory measurements, such as GHP or laser flash techniques, which often idealize boundary conditions, IR thermography enables the direct visualization of temperature fields in functioning devices. This distinction is critical because practical heat dissipation in electronics involves nonuniform heat fluxes, interfacial resistances, and transient thermal loads that are rarely captured in simplified test geometries. By imaging operational systems (e.g., high-power LEDs, processors, and transformers), IR thermography provides spatially resolved surface temperature maps that reveal hotspots, thermal gradients, and the effectiveness of interfacial heat spreading in situ [72].

Previous IR imaging studies have demonstrated precise identification of localized overheating and dynamic temperature evolution in power components, including transformers. When applied to TIM assessment, this approach enables quantitative comparison of peak temperatures (T_{max}), temperature uniformity, and temporal thermal response before and after TIM integration. Such metrics are essential for determining whether nanoscale design strategies, such as the incorporation of graphene sheets or CNTs architectures, translate into meaningful reductions in R_{th} and improved lateral heat spreading at the device level. In this context, IR thermography bridges the gap between material-level enhancements (e.g., increased intrinsic thermal conductivity) and system-level performance.

Recent advances in IR imaging hardware and data analytics have significantly improved both the spatial and temporal resolutions. Modern focal-plane array detectors generate high-definition thermal maps with rapid frame rates, enabling capture of fast transient heating and cooling events. This capability is particularly relevant for pulsed-power electronics, where temperature fluctuations occur on subsecond time scales. Moreover, quantitative temperature tracking under variable loads allows for a rigorous evaluation of TIM reliability, including thermal cycling stability and hotspot suppression efficiency [72].

Notably, IR thermography can be integrated with numerical tools, such as finite element modeling, to establish a robust experimental–computational framework. Surface temperature distributions obtained experimentally can be used to validate simulated heat flux pathways and interfacial resistance models, thereby refining predictions of device-scale thermal behavior. As electronic architectures continue to shrink while power densities increase, real-time hotspot visualization has become indispensable. In this regard, IR thermography is not merely a qualitative imaging technique but a critical validation platform for nanoengineered TIMs, ensuring that nanoscale enhancements yield tangible improvements in next-generation thermal management systems [72].

14.7 Applications in high-power electronics

Graphene and other 2DMs have attracted increasing attention for thermal management in high-power electronic systems. Their exceptional intrinsic thermal conductivity, mechanical strength, and chemical stability make them promising candidates for integration into next-generation TIMs. These materials can effectively dissipate localized heat and mitigate thermal accumulation, thereby enhancing both the performance and operational reliability of power devices operating under high thermal and electrical stress. Incorporating 2DM-based composites into device architectures enables improved heat spreading, reduced thermal resistance, and prolonged component lifetimes, which are key parameters for advancing efficient and durable electronic systems.

14.7.1 Cooling GPUs and central processing unit (CPU) dies (NVIDIA/Intel case studies)

The increasing demand for computational power from AI jobs has necessitated the adaptation of complex heat control technologies, particularly for high-performance GPUs and CPUs. With future AI expansions expected to attain greater heights, modern NVIDIA and Intel CPUs are exceeding thermal design powers of several hundred watts. Conventional air-cooling systems are insufficient. Thus advanced cooling systems and sophisticated TIMs are necessary.

Currently, NVIDIA uses liquid metals, such as Galinstan, to significantly reduce interfacial resistance and enhance cooling efficiency in AI data collection and processing centers [73]. This demonstrates that TIMs are critical for regulating the heat generated by GPUs, particularly for deep learning and generative AI applications. Likewise, Intel is also in the process of developing TIMs to address the critical issue of CPU hotspots. This clearly demonstrates the effective utilization of various materials, such as solder-based TIMs and liquid cooling, to improve thermal management [74].

Furthermore, TIMs are crucial for reducing the cooling costs of AI data centers, where thermal management generally wastes a significant amount of energy. Because future AI computers may experience an increase in power demand of up to 15,000 W, new and innovative cooling mechanisms will be required, and TIMs can provide a solution for optimal heat transport [75]. As AI workloads expand, both NVIDIA and Intel's TIM technologies are advancing to ensure the performance and reliability of high-power computing systems.

14.7.2 5G/6G infrastructure: Managing heat in radio frequency power amplifiers (GaN-on-SiC)

The erection of 5G and upcoming 6G networks is enforcing the development of more RF power amplifiers, specifically based on gallium nitride (GaN) decorated on silicon carbide (SiC). High-power RF systems prefer GaN-on-SiC over other silicon-based devices because of its greater breakdown voltage, excellent thermal conductivity, and improved electron mobility. High-power densities can create hot spots, which damage the performance and lifespan of the device. Therefore effective management solutions for governing heat are critical to ensure reliable operation in 5G and 6G infrastructure [76]. In different cases, it has been observed that the main thermal issue in GaN-on-SiC amplifiers is the case-junction thermal resistance. This can affect the manner in which heat is eliminated from the components. Conventional approaches, on the other hand, are no longer sufficient to meet modern heat dissipation requirements.

Systematic thermal characterization and sophisticated TIMs are central to accurately predicting temperature variations and device longevity. 2DMs and graphene-integrated composites are thus becoming increasingly crucial for reducing junction temperatures while increasing efficiency. GaN-on-SiC devices are preferred over GaN-on-Si devices for controlling heat; therefore, electronic packaging and device design are also important. Embedded microchannel cooling and flip-chip bonding are two strategies that reduce thermal resistance and provide uniform heat dispersion [77]. Real-time monitoring of thermal systems can provide insights into predictive maintenance. Improved temperature regulation is critical for the large-scale application of 5G and 6G networks. Well-designed solutions not only improve performance but also support infrastructure requirements for sophisticated wireless connectivity. However, inadequate temperature control can lead to reduced amplifier efficiency and service quality.

14.7.3 Electric vehicles: Battery thermal management systems using graphene-enhanced pads

The necessity of an effective battery thermal management system (BTMS) has increased in recent years owing to the growing popularity of electric vehicles (EVs) and the sensitivity of lithium-ion batteries to temperature changes. Keeping batteries within a safe temperature range (20°C–40°C) is crucial for their safety and longevity, as well as energy economy. Traditionally used cooling processes, such as liquid cooling and thermal pads, exhibit degradation with an increase in temperature to higher levels because of rapid charge/discharge cycles. Conversely, graphene-enhanced thermal pads have shown great potential as useful alternatives because of their high thermal conductivity and flexibility. These cushioning pads not only improve the heat distribution but also reduce hot spots and ensure consistent temperatures throughout battery systems [78].

Furthermore, graphene-based pads utilize less energy in EVs, simplify the system, and increase cooling efficiency. They adapt better to surface deficiencies than traditional rigid materials, resulting in improved heat transfer [79]. Studies have revealed that graphene-enhanced materials can significantly increase thermal regulation at great extent, particularly during high discharges, and support the inhibition of thermal runaway, which is a critical safety concern. Furthermore, their small weight promotes vehicle efficiency. Experimental results revealed that graphene pads can dramatically lower battery temperatures to the required limits while improving thermal profile uniformity [79]. Future developments that utilize graphene-enhanced BTMS may include the integration of sophisticated monitoring systems for active heat management and the combination of graphene with other materials for improved performance.

14.7.4 Optoelectronics: High-brightness light-emitting and laser diodes

Owing to their outstanding efficiency and extended lifespan, high-brightness LEDs and semiconductor laser diodes have revolutionized the optoelectronics industry. However, these diodes generate abundant heat during operation, which poses serious heat management challenges. The generated heat is one of the reasons for material degradation, alterations in wavelength, and decreased efficiency. This indicates that overheating is the primary cause of LED failures, highlighting the necessity of better thermal regulation to ensure dependability and brightness. The challenge is even more severe with high-power laser diodes because localized hot patches can have catastrophic effects and damage optics, necessitating careful thermal design [80]. Traditional packaging materials struggle to dissipate heat effectively, especially when applied to high-power LEDs. Innovative packaging solutions, like thermal via arrays, help reduce junction temperatures by channeling heat to external sinks, thereby enhancing performance and extending lifespan [81]. In addition, improved heat distribution using innovative TIMs increases luminous output and dependability. IR thermography tests have shown that improved thermal designs support a constant color and brightness in LED arrays.

Laser diodes have additional heat problems owing to their high power densities. Heat control techniques include reducing thermal resistance and ensuring uniform heat distribution [80]. Techniques such as flip-chip bonding and microchannel cooling are necessary for better heat removal. Effective TIMs contribute to the stabilization of the performance by reducing the temperature gradients. The increasing demand for brightness and efficiency in optoelectronics necessitates further advancements in their thermal management. Technologies such as graphene-enhanced TIMs and real-time heat monitoring are essential for preserving the performance and reliability of devices in high-power applications (Table 14.2).

14.8 Reliability, scalability, and sustainability

The transition of graphene and other 2DMs from research environments to industrial applications is fundamentally dependent on their reliability, scalability, and environmental sustainability. Reliability ensures that TIMs perform consistently and remain stable over time, even when exposed to real-world conditions, such as temperature fluctuations, mechanical stress, and environmental exposure. Scalability pertains to the capacity to reproduce the properties and performance of a material on a large scale without compromising quality or cost-effectiveness. Equally significant, sustainability involves assessing the environmental impact of the production, processing, and disposal stages, ensuring that large-scale applications adhere to ecofriendly manufacturing practices. By evaluating these three dimensions collectively, it becomes feasible to develop TIMs that not only provide excellent thermal performance but are also economically viable and environmentally sustainable. This comprehensive approach ensures that graphene-based and other 2DM technologies can progress towards practical implementation in sectors such as electronics, energy, and advanced manufacturing.

14.8.1 Thermal cycling and aging

The long-term durability of 2D material networks, such as graphene and GO composites, is essential for their use as TIMs. As discussed in earlier sections, these materials can experience frequent heating and cooling cycles because of device operation fluctuations, leading to structural alterations and a loss of thermal conductivity. Although GO degrades significantly with time, research shows that reduced GO is much more stable than GO, despite the small alterations that occur during cycling [85]. Degradation is largely a result of mechanical and chemical processes, which increase the resistance and decrease the heat conductivity. Techniques such as coating, functionalization, and hybrid composite formation can help overcome these issues by enhancing stability. According to a study, after continuous cycling, carbon fiber composites can retain ~80% of their initial conductivity [86]. For 2M-based TIMs to be widely accepted by the industry, they must exhibit consistent and long-term performance. Recycling old 2DM composites emphasizes the need for stability in developing dependable and ecofriendly heat management systems for contemporary appliances.

14.8.2 Manufacturing bottlenecks

The large-scale deployment of 2DMs in TIMs is governed not only by their intrinsic thermal transport properties but also by the techno-economic viability of their synthesis. Among the available routes,

Table 14.2 Performance benchmarks: 2DM–TIMs versus leading commercial products (Bergquist, Laird, and Honeywell).

Property/metric	2DM–TIMs (graphene, h-BN composites)	Bergquist (Gap Pad HC5000, Sil-Pad 1600S)	Laird (TPCM 7000)	Honeywell (PTM7950)
Bulk thermal conductivity ($W\ m^{-1}\bullet K$)	8–12 (graphene/h-BN composites) [3]	5.0 [82]	7.5 (TPCM 7000) [83]	8.5 [84]
Thermal Impedance ($°C\bullet cm^2\ W^{-1}$)	~2–3 (optimized graphene composites) [3]	0.61°C-in W^{-1} (~3.9°C $cm^2\ W^{-1}$) [82]	<0.2 (phase change, thin bondline) [83]	0.04–0.08 [84]
Bond line thickness (μm)	5–10 (lab composites, nanosheet alignment)	229 μm [82]	25–35 μm [83]	20–30 μm typical [84]
Interface resistance ($K\bullet cm^2\ W^{-1}$)	~1.0–1.5 [3]	~2–3 [82]	~1.8 [83]	~1.8 [84]
Reliability (thermal cycling)	High(stable Solid Electrolyte Interphase layer (SEI)-like interfacial layer) [3]	Good [82]	Proven >2000 h aging tests	1000 cycles T/C-B, 150°C bake

2DM, *two-dimensional material;* TIM, *thermal interface material.*

CVD remains the benchmark for producing high-crystallinity, large-area graphene with minimal defect density and correspondingly high in-plane thermal conductivity, often exceeding 2000 W m^{-1} K^{-1} under ideal conditions. Such structural perfection underpins the suitability of these materials for high-performance heat-spreading applications. However, the scalability of CVD-derived graphene is constrained by capital-intensive infrastructure, high-temperature operations (typically ≥1000°C), controlled atmospheres, and stringent precursor purity requirements. These factors collectively result in elevated energy consumption and limited throughput, restricting cost-effective mass production [54,87].

In contrast, liquid-phase exfoliation (LPE) offers a fundamentally different production paradigm. In LPE, graphite is dispersed in suitable solvents and exfoliated via ultrasonication or shear mixing, enabling bulk production at comparatively low cost and ambient or near-ambient processing conditions. Although LPE graphene typically exhibits smaller lateral dimensions, broader thickness distributions, and higher defect densities than CVD materials, which reduce the intrinsic thermal conductivity, its process scalability and economic accessibility render it attractive for composite-based TIMs, where percolative heat transport can compensate for reduced flake-level performance [87]. Thus the economic disparity between CVD and LPE arises from differences in thermodynamic driving forces, process complexity, and material yield, rather than solely from differences in final product quality.

Therefore the selection of CVD- or LPE-derived graphene for industrial TIM manufacturing must balance the surface chemistry, structural quality, thermal performance, and cost per unit area. CVD graphene is preferable where maximal phonon MFP and minimal grain-boundary scattering are critical, such as in high-power electronics requiring ultralow interfacial thermal resistance. In contrast, LPE graphene, despite its lower intrinsic conductivity, can achieve thermally conductive networks within polymer matrices at a substantially reduced cost, making it suitable for large-area or moderate-performance applications. Hybrid strategies that integrate CVD-grown graphene with solution-processed flakes or combine exfoliated graphene with complementary fillers have emerged as promising pathways to reconcile performance with economic feasibility [54].

Sustainability considerations further complicate these trade-offs. The energy-intensive nature of CVD, combined with high-temperature processing and gas-phase precursors, raises concerns regarding carbon footprint and lifecycle impact. Although LPE methods are generally less energy-demanding, they involve solvent handling and recovery challenges that must be addressed to ensure environmentally responsible scaling. Therefore future progress in graphene-enabled thermal management will depend not only on improving phonon transport metrics but also on reducing synthesis energy intensity, increasing yield efficiency, and implementing greener process chemistries. Overcoming these production-level constraints is essential to unlock the full potential of graphene-based TIMs in next-generation heat management technologies.

14.8.3 Sustainability

The ability to recycle 2D material-filled polymer composites, particularly those incorporated with graphene, is essential for developing highly sustainable heat management solutions. Because of their excellent thermal properties, these composites are widely used in different fields of engineering and science, even though recycling polymer composites is a huge challenge. Studies have revealed that traditional methods, which are unable to produce high-quality nanomaterials, can degrade the performance of composites through recycling. Solvent-assisted methods have shown great promise to produce graphene-based polymer composites [85]. New advancements in green chemistry can reduce environmental impact and enhance recyclability. Establishing regular recycling practices is essential to ensuring sustainability and economic feasibility as the demand for these materials in electronics and energy systems continues to increase. For environmental ethics, recyclability must be incorporated into material design.

14.9 Future outlook

The next decade will redefine TIMs through the convergence of materials informatics, nanoscale engineering, and intelligent system integration. As electronic platforms evolve toward higher power densities, driven by AI accelerators, hyperscale data centers, EVs, and 5G/6G infrastructure, the thermal load at chip–package and package–heatsink interfaces approaches critical limits. State-of-the-art AI processors have already exceeded local heat fluxes of 1 kW cm^{-2} under transient conditions, imposing unprecedented demands on interfacial thermal conductance. Therefore meeting the 2030 performance roadmap will require a paradigm shift from empirical formulation to predictive, data-driven material design.

14.9.1 Machine learning for thermal interface material design

Conventional TIM development relies heavily on iterative trial-and-error optimization of the filler loading, aspect ratio, hybridization strategy, and interfacial chemistry. This approach is intrinsically slow and resource-intensive, particularly for multicomponent systems incorporating graphene, boron nitride, MXenes, metallic particles, or ceramic nanofillers. Recent advances in ML and materials informatics offer scalable alternatives. Supervised learning model strained on curated datasets linking composition, microstructure, and measured thermal conductivity can now predict optimal filler distributions, percolation thresholds, and hybrid ratios with significantly reduced experimental overhead. For example, gradient-boosting and neural network frameworks have demonstrated accurate predictions of the composite thermal conductivity across diverse filler systems when descriptors such as intrinsic filler conductivity, interfacial thermal resistance, volume fraction, and orientation anisotropy are incorporated. Beyond property prediction, inverse design strategies enable the targeted formulation of TIMs to satisfy predefined constraints, such as thermal conductivity, viscosity, mechanical compliance, and dielectric strength. The integration of high-throughput experimentation with active learning loops further accelerates discovery cycles, positioning ML-guided optimization as a central tool in next-generation TIM engineering.

14.9.2 Eliminating the remaining interfacial bottlenecks

Current state-of-the-art TIMs typically exhibit thermal conductivities in the range of 20–60 W m^{-1} K^{-1}, with a few exceeding 100 W m^{-1} K^{-1}. According to the 2030 technology roadmap the next milestone is to surpass 200 W m^{-1} K^{-1}, which is approaching the thermal conductivity of bulk aluminum. Achieving this target will require precise control of interfacial phonon transport, particularly the minimization of interfacial thermal resistance resulting from phonon scattering and lattice mismatch. Advanced interface engineering, including functionalized surface coupling and high-density defect modulation, is likely to be central to eliminating these remaining bottlenecks.

14.9.3 Smart thermal interface materials

Beyond static composites, next-generation TIMs are engineered as adaptive systems capable of dynamically modulating thermal transport in response to operational conditions. Phase-change materials and phase-transforming 2DMs offer tunable thermal pathways. Upon reaching defined temperature thresholds, these materials can alter crystallinity, filler connectivity, or interfacial coupling, thereby modifying thermal conductivity and reducing transient hot-spot formation. In polymer-based TIMs, incorporation of thermally responsive fillers enables reversible softening and reflow under elevated temperatures, improving surface conformity and minimizing contact resistance. Emerging concepts envision thermally switchable interfaces integrated with sensor feedback loops and AI-driven control systems. In such architectures, embedded sensors detect local temperature gradients and trigger adaptive structural reconfiguration, either by activating phase transitions or modulating interfacial pressure, effectively enabling programmable heat flow. Although still in the early stages, these reconfigurable thermal interfaces could play a critical role in next-generation AI hardware and EV platforms, where rapid fluctuations in power density demand continuous thermal regulation. Between 2024 and 2034, deployment of intelligent TIM systems is expected to expand significantly in high-performance computing and electrified transportation. However, practical implementation requires simultaneous advances in scalability, long-term reliability under cyclic loading, and compatibility with automated manufacturing processes.

14.10 Conclusion

2DM–enhanced TIMs represent a significant advancement in interfacial heat management, utilizing anisotropic phonon transport and engineered filler networks to surpass the conductivity limitations of traditional polymer systems. When the dispersion, orientation, and matrix compatibility are optimized, 2DM fillers can substantially reduce the bond-line thermal resistance while maintaining mechanical compliance, which is a critical requirement for advanced electronic packaging. However, effective thermal transport in practical composites is more influenced by interfacial scattering, contact resistance, and mesoscale morphologies than by intrinsic lattice conductivity. Consequently the most promising developments have resulted from hierarchical composite architectures, surface functionalization strategies, and hybrid filler systems that stabilize conductive pathways under thermal and mechanical cycling. From a systems perspective the adoption of 2DM TIMs depends on their reliability, process compatibility, and cost parity with existing technologies. Thermal performance improvements must translate into measurable enhancements in device lifetime, junction temperature stability, and energy efficiency. Ongoing advancements in scalable synthesis, dispersion science, and interface engineering suggest that 2DM-enabled TIMs are nearing maturation, where performance optimization is increasingly linked to manufacturability. The long-term impact of these materials will depend on their integration into standardized packaging workflows and their ability to maintain thermal performance under operational stress.

References

[1] E. Pop, Energy dissipation and transport in nanoscale devices, Nano Res. 3 (3) (2010) 147–169, doi:10.1007/s12274-010-1019-z.

[2] K.M. Razeeb, E. Dalton, G.L.W. Cross, A.J. Robinson, Present and future thermal interface materials for electronic devices, Int. Mater. Rev. 63 (1) (2018) 1–21, doi:10.1080/09506608.2017.1296605.

[3] K.M.F. Shahil, A.A. Balandin, Thermal properties of graphene and multilayer graphene: applications in thermal interface materials, Solid State Commun. 152 (15) (2012) 1331–1340, doi:10.1016/j.ssc.2012.04.034.

[4] L. Lv, W. Dai, A. Li, C.-T. Lin, Graphene-based thermal interface materials: an application-oriented perspective on architecture design, Polymers (Basel) 10 (11) (2018) 1201, doi:10.3390/polym10111201.

[5] E. Pop, V. Varshney, A.K. Roy, Thermal properties of graphene: fundamentals and applications, MRS Bull. 37 (12) (2012) 1273–1281, doi:10.1557/mrs.2012.203.

[6] X.-S. Wu, W.-T. Tang, X.-F. Xu, Recent progresses of thermal conduction in two-dimensional materials, Acta Phys. Sin. 69 (19) (2020) 196602, doi:10.7498/aps.69.20200709.

[7] Y. Fu, et al., Graphene related materials for thermal management, 2D Mater. 7 (1) (2020) 012001, doi:10.1088/2053-1583/ab48d9.

[8] H. Fu, et al., High-performance thermal interface materials enabled by vertical alignment of lightweight and soft graphene foams, Nano Res. 17 (11) (2024) 9293–9299, doi:10.1007/s12274-024-6985-7.

[9] T. Zhao, B. Liu, Y. Wang, J. Liang, Z. Pei, X. Zhang, Optimizing the thermal conductivity of 2D materials/copper composites through strain-controlled electron-phonon coupling effect, Adv. Electron. Mater. 11 (16) (2025), doi:10.1002/aelm.202500133.

[10] H. Song, J. Liu, B. Liu, J. Wu, H.-M. Cheng, F. Kang, Two-dimensional materials for thermal management applications, Joule 2 (3) (2018) 442–463, doi:10.1016/j.joule.2018.01.006.

[11] W. Dai, et al., 2D materials-based thermal interface materials: structure, properties, and applications, Adv. Mater. 36 (37) (2024), doi:10.1002/adma.202311335.

[12] L. Liu, et al., Graphene-based polymer composites in thermal management: materials, structures and applications, Mater. Horiz. 12 (1) (2025) 64–91, doi:10.1039/D4MH00846D.

[13] H. Song, J. Liu, B. Liu, J. Wu, H.M. Cheng, F. Kang, Two-dimensional materials for thermal management applications, Joule 2 (3) (2018) 442–463, doi:10.1016/j.joule.2018.01.006.

[14] W. Ren, J. Chen, G. Zhang, Phonon physics in twisted two-dimensional materials, Appl. Phys. Lett. 121 (14) (2022), doi:10.1063/5.0106676.

[15] J. Li, et al., Oriented structural design of thermal interface materials: methods, properties, and challenges, Renewables 3 (4) (2025) 222–243, doi:10.31635/renewables.025.202500098.

[16] D. Ding, et al., Thermally conductive silicone rubber composites with vertically oriented carbon fibers: a new perspective on the heat conduction mechanism, Chem. Eng. J. 441 (2022) 136104, doi:10.1016/j.cej.2022.136104.

[17] Z. Wang, et al., Phonon transport in two-dimensional carbon-boron material and heterointerfaces, Surf. Interfaces 36 (2023) 102603, doi:10.1016/j.surfin.2022.102603.

[18] Z. Guo, Z. Han, A. Alkandari, K. Khot, X. Ruan, First-principles prediction of thermal conductivity of bulk hexagonal boron nitride, Appl. Phys. Lett. 124 (16) (2024), doi:10.1063/5.0210935.

[19] G. Rajasekaran, P. Narayanan, A. Parashar, Effect of point and line defects on mechanical and thermal properties of graphene: a review, Crit. Rev. Solid State Mater. Sci. 41 (1) (2016) 47–71, doi:10.1080/10408436.2015.1068160.

[20] D. Yang, F. Ma, Y. Sun, T. Hu, K. Xu, Influence of typical defects on thermal conductivity of graphene nanoribbons: An equilibrium molecular dynamics simulation, Applied Surface Science 258 (24) (2012) 9926–9931. https://doi.org/10.1016/J.APSUSC.2012.06.052.

[21] W. Blau, M. Liu, Y. Wang, Hexagonal boron nitride nanosheets: properties, preparation and applications in thermal management, Nanomaterials 16 (2) (2026) 101, doi:10.3390/nano16020101.

[22] K. Singh, et al., MXene as electrodes for energy storage: applications in batteries & supercapacitors, J. Mater. Sci. Mater. Electron. 36 (36) (2025) 2259, doi:10.1007/s10854-025-16220-w.

[23] D.M. Soares, S. Mukherjee, G. Singh, TMDs beyond MoS_2 for electrochemical energy storage, Chem. Eur. J. 26 (29) (2020) 6320–6341, doi:10.1002/chem.202000147.

[24] S.H. Lee, et al., Density-tunable lightweight polymer composites with dual-functional ability of efficient EMI shielding and heat dissipation, Nanoscale 9 (36) (2017) 13432–13440, doi:10.1039/C7NR02618H.

[25] S. Kumari, J. Dalal, V. Kumar, A. Kumar, A. Ohlan, Emerging two-dimensional materials for electromagnetic interference shielding application, Int. J. Mol. Sci. 24 (15) (2023) 12267, doi:10.3390/ijms241512267.

[26] S. Roy, et al., Structure, properties and applications of two-dimensional hexagonal boron nitride, Adv. Mater. 33 (44) (2021), doi:10.1002/adma.202101589.

[27] L.P. Yu, L. Xu, L. Lu, Z. Alhalili, X.H. Zhou, Thermal properties of MXenes and relevant applications, ChemPhysChem 23 (17) (2022) e202200203, doi:10.1002/cphc.202200203.

[28] L. Chen, X. Shi, N. Yu, X. Zhang, X. Du, J. Lin, Measurement and analysis of thermal conductivity of $Ti_3C_2T_x$ MXene films, Materials 11 (9) (2018) 1701, doi:10.3390/ma11091701.

[29] Z. Zhu, et al., Multifunctional $Ti_3C_2T_x$ MXene/boron nitride nanosheet/polydimethylsiloxane nanocomposites with integrated dielectric and electromagnetic interference shielding performance, Nanoscale 17 (38) (2025) 22368–22381, doi:10.1039/d5nr02867a.

[30] O. Samy, S. Zeng, M.D. Birowosuto, A. El Moutaouakil, A review on MoS_2 properties, synthesis, sensing applications and challenges, in: Crystals, 11, 2021, p. 355, doi:10.3390/cryst11040355.

[31] D. Gupta, V. Chauhan, R. Kumar, A comprehensive review on synthesis and applications of molybdenum disulfide (MoS_2) material: past and recent developments, Inorg. Chem. Commun. 121 (2020) 108200, doi:10.1016/j.inoche.2020.108200.

[32] A. Taube, J. Judek, A. Łapińska, M. Zdrojek, Temperature-dependent thermal properties of supported MoS_2 monolayers, ACS Appl. Mater. Interfaces 7 (9) (2015) 5061–5065, doi:10.1021/acsami.5b00690.

[33] J.H. Seol, et al., Two-dimensional phonon transport in supported graphene, Science 328 (5975) (2010) 213–216, doi:10.1126/science.1184014.

[34] I. Jo, et al., Thermal conductivity and phonon transport in suspended few-layer hexagonal boron nitride, Nano Lett. 13 (2) (2013) 550–554, doi:10.1021/nl304060g.

[35] S. Ferhat, C. Domain, J. Vidal, D. Noël, B. Ratier, B. Lucas, Flexible thermoelectric device based on $TiS_2(HA)_x$ n-type nanocomposite printed on paper, Org. Electron. 68 (2019) 256–263, doi:10.1016/j.orgel.2019.02.031.

[36] C. Chiritescu, et al., Ultralow thermal conductivity in disordered, layered WSe_2 crystals, Science 315 (5810) (2007) 351–353, doi:10.1126/science.1136494.

[37] D.D.C. Branco, G.J. Cheng, A strategy for 2D MXenes as thermal management materials by laser shock nanoshaping, J. Phys.: Mater. 6 (4) (2023) 045005, doi:10.1088/2515-7639/acef96.

[38] J. Ying, W. Dai, J. Yu, N. Jiang, C.T. Lin, Q. Yan, Rational design of graphene structures for preparing high-performance thermal interface materials: a mini review, Sci. China: Phys., Mech. Astron. 65 (11) (2022) 117005, doi:10.1007/s11433-022-2004-8.

[39] F. Lin, et al., Orientation control of graphene flakes by magnetic field: broad device applications of macroscopically aligned graphene, Adv. Mater. 29 (1) (2017), doi:10.1002/adma.201604453.

[40] H. Zhang, Ice Templating and Freeze-Drying for Porous Materials and Their Applications, 2018, pp. 1–363, doi:10.1002/9783527807390.

[41] J.S. Yeon, N. Gupta, P. Bhattacharya, H.S. Park, A new era of integrative ice frozen assembly into multiscale architecturing of energy materials, Adv. Funct. Mater. 32 (19) (2022), doi:10.1002/adfm.202112509.

[42] C.A. Banciu, F. Nastase, A.I. Istrate, L.M. Veca, 3D graphene foam by chemical vapor deposition: synthesis, properties, and energy-related applications, Molecules 27 (11) (2022) 3634, doi:10.3390/molecules27113634.

[43] L. Shi, et al., Scalable seashell-based chemical vapor deposition growth of three-dimensional graphene foams for oil–water separation, J. Am. Chem. Soc. 138 (20) (2016) 6360–6363, doi:10.1021/jacs.6b02262.

[44] Z. Zhao, L. Sun, Y. Li, W. Feng, Polymer-derived carbon materials for energy storage devices: a mini review, Carbon N. Y. 210 (2023) 118066, doi:10.1016/j.carbon.2023.118066.

[45] Z. Zhang, C. Lee, W. Zhang, Vertically aligned graphene nanosheet arrays: synthesis, properties and applications in electrochemical energy conversion and storage, Adv. Energy Mater. 7 (23) (2017), doi:10.1002/aenm.201700678.

[46] R. Nandee, M.A. Chowdhury, N. Hossain, M. Rana, M.H. Mobarak, R. Khandaker, Surface topography and surface morphology of graphene nanocomposite by FESEM, EDX and AFM analysis, Nano-Struct. Nano-Objects 38 (2024) 101170, doi:10.1016/j.nanoso.2024.101170.

[47] W. Fu, X. Zhao, W. Zheng, Growth of vertical graphene materials by an inductively coupled plasma with solid-state carbon sources, Carbon N. Y. 173 (2021) 91–96, doi:10.1016/j.carbon.2020.10.072.

[48] J. Jang, S.H. Lee, J. Kim, S.Y. Kim, S.H. Kim, Nano-bridge effect on thermal conductivity of hybrid polymer composites incorporating 1D and 2D nanocarbon fillers, Compos. B Eng. 222 (2021) 109072, doi:10.1016/j.compositesb.2021.109072.

[49] M. Aghelinejad, S.N. Leung, Thermoelectric nanocomposite foams using non-conducting polymers with hybrid 1D and 2D nanofillers, Materials 11 (9) (2018) 1757, doi:10.3390/ma11091757.

[50] Z. Shao, K. Shi, J. Wei, L. Zhou, D. Han, J. Gong, Hierarchical MOFs and derivatives toward advanced electrode materials for electrochemical energy storage, CrystEngComm 27 (41) (2025) 6697–6718, doi:10.1039/d5ce00772k.

[51] U. Hwang, et al., Structural design of hybrid fillers in a polymer matrix for advanced electromagnetic interference shielding, Small Struct. 6 (2) (2025), doi:10.1002/sstr.202400404.

[52] D. Yu, K. Park, M. Durstock, L. Dai, Fullerene-grafted graphene for efficient bulk heterojunction polymer photovoltaic devices, J. Phys. Chem. Lett. 2 (10) (2011) 1113–1118, doi:10.1021/jz200428y.

[53] S. Sinha, H. Kim, A.W. Robertson, Preparation and application of 0D-2D nanomaterial hybrid heterostructures for energy applications, Mater. Today Adv. 12 (7419) (2021) 100169, doi:10.1016/j.mtadv.2021.100169.

[54] K. Singh, et al., Panorama of physico-mechanical engineering of graphene-reinforced copper composites for sustainable applications, Mater. Today Sustain. 24 (2023) 100560, doi:10.1016/j.mtsust.2023.100560.

[55] W. Liu, X. Zhao, H. Li, Y. Ding, Research progress on the preparation and properties of graphene–copper composites, Metals 15 (10) (2025) 1117, doi:10.3390/met15101117.

[56] H. Liu, M. Zhang, S. Chen, N. Wang, Y. Zhang, J. Liu, Thermally conductive graphene film/indium/aluminum laminated composite by vacuum assisted hot-pressing, 2020 21st International Conference on Electronic Packaging Technology, ICEPT 2020, IEEE, New York, August 2020, doi:10.1109/ICEPT50128.2020.9202623.

[57] J.U. Jang, et al., Thermal percolation behavior in thermal conductivity of polymer nanocomposite with lateral size of graphene nanoplatelet, Polymers (Basel) 14 (2) (2022) 323, doi:10.3390/polym14020323.

[58] F. Kargar, et al., Thermal percolation threshold and thermal properties of composites with high loading of graphene and boron nitride fillers, ACS Appl. Mater. Interfaces 10 (43) (2018) 37555–37565, doi:10.1021/acsami.8b16616.

[59] L. Bharatham, W.S. Fong, C.J. Leong, C.-P. Chiu, A study of application pressure on thermal interface material performance and reliability on FCBGA package, in: 2006 International Conference on Electronic Materials and Packaging, IEEE, New York, December 2006, pp. 1–7, doi:10.1109/EMAP.2006.4430584.

[60] W.J. Parker, R.J. Jenkins, C.P. Butler, G.L. Abbott, Flash method of determining thermal diffusivity, heat capacity, and thermal conductivity, J. Appl. Phys. 32 (9) (1961) 1679–1684, doi:10.1063/1.1728417.

[61] R. Sundar, C. Sudha, Thermal transport and thermal diffusivity by laser flash technique: a review, Int. J. Thermophys. 46 (1) (2025) 13, doi:10.1007/s10765-024-03479-0.

[62] W. Xing, Y. Xu, C. Song, T. Deng, Recent advances in thermal interface materials for thermal management of high-power electronics, Nanomaterials 12 (19) (2022) 3365, doi:10.3390/nano12193365.

[63] S. Min, J. Blumm, A. Lindemann, A new laser flash system for measurement of the thermophysical properties, Thermochim. Acta 455 (1–2) (2007) 46–49, doi:10.1016/j.tca.2006.11.026.

[64] P. Jiang, X. Qian, R. Yang, Tutorial: time-domain thermoreflectance (TDTR) for thermal property characterization of bulk and thin film materials, J. Appl. Phys. 124 (16) (2018) 161103, doi:10.1063/1.5046944.

[65] R. Mohan, S. Khan, R.B. Wilson, P.E. Hopkins, Time-domain thermoreflectance, Nat. Rev. Methods Primers 5 (1) (2025) 55, doi:10.1038/s43586-025-00425-8.

[66] D.G. Cahill, Thermal-conductivity measurement by time-domain thermoreflectance, MRS Bull. 43 (10) (2018) 782–789, doi:10.1557/mrs.2018.209.

[67] B. Sponagle, D. Groulx, Characterization of thermal interface materials using a steady state experimental method, in: Proceedings of the ASME 2012 Summer Heat Transfer Conference, ASME Publishing, New York, July 2012.

[68] N. Yüksel, The review of some commonly used methods and techniques to measure the thermal conductivity of insulation materials, Insulation Materials in Context of Sustainability, InTech, 2016, doi:10.5772/64157.

[69] L. Fave, M.A. Pouchon, C. Hébert, A radial heat flow apparatus for thermal conductivity characterisation of cylindrical samples, J. Therm. Anal. Calorim. 130 (3) (2017) 1855–1863, doi:10.1007/s10973-017-6578-8.

[70] P. Mishra, A. Rani, R. Sekhar, R. Mulaveesala, Comparative analysis of infrared thermographic techniques for non-destructive testing and evaluation of materials, in: Lecture Notes in Mechanical Engineering, 2024, pp. 137–144, doi:10.1007/978-981-97-1036-2_13.

[71] M. Sahu, S.R. Sharma, A. Singh, R. Kumar, Y.R. Sood, An improved infrared thermography techique for hotspot temperature, per unit life and aging accelerating factor computation in transformers, in: 2020 IEEE International Conference on Computing, Power and Communication Technologies, GUCON 2020, IEEE, New York, October 2020, pp. 233–238, doi:10.1109/GUCON48875.2020.9231138.

[72] M. Zulkifli, M.Z. Muthaza, I. Safaat, S.S. Yi, N. Sahari, Thermography study on distribution transformer in substation: transformer core material, Int. J. Eng. Trends Technol. 71 (3) (2022) 9–16, doi:10.14445/22315381/IJETT-V71I3P202.

[73] T. Lei, NVIDIA Rubin Platform Thermal Management: Why Liquid Metal TIM is the Future of AI Cooling, Jiuju, 2026. https://jiujutech.com/nvidia-rubin-liquid-metal-tim-ai-cooling/ Accessed 16 Feb 2026.

[74] ServeTheHomeIntel Foundry Thermal Capabilities With TIM Options and In-Package Liquid Cooling Shown, ServeTheHome, 2026. https://www.servethehome.com/intel-foundry-thermal-capabilities-with-tim-options-and-in-package-liquid-cooling-shown/ Accessed 16 Feb 2026.

[75] Tom's HardwareFuture AI Processors Said to Consume Up to 15,360 Watts of Power—Massive Power Draw Will Demand Exotic Immersion and Embedded Cooling Tech, Tom's Hardware, 2026. https://www.tomshardware.com/pc-components/cooling/future-ai-processors-said-to-consume-up-to-15-360w-massive-power-draw-will-demand-exotic-immersion-and-embedded-cooling-tech (Accessed 16 Feb 2026).

[76] Y. Han, G. Tang, B.L. Lau, Thermal characterization and management of GaN-on-SiC high power amplifier MMIC, in: *Proceedings—Electronic Components and Technology Conference*, IEEE, New York, May 2023, pp. 1989–1993, doi:10.1109/ECTC51909.2023.00342.

[77] T. Orville, M. Tajwar, R. Bihani, P. Saha, M.A. Hannan, Enhancing thermal efficiency in power electronics: a review of advanced materials and cooling methods, Thermo 5 (3) (2025) 30, doi:10.3390/thermo5030030.

[78] R. Guo, Q. Miao, Y. Xu, Review of graphene applications in electric vehicle thermal management systems, World Electr. Vehic. J. 16 (3) (2025) 166, doi:10.3390/wevj16030166.

[79] Y. Liu, T. Thiringer, N. Wang, Y. Fu, H. Lu, J. Liu, Graphene based thermal management system for battery cooling in electric vehicles, in: *Proceedings—2020 IEEE 8th Electronics System-Integration Technology Conference, ESTC 2020*, IEEE, New York, September 2020, doi:10.1109/ESTC48849.2020.9229856.

[80] X. Liu, W. Zhao, L. Xiong, H. Liu, Thermal design and management in high power semiconductor laser packaging, in: Packaging of High Power Semiconductor Lasers, 2015, pp. 53–88, doi:10.1007/978-1-4614-9263-4_3.

[81] Y.J. Hwang, B.Y. Lee, M.J. Kim, S.C. Park, K. Won, S.U. Kim, Numerical analysis for cost-effective temperature reduction in high-power light-emitting diodes using thermal via array, Appl. Sci. 15 (12) (Jun. 2025) 6505, doi:10.3390/app15126505.

[82] Henkel AdhesivesBERGQUIST® GAP PAD® TGP HC5000, Henkel Adhesives, 2026. https://next.henkel-adhesives.com/in/en/products/thermal-management-materials/central-pdp.html/bergquist-gap-pad-tgp-hc5000/112657IB.html Accessed 21 Feb 2026.

[83] THR-DS-Tpcm 7000 Data Sheet 09022025. https://www.laird.com/products/thermal-interface-materials/phase-change/tpcm-7000.

[84] PTM7950, Honeywell. 2026. https://www.solstice.com/us/en/products/advanced-semiconductor-materials/thermal-management/phase-change-materials/ptm7950. (Accessed 21 Feb 2026).

[85] T. Hu, Y. Song, J. Di, D. Xie, C. Teng, Highly thermally conductive layered polymer composite from solvent-exfoliated pristine graphene, Carbon N. Y. 140 (2018) 596–602, doi:10.1016/j.carbon.2018.09.026.

[86] K. Dong, et al., A high-thermal-conductivity, high-durability phase-change composite using a carbon fibre sheet as a supporting matrix, Appl. Energy 264 (2020) 114685, doi:10.1016/j.apenergy.2020.114685.

[87] M. Hussain, P.K. Soni, A. Singh, Economical synthesis and optimization for few layers graphene by solid carbon sources: domestic CVD technique, Appl. Phys. A 131 (8) (2025) 636, doi:10.1007/s00339-025-08740-3.

CHAPTER

Applications of graphene and other two-dimensional materials in HVAC

15

Naveed Akram[a], Adiba Sharif[a], Afrin Jahan[b], Imdad Hussain Kalhoro[c] and Kazi Salim Newaz[d]

[a]*Department of Mechanical Engineering, Mirpur University of Science and Technology (MUST), Mirpur, Azad Jammu and Kashmir, Pakistan,* [b]*Department of Mechanical Engineering, Faculty of Engineering, Nanotechnology & Catalysis Research Centre (NANOCAT), Institute for Advanced Studies, Universiti Malaya, Kuala Lumpur, Malaysia,* [c]*Department of Electrical Engineering, Quaid-e-Awam University of Engineering, Science & Technology (QUEST), Nawabshah, Sindh, Pakistan,* [d]*Department of Mechanical Engineering, Faculty of Engineering, Universiti Malaya, Kuala Lumpur, Malaysia*

15.1 Introduction

Nanorefrigerants are an innovative category of refrigerants that involve blending nanoparticles into a pure refrigerant. Maxwell [1] proposed an initial method that entailed the dispersion of solid particles into the base fluid. This process resulted in the dispersion of numerous microparticles that enhanced the thermophysical properties of the working fluid. Despite its potential benefits, this fluid type has encountered various significant challenges, such as stability, corrosion, and erosion. Choi [2] subsequently developed a contemporary comprehension of nanoparticle distribution in a base fluid by integrating nanotechnology, resulting in the creation of nanofluids. The primary motivation behind the development of nanofluids was to enhance the efficiency of heat transfer analysis in working fluids, a concept that has now been expanded to refrigerants.

Multiple studies have reported enhanced effective thermal conductivity of base fluids by incorporating solid nanoparticles [2–4]. In particular the use of carbon nanotube nanoparticles in pure R-113 resulted in a remarkable increase of approximately 104% in effective thermal conductivity [5]. However, the analysis of nanofluids is not limited to their thermal conductivity alone. Researchers also need to explore other thermophysical properties, such as dynamic viscosity, density, specific heat capacity, and surface tension (Fig. 15.1).

Boiling heat transfer is a crucial area of study when it comes to refrigerants. With the addition of nanoparticles to pure refrigerants the process of boiling heat transfer becomes more complex. Researchers have observed various results while assessing the heat transfer rates of nanofluids and nanorefrigerants [7]. Previous studies have shown that heat transfer effectiveness with nanoparticles is higher than that of nanoparticle-based refrigerants. Although some researchers have conducted studies on both pool and flow boiling, the focus has been primarily on pool boiling of nanorefrigerants. This chapter aims to provide an overview of both pool and flow boiling of nanorefrigerants [7].

The field of nanorefrigerants can be categorized into two main areas. The first involves the incorporation of nanoparticles into a pure refrigerant, whereas the second focuses on the suspension

Graphene and 2D Materials in Heat Transfer: Fundamentals and Applications. DOI: https://doi.org/10.1016/B978-0-12-821964-5.00004-7

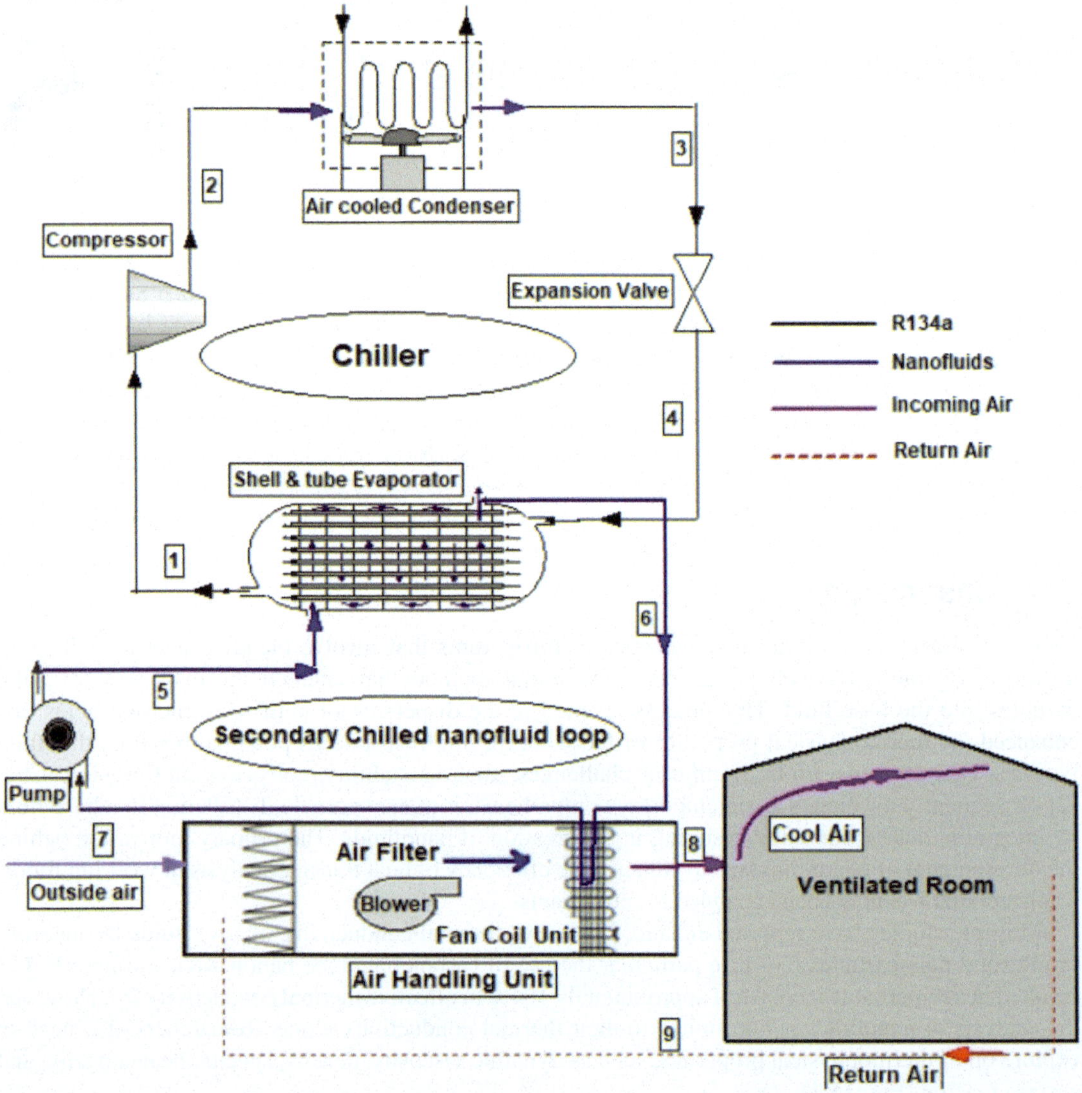

FIGURE 15.1

Schematic representation of the HVAC system [6]. *HVAC*, Heating, ventilation, and air conditioning.

of nanoparticles in a lubricant to enhance the efficiency of the cooling system. Researchers have increasingly explored the use of nanolubricants and observed a marked enhancement in cooling unit performance upon introducing nanoparticles into the coolant lubricants [8–11]. Lee et al. [12] found that the friction coefficient decreased by approximately 90% when the raw oil was replaced with the nanooil, as the latter had better lubricating qualities than the former. Fullerene C60 nanooil demonstrated an improvement of 5.26% and 5.3% in two different compressors, respectively [13]. They similarly identified a reduction of approximately 3°C–5°C in the shell compressor temperature, where reduced shell compressor temperature verified the improved lubricant stability.

Investigation of the efficiency analysis of cooling units remains in the initial phase. Every analysis, research, and experimental work, regardless of how huge or small, is essential to enhance the thermal properties of nanofluids and nanorefrigerants. The present chapter addresses essential areas of nanorefrigerants. An overview of the investigated literature on many topics, including nanorefrigerants preparation, pressure drop, thermophysical properties, and pool and flow boiling heat transfer with or without surfactants, is presented here. In addition, several experimental research studies on heating, ventilation, and air conditioning (HVAC) systems are also included in this chapter.

After Iijima [14] published his breakthrough article on CNTs in 1991, 1D nanostructures of both organic and inorganic materials began to get attention. Recent advancements in layered materials have enabled the large-scale synthesis of a variety of two-dimensional (2D) materials [5,15,16], where atoms are organized in flat layers that may be stacked on top of one another. Because of its outstanding characteristics and potential uses in many sectors, graphene is one of nature's most prevalent layered materials. The high surface area makes 2D materials ideal choices for nanofillers in heat transfer fluids. Exfoliation is a popular method of producing these multilayer nanostructures [16]. In addition to mechanical exfoliation, liquid phase techniques are essential for various applications, including but not limited to nanoelectronics, micro-/nanoelectromechanical systems, chemical sensors, and pressure sensors. One such method involves heating a metal catalyst (such as copper or silica) to elevated temperatures (ranging from 600°C to 1200°C) and introducing a carbon-containing gas (e.g., methane) over the catalyst to achieve chemical growth of individual layers or graphene sheets. This method was used by Coleman et al. to synthesize various 2D materials [MoS_2 WS_2, boron nitride (BN), Bi_2Te_3, $MoSe_2$, $MoTe_2$, etc.]. In 2D insulators such as Bi_2Te_3, Bi_2Se_3, and hexagonal boron nitride (h-BN), exfoliation would lower the residual bulk conductivity, emphasizing surface effects. As the number of layers is reduced, as predicted, electrical characteristics will also change. In addition to having intriguing physical, mechanical, and electrical characteristics, these materials have high specific surface areas, which are significant for sensing, catalysis, and energy storage applications. These layer-based materials must undergo exfoliation to unleash their full potential, much like graphene [5]. h-BN is a new 2D ceramic material with exceptional mechanical stability, chemical inertness, antiwear promotion, high electrical resistivity, and superior thermal conductivity [5,17]. It is frequently utilized in high-temperature sealing materials for aircraft engines. BN is formed by bonding boron and nitrogen atoms together. It has many identical structures to carbon but with B and N atoms alternating instead of C atoms. Many types of BN overlap with carbon structures, resulting in a wide range of characteristics and functions to suit different applications. In recent years, because of its structural closeness to graphite, h-BN has been investigated [16,17]. Each boron atom is linked to 3 nitrogen atoms by strong covalent connections, and vice versa; the B–N spacing is 1.44 Å. A van der Waals force links each boron atom to a nitrogen atom between all the layers. As a result, h-BN atomic layers are mechanically robust because of the strong B–N bond, like graphene. Single BN layers may well be separated from bulk h-BN crystals [17]. In terms of thermal conductivity, h-BN outperforms other nanofillers, making it a desirable material for composites with high thermal conductivity and electrical insulating properties [16]. It is known that high thermal conductivities (tens to millions W/m K) can only be attained from the (0 0 2) planes [18]; h-BN can be synthesized by wet exfoliation to maximize exposure to any of these (0 0 2) lattice planes. A static electric charge dissipation mechanism is required in various thermal management systems.

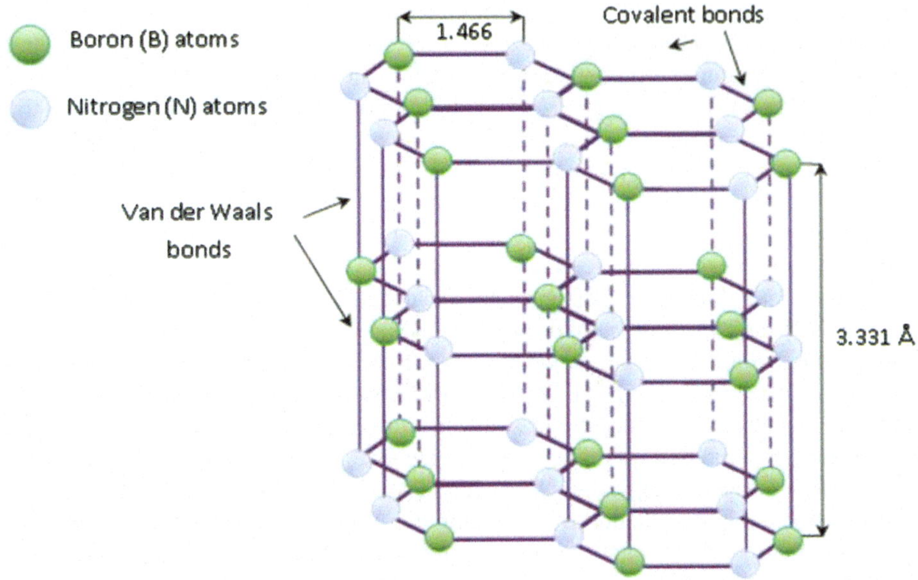

FIGURE 15.2

Scheme of h-BN and graphite structure [19]. *h-BN*, Hexagonal boron nitride.

The electrical and thermal control properties of graphene-based devices are ideal for such applications. Electrical insulation and thermal management are two areas where h-BN is very relevant and beneficial. BN is both ecologically and skin friendly, as seen by its widespread use in cosmetics. CIR 2012 found that BN is utilized in 483 cosmetic products. It is stated that BN is utilized in eye products, lipstick, powder, and perfume preparations with 25%, 2%, 16%, and 0.9%, respectively. A few layers of BN, exfoliated h-BN, can potentially be the material of the future in the thermal management area for various applications, even though little study has been done on 2D nanostructures. A diagram of h-BN and graphite structures is shown in Fig. 15.2[17].

In recent times, there has been an increased focus on enhancing the efficiency of machines, gadgets, and components by addressing cooling and antiwear issues. Researchers have taken up the task of dispersing nanostructures such as oxides, metals, ceramics, CNTs, 2D nanosheets, and others in ordinary fluids to improve their thermal conductivity, viscosity, lubrication, electrical behavior, and several other features. The effectiveness of such nanofillers is influenced by various critical parameters such as morphology, stability of dispersed nanostructures within base fluid, concentration, viscosity, sedimentation, and channel clogging. Unlike larger dispersed solid particles, nanofillers substantially impact the performance of standard fluids, allowing for smoother flow through tiny channels. The properties of nanostructure fluid surfaces are affected by different variables, including concentration, heating, chemical, and interfacial characteristics, leading to property enhancement. However, producing certain nanofluids is challenging and can be costly. The cost of nanofluids can be reduced through mass manufacturing of nanostructures and reducing the filler percentage.

15.2 Preparation and characterization

The investigation of nanofluids, a promising category of heat transfer working fluids, relies heavily on the synthesis and preparation stages. These critical steps involve the use of advanced techniques to ensure long-term stability and address issues such as agglomeration, sedimentation, and settlement. To create nanofluids a variety of solid nanoparticles can be utilized in the secondary fluid cooling unit, such as metal dispersions, metal oxides, and nonmetallic nanosized particles in base fluids [15]. Nanofluids can be synthesized through one- or two-step procedures, typically involving chemical or mechanical methods. In a one-step approach, nanoparticles are synthesized and concurrently dispersed within a liquid medium. On the other hand a two-step process involves the independent synthesis of nanoparticles in two stages, followed by their subsequent dispersion in a liquid to maintain the stability of the resulting nanofluid. Generally the two-step method is favored over the one-step technique because of its ability to ensure stability. The preparation of nanooil that might enter the refrigerator compressor was also well described in many analyses [16,20]. Several authors also summarized the strategies utilized to prepare hybrid nanofluids [21–23]. Surfactants or additives were added to produce stable nanofluids, where the stability time depends on the combination of a surfactant and a nanofluid and various other conditions, including the production technique, ultrasound control, temperature, and others.

Similar to nanofluids, single- or two-step techniques can be employed to prepare nanorefrigerants [6]. However, the boiling point of most pure refrigerants is lower than the ambient temperature, and the refrigerant is consequently not ideal for preparing nanorefrigerants in the gas phase under these circumstances. Boiling points of R-113 and R-141b are 47.6°C and 32.1°C, respectively; therefore, throughout the liquid phase, the two refrigerants were mainly utilized to produce nanorefrigerants at normal environmental temperatures [24]. Peng et al. [25] prepared a nanorefrigerant using R-113 refrigerant and CuO nanoparticles. A digital electronic balance was used to evaluate the required amount of CuO nanoparticles and then added to the coolant. Overall, 30 min of ultrasound vibration was required to disperse the nanorefrigerant mixture properly. It was noticed that the ultrasound led to an improved and dispersed solution, which was possible to maintain for 12 h without sedimentation. Low TiO_2 concentrations (0.01–0.05 vol.%) were blended by Trisaksri and Wongwises [26] in R-141b refrigerant. Ultrasound vibration was utilized to stabilize nanoparticle dispersion for 6 h. For refrigerants having a lower normal boiling point [5,27–32], the following two strategies can be utilized: (1) one- or two-step technique for liquid refrigerant at operating pressure and temperature [33,34], and (2) the preparation of a nanolubricants and blending with a refrigerant. Al_2O_3/R-141b was prepared in an orbital incubator shaker at a constant temperature of 15°C by constantly shaking the nanoparticle combination of 13 nm and 99.5% with a refrigerant for 24 h at 240 rpm. A similar nanorefrigerant preparation method for Al_2O_3/R-134a was also employed [35,36]. ZnO nanoparticles of 40 nm were mixed with oil and PAG (polyalkylene glycol). The combination was prepared in 10 h using a magnetic agitator and vibrated for 15 h with an ultrasonic homogenizer to avoid clustering of self-propelled particles. After that the mixture was utilized to form R-152a refrigerant, which minimizes energy usage by 0.5%[37].

Table 15.1 summarizes the techniques for nanorefrigerants preparation. As described earlier a nanorefrigerant can be prepared by one- or two-step methods. The two-step technique was preferred in many studies because of reasonably low-cost technology suitable for basic and mass production and requiring less preparation time. The characterization of a nanorefrigerant included the structural characteristics of nanoparticles for determining the chemical nature, agglomeration size, and morphology, and

Table 15.1 Comparison of central and local HVAC systems [39].

Criteria	Central system	Decentralized system
Temperature, humidity, and space pressure requirements	Fulfilling any or all of the design parameters	Fulfilling any or all of the design parameters
Capacity requirements	Considering HVAC diversity factors to reduce the installed equipment capacity Significant first cost and operating cost	Maximum capacity is required for each equipment Equipment sizing diversity is limited
Redundancy	Standby equipment is accommodated for troubleshooting and maintenance	No backup or standby equipment
Special requirements	• An equipment room is located outside the conditioned area or adjacent to or remote from the building • Installing secondary equipment for the air and water distribution, which requires additional cost	• Possible that no equipment room is needed • Equipment may be located on the roof and the ground adjacent building
First cost	• High capital cost • Considering longer equipment services life to compensate for the high capital cost	• Affordable capital cost
Operating cost	• Most significant energy-efficient primary equipment • A proposed operating system that saves operating cost	• Less energy-efficient primary equipment • Various energy peaks because of occupants preference • Higher operating cost
Maintenance cost	Accessible to the equipment room for maintenance and saving equipment in excellent condition, which saves maintenance cost	Accessible to equipment to be located in the basement or the living space. However, it is difficult on the roof location because of bad weather

(*continued on next page*)

Table 15.1 Comparison of central and local HVAC systems [39]—cont'd

Criteria	Central system	Decentralized system
Reliability	Central system equipment can be an attractive benefit when considering its long service life	Reliable equipment, although the estimated equipment service life may be less
Flexibility	Selecting standby equipment to provide an alternative source of HVAC or backup	Placed in numerous locations to be more flexible

HVAC, *heating, ventilation, and air conditioning.*

evaluation of the long-term stability of a nanorefrigerant. Meanwhile the characterizations of nanoparticles included transmission electron microscopy, scanning electron microscopy, X-ray diffraction, vibrating sample magnetometry, energy-dispersive X-ray spectroscopy, thermogravimetric-differential thermal analysis, ultraviolet–visible spectroscopy, Fourier transform infrared spectroscopy, infrared absorption spectroscopy, and plasma-optical spectrometry [37,38]. The selection of an appropriate surfactant is a critical factor, as it can greatly affect the stability of a solution. The amount of surfactant added to an unstable solution is a determining factor, and a reduction in surfactant volume can result in instability. On the other hand, excessive amounts of surfactants can lead to a reduction in thermal conductivity. Mechanical energy is transformed into heat energy in the sonification process, which raises the fluid temperature and varies the concentration and measurement. Sufficient sonication time and vibration frequencies supply the necessary energy to disperse nanoparticles efficiently. Ultrasonication time and nanorefrigerant temperature were significant. A nanorefrigerant might be chemically stable for about 3–4 weeks if the surfactant concentration and volume concentration could be suitably determined [38]. The behavior of TiO_2/R-141b at nanoparticle concentrations of 25–500 mg/L and particle sizes of 25, 40, 60, and 100 nm with three surfactants , sodium dodecylbenzene sulfonate, and nonylphenol ethoxylate-10 (SDBS, CTAB, and NP-10) was examined [33]. The study found that the stable hydrodynamic diameter increased with an increase in particle size and concentration, while the particle size decreased with increasing surfactant concentration. Another study by Lin et al. [38] focused on the stability of multi-walled carbon nanotubes (MWCNTs) in R-141b refrigerant (Fig. 15.3), where the addition of SDBS as a surfactant increased the relative concentration with a decrease in the length or outer diameter of MWCNTs. Because of the challenges posed by nanoparticle stability, it is crucial to distribute them appropriately in the heat transfer fluid for an accurate stability evaluation.

15.3 Physical properties of nanorefrigerants

The determination of nanorefrigerant efficiency requires consideration of both thermophysical and flow properties. To accurately calculate heat transfer, crucial thermophysical properties, such as active thermal conductivity, dynamic viscosity, specific heat capacity, latent heat, density, and surface tension, must be considered. Although previous research has mainly focused on thermal conductivity of nanofluids, more recent studies have explored the effects of viscosity. Nevertheless, a comprehensive understanding of the efficiency of nanorefrigerants can only be achieved by examining a wider range of thermophysical

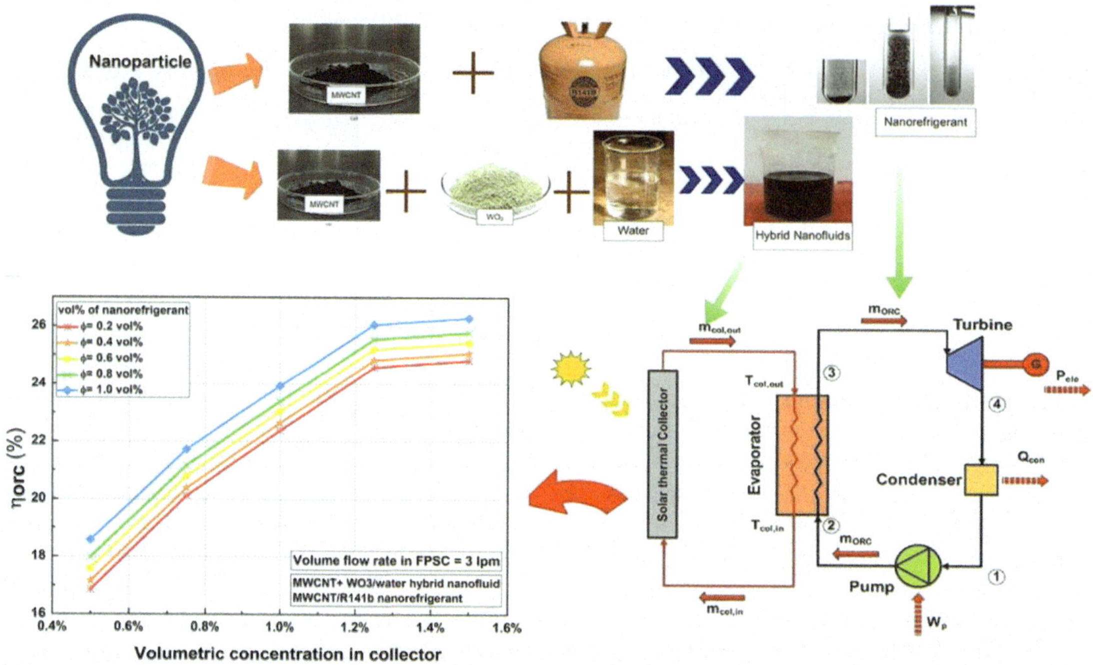

FIGURE 15.3

Synthesis and utilization of nanorefrigerants: MWCNT + R141b nanorefrigerant [40].

characteristics. Therefore it is imperative to extend research efforts to other thermophysical properties to gain a complete understanding of nanorefrigerant thermal transfer efficiency.

15.3.1 Thermal conductivity

Thermal conductivity is a vital attribute of nanofluids to enhance heat transfer efficiency. The rate of heat transfer is mainly based on the thermal conductivity value of the working fluids. Nanofluid thermal conductivity depends on various conditions, including volume fraction, temperature, particle size, particle shape, pH value, surfactant, and preparation method [41–44]. Thermal conductivity primarily influences the boiling and condensation of nanorefrigerants. The thermal conductivity of various nanorefrigerants was assessed by Jiang et al. [45]. Their research verified that the nanoparticle aggregation and the development of nanoparticle clusters had affected the improvement significantly. They also discovered a model to approximate the thermal conductivity of nanorefrigerants. CNT nanorefrigerants displayed a considerably higher thermal conductivity than CNT–water nanofluids or spherical-nanoparticle-R113 nanorefrigerants [5].

Many works explained the effect of variables on the effective thermal conductivities of various nanoparticle groups (Al_2O_3, CuO, Al, Cu, and Ni) and various pure refrigerant types (R-134a, R-113, and R-141b)-based nanotubes [31,33,36] Their findings could be outlined as follows: (1) the thermal nanorefrigerant conductivity improved with the increase of particle volume fraction, (2) the thermal nanorefrigerant conductivity increased with the decrease of particle size, and (3) the thermal nanorefrigerant conductivity improved with the increase of temperature.

15.3.2 Dynamic viscosity

Dynamic viscosity is an essential characteristic that directly affects heat transfer efficiency and drop-in pressure. Increasing dynamic viscosity reduces the convective coefficient of heat transfer, enhances the friction pressure drop, and increases the pumping power. Many authors studied water- and nanofluid-based lubricants [44,46–48]. Only a few studies have analyzed the effect of viscosity in detail, which explains why this section focuses on the analysis of nanorefrigerant viscosity.

The effects of volume fraction and temperature on the viscosity of R-134a, R-141b, and R-123 nanorefrigerants were studied by Mahbubul et al. [31,34,35]. In their research the viscosity of nanorefrigerants was improved with the nanoparticle concentration and decreased with the increase in temperature. They noticed the nanorefrigerant viscosity increased by 13.68% at the same temperature.

Alawi et al. [49,50] analyzed the effects of volume concentration and temperature on the thermophysical characteristics of CuO/R-134a and Al_2O_3/R-134a nanorefrigerants. They observed that the viscosity of nanorefrigerants improved with the increasing volume and decreasing temperature. The analysis of SWCNT/R-134a nanorefrigerant was carried out by Alawi et al. [50]. They observed that the viscosity of the nanorefrigerant was improved with the increasing particle concentrations and decreased with increasing temperature. Alawi et al. [51] also studied the effect of volume concentration and temperature on the viscosity of TiO_2/R-123 nanorefrigerant and noticed that the viscosity of nanorefrigerant was changed with the volume concentration and temperature.

15.3.3 Specific heat capacity and density

Several studies have investigated the heat transfer properties of nanofluids, particularly the influence of specific heat and density [52,53]. In one such study, Mahbubul et al. [34] examined the thermophysical characteristics of Al_2O_3/R-141b nanorefrigerant. They found that the density of nanorefrigerant was improved with increasing particle concentration and decreased with increasing temperature. Mahbubul et al. [35] also assessed the thermophysical characteristics of Al2O3/R-134a nanorefrigerant and found that the density improved by approximately 11% at the same temperature as in the pure refrigerant case. Another study by Alawi et al. [50,54] investigated the effect of particle concentration and temperature on the thermophysical properties of CuO/R-134a and SWCNT/R-134a nanorefrigerants. They observed that the specific heat increased with temperature and particle concentration, whereas the density of the nanorefrigerants improved with higher particle concentration but decreased with increasing temperature.

Surface tension is the desirable force applied upon the surface molecules of a liquid by the molecules underneath that tend to attract the surface molecules in the bulk of the liquid, besides making the liquid presumes the shape of the least surface area. This property is important in studying nanorefrigerants for flow analysis, bubble formation, dry-out, and heat transfer applications. Numerous studies have investigated the effect of surface tension on nanorefrigerants, including a study that experimentally evaluated the heat transfer characteristics of CuO/polyolester oil (POE)/R-600a nanorefrigerant during condensation in a smooth horizontal tube at various particle concentrations [55]. The addition of nanoparticles increased the surface tension of the nanorefrigerant and improved its heat transfer performance.

15.4 Nanoparticle dispersion

Choi et al. [2] explored the use of nanoparticles in suspended particle form as a thermal fluid, which they termed "nanofluid." These studies aimed to improve the heat transfer characteristics of thermal fluids by incorporating nanoparticles into host fluids at low concentrations while maintaining stable suspension and uniform dispersion. Nanofluids have been found to enhance thermophysical characteristics and heat transfer performance in a wide range of devices because of their high thermal conductivity. In fact, nanofluids have been found to exhibit improved thermophysical characteristics, such as thermal conductivity and viscosity, when compared to standard base fluids such as water and ethylene glycol. Heat transfer enhancement studies demonstrated improved heat transfer coefficients of the base fluid when nanofluids were employed.

Various applications are exploring the use of nanoparticles to achieve cleaner and more efficient energy sources. Several studies have reported using nanotechnology in areas such as the recovery of boiler discharge flue gas, cooling of engine and transmission fluid, heating and air conditioning of buildings, and biological applications [56,57]. Nanofluid can be produced using either a one- or two-step process [58].

The one-step technique involves dissolving nanophase particles that are condensed into a low-pressure vapor in a liquid, all in a single step. Nanoparticles can be produced through physical vapor decomposition or liquid chemical methods. However, the two-step process has largely replaced these methods, which include organometallic synthesis, electrochemical synthesis, salt reduction of transition metals, and ligand reduction. In the two-step process the nanoparticle and oil mixture is formed by adding nanosized powder to the oil after the procedure is complete. Various methods, such as ultrasonic agitation, high-shear mixing, magnetic force agitation, and homogenized mixing, are utilized to disperse the mixture [59].

15.5 Refrigeration system

The refrigeration system consists of four essential components: an evaporator, a compressor, a condenser, and a capillary tube. The compressor raises the pressure and temperature of the refrigerant vapor, which then flows to the condenser for cooling. During this process, heat is dissipated by natural convection to lower the refrigerant's high pressure and temperature. The refrigerant then moves through a capillary tube, which acts as a measuring device to convert the condenser pressure into evaporator pressure. This change causes the refrigerant to shift from a subcooled liquid to a mixture state as its temperature drops. The refrigerant is then circulated through a piping system within the evaporator, where it functions as a heat exchanger. The evaporator collects heat from a chilled area and transfers it to the compressor via the refrigerant. These components operate continuously in a cyclic process.

15.5.1 Heating, ventilation, and air conditioning system selection

Three primary elements influence system selection: building configuration, climatic circumstances, and customer preference. The design engineer must explore a range of systems to meet the owner's demands. Several considerations may include climatic change, building capacity, geographical needs, capital costs, operating and maintenance costs, life cycle analysis, and durability and mobility. However, limitations need to be established in choosing a system. These limitations include the capacity available according to standards, the building layout, the free space, building expenditure, the available power supply, and heating and refrigeration building loads.

15.5.2 Basic components of a heating, ventilation, and air conditioning system

The essential component of the HVAC system is used to deliver cool air to achieve thermal comfort at required buildings or spaces, such as air filters, cooling towers, air outlets, exhaust or relief fans, return air system, humidification and dehumidification equipment, cooling and heating coils, supply fan, boiler, water chiller, air control system, mixed air plenum, control system, self-regulating cooling and heating unit, ducts, water chiller, outdoor intake air, and water chiller.

15.5.3 Classification of heating, ventilation, and air conditioning systems

HVAC systems are divided into two subsystems: the central and decentralized or local systems. The type of basic work area determines whether a system is set in one place, such as cooling the entire building as a complete unit or allocating space as per room design as part of the building. Therefore distribution systems such as air conditioning should be constructed according to the system division and the main equipment area. The conditions mentioned above should also be used to choose between the two systems. In terms of the selection process, comparisons of local and central methods are given in Table 15.1[39].

15.5.4 Heating, ventilation, and air conditioning system requirements

HVAC systems require four basic components, such as main equipment, air supply, space requirement, and plumbing, as shown in Fig. 15.4[39]. Heating in buildings or spaces is accomplished using heat exchangers, specifically hot water and hot water boilers. Three types of fans are commonly used to bring cool air: axial, centrifugal, and plug or plenum. Once cool air is generated, refrigeration equipment delivers it to the building or space, creating a comfortable thermal environment. The cooling coils within the refrigeration equipment use water from water chillers or refrigerants from a refrigeration process. Whether the HVAC system is designed to be centralized or localized depends on the specific requirements of the space. In summary the five resources utilized in HVAC systems are as follows:

- Resource rooms: The total space required for machinery and electricity is between 4% and 9% of the total building area. The cooling system prefers to be installed in the center of the building to reduce the length of the pipe, to facilitate the formation of the shaft and canal cycles, and to keep sizing, operation, and adjustments in one place.
- HVAC facilities: To perform the primary function of cooling and heating, buildings must require equipment, such as heating, air temperature compressors, pumps, boiling units, and mechanical pressure control equipment. The refrigerator facility has conventional water pumps, large building cooling towers, heat exchangers, air conditioning systems, and air compressor controls. When designing equipment rooms that will accept both devices, the size and size of the equipment, the

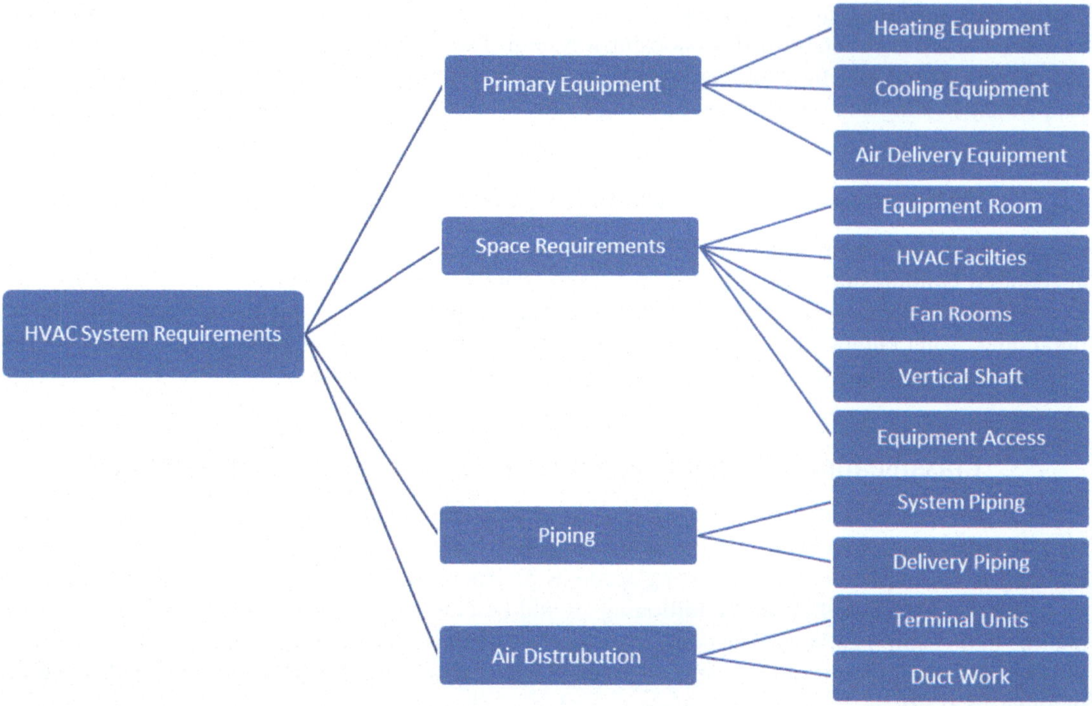

FIGURE 15.4

Horizontal hierarchy of HVAC system requirements representation [60]. HVAC, Heating, ventilation, and air conditioning.

installation and maintenance of equipment, as well as the applicable rules of air temperature and ventilation system, must be considered.

- Fan rooms: Contains HVAC fans and other various equipment. When designing rooms, it is important to consider the size required for the installation and removal of fan shafts and coils, as well as space for additional installation and storage. The size of the fans needed for air conditioning in the building will vary depending on the desired level of airflow. Factors such as cost, availability, and location will influence whether the fans are centrally located or distributed throughout the building. To ensure optimal performance, providing easy access to the outside air is advisable.
- Straight shaft: Provides space for water, air, and steam pipe distribution. The air supply system comprises an exhaust vent, an HVAC feed supply, and an exhaust air duct. Pipeline distribution includes hot water, cold water, steam delivery, condenser water, and condenser return. Some distribution of equipment and electricity, such as pipes, fire safety pipes, and power cords, is contained in the vertical shaft to provide power throughout the building.
- Availability of resources: The equipment room should be designed to remove heavy equipment during installation, repair, and replacement.

The air distribution system comprises a network of pipes that supply cold air directly, efficiently, and quietly to specific regions. The system incorporates various components, such as the air supply unit, which delivers low-velocity air to the building or space. The end-to-end air volume devices regulate the airflow by controlling the air entering the building or space. On the other hand the end-to-end fan-operated units utilize an inbuilt fan to guarantee a steady air supply to the building or space. The final ventilation units are responsible for controlling the primary air, providing fresh air, and dispersing mixed air throughout the building or space. The units also contain water intake connecting units with airflow coils. All pipes and fittings must have insulation to reduce heat loss and conserve energy. Furthermore, buildings should have adequate space to install pipes in the false roof and floor and to restore the air plenum to ensure a decrease in the return air duct.

The piping system is utilized to carry refrigerant, gas, cooled water, hot water, steam, and condensate in a direct, efficient, and cost-effective approach to and from HVAC equipment. These systems are classified into two parts, one in the supply piping and the other piping in the central plant equipment room. Insulation of HVAC pipes may or may not depend on existing code criteria.

15.5.5 Central heating, ventilation, and air conditioning systems

The central HVAC system can provide a wide range of thermal zones, and its main equipment is located outside a given area in a suitable central location to be inside, inside, or near the building [60,61]. These systems should balance the areas with their corresponding hot load. In addition, these systems will have multiple points in each location, that is, thermostats. To provide thermal power to the control system through moderation, separate the central HVAC system as shown in Fig. 15.5[62]. A thermal power transmission system can be air and water, representing all air, water, and water systems. In addition, centralized HVAC systems include water heat pumps and cooling and heating panels; the subsystems are described below. Fig. 15.6 shows that the central HVAC system has integrated units in the air handling device [61]. It comprises supply and exhaust air fans, humidifiers, cooling coils, regenerating and heating coils, filters, and outdoor air.

15.5.6 All-air systems

Air serves as a heat transfer medium in building delivery systems. These systems can be classified into single and multizone systems, with a constant airflow rate for each zone. Other classifications include air volume, variable air volume systems, terminal reheat, and dual duct systems [60,62].

15.5.7 Development in refrigeration system

At the beginning of the 20th century, home refrigerators were represented in the United States, and the demand for such devices grew in subsequent years. It was found that sales increased by more than 100% from 1919 to 1924. In 1924 the demand was 350% higher than last year. Currently, about 100 million refrigerators are operational worldwide. There has been a lot of research and development from the novel's domestic refrigerator focusing on insulation techniques with the help of balsa wood. Urethane foam was developed in the 1950s, and later, in the early 1960s, the solid urethane foam was produced with the help of hydrocarbon (HC) blower agents (i.e., R-11 and R-12). Because of

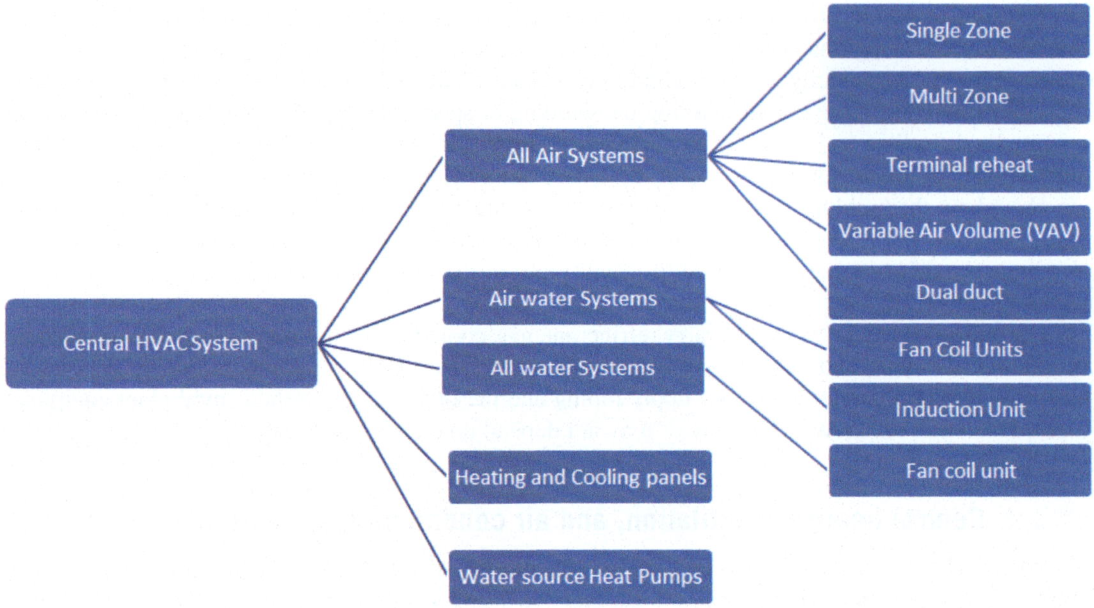

FIGURE 15.5

Horizontal hierarchy representation of the main types of central HVAC systems [62]. *HVAC*, Heating, ventilation, and air conditioning.

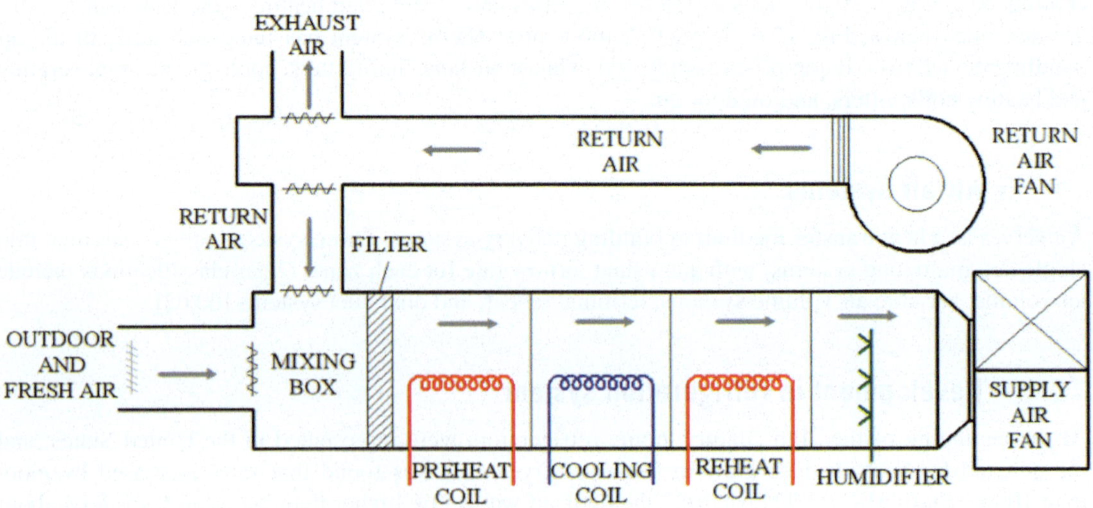

FIGURE 15.6

Equipment arrangement for a central HVAC system [62]. *HVAC*, Heating, ventilation, and air conditioning.

various environmental concerns within the framework of the Montreal Protocol on the depletion of the ozone layer by chlorofluorocarbons (CFCs), commercial potential vacuum insulations were observed as an alternate option. However, the production costs became high when vacuum insulation was made. Johnson [63] described the comparative impact of selecting suitable air-conditioning agents on climate performance in the life cycle of conventional European refrigerators and energy consumption. Two different types of blowing agents, namely pentane and hydrofluorocarbon (HFC)-245fa, were analyzed under the same product specification, assuming that 10% of the clay was in the earth's crust instead of the previously assumed 60%. The result showed that the benefit of HFC-245fa as a blasting agent, instead of n-pentane and cyclopentane in the current foam formulation, offers a different advantage in the energy consumption of refrigerators. The global warming (GWP) of HFC-245fa is also compared to a high-grade paint mixture [64]. In addition, the ingenuity of the component has been the subject of much debate among investigators. Typically, four main refrigeration components are connected, such as a condenser, capillary tube, evaporator, and compressor, respectively, to form a closed system. A compressor is an important system component that generates the basic power to distribute the refrigerator system by pressing the refrigerant. Compressor performance is measured in terms of energy efficiency in domestic refrigerators. The energy efficiency measure is the average cooling power in the refrigerator in British thermal units per hour, in the total amount of electricity (in watts) in certain specific tests. In the last few decades of research on home compressors, refrigerators have focused on lubrication, compressor valve strength, vibration and compressor efficiency, and sound effects. A wide variety of literature is available on such areas of study in compressors for domestic refrigerators. Shaffer and Lee [65] described energy consumption in the hermetic refrigerator compressors with the release of 1/4HP by analyzing multiple losses. Key performance parameters in this study are increased vehicle efficiency, reduced gas heating, and improved compressor geometry. Riffe [66] has improved the performance of nearly 40% of refrigeration compressors by introducing welded wiring harnesses and a fixed piston and cylinder construction. Schroeder suggested using solutions with good thermal conductivity and connecting them to operating capacitors to improve the efficiency of the electric motor of the refrigerator compressor from 73% to 80%. Nelson and Middleton [67] designed energy-efficient compressors and refrigerators by combining four electric motors with resistors with appropriate thermal coefficients to improve compressor design. To increase the EER of the hermetic reciprocating compressor, a four-pole electric motor was installed. Other researchers continue to work on improving the efficiency of compressor geometry, electric motors, and compressor components in subsequent years, using the same advanced techniques as previous researchers [68–70]. The evaporator and condenser are both heat exchangers, and their operation is important for all domestic refrigerators. Ding and Chen were the first to study condenser and evaporator operations in domestic refrigerators [71]. In addition, the authors reported that the weight of the evaporator also affects the energy consumption of refrigerators. Furthermore, Lee et al. [72] conducted experimental studies on frost formation effects on the evaporator of the pipes. They analyzed the effects of factors such as fin configuration, humidity, air temperature, fin spaces, and air velocity in frost growth and evaporator temperature performance. Karatas et al. [73] conducted an experimental study of the evaporator coil with a charged home refrigerator tube to determine the friction and heat transfer relationships. The two items matched Reynolds' number and fin factor, and the additions obtained were used to calculate the uneven heat transfer area. Lee et al. [74] developed a tri-tube evaporator and used it in living refrigerators. The purpose of this evaporator is to enhance the efficiency of the refrigeration system for indirect cooling, particularly during reduced and basic operations. Studies conducted in Korea have demonstrated that the implementation of this evaporator has resulted in a notable reduction in energy consumption, leading to an approximate 4% improvement in energy efficiency. The evaporator was subsequently patented in the United States in 2003 under the patent number US7726025B2 [75]. Cho et al. [76] conducted a multidisciplinary and thermal transfer study under cold conditions of successful tube overflows. With the variation of the operating conditions and the final geometry, it tasted single-stage as well as two-stage evaporators. Heat and air transfer voltages are calculated from measured data, and the tube space results and the heat transfer function adjustment are also studied. The efficiency of energy has been improved in many areas of engineering using a variety of techniques and advanced research methods. In the field of air conditioning, energy heating, and building cooling systems, increasing heat transfer is a strategy to increase energy efficiency. It is also used to manufacture petrochemical, chemicals, textile, cooking, pulp and paper, and other processing industries for cooling and heating systems.

15.5.8 Refrigerant and lubricant types

Refrigerators use a specific active ingredient, known as a refrigerant, to absorb or remove heat, which allows them to function as cooling agents. To absorb heat, the refrigerant evaporates at low temperatures and pressures, whereas it condenses at high temperatures and pressures to remove heat. A widely used refrigerant was a standard solution involving hot water during the early stages of the cooling device development. However, most early refrigerants were either toxic, hot, or both, and only a limited number proved effective. Eventually the refrigeration industry shifted from using ammonia (R-717) to propane (R-290) as a safer and more efficient alternative [77]. Propane is generally regarded as a chemically neutral substance; thus no action is taken to regulate its use. Moreover, it is not considered harmful or unpleasant, and if needed, an engineer could readily address any issues related to its moisture content. In the 1920s, carbon dioxide (R-744) was employed in refrigerated freezers and in components of centrifugal compression chillers utilized in air conditioning systems. Alternatively, refrigerants, such as ammonia and water (R-718), sulfur dioxide (R-764), dielene (1,2-dichloroethene, R-1130), and carbon tetrachloride (R-10), were employed in these types of refrigeration systems. However, only R-1130 could be utilized with centrifugal compression machines, as the others were unresponsive to various factors, including poor performance, instrument noncompliance, and safety concerns [77,78]. In 1930 fluorocarbon refrigerators were introduced, and there was a shift from concern to refrigerator safety and durability focused on removing hazardous substances and flammable structures [77]. The commercial use of dichlorodifluoromethane (R-12) as a refrigerant began the following year and was soon followed by R-11. HCs were deemed unsuitable because of their toxicity and rejected as frozen water in refrigerators. Instead, second-generation refrigerants, chlorofluorocarbons [hydrochlorofluorocarbons (HCFCs)], became the dominant choice for heat pumps and air conditioners in small businesses. However, ammonia continues to be used in large systems. Rowland and Molina [79] continue to promote the idea that certain chemicals containing chlorine and bromine, especially HCFCs and CFCs, can also be accumulated in the stratosphere (12–50 km) and greatly diminish the earth's ozone layer. Therefore, in the early 1970s, research linking CFCs with stratospheric ozone depletion was important because of their earlier status as a good refrigerator [79,80]. The properties of CFCs and HCFCs refrigerants, the discovery of ozone deflection, and their GWP lead to the Montreal Protocol (1987) [81], the Copenhagen Amendment (1992), London Amendment (1990), the Vienna Convention (1985), the Vienna Adjustment (1995), and the Montreal Amendments (1997), which set 1995 and 2030, respectively, as the end deadlines for the production, including the use of these refrigerant [82,82]. The production of a new refrigerator is very important as it affects the production of CFCs and HCFCs. Rader and Kim [78] stated that the purpose of testing the new refrigerator was reversed in the 1960s for two purposes: (1) achieving low operating temperatures with moderate compression during single-phase compression and (2) using a variety of energy-saving temperatures in the cooling process. The most popular refrigerant is HFC134a, which is expected to be used for a long time. However, these findings showed HFC134a systems tolerate less pollution than CFC12 systems. Therefore these are expected to be used for longer periods. As a result, another refrigerant was discovered, HFC152a, to replace CFC12. However, fluorochemicals, on the other hand, remain highly concentrated and focused on the long-term use of HFCs. The use of HFCs R-134a is a prevalent and attractive option as a refrigerant because there is no effect on ozone and low GWP. In 2006 the European Commission issued an order to remove R-134a from air conditioning and refrigeration systems with a GWP of less than 150 in response to the EU's commitment to the Kyoto Protocol. Because the indirect effects of GWP radiation are 1300, R-134a systems have been banned and are used in all new models since 2011 and have been used in all vehicles in European countries since 2017. Plumbing and other cooling systems are designed to replace the lubricant back compressor. Interactions between refrigeration compounds can affect properties and substances in heat transfer and system performance [83]. Sulfur dioxide was the first lubricant, which has been used in refrigeration systems. Raoul Pictet also used sulfur dioxide as a refrigerant and grease in his refrigerator system in 1875 [84]. In the present case, mineral oil is widely used in CFCs refrigerators. Systems with CFC refrigerators have incompatibility with mineral oil, which helps them return to the compressor. However, because of an increase in the ozone layer that counteracts the chlorine effects, people gradually moved to other HFC and HCFC refrigerants. HFC and HCFC refrigerators do not mislead with mineral oil, and this also affects the performance of the refrigerator system [84]. PAGs are widely used in automotive systems that use HCFC refrigerators. For HFC refrigerators such as R-134a and refrigerators based on HFC, ester-blend oil such as POE is commonly used. Because of the chemical strength and refrigeration strength of POE, it is preferred over mineral oil for use in oil applications in refrigeration systems. Similarly, with the use of nanoparticles embedded in compressors, the compressors improve efficiency and performance, which can cause systemic congestion [85]. Several factors need to be considered to decide which lubricant is to use with which refrigerator, such as solubility and viscosity with refrigerant, miscibility, stability, and lubrication and wear [83]. These structures highly affect the performance of the refrigerator.

15.6 Nanofluids in heat exchanger of cooling system

Refrigeration systems function by capitalizing on the principle that highly compressed vaporized fluids tend to become colder when allowed to expand at a specific temperature. Heat exchanger equipment is utilized to facilitate the transfer of heat between two or more fluids with different temperatures [86]. The refrigeration process involves two heat exchangers to absorb and reject heat, which is necessary to achieve its objectives. To achieve an ideal vapor compression refrigeration cycle the system uses two heat exchangers to absorb and reject heat. Improving the performance of nanofluids in heat exchangers can enhance the system's efficiency, resulting in greater energy savings. Nanofluids can increase the heat transfer coefficient in heat exchangers of refrigeration systems. Although heat transfer in straight tubes has been extensively studied, there is still a debate regarding anomalous heat transfer enhancement

[87,88]. Statistically, most previous investigations revealed only minor gains in heat transmission, where 11% of the specimens showed declining heat transfer coefficients and 3% showed no change. In another study, solutions were obtained for the hydrodynamically and thermally completely developed laminar nanofluids flow in the channels and tubes subjected to the continuous heat flux. It has been deduced that the anomalous rate of heat transfer exceeded the rate predicted by the increase in thermal conductivity, as observed in the particular cases of alumina water nanofluid in the tube and titania water nanofluid in the channel. There are restricted studies on the properties of heat transfers for multiplex geometries, such as helical tubes, heat exchangers, microchannels, and elongated tubes. Huminic and Homen [89] examined the significance of the distributed articles about improving transfer of heat in the heat exchanger using nanofluids on the two points and focused on utilizing nanofluid in various types of heat exchangers. Tiwari et al. [90] did another experiment to determine the thermal and hydraulic properties of several nanofluids using plate heat exchangers. The CeO_2/water nanofluids appear to improve heat transfer performances of the plate heat exchangers when the ideal molecular concentration has been reached [90]. Taws et al. [91] revealed that there is no prominent increase in the heat transfer for nanofluid along a volume concentration of 2.0% at similar Reynolds numbers. However, a decrease in Nusselt number was announced for the volume concentrations of about 4.65%, where CuO/water nanofluid was tested in a modern chevron-type plate heat exchanger [91]. In the heat pipes the use of nanofluids indicated a notable advancement in reducing thermal resistance and improving performance. However, former studies have demonstrated the accumulation as well as deposition of particles in small-scale channel heat sinks [92]. According to various reports, nanofluids have been found to enhance the performance of evaporators and heat exchangers in refrigeration systems. Recent experiments have shown that the application of nanofluids or a thin layer of porous nanoparticles on the evaporator's surface can improve the boiling heat transfer coefficients and critical heat flux. This enhancement is attributed to the increased capillarity and wettability of the boiling surfaces, which result from the deposition of these particles [93–95].

15.7 Potential of nanorefrigerants

Refrigeration systems can benefit from incorporating nanorefrigerants and nanolubricants, which are specialized types of nanofluids. The use of these nanofluids has the potential to enhance the mechanical and thermodynamic performance of freezing systems. Nanoparticles, when added to the base liquid, can improve the mobility characteristics, properties, and efficiency of the system. However, the influence of pressure must be systematically investigated. Nanolubricants can improve tribological properties, including softening, antiwear properties, and high-pressure conditions. Incorporating nanolubricants into refrigeration systems also has the additional advantage of enhancing compressor performance. Nanorefrigerants serve as cooling agents or heat absorbers, whereas nanolubricants lubricate compressors; thus they work with different systems. Nanorefrigerants improve heat transfer, and nanolubricants exhibit superior tribological properties, which make them excellent lubricants for moving pistons. The incorporation of nanoparticles into refrigeration and lubricating oils can result in energy savings for cooling systems. Currently, numerous applications and devices take advantage of nanofluids to enhance the performance of smoke suppression systems. Nanolubricants and nanorefrigerants have emerged as significant means to enhance the efficiency of refrigeration systems, owing to their exceptional heat transfer properties, antiaging characteristics, and improved lubricating oil properties. The inclusion of stabilized nanoparticles in refrigerants and basic lubricants boosts thermal conductivity and facilitates boiling heat transfer within the systems. Heat exchangers within the system are utilized to their full potential because of superior heat transfer in the condenser, which ensures efficient heat removal from the system. The use of nanofluids leads to minimal pumping power yet enhances heat transfer efficiency, resulting in improved evaporator performance. This is attributed to reduced critical thermal conductivity and the ideal heat transfer amount within the boiling flow and pool. The deployment of nanolubricants can also boost compressor performance by offering improved compressor properties and a lower coefficient of friction, leading to lower pumping power and better wear resistance for moving components. Nevertheless, increasing the concentration of nanoparticles can significantly increase the viscosity of the nanofluids, necessitating the development of a stable nanolubricant and nanorefrigerant solution to achieve superior vapor pressure in refrigeration systems for specific operating conditions. Redhwan et al. [24] have provided a comprehensive survey of recent advancements in nonrefractory refrigeration systems catering to a diverse range of refrigeration bases and their corresponding performance evaluations. Their investigation highlights the potential of nanorefrigerants to enhance heat transfer capacity, thereby leading to their application in various refrigeration systems. Specifically, their analysis suggests that the utilization of nanorefrigerants can result in an increase in heat transfer efficiency and subsequently improve the overall performance of the refrigeration systems.

15.8 Nanorefrigerant development

Park and Jung [96] published a study on the impact of carbon nanotubes (CNTs) on heat transfer during nucleate boiling. The experiments revealed that adding CNTs improved heat transfer rates for nucleate boiling in refrigerators. Furthermore the boiling heat transfer of nanofluids in refrigerators was investigated under different pressure conditions [26]. The findings indicated that an increase in the amount of nanorefrigerant led to a decrease in the boiling temperature of nucleates, particularly at higher temperatures. Saidur et al. [97] found no comprehensive literature on the functions of nanoparticles in standard refrigerators, such as additives and oils in the cooling system. Earlier, Jiang et al. [5] researched the thermal conduction of nanotubes (CNT) and nanorefrigerants and developed a thermal prediction model for CNT nanorefrigerants. Bi et al. [8] investigated the use of TiO_2 nanoparticle compounds in mineral oil as a solvent for domestic refrigeration systems, employing R-134a refrigerant with standard lubricant POE. The researchers assessed the performance of the nanoparticle-based cooling system using energy-efficient testing facilities, which resulted in energy savings of up to 26.1%. In addition, the team conducted experimental investigations in home refrigerators utilizing TiO_2-R600$_a$ nanorefrigerant at concentrations of 0.1 and 0.5 g/L. The cooling power was assessed, and the results showed that the power consumption could be reduced to less than 9.6% when using 0.5 g/L TiO_2-R600$_a$ nanorefrigerant [98].

15.9 Performance of the cooling system

Heat transfer enhancement is critical for improving the overall efficiency of cooling systems. Recent investigations have demonstrated that incorporating Al_2O_3 nanoparticles in refrigeration systems has led to a significant improvement in heat transfer performance, which has subsequently reduced energy consumption. This improvement can be attributed to the increased efficiency of the heat transfer process, which decreases compressor workload. The use of Al_2O_3 nanoparticles has thus proven to be a promising approach for enhancing the energy efficiency of refrigeration systems [99]. The amount of thermal energy transferred from the low-temperature reservoir to the work input of a refrigeration system can be determined by evaluating its cooling COP. Lubricants have been shown to be highly effective in enhancing mechanical durability and reducing energy consumption in cooling systems. Oil serves various purposes in cooling systems, including internal component lubrication, dissipation of heat generated during compression, structural cleaning, establishment of a water seal, and reduction of the load on the compressor [100]. Nanoparticles present in lubricants have a significant impact on the contact between different parts within the refrigerator compartment. Among all the components the condenser is widely regarded as the most susceptible to grease buildup. Consequently, there is limited documentation available regarding its maintenance. The compressor, evaporator, and plumbing systems are the most commonly discussed subjects and have been extensively tested in various scenarios.

Peng et al. [25] have investigated the effects of carbon nanotubes (CNTs) on nucleate pools that boiled and investigated the heat transfer properties of a refrigerator/oil mixture. Their experiments have used four types of CNTs with different outer diameters ranging from 15 to 80 nm and lengths ranging from 1.5 to 10 mm. The test conditions included fractions of CNTs in nanolubricant from 0 to 30 wt.% and CNT nanolubricant mass fraction from 0 to 5 wt.%. Their experimental results showed that CNTs improved the heat transfer rate of nucleate in a boiling oil tank of R113 by 61%, and the intensity increased with reducing CNTs. With a fixed body size of the CNT the enhancement increases by increasing the fraction of CNT weight in the CNT nanolubricants or by reducing the fraction of CNT weight. They suggested the adjustment to estimate the amount of nucleate dam that boils and the temperature of the oil mixture in the refrigerator and CNTs and accurately predicted its details. However, the extension of the use of their interactions with other test conditions requires verification of test details.

15.10 Conclusion

The synthesis of nanoparticles and their use as nanorefrigerants in the refrigeration loop of HVAC units have been extensively examined in this chapter. Nanofluids have shown promise in enhancing the heat transfer properties of working fluids. Consequently the use of nanorefrigerants has been explored as a means of improving the efficiency of HVAC systems. These refrigerants can be prepared using either a one- or two-step method. Numerous studies have determined that the two-step technique is more advantageous because of its affordability, shorter preparation time, and ability to facilitate mass production.

Thermophysical and flow attributes are essential for calculating the efficiency of nanorefrigerants. It is concluded that the thermal conductivity of nanorefrigerants increased with the decrease in particle size and improved with the increase in temperature and particle volume fraction. Similarly, dynamic viscosity is an essential characteristic that directly affects heat transfer efficiency and drop-in pressure. Moreover, increasing dynamic viscosity reduces the convective coefficient of heat transfer, enhances the friction pressure drop, and increases the pumping power. The specific heat was reduced with the increase in temperature and the particle concentration, whereas the density of nanorefrigerants was improved with the higher concentration of particles and decreased with the increase in temperature.

The boiling heat transfer coefficients as well as critical heat flux can be improved by using nanofluids or by enhancing the capillarity and wettability of boiling surfaces when a thin layer of porous nanoparticles deposits on the surface of the evaporator. The potential of nanorefrigerants to increase the heat transfer rate in heat exchangers will result in the use of nanorefrigerants in different systems. Moreover, nanoparticles improve the HVAC system performance and heat transfer rate in cooling systems. Lubricants play an important role in increasing mechanical strength and reducing energy consumption.

Acknowledgments

The authors gratefully acknowledge the financial support under the Fundamental Research Grant Scheme awarded by the Ministry of Higher Education Malaysia with grant number: FRGS/1/2019/TK03/UM/02/12 (FP143-2019A), research grants, ST077-2022, RMF0400-2021, Mirpur University of Engineering and Technology and Department of Mechanical Engineering, University of Malaya, for supporting the conduction of this research works. The authors also acknowledge the Centre for Energy Sciences (CES), University of Malaya, for their support.

References

[1] J.C. Maxwell, A Treatise on Electricity and Magnetism 314 (1881) 1873.

[2] S.U. Choi, J.A. Eastman, Enhancing Thermal Conductivity of Fluids With Nanoparticles, Argonne National Laboratory, IL, United States, 1995.

[3] X. Wang, X. Xu, S.U. Choi, Thermal conductivity of nanoparticle-fluid mixture, J. Thermophys. Heat Transfer 13 (4) (1999) 474–480.

[4] D.-H. Yoo, K. Hong, H.-S. Yang, Study of thermal conductivity of nanofluids for the application of heat transfer fluids, Thermochim. Acta 455 (1–2) (2007) 66–69.

[5] W. Jiang, G. Ding, H. Peng, Measurement and model on thermal conductivities of carbon nanotube nanorefrigerants, Int. J. Therm. Sci. 48 (6) (2009) 1108–1115.

[6] E.C. Okonkwo, T. Al-Ansari, Parametric investigation of a chilled water district cooling unit using mono and hybrid nanofluids, Sci. Rep. 11 (1) (2021) 19227.

[7] V. Nair, P. Tailor, A. Parekh, Nanorefrigerants: a comprehensive review on its past, present and future, Int. J. Refrig. 67 (2016) 290–307.

[8] S.-s. Bi, L. Shi, L.-l. Zhang, Application of nanoparticles in domestic refrigerators, Appl. Therm. Eng. 28 (14–15) (2008) 1834–1843.

[9] R. Wang, et al., A refrigerating system using HFC134a and mineral lubricant appended with n-TiO_2 (R) as working fluids, in: Proceedings of the 4th International Symposium on HAVC, *Tsinghua University Press*, Beijing, China, 2003.

[10] R.K. Sabareesh, et al., Application of TiO_2 nanoparticles as a lubricant-additive for vapor compression refrigeration systems–an experimental investigation, Int. J. Refrig. 35 (7) (2012) 1989–1996.

[11] X.-Q. Wang, A.S. Mujumdar, Heat transfer characteristics of nanofluids: a review, Int. J. Therm. Sci. 46 (1) (2007) 1–19.

[12] K. Lee, et al., Performance evaluation of nano-lubricants of fullerene nanoparticles in refrigeration mineral oil, Curr. Appl. Phys. 9 (2) (2009) e128–e131.

[13] M. Xing, R. Wang, J. Yu, Application of fullerene C60 nano-oil for performance enhancement of domestic refrigerator compressors, Int. J. Refrig. 40 (2014) 398–403.

[14] S. Iijima, Helical microtubules of graphitic carbon, Nature 354 (6348) (1991) 56–58.

[15] Y. Li, et al., A review on development of nanofluid preparation and characterization, Powder Technol. 196 (2) (2009) 89–101.

[16] W. Yu, H. Xie, A review on nanofluids: preparation, stability mechanisms, and applications, J. Nanomater. 2012 (2012) 1–17.

[17] J. Taha-Tijerina, L. Peña-Parás, D. Maldonado-Cortés, 2D-based nanofluids: Materials evaluation and performance, in: Two-Dimensional Materials—Synthesis, Characterization and Potential Applications, InTech, Rijeka, Croatia, 2016, pp. 153–198.

[18] O.A. Alawi, et al., Thermophysical properties and stability of carbon nanostructures and metallic oxides nanofluids, J. Therm. Anal. Calorim. 135 (2) (2019) 1545–1562.

[19] M.M. Fiume, et al., Safety assessment of boron nitride as used in cosmetics, Int. J. Toxicol. 34 (3_Suppl) (2015) 53S–60S.

[20] N. Akram, et al., A comprehensive review on nanofluid operated solar flat plate collectors, J. Therm. Anal. Calorim. 139 (2) (2019) 1309–1343.

[21] M. Goodarzi, et al., Experimental evaluation of dynamic viscosity of ZnO–MWCNTs/engine oil hybrid nanolubricant based on changes in temperature and concentration, J. Therm. Anal. Calorim. 136 (2) (2019) 513–525.

[22] S.K. Verma, et al., Performance analysis of hybrid nanofluids in flat plate solar collector as an advanced working fluid, Sol. Energy 167 (2018) 231–241.

[23] J. Sarkar, P. Ghosh, A. Adil, A review on hybrid nanofluids: recent research, development and applications, Renew. Sustain. Energy Rev. 43 (2015) 164–177.

[24] A. Redhwan, et al., Development of nanorefrigerants for various types of refrigerant based: a comprehensive review on performance, Int. Commun. Heat Mass Transfer 76 (2016) 285–293 .

[25] H. Peng, et al., Heat transfer characteristics of refrigerant-based nanofluid flow boiling inside a horizontal smooth tube, Int. J. Refrig. 32 (6) (2009) 1259–1270.

[26] V. Trisaksri, S. Wongwises, Nucleate pool boiling heat transfer of TiO_2–R141b nanofluids, Int. J. Heat Mass Transfer 52 (5–6) (2009) 1582–1588.

[27] H. Peng, et al., Nucleate pool boiling heat transfer characteristics of refrigerant/oil mixture with diamond nanoparticles, Int. J. Refrig. 33 (2) (2010) 347–358.

[28] H. Peng, G. Ding, H. Hu, Effect of surfactant additives on nucleate pool boiling heat transfer of refrigerant-based nanofluid, Exp. Therm. Fluid Sci. 35 (6) (2011) 960–970.

[29] I. Mahbubul, et al., Migration properties of TiO_2 nanoparticles during the pool boiling of nanorefrigerants, Ind. Eng. Chem. Res. 52 (17) (2013) 6032–6038.

[30] B. Sun, D. Yang, Experimental study on the heat transfer characteristics of nanorefrigerants in an internal thread copper tube, Int. J. Heat Mass Transfer 64 (2013) 559–566.

[31] I. Mahbubul, R. Saidur, M. Amalina, Thermal conductivity, viscosity and density of R141b refrigerant based nanofluid, Procedia Eng. 56 (2013) 310–315.

[32] I. Mahbubul, R. Saidur, M. Amalina, Heat transfer and pressure drop characteristics of Al_2O_3-R141b nanorefrigerant in horizontal smooth circular tube, Procedia Eng. 56 (2013) 323–329.

[33] I. Mahbubul, R. Saidur, M. Amalina, Influence of particle concentration and temperature on thermal conductivity and viscosity of Al_2O_3/R141b nanorefrigerant, Int. Commun. Heat Mass Transfer 43 (2013) 100–104.

[34] I. Mahbubul, et al., Rheological behavior of Al_2O_3/R141b nanorefrigerant, Int. J. Heat Mass Transfer 73 (2014) 118–123.

[35] I. Mahbubul, et al., Thermal performance analysis of Al_2O_3/R-134a nanorefrigerant, Int. J. Heat Mass Transfer 85 (2015) 1034–1040.

[36] D. Sendil Kumar, R. Elansezhian, ZnO nanorefrigerant in R152a refrigeration system for energy conservation and green environment, Front. Mech. Eng. 9 (2014) 75–80.

[37] A. Bhattad, J. Sarkar, P. Ghosh, Improving the performance of refrigeration systems by using nanofluids: a comprehensive review, Renew. Sustain. Energy Rev. 82 (2018) 3656–3669.

[38] L. Lin, H. Peng, G. Ding, Dispersion stability of multi-walled carbon nanotubes in refrigerant with addition of surfactant, Appl. Therm. Eng. 91 (2015) 163–171.

[39] A. Handbook, HVAC Systems and Equipment, 39, American Society of Heating, Refrigerating, and Air Conditioning Engineers, 1996 chapter.

[40] A. Kumar, et al., Performance evaluation of small scale solar organic Rankine cycle using MWCNT+ R141b nanorefrigerant, Energy Convers. Manage. 260 (2022) 115631.

[41] Z. Said, et al., Thermophysical properties of Single Wall Carbon Nanotubes and its effect on exergy efficiency of a flat plate solar collector, Sol. Energy 115 (2015) 757–769.

[42] K. Solangi, et al., A comprehensive review of thermo-physical properties and convective heat transfer to nanofluids, Energy 89 (2015) 1065–1086.

[43] S. Angayarkanni, J. Philip, Review on thermal properties of nanofluids: recent developments, Adv. Colloid Interface Sci. 225 (2015) 146–176.

[44] W.H. Azmi, et al., The enhancement of effective thermal conductivity and effective dynamic viscosity of nanofluids–a review, Renew. Sustain. Energy Rev. 53 (2016) 1046–1058.

[45] W. Jiang, et al., Experimental and model research on nanorefrigerant thermal conductivity, HVAC&R Res. 15 (3) (2009) 651–669.

[46] A.K. Sharma, A.K. Tiwari, A.R. Dixit, Rheological behaviour of nanofluids: a review, Renew. Sustain. Energy Rev. 53 (2016) 779–791.

[47] K. Bashirnezhad, et al., Viscosity of nanofluids: a review of recent experimental studies, Int. Commun. Heat Mass Transfer 73 (2016) 114–123.

[48] J.P. Meyer, et al., The viscosity of nanofluids: a review of the theoretical, empirical, and numerical models, Heat Transf. Eng. 37 (5) (2016) 387–421.

[49] O.A. Alawi, N.A.C. Sidik, Mathematical correlations on factors affecting the thermal conductivity and dynamic viscosity of nanorefrigerants, Int. Commun. Heat Mass Transfer 58 (2014) 125–131.

[50] O.A. Alawi, N.A.C. Sidik, The effect of temperature and particles concentration on the determination of thermo and physical properties of SWCNT-nanorefrigerant, Int. Commun. Heat Mass Transfer 67 (2015) 8–13.

[51] O.A. Alawi, N.A.C. Sidik, A.S. Kherbeet, Measurements and correlations of frictional pressure drop of TiO_2/R123 flow boiling inside a horizontal smooth tube, Int. Commun. Heat Mass Transfer 61 (2015) 42–48.

[52] I. Shahrul, et al., A comparative review on the specific heat of nanofluids for energy perspective, Renew. Sustain. Energy Rev. 38 (2014) 88–98.

[53] H. Riazi, et al., Specific heat control of nanofluids: a critical review, Int. J. Therm. Sci. 107 (2016) 25–38.

[54] O.A. Alawi, N.A.C. Sidik, Influence of particle concentration and temperature on the thermophysical properties of CuO/R134a nanorefrigerant, Int. Commun. Heat Mass Transfer 58 (2014) 79–84.

[55] M. Akhavan-Behabadi, et al., Experimental study on heat transfer characteristics of R600a/POE/CuO nanorefrigerant flow condensation, Exp. Therm. Fluid Sci. 66 (2015) 46–52.

[56] N.M. Rehan, Nanotechnology as a sustainable approach for achieving sustainable future, World J. Eng. Technol. 9 (4) (2021) 877–890.

[57] H.H. Al-Kayiem, S.C. Lin, A. Lukmon, Review on nanomaterials for thermal energy storage technologies, Nanosci. Nanotechnol.-Asia 3 (1) (2013) 60–71.

[58] W.S. Sarsam, et al., Synthesis, stability, and thermophysical properties of aqueous colloidal dispersions of multi-walled carbon nanotubes treated with beta-alanine, Int. Commun. Heat Mass Transfer 89 (2017) 7–17.

[59] W.S. Sarsam, et al., Stability and thermophysical properties of water-based nanofluids containing triethanolamine-treated graphene nanoplatelets with different specific surface areas, Colloids Surf. A Physicochem. Eng. Asp. 500 (2016) 17–31.
[60] R.W.H. PE, L.A.M.E.M. PE, HVAC Systems Design Handbook, McGraw-Hill Education, New York, USA, 2010.
[61] R. Bean, K.W.K. ArchD, Part 1: history of radiant heating & cooling systems, ASHRAE J. 52 (1) (2010) 40.
[62] S. Seyam, Types of HVAC systems, in: HVAC System, IntechOpen (IntechOpen Limited, London, UK), 2018, pp. 49–66.
[63] R.W. Johnson, The effect of blowing agent choice on energy use and global warming impact of a refrigerator, Int. J. Refrig. 27 (7) (2004) 794–799.
[64] G. Zyhowski, M. Spatz, S.Y. Motta, An Overview of the Properties and Applications of HFC-245fa, Honeywell International Inc., 2002.
[65] R. Shaffer, W. Lee, Energy Consumption in Hermetic Refrigerator Compressors, 1976.
[66] D. Riffe, High Efficiency Reciprocating Compressors, 1976.
[67] R.T. Nelson, M.G. Middleton, The Development of Energy Efficient Compressors for Refrigerators and Freezers, 1980.
[68] H. Kawai, et al., The Development of High Efficiency Compressors by Reducing Suction Gas Temperature, 1982.
[69] K. Ooi, S. Ng, Optimization of a Rotaprocating Compressor, 2002.
[70] N. Dreiman, et al., Two-Stage Rolling Piston Carbon Dioxide Compressor, 2004.
[71] G. Ding, Z. Chen, Effects of the Weights of the Evaporators on the Characteristics of a Household Refrigerator-Freezer, International Institute of Refrigeration (IIR), Paris, France, 1992.
[72] K.-S. Lee, et al., A numerical study on the performance of the fin-tube heat exchanger operated under the frosting condition, Int. J. Air-Cond. Refrig. 4 (1996) 23–34.
[73] H. Karatas, E. Dirik, T. Derbentli, An Experimental Study of Air-Side Heat Transfer and Friction Factor Correlations on Domestic Refrigerator Finned-Tube Evaporator Coils, International Institute of Refrigeration (IIR), Paris, France, 1996.
[74] J. Lee, et al., A Study on New Tri-Tube Type Evaporators in Domestic Refrigerator/Freezer., The Society of Air-Conditioning and Refrigerating Engineers of Korea (SAREK), 2000.
[75] J.S. Lee, Structure for Preventing Refrigerant From Leaking in Heat Exchanger and Method for Forming the Same, Google Patents, United States Patent and Trademark Office – USPTO record, 2010.
[76] H. Cho, Y. Kim, I. Jang, Performance of a showcase refrigeration system with multi-evaporator during on–off cycling and hot-gas bypass defrost, Energy 30 (10) (2005) 1915–1930.
[77] J.M. Calm, The next generation of refrigerants–historical review, considerations, and outlook, Int. J. Refrig. 31 (7) (2008) 1123–1133.
[78] R. Radermacher, K. Kim, Domestic refrigerators: recent developments, Int. J. Refrig. 19 (1) (1996) 61–69.
[79] M.J. Molina, F. Rowland, Predicted present stratospheric abundances of chlorine species from photodissociation of carbon tetrachloride, Geophys. Res. Lett. 1 (7) (1974) 309–312.
[80] M. Purvis, J. Hunt, F. Drake, Global atmospheric change and the UK refrigeration industry: redefining problems and contesting solutions, Geoforum 32 (2) (2001) 143–156.
[81] M. Protocol, in: Montreal protocol on substances that deplete the ozone layer, 26, US Government Printing Office, Washington, DC, 1987, pp. 128–136.
[82] L.N. Yu-Jose, Global environmental issues: responses from Japan, Jpn. J. Political Sci. 5 (1) (2004) 23–50.
[83] G. Short, T. Rajewski, J. Oberle, Refrigeration lubricants-Current Practice and Future Development, International Institute of Refrigeration (IIR), Paris, France, 1996.
[84] T. Mang, Encyclopedia of Lubricants and Lubrication, 1, Springer, Berlin Heidelberg, 2014.

[85] A. Celen, et al., A review of nanorefrigerants: flow characteristics and applications, Int. J. Refrig. 44 (2014) 125–140.
[86] B. Sundén, Z. Wu, Advanced heat exchangers for clean and sustainable technology, in: Handbook of Clean Energy Systems, John Wiley & Sons, Ltd. (Wiley), 2015, pp. 1–25.
[87] A. S Dalkilic, et al., Forced convective heat transfer of nanofluids-a review of the recent literature, Curr. Nanosci. 8 (6) (2012) 949–969.
[88] A.M. Hussein, et al., A review of forced convection heat transfer enhancement and hydrodynamic characteristics of a nanofluid, Renew. Sustain. Energy Rev. 29 (2014) 734–743.
[89] G. Huminic, A. Huminic, Application of nanofluids in heat exchangers: a review, Renew. Sustain. Energy Rev. 16 (8) (2012) 5625–5638.
[90] A.K. Tiwari, P. Ghosh, J. Sarkar, Heat transfer and pressure drop characteristics of CeO_2/water nanofluid in plate heat exchanger, Appl. Therm. Eng. 57 (1–2) (2013) 24–32.
[91] M. Taws, C.T. Nguyen, N. Galanis, I. Gherasim, et al., Experimental investigation of nanofluid heat transfer in a plate heat exchanger. In Heat Transfer Summer Conference, American Society of Mechanical Engineers 44786 (2012) 1–8.
[92] L. Godson, et al., Enhancement of heat transfer using nanofluids—an overview, Renew. Sustain. Energy Rev. 14 (2) (2010) 629–641.
[93] S.J. Kim, Subcooled Flow Boiling Heat Transfer and Critical Heat Flux in Water-Based Nanofluids at Low Pressure, Massachusetts Institute of Technology, Massachusetts Institute of Technology (MIT), Cambridge, Massachusetts, USA, 2009.
[94] D. Wen, et al., Boiling heat transfer of nanofluids: the effect of heating surface modification, Int. J. Therm. Sci. 50 (4) (2011) 480–485.
[95] S.B. White, A.J. Shih, K.P. Pipe, Boiling surface enhancement by electrophoretic deposition of particles from a nanofluid, Int. J. Heat Mass Transfer 54 (19–20) (2011) 4370–4375.
[96] K.-J. Park, D. Jung, Boiling heat transfer enhancement with carbon nanotubes for refrigerants used in building air-conditioning, Energy Build. 39 (9) (2007) 1061–1064.
[97] R. Saidur, K. Leong, H.A. Mohammed, A review on applications and challenges of nanofluids, Renew. Sustain. Energy Rev. 15 (3) (2011) 1646–1668.
[98] S. Bi, et al., Performance of a domestic refrigerator using TiO_2-R600a nano-refrigerant as working fluid, Energy Convers. Manage. 52 (1) (2011) 733–737.
[99] R.R. Kumar, K. Sridhar, M. Narasimha, Heat transfer enhancement in domestic refrigerator using R600a/mineral oil/nano-Al_2O_3 as working fluid, Int. J. Comput. Eng. Res. 3 (4) (2013) 42–50.
[100] G. Hundy, A. Trott, T. Welch, Chapter 14: Food refrigeration and freezing, in: Refrigeration and Air-Conditioning (Fourth Edition), Butterworth-Heinemann, Oxford, UK, 2008, pp. 177–183.
[101] L.J.M. Kuijpers, Copenhagen 1992: a revision or a landmark?: Development in international agreements and regulations, International Journal of Refrigeration 16 (3) (1993) 210–220.
[102] J.A. Mintz, Progress toward a healthy sky: an assessment of the London amendments to the montreal protocol on substances that deplete the Ozone layer, Yale J. Int'l L. 16 (1991) 571.
[103] P.H. Sand, Protecting the ozone layer: The Vienna Convention is adopted, Environment: Science and Policy for Sustainable Development 27 (5) (1985) 18–43.
[104] Boyle, K. (1995). Stock-taking on human rights: The world conference on human rights, Vienna 1993. Political Studies, 43(1), 79-95.
[105] Oberthur, S. (1997). Montreal Protocol: 10 years after. Envtl. Pol'y & L., 27, 432.update the reference at the place of [81,100-105]

CHAPTER

Environmental and economic impact of 2DM-based heat transfer fluids

16

Weng Pin Wong[a], Thachnatharen Nagarajan[b], Seitkhan Azat[c] and Mohammad Khalid[d]

[a]*Sunway Centre for Electrochemical Energy and Sustainable Technology (SCEEST), Faculty of Engineering and Technology, Sunway University, Bandar Sunway, Selangor Darul Ehsan, Malaysia,* [b]*Department of Chemical & Petroleum Engineering, Faculty of Engineering, Technology & Built Environment, UCSI University, Kuala Lumpur, Malaysia,* [c]*Laboratory of Engineering Profile, Satbayev University, Almaty, Kazakhstan,* [d]*James Watt School of Engineering, University of Glasgow, Glasgow, United Kingdom*

16.1 Introduction

16.1.1 The importance of sustainable and efficient heat transfer fluids

Heat transfer fluids (HTFs) play a central role in the thermal management of several industrial and technological systems. They are indispensable for facilitating efficient heat transport in sectors such as power generation, manufacturing, petrochemical processing, aerospace, automotive engineering, electronic cooling, food processing, and solar thermal energy harvesting. HTFs function by transferring thermal energy from a heat source to a heat sink, typically via sensible heat mechanisms (where temperature changes without a phase change) or latent heat mechanisms (involving phase transitions, such as vaporization and condensation). In modern engineering systems the effectiveness, reliability, and environmental sustainability of thermal energy management are directly dependent on the intrinsic thermophysical characteristics of the HTFs used [1,2]. In high-temperature processes, such as nuclear power generation and concentrated solar power (CSP), HTFs are employed not only for heat extraction and transfer but also for intermediate thermal energy storage, enabling load balancing and dispatchable energy delivery. This is especially crucial in CSP systems, where HTFs, such as molten salts, are used to absorb and retain solar thermal energy, mitigating the intermittency inherent to solar irradiance and enhancing grid reliability. Similarly, in the electronics industry, as the demand for high-performance microprocessors and compact integrated circuits grows, the challenge of thermal dissipation intensifies. HTFs in such systems must manage high heat fluxes while maintaining low thermal resistance, minimal volume, and compatibility with the device materials [3–5].

However, the performance of conventional HTFs, including water, ethylene glycol, silicone oils, mineral oils, and synthetic organic fluids, is fundamentally limited by their poor thermal conductivity, which typically ranges from 0.1 to 0.6 W/m K. This low conductivity hampers the rate of heat transfer, resulting in increased energy consumption, reduced operational efficiency, and elevated operating temperatures, which can lead to equipment degradation and failure [6]. In addition, many conventional fluids degrade chemically at high temperatures, exhibit poor oxidation stability, or form unwanted

Graphene and 2D Materials in Heat Transfer: Fundamentals and Applications. DOI: https://doi.org/10.1016/B978-0-12-821964-5.00015-1

deposits on heat exchange surfaces, further reducing the heat exchanger effectiveness and increasing maintenance costs. From an environmental and regulatory perspective, some widely used HTFs present substantial concerns. For instance, while ethylene glycol is readily biodegradable, its high acute toxicity poses significant environmental and safety risks in automotive and aerospace applications where operational leakage may occur. Refrigerants such as hydrofluorocarbons, while effective for thermal management, are associated with high global warming potentials (GWPs), prompting international regulatory action under frameworks such as the Kigali Amendment to phase out or restrict their use. The synthetic oils used in high-temperature systems often involve complex disposal procedures and pose long-term environmental hazards. Thus conventional HTFs are increasingly misaligned with the principles of green engineering, sustainable development, and circular economy practices [7–9].

In recent decades these limitations have catalyzed extensive research into advanced HTFs enhanced with nanotechnology, particularly nanofluid suspensions of nanoscale materials (typically 1–100 nm) in base fluids. The rationale is that nanoparticles, owing to their high surface area and intrinsic thermal conductivity, can augment the base fluid's ability to conduct and convect heat. Since the foundational work by Choi and Eastman in the 1990s, a variety of nanoparticle additives, including metal oxides (e.g., Al_2O_3, TiO_2, ZnO), metals (e.g., Cu, Ag), and carbon-based materials (e.g., carbon nanotubes, fullerenes), have been investigated [10–12]. Although these early nanofluids demonstrated moderate improvements in thermal conductivity, often in the range of 10–30% depending on the particle type and concentration, they also introduced new challenges. Particle agglomeration because of van der Waals forces, sedimentation under gravity, increased viscosity (leading to higher pumping power), erosion of internal surfaces, and incompatibility with certain system materials limit their practical viability. In addition, many traditional nanofillers exhibit isotropic geometries and low aspect ratios, which constrain the extent of interfacial contact and limit the phonon transport networks within the fluid. Moreover, stability under dynamic flow conditions remains a critical concern, especially in closed-loop and high-pressure systems, where long-term operation is essential [13,14].

These shortcomings necessitate the search for next-generation nanomaterials that offer not only improved thermal performance but also enhanced dispersion stability, lower viscosity penalties, and the potential for environmentally benign integration. In this context, two-dimensional materials (2DMs) have emerged as a transformative class of nanomaterials capable of addressing many of the limitations inherent to traditional nanofluids. 2DMs are characterized by their atomic-scale thickness and expansive lateral dimensions, resulting in ultra-high surface areas, quantum confinement effects, and unique anisotropic properties. Their sheet-like structure allows them to align parallel to fluid flow and create continuous thermal percolation networks, thereby enhancing heat transport efficiency at both the conduction and convection levels. In addition, their large surface area facilitates stronger interactions with the base fluid, improving stability without excessive surfactant use. These characteristics also reduce the particle volume fraction required to achieve significant thermal improvements, thereby mitigating viscosity and flow resistance concerns [15,16].

In summary the continuous evolution of energy systems, combined with global sustainability imperatives and advances in nanotechnology, has created an urgent need to reevaluate the design and formulation of HTFs. 2DMs, with their unprecedented combination of thermophysical properties and adaptability, represent a promising frontier for the development of high-efficiency, environmentally responsible, and economically viable HTFs. Therefore the growing body of research on 2DM-based nanofluids holds significant potential not only for improving the heat transfer performance but also for aligning thermal technologies with the broader goals of energy sustainability, resource efficiency, and climate resilience.

16.1.2 Two-dimensional materials as a solution for enhanced heat transfer

2DMs are a family of nanomaterials defined by their atomic- or molecular-level thickness (typically <10 nm) and extensive lateral dimensions, which yield an ultrahigh surface area-to-volume ratio and quantum confinement effects in the *z*-dimension. Since the groundbreaking isolation of graphene from graphite in 2004 via mechanical exfoliation, research on 2DMs has proliferated across disciplines, ranging from physics and chemistry to materials engineering and biomedical sciences [17,18]. The exceptional mechanical, thermal, electronic, and optical properties of 2DMs make them highly attractive for various applications, including electronics, energy storage, sensing, catalysis, and thermal management. Graphene, the most studied 2DM, exhibits a thermal conductivity of up to 5300 W/m K, making it one of the most thermally conductive materials known. Its planar sp^2-hybridized carbon lattice also imparts exceptional mechanical strength (Young's modulus ≈1 TPa), high electron mobility (>200,000 $cm^2/V/s$), and excellent chemical tunability through oxidation, doping, and functionalization. These attributes enable graphene to serve not only as a heat conductor but also as a flexible component that can be adapted to various fluid systems [19,20].

In addition to graphene, other 2DMs, such as hexagonal boron nitride (h-BN), transition metal dichalcogenides (TMDs) such as molybdenum disulfide (MoS_2), and MXenes (a family of transition metal carbides and nitrides), have shown remarkable potential. h-BN, often termed "white graphene," combines high thermal conductivity (up to 751 W/m K) with electrical insulation, making it ideal for applications requiring thermal transport without electrical conduction [21,22]. TMDs, such as MoS_2 and WS_2, offer moderate thermal conductivity, semiconducting properties, and tunable bandgaps ranging from 1.2 to 2.0 eV, thereby enabling multifunctional performance. MXenes, with their layered structures, high metallic conductivity, and hydrophilic surface terminations (–OH, –F, –O), exhibit significant thermal enhancement and colloidal stability when dispersed in polar fluids [23,24]. The incorporation of 2DMs into HTFs has resulted in 2DM-based nanofluids (2DM-based HTFs), which outperform traditional nanofluids in terms of thermal conductivity enhancement, stability, and energy efficiency. Unlike spherical or rod-shaped nanoparticles, 2DMs exhibit sheet-like morphologies with large aspect ratios, allowing them to form percolation networks in fluids and reduce phonon scattering. These materials facilitate more efficient heat transfer through mechanisms such as increased surface contact with base fluids, interfacial layering effects, phonon bridging, and microconvection induced by Brownian motion. Their ultrathin geometry also reduces the likelihood of sedimentation, ensuring long-term dispersion stability [25].

From a sustainability standpoint, 2DM-based HTFs can significantly reduce the volume of fluid required for effective thermal exchange, thereby lowering the pumping power and energy input. Moreover, their compatibility with environmentally benign base fluids, such as water or bioderived oils, positions them as low-impact alternatives for green engineering applications. Although the current cost of high-purity 2DMs remains relatively high owing to the complexities of synthesis, advances in scalable production methods, such as liquid-phase exfoliation (LPE), microwave-assisted synthesis, and chemical vapor deposition (CVD), are rapidly lowering the barriers to commercialization. The durability and efficiency gains achieved through 2DM-based HTFs may also offset the initial material costs by extending the system life, minimizing maintenance requirements, and improving the energy conversion efficiency.

Despite this significant progress, several challenges remain. These include achieving consistent quality and reproducibility in large-scale synthesis of 2DMs, mitigating nanoparticle aggregation under dynamic fluid conditions, elucidating the long-term environmental fate of nanomaterials, and establishing comprehensive regulatory frameworks to ensure occupational and environmental safety. Moreover, a rigorous techno-economic comparison between 2DM-based HTFs and conventional HTFs across diverse industrial applications is urgently required to support informed decision-making regarding material selection and process design.

This chapter offers a thorough and cross-disciplinary analysis of the economic and environmental impacts of 2DM-based HTFs. It brings together existing research on material properties, fluid performance, and environmental evaluations, while also identifying significant knowledge gaps and potential directions for future studies. By examining the thermophysical characteristics of 2DMs, their life cycle costs, energy efficiency benefits, and potential ecological effects, this chapter aims to guide researchers, engineers, and policymakers involved in developing sustainable thermal systems that meet global energy and climate goals.

16.2 Environmental impacts of two-dimensional-material-based heat transfer fluids

16.2.1 Life cycle assessments

Over the past decade, HTFs incorporating 2DMs have emerged as promising alternatives to conventional HTFs, primarily because of their markedly enhanced heat transfer performance. Despite the growing body of work demonstrating their thermal advantages, comparatively limited attention has been paid to the potential environmental implications associated with the use of 2DM-based HTFs. As highlighted by Pereira et al. [26], the environmental footprint of these fluids is intrinsically linked to key thermofluidic and operational parameters, including the heat transfer coefficient, thermal conductivity, pressure drop characteristics, emissions, and overall energy efficiency. From this perspective, 2DM-based HTFs, which combine a high specific surface area with superior thermal conductivity, are increasingly regarded as sustainable alternatives, as their improved heat transfer performance can translate into reduced energy consumption and lower associated environmental impacts. Consequently, size reduction of the equipment (e.g., flat-plate solar collector) can be achieved, as shown by Sundar et al. [27]. Nevertheless, Bammidi et al. [28] highlighted that a few existing concerns of nanofluids, such as disposal of nanofluids, workers' health, microbial contamination, and mist generation, need to be handled well and monitored closely to minimize the possible negative environmental impacts. Pereira et al. [26] also reported that human exposure to the nanomaterials can be harmful to the cytoskeleton, DNA repair, cell signaling, and production of inflammatory cytokines and chemokines. Toxicity can increase with decreasing size of nanomaterials; therefore in situ purifiers and water filtration equipment are required to minimize the intake via respiratory and digestive pathways. In addition, the recovery of the base fluid and nanoparticles is necessary to minimize human health risks and environmental impacts.

Overall, most of the studies in the literature review only conducted cradle-to-gate life cycle assessment (LCA) analysis for 2DMs from extraction to the manufacturing stage. Although most environmental impacts are associated with the extraction and manufacturing stages, it was pointed out that the inadequate disposal of effluents generated during the production of 2DMs and toxic emissions can also pose a threat to human health and existing ecosystems [26]. Generally, the production methods of these 2DMs can be classified into top-down (e.g., exfoliation, laser ablation, chemical reduction, ball milling, and high-temperature melting) and bottom-up (e.g., CVD, epitaxial growth, solvothermal or wet chemical synthesis, and gas-phase synthesis) approaches. Among these methods, CVD is widely utilized for the scalable synthesis of 2DMs due to its high yield and relatively low cost; however, its environmental impacts must be assessed for large-scale production. In addition, the laser ablation technique is well known for producing high-purity 2DMs without generating harmful or toxic emissions [26]. Nevertheless, most existing LCAs have focused on conventional production methods, leaving the environmental impacts of advanced synthesis techniques as a critical research gap to be addressed.

Among the LCA analyses, there were also some general limitations that reduced the accuracy of the estimated environmental impacts. First, some studies have discussed and compared the environmental impacts of the same product via different production methods; however, products synthesized using different methods have different scales, inventory databases, structures, forms, and qualities. These differences are the main obstacles in LCA comparison studies. Next, some assumptions were also made, such as no emissions of 2DMs during some stages, certain infrastructures not considered in the LCA, and recycling of the solvents in the life cycle because of data limitations. Because of the data limitations, Bobba et al. [29] have also used a mixture of data input from literature and industry partners, whereas in a simulation conducted by Hachhach et al. [30], molybdenum was also used as the substitution for the molybdenum heptamolybdate tetrahydrate (Mo precursor) during the inventory analysis. Subsequently, Deorsola et al. [31] assumed that the coproducts or byproducts produced have no associated environmental burden. In the future, data transparency and database sharing should be encouraged between industry players and research communities to develop a more accurate and reliable LCA analysis. In the following sections the environmental impacts of the 2DMs employed for HTFs are systematically reviewed and compared. The main environmental impacts, limitations, and key findings of the studies are summarized in Table 16.1.

16.2.1.1 Graphene and its derivative materials

Graphene and its derivatives, such as graphene oxide (GO), are commonly synthesized using exfoliation-based techniques, CVD, and epitaxial growth methods. Exfoliation methods can be further classified into chemical, physical (e.g., ultrasonication), and thermal approaches. Variations in processing steps across studies have led to the development of modified Hummers' methods and diverse CVD processes, as highlighted by Beloin-Saint-Pierre and Hischler [34]. The environmental implications of graphene production have been systematically evaluated by Arvidsson [32], who reviewed five major production routes for different graphene forms: chemical reduction, ultrasonication-assisted exfoliation, thermal exfoliation, CVD, and epitaxial growth. Among these methods, ultrasonication exfoliation and CVD exhibited the lowest energy consumption. The dominant contributors to environmental impact were identified as solvent production and methane feedstock supply. Consequently, strategies such as solvent reuse (e.g., diethyl ether) and the reduction or recycling of hydrocarbon gas feedstocks have been recommended to further mitigate environmental burdens.

In a subsequent study, Arvidsson et al. [33] assessed the ecotoxicity potential associated with the production of graphene and GO, incorporating projected production volumes into their evaluation

Table 16.1 Main environmental impacts, limitations, and key findings of environmental assessment studies of 2DMs.

Authors	Production methods	Feedstock	Products	Main contributor	Energy use per product (MJ/kg)	Energy use per product (MJ/cm^2)	GWP/Greenhouse gas emissions (kg CO_2/kg product)	GWP/Greenhouse gas emissions (g CO_2/cm^2 product)	Limitations/Assumptions	Key findings
Arvidsson [32]	Chemical reduction of graphite oxide	Graphite	Reduced graphene sheets in water	Reduction step/Hummers' process for water use (production of reduction agent hydrazine)	900–1000	NA	NA	NA	1. Type and quality of graphene produced are not identical 2. Assumption of no emissions of graphene occurred during production	1. Ultrasonication exfoliation has the potential for comparatively low impacts provided that there is reuse of solvents 2. Reducing or reusing the hydrocarbon gas feedstock can reduce the environmental impact of CVD-made graphene 3. Reducing the input of silicon carbide wafers and producing wafers with green electricity can lower the environmental impact of epitaxial growth
	Ultrasonication exfoliation	Graphite	Graphene sheets in diethyl ether	Production of the solvent diethyl ether	70–500	NA	NA	NA		
	Thermal exfoliation	Graphite	Pulverized graphite nanoplatelets	Microwave heating	2000	NA	NA	NA		
	CVD	Methane	Graphene on a quartz substrate	Methane feedstock production step	NA	0.007–0.2	NA	NA		

(continued on next page)

	Epitaxial growth	Silicon carbide	Graphene on a silicon carbide wafer	Electricity use for production of the silicon wafer substrate	NA	2–80	NA	NA		
Arvidsson et al. [33]	NA	NA	Graphene	NA	NA	NA	NA	NA	1. Limited number of test results 2. There are some limitations of obtaining data for graphene oxide (GO), MoS_2, and MXenes 3. Final environmental risk can be only determined with other additional parameters (e.g., emission profiles, environmental fate, exposure, and nonaquatic ecotoxicities)	1. Only MXenes exhibited high aquatic ecotoxicity (>1 L/mg), followed by graphene (~0.1 L/mg) 2. Graphene, GO, and MoS_2 have low aquatic ecotoxicity but potentially high production volumes; therefore careful monitoring of the ecotoxicological literature is recommended to ensure that aquatic toxicity not exceeding the threshold 3. MXenes with high aquatic ecotoxicity and high potential future production volumes should be considered for more detailed environmental risk assessments (e.g., exposure assessments and material flow analyses)

(*continued on next page*)

Table 16.1 Main environmental impacts, limitations, and key findings of environmental assessment studies of 2DMs—cont'd

Authors	Production methods	Feedstock	Products	Main contributor	Energy use per product (MJ/kg)	Energy use per product (MJ/cm^2)	GWP/Greenhouse gas emissions ($kg\ CO_2/kg$ product)	GWP/Greenhouse gas emissions ($g\ CO_2/cm^2$ product)	Limitations/Assumptions	Key findings
	NA	NA	GO	NA	NA	NA	NA	NA		
	NA	NA	Nanosized MoS_2	NA	NA	NA	NA	NA		
	NA	NA	MXenes	NA	NA	NA	NA	NA		
Beloin-Saint-Pierre and Hischler [34]	Baseline Hummer's process (H1)	Graphite	GO	Electricity demand	NA	NA	70–300	NA	1. More accurate data is needed on different production methods (e.g., functional properties and toxicity) 2. The infrastructure to produce graphene-based materials and wastewater treatment was not included 3. Assumptions of no graphene-based materials are released in the air, water, or soil during production	1. 3 key strategies to improve the environmental sustainability of graphene-based materials: 2. Use decarbonized energy sources 3. Increase energy efficiency of production especially for ECE and CVD 4. Use green chemistry principles for processes of chemical and thermal reduction and GO production 5. Transforming graphene-based materials solutions into dry mass can substantially increase the environmental impacts because of the energy-intensive nature of ball milling production 6. State of future electricity mixes might not affect the comparisons significantly as the conservative forecasts lead to similar changes on many production options

(continued on next page)

										7. Discrepancies between all CVD production processes are because of the substantial differences for electricity demands and amounts of methane used **8.** H2 and H3 processes have the highest climate change effect among the GO production processes **9.** EPI has higher climate change effect than the CVD processes for graphene production processes **10.** Combination of differences in electricity demands and possible evolution of impacts from the European electricity mix are not expected to modify the best and worst environmental options in the future if International Energy Agency trends are followed

(*continued on next page*)

Table 16.1 Main environmental impacts, limitations, and key findings of environmental assessment studies of 2DMs—cont'd

Authors	Production methods	Feedstock	Products	Main contributor	Energy use per product		GWP/Greenhouse gas emissions		Limitations/Assumptions	Key findings
					(MJ/kg)	(MJ/cm^2)	(kg CO_2/kg product)	(g CO_2/cm^2 product)		
	1st variant of Hummer's process (H2)	Graphite		Electricity demand	NA	NA	250–3000	NA		
	2nd variant of Hummer's process (H3)	Graphite		Electricity demand	NA	NA	500–4000	NA		
	Fugetsu variant of Hummer's process (H4)	Graphite		Chemical substance	NA	NA	15–20	NA		
	Bengal variant of Hummer's method (H5)	Graphite		Chemical substance	NA	NA	10–15	NA		
	Jeong variant of Hummer's method (H6)	Graphite		Chemical substance	NA	NA	70–90	NA		
	Staudenmaier method	Graphite		Chemical substance (nitric acid)	NA	NA	40–50	NA		
	BM	Graphite		Chemical substance (nitric acid)	NA	NA	50–70	NA		
	CVD 1st variant (CVD1)	Methane	Graphene	Carbon source	NA	NA	0.5–1.5 g CO_2 eq/cm^2	NA		
	CVD 2nd variant (CVD2)	Methane		Carbon source, chemical substance, and electricity demand	NA	NA	0.03–0.09 g CO_2 eq/cm^2	NA		

(continued on next page)

	CVD 3rd variant (CVD3)	Hexane		Energy demand	NA	NA	10 g CO_2 eq/cm^2	NA		
	CVD 4th variant (CVD4)	Methane		Energy demand	NA	NA	10–12 g CO_2 eq/cm^2	NA		
	Epitaxial growth (EPI)	Silicon carbide wafer		Electricity demand for the production of silicon carbide (i.e., carbon source)	NA	NA	NA	NA		
Diaz-Arauzo et al. [35]	LPE followed by CFF using porous ceramic membrane	Graphite	Graphene inks	Electricity consumption	0.34 MJ/L ink	NA	0.53 kg CO_2/L ink	NA	Assumptions of recycling ethyl cellulose and ethanol using rotary evaporation while recovering all the unexfoliated graphite	**1.** Higher input flow rates and concentrations enabled by porous ceramic membrane result in reduction of 97%, 96%, and 94% for greenhouse gas emissions, water consumption, and fossil fuel consumption, respectively, compared to polymer CFF membrane **2.** For polymer membrane, both microfiltration and ultrafiltration operate for longer duration because of the reduced surface area of the membrane leading to extended operation

(*continued on next page*)

Table 16.1 Main environmental impacts, limitations, and key findings of environmental assessment studies of 2DMs—cont'd

Authors	Production methods	Feedstock	Products	Main contributor	Energy use per product		GWP/Greenhouse gas emissions		Limitations/Assumptions	Key findings
					(MJ/kg)	(MJ/cm^2)	(kg CO_2/kg product)	(g CO_2/cm^2 product)		
										3. Using the ceramic membrane the ultrafiltration permeate can be directly used to produce graphene inks without rotary evaporation that is energy intensive, whereas ethanol consumption is also reduced an order of magnitude lower as feed concentration can be increased by 10 folds 4. Producing polymer membrane (29 MJ/kg) is more energy-intensive than producing ceramic membranes (0.79 MJ/kg) 5. For polymer membranes, electricity consumption is the main contributor to greenhouse gas emissions and fossil fuel consumption, whereas ethanol consumption is the main contributor to the water consumption 6. For ceramic membrane, electricity consumption is the main contributor to greenhouse gas emissions, fossil fuel consumption, and water consumption

(*continued on next page*)

	LPE followed by CFF using polymer membrane			Electricity and ethanol consumption	24 MJ/L ink	NA	16 kg CO_2/L ink	NA		
Munuera et al. [36]	Chemical oxidation with chemical reduction	Graphite	Graphene	Production and use of hydrazine and phosphoric acid Chemical inputs and the acid neutralization step	Cumulative energy demand: 3581 Process energy: 953 Abiotic depletion potential: 133	NA	243	NA	Variations in the GWP for the production processes because of different methods employed, structure of the products, and scale considered	**1.** Bengal modification of the Hummer's method has the lowest GWP because of the usage of potassium permanganate as the oxidative agent, which is lower than other oxidative agents, usage of less acid, and lower process temperature (35°C for 30 min) **2.** Biological reduction using *Escherichia coli* as the reducing agent occurs at room temperature, thus reducing the energy requirements **3.** GWP of chemical oxidation with thermal reduction is between 16% and 40% lower compared to chemical reduction as it requires less chemicals **4.** For ECE, energy efficiency increases with higher applied potential for electrolytes, but using potassium hydroxide electrolyte at 15 V resulted in the highest energy efficiency, thus having lowest GWP at 330 kg CO_2 eq/kg graphene

(*continued on next page*)

Table 16.1 Main environmental impacts, limitations, and key findings of environmental assessment studies of 2DMs—cont'd

Authors	Production methods	Feedstock	Products	Main contributor	Energy use per product		GWP/Greenhouse gas emissions		Limitations/Assumptions	Key findings
					(MJ/kg)	(MJ/cm^2)	(kg CO_2/kg product)	(g CO_2/cm^2 product)		
										5. Microwave exfoliation (3 h) consumes 5 times more energy than ultrasonic exfoliation (30 min) because of the process duration 6. For sonication LPE, 90% solvent recovery can potentially reduce 90% of the energy demand 7. Graphene derived from biomass via CVD has lower GWP compared to methane as some energy requirements were sourced internally from the thermal energy of pyrolytic gases 8. Scaling up CVD processes can reduce energy demand by around 70% 9. Smaller thickness of the SiC wafer can reduce the energy demand. Improvements in technology maturity, scaling up, and the use of renewable electricity can improve the GWP of epitaxial growth

(*continued on next page*)

	Chemical oxidation with biological reduction			Electricity consumption	NA	NA	60	NA		
	Chemical oxidation with thermal reduction			Electricity consumption	NA	NA	140	NA		
	ECE			Electricity consumption	NA	NA	439	NA		
	Ball milling LPE (melamine solution)			Electricity consumption Use of solvents	529	NA	216	NA		
	Ball milling LPE (water solution)			Electricity consumption	449	NA	183	NA		
	Ball milling LPE (dry)			Electricity consumption	210	NA	85	NA		
	Ultrasonication exfoliation			Exfoliation process	284	NA	NA	NA		
	Microwave exfoliation			Exfoliation process	1879	NA	NA	NA		
	Sonication LPE			Use of solvents	309	NA	NA	NA		
	CVD	Methane		Combustion of natural gas Use of hydrogen gas	340	4.9	594	13.2		

(continued on next page)

Table 16.1 Main environmental impacts, limitations, and key findings of environmental assessment studies of 2DMs—cont'd

Authors	Production methods	Feedstock	Products	Main contributor	Energy use per product		GWP/Greenhouse gas emissions		Limitations/Assumptions	Key findings
					(MJ/kg)	(MJ/cm^2)	(kg CO_2/kg product)	(g CO_2/cm^2 product)		
	CVD	Biomass		Electricity consumption	230	NA	235	NA		
	CVD	Hexane		Combustion of natural gas Use of hydrogen gas	NA	NA	NA	4.7		
	Epitaxial growth (50 and 350 μm)	Coke		Electricity consumption to produce SiC wafer	NA	7.9	NA	32.9		
	Epitaxial growth (500 μm)				NA	79	NA	67		
Akram [37]	Solvothermal synthesis	Ammonium molybdate tetrahydrate and sulfur	MoS_2	NA	NA	NA	NA	NA	Different inventory databases are available for different processes	1. Solvothermal method requires lower energy consumption as it operates at low temperature (180°C) compared to gas-phase synthesis (850°C) and reaction under hydrogen atmosphere (600°C) 2. Absence of intermediate phase in the solvothermal process saves energy significantly and avoids operating under harmful gaseous environment

(continued on next page)

	Gas-phase synthesis	Molybdenum trioxide and hydrogen sulfide		NA	NA	NA	NA	NA		
	Reaction under hydrogen environment	Ammonium molybdate, sulfur, and carbon disulfide		NA	NA	NA	NA	NA		
Bobba et al. [29]	Gas-phase synthesis	WO_3 H_2S	IF-like WS_2	Production of WO_3 and energy use	840 MJ LHV/kg	NA	Greenhouse: 37.98 Climate change: 39.18	NA	1. There is mixture of data input from literature and industry partners 2. High level of uncertainties at the inventory level and impact assessment level	1. Synthesis of WS_2 is highly energy-consuming, and the raw material, tungsten trioxide, is the highest contributor to the energy consumption, particularly on "Energy resources" and "Greenhouse" categories 2. Impact because of the scrubbing solution and the gas mixture (i.e., H_2, N_2, and H_2S) is very low

(*continued on next page*)

Table 16.1 Main environmental impacts, limitations, and key findings of environmental assessment studies of 2DMs—cont'd

Authors	Production methods	Feedstock	Products	Main contributor	Energy use per product (MJ/kg)	Energy use per product (MJ/cm^2)	GWP/Greenhouse gas emissions (kg CO_2/kg product)	GWP/Greenhouse gas emissions (g CO_2/cm^2 product)	Limitations/Assumptions	Key findings
Corazzari et al. [38]	Reaction under hydrogen environment at high temperatures	WO_3 H_2S	WS_2	NA	NA	NA	NA	NA	NA	1. WS_2 did not elicit a response significantly different from the negative control, indicating negligible cellular and chemical activity 2. MoS_2 showed a moderate cellular toxicity at the highest dose and was inert in all other conditions 3. Both WS_2 and MoS_2 were soluble in simulated biological fluids, indicating low durability in vivo 4. Low overall biological and chemical reactivity of WS_2 and MoS_2 indicates that tested nanomaterials are unlikely to be a hazard as far as human respiratory system is concerned
	Wet chemical synthesis	Ammonium molybdate and ammonium sulfide	MoS_2	NA	NA	NA	NA	NA	NA	

(continued on next page)

Deorsola et al. [31]	Wet chemical synthesis	Ammonium molybdate and ammonium sulfide	MoS_2	Electricity consumption	820	NA	42	NA	1. Electricity is assumed to be drawn from the grid, so the related environmental impacts are those of the Italian electricity mix 2. Ammonium citrate as a coproduct is assumed to have a market outlet, but no environmental burdens are allocated to it	1. Electricity-related impact accounts for 55% of primary energy consumption and 63% of greenhouse emissions 2. The remaining impacts are caused by the reactants (i.e., citric acid, ammonium sulfide, and ammonium molybdate)
Firouzjaei et al. [39]	Ball milling followed by tube furnace heating, etching, and delamination	Titanium pow-der/titanium dioxide Aluminum powder Graphite powder	MXenes	Laboratory electricity usage (electrical machinery in MAX phase production), mainly from tube furnace (57%)	Small laboratory scale: 153,291 Large laboratory scale: 17,364	NA	428.1	NA	MAX phase crushing to powder was not considered in LCA calculations	1. The laboratory electricity usage for the synthesis of MXenes accounts for more than 70% of the environmental impacts 2. Chemical usage is less impactful than electricity, suggesting that recycled resources and renewable energy can make MXene synthesis more sustainable 3. The demand for nonrenewable fossil energy resources contributed more than 60% of the total CED 4. Primary forest energy resources have a negligible effect on the total CED

(*continued on next page*)

Table 16.1 Main environmental impacts, limitations, and key findings of environmental assessment studies of 2DMs—cont'd

					Energy use per product		GWP/Greenhouse gas emissions			
Authors	Production methods	Feedstock	Products	Main contributor	(MJ/kg)	(MJ/cm^2)	($kg\ CO_2/kg$ product)	($g\ CO_2/cm^2$ product)	Limitations/Assumptions	Key findings
										5. The CED was reduced by ~10% when furnace holding time was cut in half from 2 to 1 h, whereas it raised by ~20% when the furnace holding time was raised from 2 to 4 h **6.** Scaling up the production of MXene can decrease the environmental impacts by about 10-fold **7.** Waste management process has a marginal effect of less than 1% in MXene production **8.** Using TiO_2 as the titanium source has greater adverse impacts in ecotoxicity, carcinogenic, acidification, and noncarcinogenic and less adverse impacts for only ozone depletion **9.** MXene synthesis in Texas has the greatest adverse effect in 6 categories, followed by China with the most adverse significant influence in 3 categories, whereas Sweden has the most negligible impact across all categories as Sweden is the global leader in the utilization of renewable energy (26.6% of energy shares)

(*continued on next page*)

Hachhach et al. [30]	Solvothermal synthesis (using LiOH)	Ammonium hepta-molybdate tetrahydrate and sulfur	MoS_2	Production process and subprocesses of lithium hydroxide	319.43 (mainly from mineral extrac-tion)	NA	391.33	NA	**1.** Data on the product life cycle steps such as manufacturing, transportation, use, and disposal should be available to conduct cradle-to-grave analysis **2.** The impact of transportation and packaging were not considered **3.** Molybdenum was used as substitution for the molybdenum heptamolybdate tetrahydrate during the inventory analysis, where 0.62 kg is used as this is the mass fraction of molybdenum in the precursor (54 wt.%)	**1.** Substitution of lithium hydroxide with sodium hydroxide can help to cope with the limited availability of resources and related processing environmental impacts, leading to decrement of estimated impacts up to 56% (nonrenewable energy and land occupation) **2.** Production process of the LiOH can lead to the aluminum emissions to water and hence contribute to the ecotoxicities and noncarcinogens impacts **3.** Land occupation impact is because of the occupation of the forest, creation of industrial and traffic areas and dumpsites, leading to carbon dioxide emission **4.** Steam production assumed to be produced from natural gas can lead to carbon dioxide and methane production and lead to climate change impact **5.** Mining exploitation of Mo can contribute to the wastes with high concentrations (up to 4000 mg/kg Mo for tailings), leading to soil, sediment, and water contamination in surrounding areas **6.** Mineral extraction has the highest impact, followed by global warming, nonrenewable energy, and land occupation

(*continued on next page*)

Table 16.1 Main environmental impacts, limitations, and key findings of environmental assessment studies of 2DMs—cont'd

Authors	Production methods	Feedstock	Products	Main contributor	Energy use per product		GWP/Greenhouse gas emissions		Limitations/Assumptions	Key findings
					(MJ/kg)	(MJ/cm²)	(kg CO_2/kg product)	(g CO_2/cm² product)		
	Solvothermal synthesis (using NaOH)				315.87 (mainly from mineral extraction)	NA	391.33	NA		
Ungureanu et al. [40]	(A) First reported synthesis: Ti_2AlC + TiC (1:1) > HF (50 wt.%) > sonication in CH_3OH	Ti_2AlC + TiC (1:1)	MXenes	MAX phase and electricity	NA	NA	1440		1. The published procedures were modeled by scaling the production of MAX phase precursor to a common amount of 200 g for all synthesis methods 2. Etching and delamination procedures were rescaled to obtain 1 g of the MXene 3. The production of the MXene discs for bulk electrical conductivity measurement was not considered in the LCA study 4. The use and end of life of MXene were neglected	1. Path A showed the lowest environmental impacts independent of the impact factors, because of the lower total reaction time (approximately 4 h) required compared to other methods (average >30 h) and etching procedure at room temperature leading to the absence of further electric energy to heat and maintain the reactants at the desired temperature 2. Paths F and G are the most environmentally impacting, with the highest percentage differences between these 2 pathways, and other pathways are observed for the impact categories of marine eutrophication, mineral source scarcity, and water consumption because of higher amounts of energy needed and specific equipment employed

(continued on next page)

									5. The contribution of transport of materials and equipment was considered by assuming 100 km as the reasonable average distance, and road freight transportation by diesel EURO 6 lorries with a capacity of 16–32 t was considered **6.** 99% of each emitted substance was considered to be retained by an aspiration system endowed with an activated carbon filter **7.** Potential release of delaminated MXene flakes and associated nanospecific impacts was not considered **8.** Hydrogen fluoride was added among the contributors to "terrestrial acidification" impact category, whereas HF emissions in air were considered contributing to the ecosystem damage category	**3.** For paths F and G, impact on the marine eutrophication is because of the release in water of nitrate waste arising from the extraction of the lignite needed in the considered energy mix **4.** For path F, impact on resource scarcity is because of the consumption of 9.27 g of nickel needed to produce the stainless steel used in the surface acoustic wave (SAW) devices employed **5.** Path G has the highest environmental score because of the electric energy needed for the vacuum pump and construction of the vacuum pump itself **6.** Path F has the short reaction time but also has relatively low yield (12 mg of MXene obtained from 100 mg of MAX phase) **7.** For paths A and G, impact on water consumption is because of the hydroelectric energy needed in the electric energy mix and employed mainly for the operation of the heating plate for more than 24 h and the vacuum pump overnight

(*continued on next page*)

Table 16.1 Main environmental impacts, limitations, and key findings of environmental assessment studies of 2DMs—cont'd

Authors	Production methods	Feedstock	Products	Main contributor	Energy use per product		GWP/Greenhouse gas emissions		Limitations/Assumptions	Key findings
					(MJ/kg)	(MJ/cm^2)	(kg CO_2/kg product)	(g CO_2/cm^2 product)		
										8. Independently, by the synthetic methods considered, the most affected damage category is human health, followed by ecosystems and resources **9.** Although path B avoids the potential hazards associated with the direct use of the HF, its environmental score is 3.7 times higher than path A because of the longer reaction time (40 h) and main contributor to the environmental impacts are because of the electric energy required for powering the aspiration system, followed by the construction of the activated carbon air filter and aspiration system **10.** Path C uses twice the amount of etching solution compared to path D, but path C has lower environmental score as lower electrical energy is required because of shorter delamination procedure duration required (4 vs. 24 h)

(*continued on next page*)

										11. Etching procedure is the step leading to higher contribution to the overall environmental impacts (except path F: delamination step) because of the long duration and high amount of electrical energy consumed **12.** Among the MAX phase synthesis the most impact contribution is always associated with the titanium-based precursors **13.** By using MAX phase E in all the pathways, environmental impacts for paths A and F may be reduced by 30%, whereas −7% to −14% can be achieved for paths B–D and path G is not affected **14.** By also considering the electrical conductivity of the MXene produced, paths B, C, and E have the low environmental impacts and high electrical conductivity, with path B identified as the excellent method with low environmental impacts, and path C has higher electrical conductivity MXene but slightly higher environmental impacts

(*continued on next page*)

Table 16.1 Main environmental impacts, limitations, and key findings of environmental assessment studies of 2DMs—cont'd

Authors	Production methods	Feedstock	Products	Main contributor	Energy use per product (MJ/kg)	Energy use per product (MJ/cm^2)	GWP/Greenhouse gas emissions ($kg\ CO_2/kg$ product)	GWP/Greenhouse gas emissions ($g\ CO_2/cm^2$ product)	Limitations/Assumptions	Key findings
	(B) Topochemical *synthesis by* LiF-HCl etching/delamination: Ti + Al + TiC (Ti:Al:C = 3:1.1:1.8) > LiF in 37 wt.% HCl > washing/centrifugation	Ti + Al + TiC *(Ti:Al:C = 3:1.1:1.8)*		MAX phase and electricity	NA	NA	4910	NA		
	(C)Al-$Ti_3C_2T_x$ synthesis from the modified Al-Ti_3AlC_2 MAX phase: TiC + Ti + Al (2:1.25:2.2) > 12M $HCl + H_2O$ + 50 wt.% HF (6:3:1 volume ratio) > 0.5M LiCl	TiC + Ti + Al (2:1.25:2.2)		MAX phase and electricity	NA	NA	7080	NA		

(continued on next page)

	(D) Scalable synthesis: TiC + Ti + Al (2:1:1) > 12M HCl + H_2O + 49 wt.% HF (6:3:1 volume ratio) > 1.18M LiCl	TiC + Ti + Al (2:1:1)		MAX phase and electricity	NA	NA	8420	NA		
	(E) $Ti_3C_2T_x$ synthesis from low-cost and environmentally friendly precursors: TiO_2 + secondary Al + tire-derived C > LiF in 9M HCl > washing/centrifugation	TiO_2 + secondary Al + tire-derived C		MAX phase and electricity	NA	NA	7540	NA		

(*continued on next page*)

Table 16.1 Main environmental impacts, limitations, and key findings of environmental assessment studies of 2DMs—cont'd

Authors	Production methods	Feedstock	Products	Main contributor	Energy use per product		GWP/Greenhouse gas emissions		Limitations/Assumptions	Key findings
					(MJ/kg)	(MJ/cm^2)	(kg CO_2/kg product)	(g CO_2/cm^2 product)		
	(F) Ultrafast, one-step salt solution-based acoustic synthesis: TiC + Ti + Al (2:1:1) > 0.5M LiF > surface acoustic waves	TiC + Ti + Al (2:1:1)		Delamination and electricity	NA	NA	15,800	NA		
	(G) First LCA-assessed synthesis of $Ti_3C_2T_x$: Ti + Al + C (3:1:2) > 37 wt.% HCl + H_2O + 48 wt.% HF (6:3:1 volume ratio) > 0.47 M LiCl	Ti + Al + C (3:1:2)		MAX phase and electricity	NA	NA	14,000	NA		

2DMs, Two-dimensional materials; BM, Brodie method; CED, cumulative energy demand; CFF, cross flow filtration; CVD, chemical vapor deposition; ECE, electrochemical exfoliation; IEA, International Energy Agency; IF, inorganic fullerene; GWP, global warming potential; LPE, liquid-phase exfoliation.

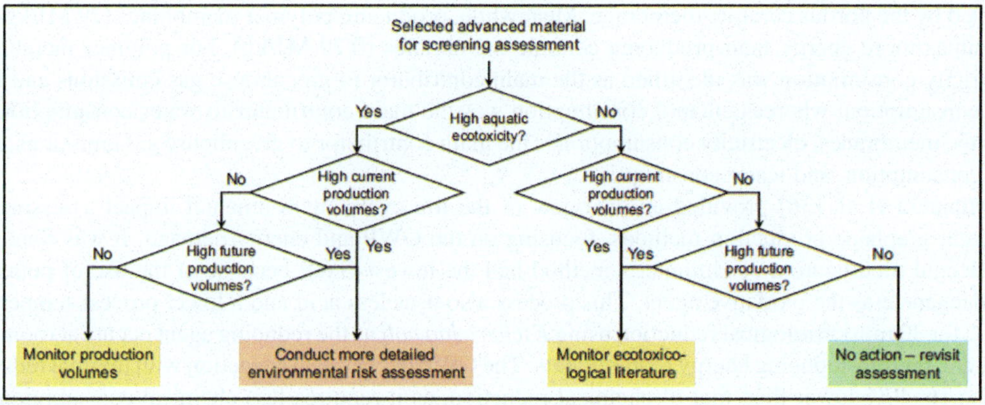

FIGURE 16.1

Recommended actions depending on the outcome of the proxy measure risk screening approach [33].

(Fig. 16.1). Although both materials exhibit low aquatic ecotoxicity, their anticipated large-scale production raises concerns regarding their cumulative environmental effects. Therefore continuous monitoring of ecotoxicological data is necessary to ensure that total aquatic ecotoxicity thresholds are not exceeded.

In addition, Beloin-Saint-Pierre and Hischler [34] investigated the principal contributors to the environmental impact associated with various graphene and GO production methods. They identified chemical inputs, carbon sources, and electricity or overall energy demand as dominant factors influencing environmental performance. In particular, the transformation of graphene-based material suspensions into dry powders was shown to substantially increase environmental impacts because of the highly energy-intensive ball-milling process. Nevertheless, the authors noted that anticipated changes in future electricity mixes are unlikely to significantly alter the comparative outcomes, as conservative energy transition scenarios lead to similar relative changes across production pathways. Their analysis indicated that the first (H2) and second (H3) variants of Hummers' method exhibited the highest climate change impacts among GO production routes, whereas epitaxial growth resulted in higher climate change impacts than CVD for graphene production. Overall, they concluded that variations in electricity demand and projected developments in the European electricity mix, following International Energy Agency trends, are not expected to alter the relative ranking of the most and least environmentally favorable production options.

Diaz-Arauzo et al. [35] have also compared the environmental impacts of the LPE followed by cross flow filtration using different membranes (i.e., porous ceramic membrane and polymer membrane) to produce the graphene inks. It was discovered that higher input flow rates and concentrations enabled by porous ceramic membranes can reduce greenhouse gas emissions, water consumption, and fossil fuel consumption by 97%, 96%, and 94%, respectively. For polymer membranes, both microfiltration and ultrafiltration operate for longer durations owing to the reduced membrane surface area, leading to extended operation. For ceramic membranes the ultrafiltration permeate can be directly used to produce graphene inks without energy-intensive rotary evaporation, whereas ethanol consumption can also be reduced by an order of magnitude because of the 10-fold increment in feed concentrations enabled by the porous ceramic membrane. Meanwhile, producing polymer membranes (29 MJ/kg) also consumes more energy than producing ceramic membranes (0.79 MJ/kg). For polymer membranes, electricity consumption was identified as the main contributor to greenhouse gas emissions and fossil fuel consumption, whereas ethanol consumption was the main contributor to water consumption. For ceramic membranes, electricity consumption is the main contributor to greenhouse gas emissions, fossil fuel consumption, and water consumption.

Munuera et al. [36] provided an overview of the life cycle environmental impact assessment of different graphene production methods, focusing on the GWP and energy demand. It was found that the Bengal modification of Hummer's method had the lowest GWP because of the use of potassium permanganate as the oxidative agent. This process also uses less acid and a lower process temperature (35°C for 30 min). Biological reduction using *Escherichia coli* as the reducing agent occurs at room temperature, thereby reducing energy requirements. The GWP of chemical oxidation with thermal reduction is also 16–40% lower than that of chemical reduction, as it requires less chemical consumption. For electrochemical exfoliation, energy efficiency increases with higher applied potential for electrolytes, but using potassium hydroxide electrolyte at 15 V resulted in the highest energy efficiency leading to lowest GWP at 330 kg CO_2 eq/kg graphene. For the other exfoliation methods, microwave exfoliation (3 h) consumes five times more energy than ultrasonic exfoliation (30 min) owing to the longer process duration, whereas 90% solvent recovery might reduce 90% of the energy demand for the sonication LPE. It was also deduced that the graphene derived from biomass via CVD has a lower GWP than methane, as some energy requirements were sourced internally from the thermal energy of pyrolytic gases. A larger production scale for CVD is also often encouraged, as scaling up can reduce the energy demand by approximately 70%. Finally, for epitaxial growth, a thinner SiC wafer can reduce the energy demand, whereas improvements in technology maturity, scaling up, and using renewable electricity can help improve the GWP of epitaxial growth.

Overall, electricity consumption is the main contributor to the environmental impact of most of the graphene production processes reviewed in this study [36]. For CVD production of graphene from methane and hexane the combustion of natural gas and use of hydrogen gas are two of the main elements contributing to environmental impacts, whereas the use of biomass as the feedstock can hinder the use of these gases in the production processes. Among the processes, chemical oxidation with chemical reduction can achieve highest cumulative energy demand (CED) of 3581 MJ/kg graphene, whereas ball milling LPE (dry) without using solvents can achieve lowest energy demand of 210 MJ/kg graphene. CVD and epitaxial growth also have higher GWP than most exfoliation methods, whereas chemical oxidation with biological reduction has good potential to reduce carbon dioxide emissions compared to other production methods because of the low operating temperature and low energy requirements.

The environmental impacts of the different production methods for graphene and GO are summarized in Table 16.2. The main contributors to environmental impact are electricity consumption for most graphene production methods, whereas chemical consumption is the main contributor to GO production methods. Subsequently, for the graphene production methods, thermal exfoliation has the largest energy consumption (2000 MJ/kg), whereas ultrasonic exfoliation can achieve an energy consumption as low as 70 MJ/kg. CVD generally has lower energy consumption than epitaxial growth. In terms of GWP, CVD has the largest impact (up to 594 kg CO_2 eq/kg), whereas chemical oxidation with biological reduction can achieve a low GWP (60 kg CO_2 eq/kg) for graphene production. Next, although Hummer's method can have the largest impact of GWP (up to 4000 kg CO_2 eq/kg), Bengal's variant of Hummer's method can achieve GWP as low as 10 kg CO_2 eq/kg for GO production. Overall, employing renewable energy

Table 16.2 Overall environmental impacts of different graphene and graphene oxide production methods [32–36].

2DMs	Production methods	Main contributors	Energy use per product	GWP	Fossil fuel consumption	Water consumption
Graphene	Chemical oxidation with chemical reduction	Production of reduction agent, chemical consumption, and acid neutralization step	900–1000 MJ/kg	243 kg CO_2 eq/kg	NA	NA
	Chemical oxidation with biological reduction	Electricity consumption	NA	60 kg CO_2 eq/kg	NA	NA
	Chemical oxidation with thermal reduction	Electricity consumption	NA	140 kg CO_2 eq/kg	NA	NA
	Ultrasonication exfoliation	Production of the solvent and exfoliation process	70–500 MJ/kg	NA	NA	NA
	Thermal exfoliation	Microwave heating (electricity consumption)	2000 MJ/kg	NA	NA	NA
	Liquid-phase exfoliation	Electricity use and ethanol consumption	0.34–24 MJ/L 210–529 MJ/kg	0.53–16 kg CO_2/L 85–216 kg CO_2 eq/kg	7.5–191 MJ/L	6.9–119 kg/L
	Microwave exfoliation	Electricity consumption	1879 MJ/kg	NA	NA	NA
	Electrochemical exfoliation	Electricity consumption	NA	439 kg CO_2 eq/kg	NA	NA
	CVD	Methane feedstock production step, use of hydrogen, and electricity consumption	0.007–4.9 MJ/cm^2 230–340 MJ/kg	0.03–13.2 g CO_2 eq/cm^2 235–594 kg CO_2 eq/kg	NA	NA
	Epitaxial growth	Electricity consumption for production of the silicon wafer substrate	2–80 MJ/cm^2	32.9–67 g CO_2 eq/cm^2	NA	NA
GO	Hummer's method	Electricity consumption and chemical consumption	NA	10–4000 kg CO_2 eq/kg	NA	NA
	Staudenmaier method	Chemical substance (nitric acid)	NA	40–50 kg CO_2 eq/kg	NA	NA
	BM	Chemical substance (nitric acid)	NA	50–70 kg CO_2/kg	NA	NA

2DMs, Two-dimensional materials; BM, Brodie method; CVD, chemical vapor deposition; GWP, global warming potential.

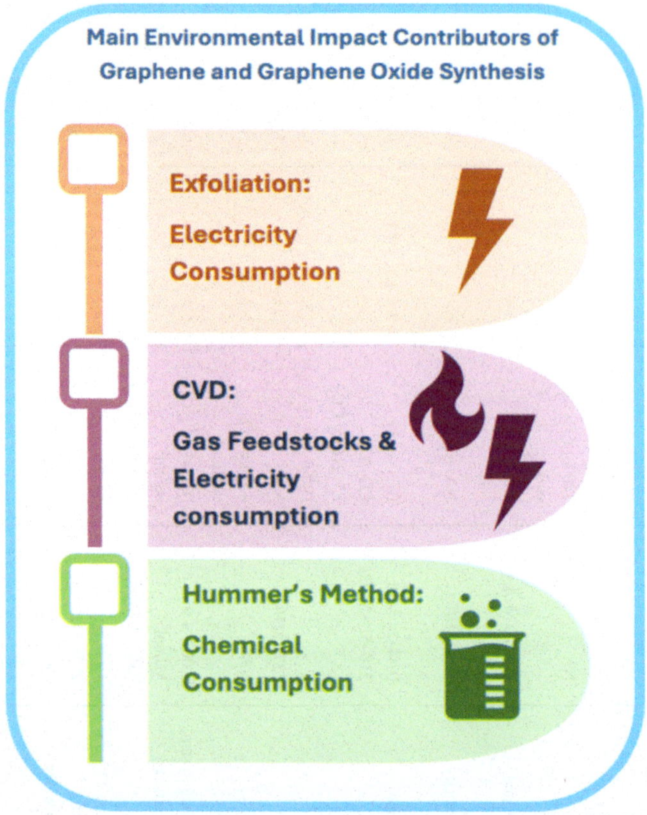

FIGURE 16.2

Main environmental impact contributors of graphene and graphene oxide synthesis. *CVD*, Chemical vapor deposition.

as the source of electricity is the key strategy to reduce environmental impacts, given that electricity consumption is the largest contributor, as shown in Fig. 16.2 and Table 16.2. Although chemical consumption and gas feedstocks are the second largest contributors to environmental impact, these production methods might be gradually replaced in the future, given that regulations and laws related to environmental protection may be strictly enforced. Alternatively, the production of gas feedstocks or chemicals from waste may be an alternative to continue employing these production methods in the future.

16.2.1.2 ***Molybdenum disulfide***

Molybdenum disulfide (MoS_2) is commonly synthesized via bottom-up routes, including solvothermal processing, wet chemical synthesis, gas-phase methods, and hydrogen-assisted reactions. Environmental and toxicological assessments indicate that, similar to graphene and GO, MoS_2 exhibits low aquatic ecotoxicity despite the potential for high production volumes, highlighting the need for continued monitoring of its ecotoxicological profile [33]. Complementary analyses have evaluated the environmental burdens associated with different MoS_2 production pathways [37]. From an energy perspective, solvothermal synthesis is comparatively advantageous because it operates at lower temperatures (~180°C) than gas-phase synthesis (~850°C) and hydrogen-based reactions (~600°C). The absence of intermediate processing stages further reduces the energy demand and minimizes exposure to hazardous gaseous environments. Toxicological investigations have shown that MoS_2 demonstrates moderate cellular toxicity only at the highest tested doses and remains largely inert under typical exposure conditions [38]. Its solubility in simulated biological fluids suggests limited in vivo persistence, and its overall low biological and chemical reactivity indicates a low likelihood of respiratory hazards.

LCA provides additional insights into the environmental footprint of MoS_2 production. A wet chemical synthesis assessment revealed that electricity consumption dominates environmental impacts, accounting for approximately 55% of primary energy demand and 63% of greenhouse gas emissions, with the remainder attributable to chemical reagents such as citric acid, ammonium sulfide, and ammonium molybdate [31]. A separate LCA comparing solvothermal synthesis using LiOH and NaOH solvents demonstrated that replacing LiOH with NaOH can substantially reduce environmental burdens, including nonrenewable energy use and land occupation, with estimated impact reductions of up to 56%[30]. These benefits arise because LiOH production is associated with aluminum emissions in aquatic systems, contributing to ecotoxicity and noncarcinogenic impacts. Additional environmental pressures stem from land occupation linked to industrial infrastructure and resource extraction, as well as greenhouse gas emissions associated with natural gas-derived steam and the release of methane. Mining and beneficiation of molybdenum ores further contribute to environmental loading through high-concentration tailings (reported up to ~4000 mg/kg Mo), which can contaminate nearby soil, sediments, and water systems. Across the evaluated categories, mineral extraction was identified as the dominant contributor, followed by GWP, nonrenewable energy depletion, and land occupation [30].

Collectively, the current evidence indicates that the environmental assessment of MoS_2 production is less comprehensive than that of graphene and GO. Existing studies consistently identify electricity demand and solvent production, particularly LiOH, as major contributors to environmental impacts [30]. Although MoS_2 itself shows low inherent aquatic ecotoxicity [33], upstream processing steps may introduce ecotoxicological risks, highlighting the importance of system-level evaluation. The variability in the reported impacts reflects the differences in methodological assumptions and process boundaries, emphasizing the need for expanded and harmonized LCAs across synthesis routes. Based on the available analyses, transitioning to renewable energy sources and substituting LiOH with lower impact solvents are practical strategies for reducing the environmental footprint of MoS_2 manufacturing [30].

16.2.1.3 ***Tungsten disulfide***

Tungsten disulfide (WS_2) is commonly synthesized through gas-phase routes and hydrogen-assisted reactions, analogous to established MoS_2 production pathways. Although WS_2 has been increasingly applied over the past decade as an additive in HTFs, LCA studies examining its environmental footprint remain limited. Available analyses indicate that the synthesis stage is energy-intensive. Bobba et al. [29] reported that tungsten trioxide, the principal precursor, dominates the overall energy demand and associated greenhouse gas emissions, particularly within the energy resource impact category. In contrast, ancillary inputs, such as scrubbing solutions and process gases (H_2, N_2, and H_2S), contribute negligibly to the total environmental burden. Complementary work by Corazzari et al. [38] compared the environmental and biological responses of WS_2 and MoS_2, showing that WS_2 produced no statistically significant deviation from the negative controls, suggesting minimal acute cellular or chemical reactivity. Consistent with the observations for MoS_2, WS_2 shows solubility in simulated biological fluids, implying limited persistence in vivo. However, broader environmental considerations extend beyond the reactivity of materials. Han et al. [41] noted that although tungsten is not widely recognized as acutely toxic, mining residues and tailings may pose environmental and human health risks. Petruzzelli and Pedron [42] further emphasized that the mobility of tungsten in soils depends strongly on physicochemical conditions, such as pH, tungstate speciation, polymerization behavior, and adsorption characteristics of soil matrices. These factors govern the transfer of tungsten species into plants and, ultimately, the food chain, highlighting the importance of regulatory oversight of tungsten emissions.

Given the strategic importance and limited substitutability of tungsten-based precursors, environmental assessments of WS_2-enabled HTFs must consider not only synthesis impacts but also the functional performance gains that improved heat transfer may deliver. Such system-level evaluations are necessary to determine whether operational efficiency offsets upstream environmental costs. Resource recovery provides additional mitigation pathways. Reprocessing tungsten tailings represents a viable secondary source of tungsten trioxide, with incomplete inventories indicating deposits containing approximately 96 kt of tungsten at an average grade of ~0.1% WO_3, compared to 0.3–1% in primary ores [41]. Established beneficiation techniques, including enhanced gravity separation, wet high-intensity magnetic separation, and flotation, can recover valuable fractions, whereas bioleaching offers a route to remove coassociated toxic elements. Therefore integrating tailings valorization into the WS_2 supply chain could reduce primary extraction pressures and improve the overall sustainability profile of tungsten-based HTF technologies.

16.2.1.4 MXenes

MXenes have attracted increasing scrutiny from an environmental perspective because their projected production scale coincides with comparatively high ecotoxicological indicators. Arvidsson et al. [33] reported that MXenes exhibit substantially higher aquatic ecotoxicity (4 L/mg) than graphene (0.1–0.5 L/mg), GO (0.001–0.05 L/mg), and MoS_2 (0.004 L/mg). Given this disparity and the anticipated growth in MXene manufacturing the authors emphasized the need for comprehensive environmental risk assessments, including exposure modeling and material flow analysis. Complementing this perspective, Firouzjaei et al. [39] performed a life cycle evaluation of MXene production involving ball milling, tube furnace heating, etching, and delamination. Their analysis identified laboratory electricity consumption as the dominant contributor to environmental impacts, accounting for more than 70% of the total burden, whereas chemical inputs were comparatively less influential. This imbalance highlights the potential sustainability gains achievable through the adoption of renewable electricity and recycled-process inputs. Energy demand analysis further revealed that nonrenewable fossil resources contribute over 60% of the CED, whereas biomass-derived energy sources exert a negligible influence. Process optimization simulations demonstrated that reducing the furnace holding time by half lowered the CED by approximately 10%, whereas doubling the holding time increased the CED by approximately 20%. Notably, scale-up modeling suggests that industrial-level production could reduce per-unit environmental impacts by nearly an order of magnitude, reflecting efficiency gains at larger production scales. Waste management contributed less than 1% to the overall impact, indicating that upstream energy use remains the primary environmental driver.

Material selection critically governs the life cycle environmental profile of MXene synthesis process. For instance, employing titanium dioxide (TiO_2) as a precursor instead of metallic titanium is associated with higher burdens in several impact categories, including ecotoxicity, carcinogenic effects, acidification, and noncarcinogenic toxicity, whereas ozone depletion is the only category showing a relative reduction [39]. Regional electricity mixes further modulated these outcomes. Process simulations indicate that MXene production in fossil-intensive grids, such as those characteristic of Texas and certain regions of China, has elevated impacts across multiple categories, whereas manufacturing in low-carbon electricity systems, exemplified by Sweden's high renewable penetration, substantially lowers overall environmental burdens. These findings collectively reinforce the central role of low-carbon energy integration in reducing the life cycle impact of MXene synthesis. Ungureanu et al. [40] systematically compared the life cycle environmental impacts of seven $Ti_3C_2T_x$ synthesis routes (paths A–G), linking each step in MAX phase preparation, etching, and delamination to specific midpoint and endpoint damage categories. Path A, the original HF-based synthesis, exhibited the lowest impact across all categories because it combined a short overall reaction time (~4 h versus more than 30 h for most alternatives) with room-temperature etching, thereby minimizing electricity use for prolonged heating. In contrast, paths F (ultrafast, salt-solution acoustic synthesis) and G (first LCA-assessed route from elemental powders) are the most environmentally intensive options, primarily because of the high electricity demand and specialized equipment requirements. The largest relative differences for these two pathways occurred in marine eutrophication, mineral resource scarcity, and water consumption, which were driven by nitrate emissions associated with lignite-based power generation, the use of nickel-containing stainless steel in surface acoustic wave devices, and the substantial electricity required to operate heating plates and vacuum pumps for extended periods. Among all routes, path G shows the highest overall burden because it relies on energy-intensive vacuum operation and the material and energy demands of vacuum pump manufacture [40].

Beyond these extremes, Ungureanu et al. [40] evaluated intermediate pathways and identified distinct trade-offs between environmental performance and product quality, as shown in Figure 16.3. Path B (LiF–HCl etching/delamination) avoids direct HF use but has an overall impact approximately 3.7-fold higher than that of path A, mainly because of its much longer reaction time (≈40 h) and the electricity required to power the aspiration system, including the production of the activated carbon filter and associated infrastructure. Paths C (Al–$Ti_3C_2T_x$ derived from a modified Al–$Ti_3C_2T_x$ MAX phase) and D (scalable $Ti_3C_2T_x$ synthesis from TiC, Ti, and Al powders) differ primarily in etchant consumption and delamination time. Although path C uses approximately twice as much etching solution as path D, it has lower overall impacts because its delamination step is considerably shorter (~4 h vs. 24 h), reducing electricity use. Across all routes except path F the etching stage is the dominant contributor to environmental impacts owing to its long duration and high-power consumption, whereas in the MAX phase synthesis step, titanium-containing precursors consistently drive the largest share of burdens. Sensitivity analysis further shows that employing the lower-impact MAX phase from path E (which incorporates tire-derived carbon) in all routes can reduce total impacts by approximately 30%, whereas only modest reductions of 7–14% are achievable for paths B–D, and no appreciable change is observed for path G. When the electrical conductivity of the resulting MXenes is considered alongside environmental metrics, paths B, C, and E offer the most favorable combinations of low impact and high conductivity. Path B is highlighted as an excellent compromise between a relatively low environmental

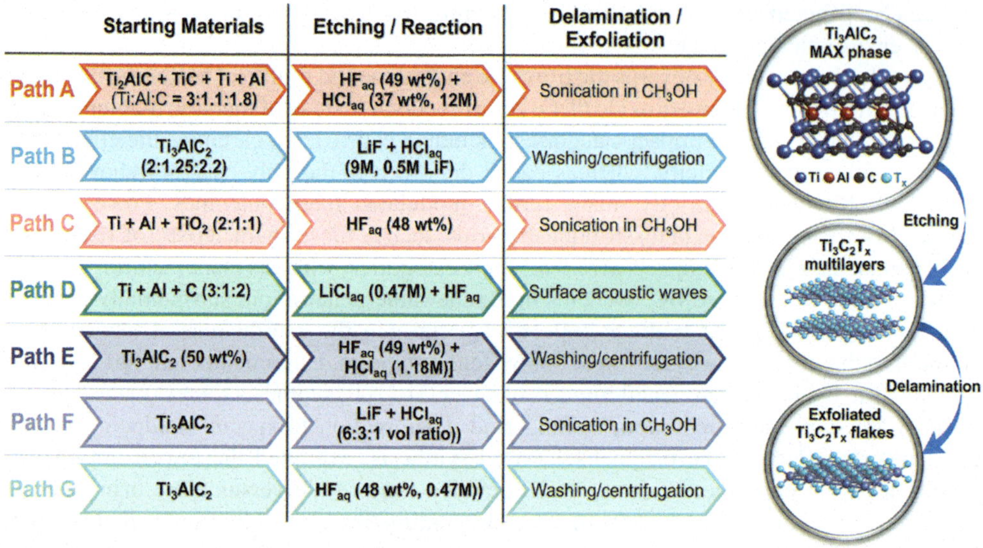

FIGURE 16.3

Graphic representation of different phases of seven synthesis methods of MXenes assessed by LCA [40]. *LCA*, Life cycle assessment.

burden and acceptable conductivity, whereas Path C provides superior conductivity at a slightly higher impact level.

Ungureanu et al. [40] comprehensively assessed the environmental impacts associated with different MXene synthesis pathways and demonstrated that most of these impacts arise from MAX-phase synthesis and electricity consumption. Consistent with the findings of Arvidsson et al. [33], MXene synthesis contributed to multiple ecotoxicity-related impact categories (Table 16.3). Based on these findings, several key characteristics can be identified for environmentally preferable MXene synthesis pathways (Fig. 16.4): (1) shorter reaction times, (2) lower reaction temperatures, (3) avoidance of specialized equipment, (4) reuse of existing equipment to minimize construction-related impacts, and (5) utilization of waste-derived materials for MAX-phase synthesis.

16.2.2 Toxicity and ecotoxicity

As addressed in the previous sections, graphene, GO, and MoS_2 generally have lower ecotoxicities than MXenes. In fact, Adhikari [43] also reported that 2D crystals of black phosphorus can be obtained using top-down synthesis, such as chemical-based intercalation or liquid-based exfoliation. Although chemical-based intercalation is much more convenient than mechanical exfoliation for large-scale 2DMs' synthesis, the use of toxic chemicals has limited its large-scale production prospects. In chemical intercalation processes, intercalating metals such as Li are considered health, physicochemical, and ecotoxicological hazards according to the National Occupational Health and Safety [44]. In addition to the intrinsic hazards associated with chemical intercalation routes the organic solvents commonly

Table 16.3 Environmental impacts of 2D nanomaterials [29–36,39,40].

2DM	GWP (kg CO_2 eq/kg)	Human toxicity (kg 1,4-DCB/kg)	Terrestrial ecotoxicity (kg 1,4-DCB/kg)	Fossil fuel consumption
Graphene	60–594	Low	Low	7.5–191 MJ/L
GO	10–4000	Low	Low	NA
MoS_2	42–391.33	1.67295×10^{-7} kg C_2H_3Cl eq. to air/kg product	1.85×10^{-15} kg TEG eq. to soil/kg	NA
WS_2	37.98	2.97	3.74×10^{-3}	11.47 kg oil eq/kg
MXenes	428.1–15,800	1500–17,500	4050–43,800	342–3610 kg oil eq/kg

2D, two-dimensional; 2DM, two-dimensional material; GO, graphene oxide; GWP, global warming potential.

employed for metal-ion disintegration and postexfoliation cleaning introduce significant occupational and environmental risk. Solvents such as hexane, benzene, acetonitrile, dimethylformamide, toluene, acetone, isopropanol, chloroform, and ethyl alcohol are widely used because of their strong solvation power and compatibility with layered materials. Toxicological assessments classify several of these solvents as posing acute oral toxicity in animal models [45], whereas others are recognized as carcinogens or substances associated with elevated long-term cancer risk [46–48]. Therefore LPE has emerged as an attractive alternative to chemically driven intercalation strategies for producing 2DMs. Unlike intercalation processes that depend on reactive chemistries, LPE is fundamentally a physical top-down technique in which external energy, most commonly ultrasonic cavitation, overcomes interlayer van der Waals forces to delaminate bulk crystals into few-layer sheets. This approach reduces the need for aggressive reagents while maintaining operational simplicity and adaptability to a range of layered solids. Notably, successful LPE protocols have been reported for materials such as black phosphorus and boron nitride, demonstrating their versatility across chemically distinct systems [43]. The drawback associated with this approach is the possible occurrence of phase transition and nonuniform size and thickness of the 2DMs because of the high external vibrations [49]. Therefore this process requires stabilization and sorting after exfoliation [50]. These processes are conducted using various organic solvents, such as polymers, copolymers, and alcohols, which are highly toxic for exfoliation [45,48,51]. Furthermore, other top-down methods such as laser-based exfoliation and ball milling-assisted exfoliation have no toxicological effects in producing 2DMs, but the quality and size of the flakes are the major issues associated with these techniques [52–55].

Meanwhile, bottom-up synthesis routes enable commercial-scale production of high-purity 2DMs, particularly through CVD, wafer-scale growth, and layer-controlled fabrication strategies [43]. In CVD processes, volatile precursors react at elevated temperatures to form atomically thin films on catalytic or insulating substrates. Although methane is widely used as a carbon source, it presents occupational and environmental hazards: it is a potent greenhouse gas, highly flammable, and capable of displacing

FIGURE 16.4

Key properties of environment-friendly MXene synthesis.

oxygen at high concentrations, creating asphyxiation risk. Consequently, strict gas handling, ventilation, and monitoring protocols are essential for mitigating operational hazards [56]. In addition to being listed as a greenhouse gas, methane is highly flammable and can be ignited at low pressure and concentrations, posing a threat to human health, as it can lead to coma and death because of oxygen deprivation if inhaled [57–59]. Hence it is necessary to handle this process safely, following all safety protocols. In addition, bottom-up approaches often deal with highly toxic metal elements, such as molybdenum. Franke and Moxon [166] demonstrated toxicity testing on Albion rats by feeding them salts of Mo, Te, and Se at levels of 25 and 50 ppm of the elements in the diets. Se salts exhibited the highest toxicity, followed by Te and Mo. Nevertheless, Mo also poses a toxic threat to living organisms, so it is necessary to safely handle the material during the bottom-up synthesis approaches of MoS_2. Other bottom-up synthesis approaches, such as molecular beam epitaxy, magnetron sputtering, and pulsed laser deposition, can produce 2DMs within a few tens of micrometers without the extensive use of toxic substances; however, they require high vacuum, complex, and expensive operating systems, so their scalable synthesis is still under development [60–62].

Numerous studies have shown that different synthetic approaches of the 2DMs can pose toxic threats to human health and the environment. However, the release of 2DMs during usage is often not considered in these studies because of data limitations and difficulties in accurately measuring the emission concentration in the workplace. Furthermore, some 2DMs, such as MXenes, are vulnerable to oxidative deterioration upon exposure to air or water. As degradation occurs, it produces carbon byproducts and titanium dioxide which can affect ecosystems [63].

16.2.3 Potential mitigation strategies

It can be foreseen that future endeavors on MXene synthesis will be focused on shortening the reaction time and lowering the reaction temperature to minimize energy consumption. Therefore microwave synthesis, which can fulfill both of these properties, is an excellent alternative to wet chemical synthesis for this purpose [64,65]. Similar synthesis methods with shorter reaction times and lower reaction temperatures should also be encouraged for the synthesis of other 2DMs. In addition, it is also encouraged to avoid the construction of new equipment, as a few studies above have shown that the construction of new equipment has contributed to most of the environmental impacts. Repurposing old equipment or proper timely maintenance of this equipment can help minimize the environmental impacts of these synthesis processes, especially for the processes with the majority of environmental impacts contributed by the chemical consumption, which is difficult to mitigate.

Waste materials can also be used to synthesize precursor materials for 2DMs' synthesis. To synthesize carbon materials, available biomass wastes, such as forest and agricultural wastes, which contain approximately 55 wt.% carbon, can be utilized [66]. Biomass resources such as rice husks, sugarcane bagasse, wheat straw, palm oil waste, fruit waste, soybeans, newspapers, populous woods, and other lignin-based biomass can also be used to synthesize monolayer and few-layer graphene [43]. In addition to using waste to derive precursors for the 2DMs' synthesis, using green synthesis techniques, such as plant-based phytochemicals or microbes, to synthesize 2DMs is also another alternative pathway [43]. Various phytochemicals derived from forest and agricultural resources and microbes, such as bacteria, fungi, and yeast, can also be used to reduce various metal salts and compounds into nanostructures [67,68]. Overall, these methods can help produce green 2DMs, which are free from toxicity. It was also found that there are few reports on the green synthesis of TMDs and other 2DMs, as there are no abundantly available natural resources for TMDs, such as carbon, whereas the reaction kinetics of transition metals with chalcogenides occur with a specific molecular structure, which is hardly achieved by utilizing natural resources. Using less toxic chemicals and environmental conditions might be a sensible approach to synthesize the 2DMs, which will release less toxic emissions during the processes [43].

Finally, it was demonstrated that energy consumption is the major contributor to the environmental impacts of most processes. Therefore it is crucial to shift the energy mix from nonrenewable resources to renewable energy sources such as solar, wind, and hydropower. As shown in a study, harvesting solar energy using a parabolic trough collector (PTC) power plant to replace coal-fired electricity production can achieve 223,100 t of carbon dioxide emissions annually, which corresponds to a 0.0204% reduction in coal-fired electricity output [69]. Currently, although photovoltaic (PV) is the trending renewable technology to harvest energy to produce electricity, PTC technology and geothermal technologies with a high capacity factor of more than 75% can also be explored further to produce 2DM-based HTFs to maximize electricity production and therefore replace a major portion of the nonrenewable portion of energy mixes [69].

16.2.4 Comparison with conventional fluids

Although many studies have been conducted to explore the enhancement of the thermal properties of 2DM-based HTFs, there are few studies on determining the environmental impacts of 2DM-based

Table 16.4 Environmental impacts of heat transfer fluids [71].

HTF	GWP (kg CO_2 eq/kg)	Human toxicity (kg 1,4-DCB/kg)	Terrestrial ecotoxicity (kg 1,4-DCB/kg)	Fossil fuel consumption (MJ/kg)
Therminol	10.35	56.0105	0.0081	171.664
Binary molten salt	4.007	0.9468	0.0012	28.272
Hitec molten salt	2.914	0.7778	0.0014	28.187

2DM, *two-dimensional material;* GWP, *global warming potential.*

HTFs. Therefore the environmental impacts of the 2DMs and conventional HTFs are tabulated in Tables 16.3 and 16.4 for comparison. According to Elsaid et al. [70], the loads of nanoparticles are generally up to 10 wt.% to limit the increment in the viscosity and density of the nanofluids. Hence based on Table 16.3, when 10% of the GWP of the 2DMs is considered, it can be observed that MXenes still contribute at least four times the GWP compared to therminol (42.81 vs. 10.35 kg CO_2 eq/kg). Although GO can contribute to a GWP of 4000 kg CO_2 eq/kg, the development of its synthesis methods may lower its GWP to as low as 10 kg CO_2 eq/kg. Considering 10% of the GWP the value is approximately three times lower than that of Hitec molten salt. Overall, graphene and GO can contribute to a very low GWP compared to HTFs, whereas more data might be required to deduce the environmental impacts of HTFs based on MoS_2 and WS_2, and MXene-based HTFs might still need further improvement in the synthesis procedures and precursors from the aspect of GWP. In terms of toxicity, it is difficult to compare these 2DMs with HTFs, as different toxicity tests were conducted for different 2DMs. Fossil fuel consumption for synthesizing 2DMs and HTFs is also difficult to compare because of the different units considered in different studies. Generally, graphene and GO have low toxicity, whereas more toxicity testing is required for MoS_2 and WS_2 to compare the values. Similar to the GWP comparison, MXenes have much higher toxicities than conventional HTFs, as shown in Tables 16.3 and 16.4.

Beyond estimating the potential CO_2 emission reductions achievable with 2DM-based HTFs, as demonstrated by Sundar et al. [27], recent studies have begun to consider broader environmental dimensions. Czaplicka et al. [72] evaluated the corrosivity and toxicity of nanofluids as indicators of environmental impact. Their analysis indicated that most 2DM additives dispersed in HTFs—predominantly water-based systems—exhibit measurable corrosive tendencies, whereas no significant toxicity was reported for the assessed materials. However, environmental assessments of 2DM-based HTFs remain sparse and fragmented. To enable a rigorous comparison with conventional HTFs and support evidence-based deployment (Fig. 16.5), future research should prioritize (1) quantifying the net environmental impact reductions attributable to 2DM-enhanced HTFs; (2) conducting comprehensive LCAs covering synthesis, formulation, use, and end-of-life stages; and (3) optimizing manufacturing pathways to balance enhanced thermal performance with minimized environmental burdens.

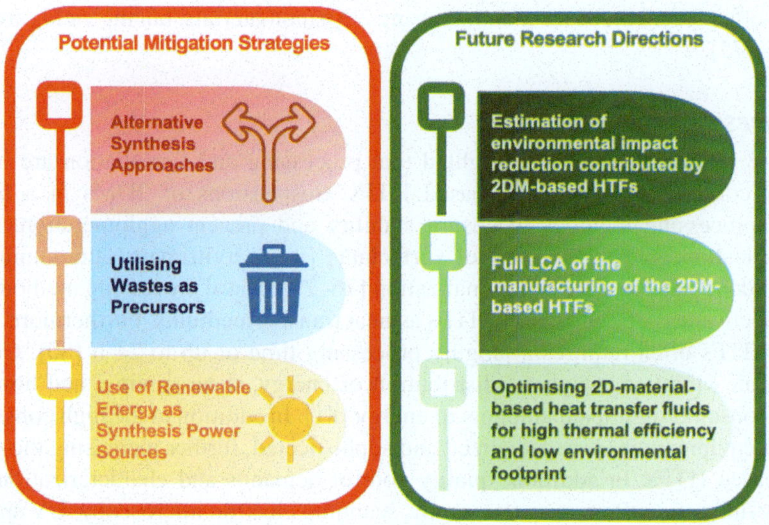

FIGURE 16.5

Potential mitigation strategies to reduce the environmental impacts and future research directions of 2DM-based HTFs. *2DM*, Two-dimensional material; HTF, heat transfer fluid; *LCA*, life cycle assessment.

16.3 Economic impacts of two-dimensional-material-based heat transfer fluids

16.3.1 Cost analysis

Although 2DM-based HTFs have improved performance for various applications, the costs associated with their entire life cycles remain an obstacle to their commercialization. Therefore it is necessary to break down the cost analysis into a few components, such as material synthesis, processing and formulation, implementation, performance and efficiency gains, and disposal, to provide a comprehensive understanding of the economic viability and identify potential areas for cost optimization throughout the life cycle of 2DM-based HTFs.

16.3.1.1 Material synthesis

The high production cost of 2DMs remains one of the primary concerns for industry stakeholders in producing 2DM-based HTFs. For instance, graphene pricing can range from \$150 to \$500/kg [73], whereas h-BN pricing can range from \$45 to \$200/kg [74], depending on the purity and production methods. In contrast, conventional HTFs, such as water, cost only \$0.01/kg, whereas silicone oils range from \$3 to \$5/kg [75]. Although 2DM-based HTFs typically require only a small percentage of 2DMs by weight, the added cost of these advanced additives can still match or exceed the total cost of conventional HTF. For example, 0.05 vol.% graphene/water has improved thermal performance, but the cost has increased up to \$0.04/kg, which is four times higher than that of water alone [72]. Studies are ongoing to develop cost-effective methods for manufacturing 2D nanomaterials, but the yields are limited to very small quantities [75].

16.3.1.2 Processing and formulation

Recent literature studies have also highlighted that processing and formulation are major economic bottlenecks for commercial-scale 2DM-based HTFs. Suspensions of 2DMs in a base fluid often require certain processing methods to ensure stability and prevent agglomeration. These methods include ultrasonication, stirring, the use of surfactants [i.e., cetyltrimethylammonium bromide and polyvinylpyrrolidone], and surface functionalization [76–79]. These processing methods are crucial for ensuring the effectiveness of 2DM-based HTFs as heat transfer mediums. Furthermore, the preparation of 2DM-based HTFs often requires a lengthy processing time of up to 24 h [80]. Generally, it also involves ultrasonic vibration, magnetic distribution of energy, shear mixing, and homogenizing ball milling, which consume significant amounts of energy [81]. In addition to the high consumption of time and energy, this equipment is also specialized and sophisticated, further increasing the final production cost of 2DM-based HTFs. In addition, quality control demands and characterizations, such as zeta potential, X-ray diffraction, Raman spectroscopy, scanning electron microscopy, UV spectroscopy, and visual stability testing, can also potentially increase labor and instrument costs [80]. Although it is difficult to quantify this cost component, future techno-economic analyses should also analyze it to facilitate quantitative comparisons with conventional HTFs.

16.3.1.3 Implementation, performance, and efficiency gains

Beyond the high synthesis and processing costs the implementation and operational phases of 2DM-based HTFs can offset these expenses via performance and efficiency gains. Various experimental studies have demonstrated that 2DM-based HTFs can significantly improve the thermal conductivity and heat transfer efficiency compared with conventional HTFs [75]. For instance, 0.05 vol% of graphene increased the thermal conductivity of water by approximately 24%[72], whereas one graphene layer improved the thermal conductivity of copper nanofluids by 3.7–18.2 times [82]. Meanwhile, a pilot-scale study has also shown that a graphene-based nanofluid applied in industrial chillers has reduced the energy consumption by up to 10% and improved the cooling performance, leading to extended lifespan of heating, ventilation, and air conditioning (HVAC) equipment and reduced maintenance costs [83].

Şen and Variyenli [84] also reported that the hybrid nanofluid of CuO/graphene nanoplatelet/distilled water applied to the concentric tube heat exchanger can improve the overall heat transfer coefficient by 5.5% and reduce the exergy losses by 55%. Therefore less energy consumption is required, which is beneficial in terms of cost savings and reduction of carbon dioxide emissions. Alawi et al. [85] have also reported that 10–13% improvements in the flat-plate solar collector efficiency can be achieved by applying the highly dispersed pentaethylene glycol-thermally treated graphene–water as the absorbing material. The improved efficiency can also translate to a proportionally higher energy capture and lower net cost per kWh of energy produced. Overall, these performance improvements have helped reduce operating costs, minimize energy consumption, and enhance the reliability of processes or applications. Improved heat dissipation has also led to less thermal fatigue, fewer maintenance cycles, and long-term economic viability of 2DM-based HTFs. Although the upfront material and processing costs remain a concern, the superior thermal properties of 2DM-based HTFs can provide a strong case for their adoption in systems where the efficiency of thermal management is paramount. Future studies should

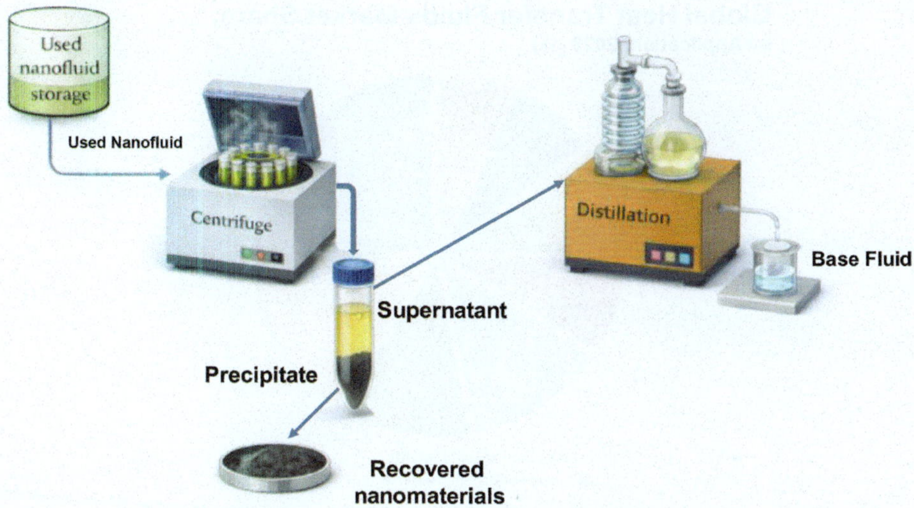

FIGURE 16.6

Schematic diagram of the potential recovery of the base fluid and nanostructures.

also quantify the potential energy and cost savings associated with 2DM-based HTFs to identify the most suitable and economically viable application fields for their deployment.

16.3.1.4 Disposal

Although the previous sections have explored the possible environmental and health effects of 2DM, the long-term environmental outcomes of many 2DMs remain unclear because of the limited toxicological data available for more advanced 2DMs, such as MXenes and TMDs [86]. The biodegradability of 2DMs is also crucial for predicting short- and long-term disposal and mitigation strategies [86]. Nanoparticle recovery processes from 2DM-based HTFs remain underdeveloped and have not yet been widely adopted because the commercial-scale production of these fluids is still in the developmental stage. Overall, the literature suggests that designing sustainable, scalable, and cost-efficient recovery processes is key to mitigating potential environmental impacts and reducing costs associated with disposal and waste management [75,87]. Shaik et al. [75] suggested a possible recovery process consisting of centrifugation, followed by evaporation of the supernatant and distillation, as shown in Fig. 16.6. Nevertheless, the economic analysis of recovery processes or waste management is still lacking in the literature, and this area needs to be explored further to quantify the economic impacts at the end of the life of 2DM-based HTFs. In addition, potentially stricter future regulations may increase end-of-life costs depending on the toxicity and safety assessments of disposed wastes [86].

16.3.2 Market analysis and projections

As of 2023 the global HTF market size was estimated at $11.06 billion, and this amount is expected to grow at a compound annual growth rate (CAGR) of 3.7% from 2024 to 2030 up to $14.22 billion. The rapid global adoption of CSP around the world is a major factor contributing to the growing market size.

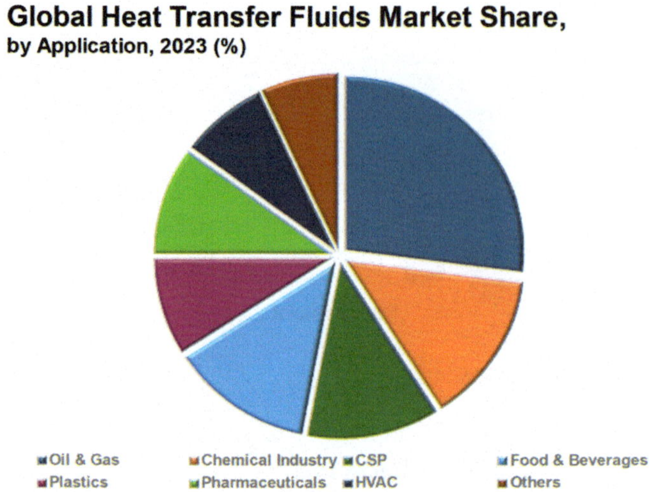

FIGURE 16.7

Global HTF market classified by segments [88]. *HTF*, Heat transfer fluid.

It was pointed out that the COVID-19 pandemic has suspended solar power plant projects and hampered the growth of the HTF market. The HTF market comprises specialized suppliers and large multinational manufacturers. Companies such as Paratherm, Hubbard Hall, Inlet Petroleum Co., and Thermic Fluids Pvt. Ltd. often involve captive consumption or distribution through third-party channels, whereas major players such as Dow Inc., ExxonMobil, BP, Royal Dutch Shell, and Eastman Chemical Company focus on in-house production, research and development (R&D), and product differentiation strategies. Overall, the HTF market was dominated by the oil and gas segment, with a revenue share of 22.2% in 2023, followed by the chemical industry, CSP, and food and beverages, as shown in Fig. 16.7. HTFs play an important role in the oil and gas segments, as specific HTFs are required to suit various desired temperatures and compatibilities. Next, the Asia-Pacific region, especially China and India, accounts for the largest revenue share (47.9%) of the HTF market. The increase is mainly attributed to the growing demand for HVAC systems owing to the increasing population, changing climatic conditions, increasing urbanization, and demographic changes [88].

Although the market for 2DM-based HTFs is not clearly shown in the current databases, an analysis of the current 2DM markets can also indirectly indicate the projections of the market for 2DM-based HTFs. Recent data have revealed significant academic and commercial momentum in the field of 2DMs. For instance, the global graphene market size is anticipated to grow from $1.32 billion in 2024 to $1.62 billion in 2025 at a CAGR of 22.8%[89]. One of the key drivers contributing to market growth is the use of thermal management materials such as HTFs. It was estimated that the CAGR could increase to 30–40% over the next 5–7 years, whereas some forecasts predicted that the market size could reach $2.5–5 billion by 2030, depending on technological breakthroughs [90]. Meanwhile, Scopus and Espacenet patent search databases have also shown an increasing trend in scientific publications and patents related to graphene in the last two decades [91]. These trends in market growth and research fields have indirectly reflected the development of graphene-based 2DM-based HTFs as next-generation

HTFs. Besides graphene, h-BN, which is also known as "white graphene," has also seen its market growth projected from $949.4 million up to $1.28 billion in 2030 with a CAGR of 5.1%. The Asia-Pacific region continues to lead the h-BN market, accounting for approximately 40.6% of the global revenue in 2024, with China representing the largest national share at approximately 41.1%. Lubricants constitute the third-largest segment of the h-BN market, following paint and coating formulations and personal care products, reflecting the established role of h-BN in friction reduction, thermal stability, and surface protection [92]. The concurrent expansion of the graphene and h-BN industries has strengthened both the material supply chain and the industrial impetus for scaling 2DM-based HTFs, particularly in sectors focused on improving energy efficiency, operational reliability, and component lifespan. From a commercialization perspective a notable example is the graphene-based nanolubricant "Infinoil," developed at Sunway University, Malaysia, which has undergone endurance validation under high-performance automotive conditions. These developments illustrate the growing translational pathway from laboratory-scale 2DM research to performance-driven industrial applications [75,93].

Overall, rising thermal management demands, especially in the Asia-Pacific region, for applications such as heat exchangers, solar energy systems, automotive cooling, and electronic device thermal regulation, are key drivers in accelerating the research, development, and potential commercialization of 2DM-based HTFs [94–96]. Advances in commercial expansion beyond lab and pilot-scale production will also determine further market projections in the coming decades. Nevertheless, the high cost of high-purity 2DMs and supply chain constraints continue to challenge their scalability and mass-scale adoption. Nanoparticle aggregation and sedimentation can be major issues in ensuring the stability of 2DM-based HTFs; therefore additional dispersion equipment is often required in the processing of 2DM-based HTFs. In addition, the high maintenance expenses and corrosion issues of nanofluid-based systems, coupled with complex system performance predictions, further diminish the interest in widespread deployment [97]. Furthermore, stringent environmental regulations and policies in different countries can significantly increase the disposal costs of 2DM-based HTFs. The strict requirements for the handling, storage, and disposal of the relevant chemicals, which are aimed at minimizing environmental and health risks, can also increase the economic burden on the manufacturers of 2DM-based HTFs to comply with the rules. Although robust market forecasts are currently limited, future economic analyses should also develop a suitable data modeling framework to understand the volumes, pricing dynamics, and uptake rates of 2DM-based HTFs and their market barriers in various application-specific sectors.

16.3.3 Economic viability and projections

The economic feasibility of deploying 2DM-based HTFs across various industrial applications should be evaluated to analyze the cost-performance trade-offs. 2DM-based HTFs provide excellent thermal properties, thereby reducing the fluid volume required to accomplish the same heat transfer as conventional fluids and resizing the equipment to a smaller size. These benefits can reduce the overall required investments [75]. Nevertheless, 2DMs can potentially increase the viscosity, requiring extra pumping power [98,99]. Hence the improvement in the heat transfer of 2DM-based HTFs must be more substantial than the additional pumping power supply to make the 2DM-based HTF system economically viable. Shaik et al. [75] proposed a price–performance factor (PPF) framework to couple thermophysical enhancement with material cost when assessing the economic feasibility of 2DM-based

HTFs. The formulation distinguishes between laminar and turbulent flow regimes because the relative influence of viscosity and thermal transport differs across the regimes. For laminar flow the PPF is defined as the ratio of the thermophysical performance index to the cumulative material price,

Laminar flow:

$$\mathrm{PPF} = \frac{C_\lambda / C_\mu \ (\text{ratio} > 0.25)}{\sum_n^{i=1} \text{price } (\$; /\text{g})}$$

whereas for turbulent flow,

$$\mathrm{PPF} = \frac{T \ (\text{for} > 1)}{\sum_n^{i=1} \text{price } (\$; /\text{g})}$$

where C_λ / C_μ and T are parameters estimating the thermal performance of nanofluids. To meet the economic needs, $C_\lambda / C_\mu > 0.25$ and $T > 1$ must be satisfied for laminar and turbulent flows, respectively, where C_λ / C_μ and T are calculated as

$$\frac{C_\lambda}{C_\mu} = \frac{(\lambda_{nf} - \lambda_{bf})/\lambda_{bf}}{(\mu_{nf} - \mu_{bf})/\mu_{bf}}$$

$$T = \left(\frac{\lambda_{nf}}{\lambda_{bf}}\right)^{0.67} \times \left(\frac{\rho_{nf}}{\rho_{bf}}\right)^{0.8} \times \left(\frac{C_{p,nf}}{C_{p,bf}}\right)^{0.33} \times \left(\frac{\mu_{nf}}{\mu_{bf}}\right)^{-0.47}$$

where μ_{nf} and μ_{bf} are the viscosity of the nanofluid and base fluid, respectively; λ_{nf} and λ_{bf} are the thermal conductivity of the nanofluid and base fluid; ρ_{nf} and ρ_{bf} are the density of nanofluid and base fluid; $C_{p,nf}$ and $C_{p,bf}$ are the specific heat of the nanofluid and base fluid, respectively.

Although a few studies have reported that nanofluids can shorten the payback period and improve the economic efficiency of thermal management systems, relevant analyses deploying 2DM-based HTFs are still lacking in the literature [75]. Although 2DM-based HTFs demonstrate strong potential for enhancing thermal performance, the evaluation of their economic feasibility should also incorporate the performance indicators and benchmarking methods used in conventional nanofluid studies. This comparative approach is essential for quantifying the viability of these fluids across various applications and justifying their adoption over conventional HTFs.

16.3.4 Policy and regulatory landscape

Evolving policy frameworks and regulatory standards that address environmental, safety, and performance concerns are crucial for shaping the commercialization of 2DM-based HTFs. These regulations can have both enabling and constraining effects on economic viability and feasibility. 2DMs such as graphene, MoS_2, and h-BN are considered nanomaterials and are therefore subjected to a few existing nanomaterial regulations in both the European Union (EU) and the United States (US). In the EU the Registration, Evaluation, Authorisation and Restriction of Chemicals (REACH) framework is applied to impose strict rules for handling and marketing nanomaterials based on environmental and human health safety concerns [100,101]. In the US the Toxic Substances Control Act was established under the Environmental Protection Agency to govern the manufacturing and use of nanoscale materials to protect against risks to human health and the environment [100]. These frameworks require comprehensive data on the toxicity, environmental persistence, and occupational exposure of 2DMs. Owing to the lack of relevant data, regulatory ambiguity remains a bottleneck for the mass-scale adoption of 2DM-based HTFs, as uncertainties can lead to longer approval timelines, additional testing and characterization costs, and extra protection strategies and plans to protect against liability risks.

Although these regulations are essential for safeguarding human health and the environment, they also introduce potential barriers to the economic viability of 2DM-based HTFs and add complexity to commercialization efforts. Many regulatory bodies have broadly classified 2DMs as nanomaterials, which can result in overly generalized risk assessments or "blanket classification issues," where the materials are treated uniformly without accounting for the unique differences between individual substances within that category. The evaluation of the potential risks of applying nanomaterials is complicated because of their distinct physical and chemical properties, such as size, distribution, agglomeration state, crystal structure, shape, surface area, surface charge, surface chemistry, chemical composition, and porosity [102]. In the guidelines on the control and safe handling of nanomaterials established by the Department of Occupational Safety and Health, it was mentioned that although two materials may have the same chemical composition, they may have significant differences in their physical and chemical characteristics; therefore the hazard data for one of the materials might not be suitable for assessing the hazards of another [103]. The Organisation for Economic Co-operation and Development (OECD) was also aware of the uncertainties in the potential risks of nanomaterials; therefore the OECD programs on manufactured nanomaterials and test guidelines were also conducting several projects to develop relevant test methods in collaboration with the EU's Gov4Nano and NanoHarmony programs [104].

Although more specific and standardized testing is paramount in assessing nanomaterial handling safety, it can potentially lead to insurmountable testing that can hamper the economic viability of producing and marketing 2DM-based HTFs [105]. Subsequently, because 2DM-based HTFs are incorporated with 2DMs labeled nanomaterials, countries may also impose additional tariffs or restrictions on 2DM-based HTFs and classify them as "chemicals of concern" to mitigate the potential risks and encourage the development of safer alternatives. Taxes or fees can be imposed on these chemicals of concern, with the possibility of using the revenues for toxic use reduction programs [106,107].

However, policies are not always barriers to the development of 2DM-based HTFs. Several global policy trends can potentially enhance the economic attractiveness of 2DM-based HTFs. Governments in Europe, the US, and Asia-Pacific countries are also offering tax credits, subsidies, and grants for clean energy, industrial decarbonization, and energy-efficient technologies. 2DM-based HTFs, which can improve thermal and energy efficiency, may be eligible for EU Green Deal energy transition programs [108], US Department of Energy's Advanced Manufacturing Office funding calls under the High Performance Computing for Manufacturing (HPC4Mfg) program [109], and Malaysia's Green Technology Financing Scheme [110]. In addition, some countries have added graphene and advanced nanomaterials to their strategic innovation roadmaps, such as China's "Made in China 2025 (MIC25)" [111] and Malaysia's "National Advanced Materials Technology Roadmap 2021–2030" [112]. These roadmaps have opened opportunities for public–private partnerships and pilot deployment funding. In addition to these strategies, other regulations and policies that promote the use of sustainable materials, low-carbon processes, and recyclability may also favor 2DM-based HTFs if ecofriendly and energy-efficient recovery and reuse methods are developed. In short, stakeholders should capitalize on policy support and mitigate regulatory risks by engaging early with regulators or policymakers and providing material-specific safety data. Globally, standardized testing and certification protocols should also be developed for emerging 2DM-based HTFs. Finally, the development of 2DM-based HTFs should be closely aligned with national goals related to energy efficiency, carbon reduction, and sustainability to ensure that emerging 2DM-based HTFs can contribute meaningfully to broader environmental and industrial policy objectives.

16.4 Sustainability and future outlook

16.4.1 Life cycle sustainability assessment

The life cycle sustainability assessment (LCSA) framework can provide a comprehensive method to evaluate overall sustainability by integrating the environmental, economic, and social dimensions across their full life cycles [113]. This approach is suitable for emerging technologies such as 2DM-based HTFs, which offer promising performance enhancements but raise concerns regarding production costs, environmental footprints, and end-of-life impacts.

In the previous sections, it was demonstrated that an LCA study examines the impacts associated with the raw material extraction, synthesis, processing, usage, and disposal of 2DM-based HTFs. For instance, the production of 2DMs often require energy-intensive processing methods such as CVD or LPE, which involve the consumption of large amounts of energy and hazardous chemicals. Furthermore, the long-term environmental impacts of these materials in terms of their toxicity in natural systems remain underexplored. Few key indicators, such as GWP, CED, water and carbon footprint, and aquatic or terrestrial toxicity, should be quantified for each component of production processes in producing 2DM-based HTFs. Although 2DM-based HTFs can improve energy efficiency and reduce operational energy use, a benefit analysis needs to be conducted to weigh against the environmental burden contributed by their synthesis and disposal risks.

Complementing the LCA the previous sections discussed the key cost elements associated with the production of 2DM-based HTFs, especially for 2DMs, over their entire lifespan. Although the production of 2DMs has high upfront costs and investments required, these costs can be offset by lower operating costs owing to improved energy efficiency, less frequent system maintenance, and extended equipment lifespan. The current literature lacks a techno-economic analysis of emerging 2DM-based HTFs. Nevertheless, future studies should also calculate suitable economic indicators, such as return on investment, net present value, payback period, and levelized cost of heat transfer, to compare the 2DM-based HTFs economic potential with that of conventional HTFs in various applications. Similarly, social considerations are limited in the literature. Future integration of social LCA can also include factors such as worker exposure during nanomaterial handling, residential area safety near manufacturing factories, and public perceptions of nanotechnology and 2DM-based HTFs. These aspects will become significantly more significant when the 2DM-based HTFs are scaled up and regulatory frameworks are developed in the later stages.

A holistic LCSA is still required in the emerging stages of 2DM-based HTFs to enable researchers, policymakers, and industry stakeholders to quantify the trade-offs, opportunities, and barriers in the development of 2DM-based HTFs. By combining environmental impacts, economic feasibility, and social dimensions, LCSA can support upcoming research, development, innovation, and strategic decision-making globally. The continuous refinement of life cycle models coupled with real-world data from lab-scale studies and pilot-scale production up to industrial deployment will ultimately ensure that

2DM-based HTFs evolve into thermal management solutions with both high performance and excellent sustainability.

16.4.2 Challenges and opportunities

Based on the previous sections a few key challenges remain associated with 2DM-based HTFs for widespread adoption and commercialization. First and foremost, scalability is a major issue limiting their commercial potential owing to costly and labor-intensive production compared to conventional HTFs. In addition, uncertainties regarding the long-term ecotoxicological behavior and biodegradability of 2DM-HTFs also contribute to regulatory bottlenecks. The lack of standardization and specification of testing protocols for environmental and safety assessments also complicates commercialization. Moreover, relevant LCA, techno-economic, and LCSA studies are lacking in the literature, making the comparison with conventional HTFs much more difficult, which can hamper the interests of stakeholders in deploying 2DM-based HTFs. However, these challenges are closely associated with potential opportunities and research directions, particularly in the sustainable design and manufacturing of 2DM-based HTFs and possible policy regulations. These opportunities are further explored in the following sections.

16.4.3 Sustainable design and manufacturing of two-dimensional-material-based heat transfer fluids

The advancement of 2DMs presents remarkable opportunities for thermal management, particularly in the development of HTFs characterized by superior thermal conductivity and stability. However, as the demand for these materials escalates, so does the need for sustainable design and manufacturing practices. This section elaborates on strategies to minimize the environmental impact of 2DM-based HTFs, focusing on the incorporation of recycled materials, renewable energy sources, and innovative practices in production and life cycle management. It also reviews recent research that highlights these methodologies.

16.4.3.1 Green synthesis of two-dimensional materials from recycled materials

The synthesis of 2DMs, such as graphene and TMDs, from recycled materials is a pivotal approach in sustainable manufacturing. Recent studies have demonstrated the feasibility of producing high-quality graphene from waste sources such as spent coffee grounds and used tires. For example, Challa et al. demonstrated the production of GO from coffee grounds, highlighting that agro-residual biomass can be effectively utilized to create and modify biomaterials [114]. This approach not only advances research on sustainable biocomposites and supports environmental conservation efforts but also conserves resources and reduces the carbon footprint associated with traditional graphite extraction and processing methods.

Another promising avenue is the upcycling of industrial byproducts, such as fly ash from coal combustion. In a recent study, Fiori et al. developed 2D nanocomposites using nanofibrillar biochar (BC) sourced from paper industry waste [115]. These nanocomposites present a sustainable and cost-effective solution by repurposing industrial byproducts for advanced material applications, including their use as additives in HTFs. This approach not only utilizes waste materials effectively, reducing

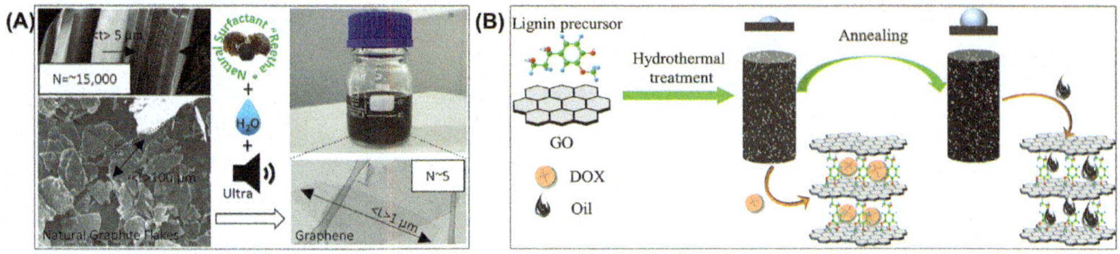

FIGURE 16.8

Schematic representation of (A) the utilization of the natural surfactant *Sapindus mukorossi* for the liquid-phase exfoliation of few-layered graphene [121] and (B) the synthesis methodology of BC–GO composite monoliths [122]. *BC–GO,* Biochar–graphene oxide.

landfill impact, but also aligns the production of 2DMs with circular economy principles, thereby enhancing environmental sustainability and resource efficiency [116,117].

16.4.3.2 Renewable energy integration in manufacturing processes

Transitioning to renewable energy sources in the manufacturing processes of 2DM-based fluids can significantly reduce their environmental impact. The traditionally energy-intensive steps of exfoliating, reducing, and functionalizing 2DMs are increasingly powered by solar, wind, and geothermal energy [118]. For example, solar-assisted techniques for producing GO have become more prevalent. Nagarajan et al. developed a solar-driven, nonhazardous, and cost-effective method to reduce GO to solar-reduced GO, which requires considerably less energy than conventional methods and eliminates the need for hazardous chemicals [119].

Integrating life cycle energy management strategies, where renewable energy sources are paired with energy storage systems, ensures a continuous and stable energy supply during 2DM production, reducing disruptions and maximizing energy efficiency. This approach is especially valuable for scaling industrial applications because it supports large-scale production while meeting sustainability standards. For example, Maka et al. investigated the performance of solar PV systems integrated with battery storage under harsh conditions [120]. Their study introduced an optimization framework that estimates the energy output of generation and storage configurations based on power potential, highlighting how such frameworks can enhance energy reliability and support sustainable industrial development.

16.4.3.3 Ecofriendly dispersants and additives

The choice of dispersants and additives in 2DM-based HTFs is essential for achieving high performance with minimal environmental impact. Recent advancements in biobased dispersants have presented effective and ecofriendly solutions. For instance, Sethurajaperumal et al. utilized the natural surfactant *Sapindus mukorossi* (SM) for the LPE of few-layered graphene, as shown in Fig. 16.8A, demonstrating improved dispersion stability in fluids [121]. This biobased surfactant not only enhances the quality of graphene but also biodegrades significantly faster than synthetic alternatives, aligning with green chemistry principles and reducing the presence of hazardous substances in HTFs. These advancements support sustainable practices in 2DM-based fluid manufacturing.

Innovative research on functionalized 2DMs is paving the way for improved thermal performance while maintaining environmental integrity. Xiong et al. introduced a simple, ecofriendly method to create robust, recyclable BC–GO composite monoliths, as shown in Fig. 16.8B, demonstrating the potential for sustainable BC–GO production across various industries [122]. These composites, notable for their recyclability and versatility, align with environmentally responsible manufacturing, demonstrating promising applications in 2DM-based systems that prioritize performance and sustainability. Recent studies on hybrid composites combining graphene with biochar have demonstrated enhanced thermal conductivity and significant adsorption capacities, adding dual functionality to HTF applications [123–125]. By incorporating waste-derived materials, these hybrids reduce the reliance on virgin resources, thereby advancing sustainable practices in 2DM-based fluid manufacturing.

16.4.3.4 Sustainable packaging and transportation

Sustainable practices in the 2DM-based fluid industry extend beyond manufacturing, embracing ecofriendly packaging and transportation to further reduce environmental impact. Transitioning to biodegradable and recyclable packaging materials helps decrease the ecological footprint of product distribution [126–128]. Many companies are adopting plant-based polymers and paper-based solutions to replace petroleum-derived plastics. Bioplastics derived from agricultural waste have shown promise as ecofriendly packaging for 2DM-based fluids, offering essential barrier properties while remaining compostable. Merino et al., for example, transformed industrial vegetable waste into bioplastic composites that biodegrade rapidly in soil [129]. These bioplastics not only serve as sustainable packaging but also improve soil health by preventing weeds and promoting plant growth when they are decomposed. Such innovations effectively repurpose waste while supporting the circular economy, making sustainable material development integral to the life cycle of 2DM-based products.

To reduce the environmental impact of 2DM-based HTFs, innovative packaging methods are essential, particularly for reducing emissions associated with transport [130,131]. Studies have shown that lightweight packaging not only reduces the total weight but also enables significant reductions in carbon emissions and overall energy costs. For instance, a study by the Flexible Packaging Association highlighted that flexible, lightweight pouches require up to 50% less energy and emit only 1/10 of the emissions compared to rigid packaging [132]. This approach aligns with sustainability goals by reducing transportation needs and supporting waste reduction during production and disposal processes.

Further advancements in flexible, lightweight films ensure that even minimal material usage does not compromise performance. Enhanced polymers and film manufacturing techniques now allow the development of tough, puncture-resistant films that preserve the functional integrity of required byproducts such as HTFs. These materials are often customizable for specific barrier needs, offering optimal protection against leaks or contamination, which are key concerns when dealing with fluids intended for heat transfer purposes [133]. In addition, by utilizing advanced design techniques, companies can transition from glass to plastic or other lighter materials, which provides further weight reduction without sacrificing the thermal properties essential for HTFs.

16.4.3.5 Future directions in sustainable two-dimensional-material-based fluid design

The future development of sustainable 2DM-based HTFs is increasingly driven by interdisciplinary innovations spanning materials science, data science, and green engineering. The integration of machine learning and artificial intelligence into materials discovery pipelines is expected to significantly accelerate the identification of novel 2DM chemistries and the optimization of fluid formulations.

These data-driven approaches enable rapid screening of compositional and structural parameters to maximize thermal transport performance while simultaneously constraining environmental burdens. In parallel, emerging predictive modeling frameworks are being developed to simulate the thermophysical behavior, dispersion stability, and life cycle impacts of multicomponent 2DM systems, thereby reducing the reliance on trial-and-error experimentation and shortening development cycles [134–136]. Another important direction involves functional hybridization, where 2DMs are combined with renewable or bioderived materials, such as nanocellulose and natural fibers. Such hybrid architectures offer opportunities to engineer synergistic interfacial interactions that enhance thermal conductivity, rheological stability, and durability while maintaining a reduced ecological footprint. Continued investigation of these biohybrid systems is expected to yield next-generation HTFs that balance high performance and sustainability, supporting broader efforts toward circular material design.

Overall, sustainable design and manufacturing principles are central to the advancement of 2DM-based HTFs. Strategies, including the use of recycled precursors, renewable energy inputs, and environmentally benign additives, collectively reduce the life cycle environmental impacts without compromising functional efficiency. Progress in this area highlights the importance of aligning thermal management technologies with global sustainability frameworks. Through coordinated research efforts and the adoption of green materials engineering practices, 2DM-enabled fluids can be developed to meet escalating thermal performance requirements while preserving ecological integrity.

16.4.4 Policy recommendations for sustainable development and adoption of two-dimensional-material-based heat transfer fluids

Effective policy frameworks are critical for enabling the sustainable development and large-scale deployment of 2DM-based HTFs. Such frameworks should integrate regulatory instruments, targeted financial incentives, coordinated research funding, and capacity-building initiatives to promote environmentally responsible practices throughout the full life cycle, spanning material synthesis, system integration, operational use, and end-of-life management. Evidence from sustainable materials governance indicates that policy alignment across these domains accelerates low-impact innovation while reducing environmental externalities and market uncertainties. In the automotive sector, well-designed policy mechanisms can catalyze technology translation, strengthen industrial adoption pathways, and align emerging thermal management solutions with broader decarbonization and circular economy objectives.

16.4.4.1 Regulatory standards and environmental certifications

Countries, such as Japan, Germany, and the Netherlands, have pioneered efforts to standardize sustainable practices in nanomaterials, providing frameworks that other nations can adopt. For instance, Japan has introduced stringent environmental certifications, such as the Eco Mark, which evaluates the environmental impact of products using LCA standards [137]. This has encouraged Japanese companies to integrate recyclable materials and sustainable synthesis methods into the production of advanced materials, including nanomaterials. Similarly, the EU's REACH regulation offers guidelines for managing the impact of chemical substances, including nanomaterials, and could serve as a regulatory model for 2DM-based HTFs [138,139].

Policy initiatives should focus on setting emission limits by implementing reduction requirements for 2DM synthesis processes, particularly those with high energy consumption. For example, Germany's

Umweltbundesamt or Federal Environment Agency has defined emission reduction strategies across industries to meet targets. Another important approach could be mandating ecolabeling and certifications based on LCA to highlight environmentally friendly products. In the Netherlands an environmental labeling scheme ensures that products with low environmental footprints are prioritized in procurement, which could extend to 2DM-based HTFs, encouraging responsible consumption [140–142].

16.4.4.2 Financial incentives and subsidies

Financial incentives are pivotal in fostering sustainable production practices in the emerging market for 2DM-based HTFs, driving innovation and making environment-friendly technologies more accessible. For instance, the US has introduced the Advanced Manufacturing Tax Credit, which provides tax incentives to companies that incorporate renewable energy sources or use recycled materials in their manufacturing processes [143,144]. This initiative helps reduce the financial burden of transitioning to more sustainable practices, making the production of 2DM-based HTFs more economically feasible for manufacturers. In addition, the EU's Horizon Europe program offers substantial funding for nanomaterial R&D [141,145]. This supports projects focused on developing green nanomaterial synthesis techniques, ecofriendly additives, and more energy-efficient manufacturing processes, all of which can significantly contribute to the growth and sustainability of 2DM-HTF technologies. This financial support is crucial for advancing innovations that minimize the environmental impact of these materials.

Moreover, in South Korea, the government has introduced low-interest loans for startups that prioritize sustainable practices in nanotechnology. These loans offer critical financial assistance to new companies, enabling them to bridge the financial gaps often associated with environmentally friendly innovations [146–148]. By supporting the development and commercialization of 2DM-based HTFs, these financial mechanisms can help accelerate the adoption of sustainable nanotechnology across industries. These incentives not only promote green manufacturing but also ensure that emerging technologies meet the growing global demand for more sustainable energy and heat transfer solutions.

16.4.4.3 Investment in research and development of sustainable alternatives

Sustainable innovation in 2DM-based HTFs requires continuous investment in targeted R&D to ensure that these materials are not only high-performing but also environmentally friendly. Both the EU and Canada have made significant strides in promoting research on sustainable nanomaterials to minimize the ecological footprint of nanotechnology [149–151]. These efforts focus on advancing cleaner production methods and creating nanomaterials that are easier to recycle and less harmful to the environment than their predecessors.

One notable example is the National Institute for Materials Science (NIMS) in Japan, which has developed biosourced materials and solar-driven exfoliation techniques for the synthesis of 2DMs. These innovations present sustainable alternatives to traditional energy-intensive processes. By harnessing renewable energy sources, such as solar power, and utilizing biobased materials, NIMS has paved the way for low-energy, ecofriendly production methods that could revolutionize the manufacturing of 2DM-based HTFs [152,153]. This shift toward greener synthesis techniques is vital for scaling up the production of 2DMs while reducing their environmental impact.

In Canada the Natural Sciences and Engineering Research Council has funded a series of projects focused on developing recyclable additives and biodegradable dispersants for nanofluids [154–156]. These projects are particularly relevant for enhancing the recyclability of 2DM-based HTFs and ensuring that the fluids do not contribute to long-term waste accumulation. By integrating ecofriendly dispersants into nanofluids, researchers have improved the sustainability of these fluids, making them easier to reuse or dispose of without harming the environment. This study is pivotal for advancing the life cycle sustainability of 2DM-based HTFs, aligning with global trends toward circular economy models in the nanomaterial sector. These efforts demonstrate how strategic investments in R&D can create more sustainable nanotechnology solutions for a variety of applications, including heat transfer.

16.4.4.4 *Education and awareness programs*

Education and awareness initiatives are critical for embedding sustainability principles within both industrial practices and consumer behavior, thereby accelerating the responsible production and use of 2DM-based HTFs. Structured programs targeting professionals, students, and the general public are increasingly being deployed to normalize life cycle thinking, environmental stewardship, and informed material selection.

Workshops and professional training: In the United Kingdom the Knowledge Transfer Network conducts workshops dedicated to sustainable nanomaterial manufacturing, with particular emphasis on LCA, resource efficiency, and environmental impact mitigation. These programs are intended to translate sustainability frameworks into actionable manufacturing strategies, covering stages from raw material sourcing to end-of-life management [157]. By promoting cross-sector knowledge exchange, such initiatives strengthen the industry's capacity to integrate sustainability into operational decision-making and innovation pathways.

University curriculum integration: Academic institutions are embedding sustainability into nanomaterial education. For example, the Technical University of Munich incorporates sustainable nanomaterial science into its curricula to train researchers to systematically evaluate environmental trade-offs during material design and development [158]. This educational model positions sustainability as a foundational design constraint rather than a secondary consideration, preparing graduates to advance 2DM-based HTFs with explicit attention to ecological performance.

Public awareness initiatives: Government-led outreach efforts broaden the sustainability discourse beyond research and industrial communities. In South Korea the Ministry of Environment has implemented awareness campaigns to communicate the environmental benefits of ecodesigned nanomaterials, including 2DM-based HTFs, to businesses and consumers [159,160]. These initiatives disseminate accessible, evidence-based information to encourage the adoption of lower impact materials and technologies. Increased public literacy on sustainability strengthens market demand for greener products and supports broader societal acceptance of environmentally responsible innovations.

16.4.4.5 *International collaboration and knowledge sharing*

International collaboration provides a mechanism for integrating resources, technical expertise, and institutional knowledge to advance sustainable innovation in 2DM-based HTFs. Cross-border partnerships between academia, industry, and regulatory bodies enable coordinated development of environmentally responsible synthesis routes, life cycle management strategies, and application frameworks. Such cooperation embeds sustainability considerations into both material design and scale-up pathways, thereby strengthening environmental stewardship within the broader nanotechnology sector.

Global standards and harmonization: The International Organization for Standardization is developing globally aligned frameworks for nanotechnology that include sustainability-oriented guidance relevant to nanomaterials such as 2DM-based HTFs. These standards outline practices for reducing environmental burdens during synthesis, improving process safety, and enhancing resource efficiency across the production chain [161,162]. Harmonized technical benchmarks support comparability across jurisdictions, reduce regulatory fragmentation, and promote adoption of environmentally sound manufacturing practices while maintaining industrial competitiveness.

Joint research initiatives: Horizon Europe supports multinational research consortia focused on sustainable materials science, including projects that address the environmental performance of advanced nanofluids [163]. These initiatives integrate expertise from materials engineering, environmental assessment, and process design to develop greener synthesis chemistries, benign stabilizing additives, and recycling or recovery strategies. Shared infrastructure, coordinated data generation, and open research networks accelerate technological maturation and strengthen the collective capacity to address the environmental risks associated with emerging nanomaterials.

Knowledge exchange programs: The Asia-Pacific Economic Cooperation facilitates regional knowledge transfer programs that promote sustainable nanomaterial production, particularly in economies with developing research and manufacturing capabilities [164,165]. Such engagement reduces technological disparities and ensures broader participation in environmentally responsible innovation, thereby strengthening the global capacity for sustainable material deployment. These activities disseminate best practices in green synthesis, resource-efficient processing, and end-of-life management. Structured training, policy dialogue, and technical cooperation enable the broader adoption of environmentally responsible production models, support capacity building, and reduce disparities in sustainable technology deployment.

The coordinated implementation of regulatory frameworks, financial incentives, targeted research investment, and international knowledge exchange establishes the institutional conditions necessary to lower the environmental footprint of 2DM-based HTFs. Integrated policy and research ecosystems encourage life cycle-oriented design, responsible manufacturing, and circular material management, positioning advanced fluids to support cleaner thermal management technologies and broader industrial sustainability objectives.

16.5 Conclusion

2DM-based HTFs represent a significant advance in thermal management engineering, introducing a class of nanofluids capable of simultaneously enhancing heat transfer performance and supporting energy efficiency goals. The incorporation of high–aspect ratio nanomaterials, including graphene, h-BN, TMDs, and MXenes, into conventional base fluids has been shown to improve the thermal conductivity, specific heat capacity, and convective heat transfer behavior. These gains arise from the intrinsic surface characteristics, tunable interfacial chemistry, and distinctive phonon transport pathways of 2DMs, enabling efficient thermal dissipation in high-density electronics, renewable energy systems, and advanced industrial processes.

Notwithstanding their performance advantages, the sustainability profiles of 2DM-based HTFs require careful evaluation. Current synthesis routes, whether top-down approaches such as mechanical exfoliation and chemical etching or bottom-up methods including CVD and solvothermal growth, are often energy-intensive and may involve hazardous reagents and byproducts. The LCA indicates that upstream environmental burdens can partially offset downstream efficiency gains if mitigation strategies are not implemented. Materials such as MXenes and MoS_2 warrant particular attention because of the potential ecological and occupational risks associated with their production and disposal. Advancing greener synthesis pathways, renewable energy–driven manufacturing, and safer chemical processing is essential to align material performance with environmental stewardship.

Economic considerations also influence the pathway to large-scale adoption. Although the initial costs of 2DM-enhanced nanofluids remain relatively high, techno-economic analyses suggest that improved thermal efficiency, lower operational energy demand, increased equipment durability, and system miniaturization can generate favorable long-term value. These benefits are especially relevant in energy-intensive sectors such as CSP, electric mobility, data infrastructure, and aerospace thermal management, where performance gains translate directly into operational savings and design optimizations.

Responsible commercialization depends on coordinated technical, environmental, and regulatory efforts. Future research should emphasize scalable, low-carbon synthesis routes supported by rigorous life cycle evaluations and materials informatics–guided optimizations. Clear occupational safety standards, nanoparticle management protocols, and product stewardship frameworks are required to minimize environmental release and exposure risks. Cross-disciplinary collaboration among materials scientists, engineers, environmental researchers, toxicologists, and policymakers will be critical to ensuring that innovation proceeds in parallel with safety and sustainability.

Overall, 2DM-based HTFs constitute a promising platform for next-generation thermal technology. When developed within a framework that integrates life cycle sustainability, economic feasibility, and regulatory oversight, these systems can enhance heat transfer performance while contributing to broader decarbonization and energy efficiency objectives. Continued research investment and policy alignment are essential for translating laboratory advances into scalable, environmentally responsible solutions for future thermal management systems.

References

[1] A.A. Minea, State of the art in PEG-based heat transfer fluids and their suspensions with nanoparticles, Nanomaterials 11 (1) (2021) 86.

[2] C.C. Kwasi-Effah, H.O. Egware, A.I. Obanor, O.O. Ighodaro, Development and characterization of a quaternary nitrate based molten salt heat transfer fluid for concentrated solar power plant, Heliyon 9 (5) (2023) e16096.

[3] A. Giaconia, et al., Assessment and perspectives of heat transfer fluids for CSP applications, Energies 14 (22) (2021) 7486.

[4] I. Arias, J. Cardemil, E. Zarza, L. Valenzuela, R. Escobar, Latest developments, assessments and research trends for next generation of concentrated solar power plants using liquid heat transfer fluids, Renew. Sustain. Energy Rev. 168 (2022) 112844.

[5] R.K. Al-Farajat, M.R. Gomaa, M.Z. Alzghoul, Comparison between CSP systems and effect of different heat transfer fluids on the performance, WSEAS Trans. Heat Mass Transf. 17 (2022) 196–205.

[6] E.A.C. Panduro, F. Finotti, G. Largiller, K.Y. Lervåg, A review of the use of nanofluids as heat-transfer fluids in parabolic-trough collectors, Appl. Therm. Eng. 211 (2022) 118346.

[7] V. Sharma, R. Patidar, K.B. Rana, A critical analysis on synthesis of nanofluids and factors affecting thermal conductivity of nanofluids for heat transfer applications: a review, Int. J. Chem. Reactor Eng. 23 (2) (2025) 143–168.

[8] W. Azmi, K.A. Hamid, N. Usri, R. Mamat, K. Sharma, Heat transfer augmentation of ethylene glycol: water nanofluids and applications—A review, Int. Commun. Heat Mass Transfer 75 (2016) 13–23.

[9] T. Marshall, C. Clark, D. Brewster, T. Henderson, Toxicological assessment of heat transfer fluids proposed for use in solar energy applications, II, Toxicol. Appl. Pharmacol. 58 (1) (1981) 31–38.

[10] R. Walvekar, et al., Deep eutectic solvents-based CNT nanofluid – a potential alternative to conventional heat transfer fluids, J. Taiwan Inst. Chem. Eng. 128 (2021) 314–326.

[11] E.C. Okonkwo, I. Wole-Osho, I.W. Almanassra, Y.M. Abdullatif, T. Al-Ansari, An updated review of nanofluids in various heat transfer devices, J. Therm. Anal. Calorim. 145 (2021) 2817–2872.

[12] B. Mehta, D. Subhedar, H. Panchal, Z. Said, Synthesis, stability, thermophysical properties and heat transfer applications of nanofluid – a review, J. Mol. Liq. 364 (2022) 120034.

[13] M.N.H. Mat, R. Saidur, Heat transfer enhancement by nanofluid coupling with surfactant in overcoming particle agglomeration for microchannel heat sinks, Microfluid. Nanofluid. 26 (9) (2022) 65.

[14] C. Huang, L. Cui, L. Qiu, Y. Zhang, G. Wei, X. Du, Impact of nanoparticle agglomeration on thermal conductivity of molten salt based nanofluids: insights from molecular dynamics and lattice Boltzmann methods, Sol. Energy Mater. Sol. Cells 266 (2024) 112691.

[15] P. Martínez-Merino, P. Estellé, R. Alcántara, I. Carrillo-Berdugo, J. Navas, Thermal performance of nanofluids based on tungsten disulphide nanosheets as heat transfer fluids in parabolic trough solar collectors, Sol. Energy Mater. Sol. Cells 247 (2022) 111937.

[16] M.H. Kalantari, X. Zhang, Thermal transport in 2D materials, Nanomaterials 13 (1) (2022) 117.

[17] V. Shanmugam, et al., A review of the synthesis, properties, and applications of 2D materials, Part. Part. Syst. Charact. 39 (6) (2022) 2200031.

[18] W. Li, X. Qian, J. Li, Phase transitions in 2D materials, Nat. Rev. Mater. 6 (9) (2021) 829–846.

[19] J. Liu, P. Li, H. Zheng, Review on techniques for thermal characterization of graphene and related 2D materials, Nanomaterials 11 (11) (2021) 2787.

[20] W. Ren, Y. Ouyang, P. Jiang, C. Yu, J. He, J. Chen, The impact of interlayer rotation on thermal transport across graphene/hexagonal boron nitride van der Waals heterostructure, Nano Lett. 21 (6) (2021) 2634–2641.

[21] S. Zafar, Multi-scale modelling and computational fluid dynamics analysis of transitional metal dichalcogenide (TMD) growth in an MOCVD chamber, Mechanical Engineering Master of Science, The Pennsylvania State University, 2020. https://etda.libraries.psu.edu/catalog/18457sxz311.

[22] H.M. Wilson, N. Jha, Chapter 6: Carbon-based and TMDs-based nanostructures for solar-driven steam generation, in: A.K. Mishra (Ed.), Carbon and TMDs Nanomaterials for Energy Applications, World Scientific Publishing Co Pte Ltd, 2024, p. 312.

[23] N. Aslfattahi, R. Saidur, M.F.M. Sabri, A. Arifutzzaman, Thermal conductivity and rheological investigation of aqueous poly(ethylene) glycol/MXene as a novel heat transfer fluid, AIP Conf. Proc. 2324 (1) (2021) 1–7.

[24] S. Sreekumar, et al., Thermo-optical characterization of novel MXene/carbon-dot hybrid nanofluid for heat transfer applications, J. Cleaner Prod. 434 (2024) 140395.

[25] P. Martínez-Merino, R. Alcántara, J. Navas, WSe_2 nanowires-based nanofluids for concentrating solar power, Mater. Today Chem. 27 (2023) 101323, doi:10.1016/j.mtchem.2022.101323.

[26] J. Pereira, A. Moita, A. Moreira, Fewer dimensions for higher thermal performance: a review on 2D nanofluids, Appl. Sci. 13 (6) (2023), doi:10.3390/app13064070.

[27] L.S. Sundar, E.V. Ramana, Z. Said, V. Punnaiah, K.V.V. Chandra Mouli, A.C.M. Sousa, Properties, heat transfer, energy efficiency and environmental emissions analysis of flat plate solar collector using nanodiamond nanofluids, Diamond Relat. Mater. 110 (2020), doi:10.1016/j.diamond.2020.108115.

[28] R. Bammidi, K.S. Prasad, P.S. Rao, Graphene nanofluids as a promising lubricant and its sustainability, Res. Rev. Int. J. Multidiscip. 5 (12) (2020) 12–25, doi:10.31305/rrijm.2020.v05.i12.004.

[29] S. Bobba, F.A. Deorsola, G.A. Blengini, D. Fino, LCA of tungsten disulphide (WS_2) nano-particles synthesis: state of art and from-cradle-to-gate LCA, J. Cleaner Prod. 139 (2016) 1478–1484, doi:10.1016/j.jclepro.2016.07.091.

[30] M. Hachhach, H. Akram, A.E. Kasmi, M. Hanafi, O. Achak, T. Chafik, Life cycle assessment of large-scale production of MoS_2 nanomaterials through the solvothermal method, J. Nanopart. Res. 24 (9) (2022), doi:10.1007/s11051-022-05563-8.

[31] F.A. Deorsola, N. Russo, G.A. Blengini, D. Fino, Synthesis, characterization and environmental assessment of nanosized MoS_2 particles for lubricants applications, Chem. Eng. J. 195–196 (2012) 1–6, doi:10.1016/j.cej.2012.04.080.

[32] R. Arvidsson, Review of environmental life cycle assessment studies of graphene production, Adv. Mater. Lett. 8 (3) (2017) 187–195, doi:10.5185/amlett.2017.1413.

[33] R. Arvidsson, G. Peters, S.F. Hansen, A. Baun, Prospective environmental risk screening of seven advanced materials based on production volumes and aquatic ecotoxicity, NanoImpact 25 (2022) 100393, doi:10.1016/j.impact.2022.100393.

[34] D. Beloin-Saint-Pierre, R. Hischier, Towards a more environmentally sustainable production of graphene-based materials, Int. J. Life Cycle Assess. 26 (2) (2021) 327–343, doi:10.1007/s11367-020-01864-z.

[35] S. Diaz-Arauzo, et al., Ultrahigh-throughput cross-flow filtration of solution-processed 2D materials enabled by porous ceramic membranes, Mater. Horiz. (2024), doi:10.1039/d4mh01205d.

[36] J. Munuera, L. Britnell, C. Santoro, R. Cuéllar-Franca, C. Casiraghi, A review on sustainable production of graphene and related life cycle assessment, 2D Mater. 9 (1) (2021), doi:10.1088/2053-1583/ac3f23.

[37] H. Akram, et al., Comparison of synthesis routes of inorganic fullerene like nano-additives for wind turbine lubrication: application of life cycle assessment approach, 2014 International Renewable and Sustainable Energy Conference (IRSEC), 2014, doi:10.1109/IRSEC.2014.7059811.

[38] I. Corazzari, et al., Hazard assessment of W and Mo sulphide nanomaterials for automotive use, J. Nanopart. Res. 16 (5) (2014), doi:10.1007/s11051-014-2401-7.

[39] M.D. Firouzjaei, S.K. Nemani, M. Sadrzadeh, E.K. Wujcik, M. Elliott, B. Anasori, Life-cycle assessment of $Ti_3C_2T_x$ MXene synthesis, Adv. Mater. 35 (31) (2023) e2300422, doi:10.1002/adma.202300422.

[40] A. Ungureanu, et al., Systematic life cycle environmental impact comparison of alternative synthetic strategies for $Ti_3C_2T_x$ MXene, ACS Sustain. Chem. Eng. 12 (15) (2024) 5893–5906, doi:10.1021/acssuschemeng.3c08491.

[41] Z. Han, A. Golev, M. Edraki, A review of tungsten resources and potential extraction from mine waste, Minerals 11 (7) (2021), doi:10.3390/min11070701.

[42] G. Petruzzelli, F. Pedron, The dynamics of tungsten in soil: an overview, Environments 8 (7) (2021), doi:10.3390/environments8070066.

[43] S. Adhikari, Toxicity of 2D materials and their future prospect, in: M.M. Rahman, J. Uddin, A.M. Asiri, M.R. Rahman Eds (Eds.), *Toxicity of Nanoparticles—Recent Advances and New Perspectives*, IntechOpen, 2023.

[44] IOHANational Govermental Links, International Occupational Hygiene Association (IOHA), 2024. https://www.ioha.net/national-governmental/ Accessed 11 January 2024.

[45] E.T. Kimura, D.M. Ebert, P.W. Dodge, Acute toxicity and limits of solvent residue for sixteen organic solvents, Toxicol. Appl. Pharmacol. 19 (1971) 699–704, doi:10.1016/0041-008x(71)90301-2.

[46] I. Rusyn, R. Bataller, Alcohol and toxicity, J. Hepatol. 59 (2) (2013) 387–388, doi:10.1016/j.jhep.2013.01.035.

[47] P.J. Brooks, DNA damage, DNA repair, and alcohol toxicity—a review, Alcohol. Clin. Exp. Res. 21 (6) (1997), doi:10.1111/j.1530-0277.1997.tb04256.x.

[48] N. Adhikari, D.R. Joshi, An overview on common organic solvents and their toxicity, J. Pharm. Res. Int. (2019) 1–18, doi:10.9734/jpri/2019/v28i330203.

[49] G.S. Bang, K.W. Nam, J.Y. Kim, J. Shin, J.W. Choi, S.Y. Choi, Effective liquid-phase exfoliation and sodium ion battery application of MoS_2 nanosheets, ACS Appl. Mater. Interfaces 6 (10) (May 28 2014) 7084–7089, doi:10.1021/am4060222.

[50] A. Zhang, et al., Recent progress of two-dimensional materials for ultrafast photonics, Nanomaterials (Basel) 11 (7) (2021), doi:10.3390/nano11071778.

[51] A. Heyndrickx, Toxicology of new synthetic organic polymers used in containers for food, pharmaceutical compounds, and drinkable water, Prog. Chem. Toxicol. 5 (1974) 273–291, doi:10.1016/b978-0-12-536505-5.50011-2.

[52] G. Wang, Y. Guan, Y. Wang, Y. Ding, L. Yang, Direct laser irradiation and modification of 2D Te for development of volatile memristor, Materials (Basel) 16 (2) (2023), doi:10.3390/ma16020738.

[53] N. Vasudevan, N.N. Nasir Ahamed, B P., A. Aravindhan, B.P. Shanmugavel, Effect of Ni addition on the densification of TiC: a comparative study of conventional and microwave sintering, Int. J. Refract. Met. Hard Mater 87 (2020), doi:10.1016/j.ijrmhm.2019.105165.

[54] B. Sadeghi, P. Cavaliere, Progress of flake powder metallurgy research, Metals 11 (6) (2021), doi:10.3390/met11060931.

[55] Z. Lin, et al., 2D materials advances: from large scale synthesis and controlled heterostructures to improved characterization techniques, defects and applications, 2D Mater. 3 (4) (2016), doi:10.1088/2053-1583/3/4/042001.

[56] H. Chen, et al., Transition-metal dichalcogenides heterostructure saturable absorbers for ultrafast photonics, Opt. Lett. 42 (21) (2017) 4279–4282, doi:10.1364/OL.42.004279.

[57] J.Y. Jo, Y.S. Kwon, J.W. Lee, J.S. Park, B.H. Rho, W.I. Choi, Acute respiratory distress due to methane inhalation, Tuberc. Respir. Dis. (Seoul) 74 (3) (2013) 120–123, doi:10.4046/trd.2013.74.3.120.

[58] I.J. Duncan, Does methane pose significant health and public safety hazards?—a review, Environ. Geosci. 22 (3) (2015) 85–96, doi:10.1306/eg.06191515005.

[59] CCOHSCCOHS: Methane, Canadian Centre for Occupational health and Safety (CCOHS), 2024. https://www.ccohs.ca/oshanswers/chemicals/chem_profiles/methane.html#section-4-hdr Accessed 1 November 2024.

[60] N. Somdock, S. Kianwimol, A. Harnwunggmoung, A. Sakulkalavek, R. Sakdanuphab, Simultaneous stoichiometric composition and highly (00l) orientation of flexible Bi_2Te_3 thin films via optimising the DC magnetron sputter-deposition process, J. Alloys Compd. 773 (2019) 78–85, doi:10.1016/j.jallcom.2018.09.216.

[61] L. Duta, I.N. Mihailescu, Advances and challenges in pulsed laser deposition for complex material applications, Coatings 13 (2) (2023), doi:10.3390/coatings13020393.

[62] Y.H. Chuai, Y.F. Wang, Y. Bai, Structural, optical, electrical, and thermoelectric properties of Bi_2Se_3 films deposited at a high Se/Bi flow rate, Nanomaterials (Basel) 13 (20) (2023), doi:10.3390/nano13202785.

[63] B. Farasati Far, N. Rabiee, S. Iravani, Environmental implications of metal-organic frameworks and MXenes in biomedical applications: a perspective, RSC Adv. 13 (49) (2023) 34562–34575, doi:10.1039/d3ra07092a.

[64] A. Kumar, Y. Kuang, Z. Liang, X. Sun, Microwave chemistry, recent advancements, and eco-friendly microwave-assisted synthesis of nanoarchitectures and their applications: a review, Mater. Today Nano 11 (2020), doi:10.1016/j.mtnano.2020.100076.

[65] E. Gabano, M. Ravera, Microwave-assisted synthesis: can transition metal complexes take advantage of this "green" method? Molecules 27 (13) (2022), doi:10.3390/molecules27134249.

[66] X. Xie, B. Goodell, Thermal degradation and conversion of plant biomass into high value carbon products, in: Deterioration and Protection of Sustainable Biomaterials, ACS Symposium Series, ACS, Washington, DC, USA, 2014, pp. 147–158.

[67] J. Singh, T. Dutta, K.H. Kim, M. Rawat, P. Samddar, P. Kumar, Green' synthesis of metals and their oxide nanoparticles: applications for environmental remediation, J. Nanobiotechnol. 16 (1) (2018) 84, doi:10.1186/s12951-018-0408-4.

[68] J. Jeevanandam, Y.S. Chan, M.K. Danquah, Biosynthesis of metal and metal oxide nanoparticles, ChemBioEng Rev. 3 (2) (2016) 55–67, doi:10.1002/cben.201500018.

[69] W.P. Wong, R. Walvekar, M. Vaka, M. Khalid, Optimisation and techno-economic assessment of parabolic trough collector system in Malaysia, China, and the United States, Int. J. Ambient Energy 44 (1) (2023) 2398–2421, doi:10.1080/01430750.2023.2237030.

[70] K. Elsaid, A.G. Olabi, T. Wilberforce, M.A. Abdelkareem, E.T. Sayed, Environmental impacts of nanofluids: a review, Sci. Total Environ. 763 (2021) 144202, doi:10.1016/j.scitotenv.2020.144202.

[71] E. Batuecas, C. Mayo, R. Díaz, F.J. Pérez, Life cycle assessment of heat transfer fluids in parabolic trough concentrating solar power technology, Sol. Energy Mater. Sol. Cells 171 (2017) 91–97, doi:10.1016/j.solmat.2017.06.032.

[72] N. Czaplicka, A. Grzegorska, J. Wajs, J. Sobczak, A. Rogala, Promising nanoparticle-based heat transfer fluids-Environmental and techno-economic analysis compared to conventional fluids, Int. J. Mol. Sci. 22 (17) (2021), doi:10.3390/ijms22179201.

[73] Graphene Price. https://www.alibaba.com/showroom/graphene-price.html. Accessed 1 July 2025.

[74] Boron Nitride Powder Price. https://www.alibaba.com/showroom/boron-nitride-powder-price.html. Accessed 1 July 2025.

[75] A.H. Shaik, et al., A review on environmental and economic impact of 2D nanomaterials-based heat transfer fluids, J. Nanomater. 2022 (1) (2022), doi:10.1155/2022/3443360.

[76] H.W. Xian, N.A.C. Sidik, R. Saidur, Impact of different surfactants and ultrasonication time on the stability and thermophysical properties of hybrid nanofluids, Int. Commun. Heat Mass Transfer 110 (2020), doi:10.1016/j.icheatmasstransfer.2019.104389.

[77] Ç.V. Yıldırım, Ş. Şirin, T. Kıvak, M. Sarıkaya, The effect of nanofluids reinforced with different surfactants on the machining and friction-wear properties of Waspaloy, Tribol. Int. 181 (2023), doi:10.1016/j.triboint.2023.108316.

[78] C. Zapata-Hernandez, G. Durango-Giraldo, D. López, R. Buitrago-Sierra, K. Cacua, Surfactants versus surface functionalization to improve the stability of graphene nanofluids, J. Dispersion Sci. Technol. 43 (11) (2021) 1717–1724, doi:10.1080/01932691.2021.1880429.

[79] K.R. Kumar, A.H. Shaik, Synthesis, thermophysical characterization and thermal performance analysis of novel Cu-MXene hybrid nanofluids for efficient coolant applications, RSC Adv. 13 (42) (2023) 29536–29560, doi:10.1039/d3ra05429b.

[80] M. Sandhya, D. Ramasamy, K. Sudhakar, K. Kadirgama, W.S.W. Harun, Ultrasonication an intensifying tool for preparation of stable nanofluids and study the time influence on distinct properties of graphene nanofluids – a systematic overview, Ultrason. Sonochem. 73 (2021) 105479, doi:10.1016/j.ultsonch.2021.105479.

[81] A.M. Ajeena, P. Víg, I. Farkas, A comprehensive analysis of nanofluids and their practical applications for flat plate solar collectors: fundamentals, thermophysical properties, stability, and difficulties, Energy Rep. 8 (2022) 4461–4490, doi:10.1016/j.egyr.2022.03.088.

[82] G.J. Olguin-Orellana, G.J. Soldano, J. Alzate-Morales, M.B. Camarada, M.M. Mariscal, Can graphene improve the thermal conductivity of copper nanofluids? Phys. Chem. Chem. Phys. 25 (7) (2023) 5489–5500, doi:10.1039/d3cp00064h.

[83] Cooling Post, Nano-fluid claims 10% chiller efficiency boost, Cooling Post (2025). https://www.coolingpost.com/world-news/nano-fluid-claims-10-chiller-efficiency-boost/#:⟨:text=6th%20March%202022,as%20water%20or%20ethylene%20glycol. Accessed 1 July 2025.

[84] Y. Şen, H.I. Variyenli, Experimental investigation of the performance of CuO-graphene nanoplatelet/water hybrid nanofluid in concentric tube heat exchanger, Int. J. Energy Stud. 6 (1) (2021) 37–52.

[85] O.A. Alawi, H.M. Kamar, H.A. Mohammed, A.R. Mallah, O.A. Hussein, Energy efficiency of a flat-plate solar collector using thermally treated graphene-based nanofluids: experimental study, Nanomater. Nanotechnol. 10 (2020), doi:10.1177/1847980420964618.

[86] H. Lin, et al., Environmental and health impacts of graphene and other two-dimensional materials: a graphene flagship perspective, ACS Nano 18 (8) (2024) 6038–6094, doi:10.1021/acsnano.3c09699.

[87] A. Kapoor, et al., Sustainable valorization of waste plastic into nanostructured materials for environmental, energy, catalytic and biomedical applications: a review, Chemosphere 364 (2024) 143279, doi:10.1016/j.chemosphere.2024.143279.

[88] Grand View ResearchHeat Transfer Fluids Market Size, Share & Trends Analysis Report by Product (Silicone Fluids, Aromatic Fluids), by Application (Oil & Gas, Chemical Industry), by Region, and Segment Forecasts, 2024–2030, Grand View Research, 2025. https://www.grandviewresearch.com/industry-analysis/heat-transfer-fluids-market Accessed 3 July 2025.

[89] The Business Research CompanyGraphene Global Market Report 2025, The Business Research Company, 2025. https://www.thebusinessresearchcompany.com/report/graphene-global-market-report#:⟨:text=What%20Is%20The%20Graphene%20Market,%2Dfriendly%20graphene%2Dbased%20products Accessed 3 July 2025.

[90] E. Zhang. What Is the Global Market Size for Graphene? https://www.linkedin.com/pulse/what-global-market-size-graphene-echo-zhang-f0gtc/. Accessed 3 July 2025.

[91] C. Galiotis, Graphene: revolutionising composite applications, Graphene 2D Mater. 9 (1–2) (2024) 1–2, doi:10.1007/s41127-024-00080-y.

[92] Grand View ResearchHexagonal Boron Nitride Market Size, Share & Trends Analysis Report by Application (Paints & Coatings, Composites, Lubricants, Personal Care & Cosmetics, Electrical Insulation), by Region, and Segment Forecasts, 2025–2030, Grand View Research, 2025. https://www.grandviewresearch.com/industry-analysis/hexagonal-boron-nitride-market-report# Accessed 3 July 2025.

[93] InfinoilMalaysian Researchers Aim to Commercialize Graphene Lubricants Within One Year, Infinoil, 2025. https://infinoil.com/malaysian-researchers-aim-to-commercialize-graphene-lubricants-within-one-year/ Accessed 3 July 2025.

[94] B. Heidarshenas, Y. Yuan, A.S. El-Shafay, Advancements in 2D nanomaterial-enhanced nanofluids: stability, thermophysical properties, and industrial applications, Powder Technol. 454 (2025), doi:10.1016/j.powtec.2025.120687.

[95] A. Nugroho, et al., Enhancing tribological performance of electric vehicle lubricants: nanoparticle-enriched palm oil biolubricants for wear resistance, Heliyon 10 (22) (2024) e39742, doi:10.1016/j.heliyon.2024.e39742.

[96] M.M. Souby, M.H.S. Bargal, H. Maher, M. Salman, J. Liu, S.C. Kim, Numerical assessment of hydrothermal and entropy generation characteristics of graphene quantum dots nanofluid in a microchannel heat sink incorporating secondary channels and ribs, Case Stud. Therm. Eng. 61 (2024), doi:10.1016/j.csite.2024.104894.

[97] H. Babar, H. Wu, W. Zhang, T.R. Shah, D. McCluskey, C. Zhou, The promise of nanofluids: a bibliometric journey through advanced heat transfer fluids in heat exchanger tubes, Adv. Colloid Interface Sci. 325 (2024) 103112, doi:10.1016/j.cis.2024.103112.

[98] D. Purbia, A. Khandelwal, A. Kumar, A.K. Sharma, Graphene-water nanofluid in heat exchanger: mathematical modelling, simulation and economic evaluation, Int. Commun. Heat Mass Transf. 108 (2019), doi:10.1016/j.icheatmasstransfer.2019.104327.

[99] Z. Said, M. Ghodbane, L.S. Sundar, A.K. Tiwari, M. Sheikholeslami, B. Boumeddane, Heat transfer, entropy generation, economic and environmental analyses of linear fresnel reflector using novel rGO-Co_3O_4 hybrid nanofluids, Renew. Energy 165 (2021) 420–437, doi:10.1016/j.renene.2020.11.054.

[100] G. Karlaganis, R. Liechti, S. Teparkum, P. Aungkavattana, R. Indaraprasirt, Nanoregulation along the product life cycle in the EU, Switzerland, Thailand, the USA, and intergovernmental organisations, and its compatibility with WTO law, Toxicol. Environ. Chem. 101 (7–8) (2019) 339–368, doi:10.1080/02772248.2019.1697878.

[101] B. Nemati, M. Kamali, T.M. Aminabhavi, R. Dewil, M.E.V. Costa, I. Capela, Sustainability assessment of carbonaceous nanostructured materials for efficient hydrogen storage, Int. J. Hydrogen Energy 135 (2025) 477–498, doi:10.1016/j.ijhydene.2025.04.307.

[102] P. Bimova, et al., Environmental risk of nanomaterials and nanoparticles and EPR technique as an effective tool to study them—a review, Environ. Sci. Pollut. Res. Int. 28 (18) (2021) 22203–22220, doi:10.1007/s11356-021-13270-5.

[103] Department of Occupational Safety and Health, Ministry of Human ResourcesGuideline on Control and Safe Handling of Nanomaterials, Department of Occupational Safety and Health, Ministry of Human Resources, Malaysia, 2018.

[104] OECDChemical Accidents Involving Nanomaterials: Potential Risks and Review of Prevention, Preparedness and Response Measures – Project Report, Series on Chemical Accidents, Organisation for Economic Co-operation and Development (OECD), Paris, France, 2022 9789264377189.

[105] R. Landsiedel, L. Ma-Hock, K. Wiench, W. Wohlleben, U.G. Sauer, Safety assessment of nanomaterials using an advanced decision-making framework, the DF4nanoGrouping, J. Nanopart. Res. 19 (5) (2017) 171, doi:10.1007/s11051-017-3850-6.

[106] OECDEconomic Instruments to Incentivise Substitution of Chemicals of Concern – A Review, OECD Series on Risk Management of Chemicals, Organisation for Economic Co-operation and Development (OECD), Paris, France, 2023 9789264444782.

[107] OECDInternational Regulatory Co-Operation in Competition Law and Chemical Safety, Organisation for Economic Co-operation and Development (OECD), Paris, France, 2022.

[108] European CommissionEnergy and the Green Deal, European Commission, 2025. https://commission.europa.eu/strategy-and-policy/priorities-2019-2024/european-green-deal/energy-and-green-deal_en#:⟨:text=The%20European%20Green%20Deal%20focuses%20on%203,the%20quality%20of%20life%20of%20our%20citizens Accessed 4 July 2025.

[109] System for Award Management (SAM.gov)Notice of Intent to Issue a Funding Opportunity From U.S. Department of Energy's (DOE) High Performance Computing for Energy Innovation (HPC4EI) Initiative, System for Award Management (SAM.gov), 2025. https://sam.gov/opp/3b5676598acf416e83e4a7d56e6a298b/view Accessed 4 July 2025.

[110] Malaysia Green Technology and Climate Change Corporation (MGTC)EMGS Certification Paves Way for Energy Efficiency, Malaysia Green Technology and Climate Change Corporation (MGTC), 2025. https://www.mgtc.gov.my/2024/07/emgs-certification-paves-way-for-energy-efficiency/#:⟨:text=Additionally%2C%20the%20government%20has%20earmarked%20RM1%20billion,(ESCOs)%20engaged%20in%20advancing%20energy%20efficiency%20solutions Accessed 4 July 2025.

[111] CKGBSMade in China 2025: Looking at a Decade of China's Self-Sufficiency Drive, CKGBS, 2025. https://english.ckgsb.edu.cn/knowledge/article/china-self-sufficiency-drive-made-in-china/#:⟨:text=According%20to%20Bloomberg%2C%20among%20the,gap%20in%20the%20other%207 Accessed 4 July 2025.

[112] Ministry of Science, Technology & Innovation (MOSTI), in: *National Advanced Materials Technology Roadmap 2021*–2030, MOSTI, Putrajaya, Malaysia, 2021, p. 2021.

[113] S. Ma, A.H. Sharaai, Z. He, N.K.a.l. Matthew, N.S. Zainordin, W. Wafa, Exploring the research frontier of life cycle sustainability assessment—A systematic literature review of applied bibliometric analysis, Chem. Eng. Trans. 106 (2023), doi:10.3303/CET23106092.

[114] A.A. Challa, et al., Graphene oxide produced from spent coffee grounds in electrospun cellulose acetate scaffolds for tissue engineering applications, Mater. Today Commun. 35 (2023) 105974.

[115] S. Fiori, et al., Nanofibrillar biochar from industrial waste as hosting network for transition metal dichalcogenides. Novel sustainable 1D/2D nanocomposites for electrochemical sensing, Chemosphere 317 (2023) 137884.

[116] A. Yadav, H. Kumar, R. Sharma, R. Kumari, Synthesis, processing, and applications of 2D (nano) materials: a sustainable approach, Surf. Interfaces 39 (2023) 102925.
[117] J. Singhal, S. Verma, S. Kumar, The physio-chemical properties and applications of 2D nanomaterials in agricultural and environmental sustainability, Sci. Total Environ. 837 (2022) 155669.
[118] M. Madani, et al., Green synthesis of nanoparticles for varied applications: green renewable resources and energy-efficient synthetic routes, Nanotechnol. Rev. 11 (1) (2022) 731–759.
[119] L. Nagarajan, A. Arulraj, G. Senguttuvan, K. Kumaraguru, Solar assisted reduced graphene oxide as adsorbent for carbon dioxide and its kinetic studies, Physica E 116 (2020) 113739.
[120] A.O. Maka, T.N. Chaudhary, Performance investigation of solar photovoltaic systems integrated with battery energy storage, J. Energy Storage 84 (2024) 110784.
[121] A. Sethurajaperumal, E. Varrla, High-quality and efficient liquid-phase exfoliation of few-layered graphene by natural surfactant, ACS Sustain. Chem. Eng. 10 (45) (2022) 14746–14760.
[122] S. Xiong, Z. Wu, Z. Li, Facile fabrication of robust, versatile, and recyclable biochar-graphene oxide composite monoliths for efficient removal of different contaminants in water, Chemosphere 287 (2022) 132418.
[123] H. Samaraweera, C.U. Pittman Jr, R.V.K. Thirumalai, F. Perez, T. Mlsna, Characterization of graphene/pine wood biochar hybrids: potential to remove aqueous Cu^{2+}, Environ. Res. 192 (2021) 110283.
[124] V. Babaahmadi, S. Pourhosseini, O. Norouzi, H.R. Naderi, Designing 3D ternary hybrid composites composed of graphene, biochar and manganese dioxide as high-performance supercapacitor electrodes, Nanomaterials 13 (12) (2023) 1866.
[125] Y. Zhao, S.A. Qamar, M. Qamar, M. Bilal, H.M. Iqbal, Sustainable remediation of hazardous environmental pollutants using biochar-based nanohybrid materials, J. Environ. Manage. 300 (2021) 113762.
[126] J.X. Chan, J.F. Wong, A. Hassan, Z. Zakaria, Bioplastics from agricultural waste, in: Biopolymers and Biocomposites From Agro-Waste for Packaging Applications, Elsevier, Duxford, UK, 2021, pp. 141–169.
[127] M. Samer, Bioplastics production from agricultural crop residues, Agric. Eng. Int. CIGR J. 21 (3) (2019) 190–194.
[128] P. Choudhary, A. Pathak, P. Kumar, N. Sharma, Commercial production of bioplastic from organic waste–derived biopolymers viz-a-viz waste treatment: a minireview, Biomass Convers. Biorefin. 14 (10) (2024) 10817–10827.
[129] D. Merino, R. Simonutti, G. Perotto, A. Athanassiou, Direct transformation of industrial vegetable waste into bioplastic composites intended for agricultural mulch films, Green Chem. 23 (16) (2021) 5956–5971.
[130] S. Ügdüler, et al., Analysis of the kinetics, energy balance and carbon footprint of the delamination of multilayer flexible packaging films via carboxylic acids, Resour. Conserv. Recycl. 181 (2022) 106256.
[131] E. Pauer, M. Tacker, V. Gabriel, V. Krauter, Sustainability of flexible multilayer packaging: environmental impacts and recyclability of packaging for bacon in block, Clean. Environ. Syst. 1 (2020) 100001.
[132] J.G. Bernal-Carrillo, F.S. Chiwo-González, A.d.C. Susunaga-Notario, M.d. Ángel-Monroy, H. Arcos-Gutiérrez, I.E. Garduño-Olvera, Development and redesign of flexible packaging under sustainability criteria, in: Rev. Cienc. Tecnol., 7, 2024, p. e253.
[133] J.G. Bernal-Carrillo, et al., Redesign and development of flexible packaging for enhanced recyclability in sustainable food applications, Preprints (2024), doi:10.20944/preprints202410.0714.v1.
[134] E. Gawronska, M. Zych, R. Dyja, G. Domek, Using artificial intelligence algorithms to reconstruct the heat transfer coefficient during heat conduction modeling, Sci. Rep. 13 (1) (2023) 15343.
[135] N. Zobeiry, K.D. Humfeld, A physics-informed machine learning approach for solving heat transfer equation in advanced manufacturing and engineering applications, Eng. Appl. Artif. Intell. 101 (2021) 104232.
[136] B. Shilpa, V. Leela, An artificial intelligence model for heat and mass transfer in an inclined cylindrical annulus with heat generation/absorption and chemical reaction, Int. Commun. Heat Mass Transfer 147 (2023) 106956.

[137] T. Nakaishi, A. Chapman, Eco-labels as a communication and policy tool: a comprehensive review of academic literature and global label initiatives, Renew. Sustain. Energy Rev. 202 (2024) 114708.
[138] M.B. Nielsen, et al., Nanomaterials in the European chemicals legislation–methodological challenges for registration and environmental safety assessment, Environ. Sci. Nano 8 (3) (2021) 731–747.
[139] E.A. Bleeker, et al., Towards harmonisation of testing of nanomaterials for EU regulatory requirements on chemical safety – a proposal for further actions, Regul. Toxicol. Pharm. 139 (2023) 105360.
[140] A. Mech, et al., Safe-and sustainable-by-design: the case of Smart Nanomaterials. A perspective based on a European workshop, Regul. Toxicol. Pharm. 128 (2022) 105093.
[141] S. Gottardo, et al., Towards safe and sustainable innovation in nanotechnology: state-of-play for smart nanomaterials, NanoImpact 21 (2021) 100297.
[142] A. Tiwari, Emerging global trends in the potential of nanotechnology for achieving the net zero goals, in: J.A.O. Barros, G. Kaklauskas, E.K. Zavadskas (Eds.), Presented at the Modern Building Materials, Structures and Techniques, MBMST 2023, Lecture Notes in Civil Engineering, 392, Springer, Cham, 2023, pp. 31–41. https://doi.org/10.1007/978-3-031-44603-0_3.
[143] D. Wen, W. Gao, F. Qian, Q. Gu, J. Ren, Development of solar photovoltaic industry and market in China, Germany, Japan and the United States of America using incentive policies, Energy Explor. Exploit. 39 (5) (2021) 1429–1456.
[144] P. Slowik, et al., Analyzing the Impact of the Inflation Reduction Act on Electric Vehicle Uptake in the United States, International Council on Clean Transportation (ICCT), 2023. https://theicct.org/publication/ira-impact-evs-us-jan23/?gad_source=1&gad_campaignid=22639629046&gbraid=0AAAAA_pFleez kZS8yEiOxTcVho2BxX_aw&gclid=CjwKCAiAkbbMBhB2EiwANbxtbTKXwg6mnCehERZqf888cvv 30GH5LJkoBeZ8NjFk9HdWauP5KmBDbxoCkVsQAvD_BwE.
[145] K. Rasmussen, R. Schoonjans, P. Jantunen, H. Rauscher, Chapter 19: European Union legislation addressing environment, health and safety aspects of nanomaterials, in: J. Szpunar, J. Jiménez-Lamana (Eds.), Environmental Nanopollutants: Sources, Occurrence, Analysis and Fate, The Royal Society of Chemistry, Cambridge, 2022, pp. 469–501. https://doi.org/10.1039/9781839166570-00469.
[146] D. Schüler, M. Suhalitca, W. Pascha, K.-y. Oh. 2020. Government policies for start-ups in Korea and its regions: motives, mechanisms and major obstacles, Working Papers on East Asian Studies, No. 127/2020, University of Duisburg-Essen, Institute of East Asian Studies (IN-EAST), Duisburg.
[147] D.U. Park, et al., The Role of Science, Technology and Innovation Policies in the Industrialization of Developing Countries, United Nations Industrial Development Organization, Vienna, Austria, 2021.
[148] S. Yusuf, Could Innovation and Productivity Drive Growth in African Countries? Lessons from Korea, Center for Global Development, Washington, DC, USA, 2023.
[149] P. Koç, A. Gülmez, Analysis of relationships between nanotechnology applications, mineral saving and ecological footprint: evidence from panel fourier cointegration and causality tests, Resour. Policy 74 (2021) 102373.
[150] L. Pokrajac, et al., Nanotechnology for a sustainable future: addressing global challenges with the international network4sustainable nanotechnology, ACS Nano 15 (12) (2021) 18608–18623.
[151] M. Aydin, P. Koc, K.I. Sahpaz, Investigating the EKC hypothesis with nanotechnology, renewable energy consumption, economic growth and ecological footprint in G7 countries: panel data analyses with structural breaks, Energy Sources, B 18 (1) (2023) 2163724.
[152] H. Kusaka, et al., Crystalline boron monosulfide nanosheets with tunable bandgaps, J. Mater. Chem. A 9 (43) (2021) 24631–24640.
[153] N. Sakai, T. Sasaki, Highly organized monolayer arrangement of 2D materials and its applications, Acc. Mater. Res. 5 (6) (2024) 752–760. https://doi.org/10.1021/accountsmr.4c00072.
[154] K. Rafiee, H. Schritt, D. Pleissner, G. Kaur, S.K. Brar, Biodegradable green composites: it's never too late to mend, Curr. Opin. Green Sustain. Chem. 30 (2021) 100482.

[155] N. Karimi, H. Fayazfar, Development of highly filled nickel-polymer feedstock from recycled and biodegradable resources for low-cost material extrusion additive manufacturing of metals, J. Manuf. Processes 107 (2023) 506–514.

[156] S.M. Parke, J.C. Lopez, S. Cui, A.M. L., G.W. Coates, Polyethylene incorporating Diels–Alder comonomers: a "Trojan Horse" strategy for chemically recyclable polyolefins, Angew. Chem. Int. Ed. 62 (30) (2023), doi:10.1002/anie.202301927.

[157] B. Walhout, Making a Difference? Nanotechnology and the Quest for Responsible Innovation, PhD Thesis, University of Twente, Enschede, 2023, 308 pp. https://doi.org/10.3990/1.9789036558778.

[158] B. Sheu, S.-K. Kao, Keynotes from nanotechnology conference [theme: keynotes from nanotechnology conference], IEEE Nanatechnol. Mag. 17 (1) (2023) 2–6.

[159] J.-H. Lee, J. Woo, Green new deal policy of South Korea: policy innovation for a sustainability transition, Sustainability 12 (23) (2020) 10191.

[160] S. Lee, et al., Eco-friendly polymer solar cells: advances in green-solvent processing and material design, Acs Nano 14 (11) (2020) 14493–14527.

[161] K. Demirbaş, S. Çevik, Regulatory policies for safety of nanomaterials, Open J. Nano 5 (1) (2020) 1–16.

[162] J. Allan, et al., Regulatory landscape of nanotechnology and nanoplastics from a global perspective, Regul. Toxicol. Pharm. 122 (2021) 104885.

[163] A. Gerber, et al., Joint declaration on mainstreaming RRI across Horizon Europe, J. Responsible Innov. 7 (3) (2020) 708–711.

[164] A. Mukherjee, A.P. Sarma, Innovation, Transfer and Diffusion of Fourth Industrial Revolution (4IR) Technologies to Catalyze Sustainable Development in Asia-Pacific, Asian and Pacific Centre for Transfer of Technology, New Delhi, India, 2022.

[165] Z. Mohammadi, S.M. Jafari, Regulatory principles on food nanoparticles legislated by Asian and Oceanian countries, in: Safety and Regulatory Issues of Nanoencapsulated Food Ingredients, Elsevier, Amsterdam, Netherlands, 2021, pp. 201–238.

[166] K.W. Franke, A.L. Moxon, The toxicity of orally ingested arsenic, selenium, tellurium, vanadium and molybdenum, The Journal of Pharmacology and Experimental Therapeutics 61 (1) (1937) 89–102.

CHAPTER

17 MXene-based nanofluids for thermal energy applications

Laveet Kumar[a], Hafiz Taimoor Ahmed Awan[b] and Rashmi Walvekar[c]

[a]*Department of Mechanical Engineering, Mehran University of Engineering and Technology, Jamshoro, Pakistan,* [b]*James Watt School of Engineering, University of Glasgow, Glasgow, United Kingdom,* [c]*Department of Chemical and Process Engineering, University of Strathclyde, Glasgow, United Kingdom*

17.1 Introduction

The rapidly growing demand for efficient thermal management and energy conversion technologies has driven intensive research on advanced heat-transfer media that outperform conventional fluids. Thermal systems used in electronics cooling, concentrated solar power (CSP), thermal energy storage, refrigeration, and waste heat recovery increasingly operate at high heat fluxes and within compact geometries. Under these conditions, classical fluids, such as water, ethylene glycol (EG), and heat-transfer oils, are fundamentally limited by their low thermal conductivity, which constrains the overall system efficiency and power density [1,2]. Nanofluids, which are engineered suspensions of nanoscale solids dispersed in base liquids, have been proposed to address these limitations by exploiting the superior thermal, interfacial, and transport properties of nanostructured additives [3,4]. Over the past two decades, nanofluids incorporating metals, metal oxides, carbon-based materials, and hybrid nanostructures have shown measurable enhancements in thermal conductivity and convective heat transfer. However, persistent challenges, such as particle agglomeration, sedimentation, increased viscosity, and limited long-term stability under realistic operating conditions, hinder their large-scale implementation [5,6].

In this context, two-dimensional (2D) materials have emerged as promising nanofluid additives owing to their high aspect ratio, large specific surface area, and tunable surface chemistry, which promote strong interfacial interactions with surrounding fluids. Among these, MXenes, a rapidly expanding family of early transition metal carbides, nitrides, and carbonitrides, have attracted significant attention for thermal applications [7–11]. MXenes are typically produced by the selective etching of the A element from the MAX phases ($M_{n+1}AX_n$), yielding atomically thin layers terminated with functional groups such as –OH, –O, and –F. These surface terminations impart hydrophilicity, electrostatic stabilization, and chemical tunability, which facilitate dispersion in both polar and nonpolar base fluids with suitable modifications [11–14]. Consequently, MXene-based nanofluids combine high intrinsic thermal and electrical conductivity with favorable interfacial heat-transfer characteristics and colloidal stability, positioning them as next-generation heat-transfer media for demanding thermal energy systems [15,16]. Recent comparative studies have demonstrated that $Ti_3C_2T_x$ MXene nanofluids exhibit thermal and electrical conductivities that are remarkably higher than those of noncovalent

Graphene and 2D Materials in Heat Transfer: Fundamentals and Applications. DOI: https://doi.org/10.1016/B978-0-12-821964-5.00016-3

graphene nanoplatelets (GNPs) at equivalent concentrations, while also offering superior stability and autofluorescence properties that are advantageous for advanced heat-transfer applications [17].

The significance of MXene nanofluids extends beyond the simple enhancement of thermal conductivity. Heat transfer in flowing systems is governed by the complex interplay between conduction, convection, Brownian motion, microconvection, and interfacial thermal resistance. The sheet-like morphology and high aspect ratio of MXenes promote the formation of percolative thermal pathways at relatively low particle loadings, thereby reducing the percolation threshold for efficient heat transport [18]. Concurrently, their surface chemistry can be tailored to generate electrostatically stabilized, sterically hindered, or chemically grafted colloids that mitigate agglomeration and sedimentation, which are key prerequisites for maintaining consistent thermophysical properties during long-term operation [19]. Experimental studies have reported substantial gains in the thermal conductivity, convective heat-transfer coefficients, and critical heat flux relative to those of conventional nanofluids. For instance, MXene-based nanofluids have demonstrated thermal conductivity enhancements of up to 30.6% compared to pure water and 27.3% in EG/water mixtures, outperforming many metal-, oxide-, and graphene-based additives without significant viscosity increases [20]. Single-layer $Ti_3C_2T_x$ nanosheets dispersed in EG have shown thermal conductivity improvements of 64.9% at a concentration of 5 vol.% while maintaining stable behavior over 30 days without sedimentation [20]. Emerging theoretical and simulation studies indicate that interfacial thermal resistance, phonon scattering, and nanoscale layering of solvent molecules at MXene surfaces play decisive roles in governing their performance [21,22]. Molecular dynamics studies have further revealed that surface functionalization critically influences colloidal stability, with oxygen-terminated MXenes exhibiting complex aggregation behavior governed by the interplay between electrostatic repulsion and strong interlayer coupling [23].

The incorporation of MXene nanofluids into multifunctional thermal energy systems is equally crucial. MXenes possess strong and wide-ranging optical absorption, which allows for effective solar-thermal conversion when mixed with suitable carrier fluids [24]. Their high electrical conductivity enables electrothermal coupling, Joule heating, and active thermal control in hybrid electronic or electrochemical devices [16]. Recent advances in surface engineering have led to the development of MXene-based nanoscale ionic materials, where organic oligomers grafted to MXene surfaces via electrostatic interactions impart room-temperature fluidity while simultaneously protecting against oxidation, demonstrating structural integrity after 570 days of air exposure [25]. These solvent-free nanofluids address the longstanding challenges of MXene stability and processability while opening new avenues for multifunctional thermal management materials [26]. In thermal energy storage, MXene-enhanced fluids and MXene-modified phase-change materials can alter the phase transition behavior, improve the thermal diffusivity, and accelerate the charging/discharging kinetics, thereby increasing the effective storage density and power capability [27,28]. For CSP applications the incorporation of multilayer $Ti_3C_2T_x$ MXenes into solar salt nanofluids has been shown to enhance decomposition temperature to 609.8°C and improve specific heat capacity, demonstrating clear potential for advancing high-temperature thermal energy storage systems [29]. These synergistic effects indicate that MXene nanofluids represent not only incremental improvements over earlier nanofluid systems but also a broader material platform that couples thermal transport with electronic, photonic, and interfacial functionalities.

Despite this rapid progress, several scientific and engineering challenges must be overcome before MXene-based nanofluids can be widely used. Key challenges include maintaining structural and colloidal stability under prolonged thermal cycling, mitigating oxidation and surface degradation in aqueous and high-temperature environments, optimizing rheological behavior to avoid excessive pumping power, and ensuring chemical compatibility with system components [30,31]. The role of surfactants and dispersants in tailoring aqueous behavior has been systematically investigated, with ionic surfactants near their critical micelle concentration demonstrating enhanced colloidal stability through complementary mechanisms: cationic surfactants yield higher supernatant concentrations and smaller hydrodynamic diameters, while anionic surfactants produce larger zeta potentials [32]. Moreover, discrepancies between property measurements on small laboratory scales and performance in full-scale devices highlight the need for standardized testing protocols, multiscale modeling, and rigorous techno-economic and life cycle assessments [33–35]. Recent advances in artificial intelligence and machine learning have enabled the development of predictive frameworks for optimizing MXene-based nanofluids, with response surface methodology models achieving high accuracy ($R^2 = 0.9997$ for thermal conductivity and 0.9984 for dynamic viscosity) and multiobjective optimization revealing optimal MXene ratios that depend on mass fraction and temperature [36]. Such intelligent frameworks offer cost-effective methodologies for reducing computational and laboratory costs while achieving high precision in nanofluid designs. A fundamental understanding of how MXene composition, surface terminations, flake size distribution, and dispersion dynamics influence macroscopic heat transfer is essential for translating promising laboratory results into robust industrial applications.

This chapter provides a comprehensive overview of MXene-based nanofluids for thermal-energy applications. It first introduces MXene chemistry, synthesis routes involving top-down and bottom-up approaches, structural features, and surface functionalization strategies that govern the dispersion behavior and stability. Subsequent sections examine the thermophysical properties of MXene nanofluids, including thermal conductivity, viscosity, specific heat, and optical characteristics, and analyze the heat-transfer mechanisms across conduction- and convection-dominated regimes. Application-oriented discussions focus on thermal energy storage and other relevant technologies. Finally, this chapter evaluates the current challenges and environmental and scalability considerations and outlines key directions for future research. By integrating perspectives from materials science, colloid chemistry, fluid dynamics, and thermal engineering, this chapter aims to elucidate the mechanistic foundations and technological potential of MXene-based nanofluids as advanced working fluids for next-generation thermal energy systems.

17.2 Synthesis of MXene

2D materials have been intensively investigated over the past decade owing to their exceptional electronic, chemical, mechanical, and optical characteristics, which collectively enable their use as modular building blocks in layered heterostructures and multifunctional composites [37,38]. Structurally, 2D materials span a wide compositional spectrum, including single-element systems such as silicene, graphene, germanene, and phosphorene; binary compounds, including transition metal dichalcogenides (TMD) and oxides; and more compositionally complex families. Among the latter, MXenes represent a rapidly expanding class of 2D transition metal carbides, carbonitrides, and nitrides, which originated from the pioneering synthesis of titanium carbide MXene (Ti_3C_2 MXene). This material was obtained by selectively etching aluminum from the layered MAX phase Ti_3AlC_2 using hydrofluoric acid under ambient conditions [39].

MXenes are generally described by the formula $M_{n+1}X_nT_x$, where M is an early transition metal (e.g., Ti, Mo, V, Nb, Ta, Cr, Zr, Hf, Sc, or Y), X denotes carbon and/or nitrogen, and n typically ranges

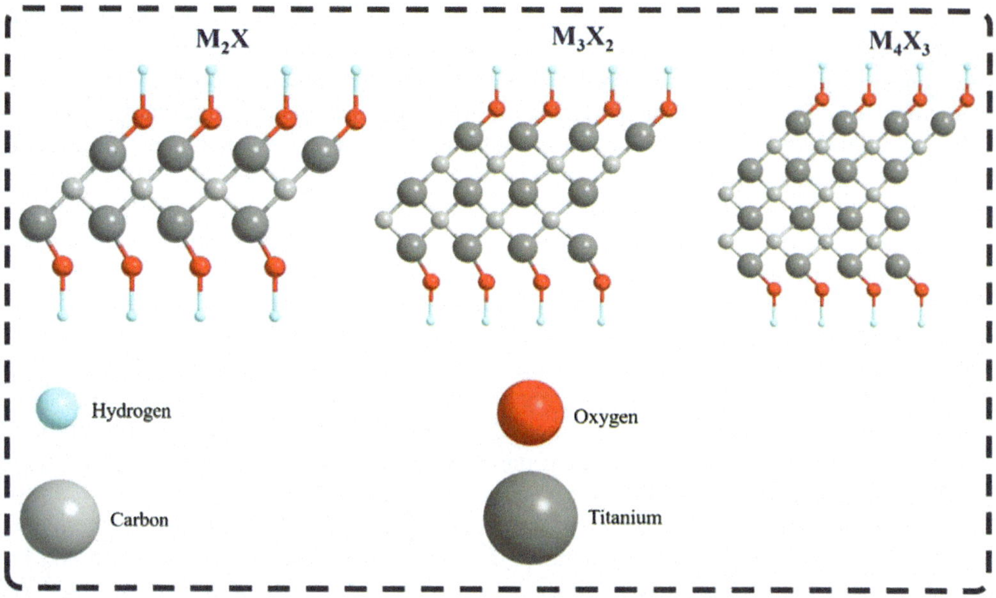

FIGURE 17.1

Various layered MXene crystal structures [40].

from 1 to 3. T_x represents surface termination groups, most commonly –OH, –F, –Cl, and –O, introduced during synthesis. These terminations are not merely chemical byproducts; they play a decisive role in governing the interlayer spacing, surface chemistry, electronic structure, hydrophilicity, and redox activity. Consequently, the compositional flexibility of the M–X backbone combined with controllable surface functionalization underpins the high tunability of MXene physicochemical properties and enables their adaptation to diverse application domains (Fig. 17.1) [40].

From a crystallographic perspective, MXenes are derived from layered MAX phases, a family of ternary carbides and nitrides with the general formula $M_{n+1}AX_n$. In these structures the A element, commonly taken from groups 12–16 (e.g., Al, Si, Ga, Ge, In, Sn), is interleaved between octahedrally coordinated M–X slabs. Strong mixed covalent–metallic–ionic bonding stabilizes the M–X layers, whereas the M–A interaction is comparatively weaker and predominantly metallic. This anisotropic bonding hierarchy is central to MXene synthesis, as selective chemical etching preferentially disrupts the M–A bonds while preserving the integrity of the M–X framework [41]. Unlike van der Waals layered solids, such as graphite or TMD, where exfoliation exploits weak interlayer forces, MAX phases require bond-selective chemical reactions to separate the layers. The differential bond strengths thus enable the controlled removal of the A layers and subsequent delamination into few- or single-layer MXene sheets, as schematically illustrated in Fig. 17.2A. The compositional landscape of the MAX precursors is shown in Fig. 17.2B.

MXenes can be synthesized using both top-down and bottom-up paradigms. The dominant top-down route involves the chemical etching and exfoliation of MAX precursors, often followed by intercalation and delamination steps to produce colloidally stable 2D flakes. Variations in etchant

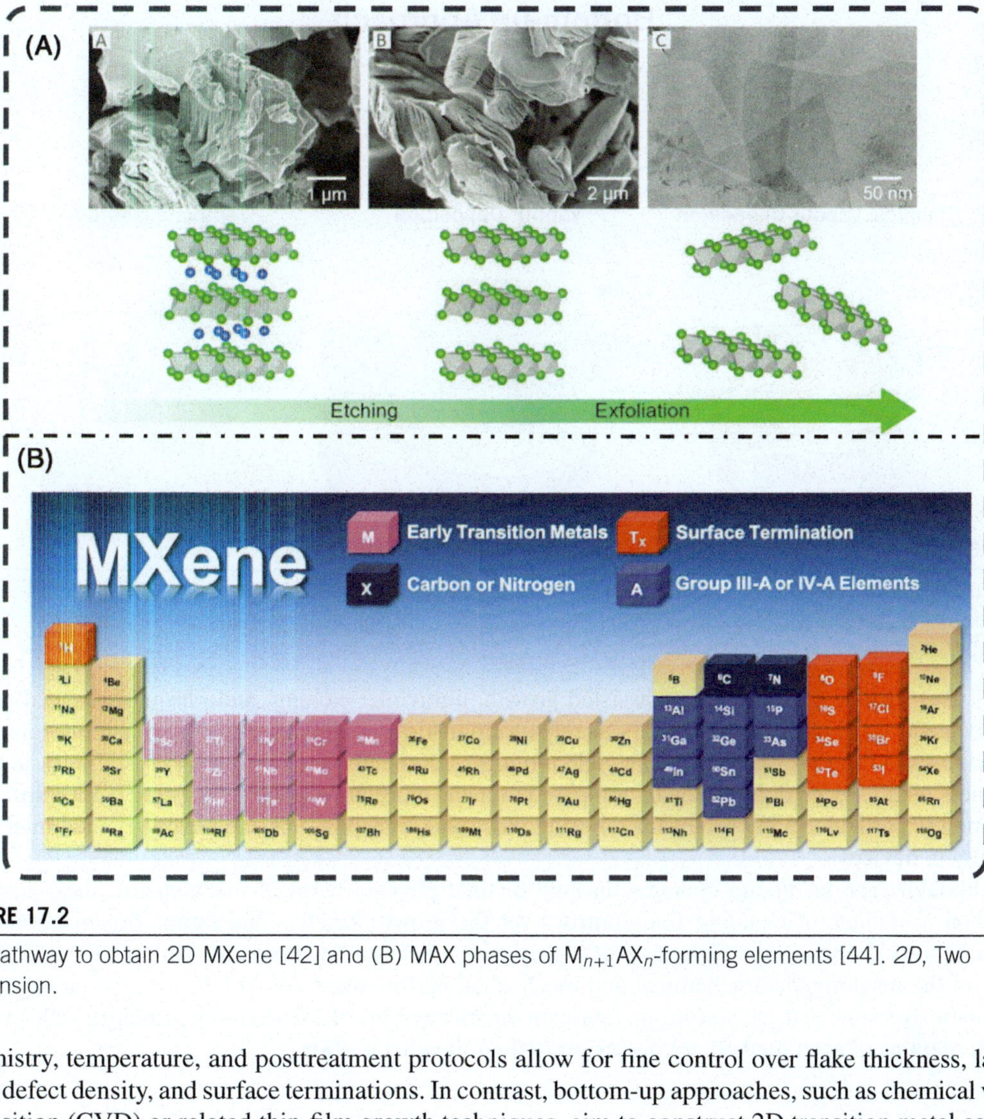

FIGURE 17.2

(A) Pathway to obtain 2D MXene [42] and (B) MAX phases of $M_{n+1}AX_n$-forming elements [44]. *2D*, Two dimension.

chemistry, temperature, and posttreatment protocols allow for fine control over flake thickness, lateral size, defect density, and surface terminations. In contrast, bottom-up approaches, such as chemical vapor deposition (CVD) or related thin-film growth techniques, aim to construct 2D transition metal carbide or nitride layers directly on substrates, offering precise thickness control and high crystallinity, albeit with greater synthetic complexity [42,43].

17.3 Bottom-up synthesis methods

Bottom-up synthesis strategies have emerged as a complementary route to conventional top-down etching for producing structurally well-defined MXene architectures with tailored properties. Representative

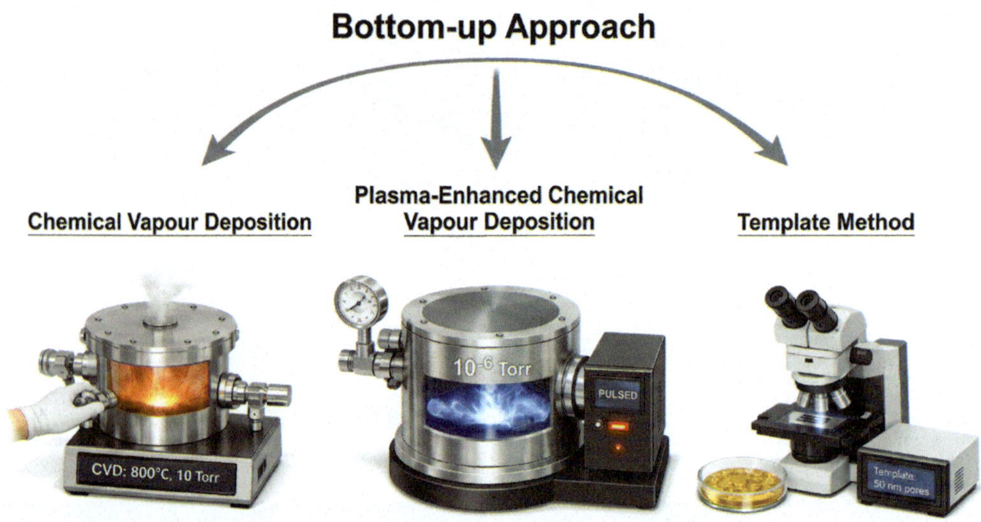

FIGURE 17.3

Various bottom-up techniques: chemical vapor deposition, plasma enhanced pulsed laser deposition, template method.

approaches include CVD, template-assisted growth, ultraviolet etching, and plasma-enhanced pulsed laser deposition (PE-PLD). These methods rely on atomistic or molecular assembly, enabling controlled nucleation and layer growth, which can yield ultrathin MXene films with superior crystallinity, uniform thickness, and reduced defect density compared to many exfoliation-based routes [42]. Such structural precision is particularly important for tailoring electronic transport, surface chemistry, and interfacial stability in devices.

Although each technique operates through distinct physicochemical mechanisms, they share the common advantage of enabling fine control over the growth kinetics and lattice organization. This controllability facilitates the formation of highly ordered crystalline domains, which are key determinants of the resulting electrochemical and mechanical performance. Fig. 17.3 provides a comparative schematic overview of these bottom-up fabrication pathways, and the underlying principles, advantages, and limitations of each method are discussed in the following sections.

17.3.1 Chemical vapor deposition

CVD has been extensively employed for the synthesis of thin films on MAX phases because of its capability to produce high-purity, dense, and well-adhered coatings. The pioneering work by Nickl et al. [17] demonstrated the feasibility of depositing MAX phase thin films via CVD, with Ti_3SiC_2 being the first successfully synthesized compound using this approach [45]. The formation of Ti_3SiC_2 typically occurs at elevated temperatures ranging from 1000°C to 1300°C, which facilitates the necessary chemical reactions and phase transformations [46]. In this method, Ti_3SiC_2 is produced through the simultaneous deposition of its constituent elements, where gaseous precursors interact with a solid

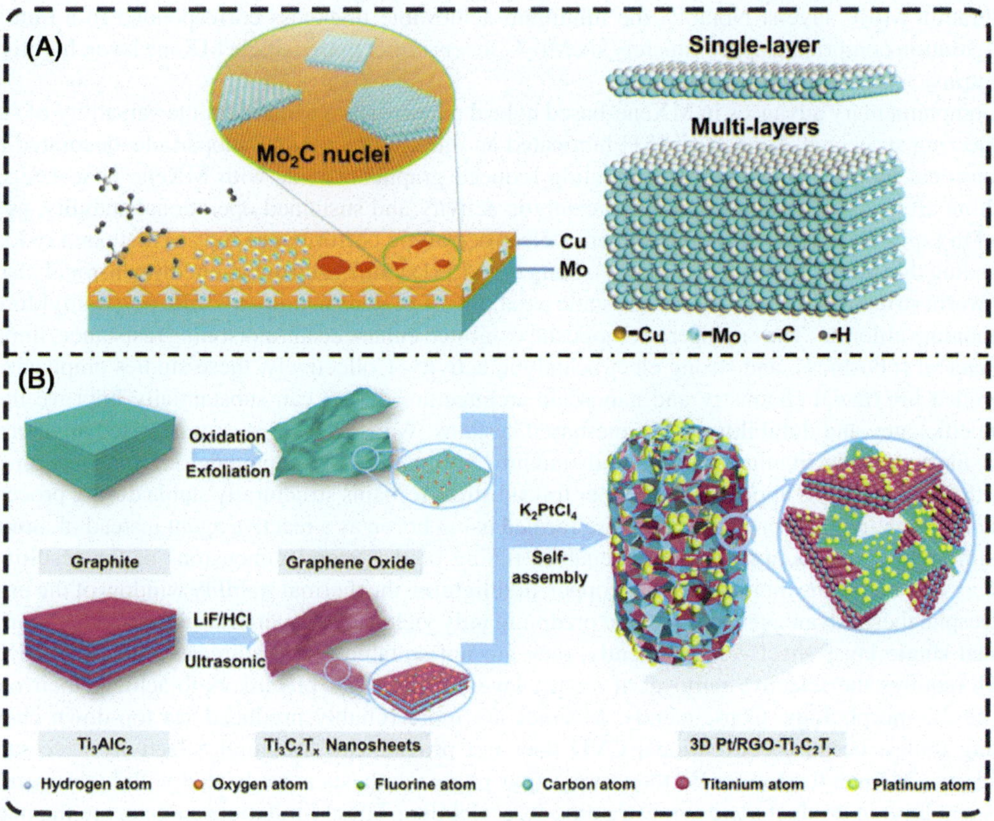

FIGURE 17.4

(A) Growth process and model of 2D crystals [50] and (B) schematic representation for the fabrication of 3D Pt/RGO–$Ti_3C_2T_x$ architecture [51]. *2D*, Two dimension.

substrate to drive the growth of the MAX phase. This gas–solid interaction enables precise control over the stoichiometry and microstructure of the resulting films, making CVD a versatile technique for tailoring MAX phase properties for high-temperature and structural applications [47,48]. Phase purity remains a critical challenge in high-temperature synthesis, where excessive thermal budgets promote parasitic reactions and surface contamination. During CVD the adsorption of reactive surface species suppresses surface diffusion, thereby influencing nucleation kinetics and film crystallinity. Leveraging this mechanism, Xu et al. [49] synthesized molybdenum carbide (Mo_2C) films using a bilayer Cu/transition-metal foil architecture. In their approach the stacked structure was heated to 1085°C under hydrogen, followed by the introduction of methane as the carbon precursor. A schematic representation of the conventional CVD process is shown in Fig. 17.4A [50]. The molten copper top layer played a decisive role in crystal growth, functioning as both a catalytic medium for methane decomposition and a diffusion pathway that enabled molybdenum atoms to dissolve and redistribute within the liquid copper phase. This coupled catalytic–diffusive process promoted the formation of ultrathin Mo_2C layers. Notably, the minimum achievable thickness corresponded to a multilayer configuration comprising approximately six Mo_2C sheets rather than a single MXene layer, highlighting the intrinsic constraints of the system.

Complementary advances in MXene-based hybrid nanostructures highlight the versatility of surface engineering strategies. Yang et al. [51] fabricated an ultrafine platinum nanoparticle–decorated three-dimensional hybrid composite by integrating reduced graphene oxide with MXene nanosheets. The resulting architecture demonstrated high catalytic activity and sustained operational stability, as illustrated in Fig. 17.4B, and exhibited a superior electrocatalytic performance toward methanol oxidation. Extending this functionalization paradigm, Yang et al. [52] synthesized quasi-one-dimensional platinum nanoworm structures through the electrostatic assembly of MXene nanosheets with polydiallyldimethylammonium chloride. The engineered composite exhibited enhanced antipoisoning resistance, improved mechanical robustness, and strong electrocatalytic activity. Collectively, these studies emphasize that controlled interfacial chemistry and nanoscale architecture design can substantially improve the catalytic efficiency and durability of MXene-based systems. Wang et al. [53] demonstrated an alternative route for producing titanium carbide and titanium nitride–based compounds through a vapor-phase growth strategy. In this approach the copper foil substrate remains structurally stable during processing, while ultrathin titanium bromide layers are formed using boron as a reactive agent instead of ammonia. Careful control of reaction conditions enables tuning of the lateral dimensions of the resulting 2D nanocrystals below the melting point of copper, highlighting the thermal stability window of the process.

Despite these advantages, this method predominantly yields multilayered structures rather than well-defined single-layer sheets. Consequently, they are not suitable for the direct synthesis of MXenes, which requires the selective removal of atomic layers from MAX precursors to achieve their characteristic 2D morphology. Consequently, MXenes are more reliably produced via top-down chemical etching routes, because conventional CVD does not provide the structural selectivity necessary for phase-pure MXene formation. Furthermore, vapor-phase synthesis is associated with high operational costs and limited product yields, constraining its scalability. These limitations underscore the need for continued advancement in MXene fabrication strategies to improve their efficiency, cost-effectiveness, and structural control.

17.3.2 Template method

The template method has emerged as a scalable alternative to CVD for producing 2D materials, particularly when a high product yield and structural uniformity are required. In this strategy, preformed 2D transition metal oxide nanosheets act as sacrificial templates that are subsequently transformed into carbides or nitrides through controlled thermal or reactive treatment. The chemical identity of the final phase is governed primarily by the precursor oxide composition and reaction atmosphere, enabling the targeted synthesis of transition metal carbides or nitrides with a preserved sheet-like morphology [26,27].

A representative example is the conversion of oxide templates into ultrathin carbide nanosheets. Jia et al. [54] demonstrated that nanosheet templates can be thermally treated at moderate temperatures (450°C–700°C) to obtain nitrogen-doped Mo_2C nanosheets with high surface area. Structural characterization revealed the presence of abundant nanopores and relatively low crystallinity, features associated with the formation of a hexagonal framework and an enhanced defect density. Such porosity is advantageous for applications that require rapid mass transport or increased active surface sites.

Similarly, nitride formation can be achieved through the ammonia-mediated nitridation of oxide templates. Joshi et al. [55] converted 2D MoO_3 nanosheets into hexagonal MoN nanosheets via high-temperature ammonia treatment at approximately 800°C. Notably, this transformation preserved the layered morphology of the original template, resulting in 2D MoN with a well-defined lamellar architecture. This retention of structural anisotropy highlights a central advantage of the template route: compositional transformation occurs without sacrificing dimensional control.

17.3.3 Plasma-enhanced pulsed laser deposition methods

Molybdenum carbide (Mo_2C) ultrathin films were fabricated using a PE-PLD strategy that integrates the features of CVD with the energetic ablation dynamics of pulsed laser deposition. Zhang et al. [56] demonstrated this hybrid route by heating a sapphire substrate to 700°C while supplying methane as the reactive carbon precursor, enabling in situ carbide formation during film growth. Although this approach successfully produces continuous ultrathin layers, structural characterization reveals limited crystallinity accompanied by a high density of stacking defects. These microstructural limitations indicate that, under the reported plasma and thermal conditions, adatom mobility and phase ordering are insufficient to promote defect-free crystal growth, thereby constraining the structural quality of the resulting films.

17.4 Top-down synthesis methods

Top-down synthesis represents the dominant manufacturing paradigm for producing 2D MXenes, relying on the controlled structural deconstruction of layered MAX precursors into atomically thin sheets. This strategy integrates a sequence of interrelated steps: precursor preparation, selective chemical etching, and subsequent exfoliation, which are designed to remove the A-layer from the parent MAX phase while preserving the integrity of the transition metal carbide or nitride framework. This process exploits the intrinsic anisotropy of MAX phases, enabling targeted bond cleavage and delamination to yield high-surface-area MXene nanosheets with tunable surface terminations and physicochemical properties.

Precursor synthesis establishes the crystallographic quality and stoichiometric uniformity of the MAX phase, which directly influences the etching kinetics and exfoliation efficiency. Selective etching then removes the A element through chemical reactions that weaken the interlayer bonding, generating multilayered MXene structures. Subsequent exfoliation, typically driven by intercalation and mechanical or chemical delamination, separates these layers into few- or single-layer nanosheets. Together, these steps form a controllable pathway for tailoring the morphology, interlayer spacing, and surface chemistry, which are critical parameters for downstream functional performance. An overview of representative top-down routes is illustrated in Fig. 17.5, where the comparative mechanisms, processing conditions, and structural transformations underlying MXene formation are summarized [43].

17.4.1 Precursor

MAX phases constitute a large family of layered ternary carbides and nitrides built from edge-sharing MX_6 octahedra separated by A-element layers, with more than 130 structurally distinct compositions reported to date [57]. Their crystallographic classification is primarily based on the stacking sequence

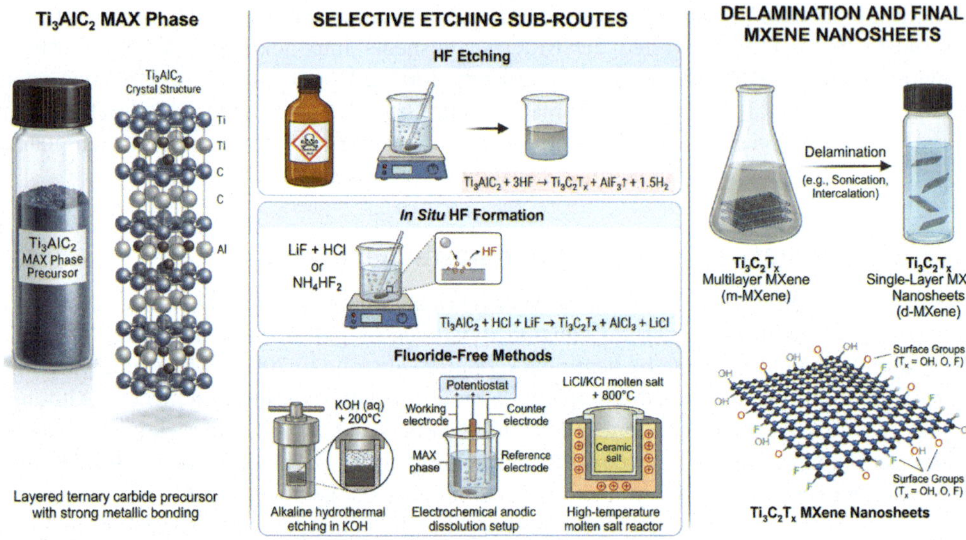

FIGURE 17.5

Various processes involve in top-down synthesis.

of the transition metal carbide/nitride slabs between adjacent A layers, commonly described as 211, 312, or 413 phases, corresponding to the number of M–X layers per repeat unit [42]. This layered architecture is central to their chemical selectivity during etching and underpins their role as precursors for 2D derivatives.

MXenes are predominantly synthesized by selectively removing the A-element, most frequently aluminum, from aluminum-containing MAX phases. Consequently, the structural and chemical characteristics of a given MXene are inherited from its parent MAX phase, including metal ordering, layer thickness, and bonding environment. Historically, slightly more than 50 MAX phases were identified in the 1960s, establishing a foundation for the subsequent exploration of layered transition metal carbides and nitrides [58]. The continued expansion of this class of materials has enabled the discovery of a growing number of MXenes, exceeding 30 experimentally realized compositions.

Recent advances have focused on chemically ordered MAX phases, which introduce additional structural complexity and tunability. Two major subclasses emerged in the mid-2010s: out-of-plane ordered (o-MAX) phases, typically associated with 312 or 413 stacking, and in-plane ordered (i-MAX) phases derived from 211 structures. In o-MAX systems, distinct transition metal layers are arranged along the *c*-axis, producing MXenes in which a single or double layer of one metal is sandwiched between layers of another, yielding anisotropic chemical environments after etching [59]. In contrast, i-MAX phases exhibit the in-plane ordering of two transition metals within the same layer, enabling compositional patterning at the atomic scale and facilitating the formation of ordered vacancy MXenes upon selective etching [60]. The recent identification of molybdenum-based i-MAX compositions has further broadened the accessible chemical space and highlighted the ongoing evolution of precursor design strategies for next-generation MXenes [61].

17.4.2 Etching

Selective etching is the foundational step in converting MAX phases into MXenes, enabling the removal of the A-layer while preserving the M–X framework. During this process the exposed surfaces are terminated with functional groups such as –OH, –O, and –F, which significantly influence the interlayer interactions, surface chemistry, and subsequent delamination behavior. The etched multilayers remain stacked through van der Waals forces, necessitating further intercalation or mechanical delamination to obtain few- or single-layer MXene sheets. Because the M–A bond exhibits strong metallic character, mechanical exfoliation alone is ineffective; therefore chemical etching remains the dominant synthesis route, as illustrated schematically in Fig. 17.6A [46,62].

Aqueous fluoride-containing acidic solutions are commonly employed to selectively dissolve A elements. The etching severity increases with the atomic number and bonding strength associated with the M component, often requiring more aggressive chemical conditions to overcome the metallic M–A bond [63]. A representative example is the synthesis of Ti_3AlC_2-derived MXene, which serves as a model system for process optimization. Early protocols involved immersing MAX powder in concentrated HF (~50 wt.%) to establish proof-of-concept selective etching [39]. Subsequent refinements demonstrated that in situ HF generation using LiF/HCl mixtures provides comparable etching efficiency while improving handling safety. For instance, Alhabeb et al. [64] reported successful etching across a range of HF concentrations (5–30 wt.%) and durations (5–24 h), yielding multilayer $Ti_3C_2T_X$ flakes. In contrast, LiF/HCl systems typically require extended reaction times (>24 h) but reduce direct exposure to concentrated HF solutions.

Process intensification strategies have aimed to shorten etching time while maintaining structural integrity. Hu et al. demonstrated rapid preparation of $Ti_3C_2T_x$ by etching Ti_3AlC_2 in 40 wt.% HF at 50°C for only 30 min under continuous stirring, compared with the ~10 h commonly required in conventional protocols [65]. The resulting MXene exhibited thermal stability up to ~300°C in air and ~800°C under vacuum, indicating preserved lattice integrity despite accelerated etching. Recently, efforts to reduce chemical hazards and energy demand have led to greener alternatives. Numan et al. [66] introduced a microwave-assisted hydrothermal route using LiF/HCl, enabling rapid Al removal and concurrent delamination in under 2 h at near-ambient temperature (~40°C), as shown in Fig. 17.6B. This approach substantially decreases the processing time relative to conventional methods, which may extend to 48 h, while minimizing the reliance on highly concentrated HF. These developments illustrate an ongoing shift toward safer, faster, and more scalable etching methodologies that preserve the MXene quality while improving environmental compatibility.

17.4.3 Exfoliation

Following selective etching, exfoliation is a critical step for delaminating multilayered MXene stacks into few- or single-layer nanosheets suitable for colloidal processing and device fabrication. Direct sonication can assist in removing residual etching byproducts, particularly when multilayers are synthesized using etchants other than HCl/LiF, thereby promoting cleaner interlayer separation and improved dispersion stability. Beyond simple mechanical agitation, liquid-phase exfoliation driven by molecular intercalation remains the principal strategy for achieving high-yield MXene colloids. In this approach the introduction of appropriately sized and chemically compatible intercalants expands the interlayer spacing, weakens electrostatic and van der Waals interactions, and facilitates delamination

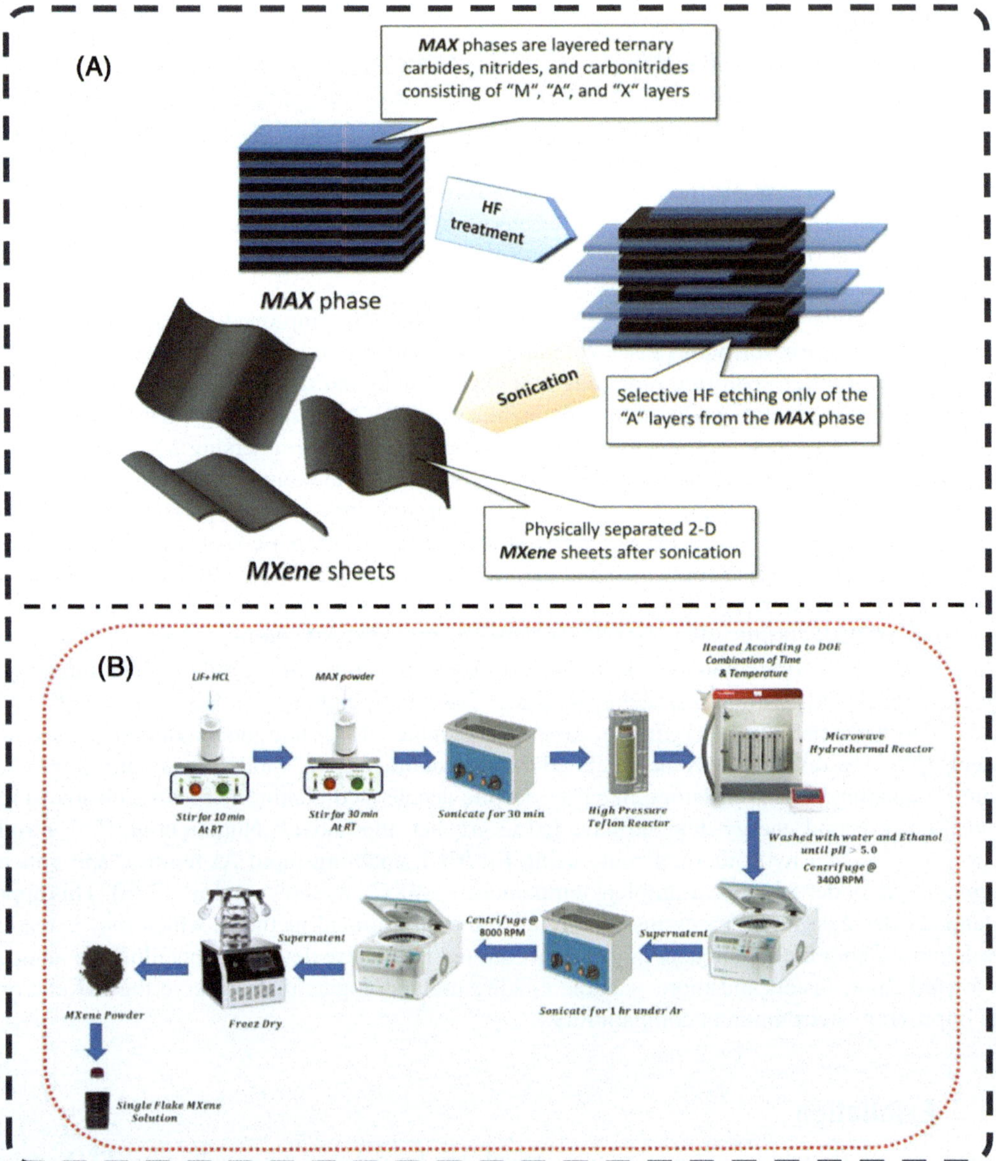

FIGURE 17.6

(A) MXene formation from MAX phases [62] and (B) synthesis scheme of MXene via etching and microwave synthesis [66].

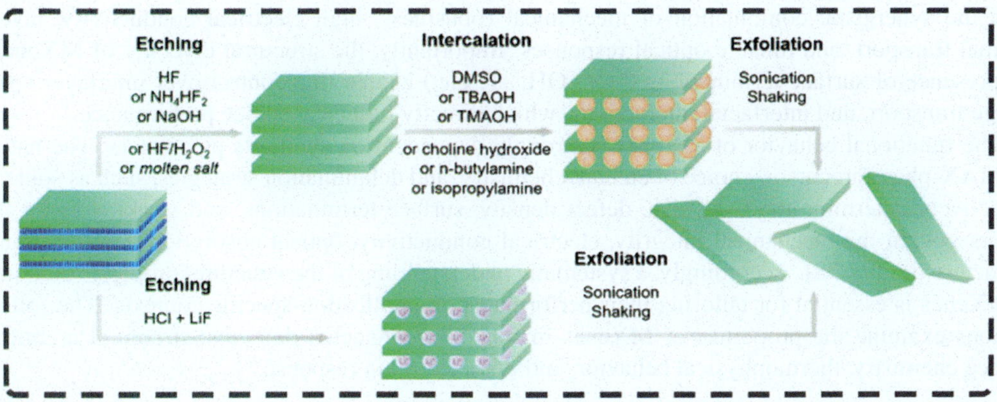

FIGURE 17.7

MXene exfoliation by etching process [42]. *DMSO*, Dimethyl sulfoxide; *TBAOH*, tetrabutylammonium hydroxide.

into individual nanosheets. Because etched multilayers typically contain intercalated cations and organic residues, the selection of intercalating species is essential to maximize the exfoliation efficiency and preserve structural integrity.

Mechanical energy supplied through controlled sonication or vigorous shaking further assists the delamination process by overcoming interlayer adhesion without inducing excessive structural damage [67–69]. Polar organic molecules are particularly effective intercalants because of their ability to penetrate hydrated interlayers and stabilize separated sheets. For example, dimethyl sulfoxide has been widely employed to exfoliate Ti_3AlC_2-derived MXenes, where solvent intercalation increases the gallery spacing and enables efficient nanosheet separation [70]. Similarly, tetrabutylammonium hydroxide promotes delamination through the intercalation of bulky TBA^+ ions, which generate steric repulsion and electrostatic stabilization within the interlayer region. This ion-assisted exfoliation pathway has become one of the most extensively adopted routes for producing high-quality MXene dispersions owing to its reproducibility and scalability [71] (Fig. 17.7).

17.5 Properties of MXene

2D nanomaterials exhibit physical and chemical characteristics that differ fundamentally from their bulk counterparts, largely because of quantum confinement, high surface-to-volume ratios, and tunable surface chemistry [72]. Representative members of this material class include graphene [73], TMDs [74], black phosphorus [75], and MXenes [76]. Among these, MXenes have emerged as a rapidly expanding family of 2D transition-metal carbides and nitrides since their first report in 2011 [39]. Their layered morphology, hydrophilic surfaces, and chemically active terminations enable property profiles that are highly responsive to synthesis and postprocessing conditions. MXenes have demonstrated strong potential across diverse technological domains, including energy conversion (e.g., photocatalysis, solar energy harvesting, and water splitting), energy storage (supercapacitors and batteries), electromagnetic interference shielding, sensing platforms, and photothermal applications. These broad capabilities arise from the synergistic combination of mechanical robustness, high electrical conductivity, favorable thermal transport, and tunable optical responses. Importantly, the structural diversity of MXenes and the presence of surface terminations (–O, –OH, –F, etc.) enable fine control over interlayer spacing, charge transport, and interfacial interactions, which directly influence device performance.

The functional behavior of MXenes is strongly dependent on synthesis parameters. The nature of the MAX-phase precursor, choice of etchant chemistry, and delamination strategies, such as sonication, collectively determine the flake size, defect density, surface terminations, and stacking order. These factors govern the mechanical integrity, electrical conductivity, optical absorption, and ion transport characteristics [77,78]. Accordingly, a systematic understanding of the structure–property relationships of MXenes is essential for tailoring their performance in application-specific contexts. The following sections examine the properties of MXenes in four interconnected domains: structural architecture, surface chemistry, thermophysical behavior, and optoelectronic response.

17.6 Morphological properties

The crystal chemistry and structural evolution from MAX precursors to MXenes have been extensively characterized to clarify how selective etching transforms a densely packed layered ceramic into a 2D material system. MAX phases belong to a family of nanolaminated carbides and nitrides that typically crystallize in hexagonal lattices. Their general stoichiometry, $M_{n+1}AX_n$ ($n = 1$–3), reflects alternating atomic layers in which M is an early transition metal (e.g., Ti, Mo, V, Zr, Nb, Ta, W, or Hf), A is a group III–VI element (e.g., Al, Si, In, Ga, or Ge), and X represents carbon and/or nitrogen [50,51]. This layered architecture yields a thermodynamically stable crystalline framework in which strong M–X bonds coexist with comparatively weaker M–A bonds, providing a structural basis for chemical exfoliation. From a crystallographic perspective, these MAX precursors are predicted and experimentally confirmed to exhibit robust structural stabilities under standard conditions [79]. MXenes are derived directly from the MAX phases, which commonly adopt a hexagonal symmetry with the P63/mmc space group, as illustrated schematically in Fig. 17.8A. The conversion process relies on selective chemical etching, which preferentially cleaves the M–A bonds while preserving the integrity of the M–X octahedral framework [80]. Exposure to strong acidic media removes the A-layer atoms, effectively delaminating the parent structure and yielding atomically thin, layered MXene sheets (Fig. 17.8B). This topochemical transformation preserves the in-plane metallic/covalent bonding network, which is essential for the mechanical and electronic properties of the resulting nanosheets.

During etching, particularly when hydrofluoric acid or fluoride-containing salts are employed, the freshly exposed MXene surfaces are terminated with functional groups such as –F, –O, and –OH [81,82]. These surface terminations are not merely byproducts; they substantially influence interlayer spacing, hydrophilicity, electronic structure, and chemical reactivity. To elucidate these effects, Khazaei et al. [83] conducted theoretical investigations of the geometric, magnetic, and electronic characteristics of representative terminated MXene systems, including M_2XF_2, $M_2X(OH)_2$, and M_2XO_2. Their analysis demonstrated that surface chemistry plays a decisive role in tuning the band structure and magnetic behavior, highlighting the coupled relationship between the termination species and functional performance. Fig. 17.8A summarizes the structural relationships between the MAX precursors and their MXene derivatives. It depicts three representative MAX architectures and the corresponding MXenes produced via selective etching, along with atomic-scale top and side views of the M_2X MXene lattice

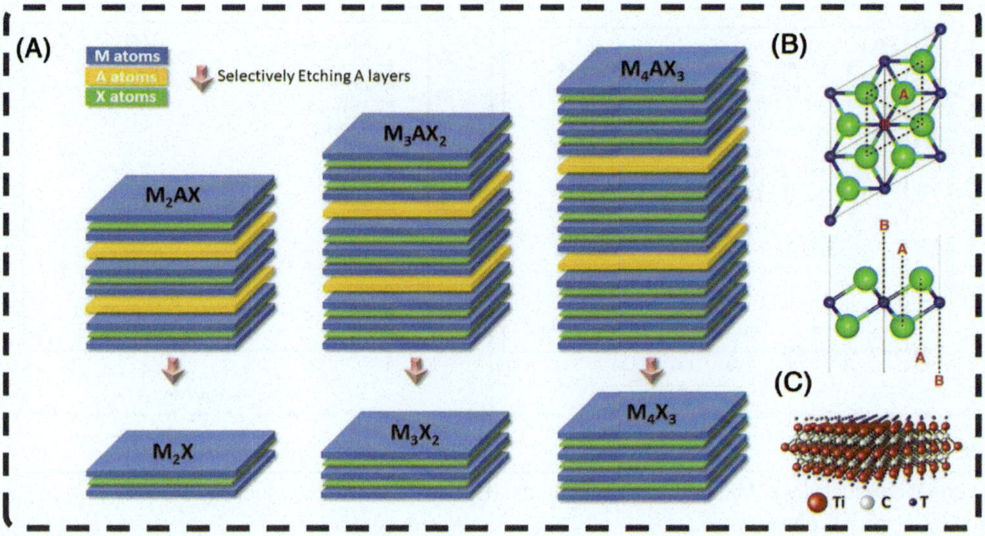

FIGURE 17.8

(A) Schematic illustrating evolution from MAX phases to MXenes, (B) top (upper) and side (lower) views of a crystal model for MXene M_2X, and (C) crystal structure of functionalized $Ti_3C_2T_x$ with surface terminations [80].

(Fig 17.8B). The labeled A and B sites in Fig 17.8B indicate distinct surface hollow positions that are relevant for termination group adsorption and subsequent physicochemical behavior. Among the over 20 different MXenes reported, $Ti_3C_2T_x$ has been the most representative and well-studied, and its crystal structure decorated with surface terminations is shown in Fig. 17.8C [80].

17.7 Thermophysical properties

MXenes derived from MAX phases have emerged as promising candidates for thermoelectric and thermal energy conversion applications because of their intrinsic ceramic-like characteristics and tunable surface chemistry. A notable feature of MXenes is their anisotropic thermal transport behavior, which arises from the structural differences along the zigzag and armchair crystallographic directions. Computational and experimental studies have reported favorable thermal conductivities in several MXenes, including Ti_2CO_2, Hf_2CO_2, Zr_2CO_2, Sc_2CF_2, Sc_2CO_2, $Sc_2C(OH)_2$ [84,85], and Mo_2C [86]. These properties make MXenes attractive materials for applications requiring efficient heat transfer and thermal management.

To exploit these characteristics in practical systems, MXenes have been incorporated into polymers and fluid matrices to enhance their thermophysical performance. For example, Jin et al. [87] fabricated poly(vinyl alcohol) (PVA)/MXene composite films using a multilayer casting strategy to create continuous MXene heat-conduction pathways. The resulting films exhibited a high in-plane thermal conductivity of 4.57 Wm^{-1} K^{-1}, representing an approximately 23-fold enhancement compared with

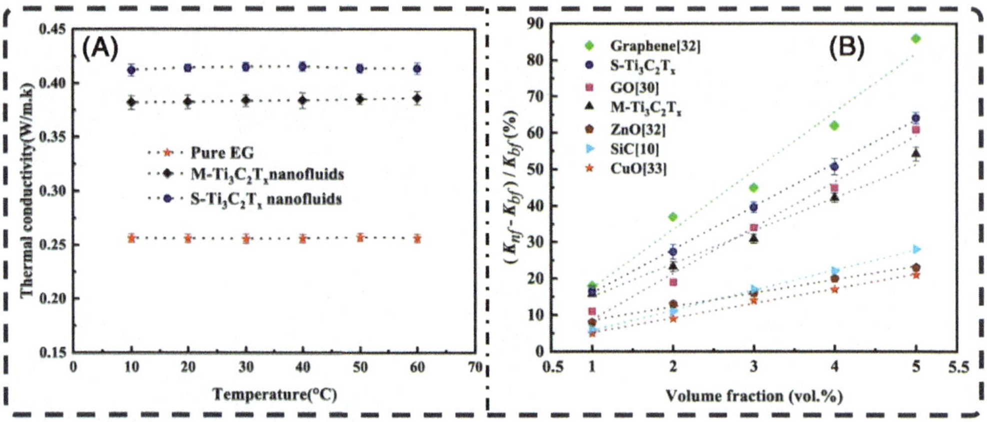

FIGURE 17.9

(A) Thermal conductivity in MXene-based nanofluids; (B) rise in thermal conductivity with increase in volume fraction [20]. *EG*, Ethylene glycol.

pristine PVA films. This improvement highlights the effectiveness of MXene alignment and percolation in promoting phonon transport within the composite structures.

Similarly, Bao et al. [20] prepared multilayer and single-layer $Ti_3C_2T_x$ MXene dispersions in EG to form nanofluids with enhanced thermal conductivity. The viscosity of EG stabilized the suspension, resulting in a nearly temperature-independent thermal conductivity over the range of 10°C–60°C (Fig. 17.9A). Furthermore, increasing the MXene volume fraction from 1% to 5% at room temperature produced a linear increase in the thermal conductivity. At 5 vol.%, nanofluids containing multilayer and single-layer MXene exhibited conductivity enhancements of 53.1% and 64.9%, respectively, relative to pure EG (Fig. 17.9B). These results demonstrate the strong potential of MXene nanofluids in heat-transfer applications.

Beyond nanofluids the hydrophilic surface chemistry, large specific surface area, and abundant functional terminations of MXenes enable strong interfacial interactions with phase-change materials. Hydrogen bonding and adsorption between polyethylene glycol (PEG) chains and MXene surfaces facilitate the formation of stable composite structures. Using a vacuum impregnation approach, Qu et al. [88] employed MXene as a thermally conductive supporting framework to fabricate form-stable phase-change materials. The resulting PEG/MXene composite exhibited a thermal conductivity of 2.052 Wm^{-1} K^{-1}, which is significantly higher than that of pure PEG. This enhancement improves the heat storage and release efficiency, which is critical for thermal energy storage systems.

Additional evidence of the thermal transport capability of MXenes was provided by Rafieerad et al. [89], who investigated water-based $Ti_3C_2T_x$ MXene nanofluids and compared them with covalently and noncovalently functionalized GNP systems. At MXene loadings of 0.1 and 0.2 mass% the thermal conductivities reached 0.732 and 0.828 Wm^{-1} K^{-1}, respectively, exceeding those of noncovalent GNP nanofluids (0.679 and 0.702 Wm^{-1} K^{-1}). These findings underscore the efficiency of MXene-based nanostructures in facilitating thermal transport at low filler concentrations.

17.8 Optical and electrical properties

First-principles calculations have predicted that certain surface terminations and stoichiometries can induce semiconducting behavior in MXenes, thereby expanding their functional space beyond metallic conduction. These theoretical insights are consistent with the growing experimental evidence that tailored MXene compositions exhibit enhanced optical absorption, carrier transport, and interfacial charge dynamics, which are attractive for optoelectronic platforms [90–92]. In particular, titanium carbide MXenes, such as $Ti_3C_2T_x$ and Ti_2CT_x, have emerged as model systems for probing the structure–property relationships relevant to transparent electrodes and photoactive devices. Experimental studies have demonstrated that controlled synthesis and postprocessing enable the simultaneous optimization of sheet conductivity and optical transparency, which are two parameters that are typically in trade-off for conventional conductive films. When deposited on substrates with precisely regulated thickness, $Ti_3C_2T_x$ and Ti_2CT_x can form transparent conductive thin films suitable for optoelectronic integration [91,93–95].

Halim et al. [95] reported transparent $Ti_3C_2T_x$ films fabricated via epitaxial growth of a precursor MAX phase followed by selective etching. In that study a Ti_3AlC_2 precursor film was magnetron-sputtered onto a sapphire substrate and subsequently etched using HF or NH_4HF_2 to remove the Al layers, yielding a continuous $Ti_3C_2T_x$ film with a thickness of approximately 19 nm. The resulting MXene film exhibited visible-to-infrared transmittance exceeding 90% while maintaining high electrical conductivity. Notably, an MAX-phase film of comparable thickness displayed only ~30% transmittance in the visible range, highlighting the substantial optical advantage of the MXene structure. The coexistence of high transparency and excellent electrical conductivity positions these films as promising alternatives to conventional transparent conductors in various applications.

Despite these advances, MXene-based optoelectronics is still an emerging field. A fundamental understanding of carrier dynamics, surface termination effects, environmental stability, and interface engineering is still developing. Therefore systematic investigations that couple theoretical modeling with device-level characterization are essential to fully elucidate the photoelectric behavior of MXenes and translate their intrinsic material advantages into robust, scalable optoelectronic technologies.

17.9 Synthesis of MXene-based nanofluids

Nanofluids are engineered colloidal suspensions in which nanometer-scale particles are dispersed within a conventional base fluid to enhance heat-transfer performance. Even at low particle loadings, such suspensions can substantially improve thermal transport, thereby increasing the overall thermal efficiency of heat-exchange systems [96]. The incorporation of nanoparticles modifies the hydrodynamic and thermophysical behaviors of the host fluid, and these changes are strongly dependent on the particle concentration. Prior studies have consistently demonstrated that nanoparticle loading governs key properties, such as viscosity, dispersion stability, and effective thermal conductivity [97]. An increase in the nanoparticle concentration typically leads to enhanced effective thermal conductivity, particularly within dilute regimes, where particle–fluid interfacial interactions and microconvection effects contribute positively to heat transport [98]. However, this benefit is accompanied by a concurrent increase in viscosity, which can elevate pumping power requirements and influence flow characteristics

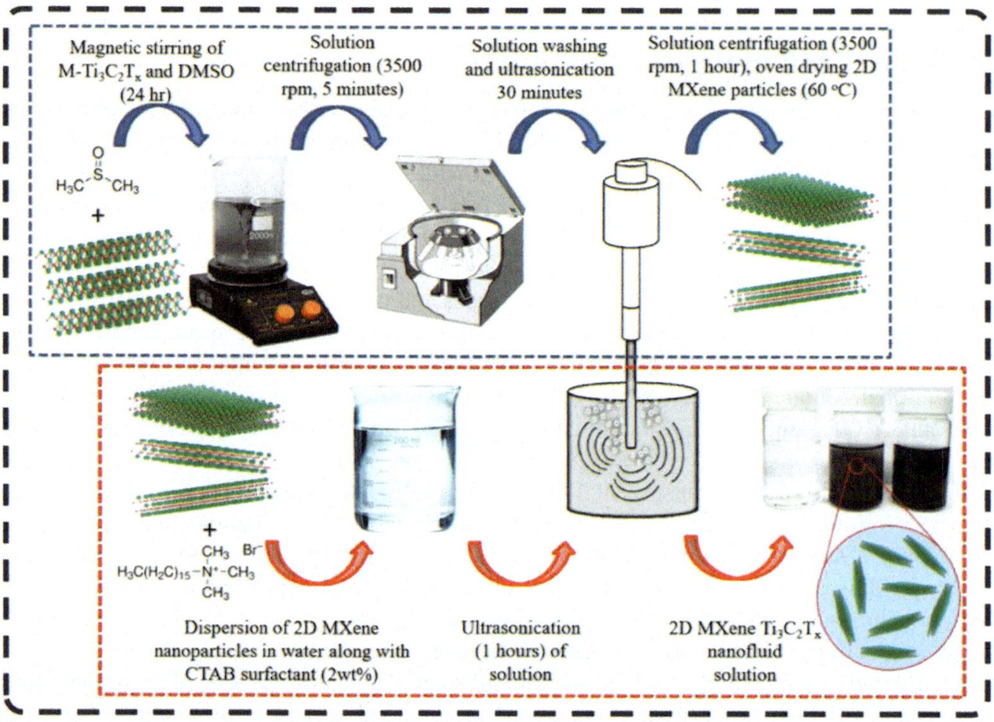

FIGURE 17.10

Synthesis of aqueous-based 2D MXene-$Ti_3C_2T_x$ nanofluid [100]. *2D*, Two dimension; *CTAB*, cetyltrimethylammonium bromide; *DMSO*, dimethyl sulfoxide.

[99]. At low concentrations, nanoparticles are more readily dispersed and maintain colloidal stability. In contrast, higher loadings increase the likelihood of particle–particle interactions, promoting aggregation and sedimentation, especially for nanoparticles with high surface area-to-volume ratios and small characteristic dimensions [96]. Surfactants are often introduced to mitigate agglomeration and improve dispersion stability, although their presence can alter interfacial thermal resistance and, in some cases, reduce the effective thermal conductivity of the nanofluid. Consequently, practical nanofluid design requires balancing enhanced thermal transport with long-term colloidal stability and manageable viscosity to ensure operational viability.

MXene-based nanofluids, particularly those incorporating $Ti_3C_2T_x$ nanosheets, are commonly synthesized using a two-step dispersion approach (Fig. 17.10). In this method a presynthesized and dried MXene nanopowder is introduced into deionized water with a stabilizing surfactant. In a representative formulation, mass fractions corresponding to nanoparticle volume fractions of 0.013% and 0.027% were prepared by dispersing dried MXene powder in deionized water containing 2 wt.% of cetyltrimethylammonium bromide (CTAB) surfactant. The mixture was then ultrasonicated for approximately 60 min to promote deagglomeration and achieve a uniform, stable suspension [100]. Ultrasonication provides sufficient acoustic energy to overcome van der Waals attractions between nanosheets, thereby facilitating homogeneous dispersion and improving short-term colloidal stability.

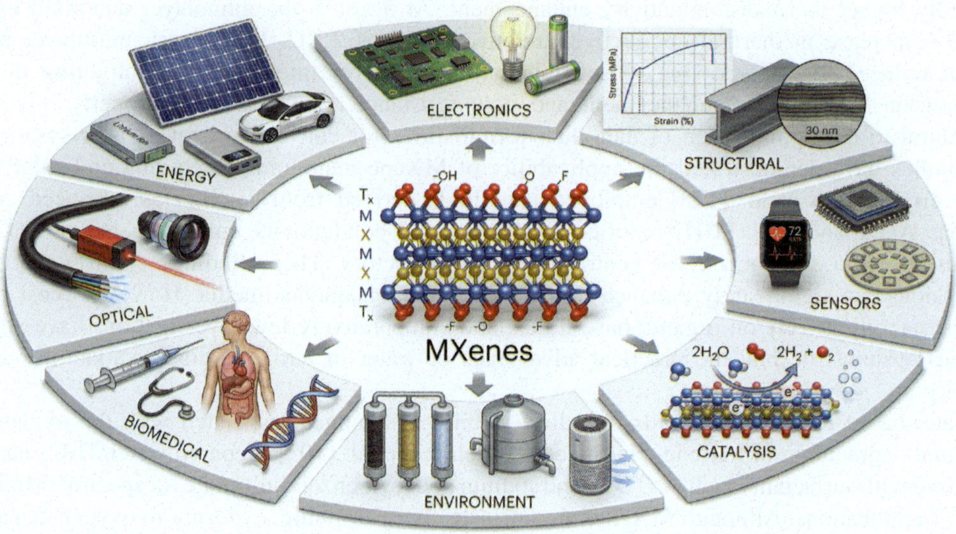

FIGURE 17.11

Application of MXens.

17.10 Applications of MXene-based nanofluids in thermal engineering

MXene-based materials have attracted sustained attention across multiple technology sectors because their 2D morphology, surface functionalization, and electronic structure enable strong coupling between thermal, optical, and electrochemical processes. As summarized in Fig. 17.11, MXenes have been explored in the electrical, thermal, electronic, optical, catalytic, electrochemical, biomedical, sensing, and environmental domains [46]. Within thermal engineering, their most compelling role arises from the combination of broadband light absorption, high intrinsic thermal conductivity, and tunable surface chemistry, which together support efficient photothermal conversion and enhanced heat transport in nanofluid systems.

A major application area is solar-thermal energy harvesting, where MXene nanosheets function as broadband absorbers and heat-conversion media. Investigations on $Ti_3C_2T_x$ MXene nanosheets demonstrated strong optical absorption spanning the near-infrared region (800–1800 nm), attributed to their high absorption coefficient and nonlinear optical response, which enables efficient light-to-heat conversion [101]. Similarly, Lin et al. [102] produced ultrathin MXene nanosheets via a two-step exfoliation route following HF etching of the parent MAX phase and reported pronounced near-infrared absorption and high photothermal conversion efficiency. These optical characteristics are critical for solar-thermal collectors, direct absorption solar collectors (DASC), and hybrid photovoltaic/thermal systems, where maximizing photon capture and rapid heat dissipation are essential.

The thermophysical performance of MXene nanofluids is strongly influenced by the nanosheet morphology, layer number, and dispersion state. Bing et al. [20] compared glycol-based nanofluids containing mono- and multilayer $Ti_3C_2T_x$ nanosheets and observed that monolayer dispersions yielded markedly higher thermal conductivity enhancement. At 5 vol.% the monolayer nanofluid exhibited a 64.9% increase in thermal conductivity relative to EG at 25°C. In contrast, multilayer MXenes exhibit reduced viscosity owing to self-lubricating interlayer interactions, highlighting the trade-off between heat-transfer enhancement and flow resistance. Such structure–property relationships are central to the optimization of nanofluid formulations for convective heat-transfer systems. Several studies have demonstrated the applicability of MXene nanofluids in cooling and solar thermal platforms. Mahesh et al. [103] employed nanofluids derived from Ti_3SiC_2 as advanced coolants, whereas Aslfattahi et al. [104] reported a 64% thermal conductivity enhancement for Ti_3C_2 MXene dispersed in silicone oil for centralized solar collectors. These findings confirm that MXene suspensions can significantly enhance heat transport in nonaqueous media. However, most reported MXene nanofluids rely on organic base fluids, and comparatively few investigations have addressed aqueous systems, despite the practical advantages of water in terms of cost, availability, and heat capacity.

Water-based MXene nanofluids introduce stability challenges associated with the oxidation and structural degradation of MXene nanosheets. Abdelrazik et al. [105] prepared water/Ti_3C_2 nanofluids stabilized with surfactants such as CTAB and sodium dodecylbenzenesulfonate for spectral management in PV/T applications. Although MXenes are intrinsically hydrophilic, exposure to oxygenated aqueous environments accelerates oxidation and diminishes their electrical and thermal performance. Strategies to mitigate degradation include low-temperature storage, oxygen-limited environments, antioxidant additives [106], and surface coupling treatments [107]. The incorporation of sodium citrate as a combined antioxidant and dispersant has been shown to suppress oxidation while improving colloidal stability [108], thereby enabling more reliable deployment in water-based DASC and related thermal systems. The intrinsic thermal conductivity of MXenes also supports their integration into composite heat-spreading materials for thermal management. Continuous MXene networks within polymer matrices can produce order-of-magnitude improvements in effective thermal conductivity; for example, PVA/MXene films exhibit nearly 23-fold enhancement relative to neat polymers. Delaminated MXenes further demonstrate stable colloidal behavior in aqueous and moderately polar solvents [109,110], reinforcing their suitability for fluid-based thermal management technologies.

From a materials perspective, only a subset of the ~70 known MAX phases can currently be converted into MXenes via chemical etching routes [104], which constrains compositional diversity. Among these, Ti_3CNT_x MXene has emerged as a particularly promising candidate for nanofluid applications because of its high aspect ratio, strong thermal conductivity, and hydrophilic surface chemistry. However, systematic studies correlating nanosheet thickness, dispersion stability, rheology, and thermal transport remain limited. Addressing these knowledge gaps is essential for predictive nanofluid design and translating MXene-enabled heat-transfer enhancements into scalable engineering systems.

17.11 Future challenges and outcomes

Although MXene-based nanofluids show promise, there are still substantial obstacles that prevent their wide-scale use in real-world thermal energy systems. The most significant concerns are colloidal and structural stability, such as particle aggregation, sedimentation, and oxidation [111]. This is especially true in water or high-temperature conditions, where MXenes are likely to degrade over time. These problems can worsen thermal conductivity and convective performance over time or as the temperature changes. In addition, higher loadings contribute to the fluid viscosity, which means that more pumping power is required. It is also important to ensure that the chemicals are compatible with the system's parts (e.g., to avoid rusting). Environmental and scalability issues make adoption more difficult. This is because many current synthesis methods use dangerous chemicals (such as HF-based etching), which makes them expensive, unsafe, and difficult to produce MXenes in large quantities that are consistent and of high quality. Environmental considerations include the potential toxicity of residual etchants or fragmented MXene pieces. This means that full life-cycle studies are needed to examine the risks throughout the production, use, and disposal of these products. Scalability issues arise from labor-intensive top-down approaches and inconsistencies in flake size, surface terminations, and quality, which affect the repeatability and performance uniformity between batches [112,113].

Future research should focus on creating ecofriendly, scalable synthesis methods (such as fluorine-free or bottom-up methods) that lower costs and risks while allowing for large-scale manufacturing. To reduce oxidation, agglomeration, and aging effects in real-world operating settings, it is important to develop better long-term stability techniques, such as enhanced surface functionalization, hybrid compositions, ionic liquid integration, and optimized surfactants. Multiscale modeling, standardized testing protocols, and AI-driven predictive frameworks (e.g., for optimizing composition, temperature, and concentration) will accelerate the design process and reduce the cost of experiments. Finally, thorough techno-economic assessments, full-system performance evaluations in devices such as heat exchangers or solar thermal systems, and environmental impact studies are necessary to connect promising laboratory discoveries with strong, long-lasting industrial technologies. MXene nanofluids can reach their full potential as multifunctional, high-performance working fluids by addressing these interconnected problems through interdisciplinary studies in materials science, colloid chemistry, fluid dynamics, and engineering.

17.12 Conclusion

MXenes have emerged as one of the most promising classes of 2DM for thermal energy systems owing to their exceptional thermophysical and surface properties. This chapter highlights the major progress in MXene synthesis, particularly the transition from hazardous HF-based etching to safer and more scalable fluorine-free routes. These advances have enhanced the control over the surface terminations, morphology, and dispersion behavior, thereby improving the performance of MXene-based nanofluids. MXene nanofluids exhibit remarkable enhancements in thermal conductivity (typically 20%–80%), convective heat transfer, and photothermal conversion efficiency, making them strong candidates for use in solar collectors, CSP systems, heat exchangers, electronic cooling, and thermal energy storage. Continued efforts toward green synthesis, hybrid formulations, and optimized dispersion strategies have further addressed the challenges of stability and scalability. Beyond thermal management, MXenes also show potential in diverse areas such as energy conversion, catalysis, sensing, biomedical devices, and environmental remediation. To fully realize their potential in sustainable thermal engineering, future research should prioritize large-scale, cost-effective production, system-level integration, and development of multifunctional MXene-based hybrid materials.

References

[1] K. Khanafer, K. Vafai, A critical synthesis of thermophysical characteristics of nanofluids, in: Nanotechnology and Energy, Jenny Stanford Publishing, Boca Raton, 2017, pp. 279–332.

[2] N.A.C. Sidik, H. Mohammed, O.A. Alawi, S. Samion, A review on preparation methods and challenges of nanofluids, Int. Commun. Heat Mass Transfer 54 (2014) 115–125.

[3] S.U. Choi, J.A. Eastman, Enhanced Heat Transfer Using Nanofluids, Argonne National Laboratory (ANL), Argonne, IL, 2000.

[4] S.K. Das, S.U. Choi, H.E. Patel, Heat transfer in nanofluids—a review, Heat Transfer Eng. 27 (10) (2006) 3–19.

[5] V. Trisaksri, S. Wongwises, Critical review of heat transfer characteristics of nanofluids, Renew. Sustain. Energy Rev. 11 (3) (2007) 512–523.

[6] M. Kole, T. Dey, Investigation of thermal conductivity, viscosity, and electrical conductivity of graphene based nanofluids, J. Appl. Phys. 113 (8) (2013).

[7] L. Kumar, M. Hasanuzzaman, N.A. Rahim, Global advancement of solar thermal energy technologies for industrial process heat and its future prospects: a review, Energy Convers. Manage. 195 (2019) 885–908, doi:10.1016/j.enconman.2019.05.081.

[8] S.F. Ahmed, M. Khalid, W. Rashmi, A. Chan, K. Shahbaz, Recent progress in solar thermal energy storage using nanomaterials, Renew. Sustain. Energy Rev. 67 (2017) 450–460, doi:10.1016/j.rser.2016.09.034.

[9] M. Naguib, et al., Two-dimensional nanocrystals produced by exfoliation of Ti_3AlC_2, in: MXenes, Jenny Stanford Publishing, New York, 2023, pp. 15–29.

[10] B. Anasori, M.R. Lukatskaya, Y. Gogotsi, 2D metal carbides and nitrides (MXenes) for energy storage, in: MXenes, Jenny Stanford Publishing, 2023, pp. 677–722.

[11] M. Ghidiu, M.R. Lukatskaya, M.-Q. Zhao, Y. Gogotsi, M.W. Barsoum, Conductive two-dimensional titanium carbide 'clay' with high volumetric capacitance, in: MXenes, Jenny Stanford Publishing, New York, 2023, pp. 379–399.

[12] P. Sati, R.C. Shende, S. Ramaprabhu, An experimental study on thermal conductivity enhancement of DI water-EG based ZnO (CuO)/graphene wrapped carbon nanotubes nanofluids, Thermochimica Acta 666 (2018) 75–81.

[13] X. Li, W. Chen, C. Zou, The stability, viscosity and thermal conductivity of carbon nanotubes nanofluids with high particle concentration: a surface modification approach, Powder Technology 361 (2020) 957–967.

[14] F. Shahzad, et al., Electromagnetic interference shielding with 2D transition metal carbides (MXenes), Science 353 (6304) (2016) 1137–1140.

[15] S.Z. Butler, et al., Progress, challenges, and opportunities in two-dimensional materials beyond graphene, ACS Nano 7 (4) (2013) 2898–2926.

[16] M.R. Lukatskaya, et al., Ultra-high-rate pseudocapacitive energy storage in two-dimensional transition metal carbides, in: MXenes, Jenny Stanford Publishing, New York, 2023, pp. 723–743.

[17] M. Rafieerad, A. Rafieerad, B. Mehmandoust, S. Dhingra, M. Shanbedi, New water-based fluorescent nanofluid containing 2D titanium carbide MXene sheets: a comparative study of its thermophysical, electrical and optical properties with amine and carboxyl covalently functionalized graphene nanoplatelets, J. Therm. Anal. Calorim. 146 (4) (2021) 1491–1504.

[18] L.B. La, C. Leatherday, Y.-K. Leong, H.P. Watts, L.-C. Zhang, Green lightweight lead-free Gd2O3/epoxy nanocomposites with outstanding X-ray attenuation performance, Compos. Sci. Technol. 163 (2018) 89–95.

[19] X. Li, H. Wang, H. Zhang, L. Yang, Comprehensive performance evaluation of Ti_3C_2 MXene/TiN composite nanofluids for photo thermal conversion, Appl. Therm. Eng. 228 (2023) 120486.

[20] Z. Bao, N. Bing, X. Zhu, H. Xie, W. Yu, $Ti_3C_2T_x$ MXene contained nanofluids with high thermal conductivity, super colloidal stability and low viscosity, Chem. Eng. J. 406 (2021) 126390.
[21] A. Aziz, A. Agha, S. Anwer, E. Abu-Nada, A. Alazzam, Numerical and experimental investigation of optimized triangular microchannels using MXene-based nanofluids for enhanced microfluidic thermal performance, Int. J. Thermofluids 26 (2025) 101074.
[22] S. Sreekumar, N. Shah, J.D. Mondol, N. Hewitt, S. Chakrabarti, Numerical investigation and feasibility study on MXene/water nanofluid based photovoltaic/thermal system, Clean. Energy Syst. 2 (2022) 100010.
[23] R. Khaledialidusti, E. Mahdavi, A. Barnoush, Stabilization of 2D graphene, functionalized graphene, and Ti_2CO_2 (MXene) in super-critical CO_2: a molecular dynamics study, Phys. Chem. Chem. Phys. 21 (24) (2019) 12968–12976.
[24] R. Li, L. Zhang, L. Shi, P. Wang, MXene Ti_3C_2: an effective 2D light-to-heat conversion material, ACS Nano 11 (4) (2017) 3752–3759.
[25] D. Wang, et al., Transforming $Ti_3C_2T_x$ MXenes into nanoscale ionic materials via an electronic interaction strategy, J. Mater. Chem. A 9 (27) (2021) 15441–15451.
[26] K. Mahesh, V. Linsha, A.P. Mohamed, S. Ananthakumar, Processing of 2D-MAXene nanostructures and design of high thermal conducting, rheo-controlled MAXene nanofluids as a potential nanocoolant, Chem. Eng. J. 297 (2016) 158–169.
[27] H. Liu, et al., MXene confined in shape-stabilized phase change material combining enhanced electromagnetic interference shielding and thermal management capability, Compos. Sci. Technol. 210 (2021) 108835.
[28] G. Qi, et al., Hierarchical graphene foam-based phase change materials with enhanced thermal conductivity and shape stability for efficient solar-to-thermal energy conversion and storage, Nano Res. 10 (3) (2017) 802–813.
[29] F. Pineda, et al., MXene-enhanced nanofluids for superior thermal energy storage in concentrated solar power plants, Sol. Energy Mater. Sol. Cells 283 (2025) 113461.
[30] C. Wang, Z. Pan, H. Chen, X. Pu, Z. Chen, MXene-based materials for multivalent metal-ion batteries, Batteries 9 (3) (2023) 174.
[31] X. Zhang, Z. Zhang, Z. Zhou, MXene-based materials for electrochemical energy storage, J. Energy Chem. 27 (1) (2018) 73–85.
[32] F. Pineda, A. Angel, F.E. Palay, A. Rosenkranz, Elucidating the role of surfactants and dispersants in tailoring the aqueous behavior of multi-layer $Ti_3C_2T_x$, Colloids Surf. A (2025) 138820.
[33] M. Dadashi Firouzjaei, S.K. Nemani, M. Sadrzadeh, E.K. Wujcik, M. Elliott, B. Anasori, Life-cycle assessment of $Ti_3C_2T_x$ MXene synthesis, Adv. Mater. 35 (31) (2023) 2300422.
[34] A. Ungureanu, et al., Systematic life cycle environmental impact comparison of alternative synthetic strategies for $Ti_3C_2T_x$ MXene, ACS Sustain. Chem. Eng. 12 (15) (2024) 5893–5906.
[35] J. Munuera, L. Britnell, C. Santoro, R. Cuéllar-Franca, C. Casiraghi, A review on sustainable production of graphene and related life cycle assessment, 2D Mater. 9 (1) (2022) 012002.
[36] M.B. Ben Hamida, et al., Optimizing MXene graphene based fluids for solar energy conversion and storage using a novel intelligent framework, Sci. Rep. 15 (1) (2025) 31702.
[37] M. Xu, T. Liang, M. Shi, H.J.C.r. Chen, Graphene-like two-dimensional materials, Chem. Rev. 113 (5) (2013) 3766–3798.
[38] M. Sajid, MXenes: are they emerging materials for analytical chemistry applications? – a review, Anal. Chim. Acta 1143 (2021) 267–280, doi:10.1016/j.aca.2020.08.063.
[39] M. Naguib, et al., Two-dimensional nanocrystals produced by exfoliation of Ti_3AlC_2, Adv. Mater. 23 (37) (2011) 4248–4253.
[40] N. Hemanth, B. Kandasubramanian, Recent advances in 2D MXenes for enhanced cation intercalation in energy harvesting applications: a review, Chem. Eng. J. 392 (2020) 123678.
[41] M. Naguib, V.N. Mochalin, M.W. Barsoum, Y. Gogotsi, 25th anniversary article: MXenes: a new family of two-dimensional materials, Adv. Mater. 26 (7) (2014) 992–1005.

[42] L. Verger, et al., Overview of the synthesis of MXenes and other ultrathin 2D transition metal carbides and nitrides, Curr. Opin. Solid State Mater. Sci. 23 (3) (2019) 149–163.
[43] F. Jamil, H.M. Ali, M.M. Janjua, MXene based advanced materials for thermal energy storage: a recent review, J. Energy Storage 35 (2021) 102322, doi:10.1016/j.est.2021.102322.
[44] M.A.A.M. Abdah, et al., Advancements in MXene-polymer composites for high-performance supercapacitor applications, J. Energy Storage 63 (2023) 106942.
[45] J.J. Nickl, K.K. Schweitzer, P. Luxenberg, Gasphasenabscheidung im system Ti · Si · C, J. Less Common Met. 26 (3) (1972) 335–353, doi:10.1016/0022-5088(72)90083-5.
[46] B. Anasori, M.R. Lukatskaya, Y. Gogotsi, 2D metal carbides and nitrides (MXenes) for energy storage, Nat. Rev. Mater. 2 (2) (2017) 1–17.
[47] Y. Zou, Z.M. Sun, S. Tada, H. Hashimoto, Synthesis reactions for Ti_3SiC_2 through pulse discharge sintering TiH_2/Si/TiC powder mixture, Mater. Res. Bull. 43 (4) (2008) 968–975, doi:10.1016/j.materresbull.2007.04.028.
[48] X. Yin, N. Travitzky, P. Greil, Three-dimensional printing of nanolaminated Ti_3AlC_2 toughened $TiAl_3$–Al_2O_3 composites, J. Am. Ceram. Soc. 90 (7) (2007) 2128–2134.
[49] C. Xu, et al., Large-area high-quality 2D ultrathin Mo_2C superconducting crystals, Nat. Mater. 14 (11) (2015) 1135–1141.
[50] C. Xu, L. Chen, Z. Liu, H.-M. Cheng, W. Ren, Bottom-up synthesis of 2D transition metal carbides and nitrides, in: 2D MetalCarbides and Nitrides (MXenes), Springer, Switzerland, 2019, pp. 89–109.
[51] C. Yang, et al., Ultrafine Pt nanoparticle-decorated 3D hybrid architectures built from reduced graphene oxide and MXene nanosheets for methanol oxidation, Chem. Mater. 31 (22) (2019) 9277–9287.
[52] C. Yang, Q. Jiang, H. Huang, H. He, L. Yang, W. Li, Polyelectrolyte-induced stereoassembly of grain boundary-enriched platinum nanoworms on $Ti_3C_2T_x$ MXene nanosheets for efficient methanol oxidation, ACS Appl. Mater. Interfaces 12 (21) (2020) 23822–23830.
[53] Z. Wang, et al., Metal immiscibility route to synthesis of ultrathin carbides, borides, and nitrides, Adv. Mater. 29 (29) (2017) 1700364.
[54] J. Jia, T. Xiong, L. Zhao, F. Wang, H. Liu, R. Hu, … S. Chen, Ultrathin N-doped Mo2C nanosheets with exposed active sites as efficient electrocatalyst for hydrogen evolution reactions, ACS nano 11 (12) (2017) 12509–12518.
[55] F. Zhang, Z. Zhang, H. Wang, C.H. Chan, N.Y. Chan, X.X. Chen, J.Y. Dai, Plasma-enhanced pulsed-laser deposition of single-crystalline M o 2 C ultrathin superconducting films, Physical Review Materials 1 (3) (2017) 034002.
[56] S. Joshi, Q. Wang, A. Puntambekar, V. Chakrapani, Facile synthesis of large area two-dimensional layers of transition-metal nitride and their use as insertion electrodes, ACS Energy Lett. 2 (6) (2017) 1257–1262.
[57] M.W. Barsoum, MAX Phases: Properties of Machinable Ternary Carbides and Nitrides, John Wiley & Sons, Germany, 2013.
[58] V.H. Nowotny, Strukturchemie einiger verbindungen der übergangsmetalle mit den elementen C, Si, Ge, Sn, Prog. Solid State Chem. 5 (1971) 27–70.
[59] B. Anasori, J. Halim, J. Lu, C.A. Voigt, L. Hultman, M.W. Barsoum, Mo_2TiAlC_2: a new ordered layered ternary carbide, Scripta Mater. 101 (2015) 5–7.
[60] Q. Tao, et al., Two-dimensional $Mo_{1.33}C$ MXene with divacancy ordering prepared from parent 3D laminate with in-plane chemical ordering, Nat. Commun. 8 (1) (2017) 1–7.
[61] Q. Tao, et al., Atomically layered and ordered rare-earth i-MAX phases: a new class of magnetic quaternary compounds, Chem. Mater. 31 (7) (2019) 2476–2485.
[62] M. Naguib, et al., Two-dimensional transition metal carbides, ACS Nano 6 (2) (2012) 1322–1331.
[63] M. Ghidiu, M.R. Lukatskaya, M.-Q. Zhao, Y. Gogotsi, M.W. Barsoum, Conductive two-dimensional titanium carbide 'clay' with high volumetric capacitance, Nature 516 (7529) (2014) 78–81.

[64] M. Alhabeb, et al., Guidelines for synthesis and processing of two-dimensional titanium carbide ($Ti_3C_2T_x$ MXene), Chem. Mater. 29 (18) (2017) 7633–7644.

[65] S. Hu, S. Li, W. Xu, J. Zhang, Y. Zhou, Z.J.C.I. Cheng, Rapid preparation, thermal stability and electromagnetic interference shielding properties of two-dimensional Ti3C2 MXene, Ceram. Int. 45 (16) (2019) 19902–19909.

[66] A. Numan, et al., Microwave-assisted rapid MAX phase etching and delamination: a paradigm shift in MXene synthesis, Mater. Chem. Phys. 288 (2022) 126429.

[67] M. Ghidiu, J. Halim, S. Kota, D. Bish, Y. Gogotsi, M.W. Barsoum, Ion-exchange and cation solvation reactions in Ti_3C_2 MXene, Chem. Mater. 28 (10) (2016) 3507–3514.

[68] M. Ghidiu, et al., Alkylammonium cation intercalation into Ti_3C_2 (MXene): effects on properties and ion-exchange capacity estimation, Chem. Mater. 29 (3) (2017) 1099–1106.

[69] O. Mashtalir, et al., Intercalation and delamination of layered carbides and carbonitrides, Nat. Commun. 4 (1) (2013) 1–7.

[70] V. Natu, M. Sokol, L. Verger, M.W. Barsoum, Effect of edge charges on stability and aggregation of $Ti_3C_2T_z$ MXene colloidal suspensions, J. Phys. Chem. C 122 (48) (2018) 27745–27753.

[71] Y. Omomo, T. Sasaki, L. Wang, M. Watanabe, Redoxable nanosheet crystallites of MnO_2 derived via delamination of a layered manganese oxide, J. Am. Chem. Soc. 125 (12) (2003) 3568–3575.

[72] C. Tan, et al., Recent advances in ultrathin two-dimensional nanomaterials, Chem. Rev. 117 (9) (2017) 6225–6331.

[73] K.S. Novoselov, et al., Electric field effect in atomically thin carbon films, Science 306 (5696) (2004) 666–669.

[74] K.F. Mak, C. Lee, J. Hone, J. Shan, T.F. Heinz, Atomically thin MoS_2: a new direct-gap semiconductor, Phys. Rev. Lett. 105 (13) (2010) 136805.

[75] Y. Sui, J. Zhou, X. Wang, L. Wu, S. Zhong, Y.J.M.T. Li, Recent advances in black-phosphorus-based materials for electrochemical energy storage, Mater. Today 42 (2021) 117–136.

[76] Y.Z. Zhang, Y. Wang, Q. Jiang, J.K. El-Demellawi, H. Kim, H.N. Alshareef, MXene printing and patterned coating for device applications, Adv. Mater. 32 (21) (2020) 1908486.

[77] M.R. Lukatskaya, et al., Cation intercalation and high volumetric capacitance of two-dimensional titanium carbide, Science 341 (6153) (2013) 1502–1505.

[78] S. Sun, C. Liao, A.M. Hafez, H. Zhu, S.J.C.E.J. Wu, Two-dimensional MXenes for energy storage, Chem. Eng. J. 338 (2018) 27–45.

[79] M. Ashton, R.G. Hennig, S.R. Broderick, K. Rajan, S.B. Sinnott, Computational discovery of stable ${M}_{2}AX$ phases, Phys. Rev. B 94 (5) (2016) 054116, doi:10.1103/PhysRevB.94.054116.

[80] H. Xu, A. Ren, J. Wu, Z. Wang, Recent advances in 2D MXenes for photodetection, Adv. Funct. Mater. 30 (24) (2020) 2000907, doi:10.1002/adfm.202000907.

[81] B. Soundiraraju, R. Raghavan, B.K. George, Chromium carbide nanosheets prepared by selective etching of aluminum from Cr_2AlC for hydrazine detection, ACS Appl. Nano Mater. 3 (11) (2020) 11007–11016, doi:10.1021/acsanm.0c02230.

[82] A.D. Dillon, et al., Highly conductive optical quality solution-processed films of 2D titanium carbide, Adv. Funct. Mater. 26 (23) (2016) 4162–4168, doi:10.1002/adfm.201600357.

[83] M. Khazaei, et al., Novel electronic and magnetic properties of two-dimensional transition metal carbides and nitrides, Adv. Funct. Mater. 23 (17) (2013) 2185–2192, doi:10.1002/adfm.201202502.

[84] X.-H. Zha, et al., The thermal and electrical properties of the promising semiconductor MXene Hf_2CO_2, Sci. Rep. 6 (1) (2016) 27971, doi:10.1038/srep27971.

[85] A.N. Gandi, H.N. Alshareef, U. Schwingenschlögl, Thermoelectric performance of the MXenes M_2CO_2 (M = Ti, Zr, or Hf), Chem. Mater. 28 (6) (2016) 1647–1652, doi:10.1021/acs.chemmater.5b04257.

[86] X.-H. Zha, et al., Intrinsic structural, electrical, thermal, and mechanical properties of the promising conductor Mo_2C MXene, J. Phys. Chem. C 120 (28) (2016) 15082–15088, doi:10.1021/acs.jpcc.6b04192.

[87] X. Jin, et al., Flame-retardant poly(vinyl alcohol)/MXene multilayered films with outstanding electromagnetic interference shielding and thermal conductive performances, Chem. Eng. J. 380 (2020) 122475, doi:10.1016/j.cej.2019.122475.

[88] X. Lu, et al., Novel light-driven and electro-driven polyethylene glycol/two-dimensional MXene form-stable phase change material with enhanced thermal conductivity and electrical conductivity for thermal energy storage, Composites, B: Eng. 177 (2019) 107372.

[89] M. Rafieerad, A.R. Rafieerad, B. Mehmandoust, S. Dhingra, M. Shanbedi, New water-based fluorescent nanofluid containing 2D titanium carbide MXene sheets: a comparative study of its thermophysical, electrical and optical properties with amine and carboxyl covalently functionalized graphene nanoplatelets, J. Therm. Anal. Calorim. 146 (4) (2021) 1491–1504, doi:10.1007/s10973-020-10088-0.

[90] H. Kim, H.N. Alshareef, MXetronics: MXene-enabled electronic and photonic devices, ACS Mater. Lett. 2 (1) (2020) 55–70, doi:10.1021/acsmaterialslett.9b00419.

[91] K. Hantanasirisakul, et al., Fabrication of $Ti_3C_2T_x$ MXene transparent thin films with tunable optoelectronic properties, Adv. Electron. Mater. 2 (6) (2016) 1600050, doi:10.1002/aelm.201600050.

[92] Y. Dong, et al., Saturable absorption in 2D Ti_3C_2 MXene thin films for passive photonic diodes, Adv. Mater. 30 (10) (2018) 1705714, doi:10.1002/adma.201705714.

[93] G. Ying, A.D. Dillon, A.T. Fafarman, M.W. Barsoum, Transparent, conductive solution processed spincast 2D Ti_2CT_x (MXene) films, Mater. Res. Lett. 5 (6) (2017) 391–398, doi:10.1080/21663831.2017.1296043.

[94] M. Mariano, et al., Solution-processed titanium carbide MXene films examined as highly transparent conductors, Nanoscale 8 (36) (2016) 16371–16378, doi:10.1039/C6NR03682A.

[95] J. Halim, et al., Transparent conductive two-dimensional titanium carbide epitaxial thin films, Chem. Mater. 26 (7) (2014) 2374–2381, doi:10.1021/cm500641a.

[96] D.K. Devendiran, V.A. Amirtham, A review on preparation, characterization, properties and applications of nanofluids, Renew. Sustain. Energy Rev. 60 (2016) 21–40, doi:10.1016/j.rser.2016.01.055.

[97] L. Yang, W. Ji, M. Mao, J.-n. Huang, An updated review on the properties, fabrication and application of hybrid-nanofluids along with their environmental effects, J. Cleaner Prod. 257 (2020) 120408, doi:10.1016/j.jclepro.2020.120408.

[98] Z. Li, S. Asadi, A. Karimipour, A. Abdollahi, I. Tlili, Experimental study of temperature and mass fraction effects on thermal conductivity and dynamic viscosity of SiO_2-oleic acid/liquid paraffin nanofluid, Int. Commun. Heat Mass Transfer 110 (2020) 104436, doi:10.1016/j.icheatmasstransfer.2019.104436.

[99] R. Prasher, D. Song, J. Wang, P. Phelan, Measurements of nanofluid viscosity and its implications for thermal applications, Appl. Phys. Lett. 89 (13) (2006) 133108, doi:10.1063/1.2356113.

[100] T. Ambreen, A. Saleem, C.W. Park, Thermal efficiency of eco-friendly MXene based nanofluid for performance enhancement of a pin-fin heat sink: experimental and numerical analyses, Int. J. Heat Mass Transfer 186 (2022) 122451, doi:10.1016/j.ijheatmasstransfer.2021.122451.

[101] X. Jiang, et al., Broadband nonlinear photonics in few-layer MXene $Ti_3C_2T_x$ (T = F, O, or OH), Laser Photonics Rev. 12 (2) (2018) 1700229, doi:10.1002/lpor.201700229.

[102] H. Lin, X. Wang, L. Yu, Y. Chen, J. Shi, Two-dimensional ultrathin MXene ceramic nanosheets for photothermal conversion, Nano Lett. 17 (1) (2017) 384–391, doi:10.1021/acs.nanolett.6b04339.

[103] K.V. Mahesh, V. Linsha, A. Peer Mohamed, S. Ananthakumar, Processing of 2D-MAXene nanostructures and design of high thermal conducting, rheo-controlled MAXene nanofluids as a potential nanocoolant, Chem. Eng. J. 297 (2016) 158–169, doi:10.1016/j.cej.2016.04.010.

[104] N. Aslfattahi, L. Samylingam, A.S. Abdelrazik, A. Arifutzzaman, R. Saidur, MXene based new class of silicone oil nanofluids for the performance improvement of concentrated photovoltaic thermal collector, Sol. Energy Mater. Sol. Cells 211 (2020) 110526, doi:10.1016/j.solmat.2020.110526.

[105] A.S. Abdelrazik, K.H. Tan, N. Aslfattahi, A. Arifutzzaman, R. Saidur, F.A. Al-Sulaiman, Optical, stability and energy performance of water-based MXene nanofluids in hybrid PV/thermal solar systems, Sol. Energy 204 (2020) 32–47, doi:10.1016/j.solener.2020.04.063.

[106] X. Zhao, et al., Antioxidants unlock shelf-stable $Ti_3C_2T_x$ (MXene) nanosheet dispersions, Matter 1 (2) (2019) 513–526, doi:10.1016/j.matt.2019.05.020.
[107] J. Ji, L. Zhao, Y. Shen, S. Liu, Y. Zhang, Covalent stabilization and functionalization of MXene via silylation reactions with improved surface properties, FlatChem 17 (2019) 100128, doi:10.1016/j.flatc.2019.100128.
[108] X. Jiang, et al., Two-dimensional MXenes: from morphological to optical, electric, and magnetic properties and applications, Phys. Rep. 848 (2020) 1–58, doi:10.1016/j.physrep.2019.12.006.
[109] S. Lim, H. Park, J. Yang, C. Kwak, J. Lee, Stable colloidal dispersion of octylated Ti_3C_2-MXenes in a nonpolar solvent, Colloids Surf. A 579 (2019) 123648, doi:10.1016/j.colsurfa.2019.123648.
[110] K. Maleski, V.N. Mochalin, Y. Gogotsi, Dispersions of two-dimensional titanium carbide MXene in organic solvents, Chem. Mater. 29 (4) (2017) 1632–1640, doi:10.1021/acs.chemmater.6b04830.
[111] S. Kulandaivel, et al., MXene nanofluids in advanced applications: an in-depth review of thermophysical characteristics and technological innovations, Mater. Today Sustain. 29 (2025) 101084.
[112] J. Chen, et al., Recent progress and advances in the environmental applications of MXene related materials, Nanoscale 12 (6) (2020) 3574–3592.
[113] A. Khosla, et al., Emergence of MXene and MXene–polymer hybrid membranes as future-environmental remediation strategies, Adv. Sci. 9 (36) (2022) 2203527.

Index

Page numbers followed by "*f*" and "*t*" indicate, figures and tables respectively.

O

P

R

S

Printed and bound by CPI Group (UK) Ltd, Croydon, CR0 4YY
14/09/2026
02216409-0015